BUILDING SCIENTIFIC APPARATUS

A Practical Guide to Design and Construction

BUILDING SCIENTIFIC APPARATUS

A Practical Guide to Design and Construction

John H. Moore • Christopher C. Davis • Michael A. Coplan
with a chapter by Sandra C. Greer
University of Maryland, College Park

PERSEUS BOOKS
Cambridge, Massachusetts

Copyright © 2003 by Westview Press, A Member of the Perseus Books Group

Westview Press books are available at special discounts for bulk purchases in the United States by corporations, institutions, and other organizations. For more information, please contact the Special Markets Department at the Perseus Books Group, 11 Cambridge Center, Cambridge MA 02142, or call (617) 252-5298.

Published in 2002 in the United States of America by Westview Press, 5500 Central Avenue, Boulder, Colorado 80301–2877, and in the United Kingdom by Westview Press, 12 Hid's Copse Road, Cumnor Hill, Oxford OX2 9JJ

Find us on the World Wide Web at www.westviewpress.com

A Cataloging-in-Publication data record for this book is available from the Library of Congress.
ISBN 0-8133-4006-3 (pbk.); 0-8133-4007-1 (hc)

The paper used in this publication meets the requirements of the American National Standard for Permanence of Paper for Printed Library Materials Z39.48–1984.

Illustrators: *James S. Kempton and Mary Ellen Didion-Cursley*
Production Services: *TIPS Technical Publishing*

10 9 8 7 6 5 4 3 2 1

To our families

CONTENTS

2 WORKING WITH GLASS

3 VACUUM TECHNOLOGY

4 OPTICS

5 CHARGED-PARTICLE OPTICS

6 ELECTRONICS

7 MEASUREMENT AND CONTROL OF TEMPERATURE

PREFACE

In the decade or more since the publication of the second edition of *Building Scientific Apparatus*, there has been a remarkable expansion in the number of technical innovations available to builders and designers of scientific equipment. These come in the form of integrated units—lasers, computer interfaces, pumping systems—making it easier and often less expensive to build an apparatus compared to the time when one was required to work with individual components such as lenses and prisms, capacitors and resistors, and valves and diffusion pumps. The third edition has grown by about a third in comparison to earlier editions, as we describe the use and specification of integrated units that can be employed in the construction of unique apparatus in the physical and engineering sciences. It is, however, still important to be able to solder together electronic components and to understand the operation of machine tools, so much discussion of fundamental principles and basic operations remains from earlier editions and, in several instances, has been expanded.

The third edition recognizes the important role of the computer and readily-available computer programs in the design, as well as the operation, of modern scientific apparatus. The use of computer-assisted drafting and design programs for mechanical, optical, and electronic systems is described. The use of computer-controlled machining is included. Ray-tracing programs for light and charged-particle optics are discussed along with electronic circuit analysis programs. There is a discussion of computer control of apparatus using programs such as LabView. We also recognize the importance that the World Wide Web has acquired as a resource for research. Although we continue to include a modest list of providers of equipment at the end of each chapter, the specifications, availability, and cost of scientific equipment and devices can be readily located using a suitable search engine.

The chapter on mechanical design and fabrication includes more information on materials, a description of electrical discharge machining, and a new section on the design of flexures. In the chapter on vacuum technology, one will find a new section on flowmeters and more information about materials, as well as a discussion of pumps for ultrahigh vacuum. The chapter on optics includes more about optical design techniques, including the use of computer software, as well as a discussion of aberrations in optical systems and their minimization. New sections discuss imaging and non-imaging systems and devices. In addition there are expanded descriptions of precision mechanical stages, new lasers, optical detectors, and fiber optics. New information on charged particle detection and magnetic field control can be found in the chapter on charged particle optics. The electronics chapter has added material on schematic capture and printed circuit board programs. Control theory and the practical implementation of computer control of experiments forms a new section and the topic of noise and timing/correlation experiments has been added. Lastly, the contribution by our colleague, Sandra C. Greer, on the measurement and control of temperature that first appeared in the second edition, is also included in the third edition.

We sincerely hope that this book will contribute to the quality and functionality of apparatus built in the laboratories of scientists and engineers. We continue to welcome the comments and suggestions of our readers and hope that our readers will recognize the incorporation of suggestions from years past.

CHAPTER 1 MECHANICAL DESIGN AND FABRICATION

Every scientific apparatus requires a mechanical structure, even a device that is fundamentally electronic or optical in nature. The design of this structure determines to a large extent the usefulness of the apparatus. It follows that a successful scientist must acquire many of the skills of the mechanical engineer in order to proceed rapidly with an experimental investigation.

The designer of research apparatus must strike a balance between the makeshift and the permanent. Too little initial consideration of the expected performance of a machine may frustrate all attempts to get data. Too much time spent planning can also be an error, since the performance of a research apparatus is not entirely predictable. A new machine must be built and operated before all the shortcomings in its design are apparent.

The function of a machine should be specified in some detail before design work begins. One must be realistic in specifying the job of a particular device. The introduction of too much flexibility can hamper a machine in the performance of its primary function. On the other hand, it may be useful to allow space in an initial design for anticipated modifications. Problems of assembly and disassembly should be considered at the outset, since research equipment rarely functions properly at first and often must be taken apart and reassembled repeatedly.

Make a habit of studying the design and operation of machines. Learn to visualize in three dimensions the size and positions of the parts of an instrument in relation to one another.

Before beginning a design, learn what has been done before. It is a good idea to build and maintain a library of commercial catalogs to be familiar with what is available from outside sources. Too many scientific designers waste time and money on the reinvention of the wheel and the screw. Use nonstandard parts only when their advantages justify the great cost of one-off construction in comparison with mass production. Consider modifications of a design that will permit the use of standardized parts. Become aware of the available range of commercial services. In most big cities, operations such as casting, plating, and heat treating are performed inexpensively by specialty job shops. In many cases it is cheaper to have others provide these services rather than attempt them yourself. The addresses of a few of the thousands of suppliers of useful services, as well as manufacturers of useful materials, are listed at the end of this chapter.

In the following sections we discuss the properties of materials and the means of joining materials to create a machine. The physical principles of mechanical design are presented. These deal primarily with controlling the motion of one part of a machine with respect to another, both where motion is desirable and where it is not. There are also sections on machine tools and mechanical drawing. The former is mainly intended to provide enough information to enable the scientist to make intelligent use of the services of a machine shop. The latter is detailed enough to allow effective communication with people in the shop.

1.1 TOOLS AND SHOP PROCESSES

A scientist must be able to make proper use of hand tools to assemble and modify research apparatus. A successful experimentalist must be able to perform elementary operations with a drill press, lathe, and milling machine in order to make or modify simple components. Even when a scientist works with instruments that are fabricated and maintained by research technicians and machinists, an elementary knowledge of machine-tool operations will allow the design of apparatus that can be constructed with efficiency and at reasonable cost. The following is intended to familiarize the reader with the capabilities of various tools. Skill with machine tools is best acquired under the supervision of a competent machinist.

1.1.1 Hand Tools

A selection of hand tools for the laboratory is given in Table 1.1. A research scientist in physics or chemistry will have frequent use for most of these tools, and if possible should have the entire set in the lab. The tool set outlined in Table 1.1 is not too expensive for any scientist to have on hand.

Table 1.1 Tool Set for Laboratory Use

Screwdrivers:

    No. 1, 2, and 3 drivers for slotted-head screws
    No. 1, 2, and 3 Phillips screwdrivers
    Allen (hex) drivers for socket-head screws, both a fractional set ($\frac{1}{16} - \frac{1}{4}$ in.) and a metric set (1.5–10 mm)
    Nut drivers for hex-head screws and nuts, both fractional ($\frac{1}{8} - \frac{1}{2}$ in.) and metric (4–11 mm)
    Set of jeweler's screwdrivers

Wrenches:

    Combination box and open-end wrenches, both a fractional set ($\frac{3}{8}$ –1 in.) and a metric set (10–19 mm)
    $\frac{3}{8}$ -in.-square-drive ratcheting socket driver with both fractional ($\frac{3}{8} - \frac{7}{8}$ in.) and metric (10–19 mm) socket sets
    6 in., 8 in., and 12 in. adjustable wrenches
    Pipe wrench

Table 1.1 Tool Set for Laboratory Use (continued)

Pliers:

    Slip joint pliers
    Channel-locking pliers
    Large and small needle-nose pliers
    Large and small diagonal cutters
    Small flush cutters
    Hemostats

Hammers:

    2 oz and 6 oz ball-peen hammers
    Soft-faced hammer with plastic or rubber inserts

Files:

    6 in. and 8 in. second-cut, flat, half-round, and round files with handles
    6 in. and 8 in. smooth-cut, flat, half-round, and round files with handles

Miscellaneous:

    Forceps
    Sheet-metal shears
    Hacksaw
    Tubing cutter
    Spring-loaded center punch
    Scriber
    Small machinist's square
    Steel scale
    Divider
    Dial caliper or micrometer
    Tap wrench with 4-40 to $\frac{1}{4}$ -20 NC, to metric, and $\frac{1}{8}$ to $\frac{1}{2}$
    NPT (pipe thread) tap sets
    Electric hand drill motor or small drill press
    Drills, $\frac{1}{16} - \frac{1}{2}$ in. in $\frac{1}{32}$ in. increments, in drill index
    Drills, Nos. 1–60, in drill index
    Small bench vise

A laboratory scientist should adopt a craftsmanlike attitude toward tools. Far less time is required to find and use the proper tool for a job than will be required to repair the damage resulting from using the wrong one.

1.1.2 Machines for Making Holes

Holes up to about 1 inch in diameter are made using a *twist drill* (see Figure 1.1) in a drill press. Large holes are made by enlarging a drilled hole using a *boring bar* in a lathe or

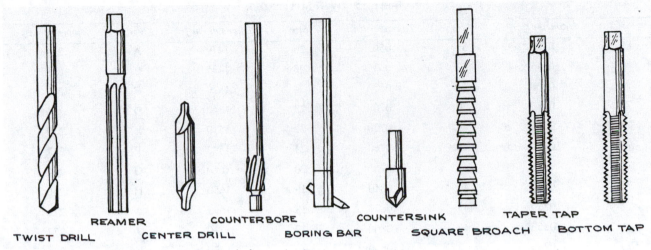

TWIST DRILL   REAMER   CENTER DRILL   COUNTERBORE   BORING BAR   COUNTERSINK   SQUARE BROACH   TAPER TAP   BOTTOM TAP

Figure 1.1   Tools for making and shaping holes.

vertical milling machine. Of course, a hole can be drilled with a twist drill in a handheld drill motor. This method, although convenient, is not very accurate and should only be employed when it is not possible to mount the work on the drill press table.

Twist drills are available in fractional inch sizes and metric sizes as well as in a numbered series of sizes at intervals of only a few thousandths of an inch. Number drill sizes are given in Table 1.2. The included angle at the point of a drill is 118°. A designer should always choose a hole size that can be drilled with a standard-size drill, and the shape of the bottom of a blind hole should be taken to be that left by a standard drill unless another shape is absolutely necessary.

Sizes designated by common fractions are available in $\frac{1}{64}$ in. increments in diameters from $\frac{1}{64}$ to $1\frac{3}{4}$ in., in $\frac{1}{32}$ in. increments in diameters from $1\frac{3}{4}$ to $2\frac{1}{4}$ in., and in $\frac{1}{16}$ in. increments in diameters from $2\frac{1}{4}$ to $3\frac{1}{2}$ in. Metric sizes are available in 0.05 increments in diameters from 1.00 mm to 2.50 mm, in 0.10 increments from 2.50 mm to 10.00 mm, and in 0.50 mm increments from 10.00 mm to 17.50 mm.

If many holes of the same size are to be drilled, it may be worthwhile to alter the drill point to provide the best performance in the material that is being drilled. In very hard materials the included angle of the point should be increased to as much as 140°. For soft materials such as plastic or fiber it should be decreased to about 90°. Many shops maintain a set of drills with points specifically ground for drilling in brass. The included angle of such a drill is 118°, but the cutting edge is ground so that its face is parallel to the axis of the drill in order to prevent the drill from digging in.

A drilled hole can be located to within about 0.010 in. by scribing two intersecting lines and making a punch mark at the intersection. The indentation made by the punch holds the drill point in place until the cutting edges first engage the material to be drilled. With care, locational accuracy of 0.001 in. can be achieved in a milling machine or jig borer. Locational error is primarily a result of the drill's flexing as it first enters the material being drilled. This causes the point of the drill to wander off the center of rotation of the machine driving the drill. The problem is alleviated by starting the hole with a *center drill* (see Figure 1.1) that is short and stiff. Once the hole is started, drilling is completed with the chosen twist drill.

A drill tends to produce a hole that is out-of-round and oversize by as much as 0.005 in. A drill point also tends to deviate from a straight line as it moves through the material being drilled. This runout can amount to 0.005 in. for a $\frac{1}{4}$ in. drill making a 1-in.-deep hole—more for a smaller diameter drill. It is particularly difficult to make a round hole when drilling material that is so thin that the drill point breaks out on the underside before the shoulder

TABLE 1.2 TWIST-DRILL SIZES

Size	Diameter (in.)	Size	Diameter (in.)	Size	Diameter (in.)	Size	Diameter (in.)
80	.0135	53	.0595	26	.1470	A	.234
79	.0145	52	.0635	25	.1495	B	.238
78	.0160	51	.0670	24	.1520	C	.242
77	.0180	50	.0700	23	.1540	D	.246
76	.0200	49	.0730	22	.1570	E	.250
75	.0210	48	.0760	21	.1590	F	.257
74	.0225	47	.0785	20	.1610	G	.261
73	.0240	46	.0810	19	.1660	H	.266
72	.0250	45	.0820	18	.1695	I	.272
71	.0260	44	.0860	17	.1730	J	.277
70	.0280	43	.0890	16	.1770	K	.281
69	.0292	42	.0935	15	.1800	L	.290
68	.0310	41	.0960	14	.1820	M	.295
67	.0320	40	.0980	13	.1850	N	.302
66	.0330	39	.0995	12	.1890	O	.316
65	.0350	38	.1015	11	.1910	P	.323
64	.0360	37	.1040	10	.1935	Q	.332
63	.0370	36	.1065	9	.1960	R	.339
62	.0380	35	.1100	8	.1990	S	.348
61	.0390	34	.1110	7	.2010	T	.358
60	.0400	33	.1130	6	.2040	U	.368
59	.0410	32	.1160	5	.2055	V	.377
58	.0420	31	.1200	4	.2090	W	.386
57	.0430	30	.1285	3	.2130	X	.397
56	.0465	29	.1360	2	.2210	Y	.404
55	.0520	28	.1405	1	.2280	Z	.413
54	.0550	27	.1440				

Note: Sizes designated by common fractions are available in $\frac{1}{64}$ in. increments in diameters from $\frac{1}{64}$ to $1\frac{3}{4}$ in., in $\frac{1}{32}$ in. increments in diameters from $1\frac{3}{4}$ to $2\frac{1}{4}$ in., and in $\frac{1}{16}$ in. increments in diameters from $2\frac{1}{4}$ to $3\frac{1}{2}$ in.

enters the upper side. The problem is alleviated by clamping the work to a backup block of similar material. When roundness and diameter tolerances are important, it is good practice in general to drill a hole slightly undersize and finish up with the correct size drill.

Before drilling in a drill press, the location of the hole should be center punched and the work should be securely clamped to the drill-press table. The drill should enter perpendicular to the work surface. When drilling curved or canted surfaces, it is best to mill a flat perpendicular to the hole axis at the location of the hole.

The speed at which the drill turns is determined by the maximum allowable surface speed at the outer edge of the bit, as well as the rate at which the drill is fed into the work. Typical tool speeds are given in Table 1.3. While drilling, the bit should be cooled and lubricated by flooding with soluble cutting oil, kerosene, or other cutting fluid. Brass or aluminum can be drilled without cutting oil if necessary.

A drilled hole that must be round and straight to close tolerances is drilled slightly undersize and then reamed using a tool such as is shown in Figure 1.1. *Reamers* are available in $\frac{1}{64}$ in. increments for a $\frac{1}{8}$ to $\frac{3}{4}$ in. diameter and in $\frac{1}{16}$ in. increments for diameters up to 3 in. The diameter tolerance on a reamed hole can be 0.001 in. or better.

A drilled hole can be threaded with a *tap* (shown in Figure 1.1). Threads in larger holes may be cut on the lathe. The head of a bolt can be recessed by enlarging the entrance of the bolt hole with a *counterbore* (also shown in Figure 1.1).

A keyway slot can be added to a drilled hole, or a drilled hole can be made square or hexagonal, by shaping the hole with a *broach* (Figure 1.1). A broach is a cutting tool with a series of teeth of the desired shape, with each successive cutting edge a few thousandths of an inch larger

TABLE 1.3 TOOL SPEEDS

Material	Speed[a] (sfpm)		
	Drill	Lathe	Mill
Aluminum	200	300	400
Brass	200	150	200
Cast iron	100	50	50
carbon steel	80	100	60
Stainless steel	30	100	60
Copper	200	300	100
Plastics	100	200	200

[a] Surface feet per minute at periphery of tool for high-speed-steel tools. For carbide-tipped tools, speeds can be twice those given.

than the one preceding. The broach can be driven through the hole by hand-driven or hydraulic presses. In some broaching machines the tool is pulled through the work. A broach can, at some expense, be ground to a nonstandard shape. The expense is probably only justified if many holes are to be broached.

1.1.3 The Lathe

A lathe (Figure 1.2) is used to produce surfaces of revolution such as cylindrical or conical surfaces. The work to be turned is grasped by a *chuck* that is rotated by the driving mechanism within the lathe *headstock*. Long pieces are supported at the free end by a center mounted in the *tailstock*. A cutting tool held atop the lathe carriage is brought against the work as it turns. As shown in Figure 1.2, the *tool holder* is clamped to the *compound rest* mounted to a rotatable table atop the *crossfeed* that in turn rests on the *carriage*. The carriage can be moved parallel to the axis of rotation along slides or ways on the lathe bed. A cylindrical surface is produced by moving the carriage up or down the ways, as in the first cut illustrated in Figure 1.3. Driving the crossfeed produces a face perpendicular to the axis of rotation. A conical surface is produced by driving the tool with the compound-rest screw.

Most lathes have a *lead screw* along the side of the lathe bed. This screw is driven in synchronization with the rotating chuck by the motor drive of the lathe. A groove running the length of the lead screw can be engaged by a clutch in the carriage *apron* to provide power to drive either the carriage or the crossfeed in order to produce a long uniform cut. The threads of the lead screw can be engaged by a split nut in the apron to provide uniform motion of the carriage for cutting threads.

A variety of attachments are available for securing work to the spindle in the headstock. Most convenient is the three-jaw chuck. All three jaws are moved inward and outward by a single control so that a cylinder placed in the chuck is automatically centered to an accuracy of about 0.002 in. A four-jaw chuck with independently controlled jaws is used for grasping workpieces that are not cylindrical or for holding a cylindrical piece off center. Large irregular work can be bolted to a face plate that is attached to the lathe spindle. Small round pieces can be

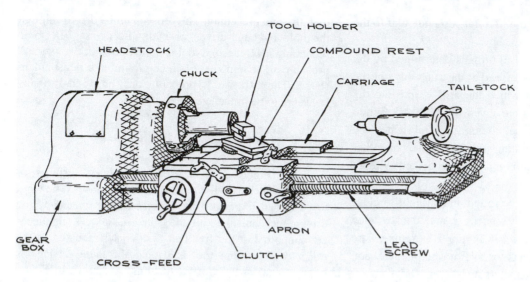

Figure 1.2   A lathe.

graspped in a *collet chuck*. A collet is a slotted tube with an inner diameter of the same size as the work and a slightly tapered outer surface. The work is clamped in the collet by a mechanism that draws the collet into a sleeve mounted in the lathe spindle.

The quality of work produced in a lathe is largely determined by the cutting tool. The efficiency of the tool bit used in a lathe depends upon the shape of the cutting edge and the placement of the tool with respect to the workpiece. A cutting tool must be shaped to provide a good compromise between sharpness and strength. The sharpness of the cutting edge is determined by the *rake angles* indicated in Figure 1.4. The indicated *relief angles* are required to prevent the noncutting edges and surfaces of the tool from interfering with the work. Placement of the tool in relation to the workpiece is illustrated in Figure 1.5. Cutting speeds for high-speed-steel tools are given in Table 1.3.

In the past, a machinist was obliged to grind tool steel stock to the required shape to make a tool bit. Virtually all work is now done with prepared tool bits. These may simply be ground-to-shape tool bits. Especially sharp, robust tools are available with sintered tungsten carbide (so-called *carbide*) or diamond tips. Many tools are made with replaceable carbide inserts. The insert that is clamped to the end of the tool is triangular or square to provide

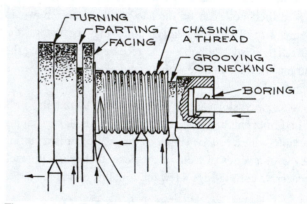

Figure 1.3   Cuts made on a lathe.

three or four cutting edges by rotating the insert in its holder. Examples are illustrated in Figure 1.6.

As in drilling, the cutting speed for turning in a lathe depends upon the material being machined. Cutting speeds for high-speed-steel tools are given in Table 1.3. Modern carbide- and ceramic-tipped cutters are much faster than tool-steel bits; they also produce a cleaner, more precise cut. Typically, a cut should be 0.003 to 0.010 in. deep, although much deeper cuts are permissible for rough work if the lathe and workpiece can withstand the stress.

Holes in the center of a workpiece may be drilled by placing a twist drill in the tailstock and driving the drill into the rotating work with the handwheel drive of the

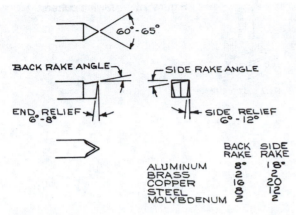

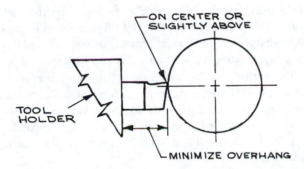

Figure 1.4   Tool angles for a right-cutting round nose tool. A right-cutting tool has its cutting edge on the right when viewed from the point end.

Figure 1.5   Placement of tool with respect to the work in a lathe.

tailstock. The hole should first be located with a *center drill* (Figure 1.1), or the drill point will wander off center.

Tolerances of 0.005 in. can be maintained with ease when machining parts in a lathe. Diameters accurate to within ±0.0005 in. can be obtained by a skilled operator at the expense of considerable time. Any modern lathe will maintain a straightness tolerance of 0.005 in./ft provided the workpiece is stiff enough not to spring away from the cutting tool.

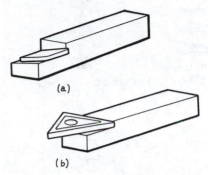

Figure 1.6   (a) Single-point carbide tip tool bit; (b) lathe tool with replaceable carbide insert (the insert has three cutting points).

1.1.4 Milling Machines

Milling, as a machine-tool operation, is the converse of lathe turning. In milling, the workpiece is brought into contact with a rotating cutter. Typical milling cutters are illustrated in Figure 1.7. A *plain milling cutter* has teeth only on the periphery and is used for milling flat surfaces. A *side milling cutter* has cutting edges on the periphery and either one or both ends so that it can be used to mill a channel or groove. An *end mill* is rotated about its long axis and has cutters on both the end and sides. There are also a number of specially shaped cutters for milling dovetail slots, T-slots, and Woodruff key slots. A *fly cutter* is another useful milling tool. It consists of a cylinder with a single, movable cutting edge and is used for cutting round holes and milling large, flat surfaces. Radial saws are also used in milling machines for cutting narrow grooves and for parting-off. Ordinary milling cutters are made of tough, hard steel known as high-speed steel. Other alloys are used as well, frequently with coatings of titanium nitride (TiN) or titanium carbonitride (TiCN) for a harder surface and improved lubricity. Both carbide-tipped cutters and tools with replaceable carbide inserts are available. Solid-carbide end milling cutters can be purchased for about twice the cost of high-speed-steel milling cutters.

There are two basic types of milling machine. The *plain miller* has a horizontal shaft, or *arbor*, on which a cutter is mounted. The work is attached to a movable bed below the

PLAIN MILLING
CUTTER

SIDE MILLING
CUTTER

END MILL

FLY CUTTER

UP MILLING

CLIMB MILLING

Figure 1.7   Milling cutters.

cutter. The plain miller is typically used to produce a flat surface or a groove or channel. It is not used much in the fabrication of instrument components. The *vertical mill* has a vertical spindle located over the bed. Milling cutters can be mounted to an arbor in the spindle of the vertical mill, or the arbor can be replaced by a collet chuck for grasping an end mill. Motion of the mill bed in three dimensions is controlled by handwheels. Big machines may have power-driven beds. An essential accessory for a vertical mill is a rotating table so that the workpiece can be rotated under the cutter for cutting circular grooves and for milling a radius at the intersection of two surfaces.

The two possible cutting operations are illustrated in Figure 1.7. *Climb milling,* in which the cutting edge enters the work from above, has the advantage of producing a cleaner cut. Also, climb milling tends to hold the work flat and deposits chips behind the direction of the cut. There is however a danger of pulling the work into the cutter and damaging both the work and the tool. *Up milling* is preferred when the work cannot be securely mounted and when using older, less rigid machines. Cutter speeds are given in Table 1.3.

Dimensional accuracy of ±0.005 in. is easily achieved in a milling operation; flatness and squareness of much higher precision are easily maintained. Both the mill operator and the designer specifying a milled surface should be aware that milled parts tend to curl after they are unclamped from the mill bed. This problem is particularly acute with thin pieces of metal. It can be alleviated somewhat if cuts are taken alternately on one side and then the other, finishing up with a light cut on each side.

The vertical milling machine is the workhorse of the *model shop* where instruments are fabricated. A scientist contemplating the design of a new instrument should become familiar with the milling machine's capability and, if possible, gain at least rudimentary skill in its operation.

Two electronic innovations have significantly increased the utility and ease of operation of the milling machine: the electronic digital position readout and computer control of the motion of the mill arbor and the mill bed.

All machine tools suffer from backlash in mechanical controls. In a milling machine, the position of the mill bed is read off vernier scales on the handwheels that drive the screws that position the bed. In addition to the inconvenience caused by this sort of readout, the operator must realize that reversing the direction of rotation of the handwheel does not instantly reverse the direction of travel of the bed owing to inevitable clearances between the threads of the drive screw and the nut that it engages. This backlash must be accounted for in even the roughest work. The problem is entirely obviated by the fitment of electronic position sensors on the bed that read out in inches or millimeters on displays mounted to the machine.

This simple innovation significantly increases the speed of operation for a skilled operator while at the same time reducing the number of errors. These displays also invariably improve the quality of work of a relatively unskilled operator.

Even modest shops now use milling machines in which the bed and arbor are driven by electric motors under computer control—Computer Numerical Control (CNC) machines. Most modern machinists, working from mechanical drawings provided by the designer, can efficiently program these machines. The time required is frequently offset by time saved by the machine operating under computer control, so in many instances it is quite reasonable to use a CNC machine for one-off production. This is particularly true when a complex sequence of bed and arbor motions are required to turn out a part. Conversely, when a CNC machine is available, the designer can contemplate many more complex shapes in a design than would be economically feasible to produce with manually controlled machines. An additional advantage for the instrument designer is that complex parts can be turned out initially in an inexpensive material, such as polyethylene, to check shape and fit before the final part is machined in a more expensive material.

The ultimate application of computer-controlled machines is to operate them using programs that are produced by software that the designer uses to prepare the engineering drawings. This is the integration of Computer-Aided Drafting (CAD) with Computer-Aided Manufacturing (CAM) in what is called a CAD/CAM system. At present, however, this mode of fabrication is not usually practical for the scientist-designer. The time involved in learning to use sophisticated CAD software, as well as its cost, cannot be justified. In addition, few model shops have machines that can be operated by the output of high-level CAD programs.

1.1.5 Electrical Discharge Machining (EDM)

A spark between an electrode and a workpiece will remove material from the work as a consequence of highly-localized heating and various electron- and ion-impact phenomena. As improbable as it may seem, the process of spark erosion has been developed into an efficient and very precise method for machining virtually any material that conducts electricity. In electrical-discharge machining (EDM), the electrode and the workpiece are immersed in a dielectric oil or deionized water, a pulsed electrical potential is applied between the electrode and the work, and the two are brought into close proximity until sparking occurs. The dielectric fluid is continuously circulated to remove debris and to cool the work. The position of the workpiece is controlled by a computer servo that maintains the required gap and moves the workpiece into the electrode to obtain the desired cut.

There are two types of electrical discharge machines: the *plunge* or *die-sinking EDM* and the *wire EDM*. The plunge EDM is used to make a hole or a well. The electrode is the male counterpart to the female concavity produced in the work. The electrode is machined from graphite, tungsten, or copper. Making the electrode in the required shape is a significant portion of the entire cost of the process. The electrode for wire EDM is typically a vertically traveling copper or copper-alloy wire 0.002 to 0.012 inch in diameter. The wire is eroded as cutting progresses. It comes off a spool to be fed through the work and is discarded after a single pass. With two-dimensional horizontal (XY) control of the workpiece, a cutting action analogous to that of a bandsaw is obtained. XYZ-control permits the worktable to be tilted up to 20° so that conical surfaces can be generated.

In an EDM process, the cavity or kerf is always larger than the electrode. This *overcut* is highly predictable and may be as small as 0.001 in. for wire EDM. As a consequence, an accuracy of ±0.001 in. is routine and an accuracy of ±0.0001 in. is possible. Furthermore, EDM produces a very high-quality finish; a surface roughness less than 12 microinches (rms) is routinely obtained. Part of the reason for the great accuracy obtained in EDM is that there is no force applied to the work and therefore no possibility of the workpiece being deflected from the cutting tool. There is no work hardening and no residual stress in the material being cut. The efficacy of EDM is independent of the hardness of the material of the work. Tool steel, conductive ceramics such as graphite and carbide, and refractory metals such as tungsten are cut with the same ease as soft aluminum. A particular advantage is that the material can be heat treated before fabrication since very little heat is generated in the cutting operation.

A precision CNC electrical-discharge machine may cost in excess of $100,000. As a consequence, these machines are seldom found in the typical university model shop. The EDM machine, however, has become the workhorse of the tool-and-die industry. Owing to the cost of the machines and the fact that, once programmed, they can run virtually 24 hours a day, job shops are usually anxious to have outside work to keep the machines running full time and welcome scientists with one-off jobs.

1.1.6 Grinders

Grinders are used for the most accurate work and to produce the smoothest surface attainable in most machine shops. A grinding machine is similar to a plain milling machine except that a grinding wheel is mounted on the rotating arbor. In most machines the work is clamped magnetically to the table and the table is raised until the work touches the grinding wheel. The table is automatically moved back and forth under the wheel at a fairly rapid rate. As an accessory, many lathes incorporate a grinder that can be mounted to the compound rest in place of the tool holder so that cylindrical and conical surfaces can be finished by grinding.

For instrument applications it is possible to produce very flat, parallel surfaces even if a surface grinder is not accessible. Clamp a drive motor with a grinding wheel above a surface plate at a height such that the wheel just touches the work resting on the surface plate. Slide the work along the surface under the grinder until no point on the work is left to touch the wheel. Invert the workpiece, shim it up 0.001 in. with brass shim stock, and finish the opposite side. Repeat the process, adding shims each time the work is inverted.

Even the hardest steel can be ground, but grinding is seldom used to remove more than a few thousandths of an inch of material. For complicated pieces to be made of hardened bronze or steel, it is advantageous to soften the stock material by annealing, machining it slightly oversize with ordinary cutting tools, rehardening the material, and then grinding the critical surfaces.

A flatness tolerance of ±0.0001 in. can be maintained in grinding operations. The average variation of a ground surface should not exceed 50 rms; a surface roughness less than 10 rms is possible.

1.1.7 Tools for Working Sheet Metal

Most machine shops are equipped with the tools necessary for making panels, brackets, and rectangular and cylindrical boxes of sheet metal. The basic sheet-metal processes are illustrated in Figure 1.8.

Sheet metal is cut in a guillotine *shear*. Shears are designed for making long, straight cuts or for cutting out inside corners. A typical shear can make a cut several feet in length in sheet metal of up to $\frac{1}{16}$ inch in thickness.

A *sheet-metal brake* is used to bend sheet stock. A typical instrument-shop brake can accommodate sheet stock 2 to 4 ft. wide and up to $\frac{1}{16}$ in. thick. The minimum bend radius is equal to the thickness of the sheet metal. Dimensional tolerances of 0.03 in. can be maintained.

Sheet can be formed into a simple curved surface on a *sheet-metal roll*. The roll consists of three long parallel rollers, one above and two below. Sheet is passed between the upper roller and the two lower rollers. The upper roller is driven. The distance between the upper roller and the lower rollers is adjustable and determines the radius of the curve that is formed.

Holes can be punched in sheet metal. A sheet-metal punch consists of a *punch*, a *guide bushing*, and a *die*. The punch is the male part. The cross section of the punch determines the shape and size of the hole. The punch is a close fit into the die, so that sheet metal placed between the two is sheared by the edge of the punch as it is driven into the die. Round and square punches are available in standard sizes. A punch-and-die set to make a nonstandard hole can be fabricated, but the cost is justified only if a large number of identical holes are required.

Sheet metal can be embossed with a *stamp and die*, similar to a punch and die, except that the die is somewhat larger than the stamp. This way, the metal is formed into the die rather than sheared off at the edge.

Sheet metal can be formed into surfaces that are figures of revolution by *spinning*. The desired shape is first turned in hard wood in a lathe. A circular sheet-metal blank is then

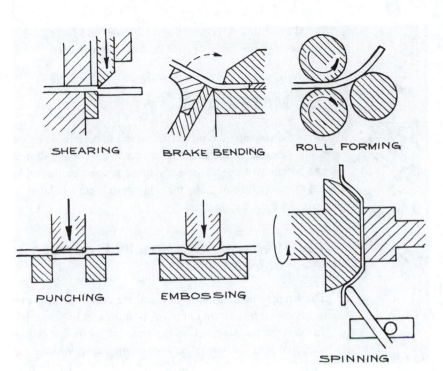

Figure 1.8   Sheet-metal shop processes.

SHEARING

BRAKE BENDING

ROLL FORMING

PUNCHING

EMBOSSING

SPINNING

clamped against the wooden form by a rubber-faced rotating center mounted in the lathe tailstock, As the wooden form is rotated, the sheet metal is gradually formed over the surface of the wood by pressing against the sheet with a blunt wooden or brass tool. Spinning requires few special tools and is economical for one-off production.

1.1.8 Casting

Sand casting is the most common process used for the production of a small number of cast parts. Although few instrument shops are equipped to do casting, most competent machinists can make the required wooden patterns that can then be sent to a foundry for casting in iron, brass, or aluminum alloy. There are both mechanical and economical advantages to producing some complicated parts by casting rather than by building them up from machined pieces. Very complicated shapes can be produced economically because the patternmaker works in wood rather than metal. Castings can be made very rigid by the inclusion of appropriate gussets and flanges that can only be produced and attached with difficulty in built-

up work. Most parts on an instrument require only a few accurately-located surfaces. In this case, the part can be cast and then only critical surfaces machined.

A designer must understand the sand-casting process in order to design parts that can be produced by this method. A sand-casting mold is usually made in two parts, as shown in Figure 1.9. The lower mold box, called the *drag,* is filled with sand, and the wooden pattern is pressed into the sand. The sand is leveled and dusted with dry parting sand. Then the upper box, or *cope,* is positioned over the lower and packed with sand. The two boxes are separated, the pattern is removed, and a filling hole (or *sprue*) is cut in the upper mold. A deep hole can be cast by placing a sand *core* in the impression (see Figure 1.9c). Note that the pattern has lugs, called *core prints,* that create a depression to support the core. The two halves of the mold are then clamped together and filled with molten metal.

It is obvious that a part to be cast must not include involuted surfaces above or below the parting plane, so the pattern can be withdrawn from the mold without damaging the impression. It is also good practice to taper all protrusions that are perpendicular to the parting plane to

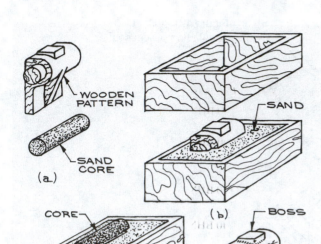

WOODEN PATTERN

SAND CORE

(a)

SAND

(b)

CORE

BOSS

(c)

CASTING

(d)

Figure 1.9 Sand casting: (a) the pattern and the core; (b) making an impression in the lower mold box; (c) inserting the sand core in the impression; (d) the finished casting.

facilitate withdrawal of the pattern from the mold. A taper or *draft* of 1 to 2° is adequate. Surfaces to be machined should be raised to allow for the removal of metal. That is the purpose of the *boss* shown in Figure 1.9d. When possible, all parts of the casting should be of the same thickness to avoid stresses that may build up as the molten metal solidifies. All corners and edges should be radiused. With care, a tolerance of 0.03 in. can be maintained.

1.2 MATERIALS

Before discussing the properties of the materials available to the designer of scientific apparatus, we shall first define the parameters used to specify these properties. In Table 1.4 the properties of many useful materials are tabulated in terms of these parameters.

1.2.1 Parameters to Specify Properties of Materials

The strength and elasticity of metal are best understood in terms of a stress-strain curve, as is shown in Figure 1.10. *Stress* is the force applied to the material per unit of cross-sectional area. This may be a stretching, compressing, or shearing force. *Strain* is the deflection per unit length that occurs in response to a stress. Most metals deform in a similar way under compression or elongation, but their response to shear is different. When a stress is applied to a metal, the initial strain is *elastic* and the metal will return to its original dimensions if the stress is removed. Beyond a certain stress however, *plastic* strain occurs and the metal is permanently deformed.

TABLE 1.4 PROPERTIES OF RESEARCH MATERIALS

Material	Density (lb/in.^{3})	Yield Strength (10^3 psi)	Tensile Strength (10^3 psi)	Modulus of Elasticity (10^6 psi)	Shear Modulus (10^6 psi)	Hardness[a]	Coefficient of Thermal Expansion ($10^{-6}\,°C^{-1}$)	Comments
Ferrous Metals								
Cast gray iron (ASTM 30)	0.253	25	30	13	5.2	200 BHN	11	
1015 steel (hot finished)	0.284	27	50	29	11	100 BHN	15	Low carbon (0.15% C)
1030 steel (hot finished)	0.284	37	68	29	11	137 BHN	15	Medium carbon (0.30% C)
1050 steel (hot finished)	0.284	50	90	29	11	180 BHN	15	High carbon (0.50% C)
Type 304 stainless steel (annealed)	0.286	35	85	28	10	150 BHN	17	Austenitic

TABLE 1.4 PROPERTIES OF RESEARCH MATERIALS (CONTINUED)

Material	Density (lb/in.^{3})	Yield Strength (10^3 psi)	Tensile Strength (10^3 psi)	Modulus of Elasticity (10^6 psi)	Shear Modulus (10^6 psi)	Hardness[a]	Coefficient of Thermal Expansion (10^{-6} °C^{-1})	Comments
Type 304 stainless steel (cold worked)	0.286	75	110	28	10	240 BHN	17	Austenitic
Type 316 stainless steel (cold worked)	0.286	60	90	28	10	190 BHN	16	Austenitic
Nickel Alloys								
Monel 400 : 25°C	0.319	25	70	26		110 BHN	14	Slightly magnetic
500°C		22	45				16	
Monel 500 : 25°C	0.306	40–150	90–180	26		140 BHN	14	Nonmagnetic
500°C		90	95				16	
Inconel 600 : 25°C	0.304	36	90	31		120 BHN	12	Strong, resists high temperature oxidation
650°C		26	65	20			16	
Invar (0–100°C)	0.294	24	70	21		140 BHN	1.3	small temperature coefficient
Aluminum Alloys								
1100-0	0.100	5	13	10	4	23 BHN	23	99% Al
2024-T4	0.098	47	68	10.6	4	120 BHN	23	3.8% Cu, 1.2% Mg, 0.3% Mn
2024-T4 (200°C)	0.098	12	18					
6061-T6	0.100	40	45	10	4	95 BHN	23	0.15% Cu, 0.8% Mg, 0.4% Si
7075-T6	0.101	73	83	10.4	4	150 BHN	23	5.1% Zn, 2.1% Mg, 1.2% Cu
Copper Alloys								
Yellow brass (annealed)	0.306	14	46	15	6		20	65% Cu, 35% Zn
Yellow brass (cold worked)	0.306	60	74	15	6	150 BHN	20	65% Cu, 35% Zn
Cartridge brass (1/2 hard)	0.308	52	70	16	6	145 BHN	20	70% Cu, 30% Zn
Beryllium copper (precipitation hardened)	0.297	140	175	19	7	380 BHN	17	98% Cu, 2% Be
Unalloyed Metals								
Copper	0.323	10	32	17	6.5	44 BHN	17	
Molybdenum	0.369	82	95	47	17	190 VHN	5.4	Refractory, nonmagnetic
Tantalum	0.600	48	67	27	10	80 VHN	6.5	Refractory, somewhat ductile
Tungsten	0.697	220	220	59	22	350 VHN	5	Refractory, very dense

TABLE 1.4 PROPERTIES OF RESEARCH MATERIALS (CONTINUED)

Material	Density (lb/in.^{3})	Yield Strength (10^3 psi)	Tensile Strength (10^3 psi)	Modulus of Elasticity (10^6 psi)	Shear Modulus (10^6 psi)	Hardness[a]	Coefficient of Thermal Expansion (10^{-6} °C^{-1})	Comments
Plastics								
Phenolics	0.049		7.5	1		125 R$_M$	81	Bakelite, Formica
Polyethylene (low density)	0.033		2	0.025		10 R$_R$	180	
Polyethylene (high density)	0.034		4	0.12		40 R$_R$	216	
Polyamide	0.040	11.8		0.410		118 R$_R$	90	Nylon
Polymethylmethacrylate	0.043		8	0.42		90 R$_M$	72	Lucite, Plexiglas
Polytetrafluoroethylene	0.077		2.5	0.060		60 R$_R$	99	Teflon
Polychlorotrifluoroethylene	0.076		6	0.25		110 R$_R$	70	Kel-F
Poly(amide-imide)	0.052		22	0.84		80 R$_R$	28	Torlon, < −200°C to 220°C
Polyimide	0.052		12	0.37		45 R$_R$	54	Vespel, −270°C to 290°C
Polycarbonate	0.043		9	0.34		70 R$_M$	70	Lexan
Ceramics								
Alumina (polycrystalline)	0.141		35	48		9 Mohs	8	99% alumina
Macor	0.091		4	9.3	3.6	400 VHN	13	Corning machinable ceramic
Steatite	0.13		9.3	15			10.6	
Wood								
Douglas fir (air dried)	0.011		9.5, 0.4[b]	1.4			6, 35[b]	Typical of softwoods
Oregon white oak (air dried)	0.026		10.2, 0.8[b]	2.3			5, 55[b]	Typical of hardwoods

[a] BHN = Brinell hardness number; R = Rockwell hardness; VHN = Vickers hardness.
[b] Parallel to fiber, across fiber.

The *tensile strength* or ultimate strength of a metal is the stress applied at the maximum of the stress-strain curve. The metal is very much deformed at this point, so in most cases it is impractical to work at such a high load.

A more important parameter for design work is the *yield strength*. This is the stress required to produce a stated, small plastic strain in the metal—usually 0.2 percent permanent deformation. For some materials the *elastic limit* is specified. This is the maximum stress that the material can withstand without permanent deformation.

The slope of the straight-line portion of the stress-strain curve is a measure of the stiffness of the material; this is the *modulus of elasticity (E)*, sometimes referred to as *Young's modulus*. It is valuable to note that E is about the same for all grades of steel (about 30×10^6 psi [pounds per square inch], [1 psi = 6900 N m^{-2}]) and about the same for all aluminum alloys (about 10×10^6 psi), regardless of the strength or hardness of the alloy. The effect of elastic deformation is specified by *Poisson's ratio (μ)*, the ratio of transverse contraction per unit dimension of a bar of unit

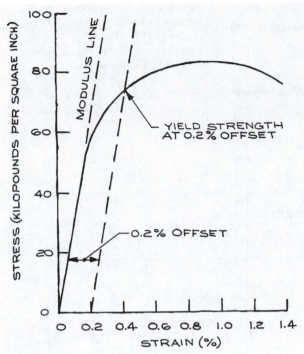

Figure 1.10   Stress-strain curve for a metal.

cross section to its elongation per unit length when subjected to a tensile stress. For most metals, $\mu = 0.3$.

The *hardness* of a material is a measure of its resistance to indentation and is usually determined from the force required to drive a standard indenter into the surface of the material or from the depth of penetration of an indenter under a standardized force. The common hardness scales used for metals are the Brinell hardness number (BHN), the Vickers hardness number (VHN), and the Rockwell C scale. Type 304 stainless steel is about 150 BHN, 160 Vickers, or 0 Rockwell C. A file is 600 BHN, 650 Vickers, or 60 Rockwell C. The Shore hardness scale is determined by the height of rebound of a steel ball dropped from a specified distance above the surface of a material under test. The Mohs hardness scale is used by mineralogists and ceramic engineers. It is based upon standard materials, each of which will scratch all materials below it on the scale. Figure 1.11 shows the approximate relation of the various hardness scales to one another.

A designer must consider the machinability of a material before specifying its use in the fabrication of a part that must be lathe-turned or milled. In general, the harder and stronger a material the more difficult it is to machine. On the other hand, soft metals such as copper and some nearly pure aluminums are also difficult to machine because the metal tends to adhere to the cutting tool and produce a ragged cut. Some metals are alloyed with other elements to improve their machinability. Free-machining steels and brass contain a small percentage of lead or sulfur. These additives do not usually affect the mechanical properties of the metal, but since they have a relatively high vapor pressure, their outgassing at high temperature can pose a problem in some applications.

1.2.2 Heat Treating and Cold Working

The properties of many metals and metal alloys can be considerably changed by heat treating or cold working to modify the chemical or mechanical nature of the granular structure of the metal.[1] The effect of heat treatment depends upon the temperature to which the material is heated relative to the temperatures at which phase transitions occur in the metal, and the rate at which the metal is cooled. If the metal is heated above a transition temperature and quickly cooled or *quenched*, the chemical and physical structure of the high temperature phase may be frozen in, or a transition to a new metastable phase may occur. Quenching is accomplished by plunging the heated part into water or oil. This process is usually carried out to harden the metal. Hardened metals can be softened by *annealing,* where the metal is heated above the transition temperature and then slowly cooled. It is frequently desirable to anneal hardened metals before machining and reharden after working, although hardening and annealing can result in distortion owing to differences in density of the high and low temperature phases. *Tempering* is an intermediate heat treatment where previously hardened metal is reheated to a temperature below the transition point in order to relieve stresses and then cooled at a rate that preserves the desired properties of the hardened material.

Repeated plastic deformation will reduce the size of the grains or crystallites within the metal. This is cold working, which accompanies bending, rolling, drawing,

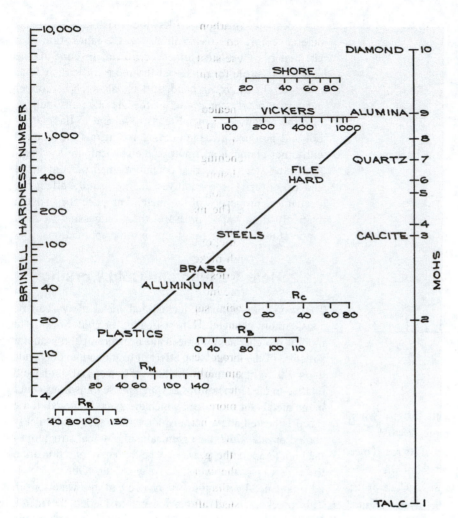

Figure 1.11 Approximate relation of Brinell (BHN), Rockwell (RR, RM, RB, RC), Vickers (VHN), Shore, and Mohs hardness scales.

hammering, and, to a lesser extent, cutting operations. Not all metals benefit from cold working, but for some the strength is greatly increased. Because of the annealing effect, the strength and hardness derived from cold working begin to disappear as a metal is heated. This occurs above 250°C for steel and above 125°C for aluminum. In rolling or spinning operations, some metals will work-harden to such an extent that the material must be periodically annealed during fabrication to retain its workability. Cold working reduces the toughness of metal. The surface of a metal part can be work-hardened, without modifying the internal structure, by peening, shot blasting, and by some rolling operations. The strength of metal

stock may depend upon the method of manufacture. For example, sheet metal and metal wire are usually much stronger than the bulk metal because of the work hardening produced by rolling or drawing.

The surface of a metal part may be chemically modified and then heat treated to increase its hardness while retaining the toughness of the bulk of the material beneath. This process is called *case hardening*. The many methods of case hardening are usually named for the chemical that is added to the surface. *Carburizing, cyaniding,* and *nitriding* are common processes for case hardening steel. Carburizing of low-carbon steel is most conveniently performed in the lab or shop. The part to be hardened is packed in bone charcoal or Kasenit in an open metal box and heated in a furnace to 900°C (salmon-red heat). The time in the furnace, from 15 minutes to several hours, determines the depth of the case. Steel will absorb carbon to a maximum depth of 0.015 in. The part is removed from the furnace and promptly quenched in water. The quenching hardens the case, but the core remains tough and ductile.

1.2.3 Iron and Steel

For instrument work, iron is used primarily as a casting material. The advantages of cast iron are hardness and a high degree of internal damping. Cast iron is harder than most steels and is used for sliding parts where resistance to wear is important. The damping capacity of iron, that is, the ability to dissipate the energy of vibration, is about ten times that of steel and for this reason the frame of a delicate instrument is often made of cast iron.

Cast iron contains a small percentage of carbon. The carbon is in a free state in gray cast iron but chemically incorporated into the structure of white cast iron. Gray iron is inexpensive and easy to machine. Class 30 gray iron has a Brinell hardness of about 200 BHN. It is, however, rather brittle and is only about half as strong as steel. White iron is stronger and harder than gray iron, and thus very wear resistant. It is very difficult to machine and is usually shaped by grinding. The machinability and ductility of cast iron can be considerably improved by various heat treatments.

Steel is an iron alloy. There are more than 10,000 different types of steel. Steels are classed as cast or wrought steels depending on the method of manufacture. *Wrought steels* are produced by rolling and are almost the only kind used for one-off machine work. Steels are also classified as either carbon steels or alloy steels. *Carbon steels* are specified by a four-digit number. Plain carbon steels are specified as 10XX, where XX is a two-digit number that indicates the carbon content in hundredths of a percent. In general, the strength and hardness of carbon steel increases with carbon content, but these properties also depend upon the nature of the heat treatment and cold working that the metal has received.

Steels with a carbon content in the 0.10 to 0.50 percent range are referred to as *low* and *medium-carbon steels* or mild steels. These steels are supplied rolled or drawn and are most suitable for machine work.

High-carbon steels are difficult to machine and weld. If they are first annealed by heating to 750°C (orange heat) and cooling slowly in air, their machinability is improved. They may be returned to their hard state by reheating to 750°C and quenching in water. This process inevitably produces some distortion.

The most common *alloy steels* are the types known as *stainless steels*. The main constituents of stainless steels are iron, chromium, and nickel. They are classified as ferritic, martensitic, or austenitic, depending on the nickel content. The high-nickel or austenitic alloys comprise the 200 and 300 series of stainless steels. The 300-series stainless steels are most useful for instrument construction because of their superior toughness, ductility, and corrosion resistance. These austenitic stainless steels are sensitive to heat treatment and cold working, although the heat-treatment processes are more complicated than those required for plain carbon steels. For example, quenching from 1000°C leaves these steels soft. Cold working, however, can more than double the strength of the annealed alloy. The heat associated with the welding of austenitic stainless steels can result in carbide precipitation at the grain boundaries in the vicinity of a weld, leaving the weld subject to attack by corrosive agents. The carbides can be returned to solution by annealing the steel after welding. There are also some special stainless steels that do not suffer from carbide precipitation. In general, stainless steels are more expensive than plain carbon steels, but their superior qualities often offset the extra cost.

Type 303 is the most machinable of the 300-series alloys. Type 316 has the greatest resistance to heat and corrosion. All of these alloys are fairly nonmagnetic—types 304 and 316 are the least magnetic. Machining and cold working tends to increase the magnetism of these alloys. Residual magnetism can be relieved by heating to 1100°C (safely above the temperature range (450–900°) where carbide precipitation occurs) and quenching in water. Thin pieces that might distort excessively may be quenched in an air blast.

1.2.4 Nickel Alloys

Nickel alloys are typically tough and strong and resistant to corrosion and oxidation. The machining characteristics of the wrought alloys are similar to those of steel, and they are readily brazed and welded. Most nickel alloys do not go by a standard specification like steel, but instead are referred to by trade names. These names often allude to an alloy or class of alloys that have been designed to optimize a particular property. Nickel alloys are expensive compared to steel alloys, but not so much so that they shouldn't be used when an instrument can exploit their special properties.

The Monel alloys are nickel-copper alloys whose properties make them a better choice than austenitic stainless steels in some applications. The most familiar is Monel 400. It is resistant to attack by both acids and bases and is sufficiently ductile for parts to be fabricated by spinning and stamping. Monel 404 is similar, but it is quite nonmagnetic, and its magnetic properties are not affected by cold working. Monel 500 is similarly nonmagnetic and retains most of its strength up to 500°C.

Inconel alloys are notable for their strength at high temperatures. The Inconel 600 series of alloys retain their strength to 700°C and are very resistant to oxidation at high temperatures. Inconel 600 is nonmagnetic. The Inconel 700 series are often called *superalloys* with tensile strengths on the order of 150,000 psi. They are used in highly stressed parts that will maintain their strength to 1000°C. They must be heat treated to obtain the maximum high-temperature strength. The Inconel 700 alloys, especially Inconel 702 and Inconel X-750, are nonmagnetic.

The Hastelloy alloys, of which Hastelloy B is most common, contain primarily nickel and molybdenum and are nearly as strong as the Inconel 700 alloys and somewhat more ductile. These alloys are notable for their resistance to corrosion.

In certain instances, the dimensions of a component of an instrument must be very stable in the face of variations of temperature. Iron-nickel alloys containing about 36 percent nickel have been found to meet this requirement. Invar is one such alloy. This material has a coefficient of expansion of 1.3×10^{-6} K^{-1} in the range 0 to 100°C and about twice this at temperatures up to 200°C. Better yet, the Super-Invar alloy (from Burleigh Instruments) has a coefficient of 0.3×10^{-6} K^{-1} in the range 0 to 100°C. By comparison, the coefficient of expansion of stainless steel is about 1.5×10^{-5} K^{-1}. Following machining, Invar may require stress relieving to achieve maximum stability. Rosebury[2] suggests the following heat treatment:

1. Heat to about 350°C for 1 hour and allow to cool in air.

2. Heat to a temperature slightly above the operating temperature and cool slowly to a temperature below the operating temperature.

3. Repeat Step 2.

One disadvantage of Invar in certain applications is that it is magnetic below 277°C.

Kovar is an iron-nickel-cobalt alloy designed for making metal-to-glass seals. Its thermal expansion characteristics match that of several hard glasses, including Corning 7052 and 7056. Unfortunately, Kovar is magnetic and easily oxidized.

1.2.5 Copper and Copper Alloys

Copper is a soft, malleable metal. Of all metals (short of the noble metals) it is the best electrical and thermal conductor. For instrument applications, oxygen-free high-conductivity (OFHC) copper is preferred because of its high purity and excellent conductivity and because it is not subject to hydrogen embrittlement. Ordinary copper becomes brittle and porous when exposed to hydrogen at elevated temperatures such as may occur in welding or brazing.

Copper alloys are classified as either brasses or bronzes. *Brass* is basically an alloy of copper and zinc. Generally, brass is much easier to machine than steel but not quite as strong, and its electrical and thermal coefficients are about half those of pure copper. Its properties vary considerably with the proportions of copper to zinc, and also with the addition of small amounts of other elements. Brass parts are readily joined by soldering or brazing. Brass is expensive, but because it is easy to work, it is well suited to instrument construction where the cost of fabrication far outweighs the cost of materials.

For general machine work, yellow brass (35% Zn) is most suitable. Free-machining brass is similar except that it contains 0.5 to 3 percent lead. Cartridge brass (30% Zn) is very ductile and is suitable for the manufacture of parts by drawing and spinning and for rivets. It work hardens during cold forming and may require periodic annealing to 600°C. Stresses that have built up in a cold-formed part during manufacture can be relieved by heating to 250°C for an hour.

The name *bronze* originally meant an alloy of copper and tin, but the term now also refers to alloys of copper and many other elements, such as aluminum, silicon, beryllium, or phosphorus. Bronzes, generally harder and stronger than brasses, often are of special value to the designer of scientific apparatus. Beryllium-copper (or beryllium- bronze) is a versatile material whose properties can be considerably modified by heat treatment. In its soft form it has excellent formability and resistance to fatigue failure. Careful heat treating will produce a material that is harder and stronger than most steels. It is useful for the manufacture of springs and parts that must be corrosion-resistant. Also, it is nearly impossible to strike a spark on beryllium bronze. This makes it useful for parts destined to be used in an explosive atmosphere. Beryllium bronze hand tools are available for use around explosive materials. Phosphor bronze cannot be heat treated, but it is very formable and has great resistance to fatigue. It is used for bellows and springs. Aluminum bronze is highly resistant to corrosion and is strong and tough. It is used in marine applications. Bronze stock can be manufactured by sintering. Sintered bronze is produced by compacting and heating a bronze powder at a temperature below the melting point. It is quite porous in this form and can be used as a filter or can be impregnated with lubricant for use as a bearing.

1.2.6 Aluminum Alloys

Aluminum alloys are valued for their light weight, good electrical and thermal properties, and excellent machinability. Aluminum is about one third as dense as steel. Its electrical conductivity is 60 percent that of copper, but it is a better conductor per unit weight.

Aluminum is resistant to most corrosive agents except strong alkalis, owing to an oxide film that forms over the surface. This layer is hard and tenacious and forms nearly instantaneously on a freshly exposed surface. Because the oxide layer is an insulator, electrical connections to aluminum parts are problematic. In some applications aluminum parts can accumulate a surface charge. This problem can be overcome by copper- or gold-plating critical surfaces. The oxide layer makes plating difficult, but reliable plating can be carried out by specialty plating shops. Aluminum parts cannot be soldered or brazed, but they can be joined by heliarc welding. Welds in aluminum tend to be porous, and welding weakens the metal in the vicinity of the weld because of the rapid conduction of heat into the surrounding metal during the operation.

Aluminum stock is produced in cast or wrought form. Wrought aluminum, produced by rolling, is used for most machine work. Aluminum alloys are specified by a four-digit number that indicates the composition of the alloy, followed by a suffix beginning with a T or H, that specifies the state of heat treatment or work hardness. The heat-treatment scale extends from T0 through T10. T0 indicates that the material is dead soft, and T10 that it is fully tempered. 1100-T0 is a soft, formable alloy that is nearly pure aluminum. It is used for extrusions. A material designated 2024-T4 or -T6 is as strong as annealed mild steel and is used for general machine work. Aluminum specified 6061-T6 is not quite as strong as 2024-T4, but is more easily cold worked. Its machinability is excellent. The strongest readily available aluminum alloy is 7075-T6. It is easily machined but its corrosion resistance is inferior to that of other alloys.

Aluminum is available as plate, bar, and round and square tube stock. Aluminum jig plate is useful for instrument construction. This is rolled plate that has been fly cut by the manufacturer to a high degree of flatness.

Aluminum may be anodized to give it a pleasing appearance and to improve its corrosion resistance. A number of proprietary formulas, such as Birchwood-Casey Aluma Black, are available to blacken aluminum to reduce its reflectivity.

1.2.7 Other Metals

Magnesium alloys are useful because their density is only 0.23 times that of steel. Magnesium is very stiff per unit weight and is completely nonmagnetic. It is very easy to machine, although there is some fire hazard. Magnesium chips can be ignited, but not the bulk material. The strength of magnesium alloys is considerably reduced above 100°C. The cost of these alloys is at least 10 times that of steel.

Titanium is about half as dense as steel, while titanium alloys are stronger than steel. Titanium is very resistant to corrosion and completely nonmagnetic. It is readily machined, if one proceeds slowly with sharp, carbide-tipped cutters, and can be welded under an inert gas atmosphere. Ninety-nine percent titanium and titanium alloys are useful for extreme-temperature service, since they maintain their mechanical properties from −250°C to well above 400°C. A titanium-manganese alloy (Ti-8Mn; 6.5–9% Mn) retains a tensile strength greater than 100,000 psi to temperatures in excess of 400°C. Its strength and hardness (up to 36 R_c) can be varied considerably by heat treatment. Titanium-aluminum-vanadium alloy (Ti-6Al-4V; 5.5–6.7% Al; 3.5–4.5% V) is a similarly strong versatile material. These alloys are more easily machined than unalloyed titanium. Titanium fasteners are available.

Molybdenum is a refractory metal. It exhibits a very low, uniform, surface potential and is used for fabricating electrodes and electron-optical elements. Molybdenum is brittle and must be machined slowly using very sharp tools.

Tungsten is the densest of the readily available elements. Its density is 2.5 times that of steel. Compact counterweights in scientific mechanisms are often made of this material. Tungsten is also used for extreme high-temperature service, such as torch nozzles and plasma electrodes.

1.2.8 Plastics

Plastics are classified as either thermoplastic or thermosetting. Within limits, *thermoplastics* can be softened by heating and will return to their original state upon cooling. *Thermosetting plastics* undergo a chemical change when heated during manufacture, and they cannot be resoftened.

Phenol-formaldehyde plastics, or phenolics, are the most widely used thermosetting plastics. Phenolics are hard, light in weight, and resistant to heat and chemical attack. They make excellent electrical insulators. Phenolics are usually molded to shape in manufacture and typically incorporate a paper or cloth filler. Reinforced in this way, the phenolic plastics are easily sawn and drilled. Bakelite is a linen-reinforced phenolic plastic.

Representative thermoplastics include polyethylene, nylon, Delrin, Plexiglas, Teflon, and Kel-F. Polyethylene is inexpensive, machinable, and very resistant to attack by most chemicals. It is soft and not very strong. Its physical properties vary with the molecular weight and the extent of chain branching within the polymer. It is produced in high-density and low-density forms that are whitish and translucent. It is also manufactured in a blackened form that is not degraded by ultraviolet radiation as is the white form. Polyethylene softens at 90°C and melts near 110°C. It can be welded with a hot wire or a stream of hot air from a heat gun. It can be cast, although it shrinks a great deal upon solidifying. Care must be exercised when heating polyethylene because it will burn. Polyethylene is an excellent electrical insulator, with a dielectric strength of 1000 volts per mil (0.001 in.).

Nylon, a polyamide, is strong, tough, and very resistant to fatigue failure. It retains its mechanical strength up to about 120°C. It does not cold flow, and is self-lubricating and thus useful for all types of bearing surfaces. Nylon is available as rod and sheet in both a white and black form. Nylon balls and gears are also readily available. There is no fire hazard with nylon, since it is self-extinguishing. It is hygroscopic and its volume increases and strength decreases somewhat when moisture is absorbed.

Delrin is a polyacetal resin with properties similar to hard grades of nylon. Delrin is not hygroscopic and therefore to be recommended in preference to nylon for use in vacuum systems.

All plastics are electrical insulators, although the hygroscopic ones are liable to surface leakage currents. DuPont produces a polyimide film known as Kapton that is widely used as an electrical insulator. This material has a high dielectric strength in thin films. It is strong and tough, resists chemicals and radiation, and is a better thermal conductor than most plastics. It is used in sheets as an insulator between layers of windings in magnets. Kapton-insulated wire is also available. DuPont supplies bulk polyimide in standard shapes under the tradename Vespel. This material is relatively hard, strong and machinable. It is probably the best plastic material for use in a vacuum because of its low vapor pressure and its high-temperature stability. Amoco produces a line of engineering plastics under the name Torlon that are based upon poly(amide-imide) copolymer. These thermoplastics are very strong and can be used up to 250°C. Torlon is also an excellent choice when plastics must be used in a vacuum.

Plexiglas, Lucite, and Perspex are all trade names for polymethylmethacrylate. These materials, also known as acrylic plastics, are strong, hard, and transparent. Because they are machinable, heat formable, and shatter resistant, polymethylmethacrylates are used in many applications as replacements for glass.

Thermoforming of acrylic sheet into curved surfaces is a fairly simple operation. Begin with a wooden or metal pattern of the desired shape. Rest the sheet on top of the pattern, warm the plastic to about 130°C, and permit gravity to pull the softened sheet down over the form. The forming may be done in an oven, or the plastic may be heated with infrared lamps or with hot air from a heat gun. It is necessary to apply the heat slowly and uniformly.

Polymethylmethacrylate dissolves in a number of organic solvents. Parts can be cemented by soaking the edges to be joined in acetone or methylene chloride for about a minute and then clamping the softened edges together until the solvent has evaporated.

General Electric manufactures a polycarbonate plastic known as Lexan. It is transparent and machinable; it has excellent vacuum properties, and is so tough that it is nearly unbreakable. Lexan is one of the most useful plastics for instrument construction.

Teflon is a fluorocarbon polymer. It is only slightly stronger and harder than polyethylene, but it is useful at sustained temperatures as high as 250°C and as low as 200°C. Teflon is resistant to all chemicals, and nothing will stick to it. It is not as good as nylon as a bearing surface in most applications since it tends to cold flow away from the area where pressure is applied. A thin film of Teflon, such as is found on modern cookware, provides dry lubrication and prevents dust and moisture adhesion. These films can be applied by specialty shops that serve the food industry. Teflon is about ten times as expensive as nylon.

Kel-F is a fluorochlorocarbon polymer. Its chemical resistance is similar to Teflon, but it is much stronger and harder. Like Teflon, Kel-F does not absorb water.

A number of rubberlike polymeric materials, known as elastomers, find application in scientific apparatus.

Materials of interest include Buna-N, Viton-A, and RTV. Buna-N is a synthetic rubber. It can be used at sustained temperatures up to 80°C without suffering permanent deformation under compression. It is available in sheet or block form and is useful as a gasket material and for vibration isolation. Viton-A is a fluorocarbon elastomer similar in appearance and mechanical properties to Buna-N. Useful up to 250°C, it tends to take a set at higher temperatures. Unlike Buna-N, Viton-A shows no tendency to absorb water or cleaning solvents. RTV, a self-curing silicone rubber produced by General Electric, comes in a semiliquid form and cures to a rubber by reaction with moisture in the air. It is useful as a sealant, a potting compound for electronic apparatus, and an adhesive. It can be cast in a plastic mold.

1.2.9 Glasses and Ceramics

Borosilicate glasses such as Corning Pyrex 7740 or Kimble KG-33 are used extensively in the lab. These glasses are strong, hard, and chemically inert, and they retain these properties to 500°C. As will be discussed in the next chapter, they are conveniently worked with a natural gas-oxygen flame.

Quartz, or fused silica, has a number of unique properties. It has the smallest known coefficient of thermal expansion of any pure material. The coefficient for quartz is 1×10^{-6} K^{-1}. By comparison, the thermal expansion coefficient of the borosilicate glasses is 3×10^{-6} K^{-1}, and that of steel is 15×10^{-6} K^{-1}. Quartz can be drawn into long, thin fibers. The internal damping in these fibers is very low, and their stress-strain relation for twisting is linear to the breaking point. Springs of quartz fibers provide the best possible realization of Hooke's law. Torsion springs and coil springs made of a quartz fiber are used in delicate laboratory balances. Quartz retains its excellent mechanical properties up to 800°C. It softens at 1500°C and therefore can only be worked with a hydrogen-oxygen flame.

Alumina (Al_2O_3) is the most durable and heat-resistant of the readily available ceramics. Only diamond and a few exotic metal carbides are harder. The compressive strength of alumina exceeds 300,000 psi, although its tensile strength and shear strength are somewhat less than those of steel. Alumina is brittle, and because of its hardness it is ordinarily shaped or cut by grinding with a diamond-grit wheel. Alumina rod, tube, thermocouple wells, and electrical insulators are available. These shapes are produced by molding a slurry of alumina powder and various binders to the desired shape and then firing at a high temperature to produce the hard ceramic. One particularly useful product is alumina rod that has been centerless-ground to a roundness tolerance of better than ±0.001 in. and a straightness of $\frac{1}{32}$ in. per ft. Polycrystalline as well as single-crystal (sapphire) balls that have been ground to a roundness tolerance of ±0.000025 in. are available. They are surprisingly inexpensive—balls of 0.125–0.500 in. diameter cost only a few dollars. These balls are used as electrical insulators and bearings, and in valves and flow meters. Alumina circuit board substrate is available as plate .010 in. to .060 in. thick and 6 in. square. Owing to the scale of the circuit board industry, the facilities for custom fabrication by laser cutting of alumina plate are widely available and inexpensive.

Corning produces a machinable glass-ceramic known as Macor, and Aremco produces machinable ceramics called Aremcolox. These materials are not nearly as strong or hard as alumina, but can be machined to intricate shapes that find wide application in instrument work. They can be drilled, tapped, turned, and milled with conventional high-speed steel tools. They resist chipping and are insensitive to thermal shock. Machinable ceramics have a dielectric strength of 1000 to 2000 volts per mil (0.001 in.).

Aluminum silicate, or lava, is another useful, machinable ceramic. Blocks of grade-A lava are available in the unfired condition. This material is readily machined into intricate shapes that are subsequently fired at 1050°C in air for half an hour. The fired ceramic should be cooled at about 150 K/hour to prevent cracking. It grows about 2 percent upon firing.

The development of so-called technical ceramics has burgeoned over the past two decades. A remarkable range of pure oxide ceramics as well as composite materials are available with mechanical, electrical, magnetic, and thermal properties to fit all sorts of special applications. The ultra-low thermal expansion glass ceramic Zerodur, produced by Schott Glass, is an example that has found use in the construction of highly stable mechanical elements and mirrors. A glass ceramic consists of a crystalline phase dispersed in a glass phase. The quartz-like crystalline phase in Zerodur has a negative thermal coefficient, while the glassy phase has a positive coefficient. In the appropriate proportion, a material with a thermal expansion coefficient of the order of 10^{-8} K^{-1} is obtained.

1.3 JOINING MATERIALS

The parts of an apparatus may be permanently joined by soldering, brazing, welding, or riveting, or they may be joined by demountable fasteners such as screws, pins, or retaining rings. We shall discuss both types of joints in this next section.

1.3.1 Soldering

In soldering and brazing, two metal surfaces are joined by an alloy that is applied in the molten state to the joint. The melting point of the alloy is below that of the base metal— that is, the metal of the parts to be joined. The two methods differ in the composition of the fusible alloy and the degree to which it is necessary to heat the base metal. Solder is a tin alloy that melts at a relatively low temperature—in the 200 to 300°C range. Brazing was originally done with a brass alloy that melts at about 900°C. There are now many different brazing alloys with melting points from 400°C to more than 2000°C.

For routine soldering of copper or brass parts, 50-50 tin-lead solder (50% tin, 50% lead) has traditionally been the alloy of choice. Because of the toxic nature of lead, tin-lead solder is being replaced by 95-5 tin-antimony and 96-4 tin-silver solders that have very similar melting points and flow properties. These solders are used when great strength and cleanliness are not required and when it is most convenient to work at low temperatures. The shear strength of a soldered joint is about 5000 psi. This strength is considerably reduced at temperatures above 100°C.

The first step in soldering a joint is to clean the surfaces to be joined with fine emery paper or steel wool. The surfaces should then be covered with soldering flux. When heated, the flux cleanses the surfaces so they will be wetted by the molten solder. A sal ammoniac (NH_4Cl) solution is used as a flux for lead-tin solders. Other proprietary formulas are available for lead-free solders. The parts are then assembled and heated. A natural gas-air flame or propane-air flame is most convenient. The assembly may also be heated on an electric or gas hot plate. Flame heating should be done indirectly if possible. Play the flame on the metal near the joint, particularly on the heavier of the two pieces. Periodically test the temperature of the two pieces by touching each with the solder. When each piece is hot enough to melt the solder, remove the flame and apply solder to the joint. The solder should flow by capillary action into and through the joint. Sufficient solder should be applied to fill the joint and to produce a buildup of solder, called a *fillet* (Figure 1.12), at the juncture of the two parts. The fillet radius should be $\frac{1}{16}$ to $\frac{1}{8}$ in. A gap of 0.003 to 0.006 in. is desirable between surfaces to be soldered. In some cases it may be necessary to insert a piece of brass shim stock to maintain this clearance. The lap between soldered surfaces should be at least four times the thickness of the thinner part.

Steel and stainless steel, as well as brass and copper, can be soldered with tin-silver alloy solders. In fact, except for plumbing, the 96-4 tin-silver alloy is almost universally applicable in the laboratory for low-temperature soldering. It melts at 221°C. The alloy is also available as a wire with a core of rosin flux for use on electronic circuits. Another useful form of tin-silver solder is available as Staintin 157 PA from Eutectic Welding Alloys. This is a slurry of tin-silver alloy powder in a liquid flux. The slurry is painted on the surfaces to be joined and heated indirectly until the solder flows. The alloy melts at 200°C and can produce a joint with a tensile strength of 15,000 psi. It does not contain lead, zinc, antimony, cadmium, or other volatile metals, and thus is suitable for high-vacuum applications.

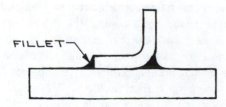

Figure 1.12   A soldered joint.

1.3.2 Brazing

There are hundreds of brazing alloys.[3] The choice of an alloy depends upon the metals to be joined and the application of the brazed part. For routine brazing of steels, borate-flux-coated brass brazing rod is available from hardware suppliers as well as specialized suppliers of brazing and welding materials. The tensile strength of joints made with brass brazing alloy is around 50,000 psi. Silver brazing alloys are preferred for most laboratory work. Silver alloys are lower-melting than brass alloys and can be used on brass as well as steel parts. The tensile strength of silver-brazed joints is in the 30,000 to 50,000 psi range, depending upon the alloy and the joint design.

In the laboratory, the alloy of approximately 60 percent silver, 30 percent copper, and 10 percent tin is generally applicable (Braze 603 from Handy and Harmon, Safety Silv 1115 from Harris, Silvaloy B-60T from Engelhard, or Cusiltin 10 from Wesgo). It melts at 600°C and flows at 720°C, and thus can be conveniently worked with the same gas-oxygen torch used for laboratory glassblowing. A fluoride flux is required. Parts made of copper, brass, steel, stainless steel, and even refractory metals can be joined with this brazing alloy. Silver-bearing brazing alloys containing cadmium or other volatile metals should be avoided, as cadmium vapor is toxic. Volatile brazing alloys may cause significant contamination in a vacuum system.

When volatile metals cannot be tolerated, the silver-copper eutectic (72% Ag, 28% Cu) makes a suitable brazing alloy for copper, stainless steel, or Kovar. EutecRod 1806 (Eutectic Welding Alloys) and Silvaloy 301 (American Platinum and Silver) are silver-copper eutectics. This alloy melts at 780°C. A lower-melting alloy with many of the same properties as the silver-copper eutectic can be made by adding indium to the silver-copper mixture. Incusil 10 (Western Gold and Platinum) is a silver-copper-indium mixture that melts at 730°C. Incusil 15 melts at 685°C.

Torch brazing is the simplest and most convenient brazing operation. An oxyacetylene torch is required for brass brazing alloys. A natural gas-oxygen flame can be used for lower-melting alloys. The flame must be nonoxidizing. The flame should be adjusted to provide a well-defined, blue, inner cone about $\frac{1}{4}$ in. long. The oxyacetylene flame should have a greenish feather at the tip of the inner cone. A fireproof working surface is also required. Most fluxes contain sodium, which emits yellow light when heated. A pair of didymium eyeglasses, such as those used in glassblowing, will filter out the sodium yellow lines and make the work much more visible while brazing. Parts to be brazed should, if possible, be designed to be self-locating. A clearance between parts of 0.001 to 0.003 in. will assure maximum strength in the finished joint. Surfaces to be wetted by the brazing alloy must be clean. Coat the joint surfaces with flux, then assemble the parts and clamp them together. Preheat the base metals in the area of the joint nearly to the melting temperature of the brazing alloy. The flux should melt from its crystalline form to a glassy fluid as the brazing temperature is approached. Then start at one end of the joint and heat a small area. Keep the flame moving over the work. Apply the brazing rod to the heated area and melt off a bead of metal. The rod should melt in contact with the work. Do not apply the flame directly to the brazing rod. With the flame moving in a circular or zigzag pattern, chase the puddle of molten alloy along the joint. The alloy should completely fill the joint and leave a fillet at the juncture of the parts being joined. It is sometimes convenient to preplace snippets of brazing wire in or near the joint before heating, rather than applying the alloy after the work is heated.

Copper, Kovar, Monel, nickel, steel, and stainless steel can be brazed without flux if the metal is heated in a hydrogen atmosphere in a furnace. The hydrogen serves as a flux by reducing surface oxides to produce a clean surface that will be wetted by the molten brazing alloy. Alloys containing chromium must be brazed in dry hydrogen. The surface oxide of chromium found on stainless steel is reduced by dry hydrogen at temperatures above 1050°C. Furnace brazing provides for more uniform heating of the work so there is usually less distortion than in torch brazing. The work is also stress-relieved to some extent. Hydrogen-brazed parts come out clean and bright over their entire surface. Although it is possible to carry out hydrogen brazing in the lab, the job is better let out to a specialty shop. Commercial heat-treating shops operate dissociated-ammonia tunnel kilns. The work travels slowly through a tunnel on a car. Ammonia introduced in the middle of the furnace dissociates to hydrogen and nitrogen to provide the flux.

Parts to be hydrogen brazed should be chemically clean. Inorganic salts from the skin are not affected by hydrogen and fingerprints will remain as stains on the finished work. Assemble the parts and place snippets of filler-metal wire near the joints to be brazed. A rough calculation of the volume to be filled will determine the amount of wire to be used. When the filler metal melts, it is drawn into the nearby joint by capillary action. If the joints to be brazed are not visible from outside the furnace, place a piece of brazing metal on top of the work where it can be observed by the operator to determine when the brazing temperature has been reached. If the walls of the brazing apparatus are opaque, a thermocouple will be required for this purpose.

The filler metal used in hydrogen brazing should melt at a temperature higher than that required for the hydrogen to effectively clean the surface of the base metal. OFHC copper has a melting point of 1083°C and is a very good filler metal for brazing steel, stainless steel, Kovar, and nickel. If low-melting alloys such as the silver-copper eutectic are used, a small amount of flux is required. Brazing alloys that contain volatile metals should never be used for hydrogen brazing. It is unwise to braze parts of ordinary electrolytic copper or use ordinary copper as a filler in a hydrogen atmosphere. Oxides in the metal are reduced to water that is trapped between the grains on the structure. This water can cause outgassing problems in vacuum applications, and, if the copper is heated, the expanding water vapor will produce internal cracks (hydrogen embrittlement). For instrument work, OFHC copper should be used for all copper parts that are to be brazed.

1.3.3 Welding

In welding, parts are fused together by heating above the melting temperature of the base metal. Sometimes a filler metal of the same type as the base metal is used. Fusion temperatures are attained with a torch, with an electric arc, or by ohmic heating at a point of electrical contact. For most instrument work, the preferred method is arc welding under an argon atmosphere using a nonconsumable electrode. This method is referred to as tungsten-inert-gas (TIG) welding or heliarc welding. All metals can be welded, although refractory metals must be welded under an atmosphere that is completely free of oxygen. Arc welding requires special skills and equipment, but most instrument shops are prepared to do TIG welding of steel, stainless steel, and aluminum on a routine basis.

Welded joints are strongest and relatively free of strain if the rate of heat dissipation is the same to each part of the work during the welding operation. When a thick piece of metal is to be joined to a thin piece, it is common practice to cut a groove in the thick piece, near the joint, to limit the rate of heat loss into the thicker piece during welding. The design of work to be welded is discussed in Section 3.6.3 (see Figure 3.40).

The effect of the high temperatures involved must be considered when specifying a welded joint. Some distortion of welded parts is inevitable. Provision should be made for remachining critical surfaces after welding. The state of heat treatment of the base metal is affected by welding. Hardened steels will be softened in the vicinity of a weld. Stainless steels may corrode because of carbide precipitation at a welded joint.

Spot welding or resistance welding is a good technique for joining metal sheets or wires. The joint is heated by the brief passage of a large electric current through a spot contact between the pieces to be joined. Steel, nickel, and Kovar, as well as refractory metals such as tungsten and molybdenum, can be spot welded. Copper and the precious metals are not easily spot welded because their electrical resistance is so low that very little power is dissipated by the welding current. Small spot-welding units with hand-held electrodes are available from commercial sources. These are quite useful in the lab.

1.3.4 Threaded Fasteners

Threaded fasteners are used to join parts that must be frequently disassembled. A thread on the outside of a cylinder, such as the thread on a bolt, is referred to as an external (or *male*) thread. The thread in a nut or a tapped hole is referred to as an internal (or *female*) thread. The terminology used to specify a screw thread is illustrated in Figure 1.13. The *pitch* is the distance between successive crests of the thread, equivalent to the axial distance traveled in one turn of a screw. In English units, the pitch is specified as the number of turns or threads per inch (tpi). The pitch of metric threads is directly specified in

millimeters. The *major diameter* is the largest diameter of either an external or an internal thread. The *minor diameter* is the smallest diameter. In the United States, Canada, and the United Kingdom, the form of a thread is specified by the Unified Standard. SI metric threads conform to ISO and DIN standards. Both the *crest* and *root* of these threads are flat or slightly rounded, as shown in Figure 1.13. The Unified Standard and SI *thread angle* is always 60°.

Two Unified Standard thread series are commonly used for instrument work. The *coarse-thread* series, designated UNC for Unified National Coarse, is for general use. Coarse threads provide maximum strength. The *fine-thread* series, designated UNF for Unified National Fine, is for use on parts subject to shock or vibration, since a tightened fine-thread nut and bolt are less likely to shake loose than a coarse-thread nut and bolt. The UNF thread is also used where fine adjustment is necessary. The pitch of standard metric fasteners is intermediate between UNC- and UNF-series threads. There is no interchangeability between Unified Standard and metric-threaded fasteners.

The fit of Unified Standard threaded fasteners may be specified by tolerances designated as 1A, 2A, 3A, and 5A for external threads and 1B, 2B, 3B, and 5B for internal threads. The fit of 2A and 2B threads is adequate for most applications, and such threads are usually provided if a tolerance is not specified. Many types of machine screws are available only with 2A threads. The 2A and 2B fits allow sufficient clearance for plating. The 1A and 1B fits leave sufficient clearance that dirty and scratched parts can be easily assembled. The 3A and 3B fits are for very

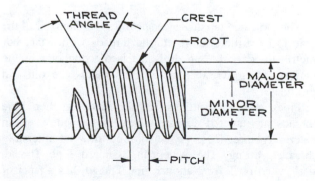

Figure 1.13   Screw-thread terminology.

precise work. The 5A and 5B are interference fits such as are used on studs that are to be installed semipermanently.

The specification of threaded parts is illustrated by the following examples:

1. An externally threaded part with a nominal major diameter of $\frac{3}{8}$ in., a coarse thread of 16 tpi, and a 2A tolerance is

$$\tfrac{3}{8}\text{-16 UNC-2A}$$

2. An internally threaded part with a nominal major diameter of $\frac{5}{8}$ in., a fine thread of 18 tpi, and a 2B tolerance is designated

$$\tfrac{5}{8}\text{-18 UNF-2B}$$

3. In the Unified Standard, major diameters less than $\frac{1}{4}$ in. are specified by a gauge number. A size 8 screw thread (with a nominal major diameter of 0.164 in.) and a coarse thread of 32 tpi is designated

$$8\text{-32 UNC}$$

4. A metric thread with a nominal major diameter of 5 mm and a pitch of 0.8 mm is designated

$$\text{M5 0.8 DIN}$$

A left-hand thread is specified by the letters LH following these designations.

Specifications for UNC and UNF thread forms are given in Table 1.5 and for metric threads in Table 1.6. The relationship between gauge number, n, in the Unified Standard and diameter, d, is given by

$$d = (0.013n + 0.060) \text{ in.}$$

The indicated tap drill size specifies the diameter of the hole to be drilled before cutting threads with a tap. For metric threads, the tap drill size is the diameter of the corresponding screw in millimeters minus the pitch in millimeters.

Pipes and pipe fittings are threaded together. Pipe threads are tapered so that a seal is formed when an externally threaded pipe is screwed into an internally threaded fitting. The American Standard Pipe Thread, designated NPT for National Pipe Thread, has a taper of 1 in 16. The diameter of a pipe thread is specified by

stating the nominal internal diameter of a pipe that will accept that thread on its outside. For example, a pipe with a nominal internal diameter of $\frac{1}{4}$ in. and a standard thread of 18 tpi is designated

$$\tfrac{1}{4}\text{-18 NPT}$$

or simply

$$\tfrac{1}{4}\text{-NPT}$$

The American Standard Pipe Thread specifications are listed in Table 1.7. The common forms of machine screws are illustrated in Figure 1.14.

Hex-head cap screws are ordinarily available in sizes larger than $\frac{1}{4}$ in. They have a large bearing surface and thus cause less damage to the surface under the head than other types of screws. A large torque can be applied to a hex-head screw, since it is tightened with a wrench.

Slotted-head screws and *Phillips* screws are driven with a screwdriver. They are available with *round, flat,* and *fillister* heads. The flat head is countersunk so that the top of the head is flush. Fillister screws are preferred to round-head screws because the square shoulders of the head provide better support for the blade of a screwdriver.

Socket-head cap screws, with a hexagonal recess, are preferred for instrument work. These are also known as *Allen* screws. Straight, L-shaped, and ball-pointed hex drivers are available that permit Allen screws to be installed in locations inaccessible to a wrench or screwdriver. An Allen screw can only be driven by a wrench of the correct size, so the socket does not wear as fast as the slot in a slotted-head screw. Socket-head screws have a relatively small bearing surface under the head and should be used with a washer wherever possible.

Setscrews are used to fix one part in relation to another. They are often used to secure a hub to a shaft. In this application it is wise to put a flat on the shaft where the setscrew will bear; otherwise the screw may mar the shaft, making it impossible for it to be withdrawn from the hub. Setscrews should not be used to lock a hub to a hollow shaft, since the force exerted by the screw will deform the shaft. In general, setscrews are suitable only for the transmission of small torque, and their use should be avoided if possible.

Machine screws shorter than 2 in. are threaded their entire length. Longer screws are only threaded for part of

TABLE 1.5 UNIFIED STANDARD THREADS

Size	Coarse (UNC)		Fine (UNF)	
(Nominal Diameter)	Threads per Inch	Tap Drill[a]	Threads per Inch	Tap Drill[a]
0 (0.060)			80	$\frac{3}{64}$
1(0.073)	64	No. 53	72	No. 53
2(0.086)	56	No. 50	64	No. 50
3 (0.099)	48	No. 47	56	No. 45
4 (0.112)	40	No. 43	48	No. 42
5(0.125)	40	No. 38	44	No. 37
6 (0.138)	32	No. 36	40	No. 33
8 (0.164)	32	No. 29	36	No. 29
10 (0.190)	24	No. 25	32	No. 21
12 (0.216)	24	No. 16	28	No. 14
1/4	20	No. 7	28	No. 3
5/16	18	Let. F	24	Let. I
3/8	16	$\frac{5}{16}$	24	Let. Q
7/16	14	Let. U	20	$\frac{25}{64}$
1/2	13	$\frac{27}{64}$	20	$\frac{29}{64}$
9/16	12	$\frac{31}{64}$	18	$\frac{33}{64}$
5/8	11	$\frac{17}{32}$	18	$\frac{37}{64}$
3/4	10	$\frac{21}{32}$	16	$\frac{11}{16}$
7/8	9	$\frac{49}{64}$	14	$\frac{13}{16}$
1	8	$\frac{7}{8}$	12	$\frac{59}{64}$

Note: ASA B1.1-1989.

[a] For approximately 75% thread depth.

TABLE 1.6 ISO (METRIC) THREADS

Diameter (mm)	Pitch (mm)	Tap Drill (mm)	Diameter (mm)	Pitch (mm)	Tap Drill (mm)
1.6	0.35	1.25	6	1.00	5.00
2	0.40	1.60	8	1.25	6.80
2.5	0.45	2.05	10	1.50	8.50
3	0.50	2.50	12	1.75	10.50
3.5	0.60	2.90	14	2.00	12.00
4	0.70	3.30	16	2.00	14.00
5	0.80	4.20	20	2.50	17.50

TABLE 1.7 AMERICAN STANDARD TAPER PIPE THREADS

Nominal Size of Pipe (in.)	Actual Outer Diameter of Pipe (in.)	Threads per Inch	Length of Engagement by Hand (in.)	Length of Effective Thread (in.)
$\frac{1}{8}$	0.405	27	0.180	0.260
$\frac{1}{4}$	0.540	18	0.200	0.401
$\frac{3}{8}$	0.675	18	0.240	0.408
$\frac{1}{2}$	0.840	14	0.320	0.534
$\frac{3}{4}$	1.050	14	0.340	0.546
1	1.315	$11\frac{1}{2}$	0.400	0.682
$1\frac{1}{4}$	1.660	$11\frac{1}{2}$	0.420	0.707
$1\frac{1}{2}$	1.900	$11\frac{1}{2}$	0.420	0.724
2	2.375	$11\frac{1}{2}$	0.436	0.756
$2\frac{1}{2}$	2.875	8	0.682	1.136
3	3.500	8	0.766	1.200

Note: ASA B2.1-1989.

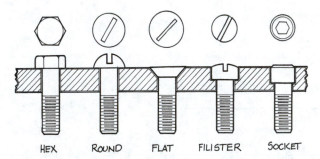

Figure 1.14 Machine screws: hex head; round head (slotted); flat head (Phillips); fillister head; socket head.

their length. Screws are usually only available with class 2A coarse or fine threads.

The type of head on a screw is usually designated by an abbreviation such as "HEX HD CP SCR" or "FIL HD MACH SCR." For example, a socket-head screw $1\frac{1}{2}$ in. long with an 8-32 thread is designated

$$8\text{-}32 \text{ UNC } 1\tfrac{1}{2} \text{ SOC HD CAP SCR}$$

The torque (T) required to produce a tension load (F) in a bolt of diameter (D) is

$$T = CDF$$

where the coefficient (C) depends upon the lubrication of the threads. In general, C may be taken as 0.2. If the threads are oiled or coated with molybdenum disulfide, a value of 0.15 may be more accurate. If the threads are very clean, C may be as large as 0.4. The tension load on the bolt is equal to the compressive force exerted by the underside of the bolt head.

Steel bolts meeting SAE specifications are identified by markings on the head:

SAE grades 0, 1, 2:   no mark

SAE grade 3:

SAE grade 5:   or

SAE grade 8:

The strength of the bolt increases with the SAE grade number. Most common steel bolts are SAE grade 2. In sizes up to 1 in., these bolts have a yield strength of about 50,000 psi. The yield strength of SAE grade 5 bolts exceeds 80,000 psi, while that of grade 8 bolts exceeds 130,000 psi.

When a bolt is to be tightened to a specified torque, the threads should first be seated by an initial tightening. The bolt should then be loosened and retightened to the computed torque with a torque wrench. The torque corresponding to the yield strength should not be exceeded during this operation.

To obtain maximum load-carrying strength, a steel bolt engaging an internal thread in a steel part should enter the thread to a distance equal to at least one bolt diameter. For a steel bolt entering an internal thread in brass or aluminum, the length of engagement should be closer to two bolt diameters.

Machine screws are readily available in a remarkable range of materials and finishes. Run-of-the-mill steel screws are often zinc plated, but nickel- and chrome-plated screws are available for enhanced corrosion resistance and better appearance. In addition to steel screws, major suppliers provide screws and nuts in types 304 and 316 stainless steel, Monel, brass, silicon bronze, 2024-T4 aluminum, titanium, Nylon, Teflon, and Fiberglass-reinforced plastic. The choice of the material for a threaded fastener depends on the desired strength in the joint to be made, the materials to be joined, and the environment of the finished assembly. The possibility of corrosion weakening a joint or preventing disassembly is often a major factor. Galvanic action between the materials of the fastener and the joined assembly is the culprit here. The material of the fastener and material of the metals being joined should be close to one another in the galvanic series. Metal screws are to be avoided when joining metals far apart in the galvanic series—brazing is preferable in this situation.

As a threaded assembly is tightened, the interlocking threaded surfaces are held together by the stress created by elastic deformation of the fastener and the clamped parts. Shock, vibration, and thermal cycling can relieve this tension and the assembly can become loose. All manner of schemes have been developed to keep threaded parts locked together. For instrument assembly, lockwashers often suffice. These come in two forms: the *spring lock* washer and the *serrated* washer. The former is simply a washer in the form of a single-turn spring that is placed under the head of a bolt or under a nut before it is screwed onto the bolt. The washer is flattened as the assembly is tightened and the spring force creates friction to resist the bolt turning. A serrated (or toothed) washer simultaneously engages the surface of the joined part and the underside of a bolt or nut resisting a bolt's loosening in a similar manner. As an aside, it is usually noted that a lockwasher should not be used with a flat washer since the lockwasher will engage the flat washer on one side but the other side of the flat washer will slip. Self-locking screws are made with a nylon insert in the threads. The nylon deforms as the screw is installed to take up the clearance between mating threaded surfaces. Nuts with nylon inserts, so-called *aircraft nuts* are readily available and quite effective in resisting the loosening from vibration. A simple means of securing a nut and bolt assembly is to install a second nut, the *locknut*, after the first nut is tightened. With the first nut held securely with a wrench to prevent its turning, the locknut is tightened down on the first. The effect is to create opposing tension in the bolt that takes up clearance in the threads and increases friction between the mating surfaces. A threaded assembly can also be secured with adhesive applied to the threads to remove clearances and stick the surfaces together. Specialized adhesives (Loctite) are available for this purpose in a range of strengths for both removable and permanent installations.

Excessive wear may result when a steel bolt is screwed into a threaded (tapped) hole in a soft material such as plastic or aluminum. Threaded inserts (Heli-Coils) can be installed in the softer material to alleviate this problem. These inserts are tightly wound helices of stainless steel or phosphor bronze wire with a diamond-shaped cross section. The insert is placed in a tapped, oversize hole and the mating bolt is screwed into the insert. Special taps are required to prepare a hole for the insert, and a special tool is required to drive the insert into the hole.

1.3.5 Rivets

Rivets are used to join sheet-metal or sheet-plastic parts. They are frequently used when some degrees of flexibility is desired in a joint, as when joining the ends of a belt to give a continuous loop. Common rivet shapes are shown in Figure 1.15. Rivets are made of soft copper, aluminum, carbon steel, and stainless steel, as well as plastic. As for threaded fasteners, the strength required in a joint and the possibility of corrosion are considered in choosing the material for a rivet. To join two pieces, a hole slightly larger than the body of the rivet is drilled or punched in each piece. The holes are aligned and the rivet is inserted. The length of the rivet should be such that it protrudes a distance equal to 1 to 1.5 times its diameter. The head of the rivet is backed-up with a heavy tool while the plain end of the rivet is *upset* by hammering to draw the two pieces together. The hammering action swells the body of the rivet to fill the hole. Viselike hand tools with appropriate die sets are available for setting rivets in a one-handed operation.

Blind rivets or *pop* rivets, illustrated in Figure 1.15, are useful in the lab. These can be installed without access to the back side of the joint. A mandrel passing through the center of the rivet is grasped by a special rivet gun, the rivet is inserted into the hole, and the mandrel is pulled back until it breaks. The head of the mandrel rolls the stem of the rivet over to form a head, and the remaining broken portion of the mandrel seals the center hole of the rivet. The installation of pop rivets in carefully positioned, reamed holes is an excellent means of aligning two thin pieces with respect to one another. Pop riveted joints can be regarded as semipermanent. The rivet is easily removed by drilling into its center with an oversize drill until the

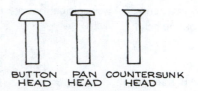

BUTTON HEAD   PAN HEAD   COUNTERSUNK HEAD

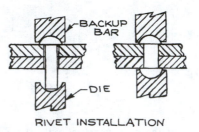

RIVET INSTALLATION

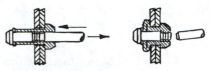

"POP" RIVET INSTALLATION

Figure 1.15   Rivets and riveted joints.

head separates from the body. With care, the drill will never touch the parts joined by the rivet.

1.3.6 Pins

Pins such as those shown in Figure 1.16 are used to precisely locate one part with respect to another or to fix a point of rotation. To install a pin, the two pieces to be joined are clamped together and the hole for the pin is drilled and then reamed to size simultaneously in both pieces.

Straight pins are made of steel that has been ground to a diameter tolerance of ±0.0001 in. They require a hole that has been reamed to a close tolerance, since they are intended to be pressed into place. If a straight pin is to serve as a pivot or to locate a part that is to be removable, the hole in the movable part is reamed slightly oversize.

A *roll pin* has the advantage of easy installation and removal and does not require a precision-reamed hole. The spring action of the walls of the pin hold it in place.

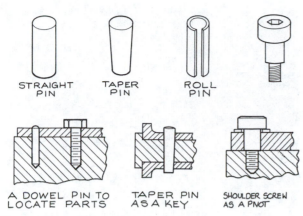

Figure 1.16   Pins and shoulder screws.

Taper pins are installed in holes that are shaped with a special reamer. The taper is 0.250 in. per ft. Plain taper pins are driven into place. Taper pins with the small end threaded are drawn into place with a nut that then secures the installed pin.

A *shoulder screw* is effectively a straight pin with a shoulder to provide location and a threaded end below for attachment. Shoulder screws are used as pivots. These screws are hardened, and the shoulder is ground to a diameter tolerance of ±0.0001 in.

1.3.7 Retaining Rings

A retaining ring serves as a removable shoulder on a shaft or in a hole to position parts assembled on the shaft or in the hole. An axially assembled external retaining ring is expanded slightly with a special pair of pliers, then slipped over the end of a shaft and allowed to spring shut in a groove on the shaft. An axially assembled internal ring is compressed, inserted in a hole, and permitted to spring open into a groove. A radially assembled external retaining ring is forced onto a shaft from the side. It springs open as it slides over the diameter and then closes around the shaft.

Some retaining rings have a beveled edge, as shown in Figure 1.17, so that the spring action of the ring on the edge of its groove produces an axial load to take up unwanted clearances between assembled parts. This scheme is frequently used to take up the end play in a ball

bearing. Another solution to end-play take up is the use of a bowed retaining ring as shown in Figure 1.17.

1.3.8 Adhesives

Epoxy resins are the most universally applicable adhesives for the laboratory. They are available in a variety of strengths and hardnesses. Epoxies that are either thermally or electrically conducting are also available. Epoxy adhesives consist of a resin and a hardener that are mixed just prior to use. The mixing proportions and curing schedule must be controlled to achieve the desired properties. A variety of epoxy adhesives are available prepackaged in small quantities in the correct proportions. Epoxies will adhere to metals, glass, and some plastics. Hard, smooth surfaces should be roughened prior to the application of the adhesive. Sandblasting is a convenient method. Parts of a surface that are not to receive the adhesive can be masked with tape during this operation. To obtain maximum strength in an epoxied joint, the gap filled by the epoxy should be 0.002 to 0.006 in. wide. To maintain this gap between flat smooth surfaces, shims can be inserted to hold them apart.

Self-curing silicone rubber such as General Electric RTV can be used as an adhesive. RTV is chemically stable and will stick to most surfaces. The cured rubber is not mechanically strong; however, this can be an advantage when making a joint that may occasionally have to be broken.

Cyanoacrylate contact adhesives, so-called *super glues*, have found wide use in instrument construction. Eastman 910 and Techni-Tool Permabond are representative of this type of adhesive. These adhesives are monomers that polymerize rapidly when pressed into a thin film between two surfaces. They will adhere to most materials, including metal, rubber, and nylon. Cyanoacrylate adhesives are not void-filling and only work to bond surfaces where contours are well matched. An adhesive film of about 0.001 in. gives the best results. A firm set is achieved in about a minute, and maximum strength is usually reached within a day. When joining metals and plastics, a shear strength in excess of 1000 psi is possible.

Sauereisen manufactures a line of ceramic cements. These are inorganic materials in powder, paste, or liquid form that are used to join pieces of ceramic or to bond

AXIALLY
ASSEMBLED
EXTERNAL RING

AXIALLY
ASSEMBLED
INTERNAL RING

RADIALLY
ASSEMBLED
EXTERNAL RING

Figure 1.17   Retaining rings and retaining installation.

FLAT RING

BEVELED RING

BOWED RING

ceramic to metal. Some are designed to be used as electrically insulating coatings and others may be used for casting small ceramic parts. The tensile strength of these materials is only on the order of 500 psi, but the materials are serviceable up to at least 1100°C.

1.4 MECHANICAL DRAWING

Mechanical drawing is the language of the instrument designer. Initial ideas for a design are expressed in terms of simple, full-scale drawings that develop in complexity and detail as the design matures. The construction of an apparatus is realized through communication with shop personnel in the form of working drawings of each part of the device. Successful design and construction relies on the designer's command of the language of mechanical drawing. Even in the event that a scientist does both the design and machine work, the completed instrument will benefit from the preparation of carefully executed mechanical drawings.

1.4.1 Drawing Tools

Earlier editions of *Building Scientific Apparatus* provided a list of the draftsman's tools—pencils, triangles, T-square, and so on—along with instructions for their use.

The traditional instruments are still appropriate for the preparation of one or two drawings, but more and more they have been replaced by Computer-Aided Drafting (CAD) programs. For the scientist whose instrument design is a small, occasional activity in the pursuit of broader scientific goals, a CAD program offers relatively simple, but important, advantages. Lines can be positioned precisely on a drawing and the line weight is consistently and accurately controlled. Scaling is done precisely so that dimensions can be determined directly from the drawing. Deletions are done at the stroke of a key—a significant timesaver compared to the difficult process of erasing pencil lines. Shapes can be duplicated and tracings can be made with ease. Files can be saved in an orderly fashion and easily retrieved for modification and updating. Lettering, once the laborious trademark of good mechanical drawing, is done at the keyboard.

In truth, CAD programs offer much more that can be effectively employed by the scientist-draftsman. With the remarkable power of modern CAD programs, modeling has evolved as a new method of design. The traditional draftsman-designer developed the plan of an instrument in two-dimensional, orthographic projections. With a powerful CAD program it is possible to model an object in a 3-D representation and extract the required 2-D projections after the fact. This approach has considerable appeal since the fit between parts can be seen graphically and interferences easily avoided. Another apparent advantage of computer-assisted design with powerful drafting programs is that the program can produce computer files that will drive the lathes and milling machines in the shop, thus obviating the need for drawings on paper, however, the integration of Computer-Aided Drafting and Computer-Aided Machining (CAD/CAM) is rarely practical for small model shops building one-off instruments. From the point of view of this book, it is hard to justify the time and effort required to learn to use a 3-D CAD program, as well as the cost of many of these programs.

The preparation of 2-D orthographic working drawings is, at least for the present, an essential step in building an instrument. The designer of scientific apparatus must learn to use the language of mechanical drawing. The most modest drafting (as opposed to *drawing*) software will ordinarily serve the requirements of the builder of scientific apparatus—Claris Cad for Mac and AutoCAD-LT for PC are examples of entirely satisfactory programs. Even if a 3-D modeling program is available, one quickly discovers that the 2-D routines at the front end of these programs will fulfill most needs. As a result, only modest computing power will be required.

A drafting program incorporates hundreds, or even thousands, of commands. The user must become facile with a significant subset of these commands to make progress without frequent reference to a manual. Many successful designers make a practice of learning at least one new command during every session at the computer.

In contrast to CPU requirements for computer drafting, it is advantageous to have the largest possible monitor—a 21 in. monitor is not too large. The conception of an instrument in the drawing phase is definitely enhanced if an entire, full-scale drawing can be viewed at once.

A printer or a plotter is an essential element of a CAD system since one ordinarily must send paper copies of working drawings to the shop. Although reduced-scale drawings are appropriate for large industrial projects, the instructions for the fabrication of the parts of an instrument are best communicated in full-scale drawings. The letter-size format of an office laser printer is usually too small for mechanical drawings. Investment in a large-format laser or ink-jet printer or a pen plotter is frequently worthwhile. Ink-jet and laser printers accepting paper up to 36 in. wide are now available. The large-format, E-size, pen plotter is the drafting-room standard.

1.4.2 Basic Principles of Mechanical Drawing

If mechanical drawing is the designer's language, then the line is the alphabet of this language. Some of the basic lines used in drawing are illustrated in the plane view shown in Figure 1.18. Three line widths are employed: *thick lines* for visible outlines, short breaks, and cutting planes (explained below); *medium lines* for hidden outlines; and *thin lines* for center, extension, dimension, and long break lines. A CAD program typically chooses a line width of 0.4 mm for thick lines, 0.2 mm for medium lines, and 0.1 mm for thin lines. If only two line weights are available, medium lines are shown thin. For pencil and paper work, H-grade lead is used for heavy lines and lettering, 2H for medium and light lines, 4H for layout.

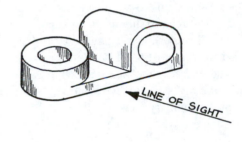

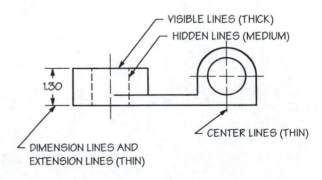

Figure 1.18   Lines used in mechanical drawing exemplified in a CAD drawing of a side view of the object at the top.

The shape of an object is described by *orthographic projection*. This is the view that one obtains when an object is far from the eye so that there is no apparent perspective. There are six principal views of an object, as illustrated in Figure 1.19. In America, the relative position of the views must always be as shown there. For example, the view seen from the right side of an object is placed to the right of the front view. Of course, for clarity or convenience, the draftsman is initially free to choose any side of the object as the *front*. As indicated in the illustration, not all views are required to completely describe an object. The draftsman should present only those views that are necessary for a complete description. Note the use of center lines in Figure 1.19 to indicate the axis of symmetry of a round hole.

In some cases an *auxiliary view* will simplify a presentation. For example, a view normal to an inclined surface may be easier to draw and more informative than one of the six principal views. Such a case is illustrated in Figure 1.20. The relationship of the auxiliary view to the front view must be as shown. That is, the auxiliary view of an inclined surface must be placed in the direction of the normal to the surface. Note the extension of the center line from the auxiliary view to the front view to show the direction of the auxiliary view.

Often the internal structure of an object cannot be clearly indicated with hidden edges indicated by dashed lines. In this case it is useful to show the view that would be seen if the object were cut open. Such a view is called a *section*. Several examples of full sections are shown in Figure 1.21. The *cutting plane* in the section is represented by cross-hatching with fine lines to suggest a saw cut. The cross-hatch lines should always be at an oblique angle to the heavy lines that indicate the outline of the section. Cross-hatch lines on two adjacent pieces should be in opposite directions. The cutting plane in a view perpendicular to the section is shown by a heavy *section line* of alternating long and two short dashes. Arrows on the section line indicate the direction of sight from which the section is viewed. Sections may be identified with letters at each end of the cutting-plane line.

It is often convenient to have the cutting plane change directions so that it passes through several features that are desired to be shown in section. Angled surfaces are then revolved into a common plane and offset surfaces are projected onto a common plane to give an *aligned section* as illustrated in Figures 1.22 and 1.23, rather than a true projection of the cutting plane.

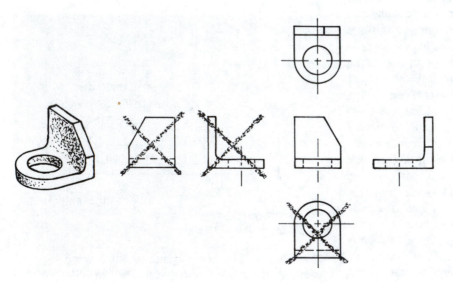

Figure 1.19 The six principal views of the bracket shown on the left. The views not required to describe the bracket in this particular case have been crossed out.

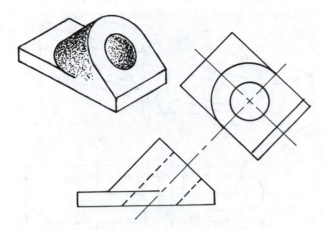

Figure 1.20   Effective use of an auxiliary view.

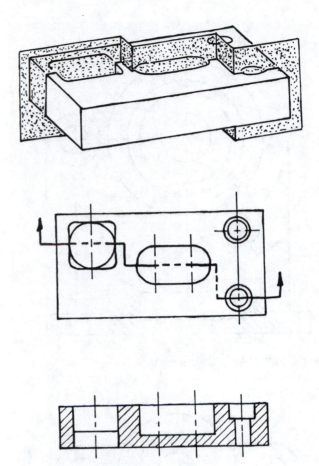

Figure 1.22   An aligned view.

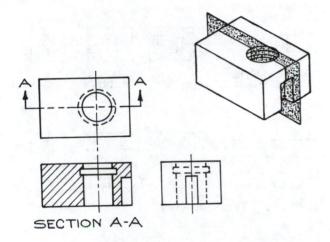

SECTION A-A

Figure 1.21   Full sections along the indicated planes.

Other convenient sectioning techniques include the *half section* of a symmetrical object as shown in Figure 1.24, and the *revolved section* of a long bar or spoke as shown in Figure 1.25. Many details, such as screw threads, rivets, springs, and welds, are so tedious to draw, and appear so often, that they are designated by symbols rather than faithful representations. Symbols for threads are illustrated in Figure 1.26.

Drawings of parts too long to be conveniently fitted on a piece of drawing paper can be represented as though a piece from the middle of the part had been broken out and discarded and the two ends moved together. Three conventional breaks are illustrated in Figure 1.27.

Mechanical drawings include dimensions, tolerances, and special descriptions and instructions in written form. Lettering on mechanical drawings employs only uppercase letters. Letters about $\frac{1}{4}$ in. high are usually appropriate.

1.4.3 Dimensions

After the shape of an object is described by an orthographic projection, its size is specified by dimensions and notes on the drawing. The dimensions to be specified depend upon the function of the object and upon the

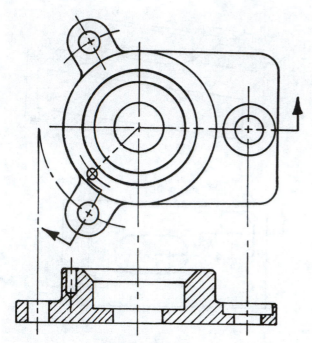

Figure 1.23   An aligned section.

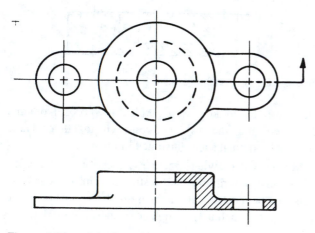

Figure 1.24   A half section.

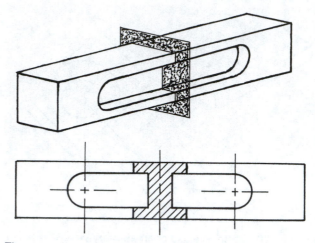

Figure 1.25   A revolved section.

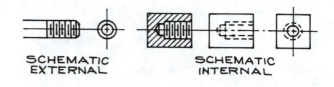

SCHEMATIC
EXTERNAL

SCHEMATIC
INTERNAL

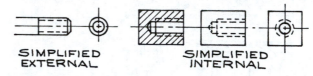

SIMPLIFIED
EXTERNAL

SIMPLIFIED
INTERNAL

Figure 1.26   Thread symbols.

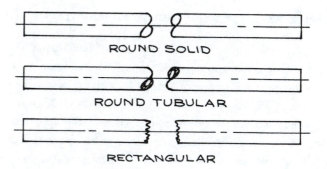

ROUND SOLID

ROUND TUBULAR

RECTANGULAR

Figure 1.27   Breaks.

machine operations to be performed by the workman when fabricating the object.

Distance on a drawing is specified by either a dimension or a note. A *dimension* indicates the distance between points, edges, or surfaces (Figure 1.28). A *note* is a written instruction that gives information on the size and shape of a part (Figure 1.29).

Dimensions may be given in *series* or *parallel* (Figure 1.30). If a combination of these two methods is used, care must be taken to ensure that the size of the object is not overdetermined. An example of an overdimensioned

drawing is given in Figure 1.31. Dimensions should not be duplicated, and a drawing should include no more dimensions than are required. When several features of a drawing are obviously identical, it is only necessary to dimension one of them.

Dimensions are often specified with respect to a *datum*. This is a single feature whose location is assumed to be exact. When choosing a datum, consider the functional importance of locating other features of an object relative to the datum, as well as the ease with which the workman can locate the datum itself.

If possible, all dimensions should be placed outside of the view with extension lines leading out from the view. As shown in Figure 1.30, the shorter dimensions are nearer

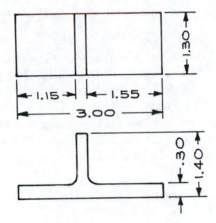

Figure 1.28   Dimensions.

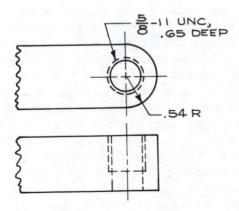

Figure 1.29   A note.

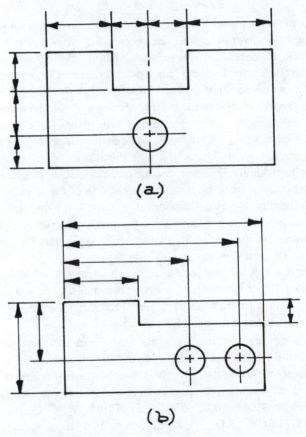

Figure 1.30   Dimensions: (a) series dimensions; (b) parallel dimensions.

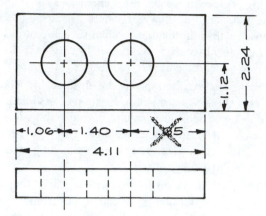

Figure 1.31 An overdimensioned drawing. One of the dimensions is superfluous.

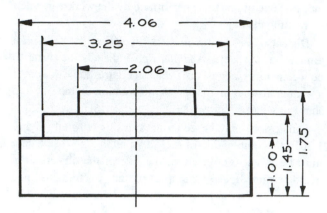

Figure 1.32 Placement of dimensions and dimension values.

to the object line. Dimension lines never cross extension lines, however extension lines may cross other extension lines if necessary. Dimensions within the view are permissible to eliminate long extension lines as long as the dimension does not obscure some detail of the view. The dimension closest to the outside of a view should be about $\frac{1}{2}$ in. from the edge. Successive dimensions should be placed at $\frac{3}{8}$ to $\frac{1}{2}$ in. intervals. Dimension values should be staggered, as in Figure 1.32, rather than stacked vertically or horizontally. Dimension values are oriented so that they can be read from the bottom or right side of the drawing. Notice that the extension lines do not touch the outline of the view, but rather stop about $\frac{1}{16}$ in. short. The units of a dimension are not ordinarily specified on a drawing. In America, all dimensions are assumed to be in inches. In Europe they are assumed to be in millimeters. Contrary to usual scientific practice, a decimal dimension less than unity is written without a zero preceding the decimal point: .256 rather than 0.256.

Notes are used to specify the size of standard parts and parts too small to be dimensioned. Notes are also used to specify a feature that is derived from a standard operation such as drilling, reaming, tapping, or spot facing. Notes such as "TAP 10-32 UNF, .75 DEEP" or "$\frac{1}{2}$ DRILL, SPOTFACE .75D .10 DEEP," that refer to a specific feature, are accompanied by a leader emanating from the beginning or end of the note and terminating in an arrowhead that indicates the location of the feature (see Figure 1.29). General notes, such as "REMOVE BURRS" or "ALL FILLETS $\frac{1}{4}$ R," that refer to the whole drawing, do not require a leader.

Many abbreviations are used in writing dimensions and notes. Common abbreviations used on mechanical drawings are given in Table 1.8.

1.4.4 Tolerances

One of the most important parts of design work is the selection of manufacturing tolerances so that a designed part fulfills its function and yet can be fabricated with a minimum of effort. The designer must carefully and thoughtfully specify the function of each part of an apparatus in order to arrive at realistic tolerances on each dimension of a drawing. It is important to appreciate the capabilities of the machines that will be used in manufacture and the time and effort required to maintain a given tolerance. As indicated in Figure 1.33, the cost of production is inversely related (approximately) to the tolerances.

Usually only one or two tolerance specifications will apply to most of the dimensions on a drawing. It is then convenient to encode the tolerance specification in the dimension rather than affix a tolerance to each dimension. Most dimensions in instrument design are written as decimals. The tolerance on a dimension can then be indicated by the number of decimal places in the

TABLE 1.8 ABBREVIATIONS USED ON MECHANICAL DRAWINGS

Word	Abbreviation	Word	Abbreviation	Word	Abbreviation
Bearing	BRG	Fillister	FIL	Right hand	RH
Bolt circle	BC	Grind	GRD	Rivet(ed)	RIV
Bracket	BRKT	Groove	GRV	Round	RD
Broach(ed)	BRO	Ground	GRD	Root mean square	RMS
Bushing	BUSH	Head	HD	Screw	SCR
Cap screw	CAP SCR	Inside diameter	ID	Socket	SOC
Center line	CL	Key	K	Space(d)	SP
Chamfer	CHAM	Keyway	KWY	Spot-face(d)	SF
Circle	CIR	Left hand	LH	Square	SQ
Circumference	CIRC	Long	LG	Stainless	STN
Concentric	CONC	Maximum	MAX	Steel	STL
Counterbore	CBORE	Minimum	MIN	Straight	STR
Counterdrill	CDRILL	Not to scale	NTS	Surface	SUR
Countersink	CSK	Number	NO	Taper(ed)	TPR
Cross section	XSECT	Opposite	OPP	Thread(ed)	THD
Diameter	DIA, D, Ø	Outside diameter	OD	Tolerance	TOL
Drawing	DWG	Pipe tap	PT	Typical	TYP
Drill(ed)	DR	Pipe thread	PT	Vacuum	VAC
Each	EA	Press	PRS	Washer	WASH
Equal(ly)	EQ	Punch	PCH	With	W/
Fillet	FIL	Reference line	REF	Without	W/O

dimension. Dimensions expressed to two decimal places have one tolerance specification, to three decimal places another, and so on. The specifications are given in a note. An example of this system is given in Figure 1.34.

Fractions may be used to give the size of some features derived from standardized operations, such as drilling or threading. A tolerance specification is usually unnecessary in these cases, since both the designer and workman should know what precision can be expected.

There are two methods of expressly stating the tolerance on a dimension. The allowable variation, plus and minus, may be stated after the dimension. The limits of the dimension can also be stated without giving a tolerance at all. Examples of these two methods are given in Figure 1.35. The trend is toward limit dimensioning.

When the absolute size of two mating parts is not important but the clearance between them is critical, specify the desired *fit*. Fits are classified as *sliding fits, clearance locational fits, transition fits, interference fits,* and *force fits*. The uses of the various classes of fits, and tables of the tolerances required to give such fits, are given in standard texts on mechanical drawing.

Beware of an undesirable accumulation of tolerances. In series dimensioning, the tolerance on the distance between two points separated by two or more dimensions is equal to the sum of the tolerances for each of the intervening dimensions. In Figure 1.30a, the tolerance on

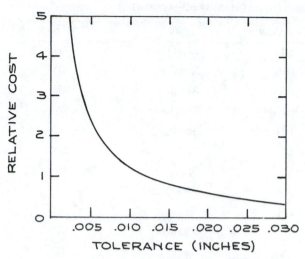

Figure 1.33   Approximate relation between tolerance and the cost of production.

1.4.5 From Design to Working Drawings

The first step in designing an instrument is to draw the entire assembly. Start with rough sketches in pencil and paper before turning to the computer. Then, with the CAD program, draw the outline of the most important parts of the instrument. Work to scale. Add subsidiary parts after the configuration of the essential elements has been established. Most dimensions, tolerances, and fine details may be omitted at this point.

Pay attention to the manner in which pieces fit together. Will the apparatus be convenient to assemble and disassemble? Are some parts much stronger or weaker than required? Does a strong part depend upon a weak one for location or support? Are all shapes as simple as possible? Are standard shapes of materials and standard bolts, couplings, bearings, and shafts used wherever possible?

Time spent at the drawing board mentally assembling and disassembling a device will save hours of frustration in the laboratory. Careful consideration of production techniques can save hundreds of dollars in the shop.

When a full-size orthographic drawing of the apparatus is complete, it is wise to discuss it with the people who will do the work.

After the design has reached its final form, the designer must make a working drawing of each piece of the apparatus. A working drawing is a fully dimensioned and toleranced drawing to be used in the shop. Work to scale if possible. The CAD program is especially useful at this point. A new layer can be superimposed on the assembly drawing and the outline of a single part transferred to the new layer without fear of error. Alternatively the outline of an individual part can be copied and transferred to a new file. The draftsman should develop a systematic way of naming drawings so they can be stored and easily retrieved.

The critical test of a set of working drawings is their usefulness. All necessary communication between scientist and machinist should appear on the drawings. The machinist should not require oral instructions, and he should not be required to make decisions affecting the performance of an instrument.

In the above discussion we have assumed that the scientific designer has the services of a model shop. This is the case in industrial laboratories and many university labs.

the lateral location of the hole with respect to the right side of the object is given by the sum of two tolerances. For parallel dimensions, the location of each feature relative to the datum depends on only one toleranc; however, the difference in the distance between two locations depends upon two tolerances. In Figure 1.30a, the tolerance on the distance between the two holes is given by the sum of the tolerances on the dimensions that locate the holes.

As much as possible, dimensions and tolerances should be chosen so that materials can be used in their stock sizes and shapes. Do not call for unnecessary finishing. The area around a hole in a casting or other rough piece of metal, for example, can be spot faced to provide a bearing surface for a bolt if the quality of the rest of the surface is not critical.

The dimensions and tolerances on a finished drawing should be carefully checked. Imagine yourself as the machinist who must use the drawing to make a part. Proceed mentally through each step of the fabrication to see that all necessary dimensions appear on the drawing and that the specified dimensions are easy to use. You should also determine that all desired tolerances are within the limits of the required machine-tool operations.

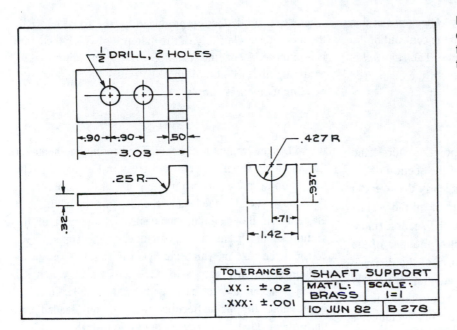

Figure 1.34   A decimal system for the specification of tolerances.

A designer, however, should follow the same procedure when he fabricates his own apparatus. All questions of sizes, tolerances, and fits must be answered before construction. If a scientist attempts to make these decisions as he proceeds with construction, the results are invariably poor.

1.5 PHYSICAL PRINCIPLES OF MECHANICAL DESIGN

No material is perfectly rigid. When any member of a machine is subjected to a force, no matter how small, it will bend or twist to some extent. Even in the absence of an external load, a mechanical element bends under its own weight (so-called *body forces*). A member subjected to a force that varies in time will vibrate. A designer must appreciate the extent of deflection of mechanical parts under load.

1.5.1 Bending of a Beam or Shaft

When a beam bends, one side of the beam experiences a tensile load and the other a compressive load. Consider a

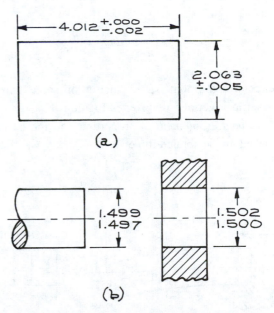

Figure 1.35   Two methods of stating the tolerance on a dimension: (a) as the allowable variation on a dimension; (b) in terms of the allowable limits of the dimension.

point in the flexed beam in Figure 1.36. The stress at this point depends upon its distance, c, from the centroid of the beam in the direction of the applied force. The *centroid* is the center of gravity of the cross section of the beam (see Figure 1.36b). The stress is given by

$$s = \frac{Mc}{I}$$

where M is the bending moment at the point of interest and I is the centroidal moment of inertia of the section of the beam containing the point of interest. Clearly the stress in a flexed beam is greatest at the outer surface of the beam.

The *centroidal moment of inertia* of the section (more correctly known as the second moment of the area of the cross section) is the integral of the square of the distance from the centroid multiplied by the differential area at that distance. The integral is taken along a line through the centroid and parallel to the applied force. For a rectangular section of height H and width B, such as that shown in Figure 1.36b, the centroidal moment of inertia is

$$I = \int_{-B/2}^{B/2} \int_{-H/2}^{H/2} h^2 \, dh \, db = \frac{BH^3}{12}$$

The centroidal moments of inertia of some common symmetrical sections are given in Figure 1.37.

The *bending moment* is proportional to the curvature produced by the applied force

$$M = EI\frac{d^2y}{dx^2}$$

where y is the deflection produced by the force and E is the modulus of elasticity of the material of which the beam is composed. For the example given in Figure 1.36, assuming the weight of the shaft can be ignored, the bending moment is simply

$$M = -F(L - x)$$

where L is the length of the beam. The bending moments for this and other simple systems are given in Figure 1.38. The deflection of a stressed beam at a point at distance x from the supported end of the beam depends upon the modulus of elasticity of the material, the centroidal moment of the section of the beam, and the nature of the support. We consider the case in which the beam is rigidly fixed to its support(s) (as in Figure 1.36) and the case in which the beam is supported at each end by a simple fulcrum. Expressions for the deflection of both point-loaded and uniformly loaded beams are given in Figure 1.38.

A sufficiently large bending moment will result in the permanent deformation of a beam. This is the *yield point*. A conservative relation between the bending moment at the yield point, M_y, and the yield strength of the material, s_y, is given by $M_y = s_y I / c_{max}$. In reality, the yield strength of a beam in bending is greater than for a beam in tension, since the inner fibers of the material must be loaded to their yield point to a considerable depth before the beam is permanently deformed. To determine the maximum allowable load on a beam or shaft, one simply sets M_y equal to the expression for the maximum value of the bending moment on the beam and solves for the force.

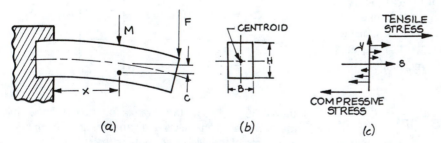

(a)   (b)   (c)

Figure 1.36   A flexed beam: (a) the beam bending under a load; (b) a cross section showing the centroid; (c) the distribution of shear forces along a parallel to the long axis of the beam along a vertical line through the centroid of the beam.

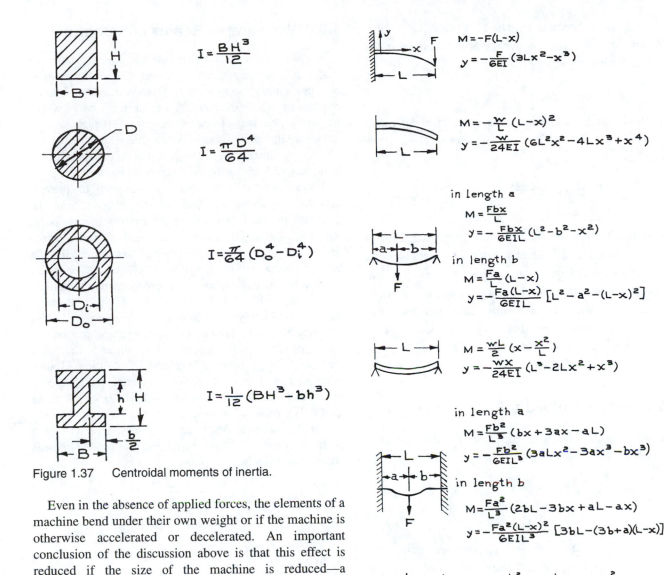

Figure 1.37   Centroidal moments of inertia.

Even in the absence of applied forces, the elements of a machine bend under their own weight or if the machine is otherwise accelerated or decelerated. An important conclusion of the discussion above is that this effect is reduced if the size of the machine is reduced—a mechanical structure becomes stiffer as it is scaled down in size. Consider a horizontal beam of rectangular cross section cantilevered from a wall. Using the deflection formula from the second panel of Figure 1.38 and setting $x = L$, the weight of the beam causes a deflection at the free end of

$$y_{\max} = \frac{wL^4}{8EI}$$

Figure 1.38   Bending formula: M = bending moment; E = modulus of elasticity; I = centroidal moment of inertia; w = weight per unit length.

where w is weight per unit length of the beam. Taking $w = \rho BH$, where ρ is the weight density of the material of the beam, and substituting for I gives

$$y_{max} = \frac{3\rho L^4}{2EH^2}$$

If the beam is scaled by some factor, the deflection decreases as the square of the scale factor. The relative deflection,

$$\frac{y_{max}}{L}$$

decreases linearly with the scale factor.

It should be noted that all of the preceding discussion assumes the tensile strength of a material to be the same as the compressive strength. For materials such as cast iron, where this is not true, the bending equations are much more complicated. The foregoing also ignores shear stresses in a loaded beam. For very short beams, shear stresses become important and the shear strength of the material must be considered.[4]

1.5.2 Twisting of a Shaft

The stress at a point in a round shaft subjected to a torsional load is

$$s = \frac{Tc}{J}$$

$$J = \frac{\pi}{32}(D_0^4 - D_i^4)$$

where c is the distance of the point of interest from the center of the shaft and T is the applied torque. (J is the *centroidal polar moment of inertia of the section* of the shaft, and D_o, and D_i are the outer and inner diameters of the shaft, respectively.

The total angle of twist in a shaft of length L is

$$\theta = \frac{57LT}{GJ} \text{ deg}$$

where G is the modulus of elasticity in shear or *shear modulus* of the material of the shaft. For metals, the shear modulus is about $\frac{1}{3}$ the elastic modulus.

1.5.3 Vibration of Beams and Shafts

In many instruments, vibration of a supporting beam or shaft can adversely affect operation. The designer must pay special attention to the natural frequencies of vibration of the parts of an instrument. A periodic disturbance with a frequency near that of one of the natural frequencies will induce a large and possibly destructive vibration. For the purpose of illustration, the response of an oscillator to the frequency of a driving force is shown in Figure 1.41.

There are two ways for the designer to prevent destructive vibrations. The critical element can be designed so that its natural frequency of vibration is far removed from the frequency of a disturbing force, or the disturbing force can be inhibited in the vicinity of the natural frequency of the critical element. The latter scheme is called *damping*. Very-low-frequency vibrations can be damped by a hydraulic or friction shock absorber such as is employed on automobile suspensions. High frequencies can be damped by coupling the objectionable source of vibration to its surroundings through a member that has a very low natural frequency of vibration. Mounts made of rubber or cork are very effective for absorbing high-frequency vibrations and sudden shocks, as these materials possess a low modulus of elasticity.

The fundamental frequency of vibration (in Hz) of a shaft or beam with one or more concentrated masses attached is given by

$$f_n^2 = \frac{\dfrac{a}{4\pi^2}\displaystyle\int_0^L \dfrac{M^2}{EI}dx}{\displaystyle\int_0^L a\rho A Y^2 dx + \sum_i F_i Y_i^2}$$

where

  $I = $ centroidal moment of inertia of the shaft

  $M = M(x) = $ *maximum* bending moment at x, i.e., the amplitude of the bending moment induced by the forces on the shaft at x

  $a = $ inertial acceleration experienced by the shaft and its weights (for a nonrotating shaft, $a = g$)

  $E = $ modulus of elasticity of shaft material

  $\rho = $ weight density of the shaft ((ρ/g) is the mass density)

  $A = $ cross-sectional area of the shaft

L = length of the shaft

$Y = y(x)$ = maximum deflection of the shaft at x, i.e., the amplitude of vibration at x

F_i = force exerted by the ith mass (for a nonrotating shaft, this is the weight of the ith mass)

Y_i = maximum deflection at the location of the ith mass

For a system similar to one of the static deflection cases treated in Figure 1.38, one need only substitute the appropriate expressions for the bending moment and deflections and solve the resulting equation to determine the natural frequency. Consider, for example, a beam of negligible mass that is rigidly supported at one end and supports a mass of weight W at its free end. This is the case illustrated in Figure 1.36. The natural frequency of vibration is

$$f_n^2 = \frac{\frac{g}{4\pi^2}\int_0^L \frac{W^2(L-x)^2}{EI}dx}{W\left\{\frac{-W}{6EI}[3Lx^2 - x^3]_{x=L}\right\}^2}$$

$$f_n^2 = \frac{1}{2\pi}\left(\frac{3gEI}{WL^3}\right)^{1/2} \text{Hz}$$

In general, the shape of the deflection curve is not known. The assumption of some reasonable deflection curve, however, will usually provide a useful result. In most cases, one can assume a sine curve for a shaft supported (but not clamped) at both ends, or a parabola for a shaft supported at only one end. More precise approaches to complex systems are given in many engineering texts on vibration.[5]

Consider Figure 1.36 once again. A parabolic deflection curve with the correct limit properties is given by

$$y = \frac{y_0}{L^2}(L-x)^2$$

where y_o is the maximum deflection of the free end. The bending moment is then

$$M = EI\frac{d^2y}{dx^2}$$

$$= EI\left(\frac{y_0}{L^2}\right)$$

Substituting in the equation for the natural frequency gives

$$f_n^2 = \frac{\frac{g}{4\pi^2}\int_0^L \frac{[-2EI(y_0/L^2)]}{EI}dx}{W(-y_0)^2}$$

$$f_n = \frac{1}{\pi}\left(\frac{gEI}{WL^3}\right)\text{Hz}$$

in reasonable agreement with the exact solution derived above.

For the special case of a heavy shaft or beam with a centrally placed load, a solution can be reached by assuming the shaft to be weightless and adding half the weight of the shaft to the concentrated load.

1.5.4 Shaft Whirl and Vibration

A shaft and rotor can never be perfectly balanced, and the axis of rotation can never be exactly located by the shaft bearings. As a shaft rotates, centrifugal force on the unbalanced mass deflects the shaft and causes it to *whirl* around its axis of rotation. This whirling motion appears as a vibration to a stationary observer, and can be treated as such. There is a *critical speed* at which the whirl becomes violent. In this unstable condition, the shaft or its bearings are likely to be damaged and intense vibration will be transmitted through the bearing mount to other parts of the machine. This critical condition occurs when the shaft speed (in rps) equals the natural frequency (in Hz) of the stationary shaft and rotor assembly. It follows that the mathematical derivations of natural frequencies of assemblies of beams and weights can be applied to the calculation of critical speeds for shafts and rotors. The actual centrifugal loads on the rotating shaft need not be determined. This is because the load F always appears as the ratio F/a and

$$\frac{F}{a} = \frac{W}{g} = m$$

where W is the weight of the element that is exerting the load.

The critical speed for a rotor-shaft assembly depends upon how rigidly the shaft bearings support the shaft. Two extreme cases are considered: a thin bearing such as a *ball bearing* for which the angle of the shaft passing through the bearing is not fixed; and a thick bearing such as a *plain sleeve bearing* that rigidly maintains the angular alignment of the shaft. The critical speed for a rotor on a weightless shaft supported on thin bearings is

$$n_c = 0.276\left(\frac{gEIL}{Wa^2b^2}\right)^{1/2} \text{ rps}$$

where W is the weight of the rotor, I is the centroidal moment of inertia of the shaft, E is the modulus of elasticity of the shaft material, L is the length of the shaft between the bearings, and a and b are the distances from the rotor to the bearings. If the ends of the shaft are rigidly supported in long bearings, the critical speed is

$$n_c = 0.276\left(\frac{gEIL^3}{Wa^3b^3}\right)^{1/2} \text{ rps}$$

Notice that the critical speed can be increased by placing the rotor near one end of the shaft so that a or b is small. In this case, the rotor acts as a gyro tending to stiffen the shaft.

For an unloaded shaft on thin bearings, the critical speed is

$$n_c = 1.57\left(\frac{EIg}{\rho AL^4}\right)^{1/2} \text{ rps}$$

and for an unloaded shaft on long bearings it is

$$n_c = 3.53\left(\frac{EIg}{\rho AL^4}\right)^{1/2} \text{ rps}$$

where ρ is the weight density of the shaft material. The units of the individual quantities appearing in brackets in the four equations above must be such that overall the quantity in brackets has units of s^{-2}. For the particular case of a solid, round, steel shaft the critical speeds are

$$n_c = 80,000\frac{D}{L^2} \text{ rps (thin bearings; D = diameter (in.);}$$

$$L = \text{length (in.))}$$

$$n_c = 180,000\frac{D}{L^2} \text{ rps (long bearings; D = diameter (in.);}$$

$$L = \text{length (in.))}$$

The special case of a centrally placed rotor on a heavy shaft can be treated by adding half the weight of the shaft to the rotor and taking the shaft to be weightless. The critical speed for a shaft carrying several rotors can be approximated by

$$n_c = \left(\frac{1}{n_s^2} + \sum_i \frac{1}{n_i^2}\right)^{-1/2}$$

where n_s is the critical speed of the shaft alone and n_i is the critical speed for the ith rotor alone on a weightless shaft.

The designer must manipulate the dimensions and material of a shaft and rotor, and the location of the rotor, so that its critical speed does not coincide with the design speed of an assembly. It is usually sufficient for these two speeds to differ by a factor of $\sqrt{2}$. The most desirable situation is for the critical speed to exceed the design speed. If this is not possible, the drive motor for the rotating assembly should be so powerful that the shaft accelerates very rapidly through the region of the critical speed. The drive motor will require an excess of power because considerable power is dissipated to vibration at the critical speed.

The formulae above give the lowest frequency of vibration and the lowest critical speed for a shaft and rotor assembly. A shaft with a number of weights has as many different fundamental modes of vibration as it has weights. Vibration will also occur at harmonics of the fundamentals.

For rotational speeds far away from a critical speed, a rotating assembly can be considered to be rigid. Imbalance can nevertheless lead to inertial (centrifugal) forces that may damage the shaft or its bearings. If the unbalanced mass lies within a plane perpendicular to the axis of rotation, as in the case of a thin rotor on a shaft, the rotor can be statically balanced. The condition for *static balance* is that the axis of rotation pass through the center of gravity. In practice this can be accomplished by placing the shaft with its rotor across a pair of knife-edge rails. The shaft will roll or rock until coming to rest with the unbalanced inertial force pointed radially downwards. Weight is then added or removed along the vertical radius until balance is achieved.

If there are unbalanced masses acting at different points along the axis, then dynamic balancing is required. The conditions for *dynamic balance* are that the axis of rotation pass through the center of gravity and that the axis of rotation coincide with a principal axis of the inertial force. Dynamic balancing is difficult without specialized equipment. Balancing is done inexpensively and accurately as a commercial service.

1.5.5 Stress Relief

An abrupt change in the cross section of a shaft produces a concentration of stresses as shown in Figure 1.39a. Stresses are increased at steps, grooves, keyways, holes, dents, and scratches. A stress concentration is an area where there is a large gradient in the stress. Failure is the result of shear forces at a stress concentration. Anyone who has ever broken a bolt is familiar with this effect—the bolt almost always fails just where the head joins the shaft. Methods of relieving stresses at a step in a shaft are suggested in Figure 1.39b, c, and d.

There are a number of means of detecting stress in a part. One of the most useful is the *photoelastic method*.[6] In this technique, a transparent plastic model is stressed in the same way as the element of interest without necessarily duplicating the magnitude of the stress. The model is illuminated with polarized, monochromatic light and viewed through a polarizer. As shown in Figure 1.40, the stress distribution appears as fringes in the image, and closely spaced fringes characterize stress concentrations. Most glassblowing shops are equipped with a polarizing device of this type for detecting residual stresses in worked glass (see Figure 2.16, next chapter).

In a similar fashion, fabricated parts can be nondestructively tested by means of holography. A hologram of the object is made; then the part is stressed and a second hologram is made. The two holograms are then superposed and placed in the holographic projector. The resulting three-dimensional image consists of patterns of fringes that are densest in the areas of greatest distortion in the stressed object. Holographic nondestructive testing (HNDT) is provided as a commercial service.

A particularly simple and inexpensive means of locating stress concentrations in a working part has been developed by Magnaflux. The part to be tested is sprayed with a substance that forms a hard, brittle layer over the surface. The part is returned to service for a time, and any flexing of the part causes the surface coating to crack. The part is then removed and dipped in a dye that permeates the cracks to reveal the areas of greatest strain.

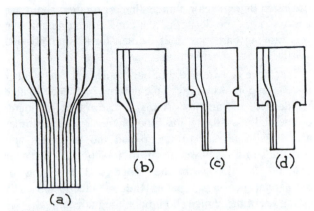

Figure 1.39   (a) Stress concentration at a step in a shaft. Lines indicate surfaces of constant stress. Parts (b), (c), and (d) show methods of relieving stress at the step; (c) and (d) are useful when a shoulder is required to locate a bearing.

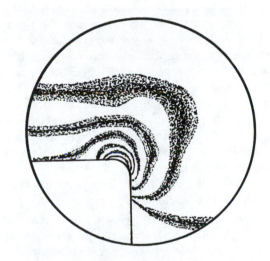

Figure 1.40   Stress distribution in transparent plastic model as observed when the part is illuminated with polarized light and observed with a crossed polarizer.

1.6 CONSTRAINED MOTION

In addition to a rigid support structure, most machines include some moving parts. The motion of these parts must usually be constrained in some fashion to obtain the required function. Most often the desired motion is translation in a plane or along a line, or rotation about one or more axes. The design of the constraining mechanism for a moving part must meet dimensional, strength, and wear-resistance specifications.

To achieve constrained motion, the designer may either create a support structure for the moving part that will give the desired alignment or make use of a standard, commercially available machine element. In the former case, it is necessary to understand the principles and limitations of geometric design. In the latter, the designer must be familiar with the range and precision of commercial bearings, shafts, sliders, and so on. The economics of the choice should not be ignored.

1.6.1 Kinematic Design

A rigid body has six independent degrees of freedom of motion. These are usually taken to be the translational motions along three orthogonal axes and the rotational motions about these axes. Any motion can be described as a linear combination of these six motions. Similarly, six coordinates define the position of a body: three position coordinates and three angle coordinates. If one of these coordinates is fixed or if some linear combination of these coordinates is fixed, motion is constrained and the number of degrees of freedom is reduced. For example, a rigid object will be constrained to move in a plane if one of its position coordinates is fixed. Kinematic design for constrained motion consists of fixing a point on a body for each degree of freedom to be constrained.[7]

In general, one degree of freedom can be removed by constraining a point on a body to remain in contact with a reference surface. The following examples illustrate this principle of kinematic design:

1. Consider a sphere constrained to maintain point contact with each of two reference planes: a ball resting in a V-groove. The ball is free to rotate about any axis but it can only translate along a line parallel to the intersection of the planes. The ball has lost two degrees of freedom.

2. A body that is constrained to maintain three points of contact with a plane surface has only three degrees of freedom: translation in two directions parallel to the

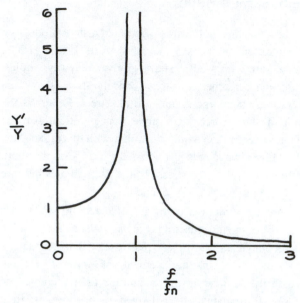

Figure 1.41   Response of a machine element to a periodically varying force. Y' is the amplitude of vibration of the element and f_n is its natural freuency of vibration. Y and f are the amplitude and frequency of the driver.

plane and rotation about an axis perpendicular to the plane.

3. A body in contact at three points with a cylindrical surface has three degrees of freedom: translation in the direction of the cylinder axis, revolution about the cylinder axis, and rotation about an axis perpendicular to the plane defined by the three points of contact.

The object of kinematic design is to permit motion to be constrained without depending upon precision of manufacture. A complication exists if two contact points are degenerate in the sense that the function of the two points can be served by a single point of contact. Consider a four-legged table *vis-à-vis* a three-legged table. One of the four legs provides a degenerate point of contact with the floor. The fourth leg may provide some much-needed stability, but it also introduces uncertainty in the location of the tabletop unless all four legs are of precisely the same length. Of course, the sort of considerations that lead to the production of four-legged tables often dictates a *semikinematic* or degenerate kinematic design for other devices as well.

Examples of kinematic design are illustrated in Figures 1.42 through 1.44. The ball feet on the carriage shown in Figure 1.42 provide five points of contact between the carriage and its base. The carriage is therefore only free to slide in one direction. One problem with this design is that it is difficult to mill a V-groove with smooth surfaces. The surface quality of the groove can be improved by lapping with carborundum, using a V-shaped brass lap that fits in the groove.

An improved version of the design in Figure 1.42 is shown in Figure 1.43. A pair of steel rods replaces the V-groove, and balls replace the feet of the carriage. This design is both more precise and more economical than the previous one. Stainless-steel shafting, hardened and centerless-ground to a diameter tolerance of $±0.00005$ in., is available at a cost of less than a dollar a foot. Stainless-steel balls with a roundness tolerance of $±0.00005$ in. are also inexpensive. The channel that contains the rods must be milled, but the surface quality in this channel is not critical. In a milling operation, the sides of the channel can easily be kept straight and parallel to within 0.0001 in./ft.

The three grooves in the platform shown in Figure 1.44 allow precise relocation of a three-legged table after it has been removed. The ball feet of the table make six points of contact with the platform, so that there are no degrees of freedom.

In applying the principles of kinematic design it must be borne in mind that all materials are more or less elastic. The deformation of the elements of a kinematic structure inevitably deflects under load, leading to a deviation from true kinematic behavior. In the absence of careful analysis of such effects, it is often possible to obtain higher precision location or movement, as well as greater stability, through the use of standard, non-kinematic elements such as bearings and ball sliders. Smith and Chetwynd, in their excellent book on the design of precision mechanisms, discuss the analysis and exploitation of elastic deformation in kinematic design.[8]

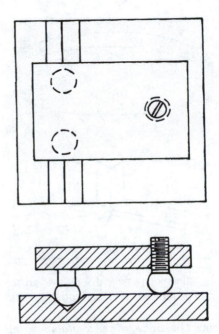

Figure 1.42   Kinematic design that constrains a carriage to move in a straight line.

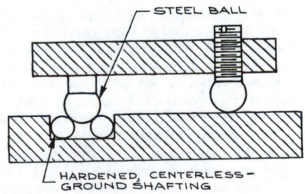

Figure 1.43   An improved version of the design shown in Figure 1.42.

1.6.2 Plain Bearings

A bearing is a stationary element that locates and carries the load of a moving part. Bearings can be divided into two categories depending upon whether there is sliding or rolling contact between the moving and stationary parts.

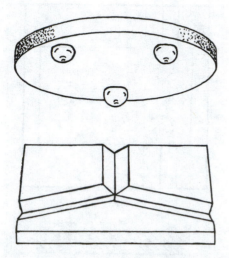

Figure 1.44   Kinematic design that permits an accurately located part to be removed and replaced in the same position.

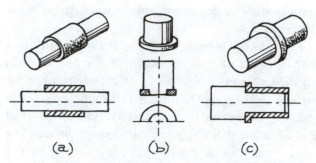

Figure 1.45   Plain bearings: (a) a journal bearing; (b) a thrust bearing; (c) a flanged journal bearing.

Sliding-contact bearings are called *plain bearings*. Rolling bearings will be discussed in the next section.

Plain bearings may be designed to carry a radial load, an axial load, or both. Different types are illustrated in Figure 1.45. A radial bearing consists of a cylindrical shaft or *journal* rotating or sliding within a shell, which is the *bearing* proper. The entire assembly is referred to as a *journal bearing*. An axial bearing consists of a flat bearing surface, like a washer, against which the end of the shaft rests. These are called *thrust bearings*. A journal bearing may incorporate a *flanged journal,* in which case it will support a radial load as well as an axial load.

A journal is usually hardened steel or stainless steel. Precision-ground shafts and shafts with precision-ground journals are available in diameters from $\frac{1}{32}$ to 1 in. Bearing shells of bronze or oil-impregnated bronze are available to fit. Nylon bearings for light loads and oil-free applications are also available. Commercial shafting and bearings are manufactured to provide a clearance of 0.0002 to 0.0010 in.

Many different methods of lubrication can be employed instead of oil impregnation. The inner surface of the bearing can be grooved, and oil or grease can be forced into the groove through a hole in the shell. If oil is objectionable, a groove on the inner surface of the bearing can be packed with molybdenum disulfide or another dry lubricant.

A plain bearing is installed by pressing it into a hole in the supporting structure. An interference of about 0.001 in. is desirable for bearings up to an inch in diameter. That is, the outer diameter of the bearing shell should be about 0.001 in. larger than the hole into which it is pressed. If the interference is too great, the inner diameter of the bearing may be significantly reduced.

A variety of *bearing housings* and *pillow blocks* (Figure 1.46) are available. These mountings are bored to accept standard bearings and in many instances the bearing is premounted. These mounts replace precision-bored bearing mounts.

It is usually necessary to provide axial location for a shaft in a journal bearing. This can be accomplished with a retaining ring in a groove on the shaft or a collar secured by a setscrew.

Plain bearings run smoothly and quietly, and have high load-carrying ability. Properly installed and lubricated, they have a very long life. Because of the close clearances between parts, they are not easily fouled by dirt in their environment. Plain bearings are limited to relatively low-speed operation. Speeds in excess of a few hundred rpm are not practical without forced lubrication. The primary disadvantage of plain bearings is their high starting friction, although, when properly installed and lubricated, their running friction can be very low.

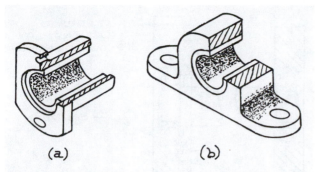

Figure 1.46   (a) A bearing housing; (b) a thrust bearing; and (c) a flanged journal bearing.

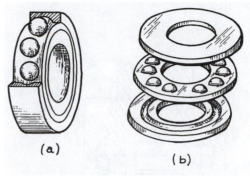

Figure 1.47   Ball bearings: (a) a radial ball bearing; (b) a thrust ball bearing.

1.6.3 Ball Bearings

The rolling element in a rolling contact bearing may be a ball, cylinder, or cone. *Ball bearings* are used for light loads and high speeds. *Roller bearings* employing cylindrical or conical rollers are suitable for very heavy loads but are not often used for instrument work. We shall discuss only ball bearings.

As with plain bearings, there are both radial and thrust ball bearings (see Figure 1.47). A *radial ball bearing* consists of an inner and an outer *race* with a row of balls between. The grooves in each race have a radius slightly larger than the radius of the balls so that there is only point contact between the balls and the race. The balls are separated by a *retainer* that prevents the balls from rubbing against one another and keeps them uniformly spaced around the bearing. A radial ball bearing can tolerate a substantial thrust load, but for pure axial loads a thrust bearing should be used. A *thrust bearing* is similar to an axial bearing except that it has upper and lower races rather than inner and outer.

Ball bearings are made of steel or stainless steel. They are graded 1, 3, 5, 7, or 9 depending on manufacturing tolerances. Grades 7 and 9 have ground races and are made to the closest tolerances. They cost little more than lower-grade ones and should be specified for instrument applications.

Proper installation is required to obtain good performance from a ball bearing. The rotating race should be given a firm interference fit, and the stationary race given a light *push fit* to permit some rotational creep. This slight movement of the stationary race helps prevent the maximum load from always bearing on the same spot.

Press fitting changes the internal clearances in a bearing. Bearing manufacturers specify the amount of interference that should be used. As shown in Figure 1.48, the press arbor used to drive a bearing onto a shaft or into a housing should be designed so that the thrust is not transmitted through the balls. Never hammer a bearing into place.

The surface quality and diameter tolerance of the shaft that is to be fitted into a bearing, or the hole that is to house a bearing, must be of the same quality as the bearing. For high-quality bearings, the mounting surfaces should be ground. Centerless-ground precision shafting is available to fit all standard bearings. Bearing mounts and pillow blocks with premounted bearings are similarly available to eliminate the need for precision machine work when installing ball bearings.

Ball bearings are designed with both radial and axial clearances. This play is intended to allow for axial misalignment and for dimensional changes that occur upon installation or because of thermal stresses. For the best locational precision, and to obtain smooth, vibration-free operation, a ball bearing should be *preloaded* to remove most of this play. A preloading force that displaces one race axially with respect to the other will remove both radial and axial play by causing the balls to roll up the sides of their grooves. The use of a shim to take up the play in a bearing is illustrated in Figure 1.49. For light-duty applications, the end play can also be taken up by installing a spring washer (Section 1.6.5) instead of a

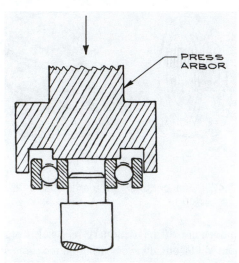

Figure 1.48   Installation of a bearing. The press arbor should bear on the race that is being fitted.

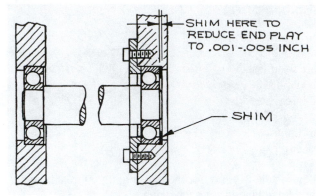

Figure 1.49   Installation of a shim to remove end play in a shaft mounted on ball bearings.

shim. With proper installation, a ball bearing will locate a revolving axis to within a few ten-thousandths of an inch.

Bearings must be protected from effects that damage the race surface on which the balls roll. Ball bearings are most likely to fail because of occasional large static loads that produce an indentation in the race. Such a dent is called a *brinell*. Dynamic loads are distributed around the race and are less likely to cause damage. A hard vibration, however, can cause brinelling in the form of a series of dents or waves on the surface of a race. Of course, a bearing is also damaged by the introduction of foreign matter that abrades or corrodes the bearing surfaces.

Bearings should be lubricated with petroleum oils or greases. For high speeds and light loads the lightest, finest grades of machine oil can be used. Ball bearings require very little oil. Lubrication is sufficient if there is enough oil to produce an observable meniscus at the point where each ball contacts a race.

Cleanliness is important. In a dirty environment, bearings with built-in side shields should be used. It is probably wise to use enclosed bearings in all instrument applications to keep the bearings clean and to prevent oil from contaminating the environment.

The chief advantages of ball bearings are their low starting friction and very low running friction. They are well suited to high-speed, low-load operation. Relative to plain bearings, ball bearings are noisy and occupy a large volume. The cost of quality ball bearings is so small that economic considerations are usually not important in choosing between rolling bearings and plain bearings for instrument use.

1.6.4 Linear-Motion Bearings

A linear-motion bearing may be of either the sliding or rolling type. A plain journal bearing can be used to locate a shaft that is to move axially. A V-shaped or dovetail groove sliding over a mating rail can serve as a bearing between a heavily loaded, slowly moving carriage and a stationary platform. This is the type of bearing used between the carriage and bed of a lathe.

Linear ball bearings are commercially available. In a bearing for use with an axially moving shaft, the balls that carry the load between the outer race and the shaft move in grooves that run parallel to the axis of the shaft. The balls are recirculated through a return track when they roll to the end of a groove. Linear-motion ball bearings will locate a shaft to within ±0.0002 in. of a reference axis. Their cost is comparable to conventional rotating ball bearings. Complete roller-slide assemblies are also available. These employ balls rolling in V-grooves. Roller slides will maintain straight-line motion to within 0.0002 in. per inch of travel.

1.6.5 Springs

In many instances it is desirable for a motion to be constrained by a flexible element such as a spring. Springs are used to hold two parts in contact when zero clearance is required, to absorb shock loads, to damp vibrations, and to measure forces.

A spring is characterized by the ratio of the magnitude of applied force to the resulting deflection, d. This is the *spring rate:*

$$k = \frac{F}{d}$$

As is the case for any flexible system, an assembly consisting of a spring and attached load has a natural frequency of vibration

$$f_n = \frac{1}{2\pi}\left(\frac{k}{m}\right)^{1/2} \text{Hz}$$

where m includes the mass of both the spring and any load that is affixed to it. Since $m = W/g = F/a$, we have upon substituting for the spring rate in this equation

$$f_n = \frac{1}{2\pi}\left(\frac{g}{d_{st}}\right)^{1/2} \text{Hz}$$

where d_{st} is the static deflection produced by the weight of the spring plus the attached load.

In most applications it is desirable to choose a spring that will not resonate with any other part of the apparatus in which it is installed. If the spring is expected to damp a vibratory motion, its natural frequency should differ from that of the disturbance by more than an order of magnitude.

A spring will exert an uneven force when it is subjected to a periodically varying load whose frequency is close to the natural frequency of the spring. When the vibration of the spring is in phase with the load, the reactive force of the spring will be less than its static force for any given deflection. When the spring is out of phase with the load, it will exert a greater force than expected. This phenomenon is called *surge.* Surge can be reduced or eliminated by using two springs with different natural frequencies. In the case of helical springs, they can be placed one inside the other.

In instrument work, helical springs are most often used. *Helical torsion, extension,* and *compression* springs are illustrated in Figure 1.50. The number of coils in a helical spring must be sufficient to ensure that the spring wire

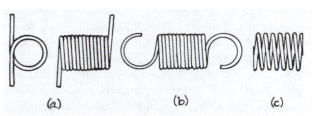

Figure 1.50 Springs: (a) torsion spring; (b) helical extension spring; (c) helical compression spring.

remains within its elastic limit when the spring is at maximum deflection. The number of coils in a compression spring also determines the minimum length that is realized when successive coils come into contact. A long compression spring may buckle under stress. This tendency is discouraged if the ends of the spring are square. It can be prevented by placing a rod through the center or by installing the spring in a hole. In general, a compression spring must be supported by some means if its length exceeds its diameter by more than a factor of five.

The spring rate of a helical spring made of round wire is

$$k = \frac{Gt^4}{8D^3N}$$

where G is the shear modulus of the spring material (about 1/3 of the elastic modulus), t is the diameter of the wire, N is the number of coils, and D is the mean diameter (the average of the inner and outer diameters) of the spring. The natural frequencies for free vibration are

$$f_n = \frac{nt}{4\pi ND^2}\left(\frac{gG}{\rho}\right)^{1/2} \text{Hz}$$

where ρ is the weight density of the spring wire and n is an integer. For steel wire

$$f_n = \frac{14000nt}{ND^2} \text{Hz}$$

when t and D are in inches.

A variety of steels and bronzes are used for spring manufacture. High-carbon steel wire works well. If necessary, the spring can be wound in the annealed state and hardened after forming. Music wire, or piano wire, is one of the best materials for one-off construction of small springs. It is available in diameters of 0.004 to 0.103 in. This wire is very strong and hard because of the drawing process used in its production, and does not need to be hardened after forming. Type-302 stainless-steel wire is useful for springs that are subject to a corrosive environment. Springs of beryllium-copper wire are especially useful in applications where spring deflection is used to gauge a force, since this material maintains a linear stress-strain relation almost to the point of permanent deformation. As mentioned earlier, quartz fiber torsion springs can also be used in these applications.

Helical coil springs are conveniently formed by winding wire on a mandrel as it is rotated in a lathe. The wire must be kept under tension as it is pulled onto the mandrel. When the tension is released, the formed spring will expand so the mandrel must be somewhat smaller than the desired inner diameter of the finished spring. The production of a small number of springs of a given size and spring rate is probably best carried out by *cut-and-try*.

Commercially manufactured springs are readily available. They are convenient to use because properties such as the spring rate and free length are specified by the supplier.

There are hundreds of possible spring configurations; however, disk springs are the only form that we shall mention other than coil springs. A disc spring (also known as a Belleville spring washer) is a cone-shaped disc with a hole in the center (Figure 1.51a). When loaded, the cone flattens. This is a very stiff spring that can absorb a large amount of energy per unit length. A spring of any desired travel can be created by stacking disc springs as in Figure 1.51b. They must be aligned by a rod passing through their centers or by stacking them in a hole slightly larger than the outer diameter of the discs. If the discs are stacked in parallel as shown in Figure 1.51c, they will provide a great deal of damping owing to friction between the faces of the discs.

1.6.6 Flexures

Flexures are beams made of a flexible spring material, such as beryllium copper or spring steel, that are intended to provide a small, controlled displacement along or about a known axis. They are basically flexible hinges. A clock pendulum is often suspended from a flexure. Owing to their uncomplicated nature and ease of manufacture, they are particularly attractive for the provision of controlled motion in small devices. They have the advantages of requiring no lubrication

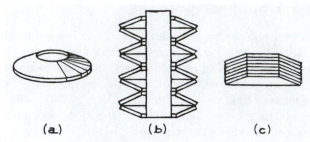

Figure 1.51 Bellville spring washers; (a) a disc spring; (b) disc springs stacked on a shaft; (c) disc springs stacked in parallel.

and having no hysteresis since there is no friction and there are no clearances. It is often possible to create a flexure by machining a thin section in a single monolithic mechanism rather than forming and mounting a separate flexible element to the machine. This is not only economical but avoids problems with alignment between the mechanism and the flexible hinge, as well as movement about the fastener that joins the hinge to the machine. A flexure is typically a simple homogeneous shape so the force-displacement relation can be accurately calculated from elementary physical principles (as described in Section 1.5.1). With careful design, a linear force-displacement relation can be achieved for small displacements

Figure 1.52 illustrates, at least conceptually, two basic motions obtainable with a flexible beam. In both cases the flexible element is rigidly mounted to a stationary base at one end and to a moveable platform at the free end. In Figure 1.52a, a force applied to the platform produces a rotation of the platform. For small displacements, the angle of rotation is proportional to the applied force. In Figure 1.52b, a bracket is attached to the platform so that the force can be applied along a line normal to the center of the flexure. The flexed spring bends near its ends and assumes an S-shape with a straight section in the middle. The bending moment, being a function of beam curvature, is zero at the midpoint in line with the applied force. The platform undergoes a linear displacement, d, that is proportional to the force for small displacements:

$$d = \frac{FL^3}{12El}$$

The spring constant, or *stiffness*, is F/d.

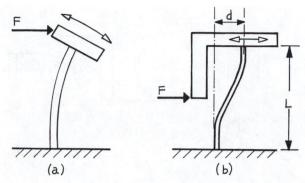

Figure 1.52   A platform mounted to a flexure attached to a solid base: (a) a transverse force applied at the end of the flexure produces a rotation of the platform; (b) a force applied normal to the center of the flexure produces a linear translation of the platform.

In practice, neither of the arrangements in Figure 1.52 is satisfactory since any misalignment of the applied force leads to pitching or twisting of the platform. A practical design of a linear translation platform is illustrated in Figure 1.53. The wide springs shown in that illustration offer resistance to twisting and provide a wider platform for the attachment of other components of the apparatus. Using two springs in a parallelogram arrangement offers resistance to pitching. In many applications of this design the force is applied directly to the platform. Maximum resistance to pitching is gained, however, if a bracket is attached as in Figure 1.52(b) so that the force acts through the midpoint of the springs. With two springs, the stiffness of the mechanism is doubled and the displacement for a given force is halved. The parallelogram arrangement results in a slightly curved path of the platform; the platform moves downward with increasing displacement. This error in the linear motion has been analyzed in detail by Jones and must be accounted for in precision applications.[9]

The vertical displacement error in the Figure 1.53 design is corrected in the compound flexure design shown in Figure 1.54. Here, the platform in the simple flexure becomes the base for a second flexure. The downward displacement of the intermediate platform is exactly compensated by an upward displacement of the lower platform. The driving force is applied to the lower platform. The overall stiffness is half that of the simple flexure in Figure 1.53. It is worth noting that the

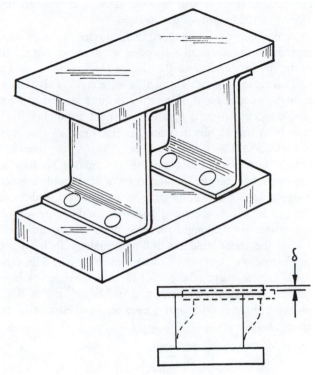

Figure 1.53   A practical design for a linear translation platform mounted on leaf springs. The inset illustrates the vertical error in the linear motion.

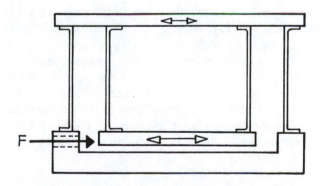

Figure 1.54   A compound flexure design to correct for the error shown in Figure 1.53.

compound design also compensates for temperature effects on the lengths of the springs.

A leaf spring with clamped ends moving in antiparallel directions bends into an S-shape as noted above. Nearly all of the flexion occurs near the clamped ends with the center section remaining straight as if it were rigid. This flexion near the clamped end of a spring is quite similar to what would occur when a bending moment is applied to a cantilever that is notched near its fixed end as shown in Figure 1.55a. This is a *notch hinge*. A leaf spring fixed to a base at one end and to a transversely-moving platform at the other end can be simulated by a beam with a notch hinge at each end. The obvious extension of this idea is a flexure in which the base, platform, and spring are all machined from a single piece of material. Figure 1.55b shows the monolithic notch-hinge analogy to the basic linear-motion flexure illustrated in Figure 1.52b. The force-displacement relation for the notch hinge has been derived analytically by Paros and Weisbord[10] and by finite element methods by Smith, Chetwynd, and Bowen[11]. The latter analysis gives

$$d = \frac{3FKRL^2}{EBt^3}$$

for the basic linear-motion flexure illustrated in Figure 1.55b, where R is the radius of each of the notches, t is the width of the web at each notch, B is the width of the beam (as in Figure 1.36), and the factor K accounts for the effective length of the flexible part of the web. K varies from about 0.2 for a relatively thin, flexible web, where t is much less than R, to about 0.7 for a relatively stiff spring with t comparable to R.

Flexures provide for relatively small displacements, especially flexures employing notch hinges. Linear force-displacement response, free of hysteresis, requires that the elastic limit of the spring material not be exceeded. For the linear-motion notch-hinge flexure in Figure 1.55(b), maximum displacement in the model of Smith *et al.* is approximated by

$$d_{\max} = 4K\left(\frac{R}{t}\right)\left(\frac{s_{\max}}{E}\right)L$$

where $s_{\max}$ is the maximum tolerable stress in the material of the spring.[11] For design purposes, $s_{\max}$ can be taken to be the yield strength of the material, or, conservatively,

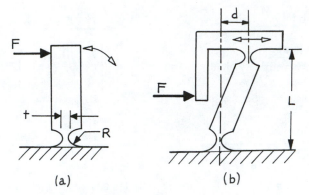

Figure 1.55 Monolithic flexures analogous to those shown in Figure 1.52.

some fraction thereof. The ratio of the yield strength to the modulus of elasticity (E) for mild steel, stainless steel, or tempered aluminum alloy is about 0.003. For hard plastics such as polyamide (Nylon) or polyimide (Vespel), the ratio is about 0.03. In a simple notch-hinge linear-motion flexure, the ratio *(R/t)* is typically about 5. This gives a maximum displacement of no more than about $0.01L$ for a simple metal flexure and $0.1L$ for a plastic flexure.

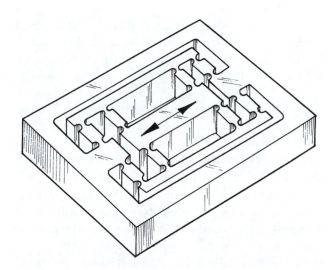

Figure 1.56 Monolithic, double-compound, rectilinear spring (ref. 11).

As suggested by the contour of the notch hinges shown in Figure 1.55, fabrication of a monolithic flexure requires only milling and boring operations. The precision of the motion, as well as the stiffness, of a monolithic flexure is determined primarily by the precision in the location of the centers of the holes that create a notch. The relative ease of manufacture of precise monolithic flexures allows for quite complex designs. An elegant, stable, and precise double-compound, rectilinear spring designed by Smith *et al.* is shown in Figure 1.56.[11]

There are many variations on the basic notch spring. On a beam, a closely-spaced pair of notches at right angles provides for two degrees of freedom. Notch springs machined into thin metal stock can serve the function of a Belleville spring to absorb shock or take up end play in a mechanism or to limit motion to a single plane. Examples are shown in Figure 1.57. In these flexures the material in the notch is stressed in torsion.

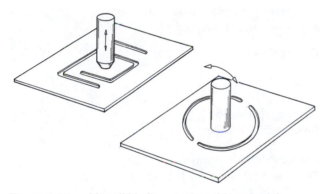

Figure 1.57   Monolithic flexures in a thin material.

CITED REFERENCES

1. *Metals Handbook, Heat Treating, Cleaning and Finishing,* Vol. 2, 8th ed., American Society for Metals, Metals Park, Ohio, 1964.

2. F. Rosebury, *Handbook of Electron Tube and Vacuum Techniques,* Addison-Wesley, Reading, Mass., 1965.

3. A comprehensive list of brazing alloys has been compiled by W. H. Kohl, in *Handbook of Vacuum Physics,* Vol. 3, *Technology,* A. H. Beck, ed., Pergamon Press, Elmsford, N. Y., 1964; *Materials and Techniques for Electron Tubes,* Reinhold, New York, 1959.

4. R. J. Roark, *Formulas for Stress and Strain,* 4th ed., McGraw-Hill, New York, 1965.

5. D. E. Newland, *Mechanical Vibration Analysis and Computation*, Longman Scientific & Technical/ Wiley, New York, 1989; S. S. Rao, *Mechanical Vibrations*, 2nd ed., Addison-Wesley, Reading, Mass., 1990; W. Weaver, Jr., S. P. Timoshenko, and D. H. Young, *Vibration Problems in Engineering*, 5th ed., Wiley, New York, 1990.

6. J. W. Dally and W. F. Riley, *Experimental Stress Analysis,* McGraw-Hill, New York, 1965, pp. 165–221.

7. A useful review of kinematic design is given by J. E. Furse, *J. Phys. E.*, **14**, 264, (1981).

8. S. T. Smith and D. G. Chetwynd, *Foundations of Ultraprecision Mechanism Design*, Gordan and Breach Science Publishers, Philadelphia, 1992.

9. R. V. Jones, *J. Sci. Instr.*, **28**, 38 (1951); *J. Sci. Instr.*, **33**, 11 (1956).

10. J. M. Paros and L. Weisbord, *Machine Design*, 151, 25 Nov. 1965.

11. S. T. Smith, D. G. Chetwynd, and D. K. Bowen, *J. Phys. E: Sci. Instrum.*, **20**, 977 (1987).

GENERAL REFERENCES

Design of Moving and Rotating Machinery

R. M. Phelan, *Dynamics of Machinery,* McGraw-Hill, New York, 1967.

Mechanical Design Texts

A. D. Deutschman, W. J. Michels, and C. E. Wilson, *Machine Design,* Macmillan, New York, 1975.

V. M. Faires, *Design of Machine Elements,* 4th ed., Macmillan, New York, 1965.

R. E. Parr, *Principles of Mechanical Design,* McGraw-Hill, New York, 1970.

R. M. Phelan, *Fundamentals of Mechanical Design,* 3rd ed., McGraw-Hill, New York, 1970.

S. T. Smith and D. G. Chetwynd, *Foundations of Ultraprecision Mechanism Design*, Gordan and Breach Science Publishers, Philadelphia, 1992.

Mechanical-Drawing Texts

T. E. French and C. J. Vierck, *Engineering Drawing and Graphic Technology,* 11th ed., McGraw-Hill, New York, 1972; (entitled *A Manual of Engineering Drawing* in its first 10 editions).

F. E. Ciesecke, A. Mitchell, H. C. Spencer, and I. L. Hill, *Technical Drawing,* 5th ed., Macmillan, New York, 1967.

Properties of Materials

A. J. Moses, *The Practicing Scientist's Handbook,* Van Nostrand Reinhold, New York, 1978. F. Rosebury, *Handbook of Electron Tube and Vacuum Techniques,* Addison-Wesley, Reading, Mass., 1965.

Goodfellow Catalog, 800 Lancaster Avenue, Berwyn, PA 19312-1780, and Goodfellow Cambridge Limited, Cambridge Science Park, Cambridge, CB4 4DJ, England, web catalog *http://www. goodfellow.com.*

Vibration Analysis

D. E. Newland, *Mechanical Vibration Analysis and Computation*, 5th ed., Longman Scientific & Technical, Essex, England and John Wiley & Sons, New York, 1989.

S. S. Rao, *Mechanical Vibrations*, 2nd ed., Addison-Wesley, Reading, Mass., 1990.

W. Weaver, Jr., S. P. Timoshenko, and D. H. Young, *Vibration Problems in Engineering*, John Wiley & Sons, New York, 1990.

MANUFACTURERS AND SUPPLIERS

Mechanical Components (general)

McMaster-Carr Supply Company
P.O. Box 440
New Brunswick, NJ 08903-0440
http://www.mcmaster.com

Support Structure Systems

80/20 Inc.
1701 South 400 East
Columbia City, IN 46725-8753
http://www.8020.net

Techno Inc.
2101 Jericho Turnpike
Box 5416
New Hyde Park, NY 11042-5416
http://www.techno-profi.com

Ceramics

Aremco Products, Inc.
P.O. Box 429
Ossining, NY 10562
(machinable ceramics)

Carborundum Company
Semiconductor Products Division
P.O. Box 664
Niagara Falls, NY 14302
Corning Glass Works (Macor)
Corning, NY 14830
(boron nitride)

Industrial Tectonics, Inc.
P.O. Box. 1128
Ann Arbor, MI 48106
(balls)

McDanel Refractory Porcelain Co.
510 Ninth Ave.
Beaver Falls, PA 15010
(rod, tube)

Glass

Corning Glass Works
Corning, NY 14830

Kimble Glass
P.O. Box 1035
Toledo, OH 43666

Metals

Burleigh Instruments, Inc.
100 Despatch Dr.
P.O. Box 388
East Rochester, NY 14445
(Super-lnvar)

Cabot Corp.
Stellite Division
1020 W. Park Ave.
Kokomo, IN 46901
(nickel alloys)

Carpenter Technology Corp.
150 W. Bern St.
Reading, PA 19603
(Invar)

H. Cross Co.
363 Park Ave.
Weehawken, NJ 07087
(refractory metal wire, ribbon)

Crucible Specialty Metals
Colt Industries
Box 977-TR
Syracuse, NY 13201
(titanium)

Goodfellow Corporation
800 Lancaster Avenue
Berwyn, PA 19312-1780
and
Goodfellow Cambridge Limited
Cambridge Science Park
Cambridge, CB4 4DJ

England
(metals, ceramics and plastics in small quantities)

Huntington Alloys, Inc.
Huntington, WV
(nickel alloys)

Kawecki Berylco Industries
220 E. 42nd St.
New York, NY
(beryllium copper)

A. D. Mackay, Inc.
104 Kings Highway-North
Darien, CT 06820
(refractory metals)

Philips Elmet Corp.
1560 Lisbon Rd.
Lewiston, ME 04240
(tungsten, molybdenum)

Plastics

Amoco Chemicals Company
Engineering Resins
200 East Randolph Dr.
Chicago, IL 60601

E. I. Du Pont de Nemours & Co.
Plastic Products and Resins Department
Wilmington, DE 19898

General Electric
Plastics Business Division
1 Plastics Ave.
Pittsfield, MA

Adhesives

Devcon Corp.
Danvers, MA 01923

Sauereisen Cements Co.
Blawnox Station
Pittsburgh, PA 15238

Techni-Tool, Inc.
5 Apollo Rd.
Plymouth Meeting, PA 19462

Tra-Con, Inc.
55-T North St.
Medford, MA 02155

Pins, Especially Spring Pins and Roll Pins

Driv-Lok, Inc.
1140 Park Ave.
Sycamore, IL 60178

Elastic Stop Nut Corp. of America
2330 Vauxhall Rd.
Union, NJ 07083

Esna Rollpin
2330 Vauxhall Rd.
Union, NJ 07083

Precision Shafts, Balls, Bearings, and Mounts

Ace Plastic Company
91-30 Van Wyck Expwy.
Jamaica, NY 11435
(nonmetallic balls)

W. M. Berg, Inc.
499 Ocean Ave.
East Rockaway, NY 11518

Industrial Tectonics, Inc.
P.O. Box 1128
Ann Arbor, MI 48106
(balls)

New Hampshire Ball Bearings, Inc.
Astro Division
Laconia, NH 03246

Northfield Precision Instrument Corp.
4400 Austin Blvd.
Island Park, NY 11558

PIC
P.O. Box 335, Benrus Center
Ridgefield, CT 06877

Product Components
30 Lorraine Ave.
Mt. Vernon, NY 10553
(nonmetallic balls)

RMB Miniature Bearings
4 Westchester Plaza
Elmsford, NY 10523

Specialty Ball Co.
P.O. Box 1128
Ann Arbor, MI 48106

Retaining Rings

Truarc Retaining Rings Div.
Waldes Kohinoor, Inc.
47-16 Austed Place
Long Island City, NY 11101

Soldering, Brazing, and Welding Supplies

Engelhard Corporation
Brazing Products
Route 152
Plainville, MA 02762

Eutectic Welding Alloys Corp.
4040 172nd St.
Flushing, NY 11358

Handy and Harmon
850 Third Ave.
New York, NY 10022

J.W. Harris, Co.
10930 Deerfield Rd.
Cincinnati, OH 45242

WESGO
477 Harbor Blvd.
Belmont, CA 94002

Spot Welders

Ewald Instruments Corp.
Route 7-T
Kent, CT 06757
Unitek Corp.
Weldmatic Div.
1820 South Myrtle
Monrovia, CA 91016

Springs

Associated Spring Corporation
Wallace Barnes Division
Bristol, CT 06010
Lee Spring Company, Inc.
30 Main St.
Brooklyn, NY 11201

Threaded Fasteners and Inserts

Heli-Coil Products
1564 Shelter Rock Lane
Danbury, CT 06810

Tridair Industries
3000 W. Lomita Blvd.
Torrance, CA 90505

Testing Services

Balance Technology
120 Enterprise Dr.
Ann Arbor, MI 48103

Jodon Engineering Associates, Inc.
145 Enterprise Dr.
Ann Arbor, MI 48103

Magnaflux Corp.
6300 W. Lawrence Ave.
Chicago, IL 60656

CHAPTER 2

WORKING WITH GLASS

Glass has been called the miraculous material. The ubiquity of glass in the modern laboratory certainly confirms this. Because glass is chemically inert, most containers are made of it. Glass is transparent to many forms of radiation, and its transmission properties can be varied by controlling its composition—all sorts of windows and lenses are made of glass. Because glass can be polished to a high degree and is dimensionally stable, most mirrors are supported on glass surfaces. Glass is strong and stiff and is often used as a structural material. Considering its mechanical rigidity and density, it is a reasonably good thermal insulator. It is an excellent electrical insulator. Perhaps the greatest virtue of this material is that most glasses are inexpensive and can be cut and shaped in the laboratory with inexpensive tools.

Thirty-five years ago, most glass laboratory apparatus were produced by the scientist or technician *in situ* by blowing molten glass or by grinding, cutting, and polishing hard glass. Today the glass industry has grown to such an extent that nearly all components of a glass apparatus are available from commercial sources at low cost. These include all sorts of containers, chemical labware, vacuum-system components, mirrors, windows, and lenses. It is often only necessary for laboratory scientists to acquaint themselves with the range of components available and to acquire the skills needed to assemble an apparatus from these components.

2.1 PROPERTIES OF GLASSES

The chemical composition of glass is infinitely variable, and therefore so are the thermal, electrical, mechanical, and chemical properties of glass. Furthermore, glass is a fluid that retains a memory of its past history. It is possible however to review the general properties of glass and to specify the properties of glasses of a particular composition and method of manufacture.

2.1.1 Chemical Composition and Properties of Some Laboratory Glasses

The chief constituent of any commercial glass is silica (SiO_2). All laboratory ware is at least three-quarters silica, with other oxides added to obtain certain thermal properties or chemical resistance.

The least expensive and, until the last quarter century, the most common glass used for laboratory ware is known as *soda-lime* glass or *soft* glass. This glass typically contains 70 to 80 percent silica, 5 to 10 percent soda (Na_2O), 5 to 10 percent potash (K_2O), and 10 percent lime (CaO). The chief advantage of this glass is that it can be softened in a natural-gas-air flame.

Soda-lime glass has been largely replaced by *borosilicate* glass for the manufacture of labware. In this glass, the alkali found in soft glass is replaced by B_2O_3 and alumina (Al_2O_3). Borosilicate glass is superior to soda-lime glass in its resistance to chemical attack and thermal or mechanical shock. It softens at a higher temperature than soft glass, however, and is more difficult to work. Borosilicate glasses of many different compositions are manufactured for various laboratory applications, but far and away the most common laboratory glass is the borosilicate glass designated by Corning as *Pyrex 7740* and by Kimble as *Kimax KG-33* (composition is: SiO_2, 80.5%; B_2O_3, 12.9%; Na_2O, 3.8%; Al_2O_3, 2.2%; and K_2O, 0.4%).

Glassware with extremely good chemical resistance is produced from borosilicate glass by heat treatment, which causes the glass to separate into two phases—one high in silica and the other rich in alkali and boric oxides. This second phase is then leached out with acid, and the remaining phase is heated to give a clear, consolidated glass that is nearly pure silica. This glass is known as *96% silica glass* and is designated by Corning as Vycor No. 7900.

Glass composed only of silica is known as *vitreous* or *fused silica* or simply as *quartz*. Because of its refractory properties and chemical durability, this material would be the most desirable glass were it not for the high cost of making it and the extremely high temperatures required to work it. The relative cost of quartz is decreasing to the extent that the market for 96% silica glass is rapidly disappearing.

Most laboratory glasses are transparent to visible light, making visual distinction between the various glass compositions impossible. For this reason it is important to label glass materials before storing and avoid mixing different kinds of glass. When necessary it is possible to distinguish different glasses from one another by differences in thermal or optical properties. The gas-air flame of a Bunsen burner will soften soda-lime glass but not borosilicate glass. A natural-gas-oxygen flame is required to soften borosilicate glass, but this flame will not affect fused silica. Fused silica must be raised to a white heat in an oxyhydrogen flame before it will soften. Glasses of different composition generally cannot be fused together successfully because of differences in the amount of expansion and contraction on heating and cooling. An unknown piece of glass may be compared with a known piece by placing the two side by side with their ends coincident. The two ends are softened together in a flame and pressed together with tweezers. The fused ends are then reheated and drawn out into a long fiber about 0.5 mm in diameter and permitted to cool. If the fiber remains straight, the two pieces have the same coefficient of expansion. If the fiber curves, the two pieces are of different composition.

A measurement of the refractive index is usually a sensitive and reliable test of glass composition. A piece of glass placed in a liquid of exactly the same refractive index will become invisible. For example, a test solution for Pyrex 7740 can be made of 16 parts by volume of methanol in 84 parts benzene. This test solution should be kept in a tightly covered container so that its composition does not change as a result of evaporation.

As can be judged from the extreme chemical environments to which glasses are routinely subjected, glass is indeed resistant to chemical action. One need only observe windows clouded by the action of rainfall in a polluted atmosphere or glassware permanently stained by laboratory chemicals, however, to confirm that glass is not entirely impervious to chemical attack. Glass is attacked most readily by alkaline solutions, and all types are affected about equally. Water has an effect. The soft glasses are most susceptible, borosilicate glass is only slightly affected, and 96% silica hardly at all, since most of its soluble components are leached out in manufacture. Acids attack glass more readily than water, although once again borosilicate glass is more resistant to acids than soda glass, and 96% silica is more resistant yet.

2.1.2 Thermal Properties of Laboratory Glasses

One of the most important properties of glass is its very low coefficient of thermal expansion. It is this property that permits glass to be formed at high temperatures in a molten state and then cooled without changing shape or breaking. The coefficients of linear expansion of several types of glass are given in Table 2.1. As can be seen from these data, borosilicate glass is much more resistant to thermal shock than soft glass. Fused silica is so stable that a white-hot piece can be immersed in liquid air without fracturing.

Glass does not have a melting point. Instead, the working properties of a glass are specified by particular points on its viscosity-temperature curve. The *softening point* is approximately the temperature at which a glass

Table 2.1 THERMAL PROPERTIES OF GLASS

Glass	Linear Expansion Coefficient $(cm\ cm^{-1}\ K^{-1})$	Strain Point $(°C)$	Annealing Point $(°C)$	Softening Point $(°C)$	Working Point $(°C)$
Soda-lime (typical)	8–10×10^{-6}	500	550	700	1000
Pyrex 7740 (borosilicate)	3.3×10^{-6}	510	555	820	1250
Vycor 7900 (96% silica)	0.75×10^{-6}	890	1020	1500	—
Fused Silica	0.55×10^{-6}	950	1100	1600	—

can be observed to flow under its own weight. The *annealing point* is the temperature at which internal stresses can be relieved (annealed) in a few minutes. The *strain point* is the temperature below which glass can be quickly cooled without introducing additional stress. The *working point* is the temperature at which the glass is formed by a skilled glassblower.

The specific heat of glass is about 0.8 J g^{-1} K^{-1}, and the thermal conductivity of glass is about 0.008 J cm sec^{-1} cm^{-2} K^{-1}. For reference, this thermal conductivity is about an order of magnitude less than that of graphite, two orders less than that of metal, and about an order of magnitude greater than that of wood.

2.1.3 Optical Properties of Laboratory Glassware

In many experiments light must be transmitted through the wall of a glass container. The transmission as a function of wavelength depends upon the composition of the glass. The composition of the soda-lime glasses varies considerably, and the optical properties of each piece should be determined by spectroscopic analysis before it is used in an experiment requiring light transmission. Spectrophotometric curves for some lab glasses of well-defined composition are given in Figure 2.1. As can be seen, fused silica is transparent over the widest wavelength range.

2.1.4 Mechanical Properties of Glass

The tensile, compressive, and shear strength of a piece of glass depends upon its shape and history and upon the time over which it is loaded. Glass appears to be much stronger in compression than tension. This is in part due to the fact that a glass surface can be made very smooth in order to uniformly distribute a compressive load. Values of 60,000 to 180,000 psi are quoted for the compression strength. For reference, 1 psi (1 pound per square inch) = 6900 N m^{-2}.

Glass is nearly perfectly elastic. When it breaks, it does so without plastic deformation. Springs made of glass or quartz fibers behave almost ideally. The modulus of elasticity of glass is high and of course depends upon composition. Young's modulus (the modulus of elasticity) for Pyrex 7740 is 9.1×10^6 psi at 0°C and 9.4×10^6 psi at 100°C. For fused silica, the corresponding values are 10.5×10^6 and 10.7×10^6 psi.

Glass fibers are very strong. A quartz fiber 0.01 inch in diameter has a tensile strength of 50,000 psi; a fiber 0.001 inch in diameter has a strength of at least 200,000 psi; and a fiber 0.0001 inch in diameter has a strength in excess of 1,000,000 psi.

2.2 LABORATORY COMPONENTS AVAILABLE IN GLASS

The list of laboratory apparatus components produced commercially in glass is nearly endless. The laboratory scientist should make a careful survey of lab suppliers' literature before embarking on the design and construction of a glass apparatus. Very often a complex device can be assembled in the lab entirely from inexpensive components without requiring the services of a skilled

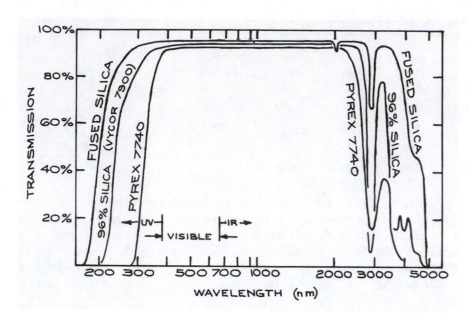

Figure 2.1 Optical-transmission curves for Pyrex 7740, Vycor 7900 (96% silica), and fused silica.

glassblower. Some of the most frequently used glass components are described below.

2.2.1 Tubing and Rod

Most laboratory supply houses carry a wide range of tubing and rod of Pyrex 7740 or Kimax KG-33 in 4 ft. lengths at a cost of only a few dollars per pound. Outer diameters between 3 and 178 mm are readily available. The standard wall thickness ranges from 0.5 mm for the smallest tubing to 3.5 mm for the largest. Heavy-wall tubing with wall thickness ranging from 2 to 10 mm is also available. In addition, heavy-wall tubing of very small bore diameter (0.5–4 mm), known as capillary tubing, is available. Finally, solid rod of 3 to 30 mm diameter is a standard item.

2.2.2 Demountable Joints

Laboratory apparatus can be quickly assembled if components are connected with joints of the type illustrated in Figure 2.2.

A gas- or liquid-tight seal between two close-fitting pieces of glass can be achieved if the mating surfaces are lightly coated with a viscous lubricant before assembly. A number of low-vapor-pressure lubricants such as Apiezon vacuum grease or Dow-Corning silicone vacuum grease are especially formulated for this purpose. The taper joint and the ball-and-socket joint in Figure 2.2a and 2.2b are assembled in this manner. The mating surfaces must fit together very well. In general, this requires that they be lapped together. This is accomplished by coating the surfaces with a fine abrasive such as Carborundum in water, fitting the pieces together, and rotating one with respect to the other. The rotation should not be continuous, but rather after half a turn or so the pieces should be pulled apart and then assembled again so as to redistribute the abrasive. Fortunately, taper and ball-and-socket joints with ground mating surfaces are now commercially produced in standard sizes with sufficient precision that lapping and grinding in the lab is seldom necessary.

The standard taper (indicated $\overline{\underline{5}}$) for ground joints is 1:10. Standard-taper joints on tubing with outer diameters between 8 and 50 mm are available. These joints are identified by figures that indicate the diameter of the large end of the taper and the length of the ground zone in millimeters. For example, $\overline{\underline{5}}$ 10/30 indicates a ground zone 10 mm in diameter at the large end and 30 mm in length. Standard-taper ground joints are available in

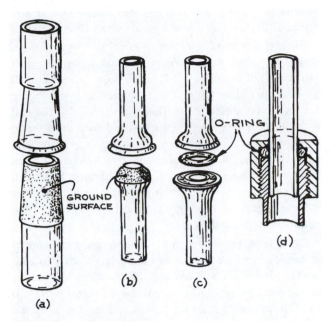

Figure 2.2   Demountable joints: (a) standard-taper joint;
(b) ball and socket; (c) O-ring joint; (d) a quick connect. (a),
(b), and (c) are used for attaching glass components to one
another; (d) is used for joining a glass tube to a metal
component.

borosilicate glass, quartz, and type-303 stainless steel.
These components are interchangeable, and thus it is often
convenient to make a transition from glass to quartz or
from glass to metal with a taper joint.

Ball-and-socket ground joints do not seal as reliably as
taper joints. Their design permits misalignment and even
some slight motion between joined parts. This type of joint
must be secured with a suitable clamp. Ball-and-socket
joints are designated by a two-number code (i.e., 28/15).
The first number gives the diameter of the ball in
millimeters, the second number the inner diameter of the
tubing to which the ball and socket are attached.
Ball-and-socket joints of standard sizes are commercially
available in borosilicate glass, quartz, and stainless steel.

The O-ring joint illustrated in Figure 2.2c is rapidly
replacing the ground joint as a means of making
demountable joints in glass vacuum apparatus. These
joints require no grease. Furthermore, the mating pieces
are sexless, since each member of the joint is grooved to a
depth of less than half the thickness of the O-ring. To date,
these joints are only available in borosilicate glass.

Quick connects of the type shown in Figure 2.2d are
also sealed with an O-ring. They are suited for joining
either glass or metal tubing to a metal container. Quick
connects are often used to join glass ion-gauge tubes to
metal vacuum apparatus.

2.2.3 Valves and Stopcocks

A glass stopcock of one of the types illustrated in Figures
2.3a through c has traditionally been used for controlling
fluid flow in glass apparatus. These consist of a hollow
tapered body and a plug with a hole bored through it that
can be aligned by rotation with inlet and outlet ports in the
body. The simplest and least expensive version uses a
glass plug with a ground surface that mates with a ground
surface on the interior of the stopcock body. The surface of
the plug must be lightly lubricated with stopcock grease to
ensure a good seal and freedom of rotation. A stopcock of
this type may be used to control liquid flow or gas flow at
pressures down to a few millitorr. For use at pressures
above 1 atm the end of the plug may be fitted with a
spring-loaded clip or collar to prevent its being blown out.
Stopcocks of this type are now available with a Teflon
plug. These are well suited for use with liquids. No
lubrication is required, and the plug seldom freezes in the
body of the stopcock.

In glass vacuum systems, the stopcock illustrated in
Figure 2.3c is far more reliable than the simple solid plug
design. The plug is hollow, and the small end of the body
is closed off by a vacuum cup so that the interior of the
stopcock can be evacuated. The plug is held securely in
place by atmospheric pressure.

Valves with threaded borosilicate glass bodies and
threaded Teflon plugs (see Figure 2.3d) are rapidly
replacing conventional stopcocks for most applications.
These valves are no more expensive than good vacuum
stopcocks. They require no lubrication; only glass and
Teflon are exposed to the interior of the system. They are
suitable for use with most liquids and gases and may be
used at pressures from 10^{-6} torr to 15 atm.

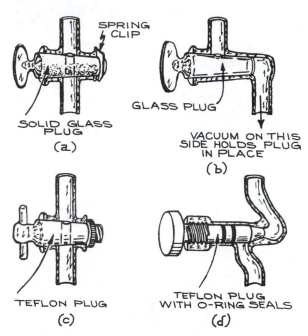

Figure 2.3 Stopcocks and valves for controlling fluid flow: (a) glass stopcock with solid glass plug; (b) high-vacuum stopcock with vacuum cup and hollow plug; (c) glass stopcock with Teflon plug; (d) threaded glass vacuum valve with O-ring-sealed Teflon stem.

2.2.4 Graded Glass Seals and Glass-to-Metal Seals

In general, glasses of different composition have different coefficients of thermal expansion. As a consequence, two glasses of different composition usually cannot be joined since differing rates of expansion and contraction produce destructive stresses as the fused joint cools. In practice, two glasses can be successfully joined if their coefficients of thermal expansion differ by no more than about $1 \times 10^{-6}\,\mathrm{K}^{-1}$. The problem of joining two glasses with significantly different coefficients of thermal expansion is solved by interposing several layers of glass, each with a coefficient only slightly different from its neighbors. When fused together, this stack of glass is called a *graded seal*. Tubing with graded seals joining different glasses is manufactured commercially. Borosilicate-to-soft-glass seals are available in tube sizes from 7 to 20 mm. Seals graded from

borosilicate glass (for example, Pyrex 7740) to quartz (Vycor 7913) are available in tube sizes from 7 to 51 mm.

Glass tubing can be joined to metal tubing with care. Ideally the glass and the metal should have the same coefficient of expansion. It has been demonstrated, however, that glass tubing can be joined to metal tubing that has a very different coefficient provided that the end of the metal tubing has been machined to a thin, sharp, feathered edge[1]. The stresses of thermal expansion and contraction can then be taken up by stretching of the metal. The stresses created by joining glass to metal can also be reduced by interposing glasses whose coefficient of thermal expansion is intermediate between that of the metal and the final glass.

Direct tube seals between borosilicate glass and copper or stainless steel are available in diameters from $\frac{1}{4}$ to 2 in. Graded glass seals between borosilicate glass and Kovar, an iron-nickel-cobalt alloy, are available in sizes from $\frac{1}{8}$ to 3 in. These graded seals are robust and far and away the most common glass-to-metal seal. Kovar is easily silver soldered or welded to most steels and brasses.

2.3 LABORATORY GLASSBLOWING SKILLS[*]

Considering the range of components now available, it is usually only necessary for the laboratory worker to be able to join commercially produced components to assemble a complete apparatus. The required skills consist primarily of joining tubes in series or at right angles, cutting and sealing-off tubing, and making simple bends. It is easiest and often necessary to make a joint between a piece of tubing held in the hand and a piece attached to an apparatus or otherwise rigidly clamped. In the following sections, only the simplest glassblowing operations with borosilicate glass are described. There are certainly many more elegant and complex manipulations that can be carried out by a skilled glassblower using only hand tools. Workers who find a need for these operations or discover

---

[*] Many of the simplified methods described in Section 2.3 were developed by John Trembly of the University of Maryland for use in his remarkably successful introductory glassblowing classes.

in themselves a flair for glassblowing should consult a more complete text on the subject.

2.3.1 The Glassblower's Tools

The necessary equipment for laboratory glasswork is illustrated in Figure 2.4. The basic tool is, of course, a torch. The National type-3A blowpipe with #2, #3, or #4 tip is the standard in the trade. A torch stand is essential to hold the lighted torch when both hands are required to manipulate a piece of glass. A source of natural gas and low-pressure oxygen is required, along with two lengths of flexible rubber or Tygon tubing for connecting the gas supplies to the torch. A rigid ring stand and appropriate clamps with fiberglass-covered jaws are needed to support glass being worked upon. The torch and stand should be set up on a fireproof bench top. Also required are a number of small hand tools, inexpensively obtained from a scientific supply house. These include a torch lighter, a wax pencil, a glass-cutting knife, a swivel, a mouthpiece (an old tobacco-pipe bit will do), glassblowing tweezers, a collection of corks and one-hole stoppers, and a variety of reaming tools. These reaming tools consist of flat triangular pieces of brass fitted with wooden file handles, and tapered round or octagonal carbon rods in various sizes. Didymium eyeglasses are necessary to filter out the intense sodium D-line emission produced when glass is exposed to the flame. Without these it is impossible to see hot glass as it is worked. Clip-on didymium lenses are available for these who wear prescription eyeglasses.

A little preliminary practice with the hand torch is advisable. The gas inlet may be connected directly to a low-pressure, natural-gas outlet in the lab. The outlet valve is turned completely open and gas flow is controlled by the valve in the body of the torch. Oxygen is usually obtained from a high-pressure cylinder. A pressure regulator for the oxygen tank is essential. The outlet pressure should be between 6 and 10 psi. Once again, the flow of oxygen gas is controlled by the valve in the body of the hand torch. The proper procedure is to open the gas control and ignite the gas first. The gas should be adjusted to give a flame 2 to 3 in. long before the oxygen is introduced. If the flame blows out, the oxygen valve should be closed and the gas permitted to flow for a few moments to purge the torch of

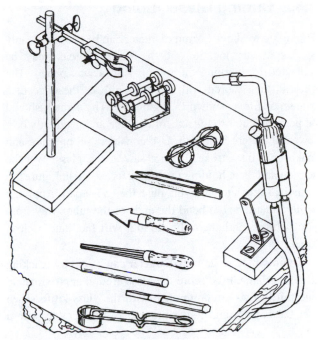

Figure 2.4   Glassblower's equipment.

the explosive gas-oxygen mixture before reignition is attempted.

With a #3 torch tip it is possible to produce a range of flames suitable for work on tubing from a few millimeters up to 50 mm in diameter. Opening the gas valve wide and adding only a little oxygen gives a large, bushy yellow flame. This is a cool, reducing flame suitable for preheating large pieces of glass and for annealing finished work. At the opposite extreme a sharp blue oxidizing flame can be obtained by restricting the gas flow. For work on fine tubing and for cutting tubing, it is possible to make a flame no more than an inch long and $\frac{1}{8}$ inch in diameter. The hottest flame is made with an excess of oxygen. The hottest part of the flame is the tip of the blue inner cone.

Some practice will be required to master even the simple operations described below. The neophyte should obtain the necessary tools and a supply of clean, dry borosilicate glass tubing of 8 to 12 mm in diameter. About 10 hours of practice is required before one can profitably embark on any serious apparatus construction.

2.3.2 Cutting Glass Tubing

Tubing up to about 12 mm diameter can be broken cleanly by hand. The location of the desired break may be indicated by marking the tubing with a wax pencil. The glass is then scored with a glass knife or the edge of a triangular file as shown in Figure 2.5. The scratch should be perpendicular to the axis of the tube and need only be a few millimeters in length. Use considerable pressure and make only one stroke. Do not saw at the glass. Wet the scratch and then, holding the tubing as shown in Figure 2.6 with the scratch toward you, push the ends apart. Applying a force that tends to bend the ends of the tube away from you, while simultaneously pulling, will facilitate a clean break.

Tubing of 12 to 25 mm diameter can be cut by cracking it with a flame. First score the glass around approximately one third of its circumference with the glass knife. Then wet the scratch and touch the end of the scratch with a very fine sharp flame that is oriented tangentially to the circumference as shown in Figure 2.7.

Large-diameter tubing can be parted by cracking it, using a resistively heated wire held in a yoke as illustrated in Figure 2.8. The tubing is placed on the resistance wire with one end butted against a stop. The wire is heated red-hot by passing current through it, and the tube is rotated to heat a narrow zone around its circumference. After a moment, the tube is quickly removed and the hot zone is brushed with a wet pipe cleaner. The thermal shock produced should result in a clean break.

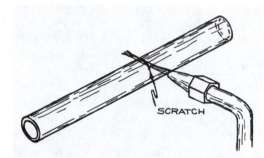

Figure 2.7   Cracking tubing with a flame.

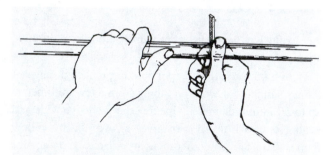

Figure 2.5   Scoring glass tubing in preparation for breaking. Only one stroke should be used with considerable force.

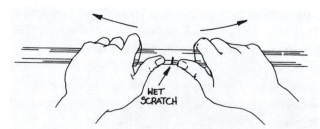

Figure 2.6   Breaking glass tubing. The tubing is bent and simultaneously pulled apart.

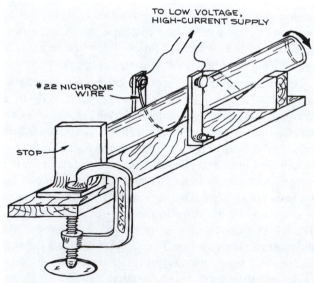

Figure 2.8   Device for cutting large-diameter glass tubing.

The edge of the glass at a fresh break is sharp and fragile. After cutting, the end of a tube can be *fire-polished* to relieve this hazardous condition. This is accomplished simply by heating the end of the tube in the flame until the glass is soft enough that surface tension rounds and thickens the sharp edge.

2.3.3 Pulling Points

Points are elongations on the end of a tube produced by heating the glass and stretching it. These points serve as handles for manipulating short pieces of tubing. Pulling a point is the first step in closing the end of a tube.

The procedure for pulling points depends upon whether the tubing can be rotated or not. If the tubing is free, then place the torch in its holder with the flame pointed away from you and hold the glass in both hands as illustrated in Figure 2.9. Adjust the torch to give a fairly full, neutral flame. Then, rotating the glass at a uniform rate, pass it through the flame to heat a zone about as wide as the tube diameter. When the glass softens, remove it from the flame and pull the heated section to a length of about 8 in. Then quickly heat the center of the stretched section and pull the glass in two, leaving a small bead at the end of each point. It is important that the points be on the axis of the original tube. This requires that the circumference of the tube be heated uniformly and that the ends of the tube be pulled straight away from one another. Some practice is required to synchronize the motion of the hands so that both ends of the tubing are rotated in such a way that the tube does not twist or bend when it becomes pliable in the flame. Beginners tend to overheat glass, thus exacerbating this problem of misalignment. When glass is sufficiently plastic for proper manipulation, it is still rigid enough to help support the end sections.

If one end of a tube is attached so that the tube cannot be rotated, then it becomes necessary to swing the torch around the tubing rather than rotate the glass in front of the stationary torch. To achieve a smooth motion with the torch so that the tube is heated uniformly around its circumference requires practice. The unattached end of the tubing is supported with the free hand until the glass is sufficiently hot to be pulled.

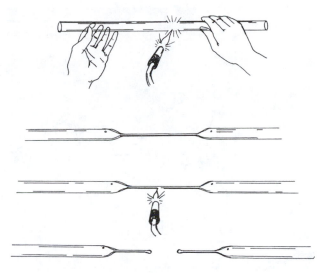

Figure 2.9   Pulling points.

2.3.4 Sealing Off Tubing: The Test-Tube End

Tubing is closed off by forming a hemispherical bubble similar to that found at the end of a test tube. In order for the *test-tube end* to form a strong closure, it must be smooth and of a uniform thickness. The hemispherical shape is formed by closing off the tubing and then softening the glass in the flame and blowing into the tube.

Begin by pulling a point on the tubing that is to be closed off. If the tubing is not rigidly mounted, place it in a clamp secured to a ring stand on the workbench. If possible, the work should be at chest height with the point upward. Press a one-hole stopper, with a piece of soft rubber tubing attached, into the open end of the tube as shown in Figure 2.10. Warm the glass at the base of the point with a soft bushy flame. Adjust the gas-oxygen mixture to the torch to give a fairly small flame with a well-defined, blue inner cone. Heat the point close to the shoulder where it joins the tube. Swing the torch around the work so that the glass is heated uniformly. When the glass melts, pull the point off the tube. Only a small bead of glass should be left behind. A large blob of glass will result in a closure that is too thick in relation to the wall of the tubing. A test-tube end that is thick in the middle and thin at the sides will develop destructive strains, since it

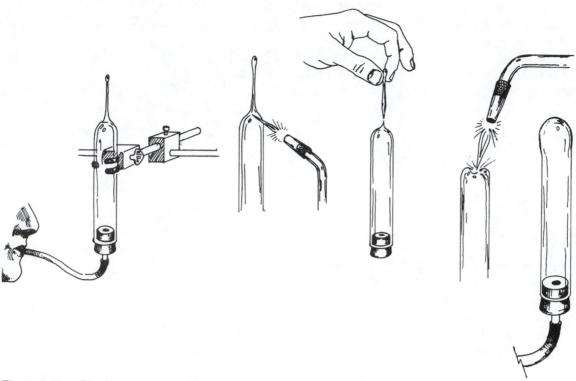

Figure 2.10   Blowing a test tube end.

does not cool at a uniform rate. The key is to heat only a narrow zone of the point near the shoulder and to perform the whole operation rather quickly. If a large globule of glass is left after pulling off the point, the excess must be removed. With a sharp flame, heat the globule until it melts. Remove the flame, touch the molten bead with the end of a piece of glass rod (*cane*), and quickly pull the rod away. Excess glass will be pulled away from the soft bead into a fiber which can be broken or melted off. The test-tube end is completed by heating the closed-off end as far as the shoulder with a somewhat larger flame until it is soft and the glass has flowed into a fairly uniform thickness. The flame should be directed downward at the end of the tube with a circular motion. Remove the flame and gently blow the end into a hemispherical shape. If a flat end is desired, reheat the round end and gently blow while pressing a flat carbon block against the end.

2.3.5 Making a T-Seal

A tube is joined at a right angle to the side of another tube with a *T-seal*. To make a T-seal, close one end of a tube with a cork and attach the blow tube to the other. If possible, mount the tube horizontally at chest height. With a sharp flame heat a spot on the top side of the tube. Using a circular motion of the torch, soften an area about the size of the cross section of the tube that is to be joined. Then move the flame away and gently blow out a hemispherical bulge as shown in Figure 2.11b. Reheat the bulge and blow out a thin-walled bubble. Finally, while blowing to pressurize the inside of the tube, touch the bubble with the tip of the flame so that a hole is blown through it. Now heat the edge of this hole so that surface tension pulls the glass back to the edges of the original bubble. The result should be a round opening slightly smaller than the diameter of the tube that is to be joined, as shown in Figure 2.11f. If the opening is too small or if the edge is

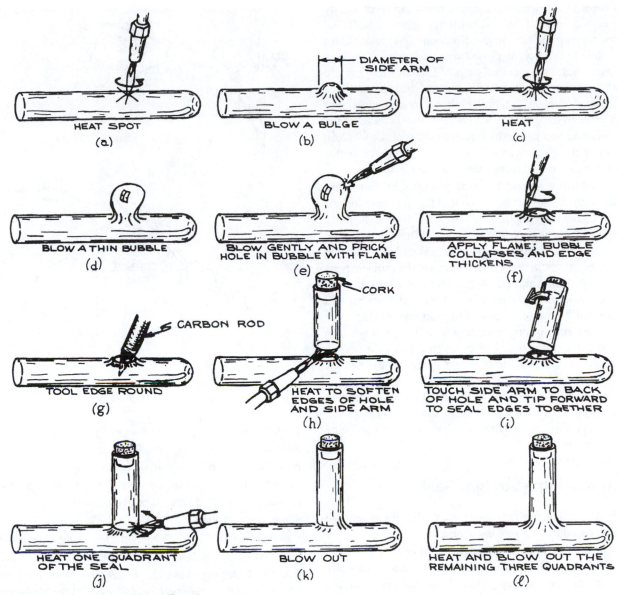

HEAT SPOT
(a)

BLOW A BULGE
(b)

DIAMETER OF SIDE ARM

HEAT
(c)

BLOW A THIN BUBBLE
(d)

BLOW GENTLY AND PRICK HOLE IN BUBBLE WITH FLAME
(e)

APPLY FLAME; BUBBLE COLLAPSES AND EDGE THICKENS
(f)

CARBON ROD

TOOL EDGE ROUND
(g)

CORK

HEAT TO SOFTEN EDGES OF HOLE AND SIDE ARM
(h)

TOUCH SIDE ARM TO BACK OF HOLE AND TIP FORWARD TO SEAL EDGES TOGETHER
(i)

HEAT ONE QUADRANT OF THE SEAL
(j)

BLOW OUT
(k)

HEAT AND BLOW OUT THE REMAINING THREE QUADRANTS
(ℓ)

Figure 2.11   Steps in making a T-seal.

irregular, the hole can be reamed and shaped by softening the glass and shaping the hole with a tapered carbon. A tool for this job can be made by sharpening a carbon rod with a diameter of a $\frac{1}{4}$ inch in a pencil sharpener.

With a cork or a point, close off the end of the tube that is to be sealed to the horizontal tube. Then, holding this tube above the hole and tipped slightly back, simultaneously heat the edges of the tube and the hole until they become soft and tacky. Bring the tube into contact with the back edge of the hole, remove the flame, and tip the tube forward until it rests squarely on the hole. The tacky edges should join together to produce an airtight seal. Blow to check the seal. Small holes can be closed by heating the seal all around until the glass is soft and tipping the vertical tube in the direction of the leak. It also is possible to *stitch* a gap closed with cane. This is done by pulling a point on a piece of glass rod. Then warm the gap which is to be repaired, heat the point of the cane until it is molten and deposit the tiny blob of molten glass thus produced into the gap. The seal at this point is quite fragile because the glass is too thick around the seal. Cracks will develop if the glass is permitted to cool before the seal is blown out in order to reduce the wall thickness.

Glassblowing around the seal proceeds in four steps. Begin with a sharp flame parallel to the axis of the horizontal tube. Swing the torch back and forth to heat one quadrant of the seal. When the glass softens, remove the flame and blow gently. Timing is important here. Blowing while the glass is in the flame will cause the thinnest parts to bulge. If, however, one waits a moment after removing the flame, the thinnest parts will cool and blowing will preferentially thin out the thick portions of the seal, yielding a more uniform wall thickness. The joint between the vertical and horizontal tubes should then be a smooth curve of glass. During the blowing operation the vertical tube must be held with the free hand to prevent its tipping or sagging into the horizontal tube. Next repeat the heating and blowing operation on the side diametrically opposite. Finally, in a third and fourth step, blow out the remaining two quadrants.

2.3.6 Making a Straight Seal

A *straight seal* joins two tubes coaxially. There are two simple procedures for making such a seal.

If the tubes are quite different in diameter, close off the larger one with a test-tube end and mount it vertically with the closed end up. Then blow a bubble of the appropriate size in the test-tube end and proceed as in making a T-seal.

Alternatively, the smaller tube can be flared out to the diameter of the larger one. The flaring operation is illustrated in Figure 2.12. With the torch in its stand, heat the edge of the tube by rotating it in a sharp flame. Remove it from the flame, and while continuing to rotate

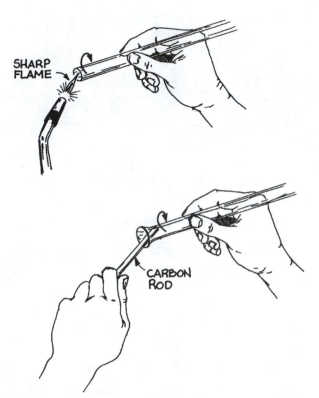

Figure 2.12   Flaring the end of a tube.

the tube, insert a tapered carbon and gently bring the soft edge of the glass up against the taper. Do not attempt to produce a flare of the desired diameter in one rotation of the tube. It will not come out round. The flaring operation is much easier if the tube is placed on a set of glassblower's rollers.

As shown in Figure 2.13a, the larger tube should be mounted vertically if possible. Suspend the smaller tube with the flare just above the larger one. With the upper tube tipped back slightly, heat the edges of the larger tube and the rim of the flare until the glass is tacky. Bring them into contact at the back of the seal, remove the flame, and tip the upper piece forward to join the two pieces. Proceed to blow out the seal in four steps as for the T-seal. Minimize the distance above and below the seal over which the glass is heated.

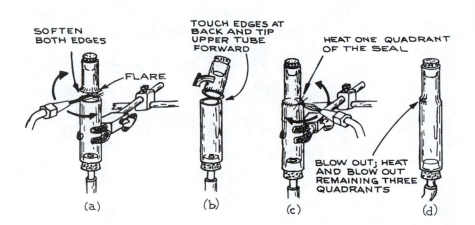

Figure 2.13   Steps in making a straight seal.

2.3.7 Making a Ring Seal

A *ring seal* is used for sealing a tube inside a larger tube as in Figure 2.14f, or for passing a tube through the wall of a container. There are many procedures for making such seals. The simplest involves bulging the smaller-diameter tube sufficiently for it to close off the hole through which the tube must pass. A bulge in a tube is known as a *maria*. Making a maria is a freehand operation and requires some practice.

Begin with the torch in its holder and the gas supply adjusted to give a long, narrow flame. Rotating the tubing in both hands, heat a very narrow zone at the desired location. As soon as the glass softens, drop the tube out of the flame, and, with your elbows braced on the bench top or against your sides, push the ends of the tube toward one another. If the tube has been heated uniformly around and the ends are held coaxial, a bulge should develop around the tube. If the heating is not uniform, the tube will bend. If the glass is too soft, the ends of the tube are liable to move out of alignment and an eccentric bulge will result. When a uniform bulge is attained, reheat it, remove it from the flame, and push again to increase the diameter of the maria. The bulge should be solid glass. If too wide a zone is heated, the maria will be hollow and dangerously fragile.

To make a coaxial ring seal between two tubes, form on the smaller tube a maria of the same diameter as the larger. Mount the larger tube vertically, fire-polish the top edge, and rest the smaller tube on top as in Figure 2.14e. With a small flame, heat the junction uniformly between the maria and the edge of the larger tube. The weight of the smaller tube should cause it to settle and become fused to the larger one. Finally blow out the seal in four steps, as in making a conventional seal. Exercise care to prevent the wall of the outer tube from sagging into the inner tube. If the inner tube is not centered, it can be realigned by softening the seal with a cool, bushy flame and manipulating its protruding end.

2.3.8 Bending Glass Tubing

Bends are best made beginning with the tubing vertical (Figure 2.15a). Close off the upper end with a cork or a point and attach the blowtube to the other end. While supporting the upper end of the tube with the free hand, apply a large bushy flame at the location of the desired bend. Heat the glass uniformly over a length equal to about four times the diameter of the tube. When the glass becomes pliable, remove the flame and quickly bend the upper end over to the desired angle. As soon as the bend is completed, blow to remove any wrinkles which may have developed and to restore the tubing to its original diameter. Some reheating may be necessary.

A sharp right-angle bend can be produced by first making a T and then pulling off one running end of the T close to the seal (Figure 2.15b). Blow out a test-tube end where the tubing is pulled off.

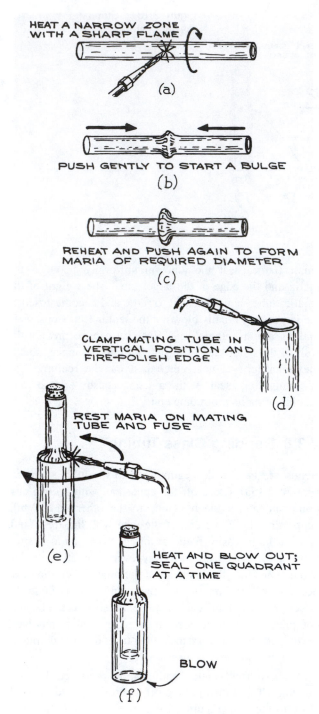

Figure 2.14   Steps in making a ring seal.

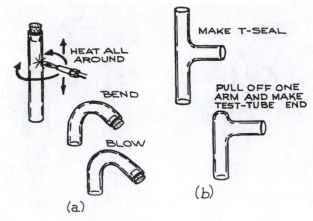

Figure 2.15   Bending tubing.

2.3.9 Annealing

Even when a seal is carefully blown out to give a uniform wall thickness, harmful stresses are inevitably frozen into the glass. Worked glass must be annealed by heating to a temperature above the annealing point (see Table 2.1) and then slowly cooling to the strain point. In a large professional glass shop, special ovens are employed for this operation. In the lab, however, glass must be annealed in the flame of the torch. After a piece of glass has been blown, it is heated uniformly over a wide area surrounding the work with a large, slightly yellow, bushy flame. For Pyrex 7740, the glass should be brought to an incipient red heat. Care must be taken to prevent the glass from becoming so hot that it begins to sag under its own weight. The heated work is then cooled by slowly reducing the oxygen to the torch until the flame becomes sooty. When soot begins to adhere to the glass, the flame can be removed.

Areas in glass which are under stress will rotate the plane of polarization of transmitted light. A *polariscope* can be used to observe strain in glass in order to test the effectiveness of the annealing operation. A polariscope can be purchased from a laboratory supplier or can be constructed with high-extinction sheet polarizers, as shown in Figure 2.16. The polarizers are crossed so that unstrained glass appears dark when placed between them. Areas of strain rotate the plane of polarization and appear bright, as in Figure 2.17. A large stationary piece of

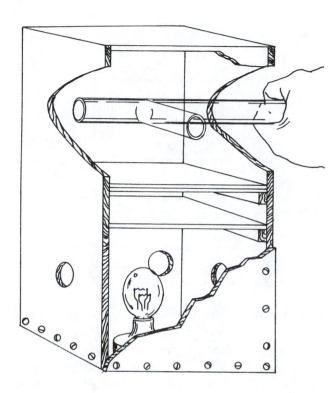

Figure 2.16   A polariscope.

Figure 2.17   The appearance of strain in unannealed glass as seen in the polariscope.

glasswork, such as a vacuum manifold, can be tested by illuminating it from behind with vertically polarized light

and then viewing the work through Polaroid sunglasses (which transmit horizontally polarized light). A suitable source can be constructed by placing a 100-watt lamp behind a diffuser in a ventilated, open-top box, and covering the box with a sheet of polarizer sandwiched between glass plates.

2.3.10 Sealing Glass to Metal

Glass-to-metal seals are fragile and difficult to produce in the lab. The simplest means of attaching metal tubing to a glass system or passing a metallic electrode through a glass wall is to make use of a commercially produced glass-to-metal seal (see Figure 3.28b). If necessary, a direct seal between a variety of metals and glasses can be produced with careful attention to design and fabrication.[2] Of these, the tungsten-to-Pyrex seal is most easily and reliably fabricated in the lab.

Although the linear coefficient of expansion of tungsten is about 50 percent greater than that of 7740 borosilicate glass, it is possible to make a strong, vacuum-tight seal between Pyrex and tungsten wire or rod of up to 0.050 inches in diameter.[3] Since the glass will wet tungsten oxide but not tungsten, it is necessary that the metal surface be properly prepared. Tungsten wire or rod usually has a longitudinal grain by virtue of the method of manufacture. The metal stock should be cut to length by grinding to prevent splintering. To prevent gas from leaking through the seal along the length of the wire, it is good practice to fuse the ends of the wire in an electric arc. Alternatively, a copper or nickel wire can be welded to each end of the tungsten wire. Tungsten electrodes with copper or nickel wires attached are also available from commercial sources. Holding the wire in a pin vise, heat the area to be sealed to a dull red and immediately rub the hot metal with a stick of sodium nitrite. The salt combines with the tungsten oxide to leave a clean metal surface. Rinse the metal with tap water to remove the salt and then rinse with distilled water. The metal must then be reoxidized in an oxygen-rich flame to produce an oxide layer of the proper thickness. If the layer is too thick it will pull away from the metal; if it is too thin it may all dissolve into the glass. The clean tungsten should be gently heated only until the metal appears blue-green when cool. A piece of 7740 glass, 2 cm long with a wall thickness of about 1 mm and an inner diameter slightly larger than the tungsten wire, is slipped over the wire and slid along to the sealing position. Be careful not to scratch the oxide layer. The glass tube is fused to the metal using a small sharp flame. Melting should proceed from one end to the other with continuous rotation to prevent the trapping of air. If the metal is properly wetted by the glass, the wire will appear oversize where it passes through the seal. A copper or amber color after the seal is cool indicates a good seal. The beaded tungsten rod is then sealed into the apparatus, as in making a ring seal (see Section 2.3.7). If necessary, a shoulder can be built up around the middle of the bead by fusing on a winding of 1 mm glass rod.

A tungsten-to-Pyrex seal makes an excellent electrical feedthrough. Wires can be brazed or spot welded to either end of the tungsten rod. Since tungsten is not a good electrical conductor, care must be taken to ensure that electrical currents do not overheat the feedthrough. The safe current-carrying capacity of the tungsten in this application is about 5000 A in.$^{-2}$.

2.3.11 Grinding and Drilling Glass

In many instances, the ends of a tube must be ground square or at an angle, or a tube must be ground to a specific length. The ends of a laser tube, for example, must be cut and ground to Brewster's angle before optical windows are glued in place. Grinding a plane surface is performed in a water slurry of abrasive on a smooth iron plate. It is done with a circular motion while applying moderate pressure. The work should be lifted from the surface frequently in order to redistribute the abrasive. If the angle of the ground surface relative to the workpiece is not critical, the work can be grasped in the hand; otherwise, a wooden jig, such as that shown in Figure 2.18, may be fabricated to orient the work with respect to the grinding surface.

Carborundum or silicon carbide grit of #30, 60, 120, or 220 mesh is used for coarse grinding, and emery of #220, 400, or 600 mesh is used for fine grinding. In any grinding operation, begin with the grit required to remove the largest irregularities and proceed in sequence through the

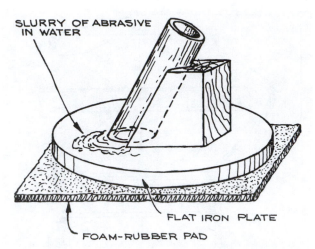

Figure 2.18   A wooden jig used in grinding the face of a piece of tubing.

finer grits. Before changing from one grit to the next finer grit, the work and the grinding surface must be scrupulously washed so that coarse abrasive does not become mixed with the next finer grade. Before changing from one grit to the next, it is useful to inspect the ground surface under a magnifying glass. If the abrasive in use has completed its job, the scratch marks left by the abrasive should be of a fairly uniform width.

Holes can be cut in glass, and round pieces of glass can be cut from larger pieces, with a slotted brass tube drill—a *cookie cutter*—in a drill press (see Figure 2.19). These drills are easily fabricated in sizes from 1/4 in. to several inches in diameter. Either flat or curved glass can be drilled. Cutting is accomplished by continuously feeding the drill with a slurry of #60 or #120 mesh abrasive. It is usually most convenient to form a dam of soft wax or putty around the hole location to hold the slurry of abrasive around the drill. Work slowly with a light touch at a drill speed of 50 to 150 rpm. The tool should be backed out frequently to permit the abrasive to flow under. Because grinding action occurs along both the leading edge and the side of the drill tube, the finished hole will be somewhat larger than the drill body. The edges of the hole will often chip where the drill breaks through on the underside of the work. This can be prevented when drilling

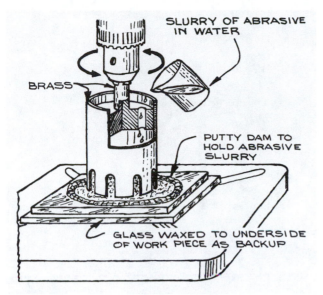

Figure 2.19 Drilling glass with a tube drill.

flat glass by backing up the workpiece with a second piece of glass waxed onto the underside.

Glass can be sawn as well as drilled. Many glass shops use a powered, diamond-grit cutting wheel for precision cutting. In the absence of such a machine, it is possible to cut glass in much the same manner as it is drilled. A hacksaw, with a strip of 20-gauge copper as a blade, is used. The work is clamped in a miter box and sawn with a light touch while feeding an abrasive slurry to the blade.

CITED REFERENCES

1. W. G. Housekeeper, Trans, A.I.E.E., **42**, 870 (1923).

2. F. Rosebury, *Handbook of Electron Tube and Vacuum Techniques,* Addison-Wesley, Reading, Mass., 1965, pp. 54–66.

3. J. Strong, *Procedures in Experimental Physics,* Prentice-Hall, Englewood Cliffs, N.J., 1938, p. 24; G. W. Green, *The Design and Construction of Small Vacuum Systems,* Chapman and Hall, London, 1968, pp. 100–102.

GENERAL REFERENCES

W. E. Barr and J. V. Anhorn, *Scientific and Industrial Class Blowing and Laboratory Techniques,* Instruments Publishing, Pittsburgh, 1959.

Corning Glass Works, *Laboratory Class Blowing with Corning's Classes,* Publ. No. B-72, Corning Glass Works, Corning, N.Y.

J.E. Hammesfahr and C. L. Strong, *Creative Class Blowing,* Freeman, San Francisco, 1968.

W. Morey, *The Properties of Class,* Reinhold, New York, 1954.

F. Rosebury, *Handbook of Electron Tube and Vacuum Techniques,* Addison-Wesley, Reading, Mass., 1965.

MANUFACTURERS AND SUPPLIERS

Ace Glass, Inc.
P.O. Box 688
1430 Northwest Blvd.
Vineland, NJ 08360

ChemGlass
3861 North Mill Rd.
Vineland, NJ 08360
Corning Glass Works
Corning, NY 14830

Kimble Glass, Owens-Illinois, Inc.
P.O. Box 1035
Toledo, OH 43666

Labglass
Northwest Blvd.
Vineland, NJ 08360

CHAPTER 3   VACUUM TECHNOLOGY

In the modern laboratory, there are many occasions when a gas-filled container must be emptied. Evacuation may simply be the first step in creating a new gaseous environment. In a distillation process, there may be a continuing requirement to remove gas as it evolves. It is often necessary to evacuate a container to prevent air from contaminating a clean surface or interfering with a chemical reaction. Beams of atomic particles must be handled *in vacuo* to prevent loss of momentum through collisions with air molecules. Many forms of radiation are absorbed by air and thus can propagate over large distances only in a vacuum. A vacuum system is an essential part of laboratory instruments such as the mass spectrometer and the electron microscope. Far-IR and far-UV spectrometers are operated within vacuum containers. Simple vacuum systems are used for vacuum dehydration and freeze-drying. Nuclear particle accelerators and thermonuclear devices require huge, sophisticated vacuum systems. Many modern industrial processes, most notably semiconductor device fabrication, require carefully controlled vacuum environments.

3.1 GASES

The pressure and composition of residual gases in a vacuum system vary considerably with its design and history. For some applications, a residual gas density of tens of billions of molecules per cubic centimeter is tolerable. In other cases, no more than a few hundred thousand molecules per cubic centimeter constitutes an acceptable vacuum: "One man's vacuum is another man's sewer."[1] It is necessary to understand the nature of a vacuum and of vacuum apparatus to know what can and cannot be done, to understand what is possible within economic constraints, and to choose components that are compatible with one another as well as with one's needs.

3.1.1 The Nature of the Residual Gases in a Vacuum System

The pressure below one atmosphere is loosely divided into vacuum categories. The pressure ranges and number densities corresponding to these categories are listed in Table 3.1. Vacuum science has been slow in adopting the SI unit of pressure—the Pascal (Pa)—preferring instead *torr* or *bar*. We will follow this preference. For the purposes of the discussion here, a pressure of 1 torr, 1 millibar, and 100 Pa can be taken as equivalent. As a point of reference for vacuum work, it is useful to remember that the number density at 1 mtorr is about 3.5×10^{13} cm^{-3} and that the number density is proportional to the pressure.

The composition of gas in a vacuum system is modified as the system is evacuated because the efficiency of a vacuum pump is different for different gases. At low pressures molecules desorbed from the walls make up the residual gas. Initially, the bulk of the gas leaving the walls is water vapor and carbon dioxide; at very low pressures, in a container that has been baked, it is hydrogen.

TABLE 3.1 AIR AT 20°C

	Pressure (torr)[a]	Number Density (cm^{-3})	Mean Free Path (cm)	Surface Collision Frequency (cm^{-2} s^{-1})	Time for Monolayer Formation[b] (s)
One atmosphere Atmosphere	760	2.7×10^{19}	7×10^{-6}	3×10^{23}	3.3×10^{-9}
Lower limit of:					
Rough vacuum	10^{-3}	3.5×10^{13}	5	4×10^{17}	2.5×10^{-3}
High vacuum	10^{-6}	3.5×10^{10}	5×10^{3}	4×10^{14}	2.5
Very high vacuum	10^{-9}	3.5×10^{7}	5×10^{6}	4×10^{11}	2.5×10^{3}
Ultrahigh vacuum	10^{-12}	3.5×10^{4}	5×10^{9}	4×10^{8}	2.5×10^{6}

[a] 1 torr = 1.33 mbarr = 133 Pa.
[b] Assumes unit sticking coefficient and a molecular diameter of 3×10^{-8} cm.

3.1.2 Gas Kinetic Theory

In order to understand mass flow and heat flow in a vacuum system, it is necessary to appreciate the immense change in freedom of movement experienced by a gas molecule as pressure decreases.

The average velocity of a molecule can be deduced from the Maxwell-Boltzmann velocity distribution law:

$$\bar{v} = \left(\frac{8kT}{\pi m}\right)^{1/2}$$

For an air molecule (molecular weight of about 30) at 20°C,

$$\bar{v} \approx 5 \times 10^{4} \, \text{cm s}^{-1} = \frac{1}{2} \, \text{km s}^{-1}$$

Each second a molecule sweeps out a volume with a diameter twice that of the molecule and a length equal to the distance traveled by the molecule in a second. As shown in Figure 3.1, this molecule collides with any of its neighbors whose centers lie within the swept volume. The number of collisions per second is equal to the number of neighbors within the swept volume. On average, the number of collisions per second (Z) is the number density of molecules (n) times the volume swept by a molecule of velocity $\bar{v}$ and diameter ξ. More precisely,

$$Z = \sqrt{2} n \pi \xi^{2} \bar{v}$$

where the $\sqrt{2}$ accounts for the relative motion of the molecules. The time between collisions is the reciprocal of this *collision frequency*. The average distance between collisions, or *mean free path*, is

$$\lambda = \bar{v} Z^{-1}$$
$$= \frac{1}{\sqrt{2} n \pi \xi^{2}}$$

It should be noted that the mean free path is independent of the temperature and the molecular weight.

The mean free path is inversely proportional to the pressure. For an N_2 or O_2 molecule, $\xi \simeq 3×10^{-8}$. The number density at 1 mtorr it is about 3.5×10^{13} cm^{-3}. The mean free path in air at 1 mtorr, therefore, is about 5 cm. Recalling the relationship between λ and P, a valuable rule of thumb is that for air at 20°C,

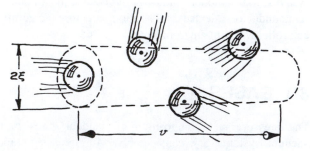

Figure 3.1   The volume swept in one second by a molecule of diameter ξ and velocity $\bar{v}$ (cm s^{-1}). The molecule will collide with any of its neighbors whose center lies within the swept volume.

$$\lambda \approx \frac{5}{P(\text{mtorr})} \text{ cm}$$

3.1.3 Surface Collisions

The frequency of collisions of gas molecules with a surface is

$$Z_{\text{surface}} = \frac{n\bar{v}}{4} \ (\text{s}^{-1}\text{cm}^{-2})$$

The sticking probability for most air molecules on a clean surface at room temperature is between 0.1 and 1.0. For water, the sticking probability is about unity for most surfaces. Assuming unit sticking probability and a molecular diameter of $\xi = 3\times10^{-8}$ cm, the time required to form a monolayer of adsorbed air molecules at 20°C is

$$t = \frac{2.5 \times 10^{-6}}{P(\text{torr})} \text{ s}$$

Thus to maintain a clean surface for a useful period of time may require a gas pressure over the surface less than 10^{-9} torr.

3.1.4 Bulk Behavior versus Molecular Behavior

The pressure of gas within a vacuum system may vary over ten or more orders of magnitude as the system is evacuated from atmospheric pressure to the lowest attainable pressure. At high pressures, when the mean free path is much smaller than the dimensions of the vacuum container, gas behavior is dominated by intermolecular interactions. These interactions, resulting in viscous forces, ensure good communication among all regions of the gas. At high pressures, a gas behaves as a homogeneous fluid. When the gas density is decreased to the extent that the mean free path is much larger than the container, molecules rattle around like the balls on a billiard table (not a pool table, where the ball density can be high), and gas behavior is determined by the random motion of the molecules as they bounce from wall to wall. Gas flow in the high pressure regime is characterized as *viscous flow*, and in the low pressure regime as *molecular flow*. In the typical laboratory vacuum system, the transition between the two flow regimes occurs at pressures in the vicinity of 100 mtorr.

In the viscous flow region, where the mean free path is relatively small, gas flow improves with increasing pressure because gas molecules tend to queue up and push on their neighbors in front of them. Gas flow is impeded by turbulence and by viscous drag at the walls of the pipe that is conducting the gas. In the viscous region, the coefficients of viscosity and thermal conductivity are independent of pressure.

In the molecular flow region, momentum transfer occurs between molecules and the wall of a container, but molecules seldom encounter one another. A gas is not characterized by a viscosity. Gas flows from a region of high pressure to one of low pressure simply because the number of molecules leaving a unit of volume is proportional to thc number of molecules within that volume. Gas flow is a statistical process. At very low pressure a molecule does not linger at a surface for a sufficient time to reach thermal equilibrium. Thus thermal conductivity at low pressures is a function of gas density, and the coefficient of thermal conductivity depends upon the pressure and the condition of the surface.

3.2 GAS FLOW

3.2.1 Parameters for Specifying Gas Flow

Before discussing vacuum apparatus, it is necessary to define the parameters used by vacuum engineers to characterize gas flow.

The volume rate of flow through an aperture or across a cross section of a tube is defined as the *pumping speed* at that point:

$$S \equiv \frac{dV}{dt} \qquad [\text{L s}^{-1}; \text{ m}^3 \text{ s}^{-1}; \text{ ft}^3 \text{ min}^{-1} \text{ (cfm)}]$$

The capacity of a vacuum pump is specified by the speed measured at its inlet:

$$S_{\text{p}} \equiv \frac{dV}{dt} \qquad (\text{at pump inlet})$$

The mass rate of flow through a vacuum system is proportional to the *throughput:*

$$Q \equiv PS$$

[(torr L s^{-1}; Pa m^3 s^{-1}; std cm^3/s (a *standard cubic centimeter* or *std cm^3* or *scc* is 1 cm^3 at 273K and 1 atm or 1.01 bar)]

To determine the throughput, it is necessary that the pressure and speed be measured at the same place, since these quantities vary throughout the system.

The ability of a tube to transmit gas is characterized by its *conductance* (C). The definition of conductance is analogous to Ohm's law for electrical circuitry. The throughput of a tube depends upon the conductance of the tube and the driving force, which in this case is the pressure drop across the tube

$$Q = (P_1 - P_2)C \qquad (P_1 > P_2)$$

Notice that conductance has the same units as pumping speed.

3.2.2 Network Equations

A complicated network of tubes can be reduced to a single equivalent conductance for the purpose of analysis. In analogy to electrical theory once again, a number of tubes in series can be replaced by a single equivalent tube with conductance C_{series} given by

$$C_{series} = \left[\frac{1}{C_1} + \frac{1}{C_2} + \frac{1}{C_3} \cdots \right]^{-1}$$

A parallel network can be replaced by an equivalent conductor with conductance

$$C_{parallel} = C_1 + C_2 + C_3 + \ldots .$$

3.2.3 The Master Equation

By use of the network equations given above, an entire vacuum system can be reduced to a single equivalent conductance leading from a gas source to a pump as shown in Figure 3.2. The net speed of the system is

$$S = \frac{Q}{P_1}$$

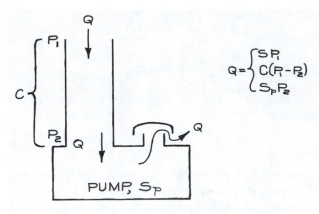

Figure 3.2   Analytical representation of a vacuum system.

$$S_p = \frac{Q}{P_2}$$

Notice that the throughput is the same at every point in this system, since there is only one gas source. The throughput of the conductance is

$$Q = (P_1 - P_2)C$$

Upon substitution for P_1 and P_2 from the previous two equations, this equation (after some rearrangement) becomes

$$\frac{1}{S} = \frac{1}{S_p} + \frac{1}{C}$$

This is the master equation that relates the net speed of a system to the capacity of the pump and the conductors leading to the pump. From this equation, we see that both the conductance of a system and the speed of the pump exceed the net speed of the system. In vacuum-system design, economics should be considered. The cost of a vacuum pump is usually greater than the cost of constructing a tube that leads from a vacuum container to the pump. In order to achieve a given net speed for a system, it is most economical to design a system whose equivalent conductance exceeds the required net speed by a factor of three or four so that the speed of the pump need only exceed the net speed by a small amount.

3.2.4 Conductance Formulae

In the viscous flow region, the conductance of a tube depends upon the gas pressure and viscosity.[2] For a tube of circular cross section, and in the absence of turbulence, the conductance from the Hagen-Poiseuille derivation is

$$C = 32,600 \frac{D^4}{\eta L} P_{av} \text{ L s}^{-1}$$

when the diameter (D) and length (L) of the tube are in centimeters, the average pressure P_{av} is in torr, and the viscosity (η) of the gas is in micropoise (μP). For air at 20°C, the conductance is

$$C = 180 \frac{D^4}{L} P_{av} \text{ L s}^{-1}$$

These equations are used for determining the dimensions required for the lines connecting to a pump that operates in the rough vacuum region. A complex system is treated as a series of simple cylindrical tubes. The conductance of each tube is calculated from the equation above and the conductance of the whole system is determined using the network equations. Joints and elbows between simple tubes will reduce the conductance of the system, so it is wise to specify tube diameters somewhat oversize to offset conductance losses in the transition between simple cylindrical shapes.

In the molecular flow region, conductance is independent of pressure. The conductance of a tube of circular cross section is

$$C = 2.6 \times 10^{-4} \bar{v} \frac{D^3}{L} \text{ L s}^{-1}$$

when the diameter and length are measured in centimeters and the average molecular velocity ($\bar{v}$) is in cm s^{-1}. For air at 20° C, the conductance is

$$C = 12 \frac{D^3}{L} \text{ L s}^{-1}$$

These equations are useful in determining the dimensions of a high-vacuum chamber and the size of the tubes leading to and from the chamber. As an example of poor design, consider the case of a vacuum chamber connected to a 100 L s^{-1} diffusion pump by a tube 2.5 cm in diameter and 10 cm long. The net speed of the pump and connecting tube is

$$S = \left(\frac{1}{100} + \frac{1}{19}\right)^{-1} = 16 \text{ L s}^{-1}$$

The pump is being strangled by the connecting tube. Consider also the pressure drop across the tube. Since the throughput is a constant, one can write

$$SP_1 = S_p P_2 = Q$$

thus

$$\frac{P_1}{P_2} = \frac{S_p}{S} = 6$$

This result indicates the importance of proper location of a pressure gauge. In this case, a gauge at the mouth of the pump would indicate a pressure six times lower than the pressure in the vacuum container.

There are frequently occasions when a vacuum system is divided into compartments at very different pressures, separated by a wall with an aperture whose diameter exceeds its length (a so-called *thin aperture*). In the viscous flow regime, the conductance of an aperture at 20°C is approximately

$$C \approx 20A \text{ L s}^{-1}$$

where A is the area in cm^2. In the molecular flow region the conductance of an aperture for a gas of molecular weight M is

$$C = 3.7\left(\frac{T}{M}\right)^{1/2} A \text{ L s}^{-1}$$

where A is the area in cm^2. These equations are useful for determining the rate of gas flow into a vacuum chamber through an aperture in a gas-filled collision chamber or ion source.

There is a *transition region* between the viscous and molecular flow regions where the mean free path approximately matches the dimensions of the container. For laboratory-scale apparatus, the transition region falls in the range of 500 mtorr down to 5 mtorr. A mathematical description of this region is difficult. Fortunately, one finds in most vacuum systems that the transition region is encountered only briefly during pumpdown from atmosphere to high vacuum. Conductance formulae, such as given above for flow in the viscous (high pressure) region compared to conductance formulae in the molecular (low pressure) region, show two important differences. Gas flow in the viscous region depends upon the gas pressure and the square of the cross section of the conductor (i.e., D^4), whereas gas flow in the molecular flow region is independent of pressure and goes as the 3/2 power of the cross section. It follows that the formulae for conductance in the molecular flow region will give a conservative estimate of the conductance in the transition region.

3.2.5 Pumpdown Time

The time required to pump a chamber of volume V from pressure P_0 to P is

$$t = \frac{V}{S}\ln\left(\frac{P_o}{P}\right)$$

assuming a constant net pumping speed and considering only the gas that resided in the container at the outset. A typical high-vacuum system is rough-pumped to about 50 mtorr with a mechanical pump, and then pumped to very low pressures with a diffusion pump or some other pump that works effectively in the molecular flow region. The pumpdown calculation is performed in two steps corresponding to these two operations. Removal of gas desorbing from the container walls must also be considered. The rate at which the pressure falls is determined by the rate of desorption.

3.2.6 Outgassing

The rate at which the pressure falls in a chamber at pressures below about 10^{-5} torr is largely determined by the rate of evolution of gas from the walls of the chamber rather than by the speed of the pump. The rate of gas evolution—*outgassing*—from solid materials depends both on the nature of the solid and the adsorbate. Water and carbon dioxide from the air are adsorbed most tenaciously. A rough, porous, or polar surface adsorbs more gas than a clean, polished surface. The outgassing rate from a surface decreases slowly, roughly as e^{-at} (a is a property of the surface and the adsorbate, as well as a strong function of temperature). Baking a vacuum chamber to 150 to 200°C under vacuum, will encourage outgassing and result in a lower ultimate pressure after the chamber cools.

Following exposure to air, stainless steel outgasses at roughly 10^{-8} torr L s^{-1} cm^{-2} of surface area after an hour of pumping at room temperature. A day of pumping may be required to reduce this rate below 10^{-9} torr L s^{-1} cm^{-2}. The outgassing rate from aluminum is more than ten times greater than from stainless steel. Outgassing from plastics is roughly ten times worse than from aluminum and outgassing from elastomers (rubber) is ten times worse yet. Plastics and rubbers have no use in a vacuum system that is to achieve pressures below about 10^{-7} torr.

Consider, for example, the problem of maintaining a base pressure of 10^{-7} torr in a stainless-steel chamber. If the outgassing rate is 10^{-9} torr L s^{-1} cm^{-2}, a pumping speed of 10^{-2} L s^{-1} is required for every cm^2 of surface. For a container 1 m in diameter and 1 m high, a pumping speed of 400 L s^{-1} is needed. This speed is typical of a high-vacuum pump (a diffusion pump or turbomolecular pump) with a nominal 4 in. diameter inlet port. Without baking to further reduce the outgassing rate, a pump ten times larger (and nearly ten times as expensive) would be required to reduce the pressure to 10^{-8} torr. Ultrahigh vacuum is out of the question.

3.3 PRESSURE AND FLOW MEASUREMENT

The pressure within a vacuum system may vary over more than thirteen orders of magnitude. No single gauge will operate over this range, so most systems are equipped with several different gauges. The pressure ranges in which the various gauges are useful are shown in Figure 3.7.

3.3.1 Mechanical Gauges

Hydrostatic gauges. The simplest pressure gauges are hydrostatic gauges such as the oil or mercury manometer. A closed-end, U-tube manometer filled with mercury is useful down to 1 torr. Diffusion pump oils such as Octoil, di-n-butyl phthalate, or Dow Corning DC704 or DC705 silicone oils, may also be used as manometer fluids. Because the density of these oils is much less than that of mercury, oil-filled manometers are useful down to about 0.1 torr. As shown in Figure 3.3, an oil manometer should be made of glass tubing of at least 1 cm diameter in order to reduce the effects of capillary depression. To prevent oil adhesion to the walls, the manometer should be cleaned with dilute hydrofluoric acid before filling. A heater is usually provided to outgas the oil prior to use.[3] A number of manometers have been designed to increase the surface area of oil exposed to the vacuum in order to hasten outgassing.[4] Both oil and mercury manometers will contaminate a vacuum system with vapor of the working fluid. If this is a problem, a cold trap is placed between the gauge and the system to condense the offending vapor. Mercury in a manometer can be isolated and protected from chemical attack by floating a few drops of silicone oil on top of the mercury column.

The McLeod gauge shown in Figure 3.4 is a sophisticated hydrostatic gauge and is sensitive to much lower pressures than a simple U-tube manometer. In the McLeod gauge, a sample of gas is trapped and compressed by a known amount (typically 1000:1) by displacing the gas with mercury from the original volume into a much smaller volume. The pressure of the compressed gas is measured with a mercury manometer and the original pressure determined from the general gas law. Of course, it is not possible to use a McLeod gauge to measure the pressure of a gas that condenses upon compression. With care, a high-quality McLeod gauge will provide accuracy of a few percent down to about 10^{-4} torr. Some gauges can be used at even lower pressures, but it is necessary to consider the error caused by the pumping action of mercury vapor streaming out of the gauge.[5]

The McLeod gauge is constructed of glass and is easily broken. If the mercury is raised into the gas bulb too quickly, it can acquire sufficient momentum to shatter the

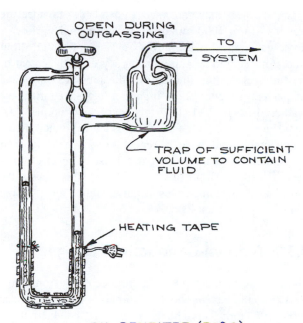

OIL DENSITES (20° C)

di-n-butyl phthalate   1.044 gm cm⁻³
Octoil   0.983
Octoil-S   0.912
DC-704   1.07
DC-705   1.09

Figure 3.3   Oil manometer. The trap is intended to contain the oil charge in the event that the low-pressure arm of the manometer is accidentally open to the atmosphere.

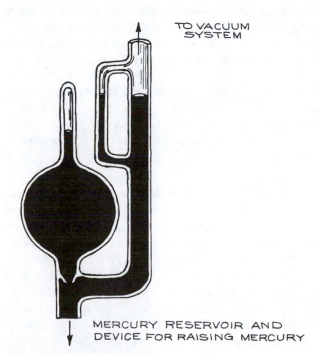

Figure 3.4   McLeod Gauge.

glass. When a gauge is broken, the external air pressure frequently drives a quantity of mercury into the vacuum system. For this reason it is wise to place a ballast bulb of sufficient volume to contain the mercury charge of the gauge between the gauge and the system.

Mechanical gauges. A number of different gauges depend upon the flexure of a metal tube or diaphragm as a measure of pressure. In the Bourdon gauge, a thin-wall, curved tube closed at one end is attached to the vacuum system. Pressure changes cause a change in curvature of the tube. A mechanical linkage to the tube drives a needle to give a pressure reading on a curved scale. Another type of mechanical gauge contains a chamber divided in two by a thin metal diaphragm. The volume on one side of the diaphragm is sealed, while the volume on the other side is attached to the system. A variation in pressure on one side relative to the other causes the diaphragm to flex, and this movement is sensed by a system of gears and levers that drives a needle on the face of the gauge. The precision of these gauges is limited by hysteresis caused by friction in the linkage. To overcome this friction, it is helpful to gently tap the gauge before making a reading. Unlike the liquid manometer, mechanical gauges are not absolute gauges. The pressure scale and zero location on these gauges must be calibrated against a McLeod gauge or U-tube manometer. Mechanical gauges are useful down to 1 torr with an accuracy of about 0.5 torr. They offer the advantage of being insensitive to the chemical or physical nature of the gas.

Capacitance manometers. A capacitance manometer is a diaphragm manometer wherein the position of the diaphragm is determined from a measurement of electrical capacitance. The gauge head is divided into two chambers by the thin metal diaphragm that serves as one plate of a capacitor; the second plate of the capacitor is fixed in one of the chambers. A change in the pressure difference between the two chambers results in a change in capacitance as the flexible diaphragm moves relative to the fixed plate. A sensitive capacitance bridge measures the capacitance between the two plates, and the capacitance is converted to a reading of the pressure difference. Capacitance manometers are inherently differential pressure gauges;

however, most are manufactured to read absolute pressure by sealing and evacuating one of the chambers. This arrangement does not provide a truly absolute gauge since the readout must be set to zero by connecting the open chamber to a true zero pressure (i.e., several orders of magnitude less than the full-scale reading). A capacitance manometer is designed so that the flexible diaphragm touches the opposing capacitor plate at a small overpressure. This protects the fragile diaphragm from a sudden inrush of gas. As might be expected, the capacitance measurement is very sensitive to the temperature of the flexible diaphragm. High-precision gauge heads incorporate a heater and temperature control that maintains the gauge head at a constant temperature.

Capacitance manometers are available to measure pressures from 1000 torr down to less than 10^{-4} torr. A single gauge head is useful over a range of three to four orders of magnitude with an accuracy of 0.01 percent of the full scale at the low end of the pressure range and 0.5 percent or better at the high end (e.g., a 1 torr gauge head reads $1\times10^{-4} \pm 1\times10^{-4}$ torr at the low end and 1.00 ±0.005 torr at the high end). Gauges are calibrated by the manufacturer and have a reputation for holding their calibration for years given careful use. The electronics have a response time of about 10 ms. This makes the capacitance manometer especially useful for pressure control in a feedback loop.

3.3.2 Thermal-Conductivity Gauges

The thermal conductivity of a gas decreases from some constant value above about 10 torr to essentially zero at about 10^{-3} torr. This change in thermal conductivity is used as an indication of pressure in the Pirani gauge and the thermocouple gauge. In both gauges, a wire filament is heated by the passage of an electrical current. The temperature of the filament depends on the rate of heat loss to the surrounding gas. In the thermocouple gauge, the pressure is determined from the voltage produced by a thermocouple in contact with the filament heated by the passage of a constant current (Figure 3.5). In the Pirani gauge, the heated filament is one arm of a Wheatstone bridge. A change in temperature of the filament produces a change of resistivity and hence a change in the voltage

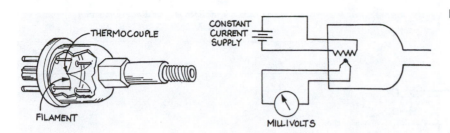

Figure 3.5   Thermocouple gauge.

across the filament. The resulting imbalance of the bridge gives an indication of pressure. Alternatively, a voltage can be applied to keep the bridge in balance and this voltage becomes the indication of pressure.

The pressure indicated by a thermocouple gauge or a Pirani gauge depends upon the thermal conductivity of the gas. They are usually calibrated by the manufacturer for use with air. For use with other gases, these gauges must be recalibrated point-by-point over their entire range, since thermal conductivity is a nonlinear function of pressure. In routine use, a thermal-conductivity gauge can only be expected to be accurate to within a factor of two. This accuracy is adequate when the gauge is used to sense the foreline pressure of a diffusion pump or to determine whether the pressure in a system is sufficiently low to begin diffusion pumping. The principal advantages of these gauges are their ease of use, ruggedness, and low cost.

3.3.3 Viscous-Drag Gauges

Above about 1 torr, the viscosity of a gas is constant, but, in the molecular flow regime, the viscous drag exerted by a gas on a moving surface depends upon the gas density and the mean velocity of the gas. It follows that below about 1 torr, a first-principles calculation yields gas pressure from a measurement of viscous drag for a gas of known temperature and composition. This approach to absolute pressure measurement is applied in the *spinning-rotor gauges* manufactured by Leybold AG and MKS Instruments.

In the gauge head of a spinning-rotor gauge, a magnetically suspended steel ball is spun up to a rotational speed of about 415 Hz by a rotating magnetic field. The drive is interrupted and the ball is permitted to decelerate to a preset lower speed—typically about 405 Hz. The

pressure is determined from a measurement of the time required for a complete revolution of the coasting ball. The rate of deceleration is very small—at 10^{-4} torr the deceleration from 415 to 405 Hz takes many minutes. The measurement of the ball rotation rate is repeated many times to obtain satisfactory precision. A microprocessor in the controller computes the statistical average of the deceleration rate after many cycles.

Spinning-rotor gauges provide absolute pressure measurement between 10^{-1} and 10^{-6} torr, with an accuracy of a few percent. At the lower end of this range, they are the only available means for making highly accurate, repeatable, absolute-pressure measurements. Spinning-rotor gauges are used as a transfer standard in calibrating other types of gauges. These instruments are expensive and must be maintained with great care.

3.3.4 Ionization Gauges

In the region of molecular flow, pressure is usually measured with an ionization gauge or *ion gauge*. In this type of gauge, gas molecules are ionized by electron impact and the resulting positive ions are collected at a negatively biased electrode. The current to this electrode is a function of pressure. There are two basic types of ion gauges, differing primarily in the mechanism of electron production. In the hot-cathode ionization gauge, electrons are produced by thermionic emission from an electrically heated wire filament. In cold-cathode gauges, electrons and ions are produced in a discharge initiated by a high-voltage electrical discharge.

The most familiar configuration of the hot-cathode ionization gauge is that devised by Bayard and Alpert and illustrated in Figure 3.6. The gauge electrodes are mounted in a glass envelope with a side arm that is attached to the

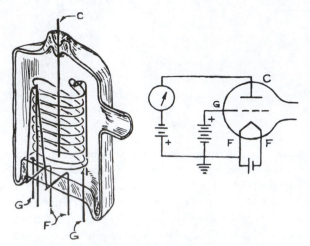

Figure 3.6   Bayard-Alpert ionization gauge.

TABLE 3.2 SENSITIVITY FACTORS FOR IONIZATION GAUGES

Gas	Formula	Multiplicative Factor[a]
Air		1.0
Nitrogen	N_2	1.0
Oxygen	O_2	0.9
Hydrogen	H_2	2.4
Carbon dioxide	CO_2	0.7
Methane	CH_4	0.7
Hydrocarbons	C_4 and up	0.2
Argon	Ar	0.9
Neon	Ne	3.5
Helium	He	6.8

[a] Calibrated for air. Recommended values based primarily upon the survey published by R. L. Summers, NASA Tech. Note TN D-5285, and an assumed nominal acceleration potential of 150 V.

vacuum system. Electrons from an electrically heated filament (F) are accelerated through the gas toward a positively biased, helical wire grid (G). Collisions between the accelerated electrons and gas molecules create ions that are collected at a central wire electrode (C), and the positive ion current is measured by a sensitive electrometer. These gauges are useful in the 10^{-3} to 10^{-10} torr range. The ion current measured with an ionization gauge is a linear function of pressure. The proportionality between ion current depends upon the electron emission current from the filament and upon the ionization efficiency of the gas in the gauge. Commercial gauge controllers are calibrated for use with air at a specified emission current—typically about 10 milliamperes (mA). An adjustment on the controller permits the user to set the emission current. For gases other than air, the indicated pressure is multiplied by a sensitivity factor that is inversely proportional to the electron-impact ionization cross section for the gas in question. Sensitivity factors for some common gases are given in Table 3.2.

Ion gauges have the sometimes useful and sometimes detrimental property of functioning as pumps. There are two mechanisms for pumping. Ionized gas molecules accelerated to and embedded in the collector are effectively removed from the system. In addition, metal evaporated from the filament deposits on the glass envelope of the gauge to produce a clean, chemically active surface that adsorbs nitrogen, oxygen, and water. At

pressures below 10^{-7} torr, this pumping action may create a significant pressure gradient between the gauge tube and the system to which it is attached, resulting in a pressure reading that is lower than that in the system. For ultrahigh vacuum systems, the pumping problem is circumvented by the use of *nude* gauges that consist of the gauge electrodes mounted to a flange that is installed so that the electrodes protrude into the system proper. There is no envelope or tubulation joining the gauge to the system.

The pressure indicated by an ionization gauge is also affected by the emission of gas adsorbed on the gauge electrodes and envelope. To overcome this problem, the gauge is periodically degassed by heating the electrodes and the envelope. This is accomplished by the gauge controller in the *degas* mode in which an ac current is passed through the grid to heat it to incandescence. Radiative heating of the surrounding components encourages outgassing. A hot-cathode ionization gauge is activated when the pressure falls below 10^{-3} torr—turn-on at higher temperatures will cause the filament to burn out. The gauge should be degassed for a few minutes when the pressure falls to about 10^{-4} torr and the degassing should be repeated with each decrease of two orders of magnitude in pressure.

The electron-emissive filament in the ion gauge is made of tungsten or thoria-coated iridium. Tungsten filaments operate at 1800°C. They provide the most stable operation but their lifetime is significantly shortened by exposure to hydrocarbons at relatively high pressures. Thoria-coated filaments operate at about 1200°C and can be expected to be longer-lived than tungsten filaments. Many gauge tubes provide two filaments with separate external connections. Only one filament is activated at a time. When the first filament burns out, the second can be used so that it is not necessary to let the system up to air pressure following the first failure. When the filaments fail in a glass-enclosed gauge, the entire gauge is replaced. Individual filaments are replaceable in nude gauges.

Cold-cathode ionization gauges in various configurations are identified as Penning gauges, magnetrons, inverted magnetrons, and Philips gauges. In these gauges, an electrical discharge is struck in a low-pressure gas between two electrodes maintained at a potential difference of several kV. The range of stability of the discharge is extended by a magnetic field provided by a permanent magnet that is an integral part of the gauge head. The magnetic field confines the electrons in the discharge so as to increase their pathlength through the gas. The discharge current measured at one of the electrodes provides a measure of the gas pressure. The measured current is relatively much larger than that obtained with a hot-cathode gauge—of the order of 1 A torr^{-1}—thus greatly simplifying the controller electronics. Cold-cathode gauges operate in the 10^{-2} to 10^{-7} torr range with an accuracy of a factor of two at best. The pressure range is extended down to less than 10^{-9} torr in some more complex designs. Without the fragile filament of the hot-cathode gauge, cold-cathode gauges are robust and economical, suffering primarily only from their limited range of applicability.

The operating pressure ranges of the vacuum gauges is shown in Figure 3.7.

3.3.5 Mass Spectrometers

A mass spectrometer can be used to determine the partial pressures of residual gases in a system and to detect leaks. Small, relatively inexpensive mass spectrometers designed specifically for these purposes are called Residual Gas Analyzers (RGAs). These instruments employ quadrupole or magnetic-deflection analyzers and typically cover the mass range from 2 to 100 amu. They are sensitive to partial pressures as low as 5×10^{-14} torr. Fixed-focus mass spectrometers set to detect helium are widely used as sensitive leak detectors. Leaks are located by monitoring the helium concentration within a vacuum system while the exterior of the system is probed with a small jet of helium. The minimum detectable leak rate of commercial helium leak detectors is less than 10^{-11} torr L s^{-1}.

3.3.6 Flowmeters

Gas flow can be measured as a throughput or as a mass flow. Flow rate as a throughput is determined from a measurement of the pressure drop across a tube or system of tubes of known conductance. The mass flow rate can be determined from a measurement of the power required to maintain a given temperature gradient across a tube through which a gas flows.

Differential pressure flowmeters. For gas flowing through a tube, the throughput depends upon the pressure difference between the entrance and exit of the tube and the conductance of the tube (from Section 3.2.1, $Q = [P_1 - P_2]C$). Obviously two pressure gauges are required to determine the throughput. In addition, the gas temperature must be measured and the dimensions of the tube chosen so that the gas flow remains in a definable flow regime (either viscous flow or molecular flow) so that the conductance of the tube can be calculated from the tube dimensions. The length of the tube should exceed the diameter by at least a factor of ten so that end effects on the conductance can be ignored. A capillary of 1 mm inner diameter and several centimeters in length would assure viscous flow conditions at downstream pressures above 1 torr and the conductance could be calculated from the Hagen-Poiseuille relation (see Section 3.2.4). To operate in the molecular flow regime, a tube of macroscopic dimensions would be only useful at very low flows and low downstream pressures. In practice, the flow rate in the molecular flow regime is measured using a parallel array of capillary tubes, the dimension of each tube being sufficiently small to assure molecular flow at pressures up to 1 torr. Capillary arrays with thousands of parallel

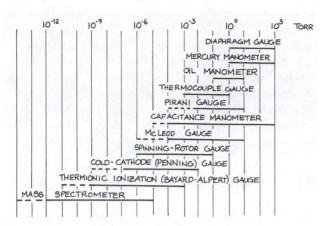

Figure 3.7   Operating ranges of pressure gauges.

channels—each 1 mm long with diameters as small as 25 micrometers (μ)—are available for this purpose.

The differential pressure flowmeter is appealing in that flow rate is accurately determinable from first-principles calculation; in practice, however, these gauges are used primarily for calibration purposes.

Thermal mass flow meters. Commercial flowmeters for routine use are of the thermal mass flow type. The mass flow of gas through a tube can be calculated from the temperature rise in the flowing gas as heat is added or from the power required to maintain a given temperature rise; the latter scheme is the basis of most commercially available gauges. Thermal mass flow gauges operate in the viscous flow regime where thermal conductivity of a gas is essentially constant. The flowmeter consists of a tube that is surrounded by a series of heaters that establish a temperature profile along the tube. Gas flowing through the tube modifies the temperature profile and this change of temperature can be used as a measure of the gas flow rate. Alternatively, the power to the heaters is adjusted by a sensitive bridge to keep the temperature profile fixed. The power to the heaters then becomes a measure of the mass flow rate. These gauges are usually calibrated for air. For other gases, a correction factor, inversely proportional to the heat capacity and density of the gas, is applied.

Thermal mass flow meters determine mass flow from a measurement of heat flow; however, these instruments are calibrated in throughput units of standard cubic centimeters per minute (sccm) of air or N_2. One standard cubic centimeter is the amount of gas contained in 1 cm^3 at 1 atm pressure and 0°C. Flowmeters are available with full-scale ranges of 10 to 50,000 sccm and an accuracy of about 1 percent of the full scale reading. Because thermal mass flow meters operate in the laminar viscous flow regime, they are always placed on the delivery side of a flow system. In choosing a gauge, one must consider the delivery pressure and the pressure drop across the gauge element. For condensable gases, it may be necessary to choose a gauge with a range larger than required to keep the delivery pressure below the vapor pressure of the sample. Thermal mass flow meters have a response time of the order of 1 sec and can be used in a feedback loop to control gas flow rate. Controllers incorporating a flow meter and an electrically driven metering valve are available commercially.

3.4 VACUUM PUMPS

Vacuum pumps are classified according to the chemical or physical phenomenon responsible for moving gas molecules out of the vacuum vessel. Like pressure gauges, every vacuum pump operates in a limited pressure range. The useful range of a pump is also limited by the vapor pressure of the materials of construction and the working fluids within the pump. In general, a pump that operates in the viscous flow region will not operate in the molecular flow region and vice versa. To achieve very low pressures, two or more pumps are used in series. In addition, some pumps are more efficient for high-molecular-weight gases than others, and some pump low boiling gases (e.g., H_2 and He) more efficiently than others. To achieve ultrahigh vacuum it is common practice to employ pumps with complimentary characteristics in parallel.

Mechanical pumps operating in the rough vacuum region with ultimate low-pressure limits of 5 torr to 5 mtorr serve two important functions in attaining high and ultrahigh vacuum. A pump that will achieve a high vacuum at its inlet will invariably have a maximum tolerable outlet pressure—this is the *critical backing pressure*. A mechanical pump is employed to maintain the high vacuum pump exhaust pressure below the critical pressure, and in this function the mechanical pump is referred to as a *backing pump* or *forepump*. In addition, it is necessary for the pressure in a vacuum chamber to be below some pressure before the high vacuum pump can be put into operation—this pressure is sometimes referred to as the *stall pressure*. A mechanical pump is used to attain this maximum starting pressure in the chamber and in this function the pump is referred to as a *roughing pump*.

3.4.1 Mechanical Pumps

Oil-sealed rotary vane pumps. The pump most commonly used for attaining pressures down to a few millitorr is the oil-sealed rotary vane pump shown schematically in Figure 3.8. In this pump a rotor turns off-center within a cylindrical stator. The interior of the pump is divided into two volumes by spring-loaded vanes attached to the rotor. Gas from the pump inlet enters one of these volumes and is compressed and forced through a one-way valve to the exhaust. The seal between the vanes and the stator is maintained by a thin film of oil. The oil used in these pumps is a good quality hydrocarbon oil from which the high-vapor-pressure fraction has been removed. These pumps are also made in a two-stage version in which two pumps with rotors on a common shaft operate in series. Rotary pumps to be used for pumping condensable vapors (water vapor in particular) are provided with a gas ballast. This is a valve that admits air to the compressed gas just prior to the exhaust cycle. This additional air causes the exhaust valve to open before the pressures of condensable vapors exceed their vapor pressure and thus prevents these vapors from condensing inside the pump.

Oil-sealed rotary vane pumps will operate for years without attention if the inner surfaces do not rust and the oil maintains its lubricating properties. It is wise to leave these pumps operating continuously so that the oil stays warm and dry. In use, an increase in the lowest attainable pressure (the base pressure) indicates that the oil has been contaminated with volatile materials. The dirty oil should be drained while the pump is warm. The pump should be filled with new oil, run for several minutes, drained, and refilled with a second charge of new oil. For storage, a pump should be filled with new oil and the ports sealed.

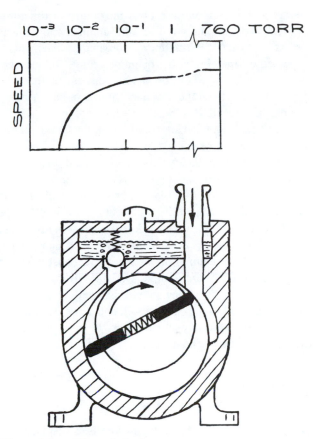

Figure 3.8   Oil-sealed rotary vane pump.

Rotary pumps are available with capacities of 1 to 500 L s^{-1}. A single-stage pump is useful down to 50 mtorr, and a two-stage pump to 5 mtorr. Typical performance of a single-stage pump is indicated by the pumping-speed curve in Figure 3.8. A two-stage pump is to be preferred as a backing pump for a diffusion pump. With a two-stage pump, a base pressure of 10^{-4} torr can be achieved after a long pumping time if the back diffusion of oil vapor from the pump is suppressed by use of a sorption or liquid-air trap on the pump inlet. This is a simple scheme for evacuating small spectrometers and Dewar flasks or other vacuum-type thermal insulators.

The exhaust gases from an oil-sealed mechanical vacuum pump contain a mist of fine droplets of oil. This oil smoke is especially dense when the inlet pressure exceeds 100 torr. The oil droplets are extremely small, usually less than 5 microns. Over the course of time, this oil settles on the pump and its surroundings and collects dirt and grime. Furthermore, breathing the finely-dispersed oil may injure the operator's lungs. In addition, over the course of weeks or months, this oil represents a significant loss from the oil charge in the pump. A pump failure may result if the oil level in the pump is not faithfully monitored. Most pump manufacturers market filters that cause the exhaust oil mist to coalesce and run back into the pump; their use is recommended. In addition, modern standards of laboratory hygiene require venting a mechanical pump into the laboratory fume hood. If toxic gases are being pumped, it is, of course, essential to pipe the exhaust out of the lab. Sometimes a pretreatment involving passage of the exhaust through a neutralizing chemical bath is required. In most cases, exhaust lines can be made of PVC drain pipe available from plumbing suppliers. The exhaust line should rise vertically from the pump so that oil in the exhaust will collect on the walls of the line and run downward back to the pump. An exception to this scheme arises in the case of a wet vacuum process. If considerable water is exhausted from the pump, there is the danger that water will condense in the exhaust line and run back into the pump.

Roots blowers.

A Roots blower is a displacement pump. As illustrated in Figure 3.9, these pumps consist of a pair of counter-rotating two-lobed rotors on parallel shafts. Rotational speeds are about 3,000 rpm. There is a clearance of a few thousandths of an inch between the rotors themselves and between the rotors and the housing. A volume of gas is trapped at the inlet and compressed as it is moved to the exhaust. There is no oil in the body of the pump to maintain a high-pressure seal, but owing to the high speed of the pumping motion a compression ratio as great as 40:1 can be achieved. The virtue of the Roots blower is its relatively high pumping speed in the 1 to 10^{-3} torr region. To achieve an ultimate vacuum of 10^{-3} torr, a Roots blower is used in series with a rotary pump at its exhaust. The required speed of the rotary pump is smaller than that of the Roots pump by a factor of the inverse of the compression ratio of the Roots pump. Roots blowers are available with speeds of a few hundred to many thousand Ls^{-1}.

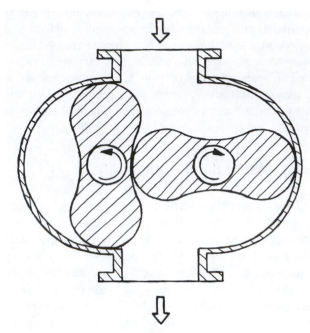

Figure 3.9   Roots blower.

Piston pumps.

Now available are piston pumps with Teflon-sealed pistons that do not contaminate the vacuum environment with oil as do oil-sealed rotary pumps. The Leybold EcoDry pump employs a novel diaphragm exhaust valve that permits a higher compression ratio than pumps with poppet valves. The low-pressure limit of these pumps is about 50 mtorr. Piston pumps are robust and reliable, but they are also relatively expensive in comparison to rotary pumps. They find use as backing pumps for pumps such as turbomolecular pumps when contamination with hydrocarbons is not permissible.

Diaphragm pumps.

Diaphragm pumps are displacement pumps employing the flexure of a thin metal or Teflon diaphragm to compress gas and drive it out through an exhaust valve. These are small pumps of low pumping speed and relatively modest ultimate vacuum. The low pressure limit is typically about 1 torr. Diaphragm pumps can be made completely free of hydrocarbons, so they are used as roughing pumps and in some applications as backing pumps in ultraclean systems.

Molecular drag pumps. A molecular drag pump incorporates a rotating cylinder within a closely fitting, stationary cylinder. The clearance between rotor and stator is of the order of 0.01 in. Pumping occurs when the surface velocity of the rotor approaches the velocity of the molecules being pumped so that molecules striking the rotor acquire a significant velocity component tangential to the rotor—they are dragged along with the rotor. Tangentially accelerated gas molecules rebounding to the stator are decelerated—they *pile up*, thus yielding compression. The process is repeated continuously around the gap between rotor and stator. Helical grooves on the surface of the rotor or stator direct the flow of continually compressing gas toward the pump outlet. Drag pumps provide efficient pumping at inlet pressures from nearly 1 torr down to about 10^{-6} torr. The *compression ratio* for air (that is, the ratio of outlet to inlet pressure) is typically 10^7 at the low-pressure end of this range. As a consequence, the exhaust pressure (that is, the critical backing pressure) falls in the 1 to 10 torr range. The practical result is that a very small pump is required to transfer the drag pump throughput to the atmosphere. Small diaphragm pumps serve well as backing pumps for molecular drag pumps.

The chamber pressure at the inlet must be below about 1 torr for a drag pump to begin to operate efficiently; therefore, a roughing pump is necessary. The backing pump, pumping through the drag pump, can serve to rough down the chamber. The gap between rotor and stator, however, is small and restrictive in the viscous flow regime. A separate roughing pump or a bypass (with a valve) from the chamber to the backing pump should be provided.

Molecular drag pumps provide modest pumping speeds at high vacuum—the inlet aperture is effectively the gap between rotor and stator. Drag pumps are currently available with pumping speeds of up to 30 L s^{-1}. The great virtue of the drag pump is the relatively high throughput in the 1 to 10^{-3} torr pressure range. Displacement pumps such as rotary-vane pumps, piston pumps, and Roots blowers, lose efficiency below 0.1 torr. High vacuum pumps such as diffusion pumps, turbomolecular pumps and ion/getter pumps, *stall* at pressures above 0.01 torr. If pumpdown time is an important issue, a molecular drag pump may be a good choice for a small vacuum system with an ultimate pressure of 10^{-6} torr. With a 1 Ls^{-1}

roughing/backing pump and a drag pump of 30 L s^{-1}, a volume of 100 L can be evacuated to less than 10^{-4} torr in a matter of minutes.

Turbomolecular pumps. Turbomolecular pumps operate in the molecular flow regime. The construction is similar to that of an aircraft-type jet turbine engine. A series of bladed turbine rotors on a common shaft turn at 20,000 to 90,000 rpm. The edge speed of a rotor approaches molecular velocities. The rotor blades are canted so that a molecule striking a blade receives a significant component of velocity in the direction of the pump exhaust (Figure 3.10). Bladed stators are interleaved between the rotors. The stator blades are canted in the opposite direction from that of the rotors in order to decelerate the molecules and compress the flowing gas before it is delivered downward to the next rotor-stator pair. These pumps provide roughly the same pumping speed for all gases; however the compression ratio depends upon the nature of the gas being pumped. A single rotor-stator pair typically provides a compression ratio of about 10 for N_2. To a first approximation, the logarithm of the compression ratio is proportional to the square root of the molecular weight of the gas. For example, a multistage pump with eight rotors may have a compression ratio of 10^8 for N_2, but only 10^2 to 10^3 for H_2. A desirable consequence of the strong dependence upon molecular weight is that the compression ratio is very high for oil vapor backstreaming from the pump bearings or from the pump exhaust. A turbo pump therefore provides an essentially oil-free vacuum. The compression ratio is extremely sensitive to the pressure at the outlet of the pump. A typical turbopump may have a compression ratio of 10^8 for air if the pressure at the exhaust is maintained below 0.1 torr, but if the exhaust pressure rises to 1.0 torr, the compression ratio falls to 10.

A turbomolecular pump is usually run in series with a conventional oil-sealed rotary pump as a backing pump. The backing pump also serves as a forepump. With the vacuum chamber at atmospheric pressure, the rotary pump is activated to evacuate the chamber, drawing gas directly through the body of the turbopump. Depending on the manufacturer's specification, the turbo is activated at the same time or somewhat later as the chamber pressure falls

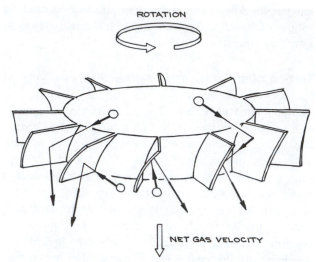

ROTATION

NET GAS VELOCITY

Figure 3.10   Rotor of turbomolecular pump.

so that the turbo is coming up to its operating speed as the chamber pressure reaches a pressure of a few millitorr.

The critical elements of a turbomolecular pump include the motor and the bearings. Typically, 400 Hz synchronous motors are used. Active electronic control is employed to maintain rotor speed over a wide range of loads and to protect the pump in the event of an overload. The motor driver and control represent a significant fraction of the total cost of a turbomolecular pumping system. Many turbopumps use water-cooled, grease-packed bearings. Some use oil as a bearing lubricant with the oil being circulated to a water-cooled heat exchanger. Magnetic suspension systems are the most recent development (Balzers, Osaka, and Varian). There is no mechanical contact in a magnetic bearing; there is essentially no friction and lubrication is unnecessary. In some pumps, the rotor is completely magnetically suspended. In others, a magnetic bearing is used at the inlet end of the rotor with a conventional grease-lubricated bearing at the outlet relying on the high pumping speeds for hydrocarbons to suppress oil backstreaming. Axial and radial magnetic fields to support the rotor are provided by electromagnets in a feedback system that senses the position of the rotor and makes corrections to offset disturbing forces. A mechanical bearing is provided to catch the rotor in the event of catastrophic imbalance and to support the rotor

when the pump is not in operation. A backup electrical system is required to protect against a sudden loss of power to the magnet system when the pump is in operation. This may simply be a battery, but in sophisticated systems the pump motor is operated as a generator when external power is lost in order to provide electricity for the magnetic bearings.

The most recent innovation is a compound pump design incorporating a multistage turbomolecular pump mounted on a common axis above a molecular drag pump, as shown in Figure 3.11. These are sometimes called hybrid pumps or compound turbomolecular/molecular drag pumps. A complete system has three pumps operating in series: a turbomolecular pump, a molecular drag pump, and a displacement-type backing pump (Balzers). The hybrid arrangement increases the compression ratio of the high vacuum pumping stage by several orders of magnitude, with a concomitant increase in the outlet pressure. The tolerable backing pump pressure of a hybrid pump is in the 1 to 10 torr range. With the increase in outlet pressure comes a corresponding decrease in the volume flow rate at the exhaust. The important consequence is that these pumps require only a small dry-diaphragm backing pump that, with the inclusion of a magnetic rotor suspension, creates an ultrahigh vacuum pumping system that is completely free of hydrocarbons. Complete turn-key systems are available that include a compound pump with a motor controller, a backing pump, and all necessary interlocks to control start up and shutdown. Pressures below 10^{-10} torr are attainable.

The high rotational speeds and close tolerances in a turbomolecular pump place a premium upon careful mounting, cleanliness, and proper maintenance. Owing to the very small clearances between moving and stationary components, twisting and bending forces that would cause distortion cannot be applied to the pump body. The pump should not be used as the mounting platform for the vacuum system. The pump should be suspended by its inlet flange. Small particles can have disastrous effects upon bearings and rotors at high speeds. The vacuum container pumped by a turbopump must be kept scrupulously clean. A protective screen over the inlet, known as a splinter shield, is essential. When the vacuum container is to be let up to atmospheric pressure, gas

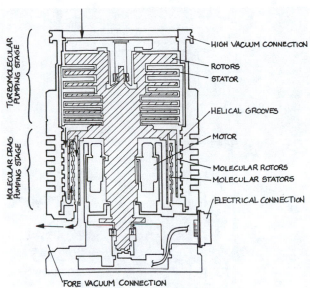

Figure 3.11 Compound turbomolecular/molecular drag pump.

should be admitted slowly so that turbulence does not carry debris into the turbopump.

Careful and systematic startup and shutdown procedures are essential to protect the turbopump and to exclude oil from the vacuum system. As mentioned above, turbos are usually operated without a valve at the inlet; however a valve is required in the *foreline* between the turbo outlet and its backing pump. At startup with the turbo and the vacuum system at atmospheric pressure, the backing pump is activated and the foreline valve is opened. The rush of dense gas into the backing pump prevents oil from backstreaming. As the pressure falls, the turbo activation is timed so that the turbo is up to at least 50 percent of its maximum speed before the pressure at the pump inlet falls below a few hundred millitorr. At this point, the turbo compression should be sufficient to prevent oil vapor from backstreaming to the vacuum system. At shutdown, the foreline valve is closed immediately after power is cut from the turbo. The vacuum system should be vented before the rotor falls below 50 percent of its maximum speed—again so that the compression ratio is sufficient to prevent backstreaming of oil from the bearings. When the pressure has risen above several hundred torr, backstreaming is essentially impossible. It follows that a pump that is not operating should not be stored under vacuum. It also should be emphasized that the pump must be vented above the outlet. Many manufacturers sell turbopumps and controllers as a complete system that incorporates the necessary valves and timing circuits to automatically perform the startup and shutdown procedures.

A turbomolecular pump offers many advantages for high- or ultrahigh-vacuum systems. Of course these must be weighed against the disadvantages, the main one being the relatively high initial cost and the cost of major servicing. A complete turbomolecular pump system may cost two or more times as much as a comparable diffusion-pump system. Turbomolecular pumps with pumping speeds from 10 to 1000 L s^{-1} are available. A turbopumped system can achieve pressures below 10^{-10} torr. Most gases can be pumped, although as noted turbos do not efficiently pump hydrogen and helium. Most corrosive gases are acceptable providing the bearing lubricant does not come under attack. Turbopumps are compact and relatively light in weight; they need not be mounted on the underside of a vacuum system. For many types of pumps, the mounting orientation is not critical—these can be mounted with the axis horizontal. Turbos can be baked to reduce outgassing and many are equipped with heating jackets for this purpose. A turbo can be used on a system that is to be baked provided the inlet temperature does not exceed 100 to 120°C. Magnetic fields can be a problem as eddy currents induced in the aluminum rotor assembly cause heating. The specification for most pumps calls for magnetic fields not to exceed 50 to 100 gauss at the inlet. A turbomolecular pump can run for one to three years without attention. The major maintenance procedure usually involves replacing the bearings and rebalancing the rotor. These procedures often require the pump to be returned to the manufacturer.

3.4.2 Vapor Diffusion Pumps

In a diffusion pump, gas molecules are moved from inlet to outlet by momentum transfer from a directed stream of oil or mercury vapor. As shown in Figure 3.12, the working fluid is evaporated in an electrically heated boiler at the bottom of the pump. Vapor is conducted upward through a tower above the boiler to an array of nozzles from which the vapor is emitted in a jet directed downward and outward toward the pump walls. The walls of the pump are cooled so that molecules of the working fluid vapor condense before their motion is randomized by repeated collisions.

The diffusion pump walls are usually water-cooled, although in some small pumps they are air-cooled. The condensate runs down the pump wall to return to the boiler. Efficient operation of a diffusion pump requires a temperature gradient from top to bottom. The pump wall temperature should be lowest at the top of the pump—less than 30°C—to effectively condense oil vapor, and highest at the bottom to drive absorbed gas out of the oil condensate. For this reason, the cooling water flows from top to bottom and the flow rate is regulated—too high a flow rate results in too low a temperature at the bottom of the pump.

Because the pumping action of a diffusion pump relies on momentum transferred in collisions between the molecules of the working fluid vapor and the molecules being pumped, the speed of the pump is greater for light gases than for heavy gases. The net speed of the pump is essentially determined by the inlet area. For air, the pumping speed is about 4 L s^{-1} cm^{-2}.

As indicated by the pumping-speed curve in Figure 3.12, the pumping action of a diffusion pump begins to fail when the inlet pressure increases to the point where the mean free path of the molecules being pumped is less than the distance from the vapor-jet nozzle to the wall. When this occurs, the net downward momentum of vapor molecules is lost and the vapor begins to diffuse upward into the vacuum system. Diffusion pumping of a vacuum chamber may be initiated at 50 to 100 mtorr, but the system pressure should quickly fall below 1 mtorr or the chamber will become significantly contaminated with the vapor of the working fluid. Oil diffusion pumps can run against an outlet pressure of 300 to 500 mtorr and mercury pumps can tolerate an outlet pressure of a few torr. A backing pump, usually an oil-sealed rotary pump, is required. The speed of the diffusion pump is insensitive to foreline pressure up to some critical pressure. If this critical pressure is exceeded, the pump is said to *stall*. This is a disaster because hot pump fluid vapor is flushed backwards through the pump into the chamber. The usual

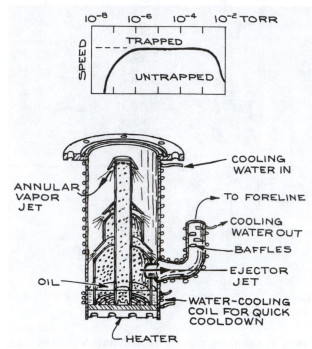

Figure 3.12   Diffusion pump.

practice is to maintain the oil diffusion pump foreline pressure below 50 mtorr.

The primary difficulty with diffusion pumps is the backstreaming of oil vapor from the pump inlet. Without appropriate precautions this can amount to as much as a microgram per minute for each cm^2 of pump inlet area. Backstreaming is much worse if the foreline pressure of a hot pump accidentally rises above the critical backing pressure of a few hundred millitorr. It is essential to maintain the pump cooling system. Most diffusion pumps are water cooled. A filter in the cooling-water line is necessary to reduce the possibility of clogging. Cooling water should enter at the top of the pump so that the uppermost part of the pump wall is at the lowest temperature. The next line of defense is a *cold cap*—typically consisting of a conical shield over the top of the jet tower. The cold cap is cooled by water flow or by solid thermal contact with the pump body. Diffusion pumps usually are supplied with a cold cap installed. In most installations, a water- or refrigerant-cooled baffle (see Figure 3.26) is mounted on top of the pump. The cold cap and baffle together reduce backstreaming by as much as three orders of magnitude, with a loss of pumping speed of perhaps 50 percent. For many purposes, the vacuum above the baffle can be considered oil-free. Oil can reach the vacuum chamber as a liquid film that creeps up the walls above the pump. To eliminate this problem, an anticreep barrier consisting of an annular ring made of Teflon (a material not wetted by oil) can be installed on top of the baffle. The ultimate performance and the cleanest vacuum are obtained with the installation of a liquid-nitrogen-cooled trap (see Figure 3.25) above the baffle.

The properties of diffusion pump fluids are given in Table 3.3; included are hydrocarbon, perfluorocarbon and silicone oils, and mercury. The various diffusion pump oils can be used interchangeably in an oil diffusion pump. The choice depends upon the desired ultimate pressure, the demands of the particular application, and the cost. The matter of economics is significant—the cost of the oils listed in Table 3.3 ranges over more than a factor of twenty. The perfluorocarbon and polyether oils are sufficiently high-priced that it is economical to reclaim them and many vacuum hardware companies offer this service. The absolute lowest pressure attainable with an untrapped diffusion pump is the room-temperature vapor pressure of the working fluid.

The silicone oils are the least expensive of the high-performance (i.e., low ultimate pressure) pump fluids. They are exceptionally resistant to oxidation and chemical attack except in the presence of BCl_3 and, to a lesser extent, CF_4 and CCl_4. The main drawback to silicone oils is that the residue of backstreaming oil found on surfaces in the vacuum system will decompose and polymerize to produce insulating films upon exposure to heat and charged-particle bombardment; silicone oils are unsuitable for diffusion pumps on instruments such as mass spectrometers or electron microscopes.

The phthalate and sebacate oils are inexpensive fluids suitable for many applications. Decomposition of these hydrocarbon oils yields carbonaceous deposits that are conductive. Polyphenyl ether oils offer exceptional chemical and thermal stability. These are well suited for use with mass spectrometers and other charged-particle beam instruments. The perfluorocarbon oils are suitable for pumping corrosive gases, with the exception of Lewis acids such as BCl_3 and AlF_3. Their performance under charged-particle bombardment is probably the best of all pump fluids.

Mercury diffusion pumps find only a few specialized uses that rely upon the chemical inertness of mercury vapor. They are occasionally used on mass spectrometers because of the simple, easily identified spectrum of mercury vapor. Mercury diffusion pumps are most often made of Pyrex glass and are intended for use with glass vacuum systems used to handle and distill small quantities of volatile chemicals. Mercury diffusion pumps tolerate inlet pressures ten times greater than the maximum inlet pressure tolerated by oil pumps. In addition, the critical foreline pressure for a mercury pump is much higher. Because the room temperature vapor pressure of mercury is about 10^{-3} torr, an inlet cold trap is required to prevent bulk migration of mercury into the vacuum system. Mercury vapor is toxic, and care must be taken to properly trap and vent the backing-pump exhaust.

Diffusion pumps with speeds of 50 to 50,000 L s^{-1} and with nominal inlet port diameters of 1 to 35 in. are available. Diffusion pumps are constructed of Pyrex glass, mild steel, or stainless steel. The jets and towers of pumps with steel barrels are aluminum. The choice of glass or steel usually depends upon the material that is used to construct the vacuum container, although it is also relatively simple to mate glass to steel with an elastomer gasket or through a metal-to-glass seal. For those systems that are intended to handle reactive gases, glass is the preferred material. Glass pumps, however, are available only in small sizes.

Diffusion pump maintenance involves assuring the quality and quantity of the pump fluid and checking that the heaters are operating. Some reaction between the pumped gas and the pump fluid is inevitable, so periodic inspection of the pump innards is essential. Moderate discoloration of the oil is not significant, but the oil should be replaced if it is opaque. The interior of the pump must be thoroughly cleaned before adding new oil, especially if a different type of oil is being installed. Scrubbing with acetone, followed by an acetone and an ethanol rinse will remove hydrocarbon oils and, with persistence, silicone oils. Deposits of badly decomposed oil can be removed with naphthalene. Polyphenyl ether oils are soluble in trichloroethylene and 1,1,1-trichloroethane. Both these solvents should be used with care in a well-ventilated environment. Fluorocarbon oils are soluble in fluorinated

TABLE 3.3 PROPERTIES OF DIFFUSION-PUMP FLUIDS

Trade Names (chemical composition)	Boiling Point at 1 torr (°C)	Approximate Vapor Pressure at 20°C (torr)
(Mercury)	120	2×10^{-3}
(Di-*n*-butyl phthalate)	140	2×10^{-5}
Octoil, Diffoil (di-2-ethyl hexyl phthalate)	200	3×10^{-7}
Octoil-S, Diffoil-S (di-2-ethyl hexyl sebacate)	210	3×10^{-8}
Convoil-20 (saturated hydrocarbon)	190	5×10^{-7}
Neovac Sy (alkyldiphenyl ether)	240	1×10^{-8}
Convalex-10 (polyphenyl ether)	280	7×10^{-10}
Santovac-5 (pentaphenyl ether)	280	5×10^{-10}
D.C. 704 (tetraphenyl tetramethyl trisiloxane)	240	3×10^{-8}
D.C. 705 (pentaphenyl trimethyl trisiloxane)	250	5×10^{-10}
Fomblin Y-VAC 25/9 (perfluoropolyether)	230	2×10^{-9}
Krytox 1625 (perfluoropolyether)	250	2×10^{-9}

alkanes such as trichlorotrifluoroethane and perfluorooctane. Severe decomposition of the oil results in black, carbonaceous deposits on the jet tower. These must be removed mechanically by scrubbing with fine abrasive or by glass-bead blasting. The tower must be carefully cleaned before reinstallation in the pump. There is a close fit between the tower and the pump body. A residue of abrasive cleaning materials will result in the tower becoming jammed in the pump body.

Diffusion pumps use flat, *pancake* heating elements bolted to the bottom of the pump or cylindrical *cartridge* heaters inserted in close-fitting holes in the bottom of the pump. As many as six cartridge heaters wired in parallel may be used. The advantage of multiple heaters is that the pump continues to operate even if one or more heaters burn out, albeit with some loss of efficiency. On the other hand, the loss of one or more heaters may go unnoticed until significant contamination of the vacuum has occurred. The individual cartridge heaters may be disconnected from one another and individually checked for electrical continuity. It is essential that diffusion pump heaters are in good thermal contact with the pump over their entire mating surfaces or the heaters will burn out prematurely. A coating of milk of magnesia (a slurry of MgO in water) will assure good contact and facilitate removal at a later time by preventing the heater from sticking to the pump because of corrosion or cold welding.

A diffusion-pumped system is the most economical route to high and even ultrahigh vacuum in chambers of moderate size. With no moving parts and no critical dimensions, diffusion pumps are robust and long-lived. Decades-old pumps can be perfectly serviceable. This is a boon for researchers on a limited budget since used pumps are available at little or no cost.

3.4.3 Entrainment Pumps

A variety of vacuum pumps remove gas from a system by chemically or physically tying up molecules on a surface or by trapping them in the interior of a solid. Two of the principal advantages of these pumps are that they require no backing pump and they contain no fluids to contaminate the vacuum.

Sorption pumps. The simplest of the entrainment pumps is the sorption pump illustrated in Figure 3.13. The sorbent material is activated charcoal or one of the synthetic zeolite materials known as molecular sieves. These materials are effective sorbents partly because of their huge surface area—on the order of thousands of square meters per gram. The most common molecular

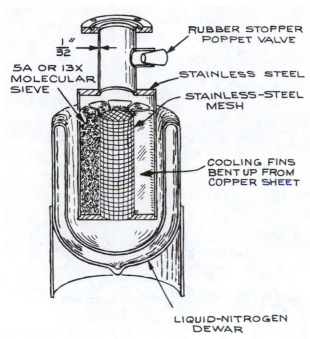

Figure 3.13   Sorption Pump.

sieves are zeolite 5A or 13X. The number in the sieve code specifies the pore size: 5A is preferred for air pumping, while 13X is used for trapping hydrocarbons. All of these materials will pump water and hydrocarbon vapors at room temperature, but they must be cooled to liquid-nitrogen temperature to absorb air. Sorbent materials do not trap hydrogen or helium at liquid-nitrogen temperature. Without some auxiliary means of removing hydrogen and helium, the partial pressures of these gases in a system will establish the lowest attainable pressure. Sorption pumps must be provided with a poppet valve because the sorbent material releases all of its absorbed air as it warms to room temperature.

The sorbent is initially activated by baking to 250°C. After several pumping cycles, the pores of the sorbent material will become clogged with water and the efficiency of the pump will deteriorate. Water is removed by baking the pump to 250°C with the poppet valve open and then cooling to room temperature with the valve closed so that moisture from the room is not reabsorbed. Baking can be accomplished by wrapping the pump with

heating tape. Custom-made heating mantles can be obtained inexpensively from the Glas-Col Company.

In a well-designed pump, 1 kg of dry Linde 5A molecular sieve will pump a 100-L volume from atmosphere to less than 10^{-2} torr in about 20 minutes. The pump must be designed to provide good thermal contact between the sieve and the coolant to obtain the maximum pumping speed. A thin-walled inlet tube will minimize heat conduction into the pump. Sorption pumps similar to that in the figure are available from suppliers of vacuum hardware. Pressures below 10^{-6} torr may be attained in a system that has been roughed down. A simple way to achieve low pressures with sorption pumping is to use two or three pumps on a manifold. Each pump is equipped with a shutoff valve. Starting at atmospheric pressure, one pump valve is opened to rough down the system. This pump should be of a size such that the pressure falls to less than 1 torr. This pump is valved off, the next pump is brought on line, and so on. To attain the lowest possible pressure with this arrangement, the vacuum chamber can be purged of hydrogen and helium before evacuation proceeds. This can be accomplished by flushing the vacuum chamber with dry air or dry nitrogen before initiating pumping with the sorption pumps.

Getter pumps.   Clean surfaces of refractory metals such as titanium, molybdenum, tantalum, or zirconium will pump most gases by chemisorption. This process is called *gettering*. In a getter pump, the active metal surface is produced *in vacuo*. Titanium or a titanium/molybdenum alloy is the preferred metal. A getter pump employing a freshly evaporated titanium surface is called a *titanium sublimation pump*. At room temperature, a titanium surface pumps H_2, N_2, O_2, CO_2, and H_2O at rates of several Ls^{-1} per cm^2 of exposed surface. Cooling the surface to liquid nitrogen temperatures may double or triple the pumping speed for N_2 and H_2. Methane and the noble gases, such as Ar and He, are not pumped at all.

A titanium sublimation pump consists of titanium sublimator and the surface surrounding the sublimator. The surface may or may not be cooled. This pump operates effectively between 10^{-3} and 10^{-11} torr, but the inert gases must be removed before ultrahigh vacuum can be achieved. This can be accomplished with a small ion pump in parallel with the getter pump. An alternative method for removing rare gases (at least initially) is to purge the system with pure dry nitrogen prior to evacuation. The titanium surface must be renewed at a rate roughly equal to the rate at which a monolayer of gas is adsorbed. Down to about 10^{-7} torr, the titanium surface is continuously deposited. Below 10^{-7} torr, the titanium need only be deposited periodically. The capacity of a titanium sublimation pump is about 30 torr L per gram Ti. A forepump is not required, but a mechanical or sorption roughing pump is needed to reach a starting pressure of about 10^{-3} torr.

Titanium sublimation pumps are quite simple in design and correspondingly inexpensive. Commercially available sublimators consist of two to four electrical feedthroughs in a flange with titanium filaments suspended between the feedthroughs, as illustrated in Figure 3.14. A controller supplies electrical current through the filaments at a rate appropriate for the vacuum conditions. The sublimator can be inserted in a side arm on the vacuum system with baffles or an elbow arranged to prevent titanium vapor from entering the main part of the vacuum chamber. The walls of the sidearm then serve as the pump. The pumping speed with this arrangement will usually correspond to the conductance of the sidearm with its baffles. Vacuum housings with water- or liquid nitrogen-cooled baffles are available to contain the sublimator. These are designed to be mounted directly atop a turbomolecular pump in an ultrahigh vacuum system. For a glass vacuum system, a simple pump can be made by wrapping titanium wire around a tungsten wire sealed in a glass bulb as shown in Figure 3.15. The tungsten filament is electrically heated to evaporate the titanium that in turn condenses on the walls of the bulb.

A new class of getter pumps employing so-called *nonevaporable* getters has recently come into use. These employ a porous, sintered, zirconium alloy that strongly absorbs and reacts with active gases. The getter is activated by heating to 700 to 900°C *in vacuo* to remove a surface oxide. In operation, the getter is maintained at a temperature of about 400°C. CO, CO_2, N_2, and O_2 are irreversibly adsorbed. Water vapor and hydrocarbons are dissociatively adsorbed. H_2 is reversibly adsorbed and there is no sorption of the rare gases. Nonevaporable getter pumps operate down to 10^{-12} torr. The pumping speed is

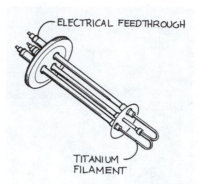

Figure 3.14   Titanium sublimator on a UHV flange.

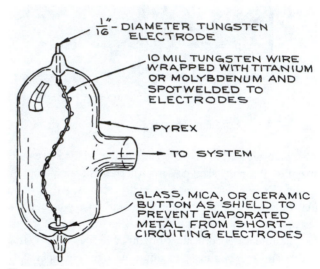

Figure 3.15   Simple titanium sublimation pump as an appendage on a glass vacuum system.

highest for H_2 and H_2O. These pumps are an excellent adjunct to a turbomolecular pump in a UHV system.

Cryopumps.   Any surface will act as a pump for a gas that condenses at the temperature of the surface. A pump that relies primarily upon condensation on a cold surface is called a cryopump. Commercial versions of these pumps employ a closed-circuit helium refrigerator to cool the active surfaces. Working temperatures are typically below 20 K. These pumps usually employ two stages. In the first stage a metal surface is maintained at 30 to 50 K to trap water vapor, carbon dioxide, and the major components of air. In some designs, this first stage is preceded by a liquid-nitrogen-cooled baffle to reduce the load of water vapor on the subsequent stages. The second stage, maintained at 10 to 20 K, is coated with a cryosorbent material such as activated charcoal to provide pumping of neon, hydrogen, and helium. Cryopumps provide high pumping speed for easily condensed gases. Their performance with helium depends critically upon the quality and recent history of the cryosorbent surface. The gas captured by a cryopump is not permanently bound. The pump must periodically be warmed to regenerate the pumping surfaces. Because of the limited capacity of the pump, a roughing pump is necessary, as is a valve between the cryopump and the vacuum chamber so that the vacuum system can be rough pumped to a pressure less than about 10^{-3} torr before cryopumping is initiated.

Ion pumps. An ion pump combines getter pumping with the pumping action exhibited by an ionization gauge. Within these pumps, a magnetically confined discharge is maintained between a stainless-steel anode and a titanium cathode. The discharge is initiated by field emission when a potential of about 7 kV is placed across the electrodes. After the discharge is struck, a current-limited power supply maintains the discharge rate at 0.2 to 1.5 A, depending on the size of the pump. Inert gas molecules as well as other molecules are ionized in the discharge and accelerated into the cathode with sufficient kinetic energy that they are permanently buried. Active gases are chemisorbed by titanium that has been sputtered off the cathode by ion bombardment and deposited on the anode.

Ion pumps function between 10^{-2} and 10^{-11} torr. A forepump is needed to reach the starting pressure. An oil-free rough pump is called for, since ion pumps are completely free of hydrocarbons. A well-trapped, oil-sealed rotary pump can be used, but roughing with sorbtion pumps will guarantee that the vacuum chamber remains clean. An ion pump has a limited lifetime of operation due largely to the rate at which the cathode erodes during operation. The operational lifetime is an inverse function of pressure. Under continuous operation

at 10^{-3} torr, the lifetime of a typical pump is only 20 to 50 hours, whereas operation at 10^{-6} torr can continue for 20,000 to 50,000 hours (6 yrs). At ultrahigh vacuum, the lifetime is essentially infinite. Ideally, an ion pump is only operated below 10^{-5} torr, but because this pressure is difficult to attain with most roughing systems, pumping can be initiated at 10^{-3} torr provided that the gas load is such that the pressure quickly falls into the desirable range. Ion pumps are frequently used in conjunction with titanium sublimation pumps. In fact, some ion pump designs include an integral sublimation pump. The sublimation pump can serve to lower the pressure from 10^{-3} torr to 10^{-5} torr before power is applied to the ion pump. At lower pressures, the titanium sublimation pump significantly increases the pumping speed for getterable gases. To attain the lowest possible pressures, an ion pump can be baked to at least 300°C. Most pumps are equipped with integral heaters for this purpose. Baking should be initiated only while the pump is in operation.

The positive-ion current to the cathode of an ion pump is a function of pressure; thus, the pump serves as its own pressure gauge. The power supply and control unit for most ion pumps include a pressure readout. This is a significant consideration in the design of a very small ultrahigh vacuum system; the cost of a 1 to 5 L s^{-1} ion pump with its control unit is little more than the cost of an ion gauge and controller.

Ion pumps require no cooling water or backing pump. They continue pumping by getter action even if the power fails. Ion pumps do not introduce hydrocarbon or mercury vapors into the vacuum. Pumps with speeds from one to many thousands of liters per second are available. The limited capacity in the 10^{-3} torr to 10^{-5} torr range can be a disadvantage. Stray magnetic and electric fields may also present a problem; ion pumps may produce a magnetic field of a few gauss at a distance of one inlet diameter. There may also be a problem with ions escaping the pump—ion emission can be suppressed by placing a suitably biased screen above the pump inlet. An ion pump with controller can cost two to three times as much as a comparable diffusion pump with an appropriately sized liquid-nitrogen-cooled baffle.

3.5 VACUUM HARDWARE

3.5.1 Materials

Glass. Two glasses are used in vacuum work: borosilicate glass or *hard glass* such as Pyrex or Kimax that is about 70 percent SiO_2, and quartz glass that is pure silica. Both materials have a low coefficient of thermal expansion and are remarkably resistant to fracture under large temperature gradients (thermal shock). Borosilicate glass has a broad softening temperature range allowing glass vacuum apparatus to be constructed and modified *in situ* by a moderately competent glassblower as described in Chapter 2. Quartz has a very high softening temperature and a narrow softening temperature range and requires a skillful glassblower. Quartz can be used in continuous operation at temperatures up to 1000°C. Quartz is used primarily when its high temperature strength or broad range of optical transmission is to be exploited. The permeability of helium through quartz is surprisingly large—about 10^{-11} torr L s^{-1} mm cm^{-2} $torr^{-1}$ (throughput [torr L s^{-1}] through unit thickness [mm] per unit area [cm^2] per unit helium pressure differential [torr]) at room temperature and increases rapidly with increasing temperature. This can be a problem with quartz windows in UHV systems. Even when the permeation of atmospheric helium is not significant, the permeation of helium through a quartz window can give rise to false signals during helium leak testing.

Hard glass tubing and glass vacuum accessories are inexpensive. Stopcocks and Teflon-sealed vacuum valves; ball joints, taper joints, and O-ring-sealed joints; traps; and diffusion pumps are all easily obtained at low cost. In most cases only the glassblowing ability to make straight butt joints and T-seals is required to make a complete system from standard glass accessories.

Glass pipe and pipe fittings are manufactured for the chemical industry. A variety of standard shapes such as elbows, tees, and crosses are available in pipe diameters of $\frac{1}{2}$ to 6 in. Glass process pipe with Teflon seals can be used to make inexpensive vacuum equipment for use down to 10^{-8} torr. Couplings are also available for joining glass pipe to metal pipe, standard metal flanges, and pipe fittings so it is easy to make a system that combines glass and metal vacuum components.

Glass and metal parts can be mated through a graded glass seal. A graded seal typically appears to simply be a section of glass tubing butt-sealed to a section of Kovar metal tube. In fact, the glass tubing consists of a series of short pieces of glass whose coefficients of thermal expansion vary in small increments from that of hard glass to that of the metal. Graded seals are useful for joining glass accessories such as ion-gauge tubes to metal systems. A graded seal can also be employed to insert metal hardware, such as a valve or an electrical feedthrough, into a glass vacuum line. Inexpensive graded seals in sizes up to 2 in. in diameter are available from commercial sources.

Windows in vacuum systems are often made of quartz. For high vacuum work, the window can be sealed to the vacuum wall with an elastomer gasket or O-ring. For ultrahigh vacuum, the edge of the quartz window is metalized so that the window can be brazed into a metal flange for installation in a vacuum wall. Window/flange assemblies are available from most vacuum equipment suppliers.

Ceramics. Ceramics are used as electrical insulators and thermal isolators. There are literally hundreds of different kinds of ceramic materials. Two find frequent use in vacuum apparatus—steatite, a magnesium silicate; and alumina ceramics containing at least 96 percent Al_2O_3. Fabrication of parts from these materials requires specialized grinding equipment ordinarily not immediately available to a laboratory scientist. There are, however, many specialized shops that provide grinding services under contract. Shapes such as rod, tubing, plate, and spheres are available in alumina. It is often possible to design an insulator that incorporates these simple shapes into a more complex assembly. Threaded steatite standoff insulators inexpensively available from electronics suppliers can be used as structural elements in vacuum apparatus.

Brass and copper. Brass has the advantage of being easily machined, and brass parts can be joined by either soft solder or silver solder. Brass unfortunately contains a large percentage of zinc, whose volatility limits the use of brass to pressures above 10^{-6} torr. Heating brass causes it to lose zinc quite rapidly. The vapor pressure of zinc is about 10^{-6} torr at 170°C and 10^{-3} torr at 300°C. Forelines for diffusion pumps are conveniently constructed of brass or copper tubing and standard plumbing elbows and tees. Ordinary copper water pipe is acceptable for forelines and other rough vacuum applications, however, Oxygen-Free High-Conductivity (OFHC) copper should be used for high-temperature and high-vacuum work. Copper can be welded by a skilled technician.

Stainless steel. The most desirable alloys for the construction of high-vacuum and ultrahigh-vacuum apparatus are the (American AISI) 300-series austenitic stainless steels, especially type-304 and type-316 stainless steels. Type-303 is not ordinarily used as it contains volatile elements added to improve machinability. Stainless steel is strong, reasonably easy to machine, bakeable, and easy to clean after fabrication. Type-304 and type-316 stainless steels are essentially nonmagnetic with magnetic permeabilities less than about 1.01. Stainless steel parts may be brazed or silver-soldered. Low-melting (230°C) silver-tin solders are useful for joining stainless-steel parts in the laboratory, however it is best if stainless parts are fused together by arc welding using a nonconsumable tungsten electrode in an argon atmosphere. This process, known commonly as heliarc fusion welding or Tungsten-Inert Gas (TIG) welding, produces a very strong joint. And because no flux or welding rod is used, such a joint is easily cleaned after welding. Most machine shops are prepared to do TIG welding on a routine basis.

The bulk of commercial vacuum fittings (flanges and shapes such as tees, crosses, and nipples) are fabricated of type-304 stainless steel. Type-304 stainless is also widely used in the milk- and food-processing industry. As a result, many stock shapes are commercially available at low cost. Elbows, tees, crosses, Ys, and many other fittings in sizes up to at least 6 in. in diameter may be purchased from McMaster-Carr.

Aluminum. Aluminum alloys, particularly the 6000 series alloys, are used for vacuum apparatus. Aluminum has the advantage (over stainless steel) of being lighter and stiffer per unit weight and much easier to machine. It is nonmagnetic. The chief disadvantages of aluminum stem from its porosity and the oxide layer that covers the surface. The rate of outgassing from aluminum is five to ten times greater than from stainless steel. Aluminum can be anodized black (black chrome) to reduce reflectivity in optical devices, but anodizing may significantly increase outgassing rates. Welds in aluminum are not as reliable as welds in stainless and tend to outgas volatile materials that have been occluded in the weld. Nevertheless, welded aluminum vacuum containers are used down to at least 10^{-7} torr. The hard oxide layer that forms instantly on a clean aluminum surface is an electrical insulator. This insulating surface tends to collect electrons or ions to produce an electric field that is undesirable in some applications. Copper or gold plating on aluminum or a coating of colloidal graphite (see Section 5.6.1) should eliminate the accumulation of surface charge.

Plastics. Plastics are used as fixtures, electrical insulators, bearings, and windows in vacuum systems at pressures down to 10^{-7} torr. Their use is limited to varying degrees because they outgas air, water, and plasticizers, and because they cannot be heated to high temperatures. Polyamide (e.g., Nylon) and acetal-resin-based plastics (e.g., Delrin) are used for mechanical elements in vacuum systems at temperatures below 100°C. These plastics are hard and machinable. Nylons are self-lubricating and make excellent bearings. Nylon is hygroscopic and outgasses water vapor after exposure to air. Outgassing is less of a problem with Delrin. Windows in vacuum systems may be made of polymethylmethacrylate (e.g., Plexiglas or Lucite) or polycarbonate (e.g., Lexan) plastics. Lexan is especially strong, tough, and machinable—it is well suited for use as a structural material or as an electrical insulator. Fluorocarbon polymers have superior high-temperature characteristics. Polytetrafluoroethylene (PTFE) (e.g., Teflon) can withstand temperatures up to 250°C. The outgassing rate falls below 10^{-8} torr L s^{-1} cm^{-2} after an initial pumpdown of a day at 100°C. Unfortunately, Teflon is relatively soft, and cold flows under mechanical pressure thus limiting its usefulness as a structural element. Various polyimides, although expensive, are finding use in vacuum systems that are to be baked. Polyetherimide (Ultem) can be used at temperatures up to 200°C. Polyamide-imide (Torlon) is the highest strength thermoplastic. Polyimide (Vespel, Kapton) is the ultimate material for both mechanical elements and electrical insulation at high temperatures. Vespel has been used continuously at temperatures approaching 300°C in high vacuum. Polyimide is somewhat hygroscopic but can be baked to reduce outgassing. Kapton is a sheet material used for electrical insulation. Kapton-insulated wire is supplied for high temperature use in high vacuum and even ultrahigh vacuum.

A variety of low vapor pressure sealers and adhesives have vacuum applications. Apiezon M grease, Dow-Corning silicone high-vacuum grease, and Fomblin perfluorinated grease are used to seal stopcocks and ground-glass taper joints and, in some instances, used as low-speed lubricants. Apiezon compounds are used to seal joints and windows as well as to fill small gaps in vacuum systems. These waxlike materials soften at temperatures of 45 to 100°C and have vapor pressures between 10^{-4} and 10^{-8} torr at room temperature. A variety of resin-based sealers in spray cans are marketed for sealing very small leaks in high-vacuum and even ultrahigh-vacuum systems while they are under vacuum.

Epoxy resins are particularly useful. Epoxy cements consist of a resin and a catalytic hardener that are combined immediately before use. The proportions of resin and hardener and subsequent curing must be carefully controlled to prevent excessive outgassing from the hardened material. Small epoxy kits with resin and catalyst prepackaged in the correct proportions are available, as are silver-filled epoxy formulations that conduct both electricity and heat.

3.5.2 Demountable Vacuum Connections

Vacuum systems require detachable joints for convenience in assembling and servicing. There are basically three means of creating a vacuum-tight seal between mating connectors: pipe threads with a sealant in between, elastomer gaskets or *O-rings*, and deformable metal gaskets.

Pipe threads. The threaded parts of a pipe joint are tapered so that the joint becomes tighter and tighter as the two are screwed together. The two parts are not made with sufficient precision to assure a vacuum-tight seal even though significant distortion of the metal threads can occur as the parts are joined. To assure a seal, the male part is covered with a thin layer of Teflon tape prior to assembly. Teflon thread tape is sold in hardware stores for use with plumbing fixtures. The tape is wrapped around the threaded end. No more than two layers are required and the tape should be stretched as it is wound on to make a very thin layer. Looking at the end of the male part, the tape should be wound in a counterclockwise direction so that the exposed tag end does not get caught up in the female threads as the joint is assembled. The tape must be replaced upon reassembly. A significant volume of gas is trapped in an assembled pipe thread joint. Pipe thread joints are not appropriate for ultrahigh vacuum systems and their use is to be avoided as much as possible in high vacuum systems.

O-Ring joints. For pressures down to about 10^{-7} torr, vacuum connections are most often sealed with rubber O-rings. Several O-ring-sealed joints are illustrated in Figure 3.16. O-rings are circular gaskets with a round cross section. They are available in hundreds of sizes from 0.125 in. ID (inner diameter) with a cord diameter of 0.070 in. (nominally $\frac{1}{16}$ in.) to 2 ft. ID with a cord diameter of 0.275 in. (nominally $\frac{1}{4}$ in.). Very large rings may be made from lengths of cord stock with the ends butted together and glued with Eastman 910 adhesive. O-rings are made of a variety of elastomers. The most common are Buna-N, a synthetic rubber, and Viton-A, a fluorocarbon polymer. Buna-N may be heated to 80°C and will not take a set after long periods of compression. Unfortunately this material outgasses badly, particularly after exposure to cleaning solvents. Viton-A has a low outgassing rate and will withstand temperatures up to 250°C. Viton-A will take a set after baking. Generally, the advantages of Viton-A O-rings offset their higher cost.

O-ring-sealed flanges and *quick connects* are easily fabricated and they are also available ready-made. Mating flanges have a groove that contains the ring after the flanges have been pulled into contact with one another. Usually, one flange is flat and a rectangular-cross-section

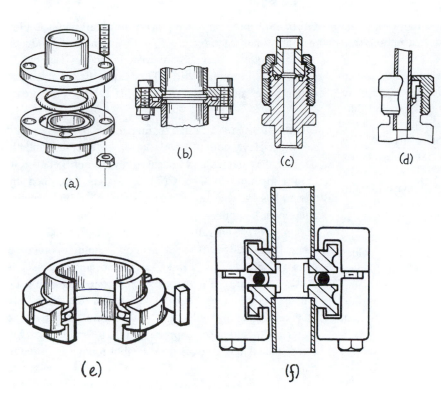

Figure 3.16 O-ring-sealed vacuum connections: (a) an exploded view of an O-ring-sealed flange joint; (b) assembled ASA-style flange joint; (c) O-ring tube coupling (Cajon VCO); (d) a quick connect; (e) QF (or KF) joint; (f) ISO joint with K-style flanges.

groove is cut into the mating flange; flanges can be made sexless by cutting a groove of half the required depth in both flanges. Groove design is critical to obtaining a reliable seal, as is the surface finish of the flange and inside the groove. Surface roughness should not exceed 32 microinches. The flange surface and groove should be cut on a lathe so that tool marks run circumferentially. Radial scratches are one of the most common causes of seal failure. To obtain the correct compression of the O-ring material, the groove depth should be about 80 percent of the actual cord diameter for static seals and 85 percent of the diameter for dynamic seals. Rubber is deformable but incompressible; the width of the groove should be such that the cross-sectional area of the groove exceeds that of the O-ring by 30 to 40 percent. The ID of the groove should match the O-ring ID; the O-ring is not to be stretched into its groove. It is sometimes convenient to undercut the sides of the groove slightly to give a closed dovetail cross section rather than a rectangular cross section, so that the ring is retained in the groove during assembly. For rings up to a foot in diameter, the nominal

cord diameter should be $\frac{1}{8}$ in. or less. Choose a ring to fit in a groove that is as close to the flange ID as possible in order to minimize the amount of gas trapped in the narrow space between the mated flanges. The bolts or clamps that pull the flanges together should be as close to the groove as possible to prevent flange distortion. For static seals, O-rings should be used dry. Before assembly, the groove should be cleaned with solvent and the ring wiped free of mold powder with a dry lint-free cloth. O-rings should not be cleaned with solvents. For rotating seals, a very light film of vacuum grease on the ring will prevent abrasion. Occasionally a film of grease may be required on a static O-ring to help make a seal on an irregular surface. Grease is a liability, however, as it tends to pick up material that disrupts the seal.

For very high temperatures, elastomer O-rings can be replaced by metal O-rings composed of a thin metal sleeve with an internal spring to provide resiliency. These are manufactured by Helicoflex in aluminum, silver, copper, nickel, and other materials suitable for temperatures from 300 to 600°C and in sizes comparable to those of standard rubber O-rings. The groove design is essentially the same as for rubber O-rings; detailed design instructions are available from the manufacturer.

There are three standard types of O-ring sealed joints: ASA, QF (or KF), and ISO. Examples are shown in Figure 3.16. The ASA design dates back to an early steamfitters standard flange. ASA design consists of two flat flanges with an O-ring groove in one flange and a circle of bolt holes well outside of the groove. These flanges are bulky and require fasteners much larger than needed to resist atmospheric pressure. The sizing of ASA flanges refers to the nominal ID of the flange or the nominal OD of the tube to which it is attached: the bore of a 4" ASA flange is about 4 in.; the OD of the flange is 9 in. Some commercial vacuum hardware, especially diffusion pumps of older design, still use ASA flanges. The QF (for Quick Flange, or KF for Klein Flange) design employs two flat, and therefore sexless, flanges with an outer diameter only slightly larger than that of the O-ring. The O-ring is mounted around an aluminum *centering ring* that prevents the O-ring from being sucked in, and whose thickness determines the extent of compression of the O-ring. The centering ring has a lip on both sides that engages the inner circumference of each flange so that the O-ring is correctly located on the flanges. A jointed, circular clamp is installed around the assembled pair of flanges and a thumbscrew is tightened to draw the two flanges together, compressing the O-ring. QF flanges are available for tube sizes from $\frac{1}{2}$ in. (10 mm) to 2 in. (50 mm). They are compact and, with only a single screw to tighten, quite convenient to use—a QF fitting can be assembled with one hand. ISO flanges are similar to QF flanges but are sized for larger tubing: from 2.5 in. (63 mm) to 24 in. (630 mm). The ISO fitting is comprised of two flat flanges with a combination centering ring and O-ring mounted between them. The centering ring has a lip on each side that engages a counterbore near the ID of the flange. ISO flanges are designed for one of two methods of joining a pair of flanges. K-style ISO flanges employ clamps that hook over the outer edges of the mating flanges. F-style flanges have holes for bolts to draw the two flanges together. K-style flanges are more compact than F-style flanges. F-style flanges make a stronger structure appropriate for joining tubing that must support large transverse forces. The centering ring/O-ring assembly in a QF or ISO joint can be replaced by an all-metal C-Flex assembly manufactured by EG&G Engineered Products. These employ a springy, tin-coated, Inconel element to replace the elastomer O-ring. They can be baked at modest temperatures and are said to be usable at UHV pressures.

Metal gaskets. Metal sealing materials are required for ultrahigh vacuum (UHV). Elastomers are unacceptable because they outgas and because they cannot be baked to high temperatures. The ConFlat design developed by Varian has become the industry standard. Many vacuum equipment suppliers now manufacture flanges of this design, now generally known as CF (for ConFlat-compatible) flanges. The CF sealing system is composed of two identical flanges and a flat OFHC copper gasket as shown in Figure 3.17. Annular knife-edge ridges in each flange cut into the gasket to make a seal as the flanges are drawn together by bolts that pass through holes in the flanges outboard of the copper gasket. The entire assembly can be baked to 450° C. A great deal of force is required to make the seal, so many high-tensile-strength bolts are used, closely spaced around the flange. The knife edge is designed so that the correct deformation of the gasket is obtained when the flanges are drawn into contact. Careful assembly is important to avoid warping the flanges. The joint is initially assembled with all the bolts just finger tight. Then, using good-fitting wrenches and working in sequence around the flange, each bolt is tightened no more than one half turn. Going round and round the flange, the

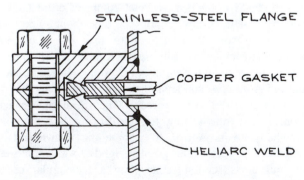

Figure 3.17   Detail of a bakeable, ultrahigh-vacuum CF flange.

bolts are drawn up just until the flange surfaces touch. The copper gaskets are permanently deformed during installation and therefore are only used once. Bolts wear and stretch and must be replaced after perhaps a dozen assembly cycles. Baking a CF flange encourages the bolts to seize in their nuts. A very thin coating of molybdenum disulfide grease helps avoid this problem. Silver-plated bolts are available to help reduce galling and seizing in systems that are frequently baked to high temperatures. CF flanges are specified by the flange OD in sizes that run from $1 \frac{1}{3}$ in. (or 34 mm) to 12 in. (305 mm) to $16 \frac{1}{2}$ in. These are used with tube sizes from .75 in. (16 mm) to 10 in. (or 250 mm) to 14 in. In UHV systems, circular cross-section tubes larger than 14 inches in diameter are joined with Wheeler flanges that employ a 2 mm diameter OFHC copper wire as a deformable gasket. Wire-sealed flanges are heavy and very large forces are required to achieve a seal. They are available from many manufacturers in sizes up to about 30 in.

3.5.3 Valves

A wide variety of valves are commercially available for use in glass systems and in high-vacuum and ultrahigh-vacuum metal systems. Characteristics to be considered in choosing a valve (roughly in order of importance) include: (i) conductance, (ii) operating temperature, (iii) leak rate around the valve-operating mechanism and the valve seat, (iv) material of construction (related to (ii)), (iv) configuration, (v) cost, (vi) reliability, and (vii) ease of maintenance.

Metal valves are always used in metal vacuum systems. The construction of glass systems is easiest if glass valves are used, however glass valves (stopcocks) have relatively low conductance. The clear passage through an open stopcock is typically only a few millimeters in diameter. When a large-bore, high-conductance valve is required in a glass system, a metal valve of suitable size can be inserted using a glass-to-metal transition on each side of valve. The two most common glass valves are illustrated in Figure 3.18. One (Figure 3.18a) is a vacuum stopcock consisting of a tapered glass plug that is fitted into a glass body by lapping. The mating parts are sealed with a film of high-vacuum grease. As shown, these stopcocks are designed so that

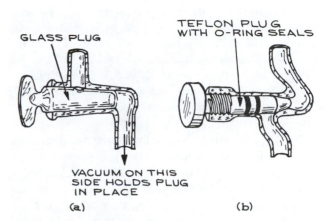

Figure 3.18   Glass vacuum valves: (a) a stopcock; (b) a glass valve with Teflon plug.

atmospheric pressure forces the plug into the body of the valve. The other more modern type of glass valve (Figure 3.18b) has a Teflon plug threaded into a glass body. The plug is sealed to the body with an O-ring. It is generally advisable to apply a light film of vacuum grease to the O-ring.

There are three basic designs for metal vacuum valves—an example of each is shown in Figure 3.19. The differences between the three designs have to do with the motion of the sealing plate as it approaches the valve seat. The valve in Figure 3.19a is a piston-type (or poppet) valve, in which an actuator moves the sealing plate axially as the valve is opened and closed. This design is inherently a *right-angle* valve with the inlet and outlet ports at 90° to one another; however the design is often modified to a *quasi-straight-through*, or *in-line*, valve by introducing a bend in the downward-facing inlet port so that it enters on the side opposite the outlet port with its centerline parallel to that of the outlet port. The conductance of inline valves is not as high as might be expected since the flow through the valve encounters two right-angle turns. The valve shown in Figure 3.19b is a swing valve or *butterfly* valve. The sealing plate rotates about an axis along its diameter as the valve is opened and closed. The actuator mechanism need only rotate one quarter of a turn. The obvious advantage of the butterfly valve is the low profile that provides the shortest possible path for gas and hence the highest possible conductance. The butterfly valve must be installed so that there is clearance for the valve plate in the open position. The valve shown in Figure 3.19c is a sliding-gate valve. The sealing plate moves radially with respect to the valve seat and the direction of gas flow. Some sort of cam mechanism is required to force the sealing plate against the valve seat once the plate is positioned above the seat. Vacuum valves of any of the three basic designs are available with bores ranging from $\frac{1}{4}$ to 8 in. or more.

Vacuum valves are fabricated of brass, cast aluminum, machined aluminum, or stainless steel. The cost increases, and the low-pressure limit of usability decreases, in the same order. Brass contains the volatile metal zinc and hence should not be used above room temperature. Small piston-type valves are often manufactured of brass. These are suitable for rough vacuum applications such as diffusion pump forelines. Cast aluminum is porous and outgases in vacuum. Inexpensive cast aluminum valves can be used in high vacuum. Valves manufactured of high-quality, vacuum-cast aluminum can be used down to at least 10^{-9} torr. Valves made of aluminum machined from solid stock, with the appropriate seals, find some use at ultrahigh vacuum. Stainless steel is the standard for valves used in high vacuum and UHV. In addition to a much lower outgassing rate (compared to aluminum), stainless steel withstands higher bakeout temperatures and is much more resistant to corrosive gases such as are used in semiconductor processing. The vacuum seal between inlet and outlet of most vacuum valves is made by an O-ring installed in the valve plate coming into contact with a smooth seating surface machined into the valve body. Viton is the standard O-ring material. A valve with Viton seals can be baked to 200°C. Valves with a polyimide seal are available for use up to 300°C. The sealing materials in a true UHV valve must be metal. Most designs employ a copper pad on the actuator plate that is driven onto a stainless-steel, knife-edge seat in the valve body. An all-metal UHV valve can be baked to 450°C. The soft copper sealing pad is effective for a limited number of cycles and must be replaced frequently. This problem, along with the relatively high cost of UHV valves, encourages UHV system designs with as few valves as possible.

All valves require an actuator mechanism that passes from the atmosphere to the evacuated innards of the valve. In butterfly valves, this is a rotating shaft; in some piston-type valves and gate valves, a rotating shaft passes through

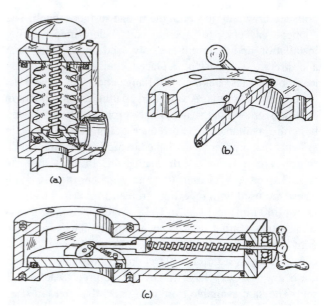

Figure 3.19 Metal high-vacuum valves: (a) vellows-sealed valve; (b) swing or *butterfly* valve; (c) sliding-gate valve.

the vacuum wall to drive a screw mechanism that provides the linear motion required to move the valve plate. In other designs, the linear motion is transmitted through the vacuum wall by a linearly translating shaft. In butterfly valves and inexpensive piston and gate valves, the actuator shaft is sealed with an O-ring. Leakage around this seal is inevitably the source of vacuum failure in these valves. In high-quality piston and gate valves, linear motion is transmitted through the vacuum wall and the drive mechanism is located outside the vacuum. The actuator shaft is sealed by a bellows that flexes to follow the driver in and out as the valve is closed and opened. The life of a bellows shaft seal ranges from 10,000 to 100,000 cycles.

The drive mechanism in a vacuum valve may be mechanical, electromagnetic, or pneumatic. The most robust actuators are driven by a rotating handwheel or crank. These designs make for slow opening and closing since the valve plate in a vacuum valve must be moved over a relatively large distance to provide good conductance when the valve is open. A large gate valve may require as many as 30 turns to open or close. Some piston and gate valves employ a toggle mechanism. The toggle handle swings through 90° or 180°. Considerable force is required at the end of the swing to effect a seal so the valve body must be mounted very securely. Valves with toggle actuators are rarely used in the laboratory. Valves with electromagnetic actuators are available for use with tubing of up to $1\frac{1}{2}$ in. (40 mm) OD. These valves are actuated by an electrical current to a solenoid that drives a plunger that opens the valve. The plunger is spring-loaded so that the valve automatically closes if power is lost. Piston-type valves and gate valves of all sizes are available with pneumatic actuators. Compressed air is admitted to a cylinder to drive a piston that is connected to the valve actuator. The motion is reversed by venting air on one side of the piston and admitting compressed air to the opposite side. An electromagnetic valve in the compressed air circuit controls the airflow. The electromagnetic valve is designed so that air is directed to the side that closes the valve when electrical power is lost. Some small valves use a simpler pneumatic circuit employing a spring to close the valve. Pneumatic valves typically require compressed air at 80 to 100 psi.

It is often necessary to admit gas at a low rate into a vacuum system. The gas source may be at a high pressure. For example, the pressure in a commercial gas cylinder may exceed 100 atm. Metering valves or *leak* valves are available for this purpose. Valves providing flow rates spanning several decades and with minimum controllable flow rates as low as 10^{-10} torr L s^{-1} are available. There are a variety of designs. Some are simply needle valves employing a long, gently tapering needle moving in a conical seat. Another design incorporates a metal or sapphire plate moving against a metal knife edge. All provide a more or less precise micrometer drive to control the moving element. In general, a leak valve should never be used to shut off the flow of gas as deformation of one of the two elements will result in a loss of low flow control. A shutoff valve, if required, should go on the upstream side of a leak valve and as close to the inlet as possible to minimize the *dead* volume of gas that must be pumped through the leak valve. A sintered metal filter should always be used at the inlet of a leak valve. The smallest piece of solid material compressed against the seat of a leak valve will certainly cause irreversible damage.

3.5.4 Mechanical Motion in the Vacuum System

Mechanical motion can be transmitted through the vacuum wall using a sliding seal, a flexible wall, or a magnetic coupling.

Rotary motion at speeds up to 100 rpm may be transmitted through an O-ring-sealed shaft. These seals are inexpensive but may fail if the O-ring becomes abraded. In addition, lubricants and the elastomer O-ring material are exposed to the vacuum.

A bellows seal to a shaft is the most common means to permit motion to pass through the vacuum wall. A simple and inexpensive linear-motion feedthrough can be made from the actuator mechanism of a small bellows-sealed vacuum valve. As shown in Figure 3.20, the valve body is truncated above the seat, a mounting flange is brazed to the body, and a fixture is brazed or screwed to the valve plate for attaching the mechanism to be driven inside the vacuum. Rotary motion may be transmitted through a bellows-sealed *wobble drive* illustrated in Figure 3.21. Relatively inexpensive rotary drives of this type are available from most vacuum hardware suppliers. Much more expensive and complex drives are also available, including drives that provide three-axis rotation and three-axis translation.

Moving parts can be magnetically coupled through a vacuum wall. This scheme is particularly useful for the transmission of linear motion through the wall of a glass vacuum system. The driven element is attached to a block of magnetic material such as iron or nickel that is placed against the inside of the vacuum wall. A magnet can then drag the driven element along the wall. A more exotic application of magnetic forces is found in the use of ferromagnetic fluids as shaft seals held in place by magnetic fields. Ferrofluidic shaft seals are commercially available.

Metal surfaces become very clean *in vacuo,* particularly after baking, and metals in close contact tend to cold weld. Because of this, unlubricated bearing surfaces within a vacuum system often become very rough after only a little use. Liquid lubricants can be used if they are thoroughly outgassed before installation. The vapor pressure of the lubricant will be the limiting factor. Diffusion pump oils can be used in diffusion-pumped vacuum systems since

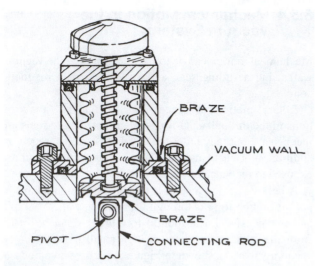

Figure 3.20   Bellows-sealed valve (Fig. 3.19a) converted to a linear-motion feedthrough.

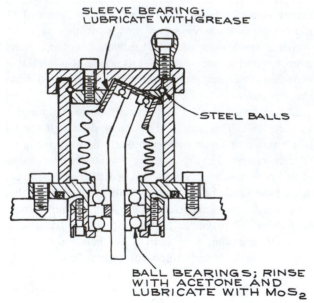

Figure 3.21   Bellows-sealed, wobble-drive, rotary-motion feedthrough.

the ultimate pressure is determined by the room temperature vapor pressure of the oil in the pump. Polyphenyl ether oils lubricate well and are among the lowest-vapor-pressure oils. Perfluorinated oils such as DuPont Krytox 143AZ are an excellent choice for liquid lubrication in a vacuum. A difficulty with liquid lubricants is their tendency to creep out of a bearing and along the surface of a shaft. A creep barrier can be placed between a lubricated bearing surface and the surroundings. This may consist of a coating or a plastic lip of a material such as Teflon that is not wetted by the lubricant.

There are a number of methods of improving bearing performance in a vacuum system without introducing high-vapor-pressure oils into the vacuum. The tendency for a bearing to gall is reduced if the two mating bearing surfaces are made of different metals. For example, a steel shaft rotating without lubrication in a brass or bronze journal will hold up better than in a steel bushing. A solid lubricant may be applied to one of the bearing surfaces. Silver, lead-indium, and molybdenum disulfide (MoS_2) have been used for this purpose. Graphite does not lubricate in a vacuum. MoS_2 is probably best. The lubricant should be burnished into the bearing surface. The part to be lubricated is placed in a lathe. As the part turns, the lubricant is applied and rubbed into the surface with the rounded end of a hardwood stick. By this means, the lubricant is forced into the pores. After burnishing, the surface should be wiped free of loose lubricant.

One component of a bearing may be fabricated of a self-lubricating material such as nylon, Delrin, Teflon, or polyimide. Teflon is good for this purpose, but its propensity to cold flow will cause the bearing to become sloppy with time. Polyimide can be used in ultrahigh vacuum after baking to 250 to 300°C. Brown, Sowinski, and Pertel have overcome the cold-flow problem in the design of a drive screw for use in a vacuum.[6] Both the screw and its nut are steel, and lubrication is accomplished by placing a Teflon key in a slot cut in the side of the screw. Teflon is then continuously wiped onto the screw threads as the screw turns.

For very precise location of rotating parts and for high rotational speeds, ball bearings are required. Precision, bakeable, low-speed ball bearings can be made using sapphire balls running in a stainless-steel race. Inexpensive precision sapphire balls are available from Industrial Tectonics. A ball retainer is required to prevent the balls from rubbing against one another. As explained in Section 1.6.1, the race must be designed so that the balls rotate without slipping against the race surface. It is helpful to burnish the race with MoS_2.

For high-speed applications, a fluid lubricant is necessary. The following process has been developed at the NASA-Goddard Space Flight Center.[7] Purchase high quality stainless-steel ball bearings with side shields and phenolic ball retainers, such as the New Hampshire PPT series or Barden SST3 series. Remove the shields and leach clean the phenolic by boiling the bearings in chloroform-acetone, and then vacuum dry at 100°C. The bearings are then lubricated by impregnating the phenolic with DuPont Krytox 143AZ, a fluorinated hydrocarbon. Impregnation is accomplished by immersing the bearing in the lubricant and heating to 100°C at a pressure of 1 torr or less until air stops bubbling out of the phenolic. After this process the bearing must be wiped almost dry of lubricant. The Texwipe Company makes ultraclean foam cubes for this process. The amount of lubricant in the bearing is determined by weighing before and after impregnation. About 25 mg of Krytox is required for an R4 ($\frac{1}{4}$ in.) bearing, and about 50 mg is needed for an R8 ($\frac{1}{2}$ in.) bearing. This lubrication process should be carried out in a very clean environment, and the bearing should be inspected under a microscope for cleanliness before the side shields are replaced. Bearings treated in this manner have been run at speeds up to 60,000 rpm *in vacuo*.

3.5.5 Traps and Baffles

Traps are used in vacuum systems to intercept condensable vapors by means of chemisorption or physical condensation. When an oil-sealed mechanical pump is used as the backing pump for a diffusion pump or a turbomolecular pump, a trap is placed in the foreline between the backing pump and the high-vacuum pump to prevent backstreaming of mechanical pump oil vapor. This is absolutely essential for the creation of a hydrocarbon-free vacuum. In some applications, such as vacuum processes employed in the semiconductor industry, the foreline trap is intended to condense corrosive gases from the process chamber to prevent harm to the forepump. In addition, a trap is placed between a diffusion pump and a vacuum chamber to pump water vapor and remove diffusion-pump fluid vapors that migrate backwards from the pump toward the chamber.

In many foreline traps, an ultraporous artificial zeolite known as molecular sieve is employed to adsorb oil vapor. These traps are similar in design to the molecular-sieve sorption pumps previously described, except, of course, that a trap must have both an inlet and an outlet. A typical foreline trap filled with zeolite 13X molecular sieve is shown in Figure 3.22. The molecular sieve is initially activated by baking to 300°C for several hours. In use, the sieve material is regenerated at intervals of about two weeks by baking under vacuum to drive out absorbed oil and water. Care must be taken to prevent the oil driven out of the trap from condensing upstream of the trap, thus doing more harm than good. A valve on the upstream side of the trap can be closed to prevent oil vapor from back-streaming, provided that the valve and the connection to the trap are themselves heated to prevent condensation while the trap is being baked. It is also possible that vapor driven from the trap will contaminate the forepump. Some trap designs provide a third connection for a separate pump to be used during the bakeout. Another trap design provides a removable basket for the sieve charge that can be baked in a separate oven. For a system that must be kept in continuous operation, two traps in parallel are used. Valves on each side of both traps allow one trap to be isolated for regeneration. An auxiliary pump (with suitable valving) is used to evacuate the trap being regenerated. Some traps are provided with a third port for connection to this auxiliary pump.

The lowest pressure attainable with a two-stage, oil-sealed mechanical pump is initially determined by the vapor pressure of the oil in the pump. By using a molecular-sieve trap in series with a mechanical pump to remove oil vapor, it is possible to achieve pressures as low as 10^{-4} torr in a small system with no gas load. This is an effective scheme for maintaining a vacuum for thermal insulation.

Molecular sieve may also be used in a high-vacuum trap over the inlet of an oil diffusion pump. As shown in Figure 3.23, these traps are designed to be optically opaque so that a molecule cannot pass through the trap in a straight line. This precaution is necessary because this type of trap is intended for use at pressures where the mean free path of a molecule is very long. To ensure high conductance,

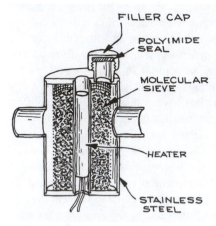

Figure 3.22 A foreline trap filled with molecular sieve.

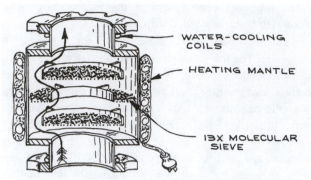

Figure 3.23 High-vacuum molecular-sieve trap for use over the inlet of a diffusion pump.

the inlet and outlet ports should have the same cross-sectional area as the inlet of the attached pump. The cross section perpendicular to the flow path through the trap (indicated by an arrow in Figure 3.23) should also be at least as large as that of the inlet and outlet ports. The molecular sieve is activated by baking under vacuum to 300°C for at least 6 hours. The trap may be heated with heavily insulated heating tape or a custom-made heating mantle such as those obtained from the Glas-Col Apparatus Company. The flanges of the trap are water cooled to prevent overheating the seals.

The maintenance of a high-vacuum trap is different from that of a foreline trap. A high-vacuum molecular-sieve trap is placed above an oil diffusion pump, and a gate valve is located above the trap to permit isolation of the trap and pump stack from the vacuum chamber. After the initial bakeout, the pump is run continuously so that the trap is always under vacuum. The isolation valve is closed whenever it is necessary to open the chamber to the atmosphere, and the chamber is rough-pumped before reopening the isolation valve. The molecular sieve cannot be regenerated after the first bakeout. The initial bake of the sieve results in the evolution of water vapor, but subsequent baking would drive out absorbed pump-fluid vapor. This oil vapor would condense on the bottom of the isolation-valve gate and be exposed to the vacuum when the valve is open. After the initial bakeout, the molecular sieve will trap oil vapor effectively for a period of several months if it is not exposed to moist air during that period.

When the sieve becomes clogged with absorbed vapor, the base pressure of the pump-and-trap combination will begin to rise. At this time, the molecular-sieve charge should be replaced. If it is absolutely necessary to stop the diffusion pump, the pump and trap should be filled with argon or dry nitrogen and isolated from the atmosphere in order to preserve the molecular sieve.

Condensation traps employing cold surfaces—cold traps—are used as foreline traps as well as inlet traps over diffusion pumps. The coolant may be simply a flow of cold water, an actively cooled refrigerant (Freon), or liquid nitrogen. In principal, a trap cooled to liquid nitrogen temperature would seem most effective; however, the vapor pressure of most pump fluids is below 10^{-12} torr at –40°C, which is a temperature easily achieved with a Freon refrigerator. A simple *thimble trap* is shown in Figure 3.24. These traps are used as foreline traps or over a small diffusion pump. For periodic operation, the thimble can be filled with liquid nitrogen. The coolant must be replenished every few hours. For continuous, long-term operation, an alternative (shown in the figure) is to fill the trap with a low-melting liquid such as isopropyl alcohol and refrigerate the liquid with an immersion cooler. These coolers consist of a refrigerator housed in a small console with a hose to carry refrigerant to a flexible probe that is inserted in the bath in the trap. A temperature of –40°C is easily attained with the smallest unit since the only heat load is thermal leakage from the surroundings into the trap. A liquid-nitrogen-cooled cryotrap for use over a large diffusion pump is illustrated in Figure 3.25. These traps are connected to a

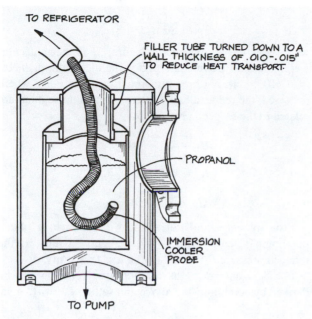

Figure 3.24   A thimble trap cooled by the flexible, cooled probe of an immersion cooler.

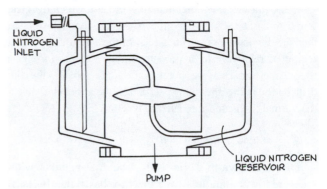

Figure 3.25   Liquid nitrogen-cooled cryotrap for a diffusion pump.

Figure 3.26   Cooled baffle.

large liquid-nitrogen Dewar with a valve operated by a sensor that detects the coolant level in the trap. When a cold trap is permitted to warm up, it must be isolated from the vacuum chamber so that condensed material does not migrate into the chamber. Also, the trap should be vented so that the evaporating condensate does not build up a dangerously high pressure.

An optically dense *chevron* baffle of the type shown in Figure 3.26 is usually placed over the inlet of a diffusion pump. These baffles may be air cooled, water cooled, or cooled by refrigerant from a small refrigerator. Baffles with thermoelectric coolers operating at −35°C have recently become available. In addition to reducing backstreaming of pump fluid vapor, a baffle serves as a radiation barrier between the hot pump and the cold trap or vacuum system above the baffle.

3.5.6 Molecular Beams and Gas Jets

Often it is necessary to introduce a gaseous sample into a vacuum system. The sample may be contained in a small chamber with holes suitably located to admit probes such

as light beams or electron beams; however, the walls of such a chamber may prove a hindrance to the proposed experiment. Alternatively, a gas can be introduced as an uncontained, but directed, beam of atoms or molecules. The absence of walls is only one of the advantages to using a gas beam. Because of the directed velocities of the particles in the beam, it is possible to maintain the sample in a collisionless environment while still obtaining useful densities. The beam can be crossed with another beam or directed at a surface to obtain collisions of a specified orientation. Under appropriate conditions, the expansion of a jet of gas results in extreme cooling to give a very narrow population distribution among quantum states of molecules in the gas. Lucas succinctly stated the case for atomic or molecular beams when he observed that "beams are employed in experiments where collisions . . . are either to be studied, or to be avoided."[8]

A gas beam is created by permitting gas to flow into a vacuum through a tube. The nature of the beam depends upon whether the flow exiting the tube is in the molecular or viscous flow regime. The shape of the beam can be established by apertures placed downstream of the channel through which the gas flows into the vacuum.

We begin with the molecular-flow case, in which the mean free path λ of the gas at the outlet of the gas channel is greater than the diameter d of the channel. For the case of flow from a region of relatively high pressure into a vacuum through a round aperture (that is, a tube of length $l = 0$), the flux distribution in the direction specified by the angle θ relative to the normal to the aperture surface is

$$I(\theta) = I(\theta = 0)\cos\theta \qquad (\text{atoms s}^{-1}\text{sr}^{-1})$$

The beam width H is 120° (the full angular width measured at the angle where the flux has fallen by half). The beam can be narrowed by increasing the length l of the tube or channel through which the gas flows into the vacuum. Any description of the gas beam issuing from such a channel must consider the molecular diameter and the nature of the scattering of molecules from the walls of the tube. Both theoretical and experimental investigations have been carried out in an effort to describe a gas beam in terms of easily measured parameters. In comparing experiment and theory, it has turned out that the observed flux and the sharpness of the beam fall short of theoretical expectations by a factor of two to five.

In general, both experimental and theoretical results imply that to obtain useful fluxes and reasonable collimation, a beam source consisting of a tube or channel must be at least ten times greater in length than its diameter and the gas pressure behind this channel should be such that

$$d < \lambda < l$$

Lucas has devised a succinct description of gas beams in the molecular flow regime in terms of a set of reduced parameters.[8] The gas pressure P behind the channel (in torr), the beam width H (degrees), the gas particle flux I (atoms s^{-1} sr^{-1} (on axis)), and throughput Q (atoms s^{-1}) are related to the corresponding reduced parameters by

$$P = \frac{P_R}{l\sigma^2}$$

$$H = \frac{H_R d}{l}$$

$$I = \left(\frac{T}{295M}\right)^{1/2} \frac{d^2 I_R}{l\sigma^2}$$

$$Q = \left(\frac{T}{295M}\right)^{1/2} \frac{d^3 q_R}{l^2 \sigma^2}$$

where T is the absolute temperature, M is the molecular weight, d and l are in cm, and σ is the molecular diameter in Å (10^{-8} cm). For pressures roughly in the range where $d < \lambda < l$, the reduced half angle, flux, and throughput are related to the reduced pressure by

$$H_R = 2.48 \times 10^2 \sqrt{P_R}$$

$$I_R = 1.69 \times 10^{20} \sqrt{P_R}$$

$$Q_R = 2.16 \times 10^{21} P_R$$

From model calculations, Lucas found that to obtain maximum intensity for a given angular width H, the reduced pressure P_R should not be less than unity.

For design work, a useful *optimum* equation can be derived by combining the above and setting $P_R = 1$ to give

$$I = 6.7 \times 10^{17} \left(\frac{T}{295M}\right)^{1/2} \frac{dH}{\sigma^2}$$

the optimum axial intensity for a given channel diameter and desired angular width. The tube length is fixed by the equations for H and $H_R(P_R = 1)$ above, and the input pressure by the equation for P. Operating at somewhat higher pressures gives some control over the flux and throughput without significantly departing from the optimum condition.

The construction of low-pressure, single-channel gas-beam sources is fairly straightforward. An excellent source can be made from a hypodermic needle cut to the appropriate length. These needles are made of stainless steel, they are available in a wide range of lengths and diameters, and they come with a mounting fixture that is reasonably gas tight.

The goal of high intensity in a sharp beam is incompatible with a single-channel source because, for a given beam width, the gas load (throughput) increases more rapidly than the flux as the channel diameter is increased. The solution to this problem is to use an array of many tubes, each having a small aspect ratio (i.e., $d/l \ll 1$).[9] Tubes with diameters as small as 2×10^{-4} cm and lengths of 1×10^{-1} cm arranged in an array 1 cm or more across are commercially available in glass (e.g. from Burle Electro-Optics or Minitubes-Grenoble).

When the gas pressure behind an aperture or nozzle leading to a vacuum is increased to the extent that the mean free path is much smaller than the dimensions of the aperture, a whole new situation arises. Not only is the density of the resultant gas jet much greater than in the molecular-flow case, but the shape of the jet changes and the gas becomes remarkably cold. This is a direct result of collisions between molecules in the gas as it expands from the orifice into the vacuum. Because of collisions, the velocities of individual molecules tend toward that of the bulk gas flow, just as an individual in a crowd tends to be dragged along with the crowd. The translational temperature of the gas, as defined by the width of its velocity distribution, decreases, while the bulk-flow velocity increases. The conversion of random molecular motion into directed motion continues until the gas becomes too greatly rarefied by expansion, at which point the final temperature is frozen in. Because the mass-flow velocity increases while the local speed of sound (proportional to the square root of the translational temperature) decreases, the Mach number rises and the flow becomes supersonic. Inelastic molecular collisions in the expanding gas also cause internal molecular energy to flow into the kinetic energy of bulk flow, with the result that there may be substantial rotational and vibrational cooling. Rotational temperatures less than 1 K and vibrational temperatures less than 50 K have been obtained. At these low temperatures only a small number of quantum states are occupied, an ideal condition for a variety of spectroscopic studies.

The low temperatures obtained in a supersonic jet can present some problems. Chief among these is that of condensation. Collisions in the high-density region of the jet cause dimer or even polymer formation. With high-boiling samples, bulk condensation may occur. Of course, if one wishes to study dimers or clusters, this condensation is desirable. To avoid dimer formation, the sample gas is mixed at low concentration with helium. This *seeded gas* sample is then expanded in a jet. The degree of clustering of the seed-gas molecules is controlled by adjusting its concentration in the helium carrier.

The cooling effect, as well as the directionality of the jet, is lost if the expanding gas encounters a significant pressure of background gas in the vacuum chamber. Unfortunately, optimum operating conditions require a large throughput of gas into the chamber. The extent of cooling depends upon the probability of binary collisions, which is proportional to the product $P_0 d$ of the pressure behind the nozzle and the diameter of the nozzle. Furthermore, to minimize condensation the ratio d/P_0 should be as large as possible. The result, in practice, is that a large-capacity vacuum pumping system is needed. In typical continuously operating supersonic jet apparatus, source pressures of 10 to 100 atm have been used with nozzle diameters of 0.01 cm down to 0.0025 cm. The conductance of an aperture under these conditions is roughly

$$C = 15d^2 \ \text{L s}^{-1}$$

when the diameter is in centimeters. This implies a throughput on the order of 10 torr L s^{-1}. To obtain a mean free path of several tens of centimeters, a base pressure of about 10^{-4} torr is required. A pump speed approaching 10,000 L s^{-1} may be necessary. Typically, several large diffusion pumps are used. Whenever possible, the jet is aimed straight down the throat of a pump.

In many experiments, the cold molecules in a supersonic jet are only probed periodically—as, for example, when doing laser spectroscopy with a pulsed laser. Operation of the jet in a synchronously pulsed mode can reduce the gas load on the pump by several orders of magnitude compared to that in the continuous mode. A number of fast valves for this purpose have been devised.[10] The simplest are based upon inexpensive automobile fuel injector valves.[11] General Valve produces a system consisting of a pulsed valve and driver that operate at pulse rates as high as 250 Hz with a pulse width as small as 160 μs.

Another efficient means of dealing with the background gas problem has been demonstrated.[12] This relies upon the fact that interaction of the expanding gas with the background gas gives rise to a shock wave surrounding the gas jet. If the pressure (P_0) behind the nozzle is increased, it is possible to achieve a mode of operation where a rarefied region is created behind the shock wave. In the region upstream of the shock front, the gas behaves like a free jet expanding into a perfect vacuum. The distance from the nozzle to the shock front is

$$l = 0.67d\left(\frac{P_0}{P}\right)^{1/2}$$

where P_0 is the pressure in the nozzle, P the background pressure, and d the nozzle diameter.

When the nozzle pressure is sufficiently high to achieve a free jet length of usable dimensions, the background pressure will increase greatly. This state of affairs has a distinct advantage, since at a high background pressure, a large throughput can be achieved with a pump of moderate speed. For example, to obtain a free length $l = 1$ cm with a 0.01 cm nozzle and a source pressure of 10 atm, the background pressure should be about 400 mtorr. The throughput in this case is about 10 torr L s^{-1}, and the required pumping speed (25 L s^{-1}) could be achieved by a large rotary mechanical pump or a Roots pump of modest capacity.

The fabrication of a nozzle is straightforward, although some difficulty may be encountered in making the necessary small hole. A skillful mechanical technician can drill a hole as small as 0.01 cm diameter. Smaller holes can be made by spark or electrolytic erosion, or by swaging a hole closed on a piece of hard wire and then withdrawing the wire. One of the first small, high-pressure nozzles was made by drilling down the axis of a stainless steel rod to within 0.1 mm of the end with a drill bit having a sharp conical point.[12] A 0.0025-cm-diameter hole was then made through the remaining metal by spark erosion.

For many experiments, the gas flowing into a vacuum system through an aperture or a channel produces a beam that is too broad or too divergent for the intended application. In this case, the beam shape can be defined by one or more apertures placed downstream from the source. It is often advantageous to build these apertures into partitions that separate the vacuum housing into a succession of chambers. Each chamber can be evacuated with a separate pump. The pressure in the first will be highest, so the large throughput obtained here will significantly reduce the gas load on the next pump. This scheme is called *differential pumping* (see Section 3.6.2).

An aperture downstream of a supersonic jet, but close enough to be in the viscous flow region, is called a *skimmer*. The design of these skimmers is critical, since they tend to produce turbulence that will destroy the directionality of the flowing gas. A skimmer is usually a cone with its tip cut off and the truncated edge ground to knife sharpness. The details of the design, location, and fabrication of a skimmer are beyond the scope of this book. An interested reader should refer to a specialized text on this subject.[13]

3.5.7 Electronics and Electricity *in Vacuo*

At low frequencies, the vacuum between electrodes or wires provides excellent insulation. At pressures below 10^{-4} torr, the breakdown voltage between smooth, gently-rounded surfaces is in excess of 1000 V for each millimeter of separation. Sparking will occasionally occur between newly made parts because of high field gradients around whiskers of metal. These whiskers quickly evaporate and sparks do not recur. The challenge in making electrical devices function in vacuum is in the design of spacers and other elements that locate electrodes and wires that are at different electrical potentials. Insulating materials must possess adequate mechanical strength and be compatible with the vacuum environment and available fabrication processes. In addition, insulators must be very clean. Electrical breakdown in the vacuum is usually initiated by a discharge traveling along the surface of an insulator.

Ceramic materials are the best insulators and, among these, alumina (at least 96% Al_2O_3) is probably the best. Simple shapes such as rod, tube, and balls are readily available to be incorporated in insulator assemblies (see Figures 5.40 and 5.41). More complex shapes require specialized grinding techniques. Alumina circuit board substrate is available as plate .010 in. to .060 in. thick and 6 in. square. Owing to the scale of the circuit board industry, the facilities for custom fabrication by laser cutting of alumina plate are widely available. The designer of vacuum apparatus can now specify quite complex shapes cut from alumina plate at a cost of just a few dollars apiece. A simple scheme employing alumina washers to locate and insulate electrically isolated parts is illustrated in Figure 3.27c.

Alumina can be metalized and subsequently joined by brazing to stainless steel or Kovar structures. It is impractical to attempt this process in the lab, but this technology is employed in many standard vacuum components, such as electrical feedthroughs to be inserted

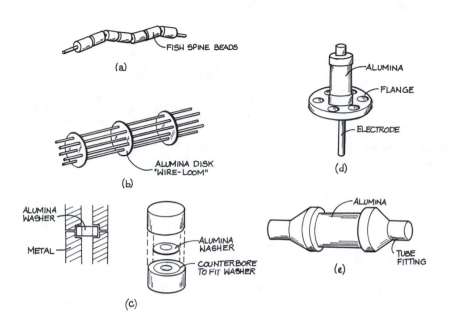

Figure 3.27 Applications of ceramics for electrical insulation in vacuum systems: (a) ceramic *fish spine* beads threaded on a wire as insulation; (b) ceramic disk with holes used as a wire-loom to keep wires in a bundle from touching one another; (c) design for electrodes located and electrically isolated with spacers laser cut from alumina circuit board substrate; (d) a ceramic-insulated electrical feedthrough; (e) a ceramic break in a vacuum line mounted between two metal flanges.

in the vacuum wall (Figure 3.27d) and ceramic *breaks* built into the middle of a length of metal vacuum tubing (Figure 3.27e).

A few precautions should be observed in designing and mounting ceramic parts. Ceramic materials are strongest in compression and cannot be plastically deformed. The best design of an assembly with ceramic components captures each ceramic part between smooth surfaces that are aligned with the face of the part so that shear forces are not applied to the ceramic material.

Ceramics are often used simultaneously for electrical insulation and thermal isolation. In this case it is important to be aware of the strong dependence of electrical resistivity on temperature. The resistivity of a ceramic material falls by roughly a factor of ten for every 100 degrees (C) increase in temperature.[14] For example, the resistivity of alumina is greater than 10^{17} ohm cm at 100°C and falls below 10^7 ohm cm at 1000°C.

The surface of a ceramic part is more or less porous and susceptible to trapping contaminates. In handling ceramic parts, inorganic salts from skin oils may be transferred from the fingers to the surface of the ceramic. Salts absorb water and encourage electrical conduction across the surface. A part should be cleaned with organic solvents to remove oils, followed by a final rinsing in hot distilled water to remove these salts. The part can be dried by rinsing with pure methanol and heating. Surgical gloves should be worn to handle a part after it has been cleaned.

If the spacing between wires in a vacuum system cannot be reliably maintained, then insulation is required. Ordinary Teflon-insulated, solid hookup wire may be used down to 10^{-7} torr, although air bleeding out from under the insulation will slow pumpdown. Stranded wire should never be used, as gas is trapped between the strands. At very low pressures and high temperatures, wires can be insulated by stringing ceramic *fish spine* beads (Figure 3.27a) or pieces of Pyrex tubing over them. The spacing between wires can be maintained with ceramic wire looms (Figure 3.27b) available from many suppliers of vacuum apparatus. Wire insulated with a Kapton (polyimide) film or a ceramic coating is available from vacuum equipment suppliers for use in UHV. The latter is better for use at elevated temperatures, but has the disadvantage of shedding bits of ceramic material when flexed. Electrical solder should not be used for electrical connections in a vacuum system because the lead in solder and soldering flux contaminate the vacuum. Mechanical connections are preferable. Connections to electrodes can be made by wrapping a wire under the head of a screw and tightening the screw. Wires may be jointed by slipping the ends into a close-fitting piece of tubing and crimping the tubing onto the wire. A variety of wire connectors are available from vacuum equipment suppliers. Tungsten, molybdenum, nichrome, or stainless steel wires may be spot welded together. In welding refractory metals, a more secure weld is obtained if a piece of nickel foil is interposed between the wires.

Electronic devices tend to overheat in a vacuum since the only cooling is by radiation. Electronic components, such as power transistors and integrated circuits that must dissipate more than about 2 watts, are particularly unreliable. Such devices can be placed in vacuum-tight, gas-filled boxes that are thermally connected to the vacuum wall.

Electrical feedthroughs for wires passing through the wall of a glass vacuum system are illustrated in Figure 3.28. As described in Section 2.3.10, an electrical feedthrough can be made by sealing a tungsten rod directly into a hard glass wall. An easier solution is to braze an electrode into the metallic end of a standard Kovar-to-glass graded seal and join the glass end of the seal to the glass vacuum wall.

3.6 VACUUM SYSTEM DESIGN AND CONSTRUCTION

Before beginning the design of a vacuum system, a number of parameters must be specified in at least a semiquantitative manner. The size and shape of the vacuum chamber must be determined. The desired ultimate pressure and the composition of the residual gas in the chamber must be specified. One must estimate the gas load on the pumps. The amount of money and time available are also important considerations. It is instructive to spend a few evenings leafing through vacuum equipment catalogs to become familiar with the specifications and cost of commercial vacuum components.

The ultimate pressure requirement dictates the choice of the primary pump. Cost and limitations on hydrocarbon contamination will determine the choice of roughing and backing pumps. An ultimate pressure down to about 1 mtorr can be achieved with a mechanical pump with an appropriate trap. Sorption pumps can be used on a system

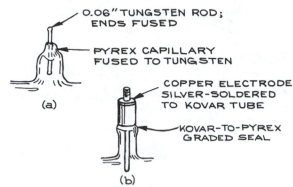

Figure 3.28   Electrical feedthroughs in a glass vacuum wall employing (a) a tungsten-to-metal seal; (b) a standard Kovar-to-glass seal.

requiring an ultimate pressure of 1 mtorr if there is not a large continuous gas load. Ultimate pressures in the high vacuum range down to about 10^{-9} torr can be achieved with a diffusion pump, a turbomolecular pump, or a cryopump. True ultrahigh vacuum systems ordinarily use ion pumps and titanium sublimation pumps. Ultrahigh vacuum can also be achieved with turbomolecular pumps. To reach UHV starting from a backing pressure near 1 torr requires a compression ratio approaching 10^{12}. This can be accomplished with two turbomolecular pumps operating in series, a turbo backed with a drag pump, or with a compound turbomolecular/drag pump.

The cost of a vacuum system is about equally divided between the cost of the pumps and the cost of the vacuum vessel and ancillary hardware such as gauges and valves. In terms of cost per unit pumping speed, mechanical pumps are most expensive—on the order of $100 per L s^{-1}. Most systems require a mechanical pump as either the primary pump or a backing pump; however, since they operate at relatively high pressures and correspondingly high throughputs, one ordinarily finds that for laboratory apparatus only modest pumping speed is required of a mechanical pump—a few L s^{-1} to a few tens of L s^{-1}. For high-vacuum systems, diffusion pumps are most cost effective although sorption or cold traps needed to prevent hydrocarbon contamination will add to the total cost. The cost of a diffusion pump amounts to $2 to $5 per L s^{-1} of pumping speed. A turbomolecular pump with its controller, a cryopump with a refrigerator, or an ion pump and controller will cost five to ten times as much as a diffusion pump of comparable pumping speed.

There are three sources of gas that must be pumped out of a vacuum vessel: molecules adsorbed on the walls of the container as well as the components therein, gas intentionally admitted as part of the process under study, and leaks. The rate of outgassing of adsorbed water, carbon dioxide and hydrocarbons, and ultimately hydrogen, from the surface of materials is difficult to measure. One finds considerable variation in outgassing rates for various materials reported in the literature. With care in the choice and processing of vacuum apparatus construction materials, it is generally possible to reduce outgassing rates below 10^{-9} torr L s^{-1} cm^{-2}, and lower yet with baking. Similarly, actual leaks can be eliminated by careful design and construction; however so-called *virtual leaks* are more or less a problem in all systems. Virtual leaks are the result of gas being tapped in small occlusions within the vacuum vessel. These typically occur where parts are fitted together. Gas will be trapped between mated surfaces and the conductance through the space between such surfaces is very small. The trapped gas may take weeks to be pumped away and thus acts as a continual gas load on the pump. Every effort should be made to vent these spaces with pumpout holes (see Figure 3.40).

3.6.1 Typical Vacuum Systems

Diffusion-pumped system. A diffusion-pumped vacuum system is shown schematically in Figure 3.29. For illustration, suppose the vacuum chamber is constructed of stainless steel and is 30 cm in diameter and 30 cm high with a surface area of 5000 cm^2. If the pump stack consists of a 2 in. oil diffusion pump with a speed of 150 L s^{-1}, and a trap and baffle with a conductance of 200 L s^{-1}, the net speed of the pump stack will be

$$\left(\frac{1}{S_p} + \frac{1}{C}\right)^{-1} = \left(\frac{1}{150\,\mathrm{L\,s^{-1}}} + \frac{1}{200\,\mathrm{L\,s^{-1}}}\right)^{-1} = 85\,\mathrm{L\,s^{-1}}$$

Assuming an initial outgassing rate of 10^{-9} torr L s^{-1} cm^{-2}, the vacuum vessel will pump down to

$$P = \frac{Q}{S} = \frac{(5000\,\mathrm{cm^2}) \times (10^{-8}\,\mathrm{torr\,L\,s^{-1}cm^{-2}})}{85\,\mathrm{L\,s^{-1}}}$$

$$= 6 \times 10^{-7}\,\mathrm{torr}$$

within a few hours. After a day of pumping and perhaps a light bake to 100°C, the outgassing rate should fall below 10^{-9} torr L s^{-1} cm^{-2} and an ultimate pressure in the 10^{-8} torr range should be achievable.

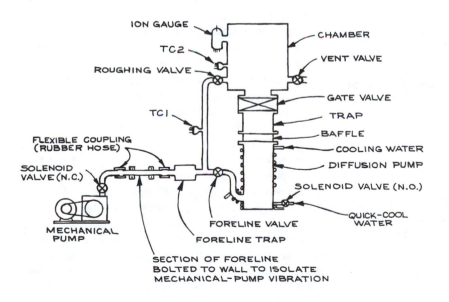

Figure 3.29   Schematic diagram of diffusion-pumped high-vacuum system.

The size of the backing pump is determined from the maximum throughput of the diffusion pump. This occurs after the vacuum vessel is rough pumped and the gate valve over the diffusion pump is first opened. The pressure in the vessel at this time should not be greater than about 50 mtorr. The maximum throughput will then be

$$Q_{max} = P_{max}S = (5 \times 10^{-2} \text{torr}) \times (85 \text{L s}^{-1})$$

$$= 4 \text{torr L s}^{-1}$$

The foreline pressure must always remain below about 200 mtorr, so the speed of the backing pump should be

$$S_{p, \text{backing}} = \frac{Q_{max}}{P_{\text{backing}}} = \frac{4 \text{torr L s}^{-1}}{2 \times 10^{-1} \text{torr}}$$

$$= 20 \text{ L s}^{-1}$$

This requirement can be met by a small, two-stage mechanical pump.

The conductance of the foreline should be at least twice the speed of the forepump so that the forepump is not strangled by the foreline. To achieve a conductance of 40 L s^{-1} at a pressure of 200 mtorr in a foreline a half meter long, the diameter (from Section 3.2.4) must be

$$D = \left(\frac{CL}{180P}\right)^{1/4} \text{cm} \qquad (L \text{ in cm})$$

$$= \left(\frac{40 \times 50}{180 \times 2 \times 10^{-1}}\right)^{1/4} = 2.7 \text{cm}$$

In this case, 1 in. copper water pipe would make an excellent foreline.

To activate a system of the type shown in Figure 3.29, starting with all valves closed and the pumps off, proceed as follows:

1. Turn on the mechanical pump.

2. When the pressure indicated by thermocouple gauge 1 (TC1) is below 200 mtorr, open the foreline valve.

3. When TC1 again indicates a pressure below 200 mtorr, turn on the cooling water and activate the diffusion pump. The pump will require about 20 minutes to reach operating temperature. When operating, it will make a crackling sound.

4. Before pumping on the chamber with the diffusion pump, the pressure in the chamber must be reduced to a rough vacuum. Close the foreline valve and open the roughing valve. When the pressure indicated by TC2 is below 50 mtorr, close the roughing valve and open the foreline valve. The diffusion pump should not be operated for more than 2 minutes with the foreline valve closed. If necessary, interrupt the roughing procedure, close the roughing valve, wait for TC1 to indicate less than 200 mtorr, and reopen the foreline valve for a moment to ensure that the diffusion-pump outlet pressure does not rise above about 200 mtorr.

5. Slowly open the gate valve that isolates the diffusion pump from the chamber. In less than 10 seconds, the pressure indicated by TC2 should fall below 1 mtorr and the ionization gauge can be turned on.

A diffusion-pumped system of the type shown in Figure 3.29 is one of the most convenient and least expensive systems for obtaining high or even ultrahigh vacuum. The system can be fabricated of metal or glass. A baffled but untrapped oil diffusion pump on this system should result in an ultimate pressure slightly below 10^{-6} torr. With a cold trap, an ultimate pressure of 10^{-9} torr is possible if the vacuum chamber is baked and only low-vapor-pressure materials are exposed to the vacuum.

To avoid contamination or damage in the event of a loss of electrical power, a number of inexpensive safeguards are incorporated in the design in Figure 3.29. A normally open (N.O.) solenoid-operated water valve, in parallel with the pump heater, admits water to the diffusion pump quick-cooling coils if the power fails. The diffusion pump is isolated from the mechanical pump by a normally closed (N.C.) solenoid valve to prevent air or oil from being sucked through the mechanical pump into the foreline. Special mechanical pump inlet valves are available for this purpose (HPS Vacuum Sentry and Leybold SECUVAC). These valves close automatically when power is lost and vent the pump to atmosphere. When power is regained, the valve does not reopen until the pressure on the pump side falls to a few hundred mtorr. The SECUVAC valve is also designed to close in the event of a sudden rise in inlet pressure. It is helpful to place a ballast volume in the foreline to maintain a low foreline pressure while the diffusion pump cools after a power outage. A relatively large foreline trap will also serve as ballast. A ballast tank of 5 to 10 liters in the foreline will permit the continued operation of the diffusion pump for a period of perhaps 10 to 20 minutes if the backing pump is turned off and the valve at the inlet to the backing pump is closed. This intermittent mode of operation can be useful in the event that vibration of the mechanical pump has an adverse effect on an instrument housed in the high vacuum chamber. For example, this mode of operation is used on some electron microscopes.

A useful fail-safe mechanism, not shown in Figure 3.29, would be a water-pressure sensor in the water line to interrupt the electrical power if the water fails. Alternatively, a heavy-duty, normally closed (N.C.), bimetallic temperature switch wired in series with the diffusion pump heater may be mounted to the bottom turn of the water-cooling coil. If the cooling water flow is inadequate, the rising temperature of the pump barrel will interrupt the electricity to the pump heater. Many diffusion pumps are equipped with a mounting plate for a temperature switch that can be obtained from the pump manufacturer.

A power interruption in the operator's absence threatens serious contamination of the vacuum vessel both at the time the power goes off and when it comes back on again. When the power goes off, it is essential that the diffusion pump stops operating before the foreline pressure rises above the critical backing pressure. This is the purpose of the quick-cool water valve described above. When power is restored in the operator's absence, there is no provision to rough out the vacuum vessel or otherwise follow the orderly initiation of pumping described above. As a minimum safeguard against a power failure, the power to the system's components should be supplied through power relays that are wired to latch *off* until reactivated by the operator. A latching relay circuit is shown in Figure 3.30.

For a vacuum system that operates for long periods of time without close supervision, it is worthwhile to incorporate a control system that monitors the pressure in the chamber and the foreline. Many high-quality commercial thermocouple-gauge and ionization-gauge controllers include relays that can be set to trip at a predetermined pressure. The relay controlled by the foreline thermocouple gauge (TC1) can be set to trip at 200 to 300 mtorr. This relay can be wired to operate a second, heavy-duty relay that removes power from the diffusion pump and simultaneously opens the quick-cool water valve and closes the foreline solenoid valve at the mechanical pump. The overpressure relay controlled by the ion gauge can be employed to isolate the high-vacuum chamber by means of an electrically controlled, pneumatically activated gate valve when the pressure in the chamber rises above a few mtorr. In lieu of a gauge

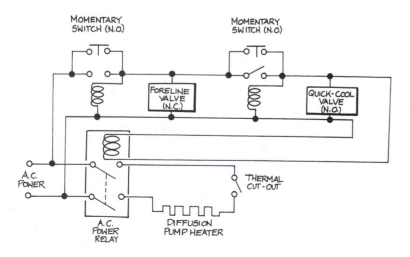

Figure 3.30   Relay circuit to latch *off* a diffusion-pumped high-vacuum system.

controller with control relays, inexpensive, modular Pirani vacuum sensors are now available with built-in pressure switches (for example, the HPS Moducell). These can be set to trip in the range 5×10^{-3} to 30 torr, making them suitable for monitoring either the high-vacuum chamber or the foreline. The design of a logical interlock system employing temperature and pressure sensors to protect a vacuum system is discussed in Sections 6.6.9 and 6.6.10.

A small, inexpensive ultrahigh-vacuum system.

A small, inexpensive system, capable of producing a vacuum of about 10^{-10} torr in a 1 L volume, is illustrated in Figure 3.31. The entire system can be made of Pyrex. Constrictions in the glass tubing joining components of the system are melted closed with a glassblower's torch to seal off each part of the system as evacuation proceeds. Rough pumping is initiated with a water aspirator or a compressed-air venturi pump. High vacuum is achieved with sorption pumps operated in sequence and ultrahigh vacuum is reached with a titanium sublimation pump, possibly with the aid of a small ion pump.[15] To achieve UHV, special precautions are taken to displace rare gases from the system and to remove water and carbon dioxide adsorbed on the walls of the system. Neither sorption pumping nor gettering efficiently pumps rare gases—helium in particular. Rare gases are swept from the system by a flow of dry nitrogen before evacuation is begun. The entire system is maintained at a moderately high temperature throughout the entire pumpdown procedure in order to drive adsorbed gases off the walls and into the pumps.

The pumpdown procedure for this vacuum system is as follows:

1. Flow pure dry nitrogen through the system from S1 to the aspirator. Do not activate the aspirator. Pure dry nitrogen gas can be obtained from the head space over liquid nitrogen in a Dewar.

2. As nitrogen flows through the system, the molecular-sieve sorption pumps are heated to 300°C and the trap is cooled with liquid nitrogen. The remainder of the system should be heated to about 100°C with heating tape or by brushing the glass with a cold flame or with hot air from a heat gun.

3. After about an hour, close pinchoff S_1 by heating the glass with a torch, and turn on the aspirator. The pressure will quickly fall to about 20 torr.

4. Seal off constriction S_2 and refrigerate the first molecular-sieve pump. The pressure will fall to about 10^{-4} torr. Continue to heat the system.

5. Seal off S_3 and cool the second sorption pump. The pressure will fall to about 10^{-7} torr. Continue heating the system for a short time to drive off the remaining adsorbed gases.

6. Pass a current through the filament of the titanium pump to evaporate a layer of titanium onto the glass envelope and activate the ion pump.

7. Seal off S_4 and deposit a fresh layer of titanium. The pressure should fall well below 10^{-9} torr.

When sealing off a constriction, it is wise to proceed slowly so that the pumps have time to take up any gas that is liberated.

A similar system can be made with the pinchoffs replaced by all-metal high-vacuum valves. In this case it would also be possible to construct the entire system of stainless steel.

An oil-free ultrahigh-vacuum system.

A large, oil-free ultrahigh-vacuum system is illustrated in Figure 3.32. The system is roughed down with sorption pumps, and the ion pump is depended upon for the removal of rare gases. The system must be baked. A number of variations on this system are possible. The rough-pumping operation can be accelerated and the gas load on the sorption pumps reduced if the system is first roughed down with one of the compressed-air aspirators (Venturi pumps) sold for this purpose. One can also employ one of the oil-free mechanical pumps that have recently appeared on the market. Varian, for example, manufactures a reciprocating piston pump with Teflon seals that would be suitable as a roughing pump. The pumping at high vacuum can be carried out with a cryopump rather than a getter pump. The choice of the high-vacuum pump or combination of pumps depends upon the composition of the residual gas as well as the

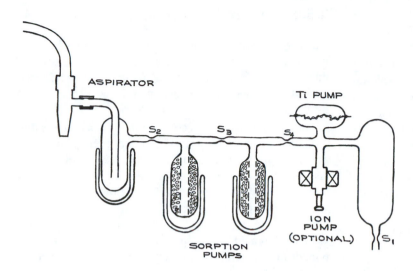

Figure 3.31 A small inexpensive ultrahigh-vacuum system made of Pyrex glass. Constrictions (labeled S) between components are heated with a glassblower's torch until they collapse and fuse under the external atmospheric pressure to seal off and isolate components as evacuation proceeds.

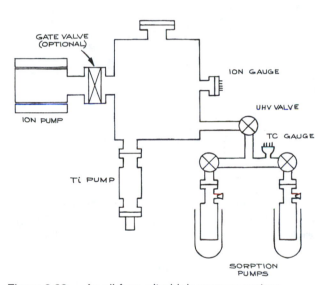

Figure 3.32 An oil-free, ultrahigh-vacuum system.

composition of gases that may be admitted to the system at high vacuum.

Small, oil-free, turbomolecular-pumped UHV system.

With appropriate precautions, hydrocarbon-free vacuum below 10^{-11} torr can be attained with a turbomolecular pump. The compression ratio of a turbomolecular pump for air is typically 10^8 to 10^9, so a foreline pressure well below 10^{-2} torr is required to achieve UHV. This is accomplished with a small high-vacuum pump between the turbomolecular pump and the mechanical foreline pump. One possibility is to back the turbo with a diffusion pump backed by a mechanical pump. The intermediate pump might just as well be another turbo or a molecular-drag pump. These intermediate pumps can be very small. In principle, the speed of these pumps need only be that of the primary turbopump divided by its compression ratio in order to deal with the throughput of the primary turbo.

The compression ratio of a turbomolecular pump exceeds 10^{16} for gases with molecular weights greater than 100, so when the pump is running, there is essentially no backstreaming of hydrocarbons from oil-lubricated bearings or from an oil-sealed mechanical forepump. Hydrocarbon contamination of the UHV chamber is possible, however, when the turbo is not operating. There are two essential precautions. When the turbo is turned off it should be vented to atmospheric pressure. Dense gas in the pump will prevent backstreaming from bearings at the bottom of the pump into the upper stages. There is usually no need for a valve between a turbo pump and the vacuum chamber, so the chamber is also vented to atmospheric pressure when the pump is vented. On the other hand, there must be a valve at the turbopump exhaust. This valve must be kept closed except when the pump is activated. In a turbomolecular-pumped system, the chamber is rough-pumped directly through the turbopump. To initiate pumping, the mechanical backing pump is turned on, the valve at the turbo exhaust is opened, and the turbo is then immediately activated so that roughing and turbo spin-up occur simultaneously. The intermediate pump can also be activated at this time. By following this procedure, the initial flow of dense gas from the chamber will prevent backstreaming while the turbo is coming up to speed. Of course, the best—albeit most expensive—means for eliminating hydrocarbons is to employ pumps that contain no hydrocarbons: turbomolecular pumps and molecular drag pumps with magnetically suspended rotors and a diaphragm pump as a forepump (see Section 3.4.1).

The ultimate pressure in a UHV chamber with a turbomolecular pump is determined by the compression ratio for H_2 and the partial pressure of H_2 in the foreline. The compression ratio for H_2 in modern pumps is in the range of 10^3 to 10^4. The exclusion of hydrocarbon oils and elastomer seals in the foreline helps reduce the hydrogen partial pressure in the foreline. In any event, it has been found that hydrogen represents 80 to 90 percent of the residual gas in a turbo-pumped chamber at pressures below about 10^{-10} torr. When this is a problem for a particular experiment, a titanium sublimation pump can be installed above the turbo. It would not be necessary for titanium to be continuously evaporated once UHV conditions are achieved.

Custom-made UHV chambers can be quite expensive, especially when many access ports are required for components such as mechanical and electrical feedthroughs and windows. When a relatively small volume is required (a few hundred cubic inches or a few thousand cubic centimeters), it is often possible to make up the chamber of standard UHV components such as tees and crosses. In the author's laboratory, a 6 in. cube cross is used as a small UHV chamber. The pump is mounted on one face and the remaining five faces provide ports for mounting components. The cube was ordered 0.5 in. oversize (i.e., 6.5" × 6.5" × 6.5"). With the additional size, it is possible to bore 0.375 in. diameter holes through the cube close to each vertical edge. Cartridge heaters inserted in the holes are used for bakeout.

3.6.2 Differential Pumping

One is occasionally required to design a system where gas is maintained at two very different pressures. Often two or more pumps with interconnecting tubing must be balanced against one another to obtain the desired pressure differential.

As an example, we will consider the problem of making a mass-spectrometric analysis of trace constituents in argon flowing through a tube at a pressure P_0 of 1.0 torr. As illustrated schematically in Figure 3.33a, the sample is to be withdrawn through a hole in the flow tube. The mass spectrometer must be operated at a pressure P_2 of 5×10^{-7} torr. To begin, let us suppose that a 6 in. diffusion pump is to be used to maintain the pressure P_2. A properly trapped 6 in. diffusion pump has a speed of about 900 L^{-1} over the range 10^{-3} to 10^{-9} torr (see Figure 3.12). The throughput is then

$$Q = P_2 S = (5 \times 10^{-7} \text{ torr})(900 \text{ L}^{-1})$$

$$= 4.5 \times 10^{-4} \text{ torr L}^{-1}$$

The conductance of the sampling aperture that provides this throughput is

$$C = \frac{Q}{P_0 - P_2} = \frac{4.5 \times 10^{-4} \text{ torr L s}^{-1}}{1.0 \text{ torr}}$$

$$= 4.5 \times 10^{-4} \text{ L s}^{-1}$$

The flow through this aperture is in the viscous flow regime. The area of the aperture that provides this conductance (from Section 3.2.4) is approximately

$$A = \frac{C(\text{L s}^{-1})}{20} \text{ cm}^2$$

$$= \frac{4.5 \times 10^{-4}}{20} = 2.3 \times 10^{-5} \text{cm}^2$$

corresponding to a diameter of 0.05 mm for a circular aperture. There are two problems associated with this small aperture. First, there is the mechanical difficulty of creating such a small hole. Second, this tiny hole may impose a severe limitation upon the sensitivity of the measurement, particularly if the analyzer samples only atoms moving along the line of sight from aperture to analyzer. A solution to these problems is to incorporate several stages of pumping—a scheme known as *differential pumping.*

A two-stage, differentially pumped system is illustrated in Figure 3.33b. An intermediate-pressure chamber with its own pump is interposed between the sample and analyzer volumes. There are two apertures now and the pressure drop occurs in two steps. For comparability with the previous example, suppose the intermediate chamber and the analyzer chamber are each evacuated with 4 in. diffusion pumps, each with a speed of 400 L s^{-1}. Take the ratio of pressures across each aperture to be the same so that $P_0/P_1 = P_1/P_2$ and the pressure P_1 in the intermediate chamber is 7×10^{-4} torr. The throughput of the pump on the lowest pressure chamber is

$$Q_2 = P_2 S$$

$$= (5\times10^{-7}\,\text{torr})(400\,\text{L s}^{-1})$$

$$= 2\times10^{-4}\,\text{torr L s}^{-1}$$

The conductance C_2 of the aperture separating the region of pressure P_1 from the region of pressure P_2 must maintain this throughput:

$$C_2 = \frac{Q_2}{P_1 - P_2} = \frac{2\times10^{-4}\,\text{torr L s}^{-1}}{7\times10^{-4}\,\text{torr}}$$

$$= 0.3\,\text{L s}^{-1}$$

The flow through this aperture is in the molecular flow regime. The area of the aperture that provides this conductance (from Section 3.2.4) for argon at 300 K is

$$A_2 = C\left[3.7\left(\frac{T}{M}\right)^{1/2}\right]^{-1}\text{cm}^2$$

$$= 0.03\,\text{cm}^2$$

corresponding to a diameter of about 2 mm for a circular aperture. Similarly, the conductance of the aperture separating the high- and intermediate-pressure regions must match the throughput of the intermediate pump:

$$Q_1 = P_1 S$$

$$= (7\times10^{-4}\,\text{torr})(400\,\text{L s}^{-1})$$

$$= 0.3\,\text{torr L s}^{-1}$$

so that

$$C_1 = \frac{Q_2}{P_0 - P_1} = \frac{0.3\,\text{torr L s}^{-1}}{1\,\text{torr}}$$

$$= 0.3\,\text{L s}^{-1}$$

The flow through this aperture is in the viscous flow regime. The area of the aperture that provides this conductance (from Section 3.2.4) is approximately

$$A_1 = \frac{C(\text{L s}^{-1})}{20}\,\text{cm}^2$$

$$= \frac{0.3}{20} = .015\,\text{cm}^2$$

corresponding to a diameter of about 1 mm for a circular aperture. In this example, we find that by pumping in stages and without an increase in total pumping speed, the required pressure differential can be maintained across apertures that are hundreds of times larger than had all the available pumping speed been applied in a single stage. Assuming that the gas flow through the first aperture is directional, the sensitivity of the measurement is also greatly increased by aligning the second aperture and analyzer with the flow through the first aperture as suggested in the figure.

3.6.3 The Construction of Metal Vacuum Apparatus

Metal vacuum chambers are often built up of one or more cylindrical sections. Tube stock is readily available and a cylindrical shape is second only to a spherical shape in strength to withstand external pressure. Even when tube stock is unavailable, a cylindrical section is easily fabricated by rolling up sheet stock. The ends of a cylindrical chamber are most conveniently closed with flat plates, as shown in Figure 3.34. A flat end plate must be quite thick to withstand atmospheric pressure, as discussed below. Domed or spherical end caps are much stronger but more expensive to fabricate.

The ability of a cylinder to resist collapsing under external pressure depends upon the length of the cylinder between supporting flanges, the diameter, the wall thickness, and the strength of the material. The American Society of Mechanical Engineers (ASME) has published standards for the wall thickness of cylindrical vessels under external pressure.[16] The data in Figure 3.35 have

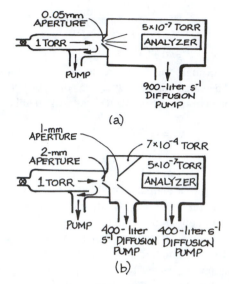

Figure 3.33 (a) Schematic representation of an analyzer in high vacuum (5×10^{-7} torr) used to sample a gas at high pressure (1 torr); (b) differential pumping arrangement that permits a much larger sampling aperture with the same total pumping speed for the analyzer.

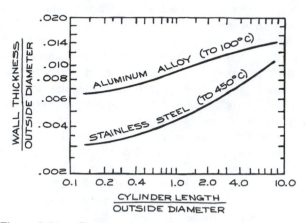

Figure 3.34 Recommended minimum wall thickness (scaled to diameter) for evacuated aluminum-alloy and stainless-steel tubes under external atmospheric pressure as a function of tube length (scaled to diameter) between supporting flanges.

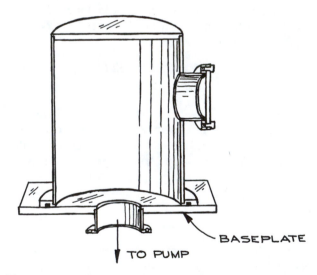

Figure 3.35 Typical aluminum or steel vacuum chamber resting on a baseplate. The pump, gauges, and electrical and mechanical feedthroughs are most conveniently mounted to the baseplate.

been abstracted from the ASME specifications for type 3003-T0 aluminum alloy and type 304 stainless steel respectively. These recommendations should be more than adequate for any of the 4000-, 6000-, or 7000-series tempered aluminum alloys (such as 2024-T4, 6061-T6 or 7075-T6) and adequate for most stainless steels. Soft aluminum such as the 1000-series alloys should not be used for vacuum chambers. Holes cut for ports in the side of a cylinder reduce the strength of the cylinder and should be avoided. If side ports are necessary, the wall thickness should be increased over the recommendations of the code. Dents and out-of-roundness in a cylinder weaken it and should be avoided.

A flat end plate will bend a surprising amount under atmospheric pressure. Assuming the plate caps a cylindrical shape, the nature of the stress in the plate and the maximum deflection depend upon whether or not the edges of the plate are clamped to the cylinder. In this case, *clamped* implies that the plate is welded to the cylinder. A removable end plate bolted in place should be considered *unclamped* when designing a vacuum chamber. The deflection at the center of a flat circular end plate clamped at its edges is

$$\delta = \frac{3PR^4(1-\mu^2)}{16Ed^2} \qquad \text{(edge clamped)}$$

where P is the external pressure, R is the radius of the plate, d the thickness, μ Poisson's ratio, and E the modulus of elasticity or Young's modulus of the material.[17] In this case, the maximum tensile stress (s_{max}) occurs at the edge of the plate:

$$s_{max} = \frac{3}{4}\left(\frac{R}{d}\right)^2 P \qquad \text{(edge clamped)}$$

The deflection of a demountable end plate is greater than that of a permanent end cap. For a circular plate with unclamped edges,

$$\delta = \frac{3PR^4(5+\mu)(1-\mu)}{16Ed^2} \qquad \text{(unclamped)}$$

The maximum tensile stress occurs at the center:

$$s_{max} = \frac{3}{8}\left(\frac{R}{d}\right)^2 (3+\mu)P \qquad \text{(unclamped)}$$

The deflection of flat end plates of steel or aluminum under atmospheric pressure is plotted in Figure 3.36. In most cases, relative deflection (δ/R) in the range 0.001 to 0.005 should be acceptable, otherwise the plate may be too heavy. Steel is about three times more dense than aluminum. For a given maximum deflection, an aluminum plate is less than half as heavy as a steel plate. A vacuum-chamber design that incorporates a stainless-steel cylinder with a demountable aluminum end plate has much to say in its favor. Welds in the stainless cylinder are much stronger and more reliable than would be the case if the cylinder were made of aluminum. A removable aluminum end plate is lighter and much less expensive to machine than one of steel.

The deflection of the walls of a vacuum chamber must be accounted for in designing the mounting for a vacuum system. If both the chamber and the pump are securely mounted, severe stresses will develop as the chamber is evacuated. The correct practice is to securely mount the vacuum chamber and leave the pump suspended from a flange on the chamber, or, alternatively, to mount the pump and permit the chamber to rest on top of the pump with no further constraint. In the latter case, it is important that forces generated by the apparatus resting on the top of the pump do not distort the pump. This is particularly important for pumps requiring a high degree of mechanical precision, such as a turbomolecular pump or a molecular drag pump. It is generally recommended that

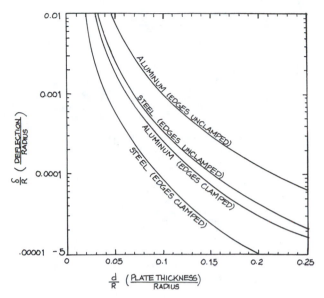

Figure 3.36   Deflection of flat circular plates under atmospheric pressure. A clamped end plate caps a cylindrical tube and is welded in place, otherwise the plate should be considered as unclamped. These curves assume a Poisson's ration of 0.3 and a Young's modulus of 3×10^7 psi for steel and 1×10^7 psi for aluminum.

turbopumps and drag pumps be suspended by their inlet flange with no other attachment.

Using standard, commercially available components saves both time and money. In fact, even quite large systems can be built entirely of standard parts. In the author's laboratory, a chamber 20 inches in diameter and 40 in. high was created from a pair of ISO 500 half nipples that were ordered with domed end caps. Ports were added to the lower half nipple for feedthroughs and for a pump as shown in Figure 3.37. The total cost was about $5000 in 1994.

Metal parts may be joined by soldering, brazing, or welding. Tungsten-inert-gas (so-called *TIG* or *heliarc*) welds produce the cleanest, strongest joints in stainless steel. Soldering and brazing result in less distortion of the joined parts and are more convenient operations in the laboratory. As shown in Figure 3.38, a joint that is to be soldered or brazed should be designed so that the metal parts take the thrust of the atmosphere and the solder or brazing alloy serves only as a sealant. If possible, the joint should be designed to provide positive location of mating

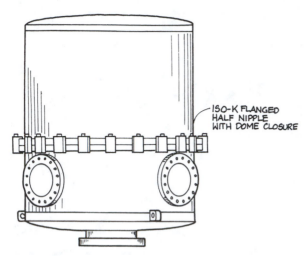

Figure 3.37   A vacuum chamber composed of a pair of ISO-5000 (500 mm diameter) half nipples with domed end caps.

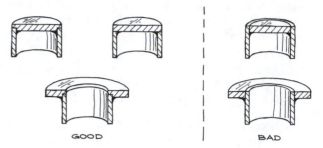

Figure 3.38   The design of joints to be soldered or brazed.

parts. The surfaces to be joined should be cleaned before assembly by sandblasting or polishing with sandpaper so that the solder or brazing metal will flow into and completely fill the joint. This is necessary to achieve maximum strength and to eliminate a narrow gap that is difficult to pump out and is liable to collect flux and other material that will contaminate the vacuum system. Tin-alloy solders (plumbing solder) should only be used for rough vacuum applications such as a diffusion-pump foreline. Brazed joints are acceptable in high vacuum. The copper-silver eutectic alloy is the preferred brazing material. Brazing is an excellent way of attaching thin metal parts, such as bellows, that can easily be burned through in a welding operation.

Welded joints are designed to minimize distortion in the welding process and avoid occlusions on the vacuum side of the joint that may trap contaminates or act as virtual leaks.

Parts to be joined by welding are designed so that the rate of heat dissipation while welding is the same to each part. This is necessary to prevent destructive thermal stresses from being frozen into the finished joint. When joining a heavy flange to a thin tube, a notch or a groove is cut in the thick part near the joint to control the rate at

which heat is dissipated, as shown in Figure 3.39. To join a very thin part to a flange, extra material such as the welding ring shown in the figure is added to back up the weld. Whenever possible, welds should be on the vacuum side—that is, on the inside of the vacuum wall. Locational welds or *tack welds* should be on the outside and should be intermittent. For material up to about $\frac{1}{8}$ in. thick, full penetration welds should be specified. A weld from the outside, often necessary when joining a small diameter tube to a flange, should be a full penetration weld if possible. When an outside weld is unavoidable, the joint should be designed to minimize the trapped volume on the inside as shown in the figure.

Metal parts inside a vacuum system may be joined by welding or brazing, but in most cases nut-and-bolt assembly is more convenient. Special care must be taken to prevent air from being trapped under a screw in a blind hole. The placement of pump-out holes for blind screw holes is illustrated in Figure 3.40.

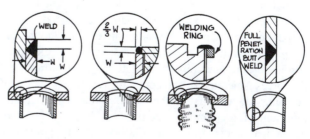

Figure 3.39   Welded joints in a vacuum system.

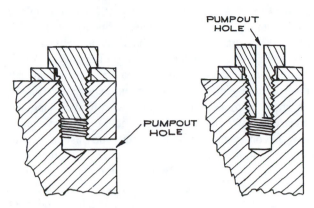

Figure 3.40   Pump-out holes for blind screw holes.

3.6.4 Surface Preparation

Surfaces that are to be exposed to a vacuum should be free of substances that have a significant vapor pressure, and they should be as smooth as possible to minimize the microscopic surface area and thus minimize the amount of adsorbed gas. The substances that must be removed from the surfaces of vacuum apparatus are mainly hydrocarbon oils and greases, and inorganic salts that are hygroscopic and outgas water vapor.

The two preferred treatments for metal vacuum system surfaces are electropolishing and bead blasting. Conventional wheel polishing and buffing is unsatisfactory, since these processes tend to flatten surface burrs and trap gas underneath. Electropolishing is conveniently carried out in the laboratory, although most machine shops are prepared to do it routinely. An electropolishing solution for stainless steel recommended by Armco Steel consists of 50 parts by volume of citric acid and 15 parts by volume of sulfuric acid plus enough water to make 100 parts of solution. The solution should be used at a temperature of 90°C. A current density of about 0.1 A cm^{-2} at 6 to 12 V is required. A copper cathode is used with the piece to be polished serving as the anode.

Blasting with 20 to 30 μm glass beads effectively reduces the adsorbing area of a metal surface and thus reduces the rate of outgassing from the surface. Glass beading can be carried out in a conventional sandblasting apparatus. The machine should be carefully cleaned of coarse grits before use.

Shop processes such as welding, plasma cutting, and high-speed turning can leave surfaces with a heavy layer of oxides and other contaminants. This layer may be removed by bead blasting, or the chemical process known as *pickling*. A pickling solution for stainless steel consists of 2 percent (by volume) concentrated nitric acid (HNO_3) and 25 percent hydrochloric acid (HCl) in distilled water at 60°C. This solution is known as *bright dip* in the shop. After dipping in this solution, the part should be thoroughly rinsed and its surface neutralized by immersion in an alkaline solution. Aluminum is pickled in a 10 percent solution of sodium hydroxide (NaOH) saturated with common salt (NaCl) at 80°C. After about half a minute in this solution the part is transferred to a 10 percent solution of hydrochloric acid to obtain a bright finish.

An ultrasonic cleaner is particularly effective for preparing vacuum parts. The cleaner should be filled with a detergent solution. If two cleaners are available, fill one with detergent solution and the other with distilled water and use them in sequence with a hot-water rinse between the two.

Diffusion-pump oil is especially difficult to remove. Nevertheless, a diffusion pump must be thoroughly cleaned when replacing old oil with new or when changing from one type of oil to another. Hydrocarbon oils, polyphenylethers, and silicone fluids can be dissolved in trichloroethylene followed by acetone and then ethyl alcohol. Trichloroethylene is somewhat toxic, and acetone and ethanol are flammable, so this cleaning should be carried out carefully in a fume hood. Extensive contact of these solvents with the skin should be avoided. Fluorinated diffusion-pump fluids, such as the perfluoropolyethers (Fomblin), are soluble in fluorinated solvents such as chlorotrifluoroethane.

Glass can usually be effectively cleaned by washing in a detergent solution followed by a rinse in hot water and immersion in a 10 to 20 percent solution of hydrofluoric acid. The glass is then rinsed with water, neutralized with an alkaline solution, and rinsed with distilled water.

After surface treatment as required for particular components, the final cleaning procedure for all components of a vacuum system should proceed as follows:

1. Scrub with a strong solution of detergent (Alconox or liquid dishwashing detergent is fine).

2. Rinse with very hot water.

3. Rinse with distilled water.

4. Rinse with pure methanol.

Do not touch clean parts with bare hands. Disposable plastic gloves (free of talc) are convenient for handling vacuum apparatus after cleaning.

3.6.5 Leak Detection

A leak that raises the base pressure of a system above about 10^{-6} torr can usually be found by probing suspected locations on the outside of the vacuum chamber, with a liquid or vapor for which the gauge sensitivity or pump speed is very different from that for air. A squeeze bottle of acetone or a spray can of a Freon cleaner is a useful tool. These liquids will usually cause a very abrupt increase in indicated pressure as they flow through a leak, but sometimes rapid evaporation of a liquid through a leak will cause the liquid to freeze and temporarily plug the leak, causing the pressure to fall. One disadvantage of this method is that the solvent may contaminate O-rings. A small jet of helium is also a useful leak probe because an ionization gauge is quite insensitive to helium. The indicated pressure will fall when helium is introduced into a leak.

In a glass system, a leak that raises the pressure into the 10 mtorr to several torr range can be located with a Tesla coil. The surface of the glass is brushed with the discharge from the Tesla coil. The discharge is preferentially directed toward the leak, and a bright white spot will reveal the location as the discharge passes through. Avoid very thin glass walls and glass-to-metal seals, as the Tesla discharge can hole the glass in these fragile areas.

For very small leaks in high-vacuum systems, a mass-spectrometer leak detector is needed (see Section 3.3.5).

3.6.6 Ultrahigh Vacuum

To achieve pressures far below 10^{-7} torr, baking is required to remove water and hydrocarbons from vacuum-system walls. Heating to 50 to 100°C for several hours will improve the ultimate pressure of most systems by an order of magnitude. A true ultrahigh-vacuum system must be baked to at least 250°C for several hours during the initial pump out. A system contaminated with hydrocarbons may require a bakeout at temperatures approaching 400°C. After a hard preliminary bake *in vacuo* to remove deeply adsorbed material, the system can be baked more gently after subsequent exposures to air.

A small system can be wrapped with heating tape and then covered with fiberglass insulation. For larger systems, an oven constructed of sheet metal and insulated with fiberglass can be erected around the vacuum chamber and heated with bar heaters. Batts of fiberglass insulation intended for use in automobile engine compartments are readily available. Do not use home-insulating fiberglass materials. Metal seals must be used in a metal vacuum system that is to be baked. It is often necessary, however, to join the vacuum chamber to its pumping station with an O-ring-sealed flange. In this case, a cooling coil should be installed around the flange (as in Figure 3.23) to prevent melting the O-ring while the chamber is being baked. Alternatively a metal O-ring can be used as described in Section 3.5.2.

Quartz lamp heaters can be installed inside a vacuum system to directly heat the surfaces that need to be outgassed. This method is frequently faster and more economical than heating from the outside. A recent innovation for removing water absorbed on the walls of vacuum apparatus employs ultraviolet radiation rather than heat. A lamp known as a Phototron, manufactured by Danielson, is inserted into the vacuum system. This lamp radiates at a wavelength absorbed by water molecules and imparts sufficient energy to break the bond holding the water to the walls. This method is not as effective as a high-temperature bakeout, since the radiation is fairly specific for water and, the radiation cannot be expected to penetrate into small involutions where water may be trapped. The method is faster than a thermal bakeout and can be used in a system that contains materials that are degraded by heat. It is claimed to be 80 to 90 percent as effective as a bakeout.

To reduce outgassing to a minimum in an ultrahigh-vacuum system, it is common practice to prefire (*stove*) components before they are assembled in order to clean the surfaces and to drive out dissolved gases. The usual shop practice is to fire components in a hydrogen atmosphere such as is provided by a hydrogen furnace or a dissociated-ammonia furnace. This has the advantage of effectively removing oxides from surfaces. The disadvantage to this procedure is that hydrogen is left to outgas at lower temperatures. Copper parts containing trapped or dissolved oxygen are also subject to embrittlement as a result of hydrogen firing. It is for this reason that oxygen-free high-conductivity (OFHC) copper should be used in constructing vacuum apparatus. High-temperature baking in vacuum or under an inert atmosphere is the preferred treatment. Iron, steel, stainless steel, and nickel alloy parts are fired at a temperature of about 900°C for up to 8 hours. Copper is fired at 500°C.

CITED REFERENCES

1. N. Milleron, *Res. Dev.*, **21**, No. 9, 40 (1970).

2. Derivations of the conductance formulae are given by C. M. VanAtta, *Vacuum Science and Engineering*, McGraw-Hill, New York, 1965, pp. 44–62; J. F. O'Hanlon, *A User's Guide to Vacuum Technology*, 2nd ed., John Wiley & Sons, New York, 1989, chap. 3.

3. M. A. Biondi, *Rev. Sci. Instr.*, **24**, 989 (1953).

4. A. T. J. Hayward, *J. Sci. Instr.*, **40**, 173 (1963); C. Veillon, *Rev. Sci. Instr.*, **41**, 489 (1970).

5. H. Ishii and K. Nakayama, *Transactions of the 2nd International Congress on Vacuum Science and Technology*, Vol. 1, Pergamon Press, Elmsford, N.Y., 1961, p. 519; R. J. Tunnicliffe and J. A. Rees, Vacuum, **17**, 457 (1967); T. Edmonds and J. P. Hobson, *J. Vac. Sci. Tech.*, **2**, 182 (1965).

6. G. R. Brown, P.J. Sowinski, and R. Pertel, *Rev. Sci. Instr.*, **43**, 334 (1972).

7. An application of this process is described by J. H. Moore and C. B. *Opal, Space Sci. Instr.*, **1**, 377 (1975).

8. C. B. Lucas, *Vacuum*, **23**, 395 (1973).

9. R. H. Jones, D. R. Olander, and V. R. Kruger, *J. Appl. Phys.*, **40**, 4641 (1969); D. R. Olander, *J. Appl. Phys.*, **40**, 4650 (1969); D. R. Olander, *J. Appl. Phys.*, **41**, 2769 (1970); D.R. Olander, R. H. Jones, and W. J. Siekhaus, *J. Appl. Phys.*, **41**, 4388 (1970); W. J. Siekhaus, R. H. Jones, *J. Appl. Phys.*, **41**, 4392 (1970).

10. M. G. Liverman, S. M. Beck, D. L. Monts, and R. E. Smalley, *J. Chem. Phys.*, **70**, 192 (1979); W. R. Gentry and C. F. Giese, *Rev. Sci. Instr.*, **49**, 595 (1978); C. M. Lovejoy and D. J. Nesbitt, *J. Chem. Phys.*, **86**, 3151 (1987).

11. D. Bassi, S. Iannotta, and S. Niccolini, *Rev. Sci. Instr.*, **52**, 8 (1981); F. M. Behlen, *Chem. Phys. Lett.*, **60**, 364 (1979).

12. R. E. Smalley, D. H. Levy, and L. Wharton, *J. Chem. Phys.*, **64**, 3266 (1976).

13. J. B. Anderson, R. P. Andres, and J. B. Fenn, in *Advances in Chemical Physics*, Vol. 10, *Molecular Beams*, J. Ross, ed., Wiley, New York, 1966, chap. 8; H. Pauly and J. P. Toennies, in *Methods of Experimental Physics*, Vol. 7, Part A, B. Bederson and W. L. Fite, eds., Academic Press, New York, 1968, chap 3.1; D. R. Miller, in *Atomic and Molecular Beam Methods*, Vol. I, G. Scoles, ed., Oxford University Press, Oxford, 1988, pp. 14–53.

14. W. D. Kingery, H. K. Bowen, and D. R. Uhlmann, *Introduction to Ceramics*, 2nd ed., John Wiley & Sons, New York, 1976, chap. 17.

15. This system is an adaptation of a design described by N. W. Robinson, *Ultra-high Vacuum*, Chapman and Hall, London, 1968, pp. 72–73.

16. *A.S.M.E. Boiler and Pressure Vessel Code*, Section VIII, Division 1, Appendix V, 1974.

17. J. F. Harvey, *Pressure Vessel Design*, Van Nostrand, Princeton, N.J., 1963, pp. 89–91.

GENERAL REFERENCES

Comprehensive Texts on Vacuum Technology

S. Dushman, *Scientific Foundations of Vacuum Technique*, 2nd ed., J. M. Lafferty, ed., Wiley, New York, 1961.

M. Hablanian, *High-Vacuum Technology*, Dekker, New York, 1990.

G. Lewin, *Fundamentals of Vacuum Science and Technology,* McGraw-Hill, New York, 1965.

J. F. O'Hanlon, *A User's Guide to Vacuum Technology,* 2nd ed., John Wiley & Sons, Inc., New York, 1989.

Foundations of Vacuum Science and Technology, edited by J. M. Lafferty, ed., John Wiley & Sons, Inc., New York, 1998.

C. M. Van Atta, *Vacuum Science and Engineering,* McGraw-Hill, New York, 1965.

"Vacuum Technology: Its Foundations, Formulae and Tables", until 1996 published as an appendix of the Leybold AG catalog, *Product and Vacuum Technology Reference Book*, Leybold, Inc., 1860 Hartog Dr., San Jose, CA 95131.

Detailed Calculation of Gas Flow

A. Roth, *Vacuum Technology,* 2nd ed., North-Holland, Amsterdam, 1982.

R. G. Livesey, "Flow of Gases through Tubes and Orifices" in *Foundations of Vacuum Science and Technology*, J. M. Lafferty, ed. John Wiley & Sons, Inc., New York, 1998.

Design of Vacuum Systems

N. T. M. Dennis and T. A. Heppell, *Vacuum System Design,* Chapman and Hall, London, 1968.

G. W. Green, *The Design and Construction of Small Vacuum Systems,* Chapman and Hall, London, 1968.

R. P. LaPelle, *Practical Vacuum Systems,* McGraw-Hill, New York, 1972.

Outgassing Data

W. A. Campbell, Jr., R. S. Marriott, and J. J. Park, *A Compilation of Outgassing Data for Spacecraft Materials,* NASA Technical Note TND-7362, NASA, Washington, D.C., 1973.

Properties of Materials Used in Vacuum Systems

W. Espe, *Materials of High Vacuum Technology:* Vol. 1, *Metals and Metaloids;* Vol. 2, *Silicates;* Vol. 3, *Auxiliary Materials,* Pergamon Press, Oxford, 1968 (a translation of the original German published in 1960).

Sealing Ceramics and Glass to Metal, Heat-Treating, Cleaning, Building Joints, and Feedthroughs

F. Rosebury, *Handbook of Electron Tube and Vacuum Techniques,* Addison-Wesley, Reading, Mass., 1969.

A. Roth, *Vacuum Sealing Techniques*, Pergamon Press, New York, 1966 and American Vacuum Society classics, American Institute of Physics, New York, 1994.

Ultrahigh Vacuum

P. A. Redhead, J. P. Hobson, and E. V. Kornelsen, *The Physical Basis of Ultrahigh Vacuum,* Chapman and Hall, London, 1968.

R. W. Roberts and T. A. Vanderslice, *Ultrahigh Vacuum and Its Applications,* Prentice-Hall, Englewood Cliffs, N.J., 1963. W. Robinson, *The Physical Principles of Ultrahigh Vacuum Systems and Equipment,* Chapman and Hall, London, 1968.

G. F. Weston, *Ultrahigh Vacuum Practice*, Butterworths, London, 1985.

J. T. Yates, Jr., *Experimental Innovations in Surface Science*, Springer-Verlag, New York, 1998.

MANUFACTURERS AND SUPPLIERS

General Vacuum Equipment and Instruments (pumps, valves, gauges, fittings)

Alcatel Vacuum Products
7 Pond St.
Hanover, MA 02339
Balzers

8 Sagamore Park Rd.
Hudson, NH 03051

CVC Products, Inc.
525 Lee Rd.
P.O. Box 1886
Rochester, NY 14603

Edwards High Vacuum, Inc.
3279 Grand Island Blvd.
Grand Island, NY 14072

Kurt J. Lesker Co.
1515 Northington Ave.
Clairton, PA 15025

Leybold Vacuum Products
5700 Mellon Rd.
Export, PA 15632

MDC Vacuum Products
23842 Cabot Blvd.
Hayward, CA 94545-1651

Nor-Cal Products, Inc.
1967 South Oregon St.
P.O. Box 518
Yreka, CA 96097

Osaka Vacuum, Ltd.
Keihan-Yodoyabashi Bldg.
2-25, Kitahama 3-chome
Chuo-ku
Osaka 541, Japan

and

911 Bern Ct.
San Jose, CA 95112

Varian Vacuum Division
611 Hansen Way
Palo Alto, CA 94303

Ceramics and Ceramic Fittings

Arenco Products
P.O. Box 429
Ossining, NY 10562
(machinable ceramics)

Ceramaseal
P.O. Box 260
New Lebanon, NY 12125
(terminal end bushings, electrical feedthroughs, vacuum breaks)

Coors Ceramics Co.
600 Ninth St.
P.O. Box 4025
Golden, CO 80401
(alumina shapes, fishspine)

Coors Ceramics Co.
2449 River Rd.
Grand Junction, CO 81505
(alumina thick film substrate)

The Materials Business
Corning Glass Works
Corning, NY 14830
(MACOR machinable ceramic)

Industrial Tectonics, Inc.
P.O. Box 1128
Ann Arbor, MI 48106
(ceramic balls)

Insulator Seal Incorporated
23874-B Cabot Blvd.
Hayward, CA 94545
(electrical feedthroughs, vacuum breaks, viewports)

Latronics Corporation
1001 Lloyd Ave.
Latrobe, PA 15650
(ceramic feedthroughs)

McDanel Refractory Porcelain Co.
510 Ninth Ave.
Beaver Falls, PA 15010
(ceramic rod and tube)

Specialty Ball Co.
951 West St.
Rocky Hill, CT 06067
(ceramic balls)

Cleaning Materials

Miller-Stephenson Chemical Co.
P.O. Box 628
Danbury, CT 06810
(freon degreasers)

Texwipe Company
Hillsdale, NJ 07642

Components

Alloy Products Company
1045 Perkins Ave.
Waukesha, WI 53186
(stainless-steel vacuum fittings)

Cajon Company
9760 Shepard Rd.
Macedonia, OH 44056
(small O-ring- and metal-sealed tube fittings)

Combination Pump Valve Company
851 Preston St.
Philadelphia, PA 19104
(small O-ring-sealed tube fittings)

Duniway Stockroom Corporation
1600 Stierlin Rd.
Mountain View, CA 94043
(relacements for OEM parts)

HPS Division of MKS
5330 Sterling Dr.
Boulder, CO 80301

Kimball Physics, Inc.
311 Kimball Hill Rd.
Wilton, NH 03086-9742
(small, monolithic UHV chambers and fittings)

Ladish Company
Cudaby, WI 53110
(stainless-steel vacuum fittings)

Tyler Griffin
46 Darby Rd.
Paoli, PA 19301

U-C Components, Inc
18700 Adams Court
Morgan Hill, CA 95037
(vented screws, vacuum-baked O-rings)

VACOA
390 Central Ave.
Bohemia, NY 11716

Glass and Glass Components

Ace Glass Inc.
P.O. Box 688
1430 Northwest Blvd.
Vineland, NJ 08360

Corning Glass Works
Corning, NY 14830

Fischer & Porter Company
Warminster, PA

Kimble Glass
P.O. Box 10350
Toledo, OH 43666

Labglass
Northwest Blvd.
Vineland, NJ 08360

Wilmad Glass Company
Route 40 & Oak Rd.
Buena, NJ 08310
(precision glass tubing)

Heating Mantles and Bakeout Lamps

Briskheat
P.O. Box 628
Columbus, OH 43216
(heating mantles)

Danielson Associates, Inc.
1989A University Ln.
Lisle, IL 60532
(bakeout lamps)

Glas-Col Apparatus Co.
709 Hulman St.
Terre Haute, IN 47802
(heating mantles)

Glass Microchannel Arrays

Burle Electro-Optics, Inc. (formerly Galileo)
Galileo Park
Sturbridge, MA 01518

Minitubes-Grenoble
7 Ave de Grand Chatelet
38100 Grenoble
France

Nonevaporable Getter Pumps

SAES Getters, S.p.A.
Group Headquarters
Viale Italia, 77
20020 Lainate (Milano) Italy

SAES Getters USA Inc.
1122 East Cheyenne Mountain Blvd.
Colorado Springs, CO 80906

O-Rings

EG&G Engineered Products
11642 Old Baltimore Pike
Beltsville, MD 20705
(metal seals to replace O-rings)

Helicoflex Company
Components Division
2770 The Boulevard
P. O. Box 9889
Columbia, SC 29290
(metal O-rings)

Parker Hannifin Corporation
2360 Palumbo Dr.
Lexington, KY 40509
(rubber O-rings)

Pressure Measuring Instruments

Granville-Phillips
5675 E. Arapahoe Ave.
Boulder, CO 80303
(ion and thermocouple gauges and controllers)

Hastings-Raydist
P.O. Box 1275
Hampton, VA 23661
(thermocouple gauge and controllers)

HPS Division of MKS
5330 Sterling Dr.
Boulder, CO 80301
(Pirani gauges)

MKS Instruments
6 Shattuck Rd.
Andover, MA 01810
tel: 978-975-2350
(capacitance manometers)

Wallace & Tiernan
25 Main St.
Belleville, NJ 07109
(mechanical gauges)

Pulsed Gas Sources

General Valve Corporation
19 Gloria Ln.
P. O. Box 1333
Fairfield, NJ 07004

Refrigerators and Refrigerated Baffles
CVC (thermoelectric baffles)

FTS Systems
P. O. Box 158
Stone Ridge, NY 12484-0158
(immersion coolers)

Naslab Instruments Inc.
871 Islington St.
P.O. Box 1178
Portsmouth, NH 03801
(immersion coolers)

CHAPTER 4

OPTICS

Physical, chemical, and biological phenomena are increasingly studied or induced optically. Such experiments can involve light absorption, light emission, or light scattering. We can characterize any experimental arrangement where light is used, produced, measured, modified, or detected as an *overall optical system*. Any such optical system will always be reducible to three parts—a source of light, a detector of that light, and everything in between. We will frequently refer to this important and varied intermediary arrangement as the *optical system*. Our discussion of overall optical system design and construction will consequently involve three key topics—sources, optical systems, and detectors. The light source may be a laser, a lamp, a light-emitting diode, or the sun. The detector may be a vacuum tube, a solid state device, or even the eye. Light intensity may vary from continuous wave (CW) to pulsed, and these pulses may have durations as short as a few femtoseconds. Passive elements in the system may transmit, reflect, combine, polarize, or separate light according to its spectral content. Nonlinear optical elements change the spectral content of light.

It is our aim in this chapter to explain the basic concepts that need to be understood by the experimentalist who uses optical techniques. We will also provide examples of useful techniques for producing, controlling, analyzing, modulating, and detecting light.

4.1 OPTICAL TERMINOLOGY

Light is electromagnetic radiation with a wavelength between 0.1 nm and 100 μm. This selection of wavelength range, of course, is somewhat arbitrary. Short wavelength, vacuum-ultraviolet light between 0.1 and 10 nm (so called because these wavelengths—and a range above up to about 200 nm—are absorbed by air and most gases) might as easily be called soft X-radiation. At the other end of the wavelength scale, between 100 and 1000 μm (the submillimeter wave region of the spectrum), lies a spectral region where conventional optical methods become difficult, as does the extension of microwave techniques for the centimeter- and millimeter-wave regions. The use of optical techniques in this region, and even in the millimeter region, is often called *quasi-optics*.[1,2] Only a few experimental techniques and devices that are noteworthy in this region will be mentioned.

Table 4.1 summarizes the important parameters that are used to characterize light and the media through which it passes. A few comments on the table are appropriate. Although the velocity of light in a medium depends both on the relative magnetic permeability μ_r and dielectric constant ε_r of the medium, $\mu_r = 1$ for all practical optical materials so that the refractive index and dielectric constant are related by

$$n = \sqrt{\varepsilon_r} \qquad (4.1)$$

When light propagates in an anisotropic medium, such as a crystal of lower than cubic symmetry, n and ε_r will

Table 4.1 Fundamental Parameters of Electromagnetic Radiation and Optical Media

Parameter	Symbol	Value	Units
Velocity of light *in vacuo*	$c_0 = (\mu_0\varepsilon_0)^{1/2}$	2.99792458×10^8	m sec^{-1}
Permeability of free space	μ_0	$4\pi \times 10^{-7}$	henry m^{-1}
Permitivity of free space	ε_0	8.85416×10^{-12}	farad m^{-1}
Velocity of light in a medium	$c = (\mu_r\mu_0\varepsilon_r\varepsilon_0)^{-1/2} = c_0/n$		m sec^{-1}
Refractive index	$n = (\mu_r\varepsilon_r)^{1/2}$		(dimensionless)
Relative permeability of a medium	μ_r	usually 1	(dimensionless)
Dielectric constant of a medium	ε_r		(dimensionless)
Frequency	$\nu = c/\lambda$		Hz
			10^9 Hz = 1 GHz(gigahertz)
			10^{12} Hz = 1 THz (terahertz)
Wavelength *in vacuo*	$\lambda_0 = c_0/\nu$		m
Wavelength in a medium	$\lambda = c/\nu = \lambda_0/n$		m
			10^{-3} m = 1mm (millimeter)
			10^{-6} m = 1 μm (micrometer)
			= 1 μ (micron)
			10^{-9} m = 1 nm (nanometer)
			= 1 mμ (millimicron)
			10^{-12} m = 1 pm (picometer)
			10^{-10} m = 1 Å(Angstrom)
Wavenumber	$\bar{\nu} = 1/\lambda$		cm^{-1} (kayser)
Wavevector	$k, |k| = 2\pi/\lambda$		m^{-1}
Photon energy	$E = h\nu$		J
			1.60202×10^{-19}J
			= 1 eV (electron volt)
Electric field of wave	**E**		V m^{-1}
Magnetic field of wave	**H**		A m^{-1}
Poynting vector	$\mathbf{P} = \mathbf{E} \times \mathbf{H}$		W m^{-2}
Intensity	$I = \langle|\mathbf{P}|\rangle_{av} = |E|^2/2Z$		W m^{-2}
Impedance of medium	$Z = \dfrac{E_x}{H_y} = \dfrac{-E_y}{H_x} = \left(\dfrac{\mu_r\mu_0}{\varepsilon_r\varepsilon_0}\right)^{1/2}$		Ω
Impedance of free space	$Z_0 = (\mu_0/\varepsilon_0)^{1/2}$	376.7	Ω

generally depend on the direction of wave propagation and its polarization state. The velocity of light *in vacuo* is currently the most precisely known of all the physical constants—its value is known[3] within 40 cm sec^{-1}. A redefinition of the meter has been adopted[4] based on the cesium-atomic clock frequency standard[5] and a velocity of light of exactly 2.99792458×10^8 m sec^{-1}. The velocity, wavelength, and wavenumber of light traveling in the air have slightly different values than they have *in vacuo*. Tables that give corresponding values of $\bar{v}$ *in vacuo* and in standard air are available.[6]

A plane electromagnetic wave traveling in the z direction can, in general, be decomposed into two independent, linearly polarized components. The electric and magnetic fields associated with each of these components are themselves mutually orthogonal and transverse to the direction of propagation. They can be written as (E_x, H_y) and (E_y, H_x). The ratio of the mutually orthogonal **E** and **H** components is called the *impedance Z* of the medium:

$$\frac{E_x}{H_y} = \frac{-E_y}{H_x} = Z = \sqrt{\frac{\mu_r \mu_0}{\varepsilon_r \varepsilon_0}} \qquad (4.2)$$

Most optical materials have $\mu_r = 1$. The negative sign in Equation 4.2 arises because (y,x,z) is not a right-handed coordinate system. The Poynting vector

$$\mathbf{P} = \mathbf{E} \times \mathbf{H} \qquad (4.3)$$

is a vector that points in the direction of energy propagation of the wave, as shown in Figure 4.1. The local direction of the Poynting vector at a point in a medium is called the *ray* direction. The average magnitude of the Poynting vector is called the *intensity* and is given by

$$I = \ <|\mathbf{P}|>_{av} \ = \frac{|E|^2}{2Z} \qquad (4.4)$$

The factor of two comes from time averaging the square of the sinusoidally varying electric field.

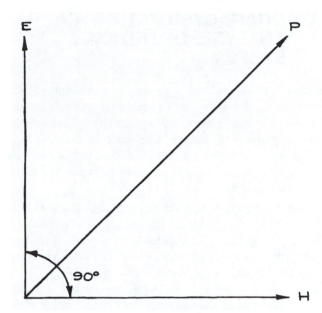

Figure 4.1   Orientation of the mutually orthogonal electric-field vector **E**, magnetic vector **H**, and Poynting vector **P**.

The wave vector **k** points in the direction perpendicular to the phase front of the wave (the surface of constant phase). In an isotropic medium, **k** and **P** are always parallel.

The photon flux corresponding to an electromagnetic wave of average intensity I and frequency v is

$$N = \frac{I}{hv} = \frac{I\lambda}{hc} \qquad (4.5)$$

where h is Planck's constant—6.6×10^{-34} J sec. For a wave of intensity 1 Wm^{-2} and wavelength 1 μm *in vacuo*, $N = 5.04 \times 10^{18}$ photons s^{-1} m^{-2}. Photon energy is sometimes measured in electron volts (eV): 1 eV = 1.60202×10^{-19} J. A photon of wavelength 1 μm has an energy of 1.24 eV. It is often important, particularly in the infrared, to know the correspondence between photon and thermal energies. The characteristic thermal energy at absolute temperature T is kT. At 300 K, $kT = 4.14 \times 10^{-21}$ J = 208.6 cm^{-1} = 0.026 eV.

4.2 CHARACTERIZATION AND ANALYSIS OF OPTICAL SYSTEMS

Before embarking on a detailed discussion of the properties and uses of passive optical components and the systems containing them, it is worthwhile to review some of the parameters that are important in characterizing passive optical systems and some of the methods that are useful in analyzing the way in which light passes through an optical system.

Optical materials have refractive indices that vary with wavelength. This phenomenon is called *dispersion*. It causes a wavelength dependence of optical system properties containing transmissive components. The change of index with wavelength is very gradual and often negligible unless the wavelength approaches a region where the material is not transparent. A description of the performance of an optical system is often simplified by assuming that the light is *monochromatic* (contains only a small spread of wavelength components).

4.2.1 Simple Reflection and Refraction Analysis

The phenomena of reflection and refraction are most easily understood in terms of plane electromagnetic waves—those in which the direction of energy flow (the ray direction) is unique. Other types of waves, such as spherical waves and Gaussian beams, are also important in optical science. The part of their wavefront that strikes an optical component can frequently be approximated as a plane wave, so plane-wave considerations of reflection and refraction still hold true.

When light is reflected from a plane mirror, or the planar boundary between two media of different refractive index, the *angle of incidence* is always equal to the *angle of reflection*, as shown in Figure 4.2. This is the fundamental *law of reflection*.

When a light ray crosses the boundary between two media of different refractive index, the *angle of refraction* θ_2 (shown in Figure 4.2) is, according to *Snell's* law, related to the angle of incidence θ_1:

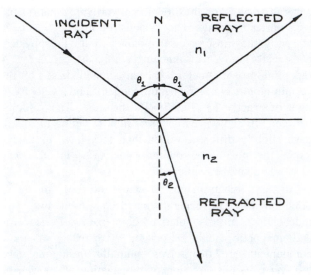

Figure 4.2 Reflection and refraction of a light ray at the boundary between two different isotropic media of refractive indices n_1 and n_2, respectively. The case shown is for $n_2 > n_1$. The incident, reflected, and transmitted rays and the surface normal N are coplanar.

$$\frac{\sin \theta_1}{\sin \theta_2} = \frac{n_2}{n_1} \qquad (4.6)$$

This result is modified if one or both of the media is anisotropic.

If $n_2 < n_1$, there is a maximum angle of incidence for which there can be a refracted wave because $\sin \theta_2$ cannot be greater than unity for a real angle. This is called the *critical* angle θ_c, and is given by

$$\sin \theta_c = n_2/n_1 \qquad (4.7)$$

as illustrated in Figure 4.3a.

If θ_1 exceeds θ_c, the boundary acts as a very good mirror, as illustrated in Figure 4.3b. This phenomenon is called *total internal reflection*. Several types of reflecting prisms discussed in Section 4.3.4 operate this way. When total internal reflection occurs, there is no transmission of energy through the boundary. The fields of the wave, however, do not go abruptly to zero at the boundary. There is an *evanescent wave* on the other side of the boundary, the field amplitudes of which decay exponentially with

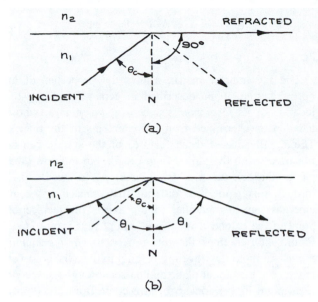

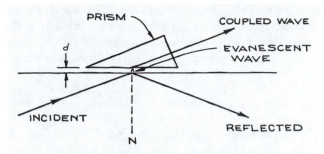

Figure 4.4   Schema of an evanescent-wave coupler. The amount of intensity reflected or coupled is varied by adjusting the spacing d; shown greatly exaggerated in the figure. This spacing is typically on the order of the wavelength. The surface boundary and adjacent prism face must be accurately flat and parallel.

Figure 4.3   (a) Critical angle ($n_1 > n_2$); (b) total internal reflection ($n_1 > n_2$). N is the surface normal.

distance. For this reason, other optical components should not be brought too close to a totally reflecting surface or energy will be coupled to them via the evanescent wave and the efficiency of total internal reflection will be reduced. With extreme care, this effect can be used to produce a variable-reflectivity, totally internal reflecting surface (as seen in Figure 4.4). Studies of total internal reflection at interfaces are important in a number of optical measurement techniques, such as attenuated total (internal) reflection[7] and photon scanning tunneling microscopy.[8]

One or both of the media in Figure 4.2 may be anisotropic, like calcite, crystalline quartz, ammonium dihydrogen phosphate (ADP), potassium dihydrogen phosphate (KDP), or tellurium. The incident wave will then generally split into two components, one of which obeys Snell's law directly (the ordinary wave) and one for which Snell's law is modified (the extraordinary wave). This phenomenon is called *double refraction*.[9–11]

If an optical system contains only planar interfaces, the path of a ray of light through the system can be easily calculated using only the law of reflection and Snell's law. This simple approach neglects diffraction effects, which become significant unless the lateral dimensions (*apertures*) of the system are all much larger than the wavelength (say 10 times larger). The simple behavior of light rays in more complex systems containing nonplanar components (but where diffraction effects are negligible) can be described with the aid of *paraxial-ray* analysis.[12,13] Transmitted and reflected intensities and polarization states cannot be determined by the above methods and are most easily determined by the *method of impedances*.

4.2.2 Paraxial-Ray Analysis

A plane wave is characterized by a unique propagation direction given by the wave vector **k**. All fields associated with the wave are, at a given time, equal at all points in infinite planes orthogonal to the propagation direction. In real optical systems, ideal plane waves do not exist, as the finite size of the elements of the system restricts the lateral extent of the waves. Nonplanar optical components will cause further deviations of the wave from planarity. The wave consequently acquires a ray direction that varies from point to point on the phase front. The behavior of the optical system must be characterized in terms of the deviations its elements cause to the bundle of rays that constitute the propagating, laterally restricted wave. This is most easily done in terms of paraxial rays. In a cylindrically symmetric optical system (for example, a coaxial system of spherical lenses or mirrors), *paraxial* *rays* are those rays whose directions of propagation occur at sufficiently small angles θ to the symmetry axis of the system that it is possible to replace $\sin\theta$ or $\tan\theta$ by θ—in other words, paraxial rays obey the small-angle approximation.

$$\sin\theta \sim \tan\theta \sim \theta \tag{4.8}$$

Matrix Formulation. In an optical system whose symmetry axis is in the z-direction, a paraxial ray in a given cross section (z = constant) is characterized by its distance r from the z-axis and the angle r' it makes with that axis. Suppose the values of these parameters at two planes of the system (an *input* and an *output* plane) are r_1, r'_1 and r_2, r'_2, respectively, as shown in Figure 4.5a. In the paraxial-ray approximation, there then is a linear relation between them of the form

$$r_2 = Ar_1 + Br'_1 \tag{4.9}$$
$$r'_2 = Cr_1 + Dr'_1$$

or, in matrix notation

$$\begin{pmatrix} r_2 \\ r'_2 \end{pmatrix} = \begin{pmatrix} A & B \\ C & D \end{pmatrix} \begin{pmatrix} r_1 \\ r'_1 \end{pmatrix} \tag{4.10}$$

Here,

$$M = \begin{pmatrix} A & B \\ C & D \end{pmatrix} \tag{4.11}$$

is called the *ray transfer matrix*. If the media where the input and output ray parameters are measured have the same refractive index, then the determinant of the ray transfer matrix is unity—i.e., $AD - BC = 1$.

Optical systems made of isotropic material are generally reversible—a ray that travels from right to left with input parameters r_2, r'_2 will leave the system with parameters r_1, r'. This leaves

$$\begin{pmatrix} r_1 \\ r'_1 \end{pmatrix} = \begin{pmatrix} A' & B' \\ C' & D' \end{pmatrix} \begin{pmatrix} r_2 \\ r'_2 \end{pmatrix} \tag{4.12}$$

where the reverse ray transfer matrix is

$$\begin{pmatrix} A' & B' \\ C' & D' \end{pmatrix} = \begin{pmatrix} A & B \\ C & D \end{pmatrix}^{-1} \tag{4.13}$$

The ray transfer matrix allows the properties of an optical system to be described in general terms by the location of its *focal points* and *principal planes*, whose location is determined from the elements of the matrix. The significance of these features of the system can be illustrated with the aid of Figure 4.5b. An input ray that passes through the *first focal point* F_1 (or would pass through this point if it did not first enter the system) emerges traveling parallel to the axis. The intersection point of the extended input and output rays (point H_1 in Figure 4.5b) defines the location of the *first principal plane*. An input ray traveling parallel to the axis, however, emerges at the output plane and passes through the second focal point F_2 (or appear to have come from this point). The intersection of the extension of these rays (point H_2) defines the location of the *second principal plane*. Rays 1 and 2 in Figure 4.5b are called the *principal rays* of the system. The location of the principal planes allows the corresponding emergent ray paths to be determined as shown in Figure 4.5b. The dashed lines in this figure, which permit the geometric construction of the location of output rays 1 and 2, are called *virtual ray paths*. Both F_1 and F_2 lie on the axis of the system. The axis of the system intersects the principal planes at the *principal points*, P_1 and P_2, as shown in Figure 4.5b. The distance f_1 from the first principal plane to the first focal point is called the *first focal length*—f_2 is called the *second focal length*.

In many practical situations, the refractive indices of the media to the left of the input plane (the *object space*) and to the right of the output plane (the *image space*) are equal. In this case, several simplifications arise:

$$f_1 = f_2 = f = -\frac{1}{C}$$

$$h_1 = \frac{D-1}{C} \tag{4.14}$$

$$h_2 = \frac{A-1}{C}$$

where h_1 and h_2 are the distances of the input and output planes from the principal planes, measured in the sense shown in Figure 4.5b.

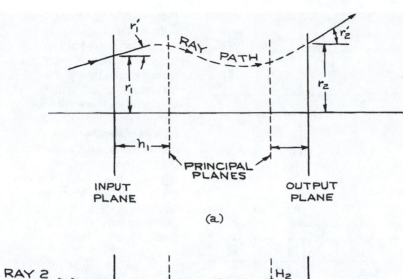

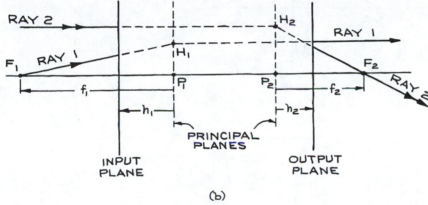

Figure 4.5   (a) Generalized schematic diagram of an optical system, showing a typical ray and its paraxial-ray parameters at the input and output planes; (b) focal points F_1 and F_2 and principal rays of a generalized optical system.

If the elements of the transfer matrix are known, the location of the focal points and principal planes is therefore determined. Graphical construction of ray paths through the system using the methods of paraxial *ray tracing* is then straightforward (see Section 4.2.2, Ray Tracing).

In using the matrix method for optical analysis, a consistent sign convention must be employed. In the present discussion, a ray is assumed to travel in the positive z-direction from left to right through the system. The distance from the first principal plane to an object is measured positively from right to left—in the negative z-direction. The distance from the second principal plane to an image is measured positively from left to right—in the positive z-direction. The lateral distance of the ray from the axis is positive in the upward direction, negative in the downward direction.

The acute angle between the system-axis direction and the ray (see r_1' in Figure 4.5a) is positive if a counterclockwise motion is necessary to go from the positive z-direction to the ray direction. When the ray crosses a spherical interface, the radius of curvature is positive if the interface is convex to the input ray. The use of ray transfer matrices in optical-system analysis can be illustrated with some specific examples.

(i) Uniform optical medium. In a uniform optical medium of length d, no change in ray angle occurs, as illustrated in Figure 4.6a; so

$$r_2' = r_1' \tag{4.15}$$

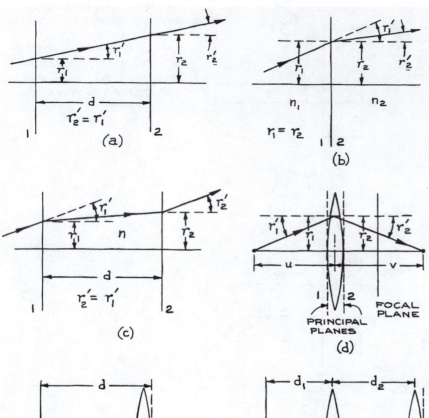

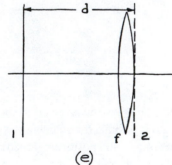

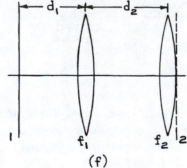

Figure 4.6   Simple optical systems for illustrating the application of ray transfer matrices, with input and output planes marked 1 and 2:
(a) uniform optical medium;
(b) planar interface between two different media;
(c) a parallel-sided slab of refractive index n bounded on both sides with media if refractive index 1; (d) thins lens (r'_2 is a negative angle; for a thin lens, $r_2 = r_1$); (e) a length of uniform medium plus a thin lens; (f) two thin lenses.

$$r'_2 = r_1 + dr'_1$$

Therefore,

$$M = \begin{pmatrix} 1 & d \\ 0 & 1 \end{pmatrix} \qquad (4.16)$$

The focal length of this system is infinite and it has no specific principal planes.

(ii) Planar interface between two different media. At the interface (as shown in Figure 4.6b) we have $r_1 = r_2$, and from Snell's law, using the approximation $\sin\theta \sim \theta$,

$$r'_2 = \frac{n_1}{n_2} r'_1 \qquad (4.17)$$

Therefore,

$$M = \begin{pmatrix} 1 & 0 \\ 0 & n_1/n_2 \end{pmatrix} \qquad (4.18)$$

(iii) A parallel-sided slab of refractive index n bounded on both sides with media of refractive index 1. In this case,

$$M = \begin{pmatrix} 1 & d/n \\ 0 & 1 \end{pmatrix} \qquad (4.19)$$

The principal planes of this system are the boundary faces of the optically dense slab (see Figure 4.6c).

(iv) Thick lens. The ray transfer matrix of the thick lens shown in Figure 4.7a is the product of the three transfer matrices:

$$
\begin{aligned}
M_3 M_2 M_1 = \;& (\text{matrix for second spherical interface}) \qquad (4.20) \\
& \times (\text{matrix for medium of length } d) \\
& \times (\text{matrix for first spherical interface})
\end{aligned}
$$

Note the order of these three matrices: M_1 comes on the right because it operates first on the column vector that describes the input ray.

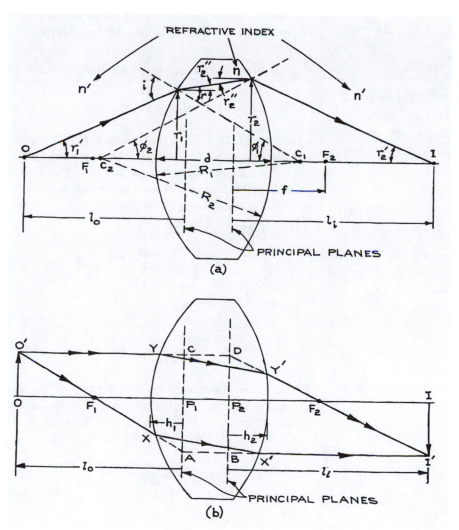

Figure 4.7   (a) Diagram illustrating ray propagation through a thick lens; (b) diagram showing the use of the principal planes to determine the principal ray paths through the lens. The dashed paths *XABX'* and *YCDY'* are the virtual-ray paths; the solid lines *XX'* and *YY'* are the real-ray paths.

At the first spherical surface,

$$n'(r_1' + \phi_1) = nr = n(r_2'' + \phi_1) \tag{4.21}$$

and, since $\phi_1 = r_1/R_1$, this equation can be rewritten as

$$r_2'' = \frac{nr_1'}{n} + \frac{(n'-n)r_1}{nR_1} \tag{4.22}$$

The transfer matrix at the first spherical surface is

$$M_1 = \begin{pmatrix} 1 & 0 \\ \dfrac{n'-n}{nR_1} & \dfrac{n'}{n} \end{pmatrix} = \begin{pmatrix} 1 & 0 \\ -\dfrac{D_1}{n} & \dfrac{n'}{n} \end{pmatrix} \tag{4.23}$$

where $D_1 = (n - n')/R_1$ is called the *power* of the surface. If R_1 is measured in meters, the units of D_1 are *diopters*.

In the paraxial approximation, all the rays passing through the lens travel the same distance d in the lens. Thus,

$$M_2 = \begin{pmatrix} 1 & d \\ 0 & 1 \end{pmatrix} \tag{4.24}$$

The ray transfer matrix at the second interface is

$$M_3 = \begin{pmatrix} 1 & 0 \\ \dfrac{n-n'}{n'R_2} & \dfrac{n}{n'} \end{pmatrix} = \begin{pmatrix} 1 & 0 \\ -\dfrac{D_2}{n'} & \dfrac{n}{n'} \end{pmatrix} \tag{4.25}$$

which is identical in form to M_1. Note that in this case both r_2' and R_2 are negative. The overall transfer matrix of the thick lens is

$$M = M_3 M_2 M_1 = \begin{pmatrix} 1 - \dfrac{dD_1}{n} & \dfrac{dn'}{n} \\ \dfrac{dD_1D_2}{nn'} - \dfrac{D_1}{n'} - \dfrac{D_2}{n'} & 1 - \dfrac{dD_2}{n'} \end{pmatrix} \tag{4.26}$$

If Equations 4.26 and 4.14 are compared, it is clear that the locations of the principal planes of the thick lens are

$$h_1 = \frac{d}{\dfrac{n}{n'}\left(1 + \dfrac{D_1}{D_2} - \dfrac{dD_1}{n}\right)} \tag{4.27}$$

$$h_2 = \frac{d}{\dfrac{n}{n'}\left(1 + \dfrac{D_2}{D_1} - \dfrac{dD_2}{n}\right)} \tag{4.28}$$

A numerical example will best illustrate the location of the principal planes for a biconvex thick lens. Suppose that

$$n' = 1(\text{air}), n = 1.5(\text{glass}) \tag{4.29}$$
$$R_1 = -R_2 = 50\text{mm}$$
$$d = 10\text{mm}$$

In this case, $D_1 = D_2 = 0.01$ mn and, from Equations 4.27 and 4.28, $h_1 = h_2 = 3.448$ mm. These principal planes are symmetrically placed inside the lens. Figure 4.7b shows how the principal planes can be used to trace the principal-ray paths through a thick lens.

From Equations 4.26 and 4.14,

$$r_2' = -\frac{r_1}{f} + \left(1 - \frac{dD_2}{n}\right)r_1' \tag{4.30}$$

where the focal length is

$$f = \left(\frac{D_1 + D_2}{n'} + \frac{dD_1D_2}{nn'}\right)^{-1} \tag{4.31}$$

If l_0 is the distance from the object O to the first principal plane, and l_i is the distance from the second principal plane to the image I in Figure 4.7b, then from the similar triangles $OO'F_1$, P_1AF_1 and P_2DF_2, $II'F_2$

$$\frac{OO'}{II'} = \frac{OF_1}{F_1P_1} = \frac{l_0 - f_1}{f_1} \tag{4.32}$$

$$\frac{OO'}{II'} = \frac{F_2P_2}{F_2I} = \frac{f_2}{l_i - f_2}$$

If the media on both sides of the lens are the same, then $f_1 = f_2 = f$ and it immediately follows that

$$\frac{1}{l_o} + \frac{1}{l_i} = \frac{1}{f} \tag{4.33}$$

This is the fundamental imaging equation.

(v) Thin Lens. If a lens is sufficiently thin that to a good approximation $d = 0$, the transfer matrix is

$$\mathbf{M} = \begin{pmatrix} 1 & 0 \\ -\dfrac{D_1 + D_2}{n'} & 1 \end{pmatrix} \tag{4.34}$$

As shown in Figure 4.6d, the principal planes of such a thin lens are at the lens. The focal length of the thin lens is f, where

$$\frac{1}{f} = \frac{D_1 + D_2}{n'} = \left(\frac{n}{n'} - 1\right)\left(\frac{1}{R_1} - \frac{1}{R_2}\right) \tag{4.35}$$

so the transfer matrix can be written very simply as

$$\mathbf{M} = \begin{pmatrix} 1 & 0 \\ -1/f & 1 \end{pmatrix} \tag{4.36}$$

The focal length of the lens depends on the refractive indices of the lens material and of the medium within which it is immersed. In air,

$$\frac{1}{f} = (n-1)\left(\frac{1}{R_1} - \frac{1}{R_2}\right) \tag{4.37}$$

For a biconvex lens, R_2 is negative and

$$\frac{1}{f} = (n-1)\left(\frac{1}{|R_1|} + \frac{1}{|R_2|}\right) \tag{4.38}$$

For a biconcave lens,

$$\frac{1}{f} = -(n-1)\left(\frac{1}{|R_1|} + \frac{1}{|R_2|}\right) \tag{4.39}$$

The focal length of any diverging lens is negative. For a thin lens, the object and image distances are measured to a common point. It is common practice to rename the distances (l_o and l_i in this case) so that the imaging Equation 4.33 reduces to its familiar form

$$\frac{1}{v} + \frac{1}{u} = \frac{1}{f} \tag{4.40}$$

Then u is called the object distance and v the image distance.

(vi) A length of uniform medium plus a thin lens. This is a combination of the systems in Section 4.2.2 (*i* and *v*, see Figure 4.6e); its overall transfer matrix is found from Equations 4.16 and 4.36 as

$$\mathbf{M} = \begin{pmatrix} 1 & 0 \\ -1/f & 1 \end{pmatrix}\begin{pmatrix} 1 & d \\ 0 & 1 \end{pmatrix} \tag{4.41}$$

$$= \begin{pmatrix} 1 & d \\ -1/f & 1 - d/f \end{pmatrix}$$

(vii) Two thin lenses. As a final example of the use of ray transfer matrices, consider the combination of two thin lenses shown in Figure 4.6f. The transfer matrix of this combination is

$$\mathbf{M} = \text{(matrix of second lens)} \tag{4.42}$$

$$\times \text{(matrix of uniform medium of length } d_2)$$
$$\times \text{(matrix of first lens)}$$
$$\times \text{(matrix of uniform medium of length } d_1)$$

which can be shown to be

$$\mathbf{M} = \begin{pmatrix} 1 - \dfrac{d_2}{f_1} & d_1 + d_2 - \dfrac{d_1 d_2}{f_1} \\[3mm] -\dfrac{1}{f_1} - \dfrac{1}{f_2} + \dfrac{d_2}{f_1 f_2} & 1 - \dfrac{d_1}{f_1} - \dfrac{d_2}{f_2} + \dfrac{d_1 d_2}{f_1 f_2} \end{pmatrix} \tag{4.43}$$

The focal length of the combination is

$$f = \frac{f_1 f_2}{(f_1 + f_2) - d_2} \tag{4.44}$$

The optical system consisting of two thin lenses is the standard system used in analyses of the stability of lens waveguides and optical resonators.[12,14]

(viii) Spherical mirrors. The object and image distances from a spherical mirror also obey Equation 4.40, where the focal length of the mirror is $R/2$—f is positive for a concave mirror, negative for a convex mirror. Positive object and image distances for a mirror are measured positive in the normal direction away from its surface. If a negative image distance results from Equation 4.40, this implies a *virtual image* (behind the mirror).

Ray Tracing

In designing an optical system, ray tracing is a powerful technique for analyzing the design and assessing its performance. Ray tracing follows the trajectories of light rays through the system. These light rays represent, at each point within the system, the local direction of energy flow, represented mathematically by the Poynting vector.

The trajectory of a light ray is governed by the laws of reflection and refraction. At a mirror surface, the angle of reflection is equal to the angle of incidence. At the boundary between two media of different refractive index n_1, n_2, respectively, the angles of incidence θ_1 and refraction θ_2 are related by Snell's Law.*

$$n_1 \sin \theta_1 = n_2 \sin \theta_2 \qquad (4.45)$$

In a medium where the refractive index varies from point to point, Equation 4.45 still holds true but it is better in this case to use the equation of light rays

$$\frac{d}{ds}\left(n\frac{dr}{ds}\right) = \text{grad } n \qquad (4.46)$$

where s is distance measured along the path of the light ray, and $n(r)$ is the index of refraction at point r, measured with respect to the origin.[13]

---

* Snell's Law must be used with some care—in anisotropic media the ray direction is not generally parallel to the wave vector (or wave normal) at the boundary separating one or more anisotropic media, which is also used to characterize the direction of the ray.[13]

Ray tracing provides an accurate description of the way light passes through an optical system unless diffraction effects become important. This happens when any apertures in the system become comparable in size to the wavelength, or when two adjacent apertures in the system subtend angles that are comparable to the diffraction angle.

At a surface of diameter $2a_1$, the diffraction angle for light of wavelength λ is

$$\theta_{diff} = \frac{1.22\lambda}{2a_1} \qquad (4.47)$$

This is the angular position of the first intensity minimum in the ring-shaped diffraction pattern that results when a plane-wave passes through a circular aperture. If a second aperture of diameter $2a_2$ is placed a distance d from the first the angle subtended is

$$\theta = \frac{2a_2}{d} \qquad (4.48)$$

Negligible diffraction requires $\theta_{diff} << \theta$, which gives

$$\frac{\lambda d}{4a_1 a_2} \ll 1 \qquad (4.49)$$

often called the Fresnel criterion.

Paraxial Ray Tracing

Practical implementation of paraxial-ray analysis in optical-system design can be very conveniently carried out graphically by *paraxial ray tracing*. In ray tracing, a few simple rules allow geometrical construction of the principal-ray paths from an object point (these constructions do not take into account the nonideal behavior, or *aberrations*, of real lenses, which will be discussed later). The first principal ray from a point on the object passes through (or its projection passes through) the first focal point. From the point where this ray, or its projection, intersects the first principal plane, the output ray is drawn parallel to the axis. The actual ray path between input and output planes can be found in simple cases—for example, the path XX' in the thick lens shown in Figure 4.7b. The second principal ray is directed parallel to the axis; from the intersection of this ray or its projection with the second principal plane, the output ray passes through (or appears to have come from) the second focal point. The actual ray path between input and output planes can again be found in simple cases—for example, the path YY' in the thick lens shown in Figure 4.7b. The intersection of the two principal rays in the image space produces the image point that corresponds to the original point on the object. If only the back projections of the output principal rays appear to intersect, this intersection point lies on a *virtual* image.

In the majority of paraxial ray tracing applications, a quick analysis of the system is desired. In this case, if all lenses in the system are treated as thin lenses, the position of the principal planes need not be calculated beforehand and ray tracing becomes particularly easy. For a thin lens, a third principal ray is useful for determining the image location. The ray from a point on the object that passes through the center of the lens is not deviated by the lens.

The use of paraxial ray tracing to determine the size and position of the real image produced by a convex lens and the virtual image produced by a concave lens are shown in Figure 4.8. Figure 4.8a shows a converging lens; the input principal ray parallel to the axis actually passes through the focal point. Figure 4.8b shows a diverging lens—the above ray now emerges from the lens so as to appear to have come from the focal point. More complex systems of lenses can be analyzed the same way. Once the image location has been determined by the use of the principal rays, the path of any group of rays (a ray *pencil*) can be found. See, for example, the cross-hatched ray pencils shown in Figure 4.8.

The use of paraxial ray-tracing rules to analyze spherical-mirror systems is similar to those described above, except that the ray striking the center of the mirror in this case reflects so that the angle of reflection equals the angle of incidence.

Imaging and Magnification

Imaging systems include microscopes, telescopes, periscopes, endoscopes, and cameras. They aim to reproduce at a particular plane in the image space—the *image plane* an exact replica in terms of relative spatial luminous intensity distribution and spectral content of an object located in the *object plane*. Each point on the object is imaged to a unique point in the image. The relative position and luminous intensity of points in the image plane preserve these qualities from their positions in the object plane. If the image is not a faithful, scaled replica of the object, then the imaging system suffers from *aberrations*.

In paraxial descriptions of an imaging system, often called the *Gaussian* description of the system, the ray transfer matrix from a point on the object to a point on the image can be written as

$$\begin{pmatrix} r_2 \\ r_2' \end{pmatrix} = \begin{pmatrix} A & B \\ C & D \end{pmatrix} \begin{pmatrix} r_1 \\ r_1' \end{pmatrix} \qquad (4.50)$$

For any ray angle of r_1' leaving a point distant r_1 from the axis on the object, the ray in the image plane must pass through a point that is distant r_2 from the axis.

It is clear from Equation 4.50 that for this to be so, the element B of the ray transfer matrix must be zero; therefore,

$$r_2 = Ar_1 \qquad (4.51)$$

The ratio of image height to object height is the *linear magnification m*, so

$$m = A = r_2/r_1. \qquad (4.52)$$

The *angular magnification* of the system is defined as

$$m' = \left(\frac{r_2'}{r_1'} \right)_{r_1 \to 0} \qquad (4.53)$$

so $D = m'$.

If we use the focal length of the imaging system $f = -1/C$, then the ray transfer matrix is

$$\begin{pmatrix} A & B \\ C & D \end{pmatrix} = \begin{pmatrix} m & 0 \\ -1/f & m' \end{pmatrix} \qquad (4.54)$$

If the media in the space where the object is located (the object space) and the image space are the same then

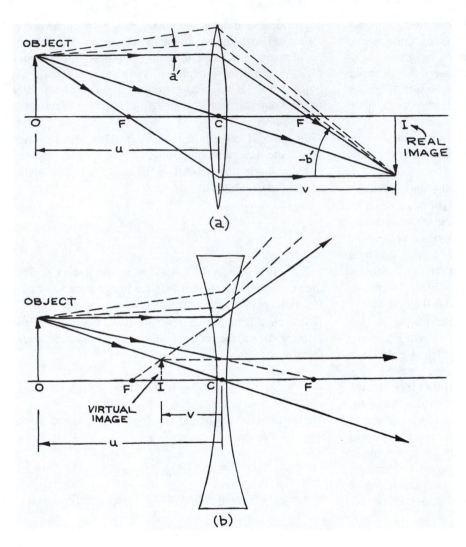

OBJECT

a'

O F C F I

REAL IMAGE

u

$-b'$

v

(a)

OBJECT

O F I C F

VIRTUAL IMAGE

v

u

(b)

Figure 4.8   Ray-tracing techniques for locating image and ray paths for (a) a converging thin lens; (b) a diverging thin lens. F = focal point; C = center of lens. The principal rays and a general ray pencil are shown in each case.

$$\det \begin{pmatrix} A & B \\ C & D \end{pmatrix} = 1 \qquad (4.55)$$

which gives $mm'=1$.

This is also a consequence of the conservation of brightness in an optical system, which will be discussed later.

In Figure 4.8a the ratio of the height of the image to the height of the object is called the *magnification m*. In the case of a thin lens,

$$m = \frac{b}{a} = \frac{v}{u} \qquad (4.56)$$

The ray transfer matrix M (in this case) includes the entire system from object O to image I. In Figure 4.8a,

$$\mathbf{M} = \text{(matrix for uniform medium of length u)} \qquad (4.57)$$

$\times$ (*matrix for lens*)

$\times$ (*matrix for uniform medium of length v*)

4.2.3 Imaging and Non-Imaging Optical Systems

The techniques of optical system analysis that we have discussed so far will provide an approximate description of simple systems. In many situations, however, a more detailed analysis is needed. Additional parameters are also introduced to characterize the system. More sophisticated methods of analysis can be illustrated through a discussion of *imaging* and *non-imaging* optical systems.

In any experimental setup that involves the collection of light from a source and its delivery to a light detection system, the properties of the optical system between source and detector should be optimized. This optimization willl take different forms depending on the application. In an experiment in which a weak fluorescent signal from a liquid in a curvette is to be detected, for example, it is generally the aim to maximize the amount of collected fluorescence and deliver this light to the active surface of the detector. This application does not require an *imaging* optical system, although an imaging system is often quite satisfactory. On the other hand, if an image of some object is being delivered to a CCD camera, then an imaging optical system is required.

Non-imaging Light Collectors

The properties of a *non-imaging* optical system can be described schematically with the aid of Figure 4.9. Light that enters the front aperture of the system within some angular range, will leave the system through the exit aperture. In such an arrangement, the *brightness* of the light leaving the exit aperture cannot exceed the brightness of the light entering the entrance aperture. The brightness of an emitting object is measured in units of W/m^2/steradian. In Figure 4.9, if the brightness of the radiation entering the apertive is B_1(w/m^2/sr), then the power entering S_1 from the source is

$$P_1 = \pi B_1 S_1 \theta_1^2 \qquad (4.58)$$

where the angular range of the entering rays is θ_1. Assuming that $R_1^2 \gg S_1$ this angular range is

$$\theta_1 \cong \frac{1}{R_1}\sqrt{\frac{S_1}{\pi}}$$

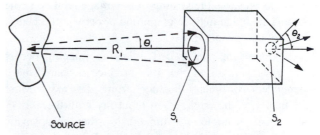

Figure 4.9   Generalized diagram of a non-imaging system. Light from the source enters the system through the aperture S_1 over an angular range defined by θ_1 and leaves through aperture S_2 over a range of angles defined by θ_2.

or

$$\theta_1^2 \cong \frac{S_1}{\pi R_1^2}$$

If all the power entering S_1 leaves through S_2, then the brightness of the emerging radiation is

$$B_2 \approx \frac{P_1}{\pi S_2 \theta_2^2} = \frac{\pi B_1 S_1 \theta_1^2}{\pi S_2 \theta_2^2} \qquad (4.59)$$

where θ_2 is the angular spread of energing rays. Therefore,

$$B_2 = \frac{B_1 S_1 \theta_1^2}{S_2 \theta_2^2} \qquad (4.60)$$

so if $B_2 \leq B_1$,

$$\theta_2 \geq \sqrt{\frac{S_2}{S_1}}\,\theta_1 \qquad (4.61)$$

Practical examples of these devices are given in Section 4.3.3, Optical Concentrators.

Imaging Systems

Generalized imaging systems. In a generalized imaging system, light from an object located in the *object plane* passes through the system until it reaches the *image plane*. In an ideal imaging system, all the light rays that leave a point on the object arrive at a single point in the imaging plane. The location of points in the image plane is also such as to preserve the scaling of the image relative to the object. The image may be magnified or demagnified but it should not be distorted.

A type of imaging called *aplanatic* imaging occurs when rays parallel to the axis produce a sharp image independent of their distance from the axis. More specifically, if the angle of an input ray with the axis θ_1 and the output angle with the axis θ_2, then an aplanatic system satisfies the *sine condition*, namely

$$\frac{\sin \theta_1}{\sin \theta_2} = constant$$

for rays at all distances from the axis that can pass through the system. *Stigmatic* imaging is sharp imaging of a point object to a point image. An object point that is imaged to an image point are together called *conjugate* points,

The numerical aperture. If an optical system is illuminated by a point source on its axis, then the amount of light collected by the system depends on the angle that the entrance pupil subtends at the source. In Figure 4.10, this angle is defined by

$$\sin \theta = \frac{D}{2u} \tag{4.62}$$

and the *numerical* aperture of the system is

$$NA = n \sin \theta \tag{4.63}$$

where n is the refractive index of the object space (which is generally 1, but could be higher—for example in microscopy when an oil-immersion objective lens is used). A good imaging system now obeys the *sine condition*, which relates the angles subtended by the exit and entrance pupils and the magnification m by

$$m = \frac{n \sin \theta}{n' \sin \theta'} \tag{4.64}$$

where the primed quantities refer to the image space. The numerical aperture in the image space is

$$(NA)' = n' \sin \theta' \tag{4.65}$$

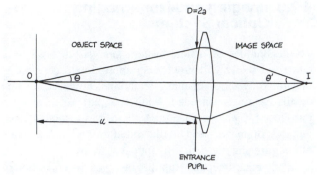

Figure 4.10 Single-lens optical system to illustrate the concepts of numerical aperture and entrance pupil.

The entrance pupil plays an important geometrical role in determining the amount of light from an object (or light source) that can enter and pass through an optical system. For an entrance pupil diameter (EPD) of D_1 located a distance d_1 from a point source of light, for example, the fraction of the light emitted by the source that will pass through the system is

$$F = \frac{1}{2}\left[1 - \left(1 + \frac{D^2}{4d^2}\right)^{-1/2}\right] \tag{4.66}$$

which for $D \ll d$ reduces to

$$F = \frac{D^2}{16d^2} \tag{4.67}$$

For a lens, the ratio of focal length to EPD is called the *f/number* (*f/#*) of the lens. An *f*/1 lens has a ratio of its focal length to aperture (the open lens diameter) equal to unity. For a point source at the focal point of such a lens, $F = 0.067$.

Apertures, Stops, and Pupils

Apertures control the ray trajectories that can pass through an optical system. The *aperture stop* is the aperture within the system that limits the angular spread, or diameter, of a cone of rays from an axial point on the object, which can pass through the system. Another aperture in the system may restrict the size or angular extent of an object that the system can image. Awareness of the position and size of these stops in an optical design is crucial in determining the light gathering power of the system, and its ability to deliver light to an image or detector. The role of the aperture and field stop can be illustrated with Figure 4.11, which shows a single lens being used to image light from a distant object. The circular aperture behind the lens clearly limits the diameter of a bundle of rays from an axial point on the object and is clearly the aperture stop. The second stop, placed in the image plane, clearly controls the maximum angle at which rays from off-axis points on the object can pass through the system. In this case, this is the *field stop*.

Whether a given aperture in an optical system is an aperture stop or a field stop is not always so easy to determine as was the case in Figure 4.11. Figure 4.12 shows a three element optical system, which actually constitutes what is called a Gullstrand opthalmoscope.[*] The first lens closest to the object, called the *objective* lens produces a real inverted image, which is then reinverted by the *erector* lens, to form an image at the focal point of the *eyepiece* lens. Aperture *A* is the aperture stop and clearly restricts the angular range of rays from an axial point on the object that can pass through the system. Aperture *B* restricts from what height on the object rays can originate and still pass through the system. This aperture is acting as a field stop.

The *entrance pupil* is the image of the aperture stop produced by all the elements of the system between the aperture stop and the object. In Figure 4.12, the two elements

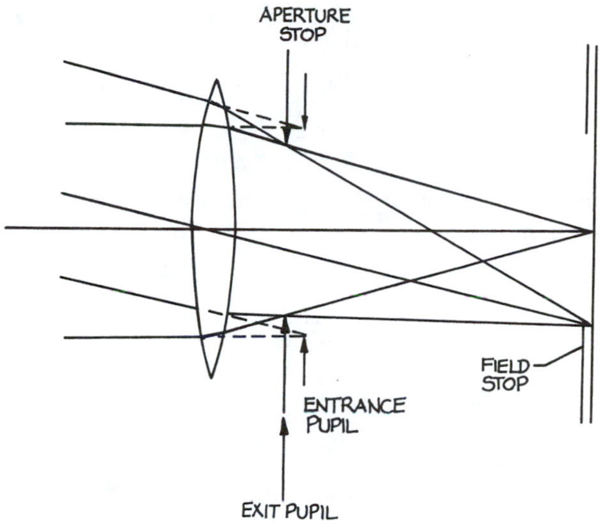

Figure 4.11   Single-lens imaging system to illustrate the concepts of aperture stop, field stop, and entrance and exit pupils.

producing this image are the erector lens and objective lens. The *exit* pupil is the image of the aperture stop produced by all the elements between the aperture stop and the image, which in Figure 4.12 is just the eye lens. The size and location of the entrance and exit pupils determines the light gathering and light delivery properties of the system. If the entrance pupil has diameter D, for example, and lies a distance R from a point source of radiant power P (watts), then the power entering the entrance pupil is

---

[*]   Used by physicians for examining the retina of a patient's eye.[15]

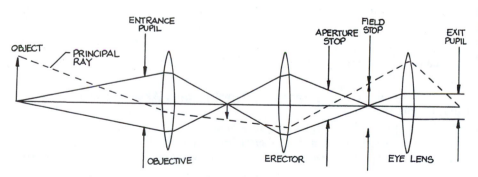

Figure 4.12   Three-lens imaging system showing aperture stop, a field stop, and the location of the entrance and exit pupils.

$$P_1 = P \sin^2 \left(\frac{\theta}{2} \right) \qquad (4.68)$$

where
$$\sin \theta = \frac{D_1}{2\sqrt{R^2 + D_1^2/4}}$$

For a point source that is far from the entrance pupil, this result becomes

$$P_1 = \frac{PD_1^2}{16R^2} \qquad (4.69)$$

The solid angle subtended by the entrance pupil in this case is

$$\Delta \omega = \frac{D_1^2}{4R^2}$$

The light that leaves the system can be characterized by the total power that leaves the exit pupil and the angular range of the rays that leave the exit pupil.

Somewhere in an imaging system there is always an *aperture stop*. This may be an actual aperture in an opaque thin screen placed on the system axis or it may be the clear aperture of a lens or mirror in the system. In either case, the aperture stop limits the range of rays that can pass through the system. This concept is illustrated in Figure 4.13. A ray of light such as A that enters the system and just passes the edge of the aperture stop is called a *marginal ray*. A ray such as B that enters the system and passes through the center of the aperture stop is called a *principal ray*. An aperture stop will always allow some light from every point on an object to reach the corresponding point on the image. If the amount of light from the edge of the object that reaches the corresponding point on the object is severely reduced, for example, this is referred to as *vignetting*.

The ray that enters the system half-way between the highest and lowest rays of an oblique beam is called the *chief ray* of the beam. In the absence of aberrations, the principal ray and chief ray are the same and both pass through the center of entrance pupil and aperture stop.

In Figure 4.12, a ray from an off-axis point on the object that passes through the center of the aperture stop is called the *principal ray* of the cone of light from the off-axis point.

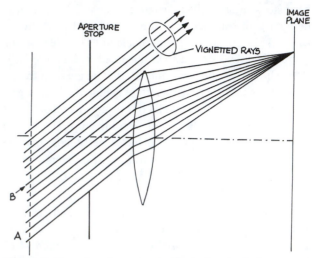

Figure 4.13 Aperture stop in front of a single lens showing vignetting of a family of rays from a distant off-axis point passing through the lens.

Vignetting

In our previous discussion, we have simplified the actual performance of a real optical system, because for oblique rays passing through the system the roles of aperture and field stops become intertwined.

This can be illustrated with Figure 4.14, which shows a simple compound lens in which the apertures of the lenses themselves restrict the rays that can pass through the system. A family of oblique rays no longer fills the aperture stop because the following lens is acting as a *stop*. This is *vignetting*. In imaging systems, it does not actually remove points of the object from the image. Parts of the image that depend on oblique rays for their illumination (especially at large angles) become less bright. Vignetting is often deliberately allowed to occur in an imaging system to remove certain rays from the system that would otherwise suffer severe aberration.

Exact Ray Tracing and Aberrations

If an imaging system is built with a series of lenses and/or mirrors, then there are inevitably various aberrations. For refraction at a spherical surface this results because the angles of incidence and refraction θ_1, θ_2, respectively, are

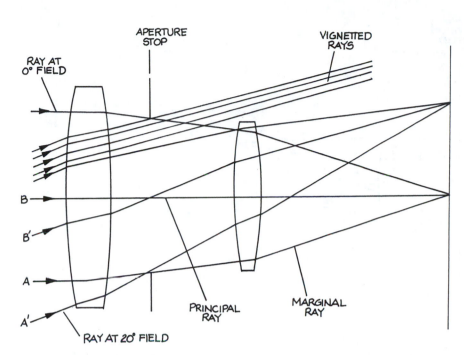

Figure 4.14 Two-lens imaging system showing aperture stop, principal and marginal rays, and vignetted rays.

not *small* angles. When light travels from a medium with refractive index n_1 to one of refractive index n_2

$$n_1 \sin \theta_1 = n_2 \sin \theta_2 \qquad (4.70)$$

The paraxial approximation, which lacks aberrations, fails because $\theta_1 \neq \sin \theta_1$, whereas in reality

$$\sin \theta = \theta - \frac{\theta^3}{3!} + \frac{\theta^5}{5!} - \frac{\theta^7}{7!}$$

The overall effect of aberrations can be greatly reduced, and some types even eliminated, by using several spherical surfaces, or in some situations by using an appropriate *aspheric* surface—a surface that is not part of a sphere but that is generally describable in terms of some other conic function or power series of the coordinates.

Analysis of imaging systems is most conveniently carried out using exact ray tracing techniques. In exact ray tracing, the small angle approximation (Equation 4.8) is not made and Snell's Law is solved exactly for each ray, at each interface.

By injecting many light rays into the system over some defining aperture or range of angles, the performance of the system in imaging, focusing, light collection, or light delivery can be evaluated precisely. In current practice, this procedure is carried out with optimal design software. The most notable software packages in this regard are Code V,[16] Zemax,[17] Oslo,[18] Solstis,[19] and Optikwerk.[20] The wide availability of these software tools, and the essentially generic way in which they are used, makes a brief description of how they work and are used worthwhile. Just as an electronic circuit designer would be likely to evaluate a current with an analog design tool such as PSpice or a digital design tool such as Magic, an optical designer can ultimately save time, and avoid mistakes, by numerical simulation of a design.

We will illustrate the use of numerical ray tracing using the set-up procedures appropriate to Code V and with a specific example—a double Gauss lens used for imaging.

The various optical elements that make up a double Gauss lens are shown in Figure 4.15. For Code V analysis, the axial position, curvature, material, and aperture size (radius of each surface perpendicular to the axis) must be specified in an *optical table* (shown in this case as Table 4.2) Note the entries that are specified in each column.

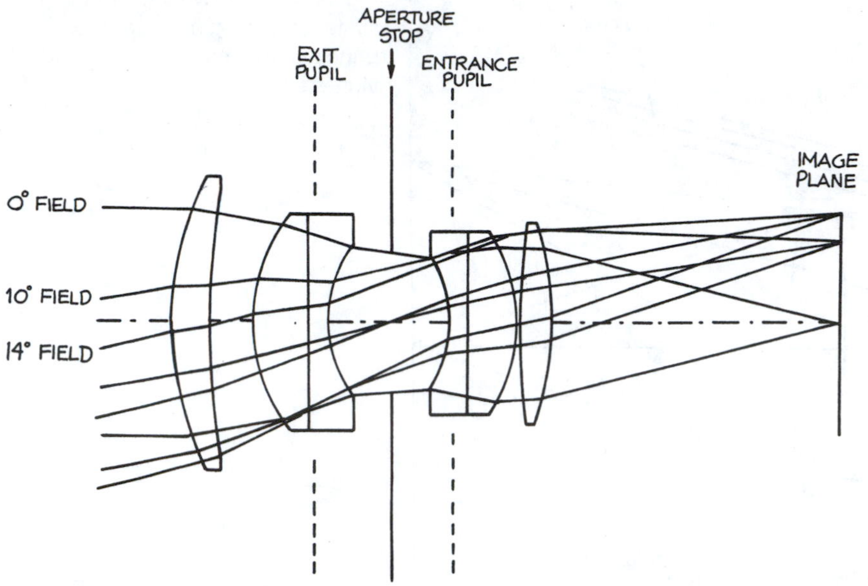

Figure 4.15   Double-Gauss lens.

With ray tracing software, the various parameters of this lens system can be simultaneously optimized so as to produce the sharpest possible image of an object.

For the double Gauss lens, for example, the curvatures of the various surfaces, their spacings, and the type of glass used for each element can be simultaneously varied. In practice, various constraints must be included in this process—lens elements should not be too thick or too thin and the glasses chosen must be available. The variable parameters for a glass are its refractive index n and its *dispersion*, $dn/d\lambda$, both of which vary with wavelength.

The dispersion is frequently characterized by the Abbe V-number, which provides a relative measure of dispersion:

$$V = \frac{n_d - 1}{n_F - n_C} \tag{4.71}$$

where the refractive index n is specified at the three wavelengths d (the helium d line at 587.6 nm), F (the hydrogen F-line at 486.1 nm), and C (the hydrogen C-line at 656.3 nm). Some of the standard wavelengths that are used in optical design are listed in Table 4.3. The different glasses that are available from a manufacturer are generally plotted on a glass chart, which plots $(n_d - 1)$ against V. Figure 4.16 shows such a chart for glasses available from Schott.

Two old classifications of glass into *crown* and *flint* glasses can be related to the glass chart. Crown glasses are glasses with a V value greater than 55 if $n_d < 1.6$, and $V > 50$ if $n_d > 1.6$. The flint glasses have V values below these limits. The rare earths glasses contain rare earth instead of SiO_2, which is the primary constituent of crown and flint glasses.

Spot Diagram

If very many rays are launched from a point on the object so as to cover the extrance pupil of the system, the resulting pattern of ray intersections with the image plane is called the *spot diagram*. A tight spot diagram indicates that several aberrations, notably spherical, astigmatism, coma, and curvature of field have been markedly reduced. Figure 4.17 shows such a spot diagram for the double Gauss lens.

Chromatic Aberrations

Because the index of refraction generally increases with decreasing wavelength, the focal length of a *singlet* lens[*] will, for example, be shorter for blue light than it is for red light. This effect leads to imperfect imaging so that in white light illumination both axial and off-axis object points will be imaged not as bright points but as regions that show blurring from red to blue. For axial object points, this blurring is spherically symmetric and results from the image being closer to the lens for blue light than for red light. This is simple *chromatic aberration*. For off-axis object points, the magnification of the system varies with wavelength, leading to what is called *lateral color* or *transverse chromatic aberration*.

Chromatic aberration can be corrected for two colors (generally red and blue) and the imaging of axial points. Such lenses take the form of an *achromatic doublet* where a doublet lens is generally fabricated from a positive (converging) crown glass element and a negative (diverging) flint glass element.

If a compound lens is designed to remove chromatic aberration at the three wavelength, it is called an *apochromat*. If corrected for four wavelengths, it is called a *superachromat*.

Achromatic doublets, because they have at least three spherical surfaces (or four in an air-spaced achromat), can also be better corrected for other aberrations than a singlet lens. Such lenses are available in a range of apertures and focal lengths from many manufacturers.

Geometrical Aberrations

In a perfect imaging system, all rays from a point on the object would pass through an identical point on the image and there would be a linear relation between the coordinates of points in the image and corresponding points on the object.

For an axially symmetric system, we can arbitrarily choose the object point 0 in Figure 4.18 to be at the (x,y) point $(0,h)$ in the object. We take the polar coordinates of a ray from this point to a point P in the entrance pupil as (ρ,θ). In the general case, this ray will intersect the image

---

[*]   A simple lens with just two spherical surfaces.

plane at point I' with coordinates in the image plane (x',y'). The aberrations $\Delta x'$, $\Delta y'$ are the displacements of (x',y') from the paraxial image point $I(x'_0, y'_0)$ for which there is a linear transformation from the coordinates of the point on the object. The lowest order aberration terms contain terms in either h^3, $h^2\rho$, or $h\rho^2$ and are referred to as *third-order aberrations*. Although there are differences in terminology concerning these terms in the literature we can write (following Born and Wolf[9])

$$\Delta y' = B\rho^3 \cos\theta - Fh\rho^2(2 + \cos 2\theta) \qquad (4.72)$$

$$+ (2C + D)h^2\rho\cos\theta - Eh^3$$

$$\Delta x' = B\rho^3 \sin\theta - Fh\rho^2\sin 2\theta + Dh^2\rho\sin\theta \quad (4.73)$$

The coefficients B, C, D, E, and F characterize the five *primary* or *Seidel* aberrations. Their magnitude in any given optical design can be calculated and visualized with ray-tracing optical software.

Spherical Aberration $B \neq 0$

When all the coefficients in Equations 4.72 and 4.73 are zero except for B, the remaining effect is spherical aberration. This aberration is strikingly observed in the imaging of an axial point source, which will be imaged to a circular bright region whose radius reveals the extent of the aberration. For an axial point $h = 0$ and from Equations 4.72 and 4.73, $\Delta y' = B\rho^3\cos\theta$, $\Delta x' = B\rho^3\sin\theta$, which can be written as

$$\Delta r = B\rho^3 \qquad (4.74)$$

For focusing of parallel light, the *best-form* singlet lens for minimum spherical aberration is close to convex-plane (with the convex suface towards the incoming parallel light).

Coma $F \neq 0$

Coma is the aberration in which the image of an off-axis point varies for rays passing through different regions of the entrance pupil. It is difficult to produce a lens where coma alone can be observed. The spot diagram in Figure

TABLE 4.2 DOUBLE GAUSS

```
     Double Gauss - U.S. Patent 2,532,751
                RDY           THI      RMD        GLA          CCY    THC    GLC
>  OBJ:       INFINITY      INFINITY                           100    100
    1:        65.81739      8.746658          BSM24_OHARA        0    100
    2:       179.17158      9.100796                             0      0
    3:        36.96542     12.424230          SK1_SCHOTT         0    100
    4:        INFINITY      3.776966          F15_SCHOTT       100    100
    5:        24.34641     15.135377                             0      0
   STO:       INFINITY     12.869768                           100      0
    7:       -27.68891      3.776966          F15_SCHOTT         0    100
    8:        INFINITY     10.833928          SK16_SCHOTT      100    100
    9:       -37.46023      0.822290                             0      0
   10:       156.92645      6.858175          SK16_SCHOTT        0    100
   11:       -73.91040     63.268743                             0    PIM
   IMG:       INFINITY     -0.279291                           100      0


SPECIFICATION DATA
    EPD       50.00000
    DIM            MM
    WL        656.30      587.60      486.10
    REF            2
    WTW            1           1           1
    XAN        0.00000     0.00000     0.00000
    YAN        0.00000    10.00000    14.00000
    VUY        0.00000     0.20000     0.40000
    VLY        0.00000     0.30000     0.40000


REFRACTIVE INDICES
    GLASS CODE              656.30      587.60      486.10
    BSM24_OHARA            1.614254    1.617644    1.625478
    SK1_SCHOTT            1.606991    1.610248    1.617756
    F15_SCHOTT           1.600935    1.605648    1.616951
    SK16_SCHOTT          1.617271    1.620408    1.627559


SOLVES
    PIM


No pickups defined in system


INFINITE CONJUGATES
    EFL        99.9866
    BFL        63.2687
    FFL       -17.7398
    FNO         1.9997
    IMG DIS    62.9895
    OAL        84.3452
    PARAXIAL IMAGE
     HT        24.9295
    ANG        14.0000
    ENTRANCE PUPIL
     DIA       50.0000
     THI       68.3242
    EXIT PUPIL
     DIA       58.0886
     THI      -52.8928
```

TABLE 4.3 STANDARD WAVELENGTHS USED IN OPTICAL DESIGN

Wavelength (nm)	Designation	Source	Wavelength (nm)	Designation	Source
312.59		mercury	632.80		He-Ne laser
334.15		mercury	643.85′	c′	cadmium
365.011	ℓ	mercury	656.27	c	hydrogen
404.66	h	mercury	706.52	r	helium
435.83	g	mercury	852.11	s	cesium
479.99′	F′	mercury	1013.98	t	mercury
486.13	F	hydrogen	1060.00		neodymium laser
546.07	e	mercury	1529.58		mercury
587.56	d	helium	1970.09		mercury
589.29	D	sodium (doublet)	2325.42		mercury

4.17, however, shows the characteristic *flaring* of the spot pattern—like a comet tail—which gives the aberration its name. Coma can be controlled by varying the curvatures of surfaces in the system.

Astigmatism $C \neq 0$

All the aberrations discussed so far are defects in the imaging of *meridional rays*—that is, rays in the plane containing the axis of the lens and the line from the object point through the center of the lens. The imaging is different for *sagittal rays*—that is, rays in the sagittal plane, which is perpendicular to the meridional plane as illustrated in Figure 4.19. The resulting aberration is called *astigmatism*. It can be controlled by lens curvature and refractive-index variations and by the use of apertures to restrict the range of angles and off-axis distances at which rays can traverse the lens.

Distortion $E \neq 0$

In distortion the magnification varies across the image plane. The height of the image above the axis relative to the height of the object various across the image.

Each object point appears as a point in the image, and the images of object points on a plane orthogonal to the axis are also on such a plane. The magnification of an object line segment may vary, however, with its distance from the axis. This is called *distortion*. It takes two common forms—*pincushion* and *barrel* distortion—as illustrated in Figure 4.20. Distortion is sensitive to the lens shape and spacing, and to the size and position of apertures in the system, which are called *stops*.

For further details of aberrations and how to deal with them, we refer the reader to specialized texts on optics, such as those by Born and Wolf,[9] Ditchburn,[21] Levi,[22] Smith,[23] and Shannon.[24]

Curvature of Field $D \neq 0$

In this aberration, the image of a plane object perpendicular to the axis is sharp over a curved surface in the image space.

Modulation Transfer Function

A common way of characterizing the performance of an imaging system is through its *modulation transfer function* or MTF. This refers to the ability of the system to replicate in the image, periodic features in the object. An optical test pattern used to show this consists of a series of white and black bands. These bands will be smeared out in the image to a greater or lesser extent because of aberrations, and ultimately if geometrical aberrations are absent, by diffraction.

The relative brightness of the object and image will appear schematically as shown in Figure 4.21. We can characterize the *visibility* or *modulation* of the image as

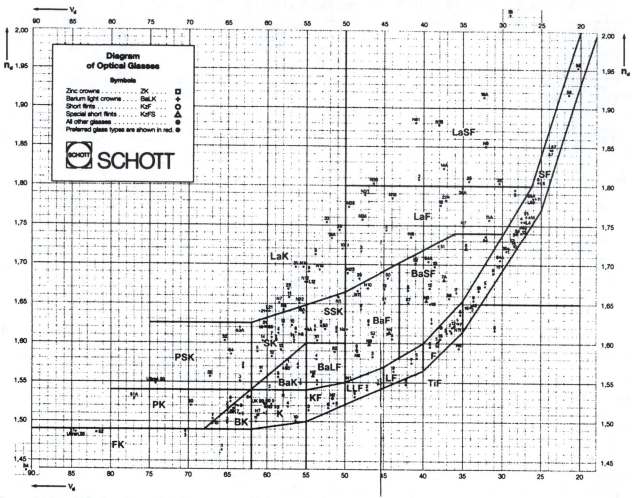

Figure 4.16 The glass chart for glasses available from Schott.

$$\text{visibility} = \frac{\max - \min}{\max + \min} \qquad (4.75)$$

If, in this case, the visibility versus the number of lines per millimeter is plotted in the object, this shows the modulation transfer function. Figure 4.22 shows an example for the double Gauss lens system shown in Figure 4.15. The MTF is ultimately limited by diffraction. In this case, for a pattern with ν lines per millimeter

$$MTF(\nu) = \frac{1}{\pi}(2\phi - \sin 2\phi) \qquad (4.76)$$

where

$$\phi = \cos^{-1}\left(\frac{\lambda_0 \nu}{2NA}\right) \qquad (4.77)$$

In Equation 4.77, λ_0 is measured in mm and NA is the numerical aperture of the system.

A quantity closely related to the MTF for charactering MTF, and also the focusing of a light beam or laser beam, is the *Strehl* ratio. The Strehl ratio is the ratio of the illumination at the center of a focused or circular image to

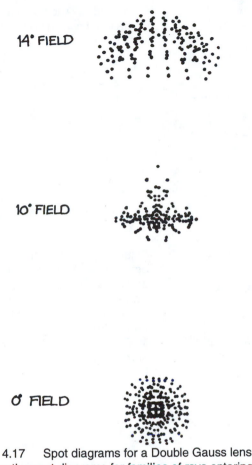

14° FIELD

10° FIELD

0° FIELD

Figure 4.17 Spot diagrams for a Double Gauss lens showing the spot diagrams for families of rays entering the lens at angles of 0°, 10°, and 14°—the *field angles*.

the illumination of an equivalent unaberrated image. The interested reader can see a more detailed discussion in Smith.[24]

4.2.4 The Use of Impedances in Optics

The method of impedances is the easiest way to calculate the fraction of incident intensity transmitted and reflected in an optical system. It is also the easiest way to follow the changes in polarization state that result when light passes through an optical system.

As we have seen in Section 4.1, the impedance of a plane wave traveling in a medium of relative permeability μ_r, and dielectric constant ε_r is

$$Z = \sqrt{\frac{\mu_r \mu_0}{\varepsilon_r \varepsilon_0}} = Z_0 \sqrt{\frac{\mu_r}{\varepsilon_r}} \tag{4.78}$$

If $\mu_r = 1$, as is usually the case for optical media, the impedance can be written as

$$Z = Z_0 / n \tag{4.79}$$

This impedance relates the transverse E and H fields of the wave

$$Z = \frac{E_{tr}}{H_{tr}} \tag{4.80}$$

When a plane wave crosses a planar boundary between two different media, the components of both E and H parallel to the boundary have to be continuous across that boundary. Figure 4.23a illustrates a plane wave polarized in the plane of incidence striking a planar boundary between two media of refractive indices n_1 and n_2. In terms of the magnitudes of the vectors involved,

$$E_i \cos\theta_1 + E_r \cos\theta_1 = E_t \cos\theta_2 \tag{4.81}$$

$$H_i - H_r = H_t \tag{4.82}$$

Equation 4.82 can be written as

$$\frac{E_i}{Z_1} - \frac{E_r}{Z_1} = \frac{E_t}{Z_2} \tag{4.83}$$

It is easy to eliminate E_t between Equations 4.81 and 4.83 to give

$$\rho = \frac{E_r}{E_i} = \frac{Z_2 \cos\theta_2 - Z_1 \cos\theta_1}{Z_2 \cos\theta_2 + Z_1 \cos\theta_1} \tag{4.84}$$

where ρ is the reflection coefficient of the surface. The fraction of the incident energy reflected from the surface is called the reflectance $R = \rho^2$. Similarly, the transmission coefficient of the boundary is

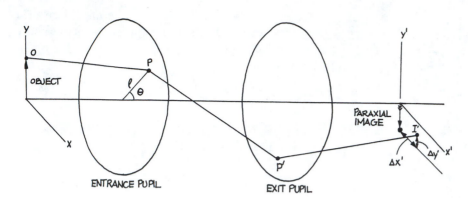

Figure 4.18 Diagram of a generalized imaging system used to discuss aberrations.

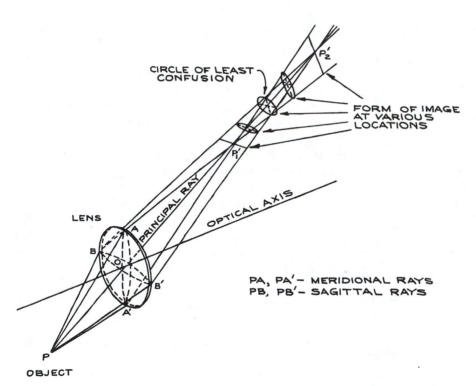

Figure 4.19 Illustrating astigmatism. Meridional rays *PA* and *PA'* are imaged at P_1', sagittal rays *PB* and *PB''* are imaged at P_2'. (After A. C. Hardy and F. H. Perrin, *Principles of Optics*, McGraw-Hill, New York, 1932; by permission of McGraw-Hill Book Company, Inc.)

$$\tau = \frac{E_t \cos \theta_2}{E_i \cos \theta_1} = \frac{2Z_2 \cos \theta_2}{Z_2 \cos \theta_2 + Z_1 \cos \theta_1} \tag{4.85}$$

By a similar treatment applied to the geometry shown in Figure 4.23b, it can be shown that for a plane wave polarized perpendicular to the plane of incidence,

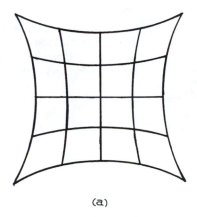

(a)

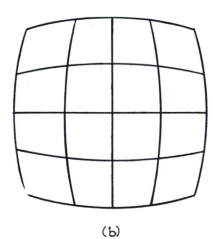

(b)

Figure 4.20   (a) *Pincushion* distortion; (b) *barrel* distortion.

$$\rho = \frac{Z_2 \sec \theta_2 - Z_1 \sec \theta_1}{Z_2 \sec \theta_2 + Z_1 \sec \theta_1} \qquad (4.86)$$

$$\tau = \frac{2Z \sec \theta_2}{Z_2 \sec \theta_2 + Z_1 \sec \theta_1} \qquad (4.87)$$

If the effective impedance for a plane wave polarized in the plane of incidence (*P-polarization*)[*] and incident on a boundary at angle θ is defined as

$$Z' = Z \cos \theta \qquad (4.88)$$

and for a wave polarized perpendicular to the plane of incidence (*S-polarization*)[†] as

$$Z' = Z \sec \theta \qquad (4.89)$$

then a universal pair of formulae for ρ and τ results:

$$\rho = \frac{Z'_2 - Z'_1}{Z'_1 + Z'_2} \qquad (4.90)$$

$$\tau = \frac{2Z'_2}{Z'_1 + Z'_2} \qquad (4.91)$$

It will be apparent from an inspection of Figure 4.23 that Z' is just the ratio of the electric-field component parallel to the boundary and the magnetic-field component parallel to the boundary. For reflection from an ideal mirror, $Z'_2 = 0$.

In normal incidence, Equations 4.84 and 4.86 become identical and can be written as

$$\rho = \frac{Z_2 - Z_1}{Z_2 + Z_1} = \frac{n_1 - n_2}{n_1 + n_2} \qquad (4.92)$$

Note that there is a change of phase of π in the reflected field relative to the incident field when $n_2 > n_1$.

Since intensity $\propto$ (electric field)^2, the fraction of the incident energy that is reflected is

$$R = \rho^2 = \left(\frac{n_1 - n_2}{n_1 + n_2}\right)^2 \qquad (4.93)$$

R increases with the index mismatch between the two media, as shown in Figure 4.24. If there is no absorption of

---

[*]   often called a TM wave.

[†]   often called a TE wave.

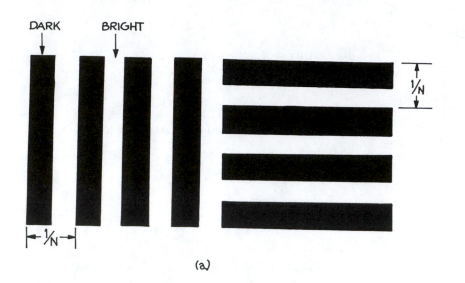

DARK   BRIGHT

$1/N$

$1/N$

(a)

Figure 4.21   (a) An object with sharp contrast between bright and dark bands; (b) the corresponding image brightness variation for differing degrees of modulation.

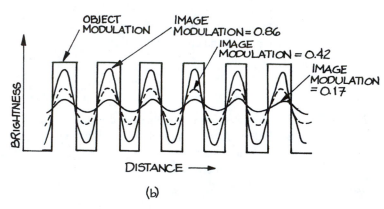

OBJECT MODULATION

IMAGE MODULATION = 0.86

IMAGE MODULATION = 0.42

IMAGE MODULATION = 0.17

BRIGHTNESS

DISTANCE →

(b)

energy at the boundary, the fraction of energy transmitted, called the *transmittance*, is

$$T = 1 - R = \frac{4n_1 n_2}{(n_1 + n_2)^2} \qquad (4.94)$$

Note that because the media on the two sides of the boundary are different, $T \neq |\tau|^2$.

Reflectance for Waves Incident on an Interface at Oblique Angles

If the wave is not incident normally, it must be decomposed into two linearly polarized components—one polarized in the plane of incidence, and the other polarized perpendicullar to the plane of incidence.

For example, consider a plane-polarized wave incident on an air glass ($n = 1.5$) interface at an angle of incidence of $30°$, with a polarization state exactly intermediate between the S-polarization and the P-polarization. The angle of refraction at the boundary is found from Snell's law:

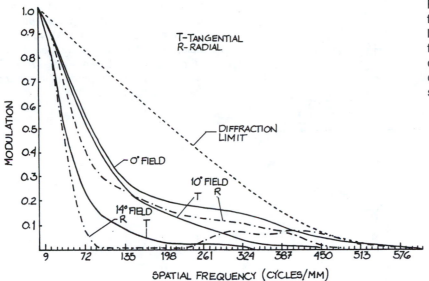

Figure 4.22   Modulation transfer function (MTF) for a double Gauss lens. The MTF for 0°, 10°, and 14° fields in both the tangential and radial directions are shown, and also the diffraction limit of an aberation-free system.

$$\sin \theta_2 = \frac{\sin 30°}{1.5} \tag{4.95}$$

and so $\theta_2 = 19.47°$.

The effective impedance of the P-component in the air is

$$Z'_{P1} = 376.7 \cos \theta_1 = 326.23 \text{ ohm} \tag{4.96}$$

and in the glass,

$$Z'_{P2} = \frac{376.7}{1.5} \cos \theta_2 = 236.77 \text{ ohm} \tag{4.97}$$

Thus, from Equation 4.90, the reflection coefficient for the P-component is

$$\rho_P = \frac{236.77 - 326.33}{236.77 + 326.23} = -0.159 \tag{4.98}$$

The fraction of the intensity associated with the P-component that is reflected is $\rho_P^2 = 0.0253$.

For the S-component of the input wave,

$$Z'_{S1} = \frac{376.7}{\cos \theta_1} = 434.98 \text{ ohm} \tag{4.99}$$

$$Z'_{S2} = \frac{376.7}{1.5 \cos \theta_2} = 266.37 \text{ ohm} \tag{4.100}$$

$$\rho_S = \frac{266.37 - 434.98}{266.37 + 434.98} = -0.240 \tag{4.101}$$

The fraction of the intensity associated with the S-polarization component that is reflected is $\rho_S^2 = 0.0578$. Since the input wave contains equal amounts of S- and P-polarization, the overall reflectance in this case is

$$R = \langle \rho^2 \rangle_{av} = 0.0416 \cong 4\% \tag{4.102}$$

Note that the reflected wave now contains more S-polarization than P, so the polarization state of the reflected wave has been changed—a phenomenon that will be discussed further in Section 4.3.6, Polarizers.

Brewster's Angle

Returning to Equation 4.84, it might be asked whether the reflectance is ever zero. It is clear that ρ will be zero if

$$n_1 \cos \theta_2 = n_2 \cos \theta_1 \tag{4.103}$$

which, from Snell's law (Equation 4.6), gives

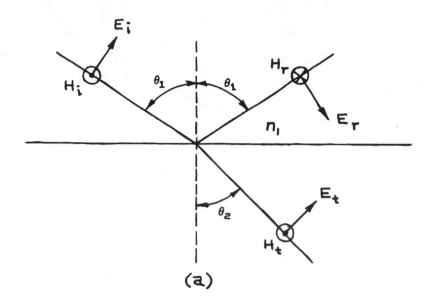

(a)

Figure 4.23   Reflection and refraction at a planar boundary between two different dielectric media: (a) wave polarized in the place of incidence (P-polarization); (b) wave polarized perpendicular to the plane of incidence (S-polarization).

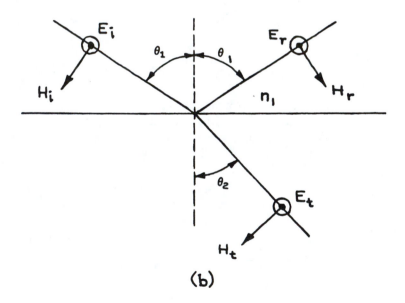

(b)

$$\cos \theta_1 = \frac{n_1}{n_2} \sqrt{1 - \left(\frac{n_1}{n_2}\right)^2 \sin^2 \theta_1} \qquad (4.104)$$

giving the solution

$$\theta_1 = \theta_B = \arcsin \sqrt{\frac{n_2^2}{n_1^2 + n_1^2}} = \arctan \frac{n_2}{n_1} \qquad (4.105)$$

The angle θ_B is called *Brewster's angle*. A wave polarized in the plane of incidence and incident on a

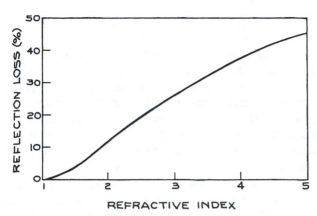

Figure 4.24 Reflection loss per surface for normal incidence as a function of refractive index.

boundary at this angle is totally transmitted. This fact is put to good use in the design of low-reflection-loss windows in laser systems, as will be seen in Section 4.6.2. If Equation 4.86 is inspected carefully, it will be seen that there is no angle of incidence that yields zero reflection for a wave polarized perpendicular to the plane of incidence.

Transformation of Impedance through Multilayer Optical Systems

The impedance concept allows the reflection and transmission characteristics of multilayer optical systems to be evaluated very simply. If the incident light is incoherent (a concept discussed in more detail later in Section 4.5.1), the overall transmission of a multilayer structure is just the product of the transmittances of its various interfaces. An air-glass interface, for example, transmits about 96 percent of the light in normal incidence. The transmittance of a parallel-sided slab is $0.96 \times 0.96 = 92$ percent. This simple result ignores the possibility of interference effects between reflected and transmitted waves at the two faces of the slab. If the faces of the slab are very flat and parallel, and if the light is coherent, such effects cannot be ignored. In this case, the method of transformed impedances is useful.

Consider the three-layer structure shown in Figure 4.25a. The path of a ray of light through the structure is shown. The angles θ_1, θ_2, θ_3 can be calculated from Snell's law. As an example, consider a wave polarized in

the plane of incidence. The effective impedances of media 1, 2, and 3 are

$$Z_1' = Z_1 \cos \theta_1 = \frac{Z_0 \cos \theta_1}{n_1} \qquad (4.106)$$

$$Z_2' = Z_2 \cos \theta_2 = \frac{Z_0 \cos \theta_2}{n_2}$$

$$Z_3' = Z_3 \cos \theta_3 = \frac{Z_0 \cos \theta_3}{n_3}$$

It can be shown that the reflection coefficient of the structure is exactly the same as for the equivalent structure in Figure 4.25b in normal incidence, where the effective thickness of layer 2 is now

$$d' = d \cos \theta_2 \qquad (4.107)$$

The reflection coefficient of the structure can be calculated from its equivalent structure using the transformed impedance concept.[25]

The transformed impedance of medium 3 at the boundary between media 1 and 2 is

$$Z''_3 = Z_2' \left(\frac{Z_3' \cos k_2 d' + i Z_2' \sin k_2 d'}{Z_2' \cos k_2 d' + i Z_3' \sin k_2 d'} \right) \qquad (4.108)$$

where $k_2 = 2\pi/\lambda_2 = 2\pi n_2/\lambda_0$. The reflection coefficient of the whole structure is now just

$$\rho = \frac{Z''_3 - Z_1'}{Z''_3 + Z_1'} \qquad (4.109)$$

the reflectance of the whole structure is $R = |\rho|^2$, and its transmittance is

$$T = 1 - R \qquad (4.110)$$

In a structure with more layers, the transformed impedance formula (Equation 4.108) can be used sequentially, starting at the last optical surface and working back to the first. More examples of the use of the technique of transformation of impedance will be given in Sections 4.3.2, Windows and Section 4.3.8, Filters.

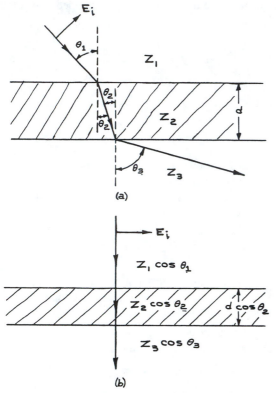

(a)

(b)

Figure 4.25 (a) Wave polarized in the plane of incidence passing through a dielectric slab of thickness d and impedance Z_2 separating two semiinfinite media of impedances Z_1 and Z_3, respectively (b) equivalent structure for normal incidence.

4.2.5 Gaussian Beams

Gaussian beams are propagating-wave solutions of Maxwell's equations that are restricted in lateral extent even in free space. They do not need any beam-confining reflective planes, as do the confined electromagnetic-field modes of waveguides.[25] The field components of a plane transverse electromagnetic wave of angular frequency ω propagating in the z-direction are of the form

$$V = V_0 e^{i(\omega t - kz)} \tag{4.111}$$

where V_0 is a constant. A Gaussian beam is of the form

$$V = \Psi(x, y, z) e^{i(\omega t - kz)} = U(x, y, z) e^{i\omega t} \tag{4.112}$$

where, for example, for a particular value of z, $\Psi(x,y,z)$ gives the spatial variation of the fields in the xy plane. $\overline{\Psi}^* \overline{\Psi}$ gives the relative intensity distribution in the plane. For a Gaussian beam, this gives a localized intensity pattern. The various Gaussian-beam solutions of Maxwell's equations are denoted as TEM_{mn} modes—a detailed discussion of their properties is given by Kogelnik and Li.[14] The output beam from a laser is generally a TEM mode or a combination of TEM modes. Many laser systems operate in the fundamental Gaussian mode, denoted TEM_{00}. This is the only mode that will be considered in detail here.

For the TEM_{00} mode, $\Psi(x,y,z)$ has the form

$$\Psi(x, y, z) = \exp\left\{-i\left[P(z) + \frac{kr^2}{2q(z)}\right]\right\} \tag{4.113}$$

where $P(z)$ is a *phase factor*, $q(z)$ is called the *beam parameter*, $k = 2\pi/\lambda$, and $r^2 = x^2 + y^2$. The beam parameter, $q(z)$, is usually written in terms of the phase-front *curvature* of the beam $R(z)$, and its *spot size*, $w(z)$, as

$$\frac{1}{q} = \frac{1}{R} - \frac{i\lambda}{\pi w^2} \tag{4.114}$$

and q and P obey the following relations:

$$\frac{dq}{dz} = 1 \tag{4.115}$$

$$\frac{dP}{dz} = -\frac{i}{q} \tag{4.116}$$

For a particular value of z, the intensity variation in the xy plane (the mode pattern) is, from Equation 4.113,

$$\Psi^* \Psi = \exp\left[\frac{-ikr^2}{2}\left(\frac{1}{q(z)} - \frac{1}{q^*(z)}\right)\right] \tag{4.117}$$

which from Equation 4.114 gives

$$\overline{\Psi^* \Psi} = e^{-2r^2/w^2} \tag{4.118}$$

Thus, $w(z)$ is the distance from the axis of the beam ($x = y = 0$) where the intensity has fallen to $1/e^2$ of its axial value and the fields to $1/e$ of their axial magnitude. The radial distribution of both intensity and field strength of the TEM_{00} mode is Gaussian, as shown in Figure 4.26.

Clearly, from Equation 4.115,

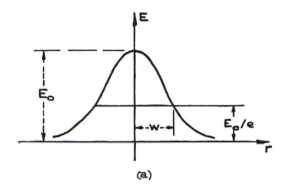

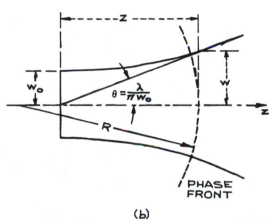

(a)

(b)

Figure 4.26   (a) Radial amplitude variation of the TEM_{00} Gaussian beam; (b) contour of a Gaussian beam.

$$q = q_0 + z \qquad (4.119)$$

where q_0 is the value of the beam parameter at $z = 0$. From Equation 4.116,

$$P(z) = -i \ln q + \text{constant} = -i \ln(q_0 + z) + \text{constant} \quad (4.120)$$

Writing the constant in Equation 4.120 as $\theta + i \ln q_0$, the full spatial variation of the Gaussian beam is

$$U = \exp\left\{-i\left[kz - i \ln\left(1 + \frac{z}{q_0}\right) + \theta + \frac{kr^2}{2}\left(\frac{1}{R(z)} - \frac{i\lambda}{\pi w^2}\right)\right]\right\} (4.121)$$

The phase angle θ is usually set equal to zero. In the plane $z = 0$,

$$U = \exp\left\{-i\frac{kr^2}{2}\left(\frac{1}{R(0)} - \frac{i\lambda}{\pi w(0)^2}\right)\right\} \qquad (4.122)$$

The surface of constant phase at this point is defined by the equation

$$\frac{kr^2}{2R(0)} = \text{constant} \qquad (4.123)$$

If the constant is taken to be zero and $R(0)$ infinite, then r becomes indeterminate and the surface of constant phase is the plane $z = 0$. In this case, $w(0)$ can have its minimum value anywhere. This minimum value is called the *minimum spot size* w_0, and the plane $z = 0$ is called the *beam waist*. At the beam waist,

$$\frac{1}{q_0} = \frac{-i\lambda}{\pi w_0^2} \qquad (4.124)$$

Using Equations 4.114 and 4.119, for any arbitrary value of z,

$$\omega^2(z) = \omega_0^2\left[1 + \left(\frac{\lambda z}{\pi w_0^2}\right)^2\right] \qquad (4.125)$$

The radius of curvature of the phase front at this point is

$$R(z) = z\left[1 + \left(\frac{\pi w_0^2}{\lambda z}\right)^2\right] \qquad (4.126)$$

From Equation 4.125 it can be seen that the Gaussian beam expands in both the positive and negative z-directions from its beam waist along a hyperbola that has asymptotes inclined to the axis at an angle

$$\theta_{\text{beam}} = \arctan \frac{\lambda}{\pi w_0} \qquad (4.127)$$

as illustrated in Figure 4.26b. The surfaces of constant phase of the Gaussian beam are in reality parabolic. For $r_2 < z$ (which is generally true, except close to the beam waist), however, they are spherical surfaces with the radius of curvature $R(z)$. Although they will not be discussed further here, the higher-order Gaussian beams—denoted TEM_{mn}—are also characterized by radii of curvature and spot sizes that are identical to those of the TEM_{00} mode and obey Equations 4.125 and 4.126.

A lens can be used to focus a laser beam to a small spot, or systems of lenses may be used to expand the beam and recollimate it (i.e., minimize the beam divergence). In such an application, a thin lens will not alter the transverse intensity pattern of the beam at the lens but it will alter its radius of curvature. Far enough from the beam waist, the radius of curvature of a Gaussian beam behaves exactly as a true spherical wave, since for

$$z \gg \frac{\pi w_0^2}{\lambda}$$

Equation 4.126 becomes

$$R(z) = z \qquad (4.128)$$

Now, when a spherical wave of radius R_1 strikes a thin lens, the object distance is clearly also R_1—the distance to the point of origin of the wave. The radius of curvature R_2 immediately after passage through the lens must therefore obey

$$\frac{1}{R_2} = \frac{1}{R_1} - \frac{1}{f} \qquad (4.129)$$

as shown in Figure 4.27. If w is unchanged at the lens, the beam parameter after passage through the lens therefore obeys

$$\frac{1}{q_2} = \frac{1}{q_1} - \frac{1}{f} \qquad (4.130)$$

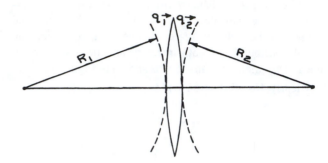

Figure 4.27   Transformation of a Gaussian beam by a thin lens.

It is straightforward to use this result in conjunction with Equations 4.114 and 4.119 to give the minimum spot size w_f of a TEM_{00} Gaussian beam focused by a lens as

$$w_f = \frac{f\lambda}{\pi w_1}\left[\left(1 - \frac{f}{R_1}\right)^2 + \left(\frac{\lambda f}{\pi w_1^2}\right)^2\right]^{1/2} \qquad (4.131)$$

where w_1 and R_1 are the laser-beam spot size and radius of curvature at the input face of the lens. If the lens is placed very close to the waist of the beam being focused, Equation 4.131 reduces to

$$w_f = f\theta_B \qquad (4.132)$$

where θ_B is the beam divergence at the input face of the lens. If θ_B is a small angle, then w_f is located almost at the focal point of the lens—in reality, very slightly closer to the lens. It is worthwhile comparing the result given by Equation 4.132 with the result obtained for a plane wave being focused by a lens. In the latter case, the lens diameter D is the factor that limits the lateral extent of the wave being focused. Diffraction theory[9,21] shows that, in this case, 84 percent of the energy is focused into a region of diameter

$$S = 2.44\lambda f/D \qquad (4.133)$$

It should be noted that the spot size to which a laser beam can be focused cannot be reduced without limit merely by reducing the focal length of the focusing lens. In the limit, the lens becomes a sphere and, of course, is then no longer a thin lens. In practice, to focus a laser beam to a small spot, the value of θ_B should be first reduced by expanding the beam and then recollimating it—generally with a Galilean telescope, as will be seen in Section 4.3.3, Laser-Beam Expanders and Spatial Filters. To prevent diffraction effects, the lens aperture should be larger than the spot size at the lens: $D = 2.8w$ is a common size used. In this case, if the focusing lens is placed at the beam waist of the collimated beam, the focal-spot diameter is

$$2w_f = \frac{5.6\lambda f}{\pi D} = \frac{1.78\lambda f}{D} \qquad (4.134)$$

In practice, it is difficult to manufacture spherical lenses with very small values of f/D (called the $f/\#$ or $f/number$) and at the same time achieve the diffraction-limited performance predicted by this equation. Commercially available spherical lenses[26] achieve values of $2w_f$ of about 10λ. Smaller spot sizes are possible with aspheric lenses.

4.3 OPTICAL COMPONENTS

4.3.1 Mirrors

When light passes from one medium to another of different refractive index there is some reflection, so the interface acts as a partially reflecting mirror. By applying an appropriate single-layer or multilayer coating to the interface between the two media, the reflection can be controlled so that the reflectance has any desired value between 0 and 1. Both flat and spherical mirrors made in this way are available—there are numerous suppliers.[27] If no transmitted light is required, high-reflectance mirrors can be made from metal-coated substrates or from metals themselves. Mirrors that reflect and transmit roughly equal amounts of incident light are often referred to as *beamsplitters*.

Flat Mirrors

Flat mirrors are used to deviate the path of light rays without any focusing. These mirrors can have their reflective surface on the front face of any suitable substrate, or on the rear face of a transparent substrate. Front-surface, totally reflecting mirrors have the advantage of producing no unwanted additional or *ghost* reflections; however, their reflective surface is exposed. Rear-surface mirrors produce ghost reflections, as illustrated in Figure 4.28, unless their front surface is antireflection-coated—but the reflective surface is protected. Most household mirrors are made this way. The cost of flat mirrors depends on their size and on the degree of flatness required.

Mirrors for high-precision applications—for example, in visible lasers—are normally specified to be flat between $\lambda/10$ and $\lambda/20$ for visible light. This degree of flatness is not required when the mirror is merely for light collection and redirection—for example to reflect light onto the surface of a detector. Mirrors for this sort of application are routinely flat to within a few wavelengths of visible light. Excellent mirrors for this purpose can be made from *float* plate glass,[28] which is flat to 1 or 2 wavelengths per inch.

Spherical Mirrors

Spherical mirrors are widely used in laser construction, where the radii of curvature are typically rather long—frequently 20 m radius or more. They can be used whenever light must be collected and focused; however, spherical mirrors are only good for focusing nearly parallel beams of light that strike the mirror close to normal. In common with spherical lenses, spherical

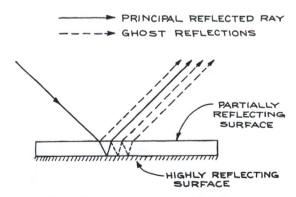

Figure 4.28   Production of ghost reflections by a rear-surface mirror whose front surface is not antireflection coated.

mirrors suffer from various imaging defects called aberrations, which have been discussed in Section 4.2.3. A parallel beam of light that strikes a spherical mirror far from normal is not focused to a small spot. When used this way, the mirror is *astigmatic*—that is, off-axis rays are focused at different distances in the horizontal and vertical planes. This leads to blurring of the image, or at best the focusing of a point source into a line image.

A useful, practical way to distinguish between a flat mirror and one with a large radius of curvature is to use the mirror to view a sharp object in grazing incidence. A curved mirror surface will reveal itself through the blurring of the image, while a flat mirror will give a sharp image.

Paraboloidal and Ellipsoidal Mirrors

Paraboloidal mirrors will produce a parallel beam of light when a point source is placed at the focus of the paraboloidal surface. These mirrors are therefore very useful in projection systems. They are usually made of polished metal, although versions using Pyrex substrates are also available. An important application of parabolic mirrors is in the off-axis focusing of laser beams, using off-axis mirrors in which the axis of the paraboloid does not pass through the mirror. When they are used in this way, there is complete access to the focal region without any shadowing, as shown in Figure 4.29. Spherical mirrors, on the other hand, do not focus well if they are used in this way. Off-axis paraboloidal mirrors, as well as metal axial paraboloidal mirrors, are available from Melles Griot, Space Optics Research Lab, Opti-Forms, Optics Plus and Oriel.[*]

Ellipsoidal mirrors are also used for light collecting. Light that passes through one focus of the ellipsoid will, after reflection, also pass through the other. These mirrors, like all-metal paraboloidal mirrors, are generally made by electroforming. The surface finish obtained in this way is quite good, although not as good as can be obtained by optical polishing. Rhodium-plated electroformed ellipsoidal mirrors are available from Hampton Controls, Melles Griot Opti-Forms, and Oriel.

---

[*] Here, and wherever suppliers are mentioned in the text, the list is intended to be representative but not necessarily exhaustive.

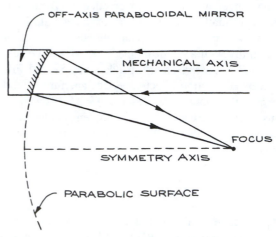

Figure 4.29   Off-axis paraboloidal mirror.

Dielectric Coatings

Several different kinds of single-layer and multilayer dielectric coatings are available commercially on the surface of optical components. These coatings can either reduce the reflectance or enhance it in some spectral region.

(i) Single-layer antireflection coatings. The mode of operation of a single-layer antireflection (AR) coating can be conveniently illustrated by the method of transformed impedances. Suppose medium 3 in Figure 4.25a is the surface to be AR coated. Apply a coating to the surface with an effective thickness

$$d' = \lambda_2/4 \tag{4.135}$$

where $\lambda_2 = c_0/v n_2$. The actual thickness of layer 2 is

$$d = d'/\cos\theta_2 \tag{4.136}$$

The transformed impedance of medium 3 at the first interface is found by substitution in Equation 4.108:

$$Z''_3 = Z'^2_2/Z'_3 \tag{4.137}$$

To reduce the reflection coefficient to zero, we need $Z''_3 = Z'_1$; so the effective impedance of the antireflection layer must be

$$Z'_2 = \sqrt{Z'_1 Z'_3} \qquad (4.138)$$

To eliminate reflection of waves linearly polarized in the plane of incidence and incident at angle θ_1, n_2 must therefore be chosen to satisfy

$$\frac{\cos \theta_2}{n_2} = \sqrt{\frac{\cos \theta_1 \cos \theta_3}{n_1 n_3}} \qquad (4.139)$$

For use in normal incidence, $n_2 = \sqrt{n_1 n_3}$ and $d = \lambda_2/4$. In normal incidence, an AR coating works for any incident polarization. To minimize reflection at an air-flint glass ($n = 1.7$) interface, we would need

$$n_2 = \sqrt{1.7} = 1.3 \qquad (4.140)$$

Magnesium fluoride with a refractive index of 1.38 and cryolite (sodium aluminum fluoride) with a refractive index of 1.36 come closest to meeting this requirement in the visible region of the spectrum. Optical components such as camera lenses are usually coated with one of these materials to minimize reflection at 550 nm. The slightly greater Fresnel reflection that results in the blue and red gives rise to the characteristic purple color of these components in reflected light, often called *blooming*.

(ii) Multilayer antireflection coatings. The minimum reflectance of a single-layer antireflection-coated substrate is $(n_2^2 - n_1 n_3)/(n_2^2 + n_1 n_3)^2$. Available robust optical coating materials such as MgF_2, however, do not have a sufficiently high refractive index to reduce the reflectance to zero with a single layer. Two-layer coatings, often called V-coatings because of the shape of their transmission characteristic, reduce reflection better than a single layer, as shown in Figure 4.30b. Multilayer coatings can be used to reduce reflectance over a broader wavelength region than a V-coating, as shown in Figure 4.31. Such broad-band coatings are usually of proprietary design—they are available commercially from many sources, as are coatings of other types.[27] Some companies offer both optical components and coatings, while

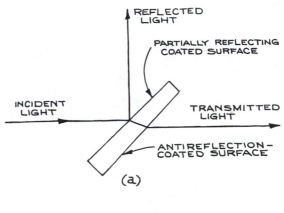

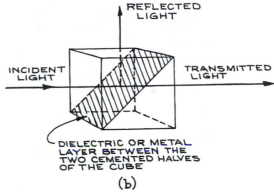

Figure 4.30 (a) Conventional planar beamsplitter (wedge angle greatly exaggerated); (b) cube beamsplitter.

others specialize in coatings and will coat customers' own materials.

(iii) High-reflectance coatings. The usually high reflectance of a metal surface can be further enhanced, as shown in Figure 4.32a, by a multiple dielectric coating consisting of an even number of layers—each a quarter wavelength thick—and of alternately high and low refractive index with a low-refractive-index material next to the metal. A very high reflectance over a narrow wavelength range can be achieved, as shown in Figure 4.32b, by using a dielectric substrate such as fused quartz, and an odd number of quarter-wavelength layers of alternately high and low refractive index with a high-index layer on the substrate. Broadband high-reflectance coatings

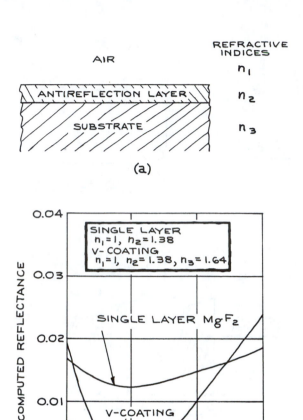

(a)

(b)

Figure 4.31 (a) Geometry of single-layer antireflection coating; (b) reflectances of single- and double-layer antireflection coatings on a substrate of refractive index 1.64. (From *Handbook of Lasers*, R. J. Pressley, ed., CRC Press, Cleveland, 1971; by permission of CRC Press, Inc.)

are also available where the thickness of the layers varies around an average value of a quarter wavelength.

Beamsplitters

Beamsplitters are semitransparent mirrors that both reflect and transmit light over a range of wavelengths. A good beamsplitter has a multilayer dielectric coating on a

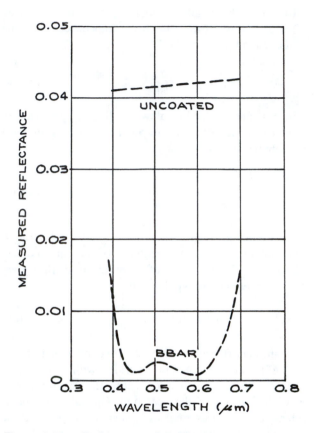

Figure 4.32 Reflectance of a typical broad-band antireflection coating (BBAR) showing considerable reduction in reflectance below the uncoated substrate. (From *Handbook of Lasers*, R. J. Pressley, ed., CRC Press, Cleveland, 1971; by permission of CRC Press, Inc.)

substrate that is slightly wedge-shaped to eliminate interference effects, and antireflection-coated on its back surface to minimize ghost images. The ratio of reflectance to transmittance of a beamsplitter depends on the polarization state of the light. The performance is usually specified for light linearly polarized in the plane of incidence (*P*-polarization) or orthogonal to the plane of incidence (*S*-polarization). Cube beamsplitters are pairs of identical right-angle prisms cemented together on their hypotenuse faces. Before cementing, a metal or dielectric semireflecting layer is placed on one of the hypotenuse faces. Antireflection-coated cube prisms have virtually no ghost image problems and are more rigid than plate type beamsplitters. The operation of both types of beamsplitter is illustrated in Figure 4.33.

Pellicles

Pellicles are beam-splitting mirrors made of a high-tensile-strength polymer stretched over a flat metal frame. The polymer film can be coated to modify the reflection-transmission characteristics of the film. The polymer generally used—nitrocellulose—transmits in the visible and near infrared to about 2 µm. These devices have some advantages over conventional coated-glass or quartz beamsplitters—the thinness of the polymer film virtually eliminates spherical and chromatic aberrations (see Section 4.3.3, Lens Aberrations) when diverging or converging light passes through them, and ghost-image problems are virtually eliminated. They will, however, produce some wavefront distortion—typically about 2 waves per in.—and are not suitable for precision applications. Pellicles are available from several suppliers,[27] such as Edmund Scientific, Newport, National Photocolor, and Melles Griot.

4.3.2 Windows

An optical window serves as a barrier between two media, for example, as an observation window on a vacuum system or on a liquid or gas cell, or as a Brewster window on a laser. When light passes through a window that separates two media, there is generally some reflection from the window and a change in the state of polarization of both the reflected and the transmitted light. If the window material is not perfectly transparent at the wavelength of interest, there is also absorption of light in the window, which at high light intensities will cause the window to heat. This will cause optical distortion of the transmitted wave, and at worst—in high-power laser applications—damage the window on its surface, internally, or both. If the two surfaces of a window are both very flat (roughly speaking, one wave per centimeter or better) and close to parallel, the window will also act as an *etalon* (see Section 4.3.8, Filters, and Section 4.7.4) and exhibit distinct variations in transmission with wavelength. To circumvent this difficulty, which is a

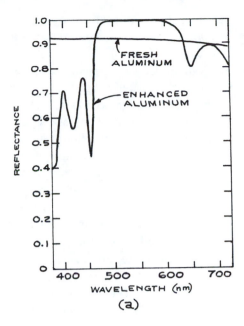

(a)

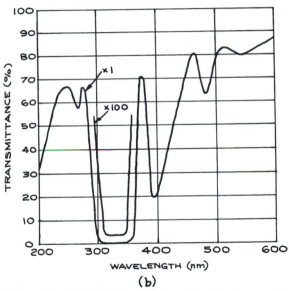

(b)

Figure 4.33   (a) Reflectance of a plain aluminum mirror and an aluminum mirror with four dielectric overlayers. (From *Handbook of Lasers*, R. J. Pressley, ed., CRC Press, Cleveland, 1971; by permission of CRC Press, Inc.); (b) typical transmittance characteristic of a narrow-band, maximum-reflectance, multilayer dielectric coating on a dielectric substrate.

particular nuisance in experiments using lasers, most precision flat optical windows are constructed as a slight wedge, with an angle usually about 30′.

The details of the reflection, refraction, and change of polarization state that occur when light strikes a window surface are dealt with in detail in Section 4.2.3 and Section 4.3.6, Polarizers. These considerations also apply when light strikes a lens or prism. The reflection at such surfaces can be reduced by a single-layer or multilayer dielectric antireflection coating. In optical systems such as multielement camera lenses, where light crosses many such surfaces, antireflection coatings are very desirable. Otherwise, a severe reduction in transmitted light intensity will result, as illustrated in Figure 4.34.

Optical cells, or cuvettes, are transparent containers—generally made of glass or fused quartz—that are used in absorption and fluorescence spectrophotometers, in light-scattering and turbidity measurements, and in many other specialized experiments. These cells are usually rectangular or cylindrical in shape and are available in many sizes and configurations, including provision for liquid circulation. Suppliers include NSG Precision Cells, Hellma, Oriel, and Esco.

4.3.3 Lenses and Lens Systems

The simplest lenses have two spherical surfaces and are referred to as *singlets*. There are 6 basic types (illustrated

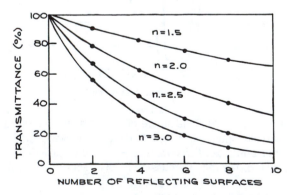

Figure 4.34 Transmittance of a multielement optical system as a function of the refractive index of the elements.

in Figure 4.35), all of which are available commercially from a number of companies such as Edmund Scientific, Melles Griot, Newport, Spindler and Hoyer, Rolyn Optics, and Opto-Sigma. Biconvex and biconcave types are frequently symmetrical. Singlet lenses are inexpensive but suffer from significant amounts of aberration, so are not a good choice in imaging situations. Compound lenses involve more than two spherical surfaces and can be designed to have reduced aberrations. Camera lenses are multi-element compound lenses especially designed for imaging applications. A wide variety of high-performance camera lenses is available commercially from companies such as Canon, Kodak, Leitz, Minolta, Nikon, and Zeiss. The choice of camera lenses for a specific application is generally guided by two parameters: (1) the *field of view* (FOV), which can be specified by the apex angle of the cone of light that can be collected by the lens, and (2) the aperture of the lens. Both these parameters influence the light gathering power of the lens.

Simple and Galilean Telescopes

The operation of a simple telescope is illustrated in Figure 4.36a. The objective is usually an *achromat* (a lens in which chromatic aberration—the variation of focal length with wavelength—has been minimized). It has a long focal length f_1 and produces a real image of a distant object, which can then be examined by the eyepiece lens of focal length f_2. This design, which produces an inverted image, is often called an *astronomical telescope*. If desired, the final image can be erected with a third lens, in which case the device is called a *terrestrial* telescope. The eyepiece of the telescope can be a singlet lens, but composite eyepiece designs such as the Ramsden and the Kellner are also common.[9] Most large astronomical telescopes use spherical mirrors as objectives, and various configurations are used such as the Newtonian, Cassegrain, and Schmidt.[7,17–19] Since most telescope development has been for astronomical applications, it is outside the scope of this book to give a detailed discussion. For further details the reader should consult Born and Wolf,[9] Levi,[22] Meinel,[29] or Brouwer and Walther.[30] Small astronomical and terrestrial telescopes are widely available at relatively low cost. Two principal manufacturers are Celestron[3] and Meade.[32]

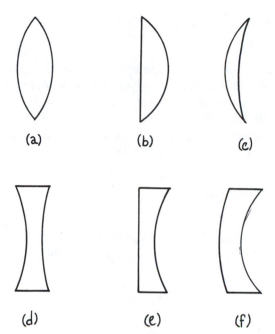

Figure 4.35 The six basic singlet lens types: (a) bi-convex; (b) plano-convex; (c) convex meniscus; (d) bi-concave; (e) plano-concave; (f) concave meniscus.

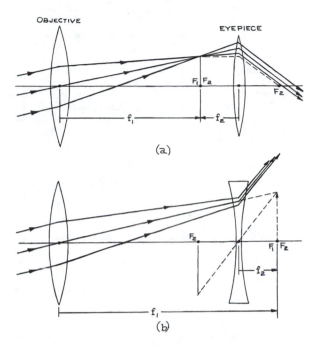

Figure 4.36 Ray paths through telescopes in *normal* adjustment (object and final image at infinity): (a) astronomical telescope; (b) Galilean telescope.

The Galilean telescope illustrated in Figure 4.36b, first constructed by Galileo in 1609, is the earliest telescope of which there exists definite knowledge. It produces no real intermediate image, but the final image is erect. In *normal* adjustment, the focal point of the diverging eyepiece is outside the telescope and coincident with that of the objective. In the simple telescope, the two focal points also coincide but are between the two. The magnification M produced by either type of telescope can be written as

$$M = -f_1/f_2 \qquad (4.141)$$

Since the simple telescope has two positive (converging) lenses, its overall magnification is negative, which indicates that the final image is inverted.

Simple and Galilean telescopes have practical uses in the laboratory that are distinct from their traditional use for observing distant objects.

Laser-Beam Expanders and Spatial Filters

Laser beams can be expanded and recollimated (or focused and recollimated) with simple or Galilean telescope arrangements, as illustrated in Figure 4.37. For optimum recollimation the spacing of the two lenses should be adjustable, as fine adjustment about the spacing $f_2 + f_1$ will be necessary for recollimating a Gaussian beam. In this application the Galilean telescope has the advantage that the laser beam is not brought to an intermediate focus inside the beam expander. Gas breakdown at such an internal focus can occur with high-power laser beams, although this problem can be solved in simple-telescope beam expanders by evacuating the telescope. Since laser beams are highly monochromatic, beam expanders need not be constructed from achromatic lenses. Attempts should be made, however, to minimize spherical aberration and beam distortion. It is best to use precision antireflection-coated lenses if possible. Laser-beam expanding telescopes that are very well corrected for

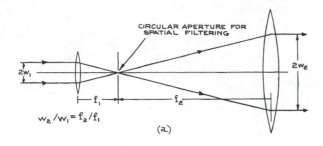

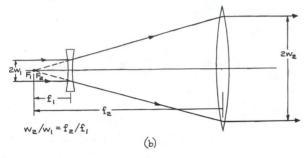

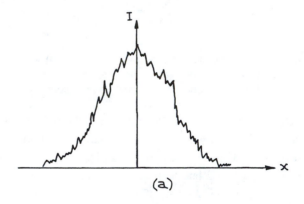

Figure 4.37 Laser-beam expanders; (a) focusing type with spatial filter; (b) Galilean type.

spherical aberration are available from Janos, Melles Griot, Oriel, and Special Optics, among others.[27] Infrared laser-beam expanders usually have ZnSe or germanium lenses. The use of biconvex and biconcave lenses distributes the focusing power of the lenses over their surfaces and minimizes spherical aberration. Better cancellation of spherical aberration is possible in Galilean beam expanders than with simple telescopes.

If a small circular aperture is placed at the common intermediate focus of a simple-telescope beam expander, the device becomes a *spatial filter* as well. Although the output beam from a laser emitting a TEM_{00} mode ideally has a Gaussian radial intensity profile, the radial profile may have some irregular structure in practice. Such beam irregularities may be produced in the laser or by passage of the beam through some medium. If such a beam is focused through a small enough aperture and then recollimated, the irregular structure on the radial intensity profile can be removed and a smooth profile restored (illustrated in Figure 4.38). This is called *spatial filtering*. The minimum aperture diameter that should be used is

$$D_{min} = \frac{2f_1 \lambda}{\pi w_1} \qquad (4.142)$$

where f_1 is the focal length of the focusing lens and w_1 is the spot size at that lens. Typical aperture sizes used in commercial spatial filters for visible lasers are on the order of 10 μm. This aperture must be accurately and symmetrically positioned at the focal point, so it is usually mounted on an adjustable XY translation stage.

Lens Aberrations

As discussed in Section 4.2.3, Exact Ray Tracing and Abberations, real lenses suffer from various forms of aberration, which lead the image of an object to be an

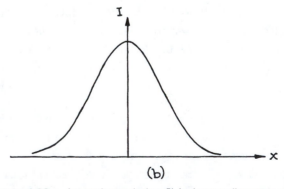

Figure 4.38 Intensity variation $I(x)$ along a diameter of a TEM_{00} laser beam possessing spatial noise structure: (a) before and (b) after spatial filtering.

imperfect reconstruction of it. If a particular aberration is judged to be detrimental in a particular experimental situation, then a special lens combination can often be obtained that minimizes that aberration—occasionally at the expense of worsening others.

(i) Chromatic aberrations. Because the refractive index of the material of a lens varies with wavelength, so does its focal length. Rays of different wavelengths from an object therefore form images in different locations. Chromatic aberration is almost eliminated with the use of an *achromatic doublet*. This is a pair of lenses, usually consisting of a positive crown-glass lens and a negative flint-glass lens, cemented together. These two lenses cancel each other's chromatic aberrations exactly at specific wavelengths in the blue and red, and almost exactly in the region in between. Good camera lenses are well corrected for chromatic aberrations.

Achromatic doublets are available from numerous suppliers, including Edmund Scientific, Klinger, Melles Griot, Oriel, and Rolyn.

(ii) Spherical aberration. This aberration can be minimized in a single lens by distributing the curvature between its two surfaces without altering the focal length. A biconvex lens, for example, will therefore produce less spherical aberration than a plano-convex one of the same focal length. Spherical aberration can be minimized in an achromatic doublet with appropriate choice of curvatures for the constituent lens elements.

(iii) Coma. When an object point is off the axis of a lens, its image is produced in different lateral positions by different zones of the lens. Coma can be controlled by an appropriate choice of lens curvatures. Other forms of aberration also occur, such as *field curvature*, in which an object plane orthogonal to the lens axis is imaged as a curved surface.

(iv) Astigmatism. This aberration can be controlled by lens curvature and index variations, and by the use of apertures to restrict the range of angles and off-axis distances at which rays can traverse the lens.

Aspheric Lenses

An aspheric lens has one nonspherical surface and a second surface that is either concave, convex, or plano (flat), as shown in Figure 4.39. With such a lens it is possible to obtain a much smaller focal length than with a conventional spherical lens of the same diameter, without increasing the spherical aberration of the lens. An ideal aspheric lens exactly cancels the spherical aberration that would otherwise be present in the optical system. Such lenses can collect and focus light rays over a much larger solid angle than conventional lenses and can be used at $f/\#$s as low as 0.6 (the $f/\#$, a frequently used lens parameter, is the ratio of the focal length to the available aperture diameter). Aspheric lenses save space and energy. They can be used as collection lenses very close to small optical sources such as high-pressure mercury or xenon arc lamps, and for collecting radiation over a large solid angle and focusing it onto a detector element. Such lenses are often referred to as *condensors*. Because aspheric lenses are generally designed only to minimize spherical aberrations, they will contribute to chromatic aberration, coma, distortion, astigmatism, and curvature of field. An aspheric lens should therefore not be used under circumstances where these aberrations will be compounded by transmission through additional components of the optical system. Glass aspheric lenses are available in a range of diameters and focal lengths from several suppliers[27]— aspheric lenses in more exotic materials can be obtained as custom items.

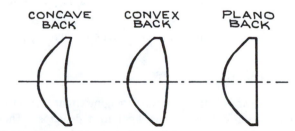

Figure 4.39   Cross sections of three common types of aspheric lens.

Fresnel Lenses

A Fresnel lens is an aspheric lens whose surface is broken up into many concentric annular rings. Each ring refracts incident rays to a common focus, so that a very large-aperture, small f/#, thin aspheric lens results. Figure 4.40 illustrates the construction and focusing characteristics of such a lens. These lenses are generally manufactured from precision molded Acrylic. Because the refractive index of Acrylic varies little with wavelength, from 1.51 at 410 nm to 1.488 at 700 nm, these lenses can be very free of chromatic and spherical aberration throughout the visible spectrum. Fresnel lenses should be used so that they focus to the plane side of the lens, and their surfaces should not be handled. They are inexpensive and available from very many suppliers, including Edmund Scientific, Fresnel Technologies, Lexitek, Oriel and Wavelength Optics. They should not be considered for use in applications where a precision image or diffraction-limited operation is required. Fresnel lenses also scatter much more light than conventional lenses.

Cylindrical Lenses

Cylindrical lenses are planar on one side and cylindrical on the other. In planes perpendicular to the cylinder axis they have focusing properties identical to a spherical lens, but they do not focus at all in planes containing the cylinder axis. Cylindrical lenses focus an extended source into a line, and so are very useful for imaging sources onto monochromator slits, although perfect matching of f/#s is not possible in this way. These lenses are also widely used for focusing the output of solid state and nitrogen lasers into a line image in the dye cell of dye lasers (see Section 4.6.3j). Cylindrical lenses are available from Esco, Bond Optics, Gould Precision Optics, Melles Griot, Rolyn, Optics for Research, Optics Plus, Oriel, and Infrared Optical Products, among various other companies.

Optical Concentrators

Optical concentrators are non-imaging light collectors that are useful in delivering light to a detector. They effectively increase the collection area of the detector, although they affect the angles of light rays.

If light enters an optical system of entrance aperture diameter $2a$ over a range of angles 2θ and leaves through an aperture of diameter $2a'$ over a range of angles $2\theta'$, then in the paraxial approximation

$$a\theta = a'\theta' \tag{4.143}$$

which follows from the brightness theorem. If the refractive indices of the input and output media are n, n', respectively, then this relation becomes

$$na\theta = n'a'\theta' \tag{4.144}$$

The quantity $n^2 a^2 \theta^2$, which is unchanged in going through the system, is called the *étendue* of the system. Figure 4.41 shows two simple concentrator designs, a linear conical concentrator, and a compound parabolic concentrator (CPC)—often called a *Winston* Cone. These devices are hollow or solid cones, which reflect entering rays one or more times before they leave the exit aperture. The range of light angles leaving such a device is greater than the range of light angle entering by the linear ratio between the entrance and exit apertures. Such a non-imaging light collector could be used to collect light from a diffusely-emitting source and direct it onto a small area photoelector. The detector's active surface, however, must be placed close to the exit aperture because of the larger range of angles of the emerging light rays.

Non-imaging light collectors are useful for collecting light whose spatial distribution may be fluctuating, yet whose overall power may remain relatively constant, as in the case of a laser beam that has passed through a fluctuating medium.

The concentration ratio is[33]

$$C = a/a' = \frac{n' \sin\theta'}{n \sin\theta} \tag{4.145}$$

In 2-D the maximum concentration ratio is

$$C_{max} = \left(\frac{n'}{n \sin\theta}\right) \tag{4.146}$$

and in 3-D for an axisymmetric concentrator

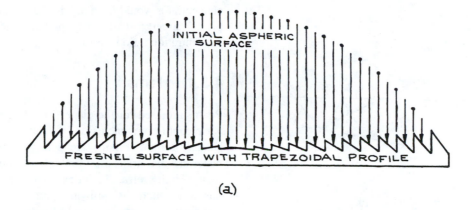

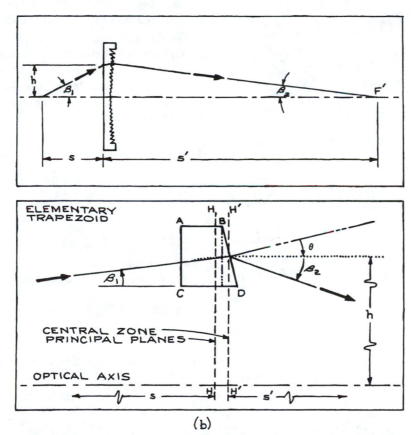

Figure 4.40   (a) A Fresnel lens, made up of trapezoidal concentric sections (it replaces a bulky aspheric lens); (b) illustration of how the focusing characteristics of a Fresnel lens result from refraction at the individual surface grooves. (Courtesy of Melles Griot, Inc.)

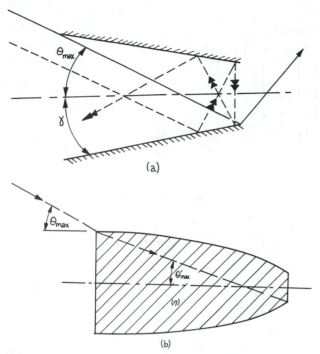

Figure 4.41 Simple optical concentrator designs: (a) linear cone; (b) compound parabolic concentrator.

$$C_{max} = \left(\frac{n'}{n\sin\theta}\right)^2 \qquad (4.147)$$

The CPC can be fabricated from a solid transparent material with refractive index n. For total internal reflection to occur for all internal rays that enter within a range of angles θ.

$$\sin\theta' \le 1 - \left(\frac{2}{n^2}\right) \qquad (4.148)$$

or

$$\sin\theta \le n - \frac{2}{n}$$

To be useful as a material for a CPC, it is clear that $n > \sqrt{2}$.

Since the rays leaving a concentrator can cover a solid angle up to $2\pi\,sr$, a detector used in conjunction with the concentrator would need to have its photosensitive surface placed directly at the exit aperture when the maximum concentration ratio is used.

4.3.4 Prisms

Prisms are used for the dispersion of light into its various wavelength components and for altering the direction of beams of light. Prisms are generally available in most materials that transmit ultraviolet, visible, and infrared light. High refractive-index semiconductor materials, however, such as silicon, germanium, and gallium arsenide, are rarely used for prisms. Consult the Manufacturers and Suppliers Section at the end of this chapter for a list of suppliers of optical materials.

The deviation and dispersion of a ray of light passing through a simple prism can be described with the aid of Figure 4.42. The various angles α, β, γ, δ satisfy Snell's Law irrespective of the polarization state of the input beam, provided the prism is made of an isotropic material. It is easy to show that

$$\sin\delta = \sin A(n^2 - \sin^2\alpha)^{1/2} - \cos A\sin\alpha \qquad (4.149)$$

and

$$D = \alpha + \delta - A \qquad (4.150)$$

The exit ray will not take the path shown if γ is greater than the critical angle.

The dispersion of the prism is defined as

$$\frac{d\delta}{d\lambda} = \frac{d\delta}{dn}\frac{dn}{d\lambda} \qquad (4.151)$$

Substituting from Equation 4.149, this becomes

$$d\delta = \frac{\sin A}{\cos\beta\cos\delta}\frac{dn}{d\lambda} \qquad (4.152)$$

When the prism is used in the position of minimum deviation,

$$\beta = \gamma = \tfrac{1}{2}A \qquad (4.153)$$

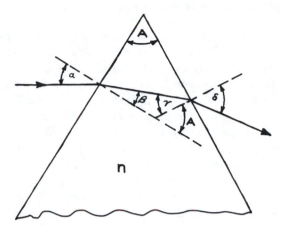

Figure 4.42   Passage of a ray of light through a prism.

and

$$\alpha = \delta = \tfrac{1}{2}(D_{\min} + A) \tag{4.154}$$

By the sine rule, Equation 4.152 can be written in the form

$$\frac{d\delta}{d\lambda} = \frac{t}{W}\frac{dn}{d\lambda} \tag{4.155}$$

where t is the distance traveled by the ray through the prism and W is the dimension shown in Figure 4.42.

A simple, yet important, application of a prism as a dispersing element is in the separation of a laser beam containing several well-spaced wavelengths into its constituent parts. Situations where this might be necessary include isolation of the 488.8 nm line in the output of an argon ion laser oscillating simultaneously at 476.5, 488.8, and 514.5 nm and other lines, or in nonlinear generation where a fundamental and second harmonic must be separated. If the spatial width of the laser beam is known, then an equation such as Equation 4.155 will determine at what distance from the prism the different wavelengths are spatially separated from each other. The advantages of prisms in this application are that they can introduce very little scattered light, produce no troublesome ghost images, and can be cut so that they operate with $\alpha = \delta = \theta_{\mathrm{B}}$, thereby eliminating reflection losses at their surfaces. Brewster-angle prisms are used in this way inside

the cavity of a laser to select oscillation at a particular frequency. A modified prism called a reflecting *Littrow prism* is frequently used in this way, as shown in Figure 4.43. A Littrow prism is designed so that for a particular wavelength the refracted ray, upon entering the prism, travels normally to the exit face. If the exit face is reflectively coated, the incident light is thus returned along its original path for the specified wavelength. Generally, a dispersing optical element used *in Littrow* has this retroreflective characteristic at a particular wavelength.

If a beam of parallel monochromatic light passes through a prism, unless the prism is used in the symmetrical minimum-deviation position, the width of the beam will be increased or decreased in one dimension. This effect is illustrated in Figure 4.44. By the use of two identical prisms in an inverted configuration, the compression or expansion, can be accomplished without an angular change in beam direction. Prism beam expanders employing this principle are used in some commercial dye lasers for expanding a small-diameter, intracavity laser beam onto a diffraction grating to achieve greater laser line narrowing. Such prism pairs can be used, in conjunction with lenses to produce *anamorphic* laser beam expanders. These devices are important for the circularization and collimation of the beams from semiconductor lasers. These lasers generally emit an elliptical shaped laser beam

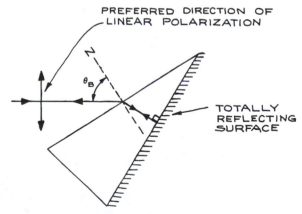

Figure 4.43   Reflecting prism used in Littrow at Brewster's angle. Only light of a specific wavelength will be refracted at the entrance face of the prism and then reflected back on itself at the coated reflecting face.

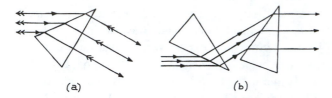

Figure 4.44   (a) One-dimensional beam expansion (or compression) with a prism; (b) one-dimensional beam expansion without beam deflection using two oppositely oriented, similar prisms.

whose beam divergence angle is different in two orthogonal planes containing the beam axis.

   Triangular prisms with a wide range of shapes and vertex angles are available. The most common types are the equilateral prism, the right-angle prism, and isosceles and Littrow prisms designed for Brewster-angle operation. Such prisms are readily available in glass and fused silica from Oriel and Melles Griot, among others.[27] Triangular and other prisms also find wide use as reflectors, utilizing total internal reflection inside the prism or a reflective coating on appropriate faces. Some specific reflective applications of prisms are worthy of special note.

Right-Angle Prisms

Right-angle prisms can be used to deviate a beam of light through 90° as shown in Figure 4.45a. Antireflection coatings on the faces shown are desirable in exacting applications. A right-angle prism can also be used to reflect a beam back parallel to its original path, as shown in Figure 4.45b. The retroreflected beam is shifted laterally. The prism will operate in this way provided the incident beam is in the plane of the prism cross section. The retroreflection effect is thus independent of rotation about the axis A in Figure 4.45b. If the angle of incidence differs significantly from zero, the roof faces of the prism need to be coated; otherwise total internal reflection at these surfaces may not result. If such a prism is rotated about the axis B, only in a single orientation does retroreflection result. Roof prisms are available from numerous suppliers, including Edmund Scientific, Rolyn, Oriel, Melles Griot, Karl Lambrecht (KLC), Opto-Sigma, and Ealing.

   Porro prisms are roof prisms with rounded corners on the hypotenuse face and beveled edges. They are widely used in pairs for image erection in telescopes and binoculars, as shown in Figure 4.46a. Amici prisms are right-angle prisms where the hypotenuse face has been replaced by a 90° roof. An image viewed through an Amici prism is left-to-right reversed (reverted) and top-to-bottom inverted as show in Figure 4.46b.

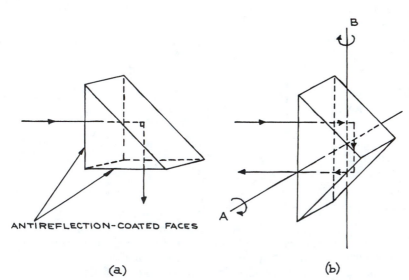

ANTIREFLECTION-COATED FACES

(a)   (b)

Figure 4.45   (a) Right-angle prism used for 90° beam deflection; (b) right-angle prism used for retroreflection. Incident and reflected rays are parallel only if the incident beam is in the plane of prism cross section; however, in this orientation, retroreflection is independent of orientation about the axis *A* within a large angular range.

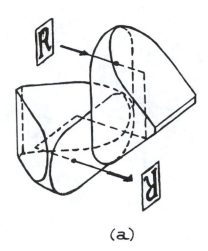

(a)

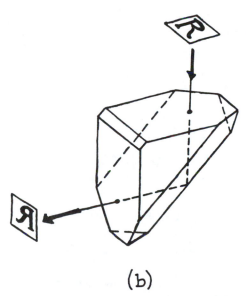

(b)

Figure 4.46 (a) Two Porro prisms used as an image-erecting element; (b) Amici prism. (Courtesy of Melles Griot, Inc.)

Dove, Penta, Polygon, Rhomboid, and Wedge Prisms

These prisms are illustrated in Figure 4.47.

Dove prisms are used to rotate the image in optical systems; rotation of the prism at a given angular rate causes the image to rotate at twice this rate. Dove prisms have a length-to-aperture ratio of about 5, so they must be used with reasonably well-collimated light. They are available from Esco, Melles Griot, Rolyn, KLC, and Edmund Scientific.

Penta prisms deviate a ray of light by 90° without inversion (turning upside down) or reversion (right-left reversal) of the image. This 90° deviation applies to all rays incident on the useful aperture of the prism, irrespective of their angle of incidence. Penta prisms do not operate by total internal reflection, so their reflective surfaces are coated. These prisms are useful for 90° deviation when vibrations or other effects prevent their alignment from being well controlled. They are available from Melles Griot, Rolyn, and KLC.

Polygon prisms, usually octagonal, are available from Rolyn and are used for high-speed, light-ray deviation—for example, in high-speed rotary prism cameras.

Rhomboid prisms are used for lateral deviation of a light ray. When used in imaging applications, there is no change of orientation of the image. Rhomboid prisms are available from Rolyn and Edmund Scientific.

Wedge prisms are used in beam steering. The minimum angular deviation D of a ray passing through a thin wedge prism of apex angle A is

$$D = (\arcsin(n\sin A) - A \cong (n-1)A) \qquad (4.156)$$

where n is the refractive index of the prism material. A combination of two identical wedge prisms can be used to steer a light ray in any direction lying within a cone of semivertical angle $2D$ about the original ray direction. Wedge prisms are available from Melles Griot and Rolyn.

Corner-Cube Prisms (Retroreflectors)

Corner-cube prisms are exactly what their name implies—prisms with the shape of a corner of a cube, cut off orthogonal to one of its triad (body-diagonal) axes. The front face of the resultant prism is usually polished into a circle, as shown in Figure 4.48. As a result of three total internal reflections, these prisms reflect an incident light ray back parallel to its original direction, no matter what

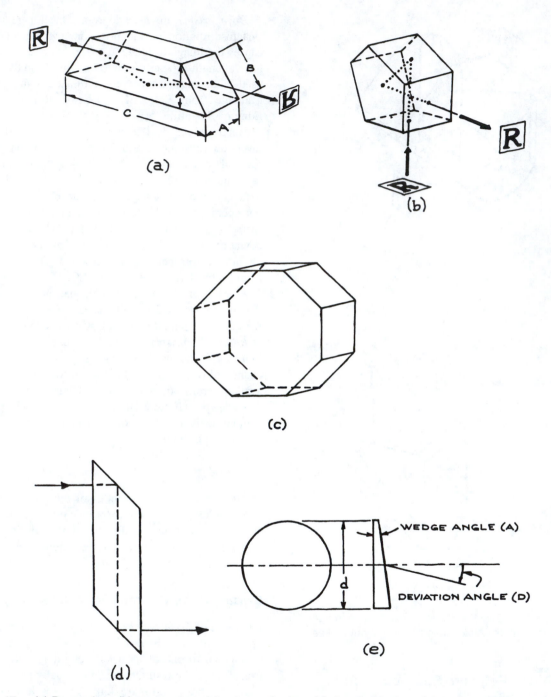

Figure 4.47   (a) Dove prism; (b) penta prism; (c) octagonal prism; (d) rhomboid prism; (e) wedge prism (a, b, and e courtesy of Melles Griot, Inc.).

the angle of incidence is. The reflected ray is shifted laterally by an amount that depends on the angle of incidence and the point of entry of the incident ray on the front surface of the prisms. These prisms are invaluable in experiments where a light beam must be reflected back to its point of origin from some (usually distant) point—in long-path absorption measurements through the atmosphere with a laser beam, for example. They are available commercially from Edmund Scientific, Melles Griot, Newport, Precision Opticals, Oriel, Rolyn, and Zygo. The angular deviation of the reflected ray from its original direction can be held to less than a half second of arc in the best prisms. For infrared applications beyond the transmission of fused silica, corner-cube prisms are not readily available, but retroreflectors can be made by using three square mirrors butted together to form a hollow cube corner. Precision hollow corner cubes of this type are available from PLX.

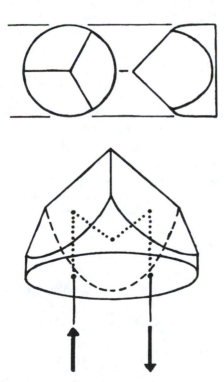

Figure 4.48   Corner-cube prism. (Courtesy of Melles Griot, Inc.)

4.3.5 Diffraction Gratings

A *diffraction grating* is a planar or curved optical surface covered with straight, parallel, equally spaced grooves. They are manufactured in three principal ways: by direct ruling on a substrate (called a *master* grating), by replication from a master grating, and holographically. Such gratings can be made for use in transmission, but are more commonly used in reflection. Light incident on the grooved face of a reflection grating is diffracted by the grooves—the angular intensity distribution of the diffracted light depends on the wavelength and angle of incidence of the incident light and on the spacing of the grooves. This can be illustrated with reference to the plane grating shown in Figure 4.49. The grooves are usually produced in an aluminum or gold layer that has been evaporated onto a flat optical substrate. For high-intensity laser applications the substrate should have high thermal conductivity. Master gratings ruled on metal are available for this purpose from Diffraction Products, Gentec, Jobin-Yvon, Richardson Grating Laboratory, and Rochester Photonics. If the light diffracted at angle β from a given groove differs in phase from light diffracted from the adjacent groove by an integral multiple m of 2π, a maximum in diffracted intensity will be observed. This condition can be expressed as

$$m\lambda = d(\sin\alpha \pm \sin\beta) \qquad (4.157)$$

The plus sign applies if the incident and the diffracted ray lie on the same side of the normal to the grating surface—m is called the order of the diffraction. For $m = 0$, $\alpha = \beta$ and the grating acts like a mirror. In any other order, the diffraction maxima of different wavelengths lie at different angles. The actual distribution of diffracted intensity among the various orders depends on the profile of the grating grooves. If the grooves are planar and cut at an angle θ to the plane of the grating, maximum diffracted intensity into a particular order results if the angle of diffraction and the angle of reflection are the same. In this case,

$$\beta - \theta = \gamma \qquad (4.158)$$

θ is called the *blaze* angle. The wavelength for which the angle of reflection from the groove face and the

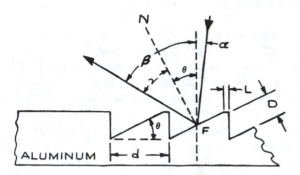

Figure 4.49 Blazed groove profile. The thickness of the aluminum is generally equal to the blaze wavelength, and the depth of the groove is about half the thickness. The dimensions shown are approximately those for a grating ruled with 1200 grooves per millimeter, blazed at 750 nm. D = depth of ruling; d = groove spacing; L = unruled land; α = angle of incidence; β = angle of diffraction; γ = angle of reflection; θ = blaze angle. The line N is normal to the groove face F. (From D. Richardson, "Diffraction Gratings," in *Applied Optics and Optical Engineering*, Vol. 5, R. Kingslake, ed., Academic Press, New York, 1969; by permission of Academic Press.)

diffraction angle are the same is called the *blaze* wavelength λ_B. It is common to use a grating in a Littrow configuration so that the diffracted ray lies in the same direction as the incident ray. In this case,

$$m\lambda = 2d\sin\beta \qquad (4.159)$$

The wavelength at which the light reflects normally from the grating groove satisfies

$$m\lambda_B = 2d\sin\theta \qquad (4.160)$$

A grating could therefore be specified to be blazed at 600 nm in the first order, for example—it would, of course, be simultaneously blazed in the second order for 300 nm, the third order for 200 nm, and so on. Gratings are also available with rectangular shaped (laminar profile) or sinusoidal groove profiles. Laminar profile grooves have low second order efficiency, which can be advantageous in vacuum ultraviolet applications. Second order rejection filters are not available at such short wavelengths. Sinusoidal gratings will operate over a broad spectral range, but their diffraction efficiency into a particular order is 33 percent at best. There are many suppliers of both plane and concave gratings, particularly Diffraction Products, Digital Optics, Jobin-Yvon, Optometrics, Oriel, Spectrogon, and TVC Jarrell-Ash.

Resolving Power

The *resolving power* of a grating is a measure of its ability to separate two closely spaced wavelengths. The resolving power depends both on the dispersion and the size of the grating. For a fixed angle of incidence, the *dispersion* is defined as

$$\left(\frac{d\beta}{d\lambda}\right)_\alpha = \frac{m}{d\cos\beta} \qquad (4.161)$$

The dispersion can be increased by increasing the number of lines per millimeter ($1/d$), by operating in a high order, and by using a large angle of incidence (grazing incidence). The incidence, however, must not be so flat as to make the projection of the groove width perpendicular to the incoming light smaller than the wavelength of the light.

The resolving power of a grating is $\Delta\lambda/\lambda$, where λ and $\lambda + \Delta\lambda$ are two closely spaced wavelengths that are just resolved by the grating. The limiting resolution depends on the projected width of the grating perpendicular to the diffracted beam. This width is $Nd\cos\beta$, where N is the total number of lines in the grating. Diffraction theory predicts that the angular resolution of an aperture of this size is $\Delta\phi \cong \lambda/Nd\cos\beta$. The angle between the two closely spaced wavelengths (from the dispersion relation) is

$$\Delta\beta = \frac{m\Delta\lambda}{d\cos\beta} \qquad (4.162)$$

In the limit of resolution, $\Delta\phi = \Delta\beta$, which gives

$$\lambda/\Delta\lambda = mN \qquad (4.163)$$

Thus, from Equation 4.157

$$\frac{\lambda}{\Delta\lambda} = \frac{Nd(\sin\alpha \pm \sin\beta)}{\lambda} \qquad (4.164)$$

and so it is clear that large gratings used at high angles give the highest resolving power. The resolving powers available from 1 cm wide gratings range up to about 5×10^5.

Concave Gratings

Concave gratings have long been used in short wavelength applications because they provide diffractive and focussing functions simultaneously. The development of holographic concave gratings has made these devices widely available. They can be used to correct for aberrations, which occur when they are used in spectrometers (see Section 4.7), and to allow compact versions of such instruments to be produced.

Efficiency

The efficiency of a grating is a measure of its ability to diffract a given wavelength into a particular order of the diffraction pattern. Blazing of the grating is the main means for obtaining efficiency in a particular order; without it, the diffracted energy is distributed over many orders. Gratings are available with efficiencies of 95 percent or more at their blaze wavelength, so they are virtually as efficient as a mirror, yet retain wavelength selectivity.

Defects in Diffraction Gratings

In principle, if an ideal grating is illuminated with a plane monochromatic beam of light, diffracted maxima occur only at angles that satisfy the diffraction equation (Equation 4.156). In practice, however, this is not so. Gratings manufactured by ruling a metal-coated substrate with a ruling engine exhibit undesirable additional maxima.

Periodic errors in the spacing of the ruled grooves produce *Rowland ghosts*. These are spurious intensity maxima, usually symmetrically placed with respect to an expected maximum and usually lying close to it. The strongest Rowland ghosts from a modern ruled grating will be less than 0.1 percent of the expected diffraction peaks. *Lyman ghosts* occur at large angular separations from their parent maximum, usually at positions corresponding to a simple fraction of the wavelength of the parent maximum—$\frac{4}{9}$ or $\frac{5}{9}$, for example. They are also associated with slow periodic errors in the ruling process. Lyman-ghost intensities from modern gratings are exceedingly weak (0.001% of the parent or less). Ghosts can be a problem when used for the detection of weak emissions in the presence of a strong laser signal, the ghosts of which can (and have been) mistakenly identified as other real spectral lines. If there is any suspicion of this, the weak signal should be checked for its degree of correlation with the laser signal—a linear correlation would be strong evidence for a ghost. The development of unruled holographic gratings, which are essentially perfect, plus the improvement of ruled gratings made by interferometrically controlled ruling engines, has considerably reduced the problem of ghosts.

There are other diffraction-grating defects. *Satellites* are misplaced spectral lines, which can be numerous, occurring very close to the parent, usually so close that they can only be discerned under conditions of high resolution—they arise from small local variations in groove spacing. *Scattering* gives rise to an apparent weak continuum over all diffraction angles when a grating is illuminated with an intense monochromatic source such as a laser. Scattering can arise from microscopic dust particles or nongroovelike, random defects on the diffraction-grating surface. In practice, it does not follow that a diffraction grating that shows apparent blemishes, such as broad, shaded bands, will perform defectively. One should never attempt to remove such apparent visual blemishes by cleaning or polishing the grating.

Specialized Diffraction Gratings

Concave gratings are frequently used in vacuum-ultraviolet spectrometers and spectrographs, as they combine the functions of both dispersing element and focusing optics. Thus, fewer reflective surfaces are required—a highly desirable feature of an instrument used at short wavelengths, where reflectances of all materials decrease markedly.

Gratings for use at long wavelengths (above 50 μm) use relatively few grooves per millimeter. Such gratings are generally ruled directly on metal and are frequently called *echelettes* (little ladders) because of the shape of their grooves.

Echelle gratings are special gratings designed to give very high dispersion, up to 10^6 for ultraviolet wavelengths, when operated in a very high order. They have the surface profile illustrated in Figure 4.50, where the dimension D is much larger than in a conventional grating, and can range up to several micrometers. Echelles are widely used as the wavelength tuning element in pulsed dye lasers because of their high dispersion and efficient operation. Other uses of diffraction gratings, such as in the production of Moiré fringes and interferometers, will not be discussed here. These subjects have been dealt with by Girard and Jacquinot.[34]

Practical Considerations in Using Diffraction Gratings

The main considerations in specifying a diffraction grating are the wavelength region where maximum efficiency is required (which will specify the blaze angle) and the number of grooves per millimeter and the size of the grating (which will determine the resolving power). Most gratings are rectangular or circular, the latter being primarily for laser-cavity wavelength selection. Diffraction gratings should be mounted with the care due all precision components. Their surfaces should *never* be touched and should be protected from dust. It is important to note that if a grating is illuminated with a given wavelength, say 300 nm, then diffraction maxima will occur in the same positions as would be found for 600 and 900 nm. This difficulty can be avoided by using appropriate filters to prevent unwanted wavelengths from passing through the system.

4.3.6 Polarizers

Polarized Light

If the electric vector of an electromagnetic wave always points in the same direction as the wave propagates through a medium, then the wave is said to be *linearly polarized*. The *direction of linear polarization* is defined as the direction of the electric displacement vector **D**, where

$$D = \varepsilon_r \varepsilon_0 E \tag{4.165}$$

Except in anisotropic media, where ε_r is a tensor, **D** and **E** are parallel and the direction of linear polarization can be taken as the direction of **E**. If a combination of two linearly polarized plane waves of the same frequency but having different phases, magnitudes, and polarization directions is propagating in the z-direction, the resultant light is said to be *elliptically polarized*. Such a pair of waves, in general, have resultant electric fields in the x- and y-directions that can be written as

$$E_x = E_1 \cos wt \tag{4.166}$$

$$E_y = E_2 \cos(\omega t + \phi) \tag{4.167}$$

where ϕ is the phase difference between these two resultant field components. These are the parametric equations of an ellipse. If $\phi = \pm \pi/2$ and $E_1 = E_2 = E_0$, then

$$E_x^2 + E_y^2 = E_0^2 \tag{4.168}$$

which is the equation of a circle. This represents *circularly polarized* light. Then the instantaneous angle that the total electric field vector makes with the x-axis, as illustrated in Figure 4.51, is

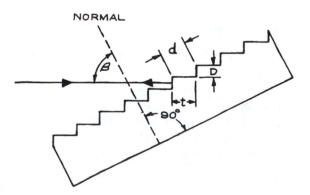

Figure 4.50   Echelle grating used in Littrow.

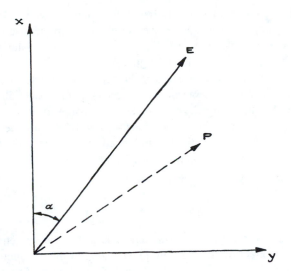

Figure 4.51 Instantaneous direction of the electric vector of an electromagnetic wave.

$$\alpha = \arctan\frac{E_y}{E_x} = \arctan(\mp\tan wt) = \mp wt \qquad (4.169)$$

For $\phi = \pi/2$, the resultant electric vector rotates counterclockwise viewed in the direction of propagation—this is *right-hand circularly polarized light*. If $\phi = -\pi/2$, the rotation is clockwise—this is *left-hand circularly polarized light*. The motion of the total electric vector as it propagates is illustrated in Figure 4.52. If $\phi = 0$ or π, we have *linearly polarized light*. Just as circularly polarized light can be viewed as a superposition of two linearly polarized waves with orthogonal polarizations, a linearly polarized wave can be regarded as a superposition of left- and right-hand circularly polarized waves. If an electromagnetic wave consists of a superposition of many independent linearly polarized waves of independent phase, amplitude, and polarization direction, it is said to be *unpolarized*.

In anisotropic media—media with lower than cubic symmetry—there is at least one direction, and at most two directions, along which light can propagate with no change in its state of polarization independent of its state of polarization. This direction is called the *optic axis*. *Uniaxial* crystals have one such axis, and *biaxial* crystals have two. When a wave does not propagate along the optic

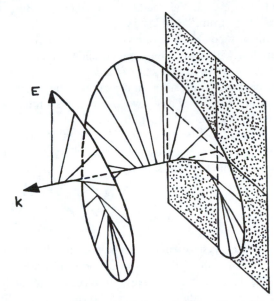

Figure 4.52 Left-hand circularly polarized light. (From E. Wahlstrom, *Optical Crystallography*, 3rd ed., Wiley, New York, 1960; by permission of John Wiley and Sons, Inc.)

axis in such crystals, it is split into two polarized components with orthogonal linear polarizations. In uniaxial crystals, these two components are called the *ordinary* and *extraordinary* waves. They travel with different phase velocities characterized by two different refractive indices—n_o and $n_e(\theta)$, respectively—where θ is the angle between the **k** vector of the wave and the optic axis. This phenomenon is referred to as *birefringence*.

D and **E** are not necessarily parallel in an anisotropic medium—**D** and **H**, however, are orthogonal to the wave vector of any propagating wave. Consequently, the Poynting vector of the wave does not generally lie in the same direction as the wave vector. The direction of the Poynting vector is the direction of energy flow—the ray direction. When a plane wave crosses the boundary between an isotropic and an anisotropic medium, the path of the ray will not, in general, satisfy Snell's law. The angles of refraction of the ordinary and extraordinary rays will be different. This phenomenon is called *double refraction*. In uniaxial crystals, however, the ordinary ray direction at a boundary does satisfy Snell's Law. For further details of these and other optical characteristics of anisotropic media, the reader should consult Born and Wolf,[9] Wahlstrom,[10] and Davis.[13]

If linearly polarized light propagates into a birefringent material, it will be split into two components polarized in these two allowed directions unless its polarization direction matches the allowed direction of the ordinary or extraordinary wave. These two components propagate at different velocities and experience different phase changes on passing through the crystal. For light of free-space wavelength λ_0 passing through a uniaxial crystal of length L, the birefringent phase shift is

$$\Delta\phi = \frac{2\pi L}{\lambda_0}[n_e(\theta) - n_0] \qquad (4.170)$$

In uniaxial crystals the allowed polarization directions are perpendicular to the optic axis (for the ordinary wave) and in the plane containing the propagation direction and the optic axis (for the extraordinary wave), as shown in Figure 4.53. If the input wave is polarized at an angle β to the ordinary polarization direction and $\Delta\phi = (2n+1)\pi$, the output wave will remain linearly polarized, but its direction of polarization will have been rotated by an angle of 2β. This rotation is always toward the ordinary polarization direction, so a roundtrip pass through the material does not affect the polarization state. It is usual to make $\beta = 45°$ in which case the crystal rotates the plane of polarization by 90°. Such a device is called a $(2n+1)$th-order *half-wave plate*. On the other hand, if $\beta = 45°$ and $\Delta\phi = (2n+1)\pi/2$, the ordinary and extraordinary waves recombine to form circularly polarized light. Such a device is called a $(2n+1)$th-order *quarter-wave plate*. If β is not 45°, a quarter-wave plate will convert linearly polarized light into elliptically polarized light.

Polarization Changes on Reflection

As shown in Section 4.2.4, if a linearly polarized wave reflects from a dielectric surface (or mirror), its state of linear polarization may be changed. For example, if a vertically polarized wave traveling horizontally striking a mirror at 45° is polarized in the plane of incidence, then after reflection it will be traveling vertically and will be horizontally polarized. A reflection off a second mirror at 45°—so oriented that the plane of incidence is perpendicular to the polarization—will produce a

horizontally traveling, horizontally polarized wave. This exemplifies how successive reflections can be used to rotate the plane of polarization of a linearly polarized wave, and is illustrated in Figure 4.54.

If circular or elliptically polarized light reflects from a dielectric surface or mirror, its polarization state will in general be changed. As an example, consider reflection at a metal mirror. From Equations 4.84 and 4.86, for the in-plane and perpendicular polarized components of the incident wave, respectively, we have

$$\rho_{\parallel} = -1 \qquad (4.171)$$
$$\rho_{\perp} = -1$$

These reflection coefficients ensure that the tangential component of electric field goes to zero at the conducting surface of the mirror. As illustrated in Figure 4.55a and 4.55b, this leads to a conversion of left- to right-hand circularly polarized light on reflection. The change of polarization state on reflection from a dielectric interface depends on the angle of incidence and whether $n_2 > n_1$. A special case illustrated in Figure 4.55c shows left-hand circularly polarized light being converted to linearly polarized light on reflection from an interface placed at Brewster's angle. Such changes of polarization state must be taken into account when designing an optical system to work with polarized light. The change in polarization of a beam of light as it passes through a series of optical components can be calculated very conveniently by the use of Jones or Mueller calculus, which use matrix methods to describe the state of polarization of the beam and its interaction with each component. Details of these techniques are given by Shurcliff.[11]

Linear Polarizers

A linear polarizer changes unpolarized light to linearly polarized light or changes polarized light to a desired linear polarization.

The simplest linear polarizers are made from dichroic materials—materials that transmit one polarization, either ordinary or extraordinary, and strongly absorb the other. Modern dichroic linear polarizers, the commonest of which is Polaroid, are made of polymer films in which long-chain molecules with appropriate absorbing side groups are

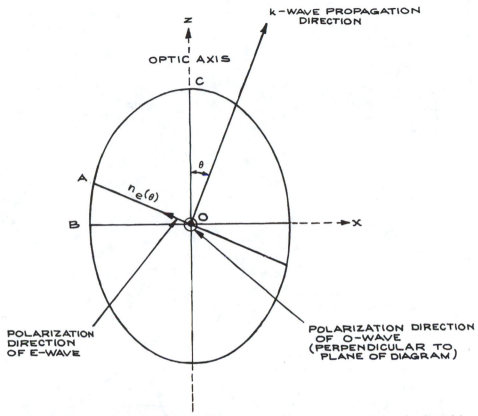

Figure 4.53 Cross section of an ellipsoidal figure called the indicatrix, which allows determination of the refractive indices and permitted polarization directions of a wave traveling in a uniaxial crystal. The equation of the ellipse shown is $x^2/n_0^2 + z^2/n_e^2 = 1$, where $OB = n_0$ and $OC = n_e$, $OA = n_e(\theta)$ is the effective extraordinary refractive index for a wave traveling at angle θ to the optic axis. The ordinary refractive index for this wave is still n_0 and is independent of θ.

oriented by stretching. The stretched film is then sandwiched between glass or plastic sheets. These polarizers are inexpensive, but cannot be used to transmit high intensities because they absorb all polarizations to some degree. They cannot be fabricated to very high optical quality, and generally only transmit about 50 percent of light already linearly polarized for maximum transmission. They do work, however, when the incident light strikes them at any angle up to grazing incidence. The extinction coefficient K that can be achieved with crossed polarizers is defined as

$$K = \log_{10}\frac{T_0}{T_{90}} \qquad (4.172)$$

where T_0 and T_{90} are the transmittances of parallel and crossed polarizers, respectively. K is a useful quantity for specifying a linear polarizer—the larger its value, the better the polarizer. For dichroic polarizers, values of K up to about 10^4 can be obtained. Polarizing beam-splitting cubes provide linearly polarized light of relatively high purity (98% or better) and can handle higher light intensities. Corning manufactures a sheet polarizer called Polacor made from glass containing billions of tiny silver crystals. This material provides high throughput in the red to near-infrared region, and extinction up to 10^4.

Higher quality, but more expensive, linear polarizers can be made by using the phenomenon of double refraction in transparent birefringent materials such as

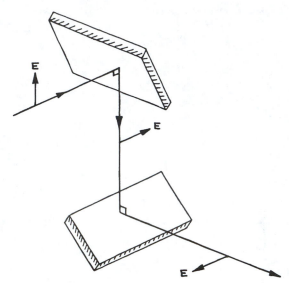

Figure 4.54 Rotation of the plane of polarization of a linearly polarized beam by successive reflection at two mirrors.

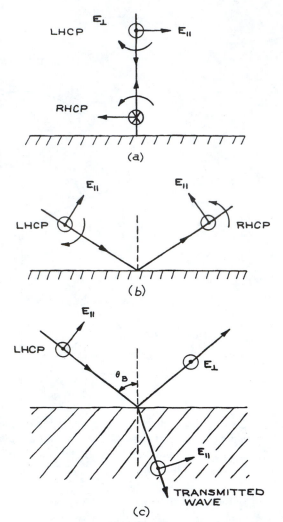

Figure 4.55 Changes of polarization state on reflection: (a) and (b) conversion of left-hand circularly polarized light (LHCP) to right-hand circularly polarized light (RHCP) on reflection at a perfect metal mirror; (c) conversion of LHCP to linear polarization on reflection at Brewster's angle.

calcite, crystalline quartz, or magnesium fluoride. Figure 4.56 schematically shows the way in which various polarizing prisms of this type operate.

In an Ahrens polarizer (a) three calcite prisms are cemented together with Canada balsam, whose refractive index is intermediate between n_0 and n_e for calcite. The O-ray suffers total internal reflection at the cement and is absorbed in the black coating on the sides. In a Glan-Foucault polarizer (b) two calcite prisms are separated with an air gap. The O-ray is totally internally reflected and either absorbed in the sides or transmitted if the sides are polished. A Glan-Thompson polarizer is similar. A Glan-Taylor polarizer (c) is similar to a Glan-Foucault polarizer except for the optic-axis orientation, which ensures that the transmitted E-ray passes through the air gap nearly at Brewster's angle; consequently, transmission losses are much reduced from those of a Glan-Foucault. For high-power laser applications, the side faces can be polished at Brewster's angle; consequently, transmission losses are much reduced from those of a Glan-Foucault. For high-power laser applications, the side faces can be polished at Brewster's angle. In a Rochon polarizer (d) two calcite prisms with orthogonal optic axes are cemented

together with Canada balsam. ϕ depends on the interface angle. The intensities of the transmitted O- and E-rays are different. Other birefringent materials can be used; in the case of quartz, for example, the E-ray would exit below the O-ray in Figure 4.56. In a Senarmont polarizer (e) two calcite prisms are cemented together with Canada balsam.

The angle ϕ depends on θ. This type of polarizer is less commonly used than the Rochon polarizer. A Wollaston polarizer (*f*) has two calcite prisms cemented together with Canada balsam. The O- and E-ray transmitted intensities are different. The beam separation angle ϕ depends on the interface angle θ.

Some of these polarizers generate two orthogonally polarized output beams separated by an angle, while others (the Glan-Taylor, for example) reject one polarization state by total internal reflection at a boundary. Extinction coefficients as high as 10^6 and good optical quality can be obtained from such polarizers. However, the acceptance angle of these polarizers is generally quite small, although it can range up to about 38° with an Ahrens polarizer. Suppliers of polarizing prisms include Karl Lambrecht Corporation, Conoptics, Inrad, Meadowlark Optics, Newport, Oriel, and Precision Optical.

Infrared polarizers (beyond about 7 μm) are often made in the form of very many fine, parallel, closely spaced metal wires. Waves with their electric vector perpendicular to the wires are transmitted—the parallel polarization is reflected. Such polarizers are available from Coherent Inc., Molectron, Rochester Photonics, and Specac.

Linear polarizers for any wavelength where a transparent window material is available can be made using a stack of plates placed at Brewster's angle. Unpolarized light passing through one such plate becomes slightly polarized, since the component of incident light polarized in the plane of incidence is completely transmitted, whereas only about 90 percent of the energy associated with the orthogonal polarization is transmitted. A pile of 25 plates gives a high degree of polarization. Stacked-plate polarizers are rather cumbersome, but they are most useful for polarizing high-energy laser beams where prism polarizers would suffer optical damage. They are available commercially from Inrad.

Retardation Plates

Quarter–wave plates are generally used for converting linear to circularly polarized light. They should be used with the optic-axis direction at 45° to the incident linear polarization direction. They are available commercially in two forms: multiple- and single-order. Multiple-order plates produce a phase change (retardation) between the ordinary and extraordinary waves of $(2n+1)\pi/2$. This phase difference is very temperature-sensitive—a thickness change of one wavelength will produce a substantial change in retardation. Single order plates are very thin, their thickness being

$$L = \frac{\lambda_0}{4(n_e - n_o)} \qquad (4.173)$$

With calcite, for example, which has $n_o = 1.658$, $n_e = 1.486$, and with $\lambda_0 = 500$ nm, this thickness is only 726.7 nm. Most single-order plates are therefore made by stacking a $(2n + 1)$-order quarter-wave plate on top of a $2n$-order plate whose optic axis is orthogonal to that of the first plate. The net retardation is just $\pi/2$ and is very much less temperature-sensitive.

Direct single-order quarter-wave plates for low-intensity applications can be made from mica, which cleaves naturally in thin slices. The production of a plate for a particular wavelength is a trial-and-error procedure.[11,35] Quarter-wave plates are invaluable, in combination with a linear polarizer, for reducing reflected glare and for reducing back reflections into an optical system as shown in Figure 4.57.

Half-wave plates are generally used for rotating the plane of polarization of linearly polarized radiation. Multiple- and zero-order plates are available. As in the case of quarter-wave plates, the operating wavelength must be specified in buying one. This operating wavelength can be tuned to some extent by tilting the retardation plate. Half-wave plates cannot be used to make an optical isolator, as the polarization change that occurs on one pass through the plate is reversed on the return path.

Retardation plates only produce their specified retardation at a particular wavelength, and are usually specified for normal, or almost normal incidence. Retardation plates are available from several suppliers, including Karl Lambrecht Corporation, Optics for Research, Inrad, II-VI Inc., Esco, Melles Griot, Newport, Precision Optical, and Oriel. Continuously adjustable retardation plates, called Babinet-Soleil compensators, are available from Karl Lambrecht Corporation, Melles Griot,

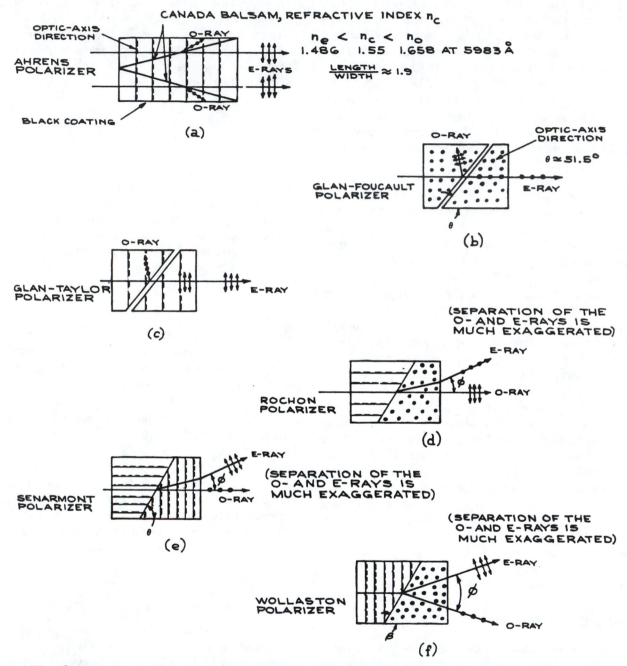

Figure 4.56   Construction of various polarizers using birefringent crystals: (a) Ahrens polarizer, (b) Glan-Foucault polarizer, (c) Glan-Taylor polarizer, (d) Rochon polarizer, (e) Senarmont polarizer, and (f) Wollaston polarizer.

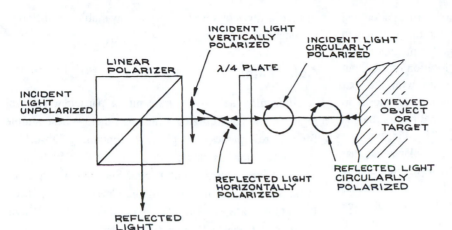

Figure 4.57   An optical isolator constructed from a linear polarizer and a quarter-wave plate.

II-VI Inc., New Focus, Continental Optical, and Optics for Research, among others.[27]

Retardation Rhombs

Because there is generally a different phase shift on total internal reflection for waves polarized in and perpendicular to the plane of incidence, a retarder that is virtually achromatic can be made by the use of two internal reflections in a rhomboidal prism of appropriate apex angle and refractive index. Two examples are shown in Figure 4.58. The Fresnel rhomb, for example, uses glass of index 1.51, and with an apex angle of approximately 54.6° produces a retardation of exactly 90°. The optimum angle must be determined by trial and error, however, and the desired retardation will only be obtained for a specific angle of incidence. Fresnel rhombs are available from KLC, Oriel, and II-VI.

4.3.7 Optical Isolators

Optical isolators are devices that allow light to pass through in one direction, but not in the reverse direction. These devices contain two linear polarizers and between them a medium (generally a special glass) that exhibits the *Faraday effect*. The medium is placed in an axial magnetic field provided by a permanent magnet or magnets. A Faraday-active medium rotates the plane of an input linearly polarized wave, and the direction of rotation

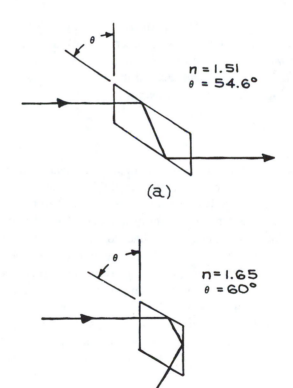

Figure 4.58   Rhomb retarders: (a) Fresnel; (b) Mooney.

depends on the relative direction of light propagation and the magnetic field direction. In an isolator, the optical path length through the Faraday material and the magnetic field strength are selected to provide a 45° rotation of the input linear polarization, which then lines up with the exit polarizer set at 45° to the input polarizer as shown in Figure 4.59. Light passing in the reverse direction has its plane of linear polarization also rotated by 45°, but in a sense that makes it orthogonal to the next polarizer so that negligible light emerges. Commercially available Faraday isolators typically provide 30 dB isolation and must be selected for the specific wavelength desired. Both free space and optical fiber-based devices are available from companies such as Electro-optics Technology, New Focus, and Optics for Research.

Optical isolators are particularly useful in certain precision experiments using lasers to avoid *feedback instability*. If some of the light from a laser re-enters the laser, the amplitude and phase of the laser will fluctuate—this can sometimes be very marked.

4.3.8 Filters

Filters are used to select a particular wavelength region from light containing a broader range of wavelengths than is desired. They therefore allow red light to be obtained from a white light source, for example, or they allow the isolation of a particular sharp line from a lamp or laser in the presence of other sharp lines or continuum emission. Although filters do not have high resolving power (see Section 4.3.5), they have much higher optical efficiency at their operating wavelength than higher resolving power, wavelength-selective instruments such as prism or grating monochromators. That is, they have a high ratio of transmitted flux to input flux in the wavelength region

desired. Filters of several types are available for different applications.

Color Filters

A *color filter* selectively transmits a particular spectral region while absorbing or reflecting others. These simple filters are glasses or plastics containing absorbing materials such as metal ions or dyes, which have characteristic transmission spectra. Such color filters are available from Chance-Pilkington, Corning, CVI, Hoya, Kopp Glass, Melles-Griot, Omega Optical, Schott, Rolyn, Oriel, and Kodak (Wratten filters). Transmission curves for these filters are available from the manufacturers and in tabulations of physical and chemical data.[5,36]

Care must be taken when color filters are being used to block a strong light signal, such as a laser. These filters may fluoresce, so although the primary incident light is blocked, a weak, longer wavelength fluorescence can occur. We have seen this phenomenon quite strongly when using orange plastic to block a blue/green laser.

Band-Pass Filters

A *band-pass filter* is usually a *Fabry-Perot etalon* of small thickness. Although etalons will be discussed in further detail in Section 4.7.4, a brief discussion is in order here.

In its simplest form, the Fabry-Perot etalon consists of a plane, parallel-sided slab of optical material of refractive index n and thickness d, with air on both sides. In normal incidence, the device has maximum transmission for wavelengths for which the thickness of the device is an integral number of half wavelengths. In this case, from Equation 4.108,

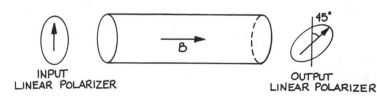

Figure 4.59 Optical isolator that uses two linear polarizers and a Faraday material producing a 45° polarization rotation.

$$Z''_3 = Z'_3 = Z'_1 \qquad (4.174)$$

and there is zero reflection. For incidence at angle θ, regardless of the polarization state of the light, there is maximum transmission for wavelengths (in vacuo) that satisfy

$$m\lambda_0 = 2d\cos\theta_2 \qquad (4.175)$$

where θ_2 is the angle of refraction in the etalon.

A Fabry-Perot etalon is a *comb* filter (Figure 4.148 shows an example of a transmission characteristic of such a device). If the thickness of the etalon is small, the transmission peaks are broad. In a typical band-pass filter operated as an etalon, all the transmission peaks but the desired one can be suppressed by additional absorbing or multilayer reflective layers. The spectral width of the transmission peak of the filter can be reduced by stacking individual etalon filters in series. A single etalon filter might consist, for example, of two multilayer reflective stacks separated by a half-wavelength-thick layer as shown in Figure 4.60a, while a two-etalon filter would have two half wavelength-thick layers bounded by multilayer reflective stacks. Commercially available band-pass filters for use in the visible usually have transmission peaks in a range from 1 nm to 50 nm (full width at half maximum transmission). The narrowest band-pass filters are frequently called spike filters. Filters to transmit particular wavelengths such as 632.8 nm (He-Ne laser), 514.5 and 488 nm (argon ion laser), 404.7, 435.8, 546.1, 577.0, and 671.6 nm (mercury lamp), 589.3 nm (sodium lamp), and other wavelengths are available as standard items.[27] Filters at nonstandard wavelengths can be fabricated as custom items. Bandpass filters for use in the ultraviolet usually have lower transmittance (typically 20 percent) than filters in the visible, whose transmittance usually ranges from 40 to 70 percent, being lowest for the narrowest passband. Infrared band-pass filters are also readily obtained, at least out to about 10 µm, and have good peak transmittance. The narrowest available band pass for a filter whose peak transmission is λ_{peak} can usually be estimated as $\lambda_{peak}/50$.

Band-pass filters are usually designed for use in normal incidence. Their peak of transmission can be moved to shorter wavelengths by tilting the filter, although the sharpness of the transmission peak will be degraded by this procedure. If a filter designed for peak wavelength λ_0 is tilted by an angle θ, its transmission maximum will shift to a wavelength

$$\lambda_s = \lambda_0 \sqrt{1 - \frac{\sin^2\theta}{n^2}} \qquad (4.176)$$

where n is the refractive index of the half-wavelength-thick layer of the filter. For small tilt angles,

$$\lambda_0 - \lambda_s = \frac{\lambda_0 \sin^2\theta}{2n^2} \qquad (4.177)$$

Thus, a band-pass filter whose peak-transmission wavelength is somewhat longer than desired can be optimized by tilting.

Long- and Short-Wavelength-Pass Filters

A *long wavelength-pass filter* is one that transmits a broad spectral region beyond a particular cutoff wavelength, λ_{min}. A *short wavelength-pass filter* transmits in a broad spectral region below its cutoff wavelength λ_{max}. A combination of a long wavelength-pass and a short wavelength-pass filter can be used to produce a bandpass filter, as shown in Figure 4.61. Although long and short wavelength-pass filters usually involve multilayer dielectric stacks, the inherent absorption and transmission characteristics of materials can be utilized. Semiconductors exhibit fairly sharp long-wavelength-pass behavior beginning at the band gap energy. Germanium is a long-pass filter, for example, with $\lambda_{min} = 1.8$ µm.

Holographic Notch Filters

These filters are special diffractive structures that reflect a narrow range of wavelengths (~ 5 nm) very strongly, and consequently they reduce transmitted light substantially (40–60 dB) over the same narrow range. These filters are ideal for rejection of laser light in experiments where a weaker light signal—fluorescence or Raman scattering—is being studied, and some scattering or reflection of the

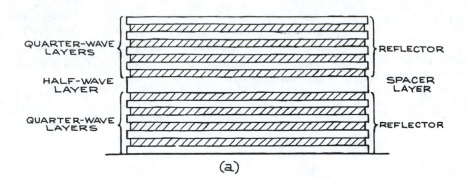

QUARTER-WAVE LAYERS

HALF-WAVE LAYER

QUARTER-WAVE LAYERS

REFLECTOR

SPACER LAYER

REFLECTOR

(a)

Figure 4.60 (a) Construction of an all-dielectric single-cavity Fabry-Perot interference filter; (b) theoretical transmittance characteristics of single- and double-cavity filters. (From *Handbook of Lasers*, R. J. Pressley, ed., CRC Press, Cleveland, 1971; by permission of CRC Press, Inc.)

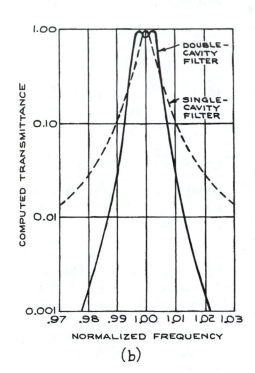

(b)

excitation laser light is occuring. These filters are available from Kaiser Optical Systems, Rochester Photonics, CVI, and Oriel.

Reststrahlen Filters

When ionic crystals are irradiated in the infrared, they reflect strongly when their absorption coefficient is high and their refractive index changes sharply. This high reflection results from resonance between the applied infrared frequency and the natural vibrational frequency of ions in the crystal lattice. The characteristic of reflected light from different crystals, termed *Reststrahlen*, allows specific broad spectral regions to be isolated by reflection from the appropriate crystal. Some examples of Reststrahlen filters are given in Figure 4.62.

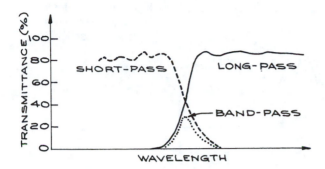

Figure 4.61   Schematic transmission of long- and short-pass optical filters showing the band-pass transmission characteristics of the two in combination,

Christiansen Filters

In the far infrared, the alkali halides exhibit *anomalous* dispersion—at certain wavelengths their refractive indices pass through unity (normal dispersion involves an increase of refractive index with wavelength). A powdered alkali halide, which would generally severely attenuate a transmitted wave because of scattering, will therefore transmit well at wavelengths where its refractive index is the same as that of air. Filters using such an alkali halide powder held between parallel plates are called Christiansen filters and provide sharp transmission peaks at certain wavelengths listed in Table 4.4.

Neutral-Density Filters

Neutral-density (ND) filters are designed to attenuate light uniformly over some broad spectral region. The ideal neutral-density filter should have a transmittance that is independent of wavelength. These filters are usually made by depositing a thin metal layer on a transparent substrate. The layer is kept sufficiently thin that some light is transmitted through it. The optical density D of a neutral-density filter is defined as

$$D = \log_{10}(1/T) \qquad (4.178)$$

where T is the transmittance of the filter. T is controlled both by reflection from and by absorption in the metal film. If such filters are stacked in series, the optical density of the combination is the sum of the optical densities of the individual filters, provided the filters are positioned so that multiple reflection effects do not occur between them in the direction of interest.

Neutral-density filters are used for calibrating optical detectors and for attenuating strong light signals falling on detectors to ensure that they respond linearly. Because neutral-density filters usually reflect and transmit light, they can be used as beamsplitters and beam combiners for any desired intensity ratio. There are many suppliers.[27] Variable ND filters are available from Cambridge Research and Instrumentation, CVI, OCLI, and Reynard.

Light Traps

If a beam of light must be totally absorbed (for example, in an application where any reflected light from a surface could interfere with some underlying weak emission), a light trap can be constructed. The two types illustrated in Figure 4.63 work well—the Wood's horn, in which a beam of light entering the trap is gradually attenuated by a series of reflections at an absorbing surface made in the form of a curved cone (usually made of glass), and a stack of razor blades, which absorb incident light very efficiently. Light traps are commercially available from Klinger.

4.3.9 Fiber Optics

As is well known, the use of optical fibers has become widespread in the telecommunications field. There are, therefore, numerous suppliers of the various components and subsystems involved.[15] The use of optical fibers in scientific research can be a very valuable technique, and one that is not particularly complicated. In experiments where radio frequency (rf) interference, pickup, and ground loops are a problem, for example, an experimental signal can be used to modulate a small light-emitting diode or laser. The modulated optical signal is then passed along an optical fiber to the observation location, where the original electrical signal can be recovered with a photodiode. Complete analog transmission systems of this kind are available from Alcatel, Analog Modules, Force, Hewlett Packard, JDS Uniphase, Litton Poly-Scientific, and Metrotek Industries. A schematic diagram of the various

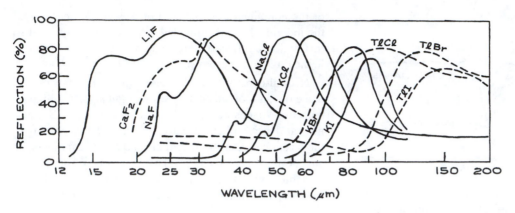

Figure 4.62 *Restrahlen* filters. (Courtesy of Harshaw Chamical Co.)

TABLE 4.4 CHRISTIANSEN FILTERS

Crystal	Wavelength of Maximum Transmission (μm)
LiF	11.2
NaCl	32.0
NaBr	37.0
KCl	37.0
RbCl	45.0
NaI	49.0
CsCl	50.0
KBr	52.0
CsBr	60.0
TlBr	64.0
KI	64.0
RbBr	65.0
RbI	73.0
TlI	90.0

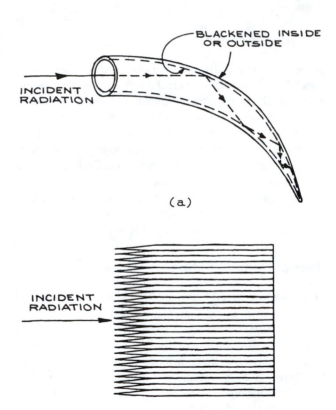

(a)

(b)

Figure 4.63 Light traps: (a) Wood's horn; (b) stacked razor blades.

components of such a system is shown in Figure 4.64a. The source may be supplied fabricated directly onto a fiber. The modulator is frequently unnecessary, as the source itself can be directly amplitude-modulated. The detector may also be supplied fabricated directly onto the end of the fiber. A few specific points are worthy of note for the experimentalist who may wish to utilize this technology.

For certain laboratory applications, the construction of a fiber system without any connectors is straightforward. Light from the source is focused into the end of the fiber with a microscope objective. Light emerging from the other end of the fiber is focused onto the detector in a similar way. A convenient range of microscope objectives for this purpose is available from Newport, which also supplies a wide range of components for holding and positioning fibers. The choice of lens focal length and placement is governed by the numerical aperture (NA) of the fiber. The meaning of this parameter can be understood with reference to Figure 4.64b, which shows a meridional section through a *step-index* fiber. In this composite fiber, the cylindrical core has refractive index n_1, and the surrounding cladding has index n_2, where for total internal reflection of rays to occur in the core, $n_1 > n_2$. Light entering the fiber at angles $\leq \theta_0$ will totally internally reflect inside the core. The NA is

$$\sin \theta_0 = \sqrt{n_1^2 - n_2^2} \qquad (4.179)$$

Commercially available graded index fibers typically have a smoothly varying radial index profile, but the NA is still the appropriate parameter for determining the acceptance angle. Two specific types of fiber are in most common use: single mode and multimode. Single-mode fibers typically have core diameters on the order of 10 μm and cladding diameters of 125 μm. They require very precise connectors and are only needed in specialized experiments where the ability of the fiber to support only a single propagating mode is important. Multimode fibers have larger core diameters, from 50 μm to above 1 mm, and cladding diameters somewhat larger than their respective core diameters. These fibers are easy to use. The larger sizes are frequently used to channel light for illumination into positions that are difficult to access, or to collect light from one location and channel it to a detector somewhere else. Large fibers that are suitable for these purposes are available from Edmund Scientific, Fiberguide Industries, Newport, Spectron, and 3M Speciality Optical Fibers.

Fibers can be cut to provide a flat end face with the aid of specialist cleaving tools of varying degrees of precision and complexity, available from suppliers such as Newport and York. In simple experimental setups, an adequate cleaved flat face can be obtained in the following way:

1. Remove the outer plastic protective coating from the cladding in the region that is to be cut by using a fiber stripping tool or by immersing this part of the fiber in methylene chloride (the active ingredient in most proprietary paint and varnish strippers).

2. Lightly scratch the cladding with a cleaving tool (slightly more precise than a conventional glass cutter).

3. Fasten one half of the fiber to a flat surface with masking tape, and pull the other half of the fiber axially away, keeping the fiber flat on the surface.

This procedure usually gives a flat enough face for use with a microscope objective. If fibers are to be cut and connectorized, however, they must be cut and polished according to the instructions supplied with the connector.

Fiber Optic Connectors

If two identical fibers are to be joined together, this can be done with mechanical coaxial connectors, with a mechanical splice, or the fibers can be fused together with a fusion splicer. In all cases, the flat, cleaved faces of the two fibers to be joined together must be brought together so that the two faces are in close proximity and parallel, and that the fibers are coaxial. Proper alignment for orthogonal and angle cleaves is shown in Figures 4.65a and b, respectively. Any lateral or tilt offset of the fibers when they are joined will reduce coupling efficiency. Precision mechanical optical fiber connectors are able to couple two fibers together with losses in the range 0.1–0.2 dB. Sometimes the connector contains a small amount of index-matching gel that is squeezed in the gap between the fibers. Mechanical fiber splices are an inexpensive way to joint fibers. They generally have a sleeve-like construction into which the two fibers are pushed and then glued or clamped in place. When fibers are fused together, a fusion splicer is used (it can be an expensive device). A good fusion splice can have a loss below 0.1 dB. After two fibers are fused together, a *splice protector* should be included if the joined fiber is to be protected from failure of the splice.

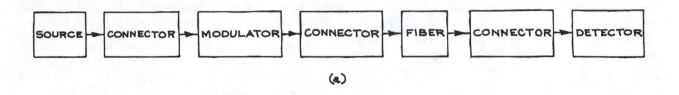

(a)

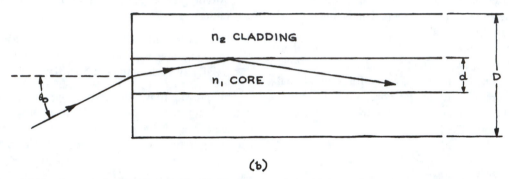

(b)

Figure 4.64   (a) Fundamental components of a fiber-optic data link; (b) Meridional section through a step-index optical fiber. The core refractive index in n_1; the cladding refractive index in n_2. The core and cladding diameters are d and D, repectively. A ray entering the fiber will be totally internally reflected provided its angle of incidence is less than θ_0, where $\sin \theta_0 = NA$.

Polarization-maintaining (PM) fiber will preserve the polarization state of launched light (up to about 1% purity). Special connectors and splicing techniques are needed for PM fiber, as the azimuthal orientation of the two fibers is now important.

The use of optical fiber devices is much simplified if these devices are purchased with appropriate fiber optic connectors already attached. There are many different types of fiber optic connectors such as SMA, FC, SC, ST, and Optoclips. For interconnections involving different kinds of connectors, fiber optic jumpers are available from a number of suppliers such as Fiberall. For special applications, angle-cleaved connectors are available, which reduce the inevitable small reflection that occurs at a fiber connector. The trade journal *Lightwave*[37] is a useful source of information about suppliers and products.

4.3.10 Precision Mechanical Movement Systems

It is frequently necessary to be able to position components in an optical system with high precision. The type of stage with which this is done depends on the scale of movement, the accuracy, repeatability resolution, and freedom from backlash that will be required. There are a large number of suppliers of stages, including translation, rotation, tilting, multi-axis positioning, and combinations of these, including Newport, New Focus, Melles Griot, Optosigma, Thor Labs, Aerotech, and Velmex. Both manual and motor-driven stages are available. It is convenient to divide the movement scales available into three broad categories.

Medium. Medium is the scale of movement provided by manual micrometer drives or precision lead screws. Typical micrometer drives provide up to 50 mm travel

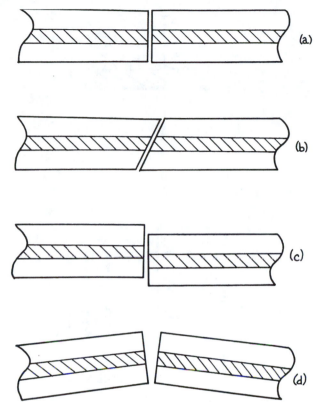

Figure 4.65   Alignment of optical fibers for optical connection: (a) correct alignment for perpendicular-cleaved fibers; (b) correct alignment for angle-cleaved fibers; (c) lateral misalignment; (d) angle misalignment.

with accuracy on the order of 1 μm. Some commercial drives claim a resolution below 0.1 μm, but in practice such claims must be reviewed with skepticism, since the backlash, creep, and repeatability of all mechanical, screw-based stages is likely to be worse than this.

The accuracy of adjustment is the difference between the actual and desired positions. *Repeatability* refers to the ability to reproduce a desired position after a movement of a stage to a different position. In a fair comparison of stages, this should involve significant, comparable movements away from—and then back to—a desired position. Resolution is the smallest motion that a stage can make, which for a screw-driven device will be limited by factors such as the number of threads per inch in the lead screw, and friction. A precision micrometer might, for example, have a resolution of about 1/10 of the smallest scale graduation around the micrometer head. Another factor of importance is creep. After a stage has been set to a desired position does it stay there, or does removal of stress from a positioning screw cause it to relax? Absence of creep means that a mount, once positioned, holds its position for a long period of time. It is our experience that stainless steel mounts are less susceptible to creep than aluminum mounts.

Resolution alone as a specification does not mean much without the additional specifications of accuracy, repeatability, and stability. The long term stability of a mount is also influenced by the thermal stability and expansion of the materials from which the stage is made. The convenient materials of construction are aluminum alloy and stainless steel. Aluminum has a higher coefficient of thermal expansion (24×10^{-6}/°C) compared with stainless steel (10–16×10^{-6}/°C), but aluminum has a higher thermal conductivity and reaches thermal equilibrium more quickly, which minimizes thermally-induced strain. Aluminum takes a poor thread, however, and brass threaded inserts must usually be used for brass or stainless steel positioning screws.

The thermal stability of a optical mount can be improved by using mounting assemblies constructed with two dissimilar metals whose thermal expansions off set each other. Commercially available mounts of the highest precision may include this technology. The manufacturer's technical specifications should be consulted.

Fine. The scales of movement, precision, and repeatably are provided by motor-driven stages, which may be driven either by a stepping motor or dc motor. For repeatability, either some form of encoding of the position or feedback based on the characteristics of the state is needed. Position encoding is usually done with either a shaft encoder, which tells the drive motor on the lead screw exactly what its angular position is, or a linear encoder. A linear encoder gives a more precise, repeatable measure of the actual location of the object being moved. Precision drives incorporating encoders are available from Newport. Precise motor driven stages are available from Melles-Griot and New Focus.

Nanoscale. Nanoscale is positioning down to 1 nm resolution with varying degrees of repeatability. Positioning on this scale requires the use of piezoelectric or electrostatic transducers. Piezolectric transducers generally use a ceramic material such as lead zirconate titanate (PZT), which changes its shape in an applied electric field. Typical PZT transducers provide mechanical motion on the order of a few hundreds of nanometers for applied voltages on the order of 1 kV. They provide resolution of 10 nm or less but do exhibit hysteresis, as shown in Figure 4.66, unless some absolute position encoding scheme is used. Individual PZT transducers are generally available either in the form of tubes or disks, as shown in Figure 4.67. Larger motions can be obtained by stacking transducers in either a bimorph or polymorph arrangement. In a bimorph, two slabs of PZT are driven so that one slab extends while the other gets shorter, which causes the bimorph to bend, as shown in Figure 4.68.

In a slab transducer, voltage is applied across the narrow dimension of the slab, which will then change its thickness with applied voltage. In a cylindrical transducer, voltage is applied between the inner and outer metallized cylindrical surface of the tube, which then changes its length in the axial direction. Nanoscale precision piezoelectric drives are commercially available from Melles Griot, Queensgate Instruments, Polytec, Physik Instruments, and Thor Labs, among others. Bimorphs are available from Sensor Technology.

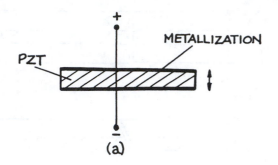

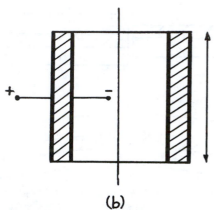

Figure 4.67   Simple piezoelectric elements: (a) slab; (b) cylinder.

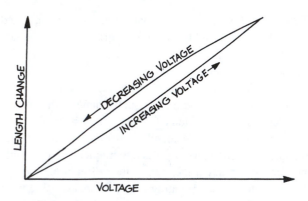

Figure 4.66   Hysteresis in the length change of a piezoelectric element with applied voltage.

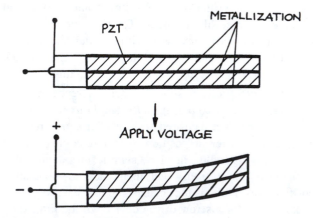

Figure 4.68   Operation of a bimorph piezoelectric element in a *parallel* configuration.

Precision mechanical stages can be built in the laboratory, but are widely available commercially.

Small scale (~ 10 µm) precision motion in each of three orthogonal axes can be accomplished inexpensively with a modified piezotube. A typical piezotube is on the order of 30 mm long by 10 mm, and is generally supplied with a metal coating on its inner and outer cylindrical surfaces. The outer metallization should be removed in four segements approximately 1 mm wide, spaced 90° apart on the outer cylindrical surface, as shown in Figure 4.69. The four segments are then driven by a *bridge* amplifier current, in which opposite pairs of electrodes are driven in a push-pull configuration as shown schematically in Figure 4.70. Driving electrodes 1 and 2 produces an *x*-directed tilt of the tube— electrodes 3 and 4 produce a *y*-directed tilt of the tube. An

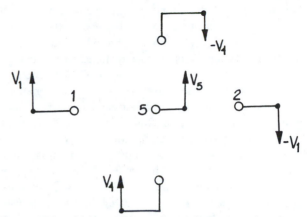

Figure 4.70   Voltage drive scheme for a piezoelectric element such as the one shown in Figure 4.69. Electrodes 1–4 are the quadrant electrodes. Electrode 5 is the inner electrode.

additional amplifier drives electrode 5 for *z*-directed extension of the tube. Because inter-electrode voltages used need to be as large as 1000 V, special op-amps are required. Appropriate amplifers are available from Apex Microtechnology Corporation. Piezo tubes are available from Staveley Sensors, Inc.

4.3.11 Devices for Positional and Orientational Adjustment of Optical Components

For positioning any given component, the maximum number of possible adjustments is six: translation along three mutually orthogonal axis and rotation about these (or other) axes. Mounts can be made that provide all six of these adjustments simultaneously. Usually, however, they are restricted and have no more degrees of freedom than are necessary. Thus, for example, a translation stage is usually designed for motion in only one dimension— independent motion along other axes can then be obtained by stacking one-dimensional translation stages. Orientation stages are usually designed for rotation about one axis (polar rotation) or two orthogonal axes (mirror mounts or tilt tables). Once again, such mounts can be stacked to provide additional degrees of freedom. It is

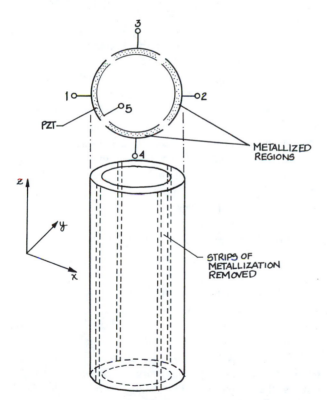

Figure 4.69   A cylindrical piezoelectric element with outer quadrant electrodes. The element can be driven to provide axial motion as well as tilt in two orthogonal axes.

often unnecessary for angular adjustment to take place about axes that correspond to translation directions.

The ideal optical mount provides independent adjustment in each of its degrees of freedom without any interaction with other potential degrees of freedom. It should be sturdy, insensitive to extraneous vibrational disturbances (such as air or structure-borne acoustic vibrations), and free of backlash during adjustment. It should generally be designed on kinematic principles (see Section 1.6.1). The effect of temperature variations on an optical mount can also be reduced, if desired, by constructing the mount from low coefficient-of-expansion material (such as Invar), or by using thermal-expansion compensation techniques. Very high ridigity in an optical mount generally implies massive construction. When weight is a limitation, honeycombed material or material appropriately machined to reduce excess weight can be used. Most commercially available mounts are not excessively massive but represent a compromise between size, weight, and rigidity. Mounts generally do not need to be made any more rigid than mechanical considerations involving Young's modulus and likely applied forces on the mount would dictate.

The design of optical mounts on kinematic principles involves constraining the mount just enough so that, once adjusted, it has no degrees of freedom. Some examples will illustrate how kinematic design is applied to optical mounts.

Two-Axis Rotators

Within this category, the commonest devices are adjustable mirror or grating mounts and tilt tables used for orienting components such as prisms. Devices that adjust orientation near the horizontal are usually called tilt adjusters. Mirror holders, on the other hand, involve adjustment about a vertical or near-vertical axis. A basic device of either kind involves a fixed stage equipped with three spherical or hemispherical ball bearings, which locate in a V-groove, in a conical hole, and on a plane surface machined on a movable stage, respectively. A typical device is shown in Figure 4.71. The movable stage makes contact with the three balls on the fixed stage at just six points—sufficient to prevent motion in its six degrees of freedom—and is held in place with tension springs between the fixed and movable stages. Adjustment is

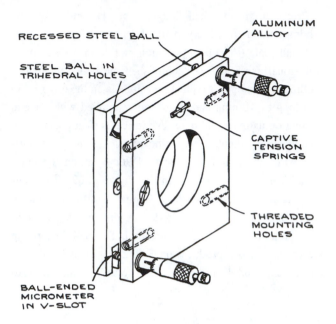

Figure 4.71 Kinematic double-hinge mirror mount.

provided by making two of the locating balls the ends of fine-thread screws or micrometer heads. Suitable micrometer heads for this purpose are manufactured by Starrett, Brown and Sharpe, Moore and Wright, Shardlow, and Mitutoyo, and are available from most wholesale machine-tool suppliers. The ball of the micrometer should be very accurately centered on the spindle. The adjustment of the mount shown in Figure 4.76 is about two almost orthogonal axes. There is some translation of the center of the movable stage during angular adjustment, which can be avoided by the use of a true gimbal amount.

A few of the constructional details of the mount in Figure 4.71 are worthy of note. Stainless steel, brass, and aluminum are all suitable materials of construction—an attempt to reduce the thermal expansion of the mount by machining it from a low-expansion material such as Invar may not meet with success. Extensive machining of low expansion-coefficient alloys generally increases the coefficient unless they are carefully annealed afterward. Putting the component in boiling water and letting it gradually cool will lead to partial annealing. True kinematic design requires ball location in a trihedral hole, in a V-slot, and on a plane surface (or in three V-slots, but this no longer permits angular adjustment about orthogonal axes; see also Section 1.6.1). If the mount is made of aluminum, the locating hole, slot, and plane on the movable stage should be made of stainless steel, or some other hard material, shrunk-fit into the main aluminum body. A true trihedral slot can be made with a punch as shown in Figure 4.72a, or it can be machined with a 45° milling cutter as shown in Figure 4.72b. A

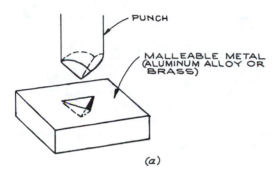

(a)

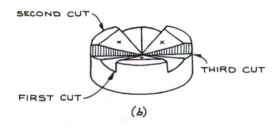

(b)

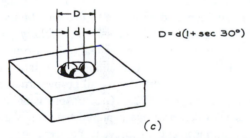

(c)

Figure 4.72 Methods for making trihedral locating holes for kinematic design; (a) tridhedral steel punch; (b) trihedral hollow machined with 45° milling tool; (c) steel balls fitting tightly in a hole.

conical hole is not truly kinematic—a compromise design is shown in Figure 4.73. Although motion of the mount in Figure 4.71 results from direct micrometer adjustment, various commercial mounts incorporate a reduction mechanism to increase the sensitivity of the mount. Typical schemes that are used involve adjustment of a wedge-shaped surface by a moving ball, as shown in Figure 4.74, or the use of a differential screw drive. The two or more movable stages of an adjustable mount can be held together with captive tension springs as in Figure 4.71. Designs involving only small angular adjustment, however, can utilize the bending of a thin section of material as shown in Figure 4.75. These are called *flexure* mounts, also discussed in Section 1.6.6.

The basic principle of a gimbal mount for two-axis adjustment is shown in Figure 4.76. True orthogonal motion about two axes can be obtained in such a design without translation of the center of the adjusted component. A kinematically designed gimbal mount is much more complicated in construction, however, than the mount shown in Figure 4.76. Precision gimbal mounts

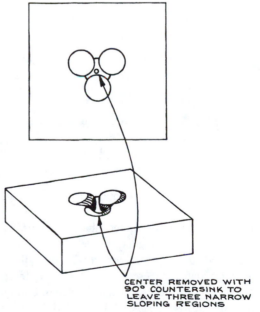

CENTER REMOVED WITH 90° COUNTERSINK TO LEAVE THREE NARROW SLOPING REGIONS

Figure 4.73 Compromise design for kinematic trihedral hole.

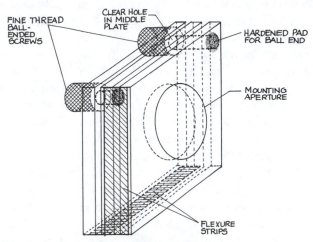

Figure 4.74   Cutaway views of a kinematic, single-axis rotator incorporating a reduction mechanism involving a ball bearing on an angled surface. Two rotators of this kind could be incorporated in a single mount to give two-axis gimbal adjustment.

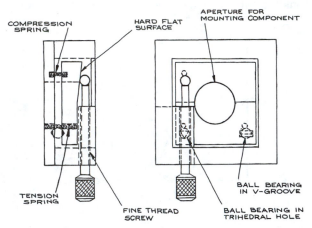

Figure 4.75   Flexure type double-hinge mirror design.

should be purchased rather than built. Very many companies manufacture adequate devices with a variety of designs, and their prices are lower than the cost of duplicating such devices in the laboratory. Suppliers of gimbal mounts include Aerotech, Burleigh, Ealing Optics, New Focus, Newport, J. A. Noll, and Oriel. Most of these

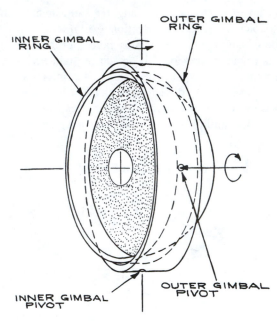

Figure 4.76   Schematic diagram of a true two-axis gimbal mount. The center of the mounted component does not translate during rotation.

manufacturers also manufacture good non-gimbal mounts similar to the one shown in Figure 4.71.

Translation Stages

Translation stages are generally designed to provide linear motion in a single dimension and can be stacked to provide additional degrees of freedom. Mounts for centering optical components, particularly lenses, incorporate motion in two orthogonal (or near-orthogonal) axes in a single mount. The simplest translation stages usually involve some sort of dovetail groove arrangement, as shown in Figure 4.77a. Greater precision can be obtained with ball bearings captured between V-grooves with an arrangement to prevent adjacent balls from touching as shown in Figure 4.77b, or with crossed roller bearings as shown in Figure 4.77c. A simple ball-bearing groove design is shown in Figure 4.78, although variants are possible. Suppliers of precision, roller bearing translators include Aerotech, Ealing, Melles Griot, Newport, OptoSigma, and Oriel. Vertical translators using

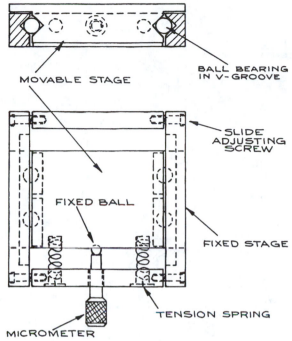

Figure 4.77 Section drawings of a simple, one-dimensional translation stage.

a fine-thread screw drive are available from Oriel. Stages can be built to provide short range, highly stable, frictionless precision movement by using flexures, as shown in Figure 4.79. Typical travel of a flexure stage is on the order of 10 percent to 15 percent of the dimensions of the stage.

Less precise, but generally adequate, nonball-bearing translation stages are available from Newport and Velmex. Mounts for centering optical components such as lenses are available from Aerotech, Ealing, and Newport, among others. A simple kinematic design for a centering mount is illustrated in Figure 4.80.

Rotators and Other Mounts

Rotators are used, for example, for orienting polarizers, retardation plates, prisms, and crystals. They provide rotation about a single axis. Precision devices using ball bearings are available from Aerotech, New Focus, Newport, Oriel, OptoSigma, and Thor Labs, among various other suppliers.

There are numerous more complicated optical mounts than those described in the previous sections that provide in a single mount any combination of degrees of freedom desired. Motor-driven and computer-controllable versions of all these mounts are generally available. For details, the catalogs of the manufacturers previously mentioned should be consulted.

Optical Benches and Components

For optical experiments that involve the use of one or more optical components in conjunction with a source and/or a detector in some sort of linear optical arrangement, construction using an optical bench and components is very convenient, particularly in breadboarding applications. The commonest such bench design is based on the equilateral triangular bar developed by Zeiss and typically machined from lengths of stabilized cast iron. Aluminum benches of this type are also available, but are less rigid. Components are mounted on the bench in rod-mounted holders that fit into carriers that locate on the triangular bench. A cross-sectional diagram of a typical bench and carrier is shown in Figure 4.81.

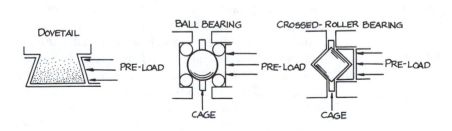

Figure 4.78 Groove mounting arrangements for linear translation stages: (a) dovetail V-groove; (b) ball bearing; (c) crossed roller bearing.

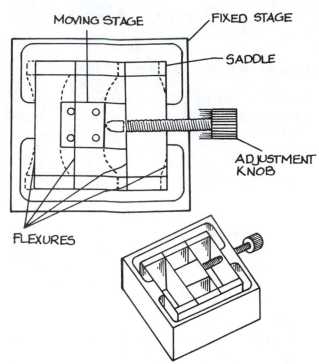

Figure 4.79   Linear translation stage using flexure mounts.

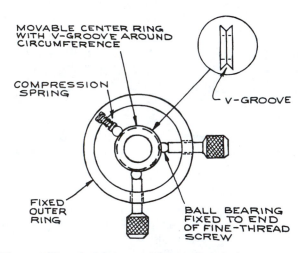

Figure 4.80   A simple precision centering stage.

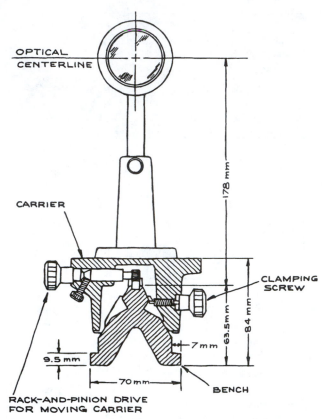

Figure 4.81   Typical dimensions of a Zeiss triangular optical bench with carrier and mounted optical component. (Courtesy of Ealing Corporation.)

Coarse adjustment along the length of the bench is provided by moving the carrier manually. On many benches this is accomplished with a rack-and-pinion arrangement. A wide range of mounts and accessories are available, providing for vertical, lateral, and longitudinal adjustment of mounted-component positions. Holders for filters, for centering lenses, and for holding and rotating prisms and polarizers are readily obtained. Precision gimbal mounts, translators, and rotators can be readily incorporated into the structure. Examples of devices that lend themselves well to optical-bench construction are telescopes, collimators, laser-beam expanders and spatial filters, optical isolators, and electro-optic light modulators. Triangular cross-section optical benches and accessories are available from Ealing, Edmund Scientific, Melles

Griot, Newport, Oriel, and Spindler and Hoyer, among others. Other designs of optical benches are also available, but the widespread use of the Zeiss equilateral design offers the convenience that mounts made by different manufacturers can be used interchangeably. For very high precision optical-bench arrangements, precision-machined lathe bed optical benches are recommended. Ealing guarantees the straightness of its best-quality bench to 12.5 μm or better over a 2.4 m length.

Piezoelectric Transducers

Piezoelectric transducers (PZTs) are made from materials whose physical dimensions change when a voltage is applied between two electrodes deposited on the material. Most PZTs are made from ferroelectric ceramics such as lead zirconate titanate or barium titanate, but polymers such as polyvinylidene fluoride (PVF) are also used. Simple ceramic transducers are usually supplied as flat discs with a metalized layer on each flat face, or as short hollow cylinders, in which case the metalized electrodes are deposited on the inner and outer cylindrical surfaces. The flat discs change their thickness with applied voltage, the cylindrical types change their axial length. Typical length changes are on the order of 10^{-9} meters per volt. These devices, therefore, are particularly suitable for very fine (submicron) positioning. Much larger motions can be obtained with bimorph PZTs. These are fabricated from two thin sheets of transducer material that are oriented differently and then sandwiched together. Application of a voltage causes one half of the sandwich to shrink, and the other to expand, so that the bimorph bends. Bimorphs can be stacked to provide macroscopic motions for modest applied voltages. Excursions of several millimeters can be obtained for voltages of 1 kV. Even larger motions can be obtained with Inchworm transducers manufactured by Burleigh. Basic ceramic transducer elements are available from EDO Electro-Ceramic Products, Physik Instruments (PI), Piezo Systems, CTS Piezoelectric Products, Polytec, and Queensgate Instruments. PVF transducers are available from Atochem North America (formerly Pennwalt). Complete transducer assemblies are available from Burleigh, Polytec Optronics, and Klinger.

4.3.12 Optical Tables and Vibration Isolation

A commercially made optical table provides the perfect base for construction of an optical system. Such tables have tops made of granite, steel, invar, superinvar (very expensive), or ferromagnetic (or nonmagnetic) stainless steel. When vibrational isolation of the table top is required, the top is mounted on pneumatic legs. Such an arrangement effectively isolates the table top from floor vibrations above a few Hertz. To minimize acoustically driven vibrations of the table top itself, the better metal ones employ a laminated construction with a corrugated or honeycomb metal cellular core bonded with epoxy to the top and bottom table surfaces. Granite table tops are much less intrinsically resonant than metal tops, and represent the ultimate in dimensional stability and freedom from undesirable vibrational modes. Complete pneumatically isolated optical table systems, nonisolated tables, and table tops alone are available from Ealing, TMC, Melles Griot, Newport, Oriel, and Modern Optics. In selecting a table, points of comparison to look for include the following:

- The overall flatness of the table top (flatness within 50 μm is good, although greater flatness can be obtained)
- The low-frequency vibration isolation characteristics of the pneumatic legs
- The rigidity of the table top (how much it deforms with an applied load at the center or end)
- The isolation from both vertical and horizontal vibrations
- The frequencies and damping of the intrinsic vibrational modes
- The threaded holes in the table top sealed so that liquid spills do not disappear into the interior honeycombs of the table

Other criteria such as cost and table weight may be important. Typical sizes of readily available tops range up to 5 × 12 ft. (larger tables are also available) and in thickness from 8 in. to 2 ft. The thicker the table, the greater its rigidity and stability. Most commercial tables are available with $\frac{1}{4}$-20 threaded holes spaced on 1 in. centers over the surface of the table. Better designs use blind holes of this type so that liquids spilt on the table do not vanish into its interior. Ferromagnetic table tops permit the use of magnetic hold-down bases for mounting optical components. Such mounts are available from numerous optical suppliers. It is cheaper and just as satisfactory, however, to buy magnetic bases designed for machine tool use from a machine tool supply company.

Steel and granite surface plates that can be used as small, nonisolated optical tables are available from machine tool supply companies. If it is desired to construct a vibration-isolated table in the laboratory, a 2 in. thick aluminum plate resting on underinflated automobile tire inner tubes works rather well, although the plate will have resonant frequencies. Very flat aluminum plate, called *jig* plate, can be obtained for this application at a cost not much greater than ordinary aluminum plate. Just mounting a heavy plate on dense polyurethane foam also works surprisingly well. In fact, the vibrational isolation of a small optical arrangement built on an aluminum plate and mounted on a commercial optical table is increased by having a layer of thick foam between plate and table top.

Even the best-constructed precision optical arrangement on a pneumatic isolation table is subject to airborne acoustic disturbances. A very satisfactory solution to this problem is to surround the system with a wooden particle-board box lined with acoustic absorber. Excellent absorber material (a foam-lead-foam-lead-foam sandwich) called Hushcloth DS is available from American Acoustical Products.

4.3.13 Alignment of Optical Systems

Small helium-neon lasers are now so widely available and inexpensive that they must be regarded as the universal tool for the alignment of optical systems. Complex systems of lenses, mirrors, beamsplitters, and so on can be readily aligned with the aid of such a laser—even visible-opaque infrared components can be aligned in this way by the use of reflections from their surfaces. Fabry-Perot interferometers, perhaps the optical instruments most difficult to align, are easily aligned with the aid of a laser. The laser beam is shone orthogonally through the first mirror of the interferometer (easily accomplished by observing part of the laser beam reflect back on itself), and the second interferometer mirror is adjusted until the resultant multiple-spot pattern between the mirrors coalesces into one spot.

4.3.14 Mounting Optical Components

A very wide range of optical devices and systems can be constructed using commercially available mounts designed to hold prisms, lenses, windows, mirrors, light sources, and detectors. Some commercially made optical mounts merely hold the component without allowing precision adjustment of its position. Many of the rodmounted lens, mirror, and prism holders designed for use with standard optical benches fall into this category, although they generally have some degree of coarse adjustability. A wide range of commercially available optical mounts, however, allow precision adjustment of the mounted component.

The construction of a complete optical system from commercial mounts can be very expensive while providing, in many cases, unnecessary and consequently undesirable degrees of freedom in the adjustment and orientation of the various components of the system. Although extra adjustability is often desirable when an optical system is in the *breadboard* stage, it reduces overall stability in the final version of the system. An adjustable mount can never provide the stability of a fixed mount of comparable size and mass. Stability in the mounting of optical components is particularly important in the construction of precision optical systems for use—for example, in interferometry and holography.

A custom-made optical mount must allow the insertion and stable retention of the optical component without damage to the component, which is usually made of some fragile (glassy or single crystal) material. If possible, the precision surface or surfaces of the component should not contact any part of the mount. The component should be clamped securely against a rigid surface of the mount. It should not be clamped between two rigid metal parts, as thermal expansion or contraction of the metal may crack the component or loosen it. The clamping force should be applied by an intermediary rubber pad or ring. Some examples of how circular mirrors or windows can be mounted in this way are shown in Figure 4.82. If a precision polished surface of the component must contact part of the mount (for example, when a precision lens, window, mirror, or prism is forming a vacuum-tight seal), the metal surface should be finely machined—the edges of the mounting surface must be free of burrs. The metal with which the component is in contact should ideally have a lower hardness than the component. A thin piece of paper placed between the component and metal will protect the surface of the component without significantly detracting from the rigidity of the mounting. Generally speaking, unless an optical component has one of a few standard shapes and sizes, it will not fit directly in a commercial mount and a custom adaptor will need to be made.

It is often necessary to mount a precision optical component (usually a flat window or mirror, but occasionally a lens or spherical mirror) so that it provides a vacuum or pressure-tight seal. There are two important questions to be asked when this is done: (1) Can the component of a certain clear aperture and thickness withstand the pressure differential to which it will be subjected, and (2) will it be distorted optically either by this pressure differential or the clamping forces holding it in place?

There are various ways of mounting a circular window so that it makes a vacuum- or pressure-tight seal. Demountable seals are generally made by the use of rubber O-rings (see Section 3.5.2) for vacuum or flat rubber gaskets for pressure. More or less permanent seals are made with the use of epoxy resin, by running a molten bead of silver chloride around the circumference of the window where it sits on a metal flange, or in certain cases by directly soldering or fusing the window in place. Some crystalline materials, such as germanium, can be soldered to metal and many others can be soldered with indium if they, and the part to which they are to be attached, are appropriately treated with gold paint beforehand.[38] If a window is fused in place, its coefficient of expansion must be well matched to that of the part to which it is fixed. Suitably matched glass, fused silica, and sapphire windows on metal are available. Either the glass and metal are themselves matched, or an intermediate graded seal is used (see Section 2.2.4). To decide what thickness-to-unsupported-diameter ratio is appropriate for a given pressure differential, it must be determined whether the method of mounting corresponds to Figure 4.83a or b. In either case, the maximum stress in the window must be

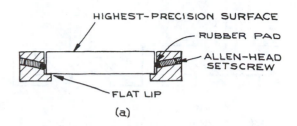

(a)

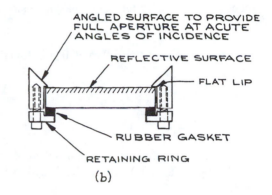

(b)

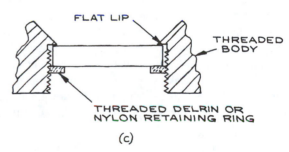

(c)

Figure 4.82   Ways for mounting delicate optical components; (a) no contact between precision surface of compnent and mount; (b) precision surface in minimal contact with mount and held in place with retaining ring; (c) precision surface in minimal contact with bottom of recess in mount and held in place with threaded plastic ring.

kept below the modulus of rupture Y of the window by an appropriate safety factor S_f. A safety factor of four is generally adequate for determining a safe window thickness, but a larger factor may be necessary to avoid deformation of the window that would lead to distortion of transmitted wavefronts. The desired thickness-to-free-

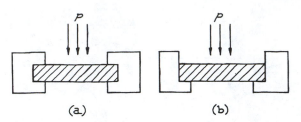

Figure 4.83   Window mounting arrangements: (a) clamped or fixed edge (maximum stress at edge); (b) unclamped or freely supported edge (maximum stress at center).

diameter ratio for a pressure differential p is found from the formula

$$\frac{t}{D} = \sqrt{\frac{KpS_f}{4Y}} \qquad (4.180)$$

where K is a constant whose value can be taken as 0.75 for a clamped edge and 1.125 for a free edge (see also Section 3.6.3).

A window mounted with an O-ring in a groove, as shown in Figure 4.84a, may not provide an adequate seal at moderately high pressures (above 1000 psi) unless the clamping force of the window on the ring is great enough to prevent the pressure differential deforming the ring in its grooves and creating a leak. A better design for high-pressure use, up to about 7000 psi, is shown in Figure 4.84b: the clamping force and internal pressure act in concert to compress the ring into position. Clamping an optical-quality window or an O-ring seal of the sort shown in Figure 4.84a may cause slight bending of the window. If an optical window has been fixed *permanently* in place with epoxy resin, it may be possible to recover the window—if it is made of a robust material—by heating the seal to a few hundred degrees or by soaking it in methylene chloride or a similar commercial epoxy remover.[39]

If optical windows are to be used in pressure cells where there is a pressure differential of more than 500 bars (~7000 psi), O-rings are no longer adequate and special window-mounting techniques for very high pressures must be used.[40–42] Optical cells for use at pressures up to 3 kbar are available from SLM and ISS.

4.3.15 Cleaning Optical Components

It is always desirable for the components of an optical system to be clean. At the very least, particles of dust on mirrors, lenses, and windows increase scattering. Organic films lead to unnecessary absorption, possible distortion of transmitted wavefronts, and—in experiments with ultraviolet light sources—possible unwanted fluorescence. In optical systems employing lasers, dust, and films will frequently lead to visible, undesirable interference and diffraction effects. At worst, dirty optical components exposed to high-intensity laser beams can be permanently damaged, as absorbing dust and films can be burnt into the surface of the component.

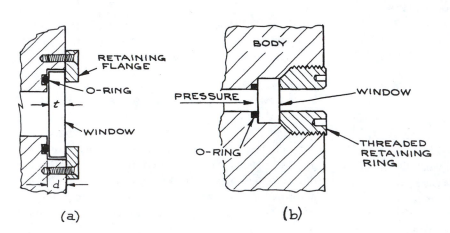

Figure 4.84   Mounting arrangements for optical windows in vacuum and pressure applications: (a) simple vacuum window; (b) high-pressure window.

The method used for cleaning optical components, such as mirrors, lenses, windows, prisms, and crystals, will depend on the nature of both the optical component and the type of contaminant to be removed. For components made of hard materials such as borosilicate glass, sapphire, quartz, or fused silica, gross amounts of contamination can be removed by cleaning in an ultrasonic tank. The tank should be filled with water and the components to be cleaned placed in a suitable pure solvent—such as trichloroethylene or acetone—contained in a clean glass or plastic beaker. The ultrasonic cleaning is accomplished by placing the beaker in the tank of water. Dust can be removed from the surface of hard materials, including most modern hard dielectric reflecting and antireflecting coatings, by brushing with a soft camelhair brush. Suitable brushes are available from most photographic stores. Lens tissue and solvent can be used to clean contaminants from both hard and medium-hard surfaces, provided the correct method is used. The category of medium-hard materials includes materials such as arsenic trisulphide, silicon, germanium, gallium arsenide, KRS-5, and tellurium. Information about the relative hardness of these and other materials is contained in Table 4.5. Dust should first be removed from the component by wiping, very gently, in one direction only, with a folded piece of dry lens tissue. The tissue should be folded so that the fingers do not come near any part of the tissue that will contact the surface. Better still, if a cylinder of dry nitrogen is available, dust can be blown from the surface of the component. Because the gas can itself contain particles, however, it should be passed through a suitable filter such as fine sintered glass before use. Clean gas can be used in this way to remove dust from the surface of even very delicate components, such as aluminum and gold-coated mirrors and diffraction gratings. Commercial *dust-off* sprays are available that use a liquified propellant gas that vaporizes on leaving the spray-can nozzle. These sprays are not recommended for cleaning delicate or expensive components. They can deposit unvaporized propellant on the component or condense water vapor into the surface of a hygroscopic material.

There are two good methods for cleaning hard and medium-hard components with lens tissue and solvent. The lens tissue should be folded as indicated previously, and a few drops of solvent placed on the tissue in the area to be used for the cleaning. The tissue should be wiped just once in a single direction across the component and then discarded. Fingers should be kept away from the solvent-impregnated part of the tissue, and it may be desirable to hold the component to be cleaned with a piece of dry tissue to avoid the transmission of finger grease to its surface. Hemostats, available from surgical supply stores, are very convenient for holding tissue during cleaning operations. To clean a single, fairly flat optical surface, such as a dielectric coated mirror or window, place a single lens tissue flat on the surface. Put one drop of clean solvent (spectroscopically pure grades are best) on the tissue over the component. Then draw the tissue across the surface of the component so that the dry portion of the tissue follows the solvent-soaked part and dries off the surface.

Several suitable solvents exist for these cleaning operations, such as acetone, trichloroethylene, and methyl, ethyl, and isopropyl alcohols. Diethyl ether is a very good solvent for cleaning laser mirrors, but its extreme flammability makes its use undesirable.

If there is any doubt as to the ability of an optical component to withstand a particular solvent, the manufacturer should be consulted. This applies particularly to specially coated components used in ultraviolet (and some infrared) systems. Some delicate or soft components cannot be cleaned at all without running the risk of damaging them permanently, although it is usually safe to remove dust from them by blowing with clean, dust-free gas. It is worth noting that diffraction gratings frequently look as though they have surface blemishes. These blemishes, however, frequently look much worse than they actually are and do not noticeably affect performance. Never try to clean diffraction gratings. It is also very difficult to clean aluminum- or gold-coated mirrors that are not overcoated with a hard protective layer such as silicon monoxide. Copper mirrors are soft and should be cleaned with extreme care.

Most optical components used in the visible and near ultraviolet regions of the spectrum are hard—soft materials are often used in infrared systems. In this region, several soft crystalline materials such as sodium chloride, potassium chloride, cesium bromide and cesium iodide are commonly used. These materials can be cleaned, essentially by repolishing their surfaces. For sodium and potassium chloride windows, this is particularly straightforward. Fold a piece of soft, clean muslin several times to form a large pad. Place the pad on a flat surface— a piece of glass is best. Stretch the pad tight on the surface and fix it securely at the edges with adhesive tape. Dampen an area of the pad with a suitable solvent such as trichloroethylene or alcohol. Place the window to be polished on the solvent-soaked area, and work it back and forth in a figure-eight motion. Gradually work the disc onto the dry portion of the pad. This operation can be repeated as many times as necessary to restore the surface of the window. Do not attempt to clean windows of this kind, which are hygroscopic, by wiping them with solvent-soaked tissue. The evaporation of solvent from the surface will condense water vapor onto it and cause damage. The above recommended cleaning procedure is quite satisfactory for cleaning even laser windows. Small residual surface blemishes and slightly foggy areas do not detract significantly from the transmission of such windows in the middle infrared and beyond—which, in any case, is the only spectral region where they should be used. For removing slightly larger blemishes and scratches from such soft windows, very fine aluminum oxide powder (available from Adolf Meller) can be mixed with the cleaning solvent on the pad in the first stages of polishing. Perhaps the best way of all for cleaning hard and medium-hard optical components is to vapor-degrease them (see Figure 3.35). This is a very general procedure for cleaning precision components. Place some trichloroethylene or isopropyl alcohol in a large beaker (500 ml or larger) to a depth of about 1 cm. Suspend the components to be cleaned in the top of the beaker so that their critical surfaces face downward—an improvised wire rack is useful for accomplishing this. Cover the top of the beaker tightly with aluminum foil. Place the beaker on an electric hot plate, and heat the solvent until it boils. The solvent vapor thus produced is very pure. It condenses on and drips off the suspended item being cleaned, effectively removing grease and dirt. If large items are being cleaned, it may be necessary to blow air on the aluminum foil with a small fan. At the end of several minutes turn off the heat and remove the clean items while they are still warm. This cleaning procedure is recommended for components that must be extremely clean, such as laser Brewster windows.

4.4 OPTICAL MATERIALS

The choice of materials for the various components of an optical system such as windows, prisms, lenses, mirrors, and filters is governed by several factors:

- Wavelength range to be covered
- Environment and handling that components must withstand
- Refractive-index considerations
- Intensity of radiation to be transmitted or reflected
- Cost

4.4.1 Materials for Windows, Lenses, and Prisms

For the purposes of classification, the characteristics of various materials for use in transmissive applications will be considered in three spectral regions:

1. Ultraviolet, 100–400 nm

2. Visible and near infrared, 400 nm–2 μm

3. Middle and far infrared, 2–1000 μm

There are, of course, many materials that can be used in part of two or even three of these regions. Table 4.5 summarizes the essential characteristics of all the common—and most of the rarely used—materials in these three regions. The useful transmission range given for each material is only a guide—the transmission at the ends of this range can be increased and the useful wavelength range of the component extended by using thinner pieces of material, if possible. The refractive index of all these materials varies with wavelength, as is illustrated for several materials in Figure 4.85. The Knoop hardness[43] is a static measure of material hardness based on the size of impression made in the material with a pyramidal diamond indenter under specific conditions of loading, time, and temperature. Roughly speaking, a material with a Knoop hardness above about 60 is hard enough to withstand the cleaning procedures described in Section 4.3.13.

Transmission curves for many of the useful optical materials summarized in Table 4.5 are collected together in Figures 4.86 through 4.95.[15] Most of these materials are

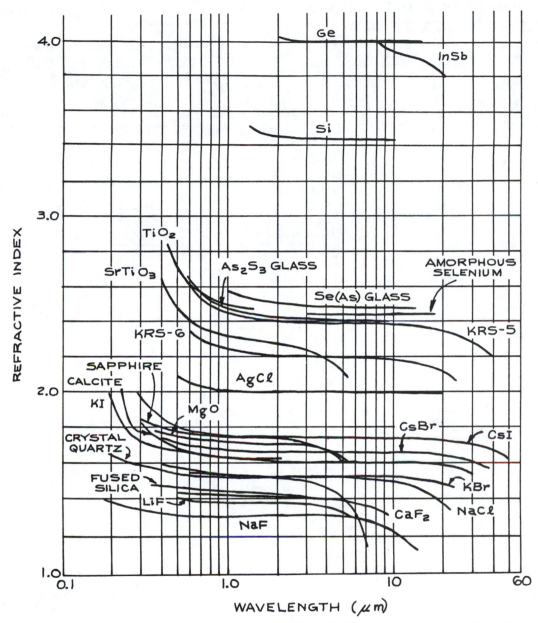

Figure 4.85   The refractive indices of various optical materials. (Adapted with permission from W. C. Wolfe and S. S. Ballard, "Optical Materials, Films, and Filters for Infrared Instrumentation," *Proceedings of the IRE*, Vol. 47, pp. 1540–1546, 1959. © 1959 IRE [now IEEE].)

TABLE 4.5 CHARACTERISTICS OF OPTICAL MATERIALS

Material	Useful Transmission Range (≥ 10% transmission) in 2 mm Thickness	Index of Refraction [wavelength (μm) in parentheses]	Knoop Hardness	Melting Point (°C)
LiF	0.104–7	1.60 (0.125), 1.34 (4.3)	100	870
MgF$_2$	0.126–9.7	$n_o = 1.3777$, $n_e = 1.38950$ (0.589)[f]	415	1396
CaF$_2$	0.125–12	1.47635 (0.2288), 1.30756 (9.724)	158	1360
BaF$_2$	0.1345–15	1.51217 (0.3652), 1.39636 (10.346)	65	1280
Sapphire(Al$_2$O$_3$)	0.15–6.3	$n_o = 1.8336$ (0.26520),[f] $n_o = 1.5864$ (5.577),[f] n_e slightly less than n_o	1525–2000[c]	2040 ±10
Fused silica (SiO$_2$)	0.165–4[d]	1.54715 (0.20254), 1.40601 (3.5)	615	1600
Pyrex 7740	0.3–2.7	1.474 (0.589), ≈ 1.5 (2.2)	≈ 600	820[g]
Vycor 7913	0.26–2.7	1.458 (0.589)	—	1200
As$_2$S$_3$	0.6–0.13	2.84 (1.0), 2.4 (8)	109	300
RIR 2	≈ 0.4–4.7	1.75 (2.2)	≈ 600	≈ 900
RIR 20	≈ 0.4–5.5	1.82 (2.2)	542	760
NaF	0.13–12	1.393 (0.185), 0.24 (24)	60	980
RIR 12	≈ 0.4–5.7	1.62 (2.2)	594	≈ 900
MgO	0.25–8.5	1.71 (2.0)	692	2800
Acrylic	0.340–1.6	1.5066 (0.4101), 1.4892 (0.6563)	—	Distorts at 72
Silver chloride (AgCl)	0.4–32	2.134 (0.43), 1.90149 (20.5)	9.5	455
Silver bromide (AgBr)	0.45–42	2.313 (0.496), 2.2318 (0.671)	≥ 9.5	432
Kel-F	0.34–3.8	—	—	—
Diamond (type IIA)	0.23–200+	2.7151 (0.2265), 2.4237 (0.5461)	5700–10400[c]	—
NaCl	0.21–25	1.89332 (0.185), 1.3403 (22.3)	18	803
KBr	0.205–25	1.55995 (0.538), 1.46324 (25.14)	7	730
KCl	0.18–30	1.78373 (0.19), 1.3632 (23)	—	776
CsCl	0.19– ≈ 30	1.8226 (0.226), 1.6440 (0.538)	—	646
CsBr	0.21–50	1.75118 (0.365), 2.55990 (39.22)	19.5	636
KI	0.25–40	2.0548 (0.248), 1.6381 (1.083)	5	723
CsI	0.235–60	1.98704 (0.297), 1.61925 (53.12)	—	621
SrTiO$_3$	0.4–7.4	2.23 (2.2), 2.19 (4.3)	620	2080
SrF$_2$	0.13–14	1.438 (0.538)	130	1450
Rutile (TiO$_2$)	0.4–7	$n_o = 2.5$ (1.0), $n_e = 2.7$ (1.0)[f]	880	1825
Thallium bromide (TlBr)	0.45–45	2.652 (0.436), 2.3 (0.75)	12	460
Thallium bromoiodide (KRS-5)	0.56–60	2.62758(0.577), 2.21721(39.38)	40	414.5
Thallium chlorobromide (KRS-6)	0.4–32	2.3367(0.589), 2.0752(24)	39	423.5
ZnSe	0.5–22	2.40(10.6)	150	—
Irtran 2 (ZnS)	0.6–15.6	2.26(2.2), 2.25(4.3)	354	800
Si	1.1–15[e]	3.42(5.0)	1150	1420
Ge	1.85–30[e]	4.025(4.0), 4.002(12.0)	692	936
GaAs	1–15	3.5(1.0), 3.135(10.6)	750	1238
CdTe	0.9–16	2.83(1.0), 2.67(10.6)	45	1045
Te	3.8–8+	$n_o = 6.37(4.3)$, $n_e = 4.93(4.3)$[f]	—	450
CaCO$_3$	0.25–3	$n_o = 1.90284(0.200)$, $n_e = 1.57796(0.198)$[f] $n_o = 1.62099(2.172)$, $n_e = 1.47392(3.324)$	135	894.4[h]

[a] Parallel to c-axis.
[b] Perpendicular to c-axis.
[c] Depends on crystal orientation.
[d] Depends on grade.

TABLE 4.5 CHARACTERISTICS OF OPTICAL MATERIALS (CONTINUED)

Material	Thermal-Expansion Coefficient $(10^{-6}/°C)$	Solubility in Water $[g/(100\ g),\ 20°C]$	Soluble in	Comments
LiF	9	0.27	HF	Scratches easily
MgF$_2$	16	7.6×10^{-3}	HNO$_3$	—
CaF$_2$	25	1.1×10^{-3}	NH$_4$ salts	Not resistant to thermal or mechanical shock
BaF$_2$	26	0.12	NH$_4$Cl	Slightly hygroscopic, sensitive to thermal shock
Al$_2$O$_3$	6.66,[a] 5.0[b]	9.8×10^{-5}	NH$_4$ salts	Very resistant to chemical attack, excellent material
SiO$_2$	0.55	0.00	HF	Excellent material
Pyrex	3.25	0.00	HF, hot H$_2$PO$_4$	Excellent mechanical, optical properties
Vycor	0.8	0.00	HF, hot H$_2$PO$_4$	Excellent mechanical, optical properties
As$_2$S$_3$	26	5×10^{-5}	Alcohol	Nonhygroscopic
RIR 2	8.3	0.00	1% HNO$_3$	Good mechanical, optical properties
RIR 20	9.6	—	—	Good mechanical, optical properties
NaF	36	4.2	HF	Lowest ref. index of all known crystals
RIR 12	8.3	—	—	Good mechanical and optical properties
MgO	43	6.2×10^{-4}	NH$_4$ salts	Nonhygroscopic; surface scum forms if stored in air
Acrylic	110–140	0.00	Methylene chloride	Easily scratched, available in large sheets
AgCl	30	1.5×10^{-4}	NH$_4$OH, KCN	Corrosive, nonhygroscopic, cold flows
AgBr	—	12×10^{-6}	KCN	Cold flows
Kel–F	—	—	—	Soft, easily scratched
Diamond	0.8	0.00	—	Hardest material known; thermal conductivity 6 × Cu at room temp., chemically inert
NaCl	44	36	H$_2$O, glycerine	Corrosive, hygroscopic
KBr	—	65.2	H$_2$O, glycerine	Hygroscopic
KCl	—	34.35	H$_2$O, glycerine	Hygroscopic
CsCl	—	186	Alcohol	Very hygroscopic
CsBr	48	124	H$_2$O	Soft, easily scratched, hygroscopic
KI	—	144.5	Alcohol	Soft, easily scratched, hygroscopic, sensitive to thermal shock
CsI	50	160	Alcohol	Soft, easily scratched, very hygroscopic
SrTiO$_3$	9.4	—	—	Refractive index $\simeq \sqrt{5}$
SrF$_2$	—	1.17×10^{-2}	Hot HCl	Slightly sensitive to thermal shock
TiO$_2$	9	0.00	H$_2$SO$_4$	Nonhygroscopic, nontoxic
TlBr	—	0.0476	Alcohol	Flows under pressure, toxic
KRS-5	51	<0.0476	HNO$_3$, aqua regia	Cold-flows, nonhygroscopic, toxic
KRS-6	60	0.32	HNO$_2$, aqua regia	Cold-flows, nonhygroscopic, toxic
ZnSe	8.5	0.001	—	Very good infrared materials, also transparent to some visible light
ZnS	—	6.5×10^{-5}	HNO$_3$, H$_2$SO$_4$	—
Si	4.2	0.00	HF + HNO$_3$	Resistant to corrosion, must be highly
Ge	5.5	0.00	Hot H$_2$SO$_4$, aqua regia	polished to reduce scattering losses at surface
GaAs	5.7	0.00	—	Very good high-power infrared-laser window material
CdTe	4.5	—	—	—
Te	16.8	0.00	H$_2$SO$_4$, HNO$_3$	Poisonous, soft, easily scratched
CaCO$_3$	—	1.4×10^{-3}	Acids, NH$_4$Cl	—

[e] Long-wavelength limit depends on purity of material.
[f] Birefringent.
[g] Softening temperature.
[h] Decomposition temperature.

available from several suppliers. Some of the materials listed in Table 4.5 are worthy of brief extra comment.

Ultraviolet Transmissive Materials

The following are suitable for use in the UV region.

(i) Lithium fluoride. Lithium fluoride (LiF, see Figure 4.86) has useful transmittance further into the vacuum ultraviolet than any other common crystal. The transmission of vacuum-ultraviolet quality crystals is more than 50 percent at 121.6 nm for 2 mm thickness, and thinner crystals have useful transmission down to 104 nm. The short wavelength transmission of the material, however, deteriorates on exposure to atmospheric moisture or high-energy radiation. Moisture does not affect the infrared transmission, which extends to 7 μm. Several grades of LiF are available. Vacuum ultraviolet-grade material is soft, but visible-grade material is hard. LiF can be used as a vacuum-tight window material by sealing with either silver chloride or a suitable epoxy.

(ii) Magnesium fluoride. Vacuum-ultraviolet-grade magnesium fluoride (MgF_2, see Figure 4.86) transmits farther into the ultraviolet than any common material except LiF. The transmission at 121.6 nm is 35 percent or more for a 2 mm thickness. MgF_2 is recommended for use as an ultraviolet transmissive component in space work, as it is only slightly affected by ionizing radiation. MgF_2 is birefringent and is used for making polarizing components in the ultraviolet. Irtran 1 is a polycrystalline form of magnesium fluoride manufactured by Kodak, which is not suitable for ultraviolet or visible applications. Its useful transmission extends from 500 nm to 9 μm.

(iii) Calcium fluoride. Calcium fluoride (CaF_2, see Figure 4.86) is an excellent, hard material that can be used for optical components from 125 nm to beyond 10 μm. Vacuum ultraviolet-grade material transmits more than 50 percent at 125.7 nm for a 2 mm path length. Calcium fluoride is not significantly affected by atmospheric moisture at ambient temperature. Calcium fluoride lenses are available from Argus, Janos Technology, Linos Photonics, Meller Optics, Melles Griot, Newport,

Optovac, Oriel, and Precision Optical. Irtran 3 is a polycrystalline form of CaF_2, which is not suitable for ultraviolet or visible applications but transmits in the infrared to 11.5 μm.

(iv) Barium fluoride. Although barium fluoride (BaF_2, see Figure 4.87) is slightly more water soluble than calcium or magnesium fluoride, it is more resistant than either of these to ionizing radiation. It is a good general-purpose optical material from ultraviolet to infrared. It has excellent transmission to beyond 10 μm, and windows that are not too thick (< 2mm) can be used in CO_2-laser applications except at very high energy densities. Barium fluoride lenses are available from Infrared Optical Products and International Scientific Products.

(v) Synthetic sapphire. Synthetic sapphire (corundum, Al_2O_3; see Figure 4.87) is probably the finest optical material available in applications from 300 nm to 4 μm. It is very hard, strong, and resistant to moisture and chemical attack (even HF below 300°C). The useful transmissive range extends from 150 nm to 6 μm, the transmission of a 1 mm thickness being 21 percent and 34 percent, respectively, at these two wavelengths. Sapphire Optical components are available from many suppliers.[27]

Sapphire is also resistant to ionizing radiation, has high thermal conductivity, and can be very accurately fabricated in a variety of forms. It has low dispersion, however, and is not very useful as a prism material. Chromium-doped Al_2O_3 (ruby) is used in laser applications.

(vi) Fused silica. Fused silica (SiO_2, see Figure 4.88) is the amorphous form of crystalline quartz. It is almost as good an optical material as sapphire and is much cheaper. Its useful transmission range extends from about 170 nm to about 4.5 μm. Special ultraviolet-transmitting grades are manufactured (such as Spectrosil and Suprasil), as well as infrared grades (such as Infrasil) from which undesirable absorption features at 1.38, 2.22, and 2.72 μm due to residual OH radicals have been removed. Fused silica has a very low coefficient of thermal expansion, $5\times10^{-7}/K$ in the temperature range from 20 to 900°C, and is very useful as a spacer material in applications such as

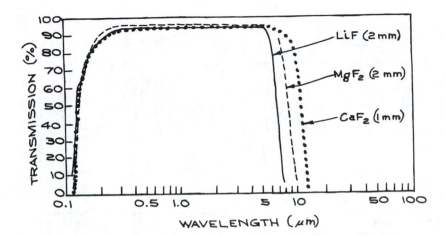

Figure 4.86   Transmission curves of lithium flouride, magnesium flouride, and calcium flouride windows of specified thickness.

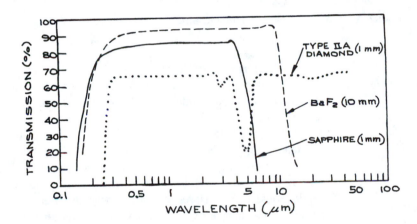

Figure 4.87   Transmission curves of sapphire, barium flouride, and type-IIA diamond windows of specified thickness.

Fabry-Perot etalons and laser cavities. Fused-silica optical components such as windows, lenses, prisms, and etalons are readily available. Crystalline quartz is birefringent and is widely used in the manufacture of polarizing optics, particularly quarter-wave and half-wave plates.

Very pure fused silica is also the material of which most optical fibers are made. In fiber optic applications, the index gradient inside the fiber is generally produced by germania (GeO_2) doping. The pure silica used in optical fibers is most transparent in the infrared, generally near 1.55 μm.

(vii) Diamond.   Windows made of type-IIA diamond (see Figure 4.87) have a transmittance that extends from 230 nm to beyond 200 μm, although there is some absorption between 2.5 and 6.0 μm. The high refractive index of diamond ($n > 2.4$) over a very wide band leads to a reflection loss of about 34 percent for two surfaces. Diamond windows are extremely expensive, ($10,000–20,000 for a 1 cm-diameter, 1 mm-thick window). Some properties of diamond, however, are unique—its thermal conductivity is six times that of pure copper at 20°C, and it is the hardest material known, very resistant to chemical attack, with a low thermal-expansion coefficient, and high resistance to radiation damage.

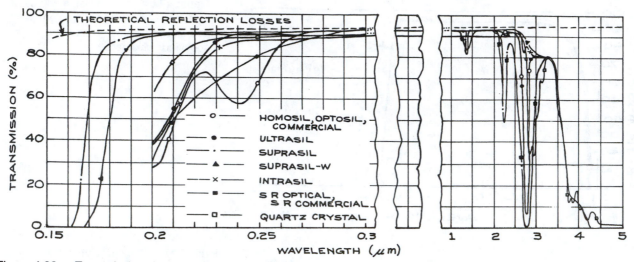

Figure 4.88 Transmission characteristics of various grades of fused silica; all measurements made on 10 mm thick materials. (Courtesy of Heraeus-Amersiol, Inc.)

Visible- and Near Infrared-Transmissive Materials

All the materials so far considered as ultraviolet-transmitting are also excellent for use in the visible and near infrared. There are several materials, however, that transmit well in the visible and near infrared but are not suitable for ultraviolet use.

(i) Glasses.

There is an extremely large number of types of glass used in the manufacture of optical components. They are available from many manufacturers, such as Chance, Corning, Hoya, O'Hara, and Schott.

Two old classifications of glass, *crown* and *flint* glass, can be related to the glass chart. Crown glasses are glasses with a V value greater than 55 if $n_d < 1.6$ and $V > 50$ if $n_d > 71.6$. The flint glasses have V values below these limits. The rare earth glasses contain rare earths instead of SiO_2, which is the primary constituent of crown and flint glasses. Glasses containing lanthanum contain the symbol La.

A few common glasses are widely used in the manufacture of lenses, windows, prisms, and other components. These include the borosilicate glasses Pyrex, BK7/A, and crown (BSC), as well as Vycor, which is 96 percent fused quartz. Transmission curves for Pyrex No. 7740 and Vycor No. 7913 are shown in Figure 4.92. These glasses are not very useful above 2.5 μm—their transmission at 2.7 μm is down to 20 percent for a 10 mm thickness. In the spectral region between 350 nm and 2.5 μm, however, they are excellent transmissive materials. They are inexpensive, hard, chemically resistant, and can be fabricated and polished to high precision. Special glasses for use further into the infrared are available from Hoya and Corning. In window applications, these glasses do not appear to offer any particular advantages over more desirable materials such as infrared-grade fused silica or sapphire. In specialized applications such as infrared lens design, however, their higher refractive induces are useful.

For special applications, many colored filter glasses can be obtained. Glasses that transmit visible but not infrared, or vice versa, or that transmit some near ultraviolet but no visible (and many other combinations) are available. The reader should consult the catalogs published by the various manufacturers and suppliers such as Corning, Schott, Chance-Pilkington, Kodak, Oriel, or Rolyn.

Special glasses whose expansion coefficients match selected metals can be made because the composition of glass is continuously variable. These glasses are useful for sealing to the corresponding metals in order to make windows on vacuum chambers. Corning glass No. 7056 is often used in this way, as it can be sealed to the alloy Kovar. Vacuum chamber windows of quartz or sapphire are also available,[44] but their construction is complex, which makes them costly. Various grades of flint glass are available that have high refractive indices and are therefore useful for prisms. For a tabulation of the types available the reader is referred to Kaye and Laby.[5] The refractive indices available range up to 1.93 for Chance-Pilkington Double Extra Dense Flint glass No. 927210, but this and other high-refractive-index glasses suffer long-term damage such as darkening when exposed to the atmosphere.

(ii) Plastics. In certain noncritical applications, such as observation windows, transparent plastics such as polymethylmethacrylate (Acrylic, Plexiglas, Lucite, Perspex) or polychlorotrifluoroethylene (Kel-F) are cheap and satisfactory. They cannot be easily obtained in high optical quality (with very good surface flatness, for example) and are easily scratched. The useful transmission range of Acrylic windows 1 cm thick runs from about 340 nm to 1.6 μm. A transmission-versus-wavelength curve for this material is shown in Figure 4.89. One important application of Acrylic is in the manufacture of Fresnel lenses (see Section 4.3.3). Kel-F is useful up to about 3.8 μm, although it shows some decrease in transmission near 3 μm. It is similar to polytetrafluoroethylene (Teflon) and is very resistant to a wide range of chemicals—even gaseous fluorine.

(iii) Arsenic trisulfide. Arsenic trisulfide (As_2S_3, see Figure 4.90) is a red glass that transmits well from 800 nm to 10 μm. Its usefulness stems from its relatively low price, ease of fabrication, nontoxicity, and resistance to moisture. Although primarily an infrared-transmissive material, it is also transmissive in the red, which facilitates the alignment of optical systems containing it. As_2S_3 is relatively soft and will cold-flow over a long period of time. A wide range of As_2S_3 lenses are available from Amorphous Materials, Pacific Optical, and Reynard.

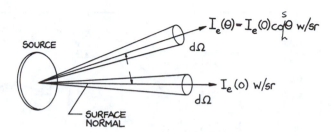

Figure 4.89　Transmission versus wavelength of 3.175 mm thick acrylic. (Courtesy of Melles Griot, Inc.)

Middle- and Far Infrared-Transmissive Materials

The following are useful well into the IR region.

(i) Sodium chloride. Sodium chloride (NaCl, see Figure 4.91) is one of the most widely used materials for infrared-transmissive windows. Although it is soft and hygroscopic, it can be repolished by simple techniques and maintain exposure to the atmosphere for long periods without damage simply by maintaining its temperature higher than ambient. Two simple ways of doing this are to mount a small tungsten-filament bulb near the window, or to run a heated wire around the periphery of small windows. High-precision sodium chloride windows are in widespread use as Brewster windows in high-energy, pulsed CO_2 laser systems.

(ii) Potassium chloride. Potassium chloride (KCl, see Figure 4.91) is very similar to sodium chloride. It is also an excellent material for use as a Brewster window in pulsed CO_2 lasers, as its transmission at 10.6 μm is slightly greater than that of sodium chloride.

(iii) Cesium bromide. Cesium bromide (CsBr, see Figure 4.91) is soft and extremely soluble in water, but has useful transmission to beyond 40 μm. It will suffer surface damage if the relative humidity exceeds 35 percent.

(iv) Cesium iodide. Cesium iodide (CsI, see Figure 4.93) is used for both infrared prisms and windows—it is similar to CsBr, but its useful transmission extends beyond 60 μm.

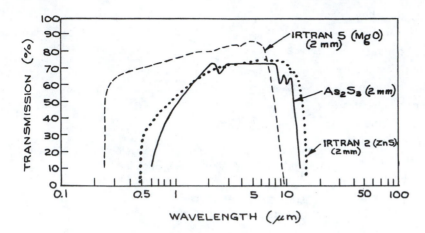

Figure 4.90   Transmission curves for Irtran 2 (ZnS), Irtran 5 (MgO), and As$_2$S$_3$ windows of specified thickness.

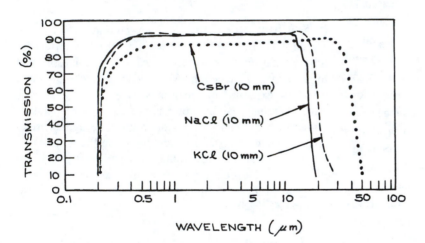

Figure 4.91   Transmission curves for cesium bromide, potassium chloride, and sodium chloride windows of specified thickness.

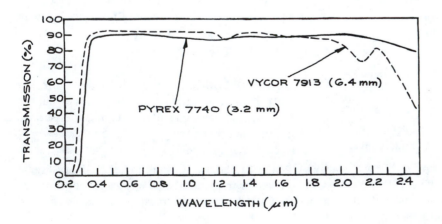

Figure 4.92   Transmission curves for Pyrex 7740 and Vycor 7913.

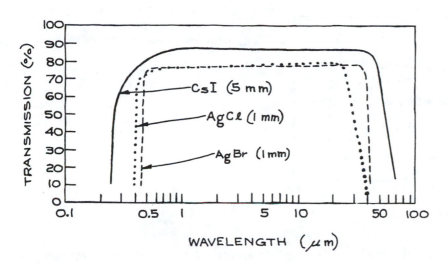

Figure 4.93   Transmission curves for cesium iodide, silver bromide, and silver chloride windows of specified thickness.

(v) Thallium bromoiodide (KRS-5).
Thallium bromoiodide (see Figure 4.94), widely known as KRS-5, is an important material because of its wide transmission range in the infrared—from 600 nm to beyond 40 μm for a 5 mm thickness—and its small solubility in water. KRS-5 will survive atmospheric exposure for long periods of time and can even be used in liquid cells in direct contact with aqueous solutions. Its refractive index is high—reflection losses at 30 μm are 30 percent and it has a tendency to cold-flow. KRS-5 lenses are available from Bicron, Infrared Optical Products, International Scientific Products, and Janos Technology.

(vi) Zinc selenide.
Chemical-vapor-deposited zinc selenide (ZnSe, see Figure 4.94) is used widely in infrared-laser applications. It is transmissive from 500 nm to 22 μm. Zinc selenide is hard enough that it can be cleaned easily. It is resistant to atmospheric moisture and can be fabricated into precision windows and lenses. Its transmissive qualities in the visible (it is orange-yellow in color) makes the alignment of infrared systems that use it much more convenient than in systems using germanium, silicon, or gallium arsenide. Irtran 4 is a polycrystalline form of zinc selenide manufactured by Kodak.

(vii) Plastic films.
Several polymer materials are sold commercially in film form, including polyethylene (Polythene, Polyphane, Poly-Fresh, Dinethene, etc.), polyvinylidene chloride copolymer (Saran Wrap), polyethylene terephthalate (Mylar, Melinex, Scotchpar), and polycarbonate (Lexan). These plastic materials can be used to wrap hygroscopic crystalline optical components to protect them from atmospheric moisture. If the film is wrapped tightly, it will not substantially affect the optical quality of the wrapped components in the infrared. Care should be taken to use a polymer film that transmits the wavelength for which the protected component is to be used. Several commercial plastic food wraps, for example, can be used to wrap NaCl or KCl windows for use in CO_2-laser applications. Not all will prove satisfactory, however, and a particular plastic film should be tested for transmission before use. A note of caution: Plastic films contain plasticizers, from which the vapor may be a problem.

(viii) Plastics and other materials for far-infrared use.
Plastics are widely used for windows, beamsplitters, and light guides in the far infrared. A Mylar film is widely used as the beam-splitting element in far-infrared Michelson interferometers, and polyethylene, polystyrene, nylon, and Teflon are used as windows in far-infrared spectrometers and lasers. Black polyethylene excludes visible and near-infrared radiation but transmits far-infrared. Lenses and stacked-sheet polarizers for far-infrared use can be made from polyethylene. Several crystalline and semiconducting materials, although they

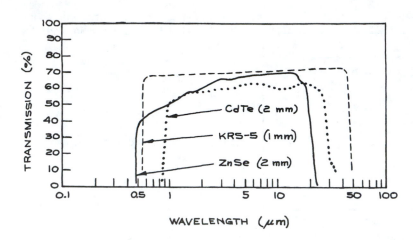

Figure 4.94 Transmission curves for zinc selenide, cadmium telluride, and KRS-5 windows of specified thickness.

absorb in the middle infrared, begin to transmit again in the far infrared. A 1 mm thick quartz window has 70 percent transmission at 100 μm and more than 90 percent transmission from 300 to 1000 μm. The alkali and alkaline-earth halides such as NaF, LiF, KBr, KCl, NaCl, CaF_2, SrF_2, and BaF_2 show increasing transmission for wavelengths above a critical value that varies from one crystal to another but is in the 200 to 400 μm range. Further details of the specialized area of far-infrared materials and instrumentation are available in the literature.[45–47]

Semiconductor Materials

Several of the most valuable infrared optical materials are semiconductors. In very pure form, these materials should be transmissive to all infrared radiation with wavelength longer than that corresponding to the band gap. In practice, the long-wavelength transmission will be governed by the presence of impurities—a fact that is put to good use in the construction of infrared detectors. Semiconductors make very convenient *cold mirrors*, as they reflect visible radiation quite well but do not transmit it.

(i) Silicon. Silicon (Si, see Figure 4.95) is a hard, chemically resistant material that transmits wavelengths beyond about 1.1 μm. It does, however, show some absorption near 9 μm, so it is not suitable for CO_2-laser windows. It has a high refractive index (the reflection loss

for two surfaces is 46% at 10 μm), but can be antireflection-coated. Because of its high refractive index, it should always be highly polished to reduce surface scattering. Silicon has a high thermal conductivity and a low thermal expansion coefficient, and it is resistant to mechanical and thermal shock. Silicon mirrors can therefore handle substantial infrared-laser power densities. Silicon lenses are available from Argyl International, Hampton Controls, International Scientific Products, Lambda Research Optics, Laser Research Optics, and V-A Optical Labs.

(ii) Germanium. Germanium (Ge, see Figure 4.95) transmits beyond about 1.85 μm. Very pure samples can be transparent into the microwave region. It is somewhat brittle, but is still one of the most widely used window and lens materials in infrared-laser applications. It is chemically inert and can be fabricated to high precision. Germanium has good thermal conductivity and a low coefficient of thermal expansion; it can be soldered to metal. In use its temperature should be kept below 40°C, as it exhibits thermal runaway—its absorption increases with increasing temperature. A wide range of germanium lenses are available from Argyle International, Hampton Controls, Lambda Research Optical, Laser Research Optics, and Oriel.

(iii) Gallium arsenide. Gallium arsenide (GaAs, see Figure 4.95) transmits from 1 to 15 μm. It is a better,

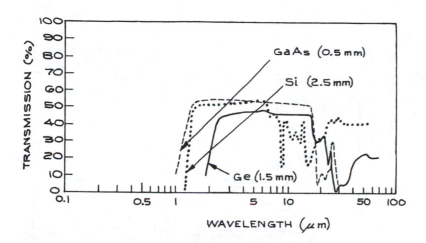

Figure 4.95 Transmission curves for the semiconductor materials germanium, gallium arsenide, and silicon.

although more expensive, material than germanium, particularly in CO- and CO_2-laser applications. It is hard and chemically inert, and maintains a very good surface finish. It can handle large infrared power densities because it does not exhibit thermal runaway until it reaches 250°C. It is also used for manufacturing infrared-laser electro-optic modulators. Gallium arsenide lenses are available from Infrared Optical Products, International Scientific Products, and REFLEX Analytical.

(iv) Cadmium telluride. Cadmium telluride (CdTe, see Figure 4.94) is another excellent material for use between 1 and 15 μm. It is quite hard, chemically inert, and takes a good surface finish. It is also used for manufacturing infrared-laser electro-optic modulators. Irtran 6 is a polycrystalline form of cadmium telluride available from Kodak—it transmits to 31 μm. Cadmium telluride lenses are available from II-VI, International Scientific Products, and REFLEX Analytical.

(v) Tellurium. Tellurium is a soft material that is transparent from 3.3 μm to beyond 11 μm. It is not affected by water. Its properties are highly anisotropic. It can be used for second harmonic generation with CO_2 laser radiation. Tellurium should not be handled, as it is toxic and can be absorbed through the skin.

4.4.2 Materials for Mirrors and Diffraction Gratings

Metal Mirrors

The best mirrors for general broadband use have pure metallic layers, vacuum-deposited or electrolytically-deposited on glass, fused silica, or metal substrates. The best metals for use in such reflective coatings are aluminum, silver, gold, and rhodium. Solid metal mirrors with highly polished surfaces made of metals such as stainless steel, copper, zirconium-copper, and molybdenum are also excellent, particularly in the infrared. The reflectance of a good evaporated metal coating exceeds that of the bulk metal. Flat solid-metal mirrors can be made fairly cheaply by machining the mirror blank, surface-grinding it, and finally polishing the surface. If precision polishing facilities are not available, there are numerous optical component suppliers who will undertake the polishing of such surface-ground blanks to flatnesses as good as $\lambda/10$ in the visible. Finished flat and spherical solid-metal mirrors are available from AOET, Opti-Forms, Oriel, Surface Finishes, Space Optics Research Labs (SORL), and Spawr. Solid-metal mirrors are most valuable for handling high-power infrared laser beams when metal-coated mirrors cannot withstand the power dissipation in the slightly absorbing mirror surface.

The spectral reflectances in normal incidence of several common metal coatings are compared in Figure 4.96.

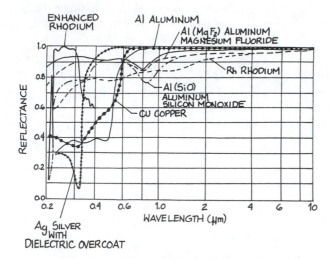

Figure 4.96 Reflectance of freshly deposited films of aluminum, copper, gold, rhodium, and coated silver as a function of wavelength from 0.2 to 10 μm.

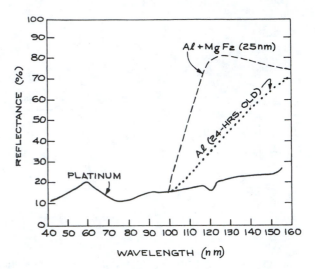

Figure 4.97 Normal-incidence reflectance of platinum, aluminum with a 25 nm thick overcoat of MgF$_2$, and unprotected aluminum 24 hours after film deposition.

Several points are worthy of note. Gold and copper are not good for use as reflectors in the visible region. Unprotected gold, silver, copper, and aluminum are very soft. Aluminum rapidly acquires a protective layer of oxide after deposition; this significantly reduces the reflectance in the vacuum ultraviolet and contributes to increased scattering throughout the spectrum. Aluminum and gold are often commercially supplied with a thin protective dielectric layer, which increases their resistance to abrasion without significantly affecting their reflectance and also protects the aluminum from oxidation. The protective layer is usually a $\lambda/2$ layer (at 550 nm) of SiO for aluminum mirrors used in the visible region of the spectrum. Aluminum coated with a thin layer of MgF$_2$ can be used as a reflector in the vacuum ultraviolet, although the reflectance is substantially reduced below about 100 nm, as shown in Figure 4.97. Many optical instruments operating in the vacuum ultraviolet, including some spectrographs, have at least one reflective component. Above 100 nm, coated aluminum is almost always used for this reflector. Below 100 nm, platinum and indium have superior reflectance to aluminum. At very short wavelengths, below about 40 nm, the reflectance of platinum in normal incidence is down to only 10 percent, as shown in Figure 4.97. Reflective

components in this wavelength region are therefore generally used at grazing incidence, as the reflectance under these conditions can remain high. At 0.832 nm, for example, the reflectance of an aluminum film at a grazing angle of $\frac{1}{2}°$ is above 90 percent.

Freshly evaporated silver has the highest reflectance of any metal in the visible, but it tarnishes so rapidly that it is rarely used except on internal reflection surfaces. In this case, the external surface is protected with a layer of Inconel or copper and a coat of paint.

Multilayer Dielectric Coatings

Extremely high-reflectance mirrors (up to 99.99%) can be made by using multilayer dielectric films involving alternate high- and low-refractive index layers, ranging around $\lambda/4$ thick, deposited on glass, metal, or semiconductor substrates. The layers are made from materials that are transparent in the wavelength region where high reflectance is required. Both narrow-band and broad-band reflective coated mirrors are commercially available. Figure 4.32 shows an example of each. Multilayer or single-layer dielectric-coated mirrors having almost any desired reflectance at wavelengths from 150 nm to 20 μm are also commercially available.[27]

Substrates for Mirrors

The main factors influencing the choice of substrate for a mirror are (1) the dimensional stability required, (2) thermal dissipation, (3) mechanical considerations such as size and weight, and (4) cost. Glass is an excellent substrate for most totally reflective mirror applications throughout the spectrum, except where high thermal dissipation is necessary. Glass is inexpensive and strong, and takes a surface finish as good as $\lambda/200$ in the visible. Fused silica and certain ceramic materials such as Zerodur or Cervit (both available from Corning Glass Works, Optical Products Department) are superior—but more expensive—alternatives to glass when slightly greater dimensional stability and a lower coefficient of thermal expansion are required. Fused silica has the lowest coefficient of thermal expansion of any readily available material. Some ceramics and the alloy Superinvar have lower values, but generally over a restricted temperature range. Machining or polishing of these materials can also produce stresses in them and increase their expansion coefficients. Metal mirrors are frequently used when a very high light flux must be reflected but even the small absorption loss in the reflecting surface necessitates cooling of the substrate. Metals are, however, not as dimensionally stable as glasses or ceramics, and their useas mirrors should be avoided in precision applications—particularly in the visible or ultraviolet.

Partially reflecting, partially transmitting mirrors are used in many optical systems, such as interferometers and lasers. In this application, the substrate for the mirror has to transmit in the spectral region being handled. The substrate material should possess the usually desirable properties of hardness and dimensional stability, plus the ability to be coated. Germanium, silicon, and gallium arsenide, for example, are frequently used as partially transmitting mirror substrates in the near infrared. Only the least expensive partially reflecting mirrors use thin metal coatings. Such coatings are absorbing, and multilayer, dielectric-coated reflective surfaces are much to be preferred.

There are numerous suppliers of totally and partially reflecting mirrors.[27] The experimentalist need only determine the requirements and specify the mirror appropriately. Mirrors that reflect visible radiation and transmit infrared *(cold mirrors)* or transmit visible radiation and reflect infrared *(hot mirrors)* are available as standard items from several suppliers.[27]

Diffraction Gratings

Diffraction gratings are generally ruled on glass, which may then be coated with aluminum or gold for visible or infrared reflective use, respectively. Many commercial gratings are replicas made from a ruled master, although high quality holographic gratings are available from several suppliers.[27] For high-power infrared laser applications, gold-coated replica gratings are unsatisfactory and master gratings ruled on copper should be used. Such gratings are available from American Holographic, Gentec, Jobin-Yvon, Richardson Grating Laboratory, and Rochester Photonics.

4.5 OPTICAL SOURCES

Optical sources fall into two categories—incoherent sources such as mercury or xenon arc, tungsten filament, and sodium lamps, and coherent sources (lasers). There are many different types of laser, some of which exhibit a high degree of coherence while others are not significantly more coherent than a line source—such as a low-pressure mercury lamp.

Incoherent sources fall into two broad categories—line sources and continuum sources. Line sources emit most of their radiation at discrete wavelengths, which correspond to strong spectral emission features of the excited atom or ion that is the emitting species in the source. *Continuum sources* emit over some broad spectral region, although their radiant intensity varies with wavelength.

4.5.1 Coherence

There are two basic coherence properties of an optical source. One is a measure of its relative spectral purity. The other depends on the degree to which the wavefronts coming from the source are uniphase or of fixed spatial phase variation. A wavefront is said to be uniphase if it has the same phase at all points on the wavefront. The Gaussian TEM$_{00}$ mode emitted by many lasers has this property.

The two types of coherence are illustrated by Figure 4.98. If the phases of the electromagnetic field at two points, A and B, longitudinally separated in the direction of propagation, have a fixed relationship, then the wave is said to be temporally coherent for times corresponding to the distance from A to B. The existence of such a fixed phase relationship could be demonstrated by combining waves extracted at points separated like A and B and producing an interference pattern. The Michelson interferometer (see Section 4.7.6a) demonstrates the existence of such a fixed phase relationship. The maximum separation of A and B for which a fixed phase relationship exists is called the *coherence length* of the source l_c. The coherence length of a source is related to its *coherence time* τ_c by

$$l_c = cd\tau_c \qquad (4.181)$$

Conceptually, l_c is the average length of the uninterrupted wave trains from the source between random phase interruptions. Fourier theory demonstrates that the longer in time a sinusoidal wave train is observed, the narrower is its frequency spread. The spectral width of a source can therefore be related to its coherence time by writing

$$\Delta \nu \sim 1/\tau_c \qquad (4.182)$$

The most coherent conventional lamp sources are stabilized, low-pressure mercury lamps, which can have τ_c ranging up to about 10 ns. Lasers, however, can have τ_c ranging up to at least 1 ms.

Spatial coherence is a measure of phase relationships in the wavefront transverse to the direction of wave propagation. In Figure 4.98, if a fixed phase relationship exists between points C and D in the wavefront, the wave is said to be spatially coherent over this region. The *area of coherence* is a region of the wavefront within which all points have fixed phase relationships. Spatial coherence can be demonstrated by placing pinholes at different locations in the wavefront and observing interference

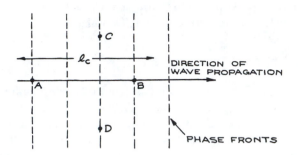

Figure 4.98   Temporal and spatial coherence.

fringes. Temporally incoherent sources can exhibit spatial coherence. Small sources (*point sources*) fall into this category. The light from a star can be spatially coherent.

Extended, temporally incoherent sources have a low degree of spatial coherence because light coming from different parts of the source has different phases. Lasers emitting a TEM$_{00}$ (fundamental) mode have very good spatial coherence over their whole beam diameter.

4.5.2 Radiometry: Units and Definitions

Radiometry deals with the measurement of amounts of light. In radiometric terms, the characteristics of a light source can be specified in several ways.

Radiant power W, measured in watts, is the total amount of energy emitted by a light source per second. The spectral variation of radiant power can be specified in terms of the radiant power density per unit wavelength interval W_λ. Clearly,

$$W = \int_0^\infty W_\lambda d\lambda \qquad (4.183)$$

If a light source emits radiation only for some specific duration—which may be quite short in the case of a flashlamp—it is more useful to specify the source in terms of its radiant energy output Q_e, measured in joules. If the source emits radiation for a time T, we can write

$$Q_e = \int_0^T W(t)dt \qquad (4.184)$$

The amount of power emitted by a source in a particular direction per unit solid angle is called *radiant intensity* I_e, and is measured in watts per steradian. In general,

$$W = \oint I_e(\omega)d\omega \qquad (4.185)$$

where the integral is taken over a closed surface surrounding the source. If I_e is the same in all directions, the source is said to be an isotropic radiator. At a distance r from such a source—if r is much greater than the dimensions of the source—the radiant flux crossing a small area ΔS is

$$\phi_e = \frac{I_e \Delta S}{r^2} \qquad (4.186)$$

The *irradiance* at this point, measured in W m^{-2}, is

$$E_e = \frac{I_e}{r^2} \qquad (4.187)$$

which is equal to the average value of the Poynting vector measured at the point. The radiant flux emitted per unit area of a surface (whether this be emitting or merely reflecting and scattering radiation) is called the *radiant emittance* M_e, and it is measured in W m^{-2}. For an extended source, the radiant flux emitted per unit solid angle per unit area of the source is called its *radiance* or *brightness* L_e:

$$L_e = \frac{\delta I_e}{\delta S_n} \qquad (4.188)$$

where the area δS_n is the projection of the surface element of the source in the direction being considered. When the light emitted from a source or scattered from a surface has a radiance that is independent of viewing angle, the source or scatterer is called a perfectly diffuse or *Lambertian radiator*. For such a source, the radiant intensity at an angle θ to the normal to the surface is clearly

$$I_e(\theta) = I_e(0)\cos\theta \qquad (4.189)$$

as illustrated in Figure 4.99.

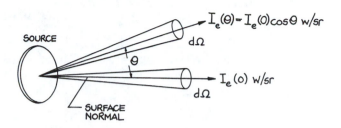

Figure 4.99   Radiant intensity characteristics of a Lambertian radiator.

The total flux emitted per unit area of such a surface is its radiant emittance, which in this case is

$$M_e = \pi I_e(0) \qquad (4.190)$$

Illuminated diffusing surfaces made of finely ground glass or finely powered magnesium oxide will behave as Lambertian radiators.

For plane waves, since all the energy in the wave is transported in the same direction, the concepts of radiant intensity and emittance are not useful. It is customary to specify the radiant flux crossing unit area normal to the direction of propagation, and call this the *intensity I* of the plane wave. Because lasers emit radiation into an extremely small solid angle, they have very high radiant intensity, and it is once again more usual to refer simply to the *intensity* of the laser beam at a point as the energy flux per second per unit area. The total power output of a laser is

$$W = \int_{\text{beam}} I dS \qquad (4.191)$$

4.5.3 Photometry

The response of the human eye gives rise to a nonlinear and wavelength-dependent subjective impression of radiometric quantities. The response function of the human eye extends roughly from 400 to 700 nm, with a peak at 555 nm for the photopic (light-adapted) eye. Measures in photometry take into account the *relative spectral luminous efficiency* $V(\lambda)$. In physiological photometry, for example, the *luminous flux F* is related to radiant flux $\phi_e(\lambda)$ by

$$F = K \int_0^\infty V(\lambda)\phi_e(\lambda)d\lambda \qquad (4.192)$$

where K is a constant. When F is measured in lumens and $\phi_e(\lambda)$ in watts, $K = 679.6$ lumen W^{-1}. Other photometric quantities that may be encountered in specifications of light sources are as follows:

1. The *luminous intensity* I_v measured as candela (Cd), where

   1 candela = 1 lumen str^{-1}

   This is the photometric equivalent of the radiometric quantity radiant intensity.

2. The *illuminance*, E_v measured in lumen/m^2 ≡ lux; lumen/cm^2 ≡ phot; or lumen/ft^2 ≡ footcandle.

3. The *luminance* or *brightness*, L_v measured in candela/m^2 ≡ nit; candela/cm^2 ≡ stilb; candela/π ft^2 ≡ footlambert; candela/π m^2 ≡ apostilb; or candela/π cm^2 ≡ lambert.

4. The *luminous energy*, Q_v, measured in lumen seconds ≡ talbot.

For a Lambertian source, the luminance is independent of the observation direction. Photometric description of the characteristics of light sources should be avoided in strict scientific work, but some catalogs of light sources use photometric units to describe lamp performance. For further details of photometry and other concepts such as color in physiological optics, the reader should consult Levi[22] or Fry.[48]

Calibration standards and devices for radiometric and photometric calibration of light sources are available from Gamma Scientific, Graseby, and International Light.

4.5.4 Line Sources

Line sources are used as wavelength standards for calibrating spectrometers and interferometers—as sources in atomic absorption spectrometers, in interferometric arrangements for testing optical components such as Twyman-Green interferometers (see Section 4.7.4, in Confocal Fabry-Perot Interferometers), in a few special cases for optically pumping solid-state and gas lasers, and for illumination.

The emission lines from a line source are not infinitely sharp. Their shape is governed by the actual conditions and physical processes occurring in the source. The variation of the radiant intensity with frequency across a line whose center frequency is v_0 is described by its *lineshape function* $g(v,v_0)$, where

$$\int_{-\infty}^{\infty} g v, v_0 \, dv = 1 \qquad (4.193)$$

The extension of the lower limit of this integral to negative frequencies is done for formal theoretical reasons connected with Fourier theory and need not cause any practical problems, since for a sharp line the major contribution to the integral in Equation 4.193 comes from frequencies close to the center frequency v_0. There are three main types of lineshape function: Lorentzian, Gaussian, and Voigt.[49]

The *Lorentzian* lineshape function is

$$g_L(v, v_0) = \frac{2}{\pi \Delta v} \frac{1}{1 + [(v - v_0)/\Delta v]^2} \qquad (4.194)$$

where Δv is the frequency spacing between the half-intensity points of the line (the full width at half maximum height, or FWHM). Spectral lines at long wavelengths (in the middle and far infrared) and lines emitted by heavy atoms at high pressures and/or low temperatures frequently show this type of lineshape.

The *Gaussian* lineshape function is

$$g_D(v, v_0) = \frac{2}{\Delta v_D}\left(\frac{\ln 2}{\pi}\right)^{1/2} \exp\left\{-\left[2\frac{v-v_0}{\Delta v_D}\right]^2 \ln 2\right\} \qquad (4.195)$$

where Δv_D is the FWHM. Spectral calibration lamps are available from Cathodeon, Gamma Scientific, Oriel, and Resonance. Gaussian lineshapes are usually associated with visible and near-infrared lines emitted by light atoms in discharge-tube sources at moderate pressures. In this case, the broadening comes from the varying Doppler shifts of emitting species, whose velocity distribution in the gas is Maxwellian. Emitting ions in real crystals sometimes have this type of lineshape because of the random variations of ion environment within a real crystal produced by dislocations, impurities, and other lattice defects. A Lorentzian and Gaussian lineshape are compared in Figure 4.100.

The broadening processes responsible for Lorentzian and Gaussian broadening are often simultaneously operative, in which case the resultant lineshape is a convolution of the two and is called a *Voigt profile.*[13]

The low-pressure mercury lamp is the most commonly used narrow-line source. These lamps actually operate with a mercury-argon or mercury-neon mixture. The principal lines from a mercury-argon lamp are listed in Table 4.6. Numerous other line sources are also available. Hollow-cathode lamps in particular emit the strongest spectral line of any desired element for use in atomic absorption spectrometry. Such lamps are available from Bulbtronics, GBC Scientific Equipment, Hamamatsu, and Vitro Technology.

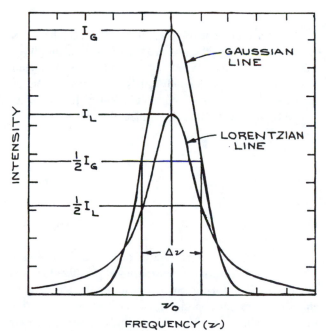

Figure 4.100 Gaussian and Lorentzian lineshapes of the same FWHM, Δv.

4.5.5 Continuum Sources

A continuum source in conjunction with a monochromator can be used to obtain radiation whose wavelength is tunable throughout the emission range of the source. If the wavelength region transmitted by the monochromator is made very small, however, not very much energy will be available in the wavelength region selected. Even so, continuum sources find extensive use in this way in absorption and fluorescence spectrometers. Certain continuum sources, called *blackbody sources*, have very well characterized radiance as a function of wavelength and are used for calibrating both the absolute sensitivity of detectors and the absolute radiance of other sources.

Blackbody Sources

All objects are continuously emitting and absorbing radiation. The fraction of incident radiation in a spectral band that is absorbed by an object is called its *absorptivity*. When an object is in thermal equilibrium with its surroundings, it emits and absorbs radiation in any spectral interval at equal rates. An object that absorbs all radiation incident on is called a *blackbody*—its absorptivity α is equal to unity. The ability of a body to radiate energy in a particular spectral band compared to a blackbody is its *emissivity*. A blackbody is also the most efficient of all emitters—its emissivity ε is also unity. For any object emitting and absorbing radiation at wavelength λ, $\varepsilon_\lambda = \alpha_\lambda$. Highly reflecting, opaque objects, such as polished metal surfaces, do not absorb radiation efficiently—nor do they emit radiation efficiently when heated.

The simplest model of a blackbody source is a heated hollow object with a small hole in it. Any radiation entering the hole has minimal chance of reemerging. The radiation leaving the hole will therefore be characteristic of the interior temperature of the object. The energy density distribution of this *blackbody radiation* in frequency is

$$\rho(v) = \frac{8\pi h v^3}{c^3} \frac{1}{e^{hv/kT} - 1} \qquad (4.196)$$

TABLE 4.6 CHARACTERISTIC LINES FROM
A MERCURY LAMP

Wavelengtha (μm)
0.253652
0.313156
0.313184
0.365015
0.365438
0.366328
0.404656
0.435835
0.546074
0.576960
0.579066
0.69075
0.70820
0.77292
1.0140
1.1287
3.9425

Note: Extensive listings of calibration lines from other sources can be found in C.R. Harrison, *M.I.T. Wavelength Tables*, M.l.T. Press, Cambridge, Mass, 1969; and in A. R. Striganov and N. S. Sventitskii, *Tables of Spectral Lines of Neutral and Ionized Atoms*, IFT/Plenum Press, New York, 1968.

a *In vacuo.*

where $\rho(v)dv$ is the energy stored (J m^{-3}) in a small frequency band dv at v. The energy density distribution in wavelength is

$$\rho(\lambda) = \frac{8\pi hc}{\lambda^5} \frac{1}{e^{hc/\lambda kT} - 1} \qquad (4.197)$$

This translates into a spectral emittance (the total power emitted per unit wavelength interval into a solid angle 2π by unit area of the blackbody) given by

$$M_{e\lambda} = \frac{C_1}{\lambda^5(e^{C_2/\lambda T} - 1)} \qquad (4.198)$$

where $C_1 = 2\pi hc^2$, called the *first radiation constant*, has the value 3.7418×10^{-16} W m^2 s^{-1}, and $C_2 = ch/k$, called the *second radiation constant*, has the value 1.43877×10^{-2} mK.

A true blackbody is also a diffuse (Lambertian) radiator. Its radiance is independent of the viewing angle. For such a source

$$M_{e\lambda} = \pi L_{e\lambda} \qquad (4.199)$$

The variation of $L_{e\lambda}$ with wavelength for various values of the temperature is shown in Figure 4.101. The wavelength of maximum emittance, λ_m at temperature T obeys Wien's displacement law,

$$\lambda_m T = 2.8978 \times 10^{-3} \text{mK} \qquad (4.200)$$

The total radiant emittance of a blackbody at temperature T is

$$M_e = \int_0^\infty M_{e\lambda} d\lambda = \frac{2\pi^5 k^4}{15c^2 h^3} T^4 = \sigma T^4 \qquad (4.201)$$

This is a statement of the Stefan-Boltzmann law. The coefficient σ, called the *Stefan-Boltzmann constant*, has a value of 5.6705×10^{-8} W m^{-2} K^{-4}. The known parameters $M_{e\lambda}$ and M_e of a blackbody allow it to be used as an absolute calibration source in radiometry. If a detector responds to photons, the spectral emittance in terms of photons N_λ may be useful:

$$N_\lambda = \frac{M_{e\lambda}}{hc/\lambda} \qquad (4.202)$$

Curves of N_λ are given by Kruse, McGlauchlin, and McQuistan.[50]

A source whose spectral emittance is identical to that of a blackbody apart from a constant multiplicative factor is called a *graybody*. The constant of proportionality ε is its emissivity. Several continuum sources, such as tungsten-filament lamps, carbon arcs, and flashlamps, are approximately graybody emitters within certain wavelength regions.

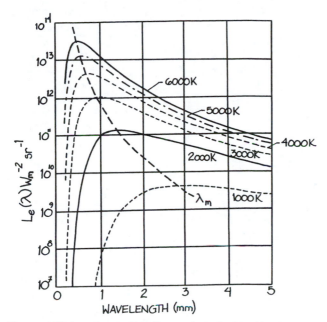

Figure 4.101 Spectral radiance $L_e\lambda$ of a blackbody source at various temperatures.

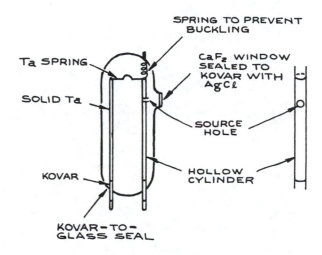

Figure 4.102 Construction details of a simple high-temperature blackbody source. (From P. W. Kruse, L. D. McGlauchlin, and R. B. McQuistan, *Elements of Infrared Technology: Generation, Transmission, and Detection*, Wiley, New York, 1962; by permission of John Wiley and Sons, Inc.)

Practical Blackbody Sources

The radiant emittance of a blackbody increases at all wavelengths as the temperature of the blackbody is raised, so a practical blackbody should, ideally, be a heated body with a small emitting aperture that is kept as small as possible. Kruse, McGlauchlin, and McQuistan describe such a source, illustrated in Figure 4.102, that can be operated at temperatures as high as 3000 K.[50] A 25 µm thick tungsten ribbon 2 cm wide is rolled on a 3 mm diameter copper mandrel and seamed with a series of overlapping spot welds. A hole about 0.75 mm in diameter is made in the foil, and the copper dissolved out with nitric acid under a fume hood. The resulting cylinder is mounted on 1 mm diameter Kovar or tungsten rod feedthroughs in a glass envelope and heated from a high-current, low-voltage power supply. The glass envelope should be fitted with a window that is transmissive to the wavelength region desired from the source.

Another design of blackbody source is shown in Figure 4.103. This design is based on a heated copper cylinder, containing a conical cavity of 15° semivertical angle, that is allowed to oxidize during operation (so that it becomes nonreflective and consequently of high emissivity). The cylinder is heated by an insulated heater wire wrapped around its circumference. If nichrome wire is used, the cylinder can be heated to about 1400 K. This assembly is mounted in a ceramic tube (alumina is quite satisfactory) or potted in high-temperature ceramic cement. For high-temperature operation the whole assembly can be mounted inside a water-cooled block.

A popular blackbody source uses a *Globar*—a rod of bonded silicon carbide—although in the high temperature blackbody source supplied by Oriel this has been replaced by pyrolytic graphite. For further details of the advantages and disadvantages of this and various other blackbody sources, the reader is referred to Hudson.[51] Blackbody sources are available commercially from EDO/Barnes Engineering, Eppley, Infrared Industries, Micron, and Oriel.

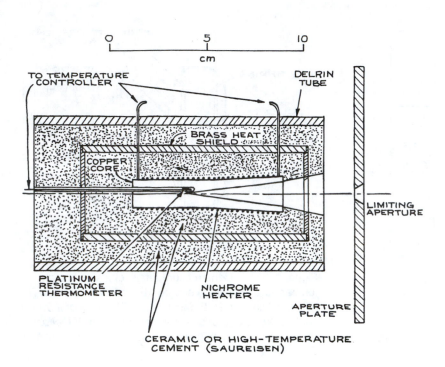

Figure 4.103   Construction details of an NBS-type blackbody source.

Tungsten-Filament Lamps

Tungsten-filament lamps are approximately graybodies in the visible with an emittance between 0.45 and 0.5. Such lamps are frequently described in terms of their *color temperature* T_c, which is the temperature at which a blackbody would have a spectral emittance closest in shape to the lamp's. The color temperature will depend on the operating conditions of the lamp.

Tungsten-filament lamps can most conveniently be operated in the laboratory with a variable transformer. For best stability and freedom from ripple on their output, however, they should be operated from a stabilized dc supply. Typical supply requirements range up to a few hundred volts. Lamps with wattage ratings up to 1 kW are readily available.

Small tungsten-filament lamps that can be used as point sources are available from Oriel. Very long life, constant efficiency tungsten-halogen lamps are available, in which the lamp envelope usually contains a small amount of iodine. In operation, the iodine vaporizes and recombines with tungsten that has evaporated from the filament and deposited on the inside of the lamp envelope. The tungsten iodide thus formed diffuses to the hot filament, where it decomposes, redepositing tungsten on the filament. The constant replacement of the filament in this way allows it to be operated at very high temperature and radiant emittance. Because the lamp envelope must withstand the chemical action of hot iodine vapor and high temperatures, it is made of quartz. Such lamps are therefore frequently called quartz-iodine lamps. Such lamps can be quite compact. A 1 kW lamp will have a filament length of about 1 cm. The NBS standard of spectral irradiance consists of a quartz-iodine lamp with a coiled-coil tungsten filament operating at about 3000 K and calibrated from 250 nm to 2.6 µm against a blackbody source. Such calibrated lamps are available from EG&G and Oriel. Because they are intense sources of radiant energy, these lamps can also be used for heating. In particular, they are often placed inside complex vacuum systems to bake out internal components that are well insulated thermally from the chamber walls.

Continuous Arc Lamps

High-current electrical discharges in gases, with currents that typically range from 1 to 100 A, can be intense sources of continuum or line emission, and sometimes both at the same time. For substantial continuum emission, the most popular such lamps are the high-pressure xenon, high-pressure mercury, and high-pressure mercury-xenon lamps. The arc in these lamps typically ranges up to about 5 cm long and 6.2 mm in diameter (for a 10 kW lamp– 100 V, 100 A input). Because of their small size, arc lamps have much higher spectral radiance (brightness) than quartz-iodine lamps of comparable wattage. In the visible region at 500 nm, a typical xenon arc lamp shows 1.9 times the output of a quartz-iodine lamp; at 350 nm, 14 times; and at 250 nm, 200 times. Because of their small size, high-pressure arc lamps also work well in the illumination of monochromator slits in spectroscopic applications. Lower-wattage arc lamps come close to being point sources and are ideal for use in projection systems and for obtaining well-collimated beams.

There are two different kinds of high-pressure arc lamps—those where the discharge is confined to a narrow quartz capillary (which must be water cooled), and those where the discharge is not confined (which usually operate with forced-air cooling). The former are available from Flashlamps (Verre et Quartz), ILC Technology, Xenon Corp., and Ushio, and are used for pumping CW solid-state lasers. (Krypton arc lamps are better than xenon arc lamps for pumping Nd^{3+} lasers, as their emission is better matched to the absorption spectrum of Nd^{3+} ions.)

Because high-pressure arc lamps operate at very high pressures when hot (up to 200 bars), they must be housed in a rugged metal enclosure to contain a possible lamp explosion. The mounting must be such as to allow stress-free expansion during warmup. Generally speaking, commercial lamp assemblies should be used. The power-supply requirements are somewhat unusual. An initial high-voltage pulse is necessary to strike the arc, and then a lower voltage, typically in the range 70 to 120 V, to establish the arc. When the arc is fully established, the operating voltage will drop to perhaps as low as 10 V. Arcs containing mercury need a further increase in operating voltage as they warm up and their internal mercury pressure increases. Complete lamp assemblies and power supplies are available from several companies, among them, EG&G, ORC Lighting Products, ILC Technology, Oriel, and Spectral Energy.

High-pressure arc amps give substantial continuum emission with superimposed line structure, as can be seen in Figure 4.104. These lamps are not efficient sources of infrared radiation. They give substantial UV emission, however, and care should be taken in their use to avoid eye or skin exposure. Their UV output will also generate ozone, and provision should be made for venting this safely from the lamp housing.

Deuterium lamps are efficient sources of ultraviolet emission with very little emission at longer wavelengths, as shown in Figure 4.104a. They are available from Catheodeon, Hamamatsu, and Oriel.

Discharges in high-pressure noble gases can also be used as intense continuum sources of vacuum-ultraviolet radiation. This radiation arises from noble-gas excimer emission, which in the case of helium, for example, arises from a series of processes that can be represented as

$$He + e \rightarrow He^* \tag{4.203}$$

$$He^* + He \rightarrow He_2^*$$

$$He_2^* \rightarrow He + He + h\nu$$

The spectral regions covered by the excimer continua are He, 105 to 400 nm; Ar, 107 to 165 nm; Kr, 124 to 185 nm; and Xe, 147 to 225 nm. Because there are no transmissive materials available for wavelengths below about 110 nm, sources below this wavelength are used without windows. Radiation leaves the lamp through a small slit, or through a multicapillary array. The latter is a close-packed array of many capillary tubes, which presents considerable resistance to the passage of gas but is highly transparent to light. Multicapillary arrays can be obtained from Burle Electro-Optics. To maintain lamp pressure, gas is continuously admitted. Gas that passes from the lamp into the rest of the experiment (a vacuum-ultraviolet monochromator, for example) is continuously pumped away by a high-speed vacuum pump as shown in Figure 4.105. This technique is called *differential pumping* (see Section 3.6.2). For further details of vacuum-ultraviolet sources and technology, the reader is referred to Samson.[52]

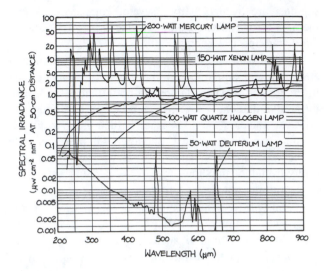

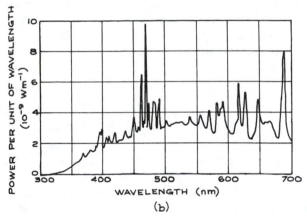

(b)

Figure 4.104   (a) Spectral irradiance of various moderate power, gas-discharge lamps (courtesy of Oriel Corporation, Stamford, Conn.); (b) spectral energy distribution for a high-power (10 kW) xenon arc lamp with a long discharge column. (From A. A. Kruithof, "Modern Light Sources," in *Advanced Optical Techniques*, A. C. S. Van Heel, ed., North-Holland, Amsterdam, 1967; by permission of North-Holland Publishing Company.)

Microwave Lamps

Small microwave-driven discharge lamps are very useful where a low-power atomic emission line source is desired, particularly if the atomic emission is desired from some reactive species such as atomic chlorine or iodine. A small cylindrical quartz cell containing the material to be excited, usually with the addition of a buffer of helium or argon, is excited by a microwave source inside a small tunable microwave cavity as shown in Figure 4.106. Suitable cavities for this purpose, designed for operation at 2450 MHz, are available from Opthos Instruments—power supplies for these lamps are available from Opthos and Cathodeon.

Flashlamps

The highest-brightness incoherent radiation is obtained from short-pulse flashlamps. By discharging a capacitor through a gas-discharge tube, much higher discharge currents and input powers are possible than in dc operation. As the current density or pressure of a flashlamp is increased, the emission from the lamp shifts from radiation that is characteristic of the fill gas with lines superimposed on a continuum, to an increasingly close approximation to blackbody emission corresponding to the temperature of the discharge gas. Commercial flashlamps can be roughly divided into two categories. In long-pulse lamps, fairly large capacitors (100–10,000 μF), charged to moderately high voltages (typically up to about 5 kV), are discharged slowly (on time scales from 100 μs to 1 ms) through high-pressure discharge tubes. In short-pulse, high peak-power lamps, smaller, low-inductance capacitors (typically 0.1–10 μF), charged to high voltages (10–80 kV) are discharged rapidly (on time scales down to 1 μs) through lower-pressure discharge tubes.

Long-Pulse Lamps

Long-pulse lamps are generally filled with xenon,[53] but krypton lamps[54] (for pumping Nd^{3+} lasers) and alkali-metal lamps are also available. Fill pressures are typically on the order of 0.1 to 1 bar. Although the discharge current in such lamps can run to tens of thousands of A cm^{-2}, at low repetition frequencies (< 0.1 Hz) ambient cooling is all that is necessary. Three of the most commonly used circuits for operating such lamps are shown in Figure 4.107. In all three of these circuits, the capacitor—or pulse forming network shown in Figure 4.107c—is charged through a resistor R. The capacitor is discharged through the lamp by triggering the flashlamp with one of the

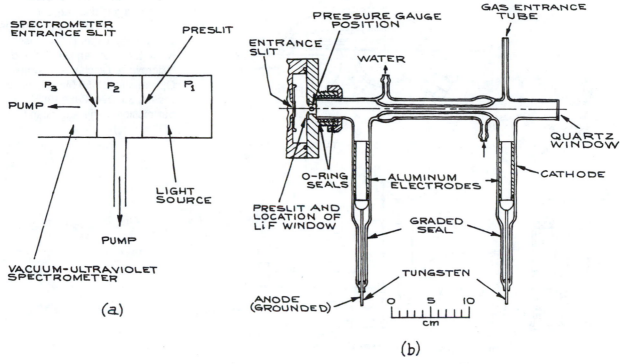

(a)

(b)

Figure 4.105 (a) Differential pumping arrangement (P1, P2, P3 are the light source, differential pumping unit, and spectrometer operating pressures, respectively); (b) light source for the production of the noble gas continua—in a differentially pumped mode, no LiF window would be used. (From R. E. Huffman, Y. Tanaka, and J. C. Larrabee, "Helium Continuum Light Source for Photoelectric Scanning in the 600–1100 Å Region," *Appl. Opt.*, 2, 617–623, 1963; by permission of the Optical Society of America.)

trigger circuits shown in Figure 4.108. The lamp itself behaves both resistively and inductively when it is fired. If the inductance and resistance of the lamp are not sufficiently large, the capacitor may discharge too rapidly, which can lead to damage of both lamp and capacitor. Generally speaking, an additional series inductor will be desirable to control the discharge. The problem of selecting the appropriate inductor for a particular capacitor size and lamp has been dealt with in detail by Markiewicz and Emmett.[55] Slow flashlamps have a nonlinear V-I characteristic, which can be approximated by

$$V = K_0 \sqrt{|I|} \qquad (4.204)$$

where the sign of V is taken to be the same as the sign of I. The value of K_0, measured in $\Omega \, A^{1/2}$, is a parameter

specified by a manufacturer for a given lamp. Given this value and the capacitor size to be used, the calculations of Markiewicz and Emmett allow a suitable value of series inductor to be chosen. Other factors must be taken into account in designing the flashlamp circuit: the maximum power loading, usually specified as the explosion energy of the lamp (which will depend on the discharge-pulse duration), and the maximum repetition frequency of the lamp. The larger the capacitor stored energy, the lower the permitted repetition frequency will usually be. As lamps and discharge energies get smaller, repetition frequencies can be extended, to several kilohertz for the smallest low-energy lamps such as those used in stroboscopes. The explosion energy of the lamp is the minimum input required to cause lamp failure in one shot. To obtain long flashlamp life, a lamp should be operated only at a fraction

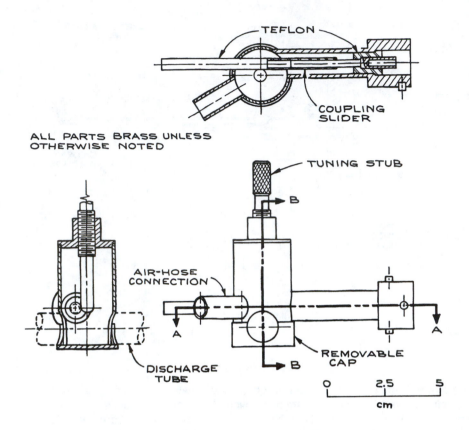

ALL PARTS BRASS UNLESS OTHERWISE NOTED

Figure 4.106   Microwave cavity for exciting a gas discharge in a cylindrical quartz tube (design used by Opthos). (After F. C. Fehsenfeld, K. M. Evenson, and H. P. Broida, "Microwave Discharge Cavities Operating at 2450 MHz," *Rev. Sci. Instru.*, **36**, 294–298 (1965); by permission of the American Institute of Physics.)

of its explosion energy. For example, at 50 percent of the explosion energy the expected lifetime is from 100 to 1000 flashes, while at 20 percent it is from 10^5 to 10^6 flashes. The lamp should be mounted freely and not clamped rigidly at its ends when it is operated.

The spectral output of a flashlamp varies slightly during the flash. For moderate- and high-power lamps (> 100 J in 1 ms), the spectral output approaches that of a blackbody as shown in Figures 4.109 and 4.110.[56] Small-power lamps exhibit the spectral feature of their fill gas, as shown in Figure 4.111. Figure 4.109 shows the transition from a spectrum with some line structure to a blackbody continuum as the current density through the lamp is increased.

A word about the triggering schemes shown in Figure 4.108—external triggering, accomplished by switching a high-voltage pulse from a transformer to a trigger wire wrapped around the lamp, is perhaps the simplest scheme. It is used only with long-pulse lamps and does not give

quite as good time synchronization as series-spark-gap or thyratron-switched operation. The other methods are more complex but can be also used with short-pulse, high-power lamps. Series triggering allows the incorporation of triggering and lamp series inductance in a single unit. Trigger transformers of various types are available from EG&G and ILC. EG&G and EEV supply a range of excellent thyratrons.

Short-Pulse High-Power Lamps

Flashlamps that can handle the rapid discharge of hundreds of joules on time scales down to a few microseconds are commercially available from ILC, EG&G, Flashlamps (Verre et Quartz), and Xenon Corporation, among others. To achieve such rapid discharges, these lamps are generally operated in a low-inductance discharge circuit. When these lamps are fired at

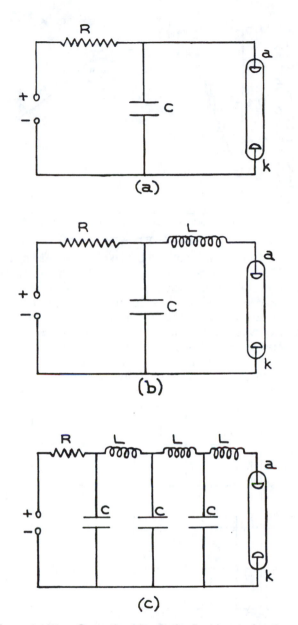

Figure 4.107 Operating circuits for flashlamps (a = lamp anode; k = lamp cathode): (a) RC discharge; (b) RLC criticallly damped discharge; (c) pulse forming network.

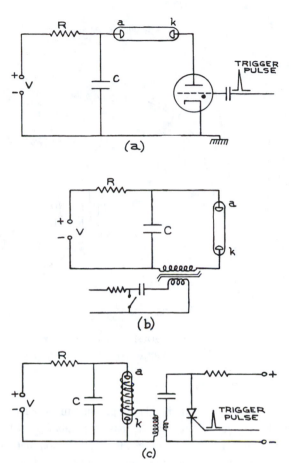

Figure 4.108 Flashlamp triggering schemes (a = lamp anode; k = lamp cathode): (a) overvoltage triggering ($V >$ self-flash voltage of flashlamp; switching is accomplished in this case with a thyratron, but a triggered spark gap or ignitron can also be used); (b) series triggering (a saturable transformer is used with a fast risetime 10–20 kV pulse with sufficient energy to trigger the lamp and saturate the core); (c) external triggering (the flashlamp is ionized by a trigger wire wrapped around the outside of the lamp and connected to the 5–15 kV secondary of a high-voltage pulse transformer).

high peak power inputs, a severe shock wave is generated in the lamp. To withstand the shock wave, the lamps are designed with special reentrant electrode seals. If lamps designed for long-pulse operation are discharged too rapidly, their electrodes are quite likely to pop off because the electrode seals are not shock resistant.

The spectral emission of short-pulse, high-energy lamps is generally quite close to that of a high-temperature blackbody,

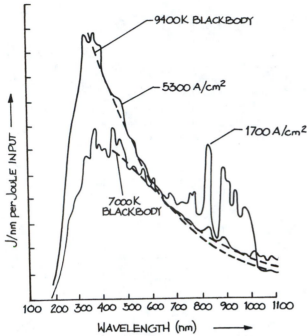

Figure 4.109   Output spectrum of a Perkin-Elmer FX-47A xenon flashlamp (0.4 atm) at two current densities: 1700A/cm² and 5300 A/cm².

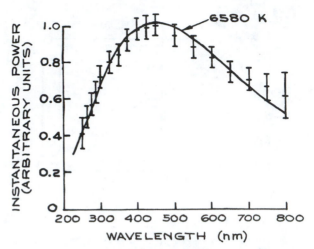

Figure 4.110   Spectral distribution of power radiated by an EG&G FX42 xenon-filled flashlamp (76 mm long x 7 mm bore) operated in a critically damped mode with 500 J discharged in 115 ms. The spectral distribution was observed 0.7 ms from flash initiation. The line is the blackcbody radiation curve of best fit. (After J.G. Edwards, "Some Factors Affecting the Pumping Efficiency of Optically Pumped Lasers," *Appl. Opt.*, **6**, 837–843 (1967); by permission of the Optical Society of America.)

perhaps as hot as 30,000 K. The fill pressure in these lamps is often lower than in long-pulse lamps— typically from 0.1 to a few tens of millibars. The fill gas is usually xenon, but other noble gases such as krypton or argon can be used. If a short-pulse lamp has a narrow discharge-tube cross section, at high power inputs material can ablate from the wall and a substantial part of the emission from the lamp can come from this material. Such lamps can be made with heavy-walled discharge-tube bodies of silica, glass, or even Plexiglas. These lamps can be made fairly simply. A good design described by Baker and King is illustrated in Figure 4.112.[57] It can be operated in nonablating or ablating modes (at high or low pressure, respectively). When operated in the ablating regime, such lamps can be filled with air as the nature of the fill gas is unimportant. Figure 4.112 also shows the use of a field distortion-triggered spark gap for firing the lamp—a spark-gap design that offers quiet, efficient switching of rapid discharge capacitors. Rapid-discharge (low-inductance) capacitors suitable for fast flashlamp and other applications

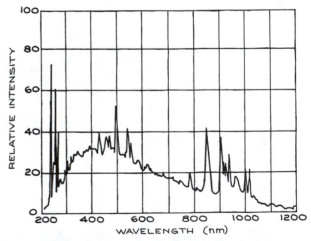

Figure 4.111   Spectral distribution of intensity from an ILC 4L2 xenon flashlamp (51 mm long by 4 mm bore) operated in a critically damped mode with 10 J discharged in 115 µs. (Courtesy of ILC.)

are available from Maxwell Energy Products, Hi Voltage Components, and CSI Technologies, among others.

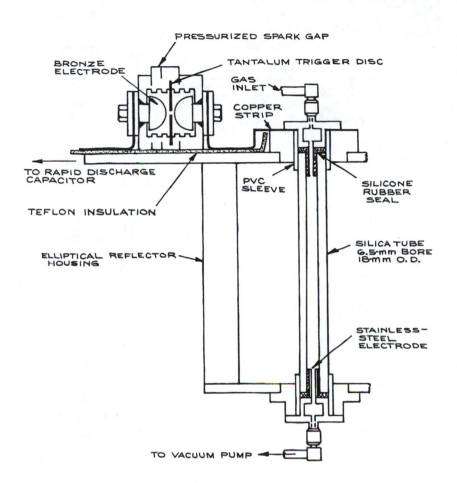

Figure 4.112   High pulse-energy, fast flashlamp design for conventional or ablation-mode operation. (From H. J. Baker and T. A. King, "Optimization of Pulsed UV Radiation from Linear Flashtubes," *J. Phys. E.: Sci. Instr.*, **8**, 219–223 (1975). © 1975 by the Institute of Physics, used with permission.)

4.6 LASERS

Lasers are now so widely used in physics, chemistry, the life sciences, and engineering that they must be regarded as the experimentalist's most important type of optical source. Generally speaking, anyone who wants a laser for an experiment should buy one. This is certainly true in the case of helium-neon, argon-ion, and helium-cadmium gas lasers and all solid-state crystalline, glass, or semiconductor lasers. It's true that these lasers can be built in the laboratory, but this is the province of the laser specialist and will generally be found time-consuming and unproductive for scientists in other disciplines. On the other hand, there are some lasers—such as nitrogen, exciplex, CO_2, and CO gas lasers and dye lasers—that it

might occasionally pay to build for oneself even though models are commercially available. A detailed discussion of how to construct all these lasers will not be given here, but to illustrate how it is accomplished, design features of some specific systems will be discussed.

Tables 4.7 through 4.12 highlight some of the characteristic features of the more commonly used lasers presently available on the commercial market. Before briefly describing some of these laser systems, some background material will be presented that is pertinent to a discussion of lasers in general. More detailed information regarding the physics of laser operation is given in the books by Davis,[13] Silfvast,[58] Saleh and Teich,[59] Yariv,[12] and Siegman.[60]

TABLE 4.7 CHARACTERISTICS OF CW GAS LASERS

Type	Principal Operating Wavelengths (μm)	Output Power (W)		Suppliers[a]
		TEM$_{00}$	Multimode	
Ar ion	0.229[b]			Lexel
	0.244[b]	medium		
	0.248[b]			
	0.257[b]			
	0.275[b]	high		Spectra-Physics
	0.3336 ⎫			
	0.3344 ⎪			Coherent
	0.3358 ⎬		medium-high	Spectra-Physics
	0.3511[c] ⎪			Lexel
	0.3638[c] ⎭			
	0.4545 ⎫			
	0.4579 ⎪			
	0.4658 ⎪			Coherent
	0.4765 ⎪			Spectra-Physics
	0.4880[c] ⎪			Lexel
	0.4965 ⎬	medium-high	high	Melles Griot
	0.5017 ⎪			American Laser
	0.5145[c] ⎪			JDS Uniphase
	0.5287 ⎪			
	1.0923 ⎭			
Kr ion	0.3375 ⎫			Coherent
	0.3507 ⎬	medium	high	Spectra-Physics
	0.3564 ⎭			
	0.4067 ⎫			
	0.4131 ⎪			
	0.4619 ⎪			Coherent
	0.4680 ⎪			Spectra-Physics
	0.4762 ⎪			Lexel
	0.4825 ⎬	medium	medium-high	Melles Griot
	0.5309[c] ⎪			American Laser
	0.5681[c] ⎪			JDS Uniphase
	0.6470 ⎪			Laser Physics
	0.6764 ⎪			
	0.7993 ⎭			

TABLE 4.7 CHARACTERISTICS OF CW GAS LASERS (CONTINUED)

Type	Principal Operating Wavelengths (μm)	Output Power (W)		Suppliers[a]
		TEM$_{00}$	Multimode	
He-Ne	0.543	low		Melles Griot, Coherent,
	0.5941	low	low	JDS Uniphase,
	0.612	low	low	Research Electro-Optics
	0.6328[d]	low-medium		Spectra-Physics, Coherent, Melles Griot, JDS Uniphase
	1.15	low		Research Electro-Optics
He-Ne	3.39	low		Research Electro-Optics
He-Cd	0.3250	low	low-medium	Melles Griot, Kimmon
	0.4416	low	low-medium	Melles Griot, Kimmon
CO	5–6.5[e]	medium-high		Edinburgh Inst., Laser Systems Devices
CO$_2$	9–11[e] 10.6	high-v. high	v. high–industrial[f]	Synrad, Spectra Lasers, De Maria ElectroOptics, Edinburgh Inst., Rofin Synar, A-B Lasers, Trumpf, Howden Laser

[a] For a more extensive listing, see Reference 27.
[b] Produced by second harmonic generation of blue-green argon ion laser lines.
[c] Strongest lines.
[d] Most readily available wavelength.
[e] Molecular lasers that offer discrete tunability over several lines.
[f] Industrial CO$_2$ lasers generally operate only at 10.6 μm.

Some comments on Tables 4.7 through 4.12 are in order. It is not meant to be an exhaustive compilation of all the laser types, wavelengths, and power outputs that can be obtained. The lasers listed in the table fall into two categories—pulsed and CW (continuous wave).

For pulsed lasers, the available energy outputs per pulse are classified as low < 10mJ) medium (10mJ–1J), high (1J–100J), and very high (> 100J). Pulse lengths depend on the type of laser and its mode of operation. Almost all pulsed solid state lasers operate in a *Q-switched* mode, which typically provides a pulse length in the 10 ns range, but varies somewhat from manufacturer to manufacturer.[13] These lasers are designated with the letter Q in the table. Non-Q-switched, long pulse (LP), solid-state lasers have pulse lengths > 0.1 ms. Solid-

TABLE 4.8 CHARACTERISTICS OF PULSED GAS LASERS

Type	Operating Wavelengths (μm)	Output Energy (J/pulse)		Pulses per Second	Pulse Length (ns)	Suppliers
		TEM$_{00}$	Multimode			
ArF	0.193		med-v. high	0–1000	25–30	Lambda Physik, MPB Technologies, JPSercel Associates (JPSA)
CO_2	10.6	low-v. high	low-v. high	0–500,000	50–CW	Edinburgh Inst., Coherent, Laser Systems Devices, Spectron Lasers, Parselax Technology, Laser Engineering, Howden Laser
	9–11	low-high	low-high	0–1000	> 200	Laser Systems Devices, Spectron Lasers, Parselax Technology, Laser Engineering, Howden Laser
Cu vapor	0.5105 0.5782	low	low-medium	0–5000	20–50	Oxford Lasers
F_2	0.157	—	low-medium	0–1000	2.5–20	Lambda Physik, MPB Technologies, GAM Laser
KrF	0.248		low-v. high	0–2000	5–45	Lambda Physik, Sopra, Potomac Photonics, GAM Laser, MPB Technologies (JPSA)
N_2	0.3371	—	low	0–1000	0.5–10	Potomac Photonics, Lambda Physik, GAM Laser, Sopra, MPB Technologies
XeCl	0.308	—	low-v. high	0–1000	1.5–120	Potomac Photonics, Lambda Physik, GAM Laser, Sopra, MPB Technologies
XeF	0.351	—	low-high	0–1000	2.5–30	GAM Laser, Lambda Physik, MPB Technologies

TABLE 4.9 CHARACTERISTICS OF CW SOLID-STATE LASERS

Type	Operating Wavelengths (μm)	Output Power (W)		Suppliers
		TEM_{00}	Multimode	
Alexandrite	0.193	low		Light Age
	0.375	low		
	0.75	medium		
Cr:LiSAF	0.42–0.44 (tunable)	low		B&W Tek
Nd:YAG	0.266[a]	med-high		
	0.355[b]	med-high		
	0.473	med-high		Intelite
	0.530[c]	low-v. high	high-v. high	CVI Laser, Quantronix
	0.946	medium		B&W Tek
	1.06	low-v. high	low-industrial	Quantronix, Lee Laser, Spectron, MeshTel, Trumpf Electo
	1.318	high	high-v. high	Lee Laser
Ti:Sapphire	0.7–1.02 (tunable)	med-high		Laser Systems Devices, Lexel

[a] Fourth harmonic of 1.06 μm Nd:YAG wavelength.
[b] Third harmonic of 1.06 μm.
[c] Second harmonic of 1.06 μm.

TABLE 4.10 CHARACTERISTICS OF PULSED SOLID-STATE LASERS

Type	Operating Wavelengths (μm)	Output Energy (J)		Pulses per Second	Pulse Length[a] (ns)	Suppliers
		TEM_{00}	Multimode			
Alexandrite	0.72–0.78 (tunable)		medium-high	0–50	$100–2\times10^6$	Light Age
	0.755					
Er:YAG	2.94		high	2–50	70,000–500,000	Fotano, Equilasers
Er:fiber lasers	0.76–0.79	v. low		$5–100\times10^6$	0.15	Calmar Optron,
	1.52–1.58	v. low		$5–100\times10^6$	0.15	Laser Systems Devices
	2.94	medium		1–40		
Er:glass	1.534		low	1	25	Kigre

TABLE 4.10 CHARACTERISTICS OF PULSED SOLID-STATE LASERS (CONTINUED)

Type	Operating Wavelengths (μm)	Output Energy (J) TEM$_{00}$	Multimode	Pulses per Second	Pulse Length[a] (ns)	Suppliers
Ho:YAG	2–1	medium	medium	1–40	50–10^6	Laser Systems Devices
Nd:glass	0.266[b]	low	med-high	1/3	13–20	Quantel
	0.355[c]	medium	med-high	1/3	< 13–20	Quantel
	0.532[d]	medium	med-high	< 1/3	13–20	Quantel
	1.06[e]	medium[f]	med-v. high	0–0.1	25–6×10^4	Quantel, Thomson CSF
Ti:sapphire	0.68–1 (tunable)		low-high	up to 10^8	mode locked /35 fs	Spectral-Physics, Coherent, Continuum, MPB Technologies, Light Age
Nd:YAG	0.213[g]	low	med-high	0–1000	2–7000	Quantel, Lamdba Physik
	0.265[b]	low-med	low	0–1000	4–25,000	Spectron, Spectra Physics, Quantel, Thomson CSF, Laser
	0.355[c]	low-med	low-med	0–1000	4–25,000	Quantel, Spectral Physics, Spectron, Eksma
	0.53[d]	low-high	low-high	0–3000	2–25,000	Spectra-Physics, Quantel, Lambda Physik, Spectron, New Wave Research, Thompson CSF Laser, Polytec PI
	1.064	low-v. high	low-v. high	0–10^5	2 CW	Thomson CSF Laser, Quantel, Spectra-Physics, Eksma, Lasag, Sopra, Kigre, Trumpf, Lambda Physik, Solar Laser Systems, Opton Laser, Photonics Industrial Intl.
Ruby	0.6943	low-high	high	0–5	25–250	Spectron, Thomson CSF, Laser

[a] Shorter pulses can be obtained by mode locking.
[b] Frequency quadrupled from 1.06 μm output. Phosphate-glass gives 0.263 μm.
[c] Frequency tripled from 1.06 μm output. Phosphate glass gives 0.35 μm.
[d] Frequency doubled from 1.06 μm output. Phosphate glass gives 0.525 μm.
[e] Phosphate glass gives 1.05 μm.
[f] Much higher energies can be obtained from oscillator amplifier configurations.
[g] Fifth harmonic from 1.06 μm output.

TABLE 4.11 CHARACTERISTICS OF CW TUNABLE DYE LASERS (PUMPED WITH ION LASERS, FREQUENCY DOUBLED TI:SAPPHIRE, AND FREQUENCY DOUBLED DIODE LASERS)

Tuning Range[a]	Power (W)	Linewidth	Supplier	Tuning Range[a]	Power (W)	Linewidth	Supplier
0.225–1.1	+3	< 500 GHz	Coherent	0.38–1100	1	< 40 GHz	Spectra-Physics

[a] Nd:YAG pump sources are frequency doubled to 530 nm or tripled to 344 nm to pump the dye.

TABLE 4.12 CHARACTERISTICS OF PULSED TUNABLE DYE LASERS

Tuning Range (μm)	Energy (J)	Pulses per Second	Pulse Length (ns)	Linewidth (nm)	Pumping Method	Supplier
0.197–0.42	low	0–50	5–100	.001	Nd:YAG	Lambda Physik, Spectron
0.205–1.0	medium	0–50	8.30	.0015	Nd:YAG or excimer laser	Polytec PI
0.225–4.5	high	10–50	5	.001	Nd:YAG	Continuum
0.32–1.036	low	0–500	4–20	> 900 MHz	Nd:YAG excimer 308 nm excimer 351 nm	Lambda Physik
0.4–0.9	low	50	< 0.5	1–8	Nd:YAG	LTB Laser Technik
0.55–1	low	50	5–10	.001	Nd:YAG	Spectron

state and dye lasers are frequently operated in a mode-locked (ML) configuration, which provides pulse lengths in the 0.1 to 10 ps range. A given laser can often be operated in ML, Q, and LP modes. The energy output per pulse will decrease as LP > Q > ML.

CW lasers are classified by output power as low (< 10 mW), medium (10 mW–1 W), high (1 W–10 W), very high (10 W–100 W), and industrial (> 100 W). CW gas lasers, in particular, provide power outputs at many different wavelengths and the available power varies from line to line. These lasers are generally supplied to operate in a specific wavelength range with discrete tunability within that range. Ultraviolet gas lasers, for example, are equipped with special optics, so do not generally operate in the visible range as well.

The energy outputs listed for the pulsed lasers, together with the available pulse repetition rates and pulse lengths, represent the ranges that are readily available commercially. They are not intended to imply that all possible combinations can be obtained simultaneously. Indeed, in the case of pulsed solid-state lasers, the highest available energy outputs are generally available only at the lowest pulse repetition rates. In the case of the CW lasers, available power outputs vary from line to line and from manufacturer to manufacturer.

To the intended purchaser of a laser system, we recommend the following questions:

1. What fixed wavelength or wavelengths must the laser supply?

2. Is tunability required?

3. What frequency and amplitude stability are required?

4. What laser linewidth is required?

5. What power or pulse energy is required?

6. Are high operating reliability and long operating lifetime required?

7. For time-resolved experiments, what pulse length is needed?

8. Is the spatial profile of the output beam important?

The desirable attributes in each of these areas are unlikely to occur simultaneously. Frequency tunability, for example, is not readily available throughout the spectrum without resorting to the specialized techniques of nonlinear optics.[12,13,61–65]

In Tables 4.7–4.12, the various categories of laser—CW gas, pulsed gas, CW solid-state, pulsed solid state, CW dye, and pulsed dye—frequently show tuning ranges and wavelengths that are obtained from these lasers by the use of nonlinear optics. Shorter wavelengths from argon ion, Nd:YAG, Ti:sapphire, and CO_2, for example, can be obtained by frequency doubling (tripling or even quadrupling) by the use of a nonlinear crystal. The nonlinear crystal is generally external to the primary laser, although it may be housed in its own resonant structure. Generating second or third harmonic light from a sufficiently powerful laser is not especially difficult, although this is still a relatively specialized endeavor.

New frequency generation from a primary pump laser can also be obtained by the use of an optical paramatic oscillator (OPO). In such a device, an appropriate nonlinear crystal is pumped with a frequency ν_P and in this process generates two new frequencies ν_S and ν_i, called the *signal* and *idler* frequencies, respectively. In this process photon energy must be conserved, so

$$\nu_P = \nu_S + \nu_i \qquad (4.205)$$

High-energy and high-power lasers are generally less stable both in amplitude and in frequency, and will generally be less reliable and need more *hands-on* attention than their low-energy or low-power counterparts. When a decision is made to purchase a laser system, compare the specifications of lasers supplied by different manufacturers. Discuss the advantages and disadvantages of different systems with others who have purchased them previously. Reputable laser manufacturers are generally very willing to supply the names of previous purchasers.

To some experimental scientists, a laser is merely a rather monochromatic directional lightbulb. For others, detailed knowledge of its operating principles and characteristics is essential. Certain aspects of laser designs and operating characteristics, however, are very general and are worthy of some discussion.

4.6.1 General Principles of Laser Operation

A laser is an optical-frequency oscillator—in common with electronic circuit oscillators, it consists of an amplifier with feedback. The optical frequency amplifying part of a laser can be a gas, a crystalline or glassy solid, a liquid, or a semiconductor. This medium is maintained in an amplifying state, either continuously or on a pulsed basis, by pumping energy into it appropriately. In a gas laser, the input energy comes from electrons (in a gas discharge or an electron beam), or from an optical pump (which may be a lamp or a laser). Solid-state crystalline or glassy lasers receive their pumping energy from continuous or pulsed lamps, and in more and more cases from semiconductor lasers. Liquid lasers can be pumped with a flashlamp or—on a continuous or pulsed basis—by another laser. Semiconductor lasers are *p-n* junction devices and are operated by passing pulsed or continuous electrical currents though them.

The amplifying state in a laser medium results if a population inversion can be achieved between two sublevels of the medium. The energy *sublevels* of a system are single states with their own characteristic energy. A set of sublevels having the same energy is called an energy *level*. This can be seen with reference to Figure 4.113, which shows a schematic partial energy-level diagram of a typical laser system. Input energy excites ground-state particles (atoms, molecules, or ions) of the medium into the state (or states) indicated as 3. These particles then transfer themselves or excite other particles preferentially to sublevel 2. In an ideal system, negligible excitation of particles into sublevel 1 should occur. If the populations of the sublevels 2 and 1 are n_2 and n_1, respectively, and $n_2 > n_1$, this is called a *population inversion*. The medium then becomes capable of amplifying radiation of frequency

$$\nu = \frac{E_2 - E_1}{h} \qquad (4.206)$$

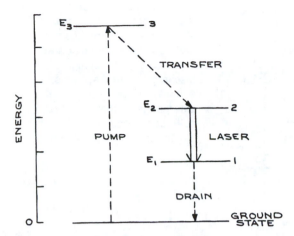

Figure 4.113   Schematic partial energy-level diagram of a laser system.

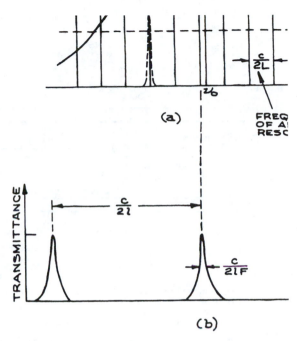

Figure 4.114   (a) Laser gain profile showing position of cavity resonances and potential laser oscillation frequencies near cavity resonances where gain lies above loss (*L* is length of laser cavity); (b) transmission characteristic of intracavity etalon, on same frequency scale as (a), for single-mode operation (*l* is the etalon thickness and *F* its finesse).

where h is Planck's constant. In practice, energy levels of real systems are of finite width, so the medium will amplify radiation over a finite bandwidth. The gain of the amplifier varies with frequency in this band and is specified by a gain profile $g(v)$, as shown in Figure 4.114. Actual laser oscillation occurs when the amplifying medium is placed in an optical resonator, which provides the necessary positive feedback to turn the amplifier into an oscillator. Most optical resonators consist of a pair of concave mirrors—or one concave and one flat mirror—placed at opposite ends of the amplifying medium and aligned parallel. A laser resonator is, in essence, a Fabry-Perot (see Section 4.7.4), generally of large spacing. The radii and spacing of the mirrors will determine to a large degree the type of Gaussian beam that the laser will emit. At least one of the mirrors is made partially transmitting, so that useful output power can be extracted. The choice of optimum mirror reflectance depends on the type of laser and its gain.

Low-gain lasers such as the helium-neon have high-reflectance mirrors, (98–99%), while higher-gain lasers such as pulsed nitrogen, CO_2, or Nd^{3+}:YAG can operate with much lower reflectance values. In certain circumstances, when their gain is very high, lasers will emit laser radiation without any deliberately applied feedback. Lasers that operate in this fashion are generally operating in an *amplified spontaneous emission* (ASE) mode. Such lasers are just amplifying their own spontaneous emission very greatly in a single pass.

4.6.2 General Features of Laser Design

Even though the majority of laser users do not build their own, an awareness of some general design features of laser systems will help the user to understand what can and cannot be done with a laser. It will also assist the user who wishes to make modifications to a commercial laser system to increase its usefulness or convenience in a particular experimental situation. Figure 4.115 shows a schematic diagram of a typical gas laser, which incorporates most of the desirable design features of a precision system. The amplifying medium of a gas laser is generally a high- or low-pressure gas excited in a

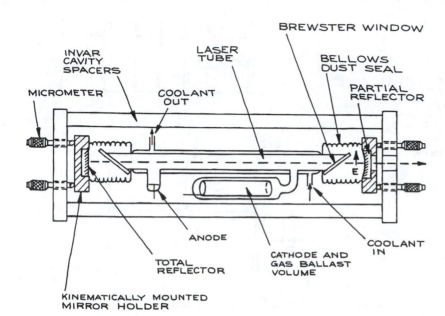

Figure 4.115 Schematic diagram of a typical gas laser showing some of the desirable features of a well-engineered research system. In high discharge-current systems, some form of internal or external gas return path from cathode to anode must be incorporated into the structure.

discharge tube. The excitation may be pulsed (usually by capacitor discharge through the tube), dc, or occasionally ac. Some gas lasers are also excited by electron beams, by radiofrequency energy, or by optical pumping with either a lamp or another laser. Far infrared lasers in particular are frequently pumped with CO_2 lasers. Depending on the currents that must be passed through the discharge tube of the laser, it can be made of Pyrex, quartz, beryllium oxide, graphite, or segmented metal. These last three have been used in high current-density discharge tubes for CW ion lasers. Several types of pulsed high-pressure or chemical lasers, such as CO_2, HF, DF, and HCl, can be excited in structures built of Plexiglas or Kel-F.

Because most gas lasers are electrically inefficient, a large portion of their discharge power is dissipated as heat. If ambient or forced-air cooling cannot keep the discharge tube cool enough, water cooling is used. This can be done in a closed cycle, using a heat exchanger. Some lasers need the discharge tube to run at temperatures below ambient. In such cases, a refrigerated coolant such as ethylene glycol can be used. Many gas lasers use discharge tubes fitted with windows placed at Brewster's angle. This permits linearly polarized laser oscillation to take place in the direction indicated in Figure 4.115—light bouncing back and forth between the laser mirrors passes through the windows without reflection loss. Mirrors fixed directly on the discharge tube can also be used—this is common in commercial He-Ne lasers. In principle, such lasers should be unpolarized, although in practice various slight anisotropies of the structure often lead to at least some polarization of the output beam.

The two mirrors that constitute the laser resonator, unless these are fixed directly to the discharge tube, should be mounted in kinematically designed mounts and held at fixed spacing l by a thermally stable resonator structure made of Invar or quartz. In this structure, an important parameter is the *optical length L* of the structure. For a laser in which the whole space between the two resonator mirrors is filled with a medium of refractive index n, $L = nl$. For a composite structure with regions of different refractive index, this result is easily generalized. Because the laser generates one or more output frequencies which are close to integral multiples of $c/2L$, any drift in mirror spacing L causes changes in output frequency v. This frequency change Δv satisfies

$$\frac{\Delta v}{v} = \frac{\Delta L}{L} \qquad (4.207)$$

For a laser 1 m long, even with an Invar-spaced resonator structure, the temperature of the structure would need to be held constant within 10 mK to achieve a frequency stability of only 1 part in 10^8. The discharge tube should be thermally isolated from the resonator structure unless the whole forms part of a temperature-stabilized arrangement.

If single-frequency operation is desired, the laser should be made very short so that $c/2L$ becomes larger than Δv in Figure 4.114—or operated with an intracavity etalon (see Sections 4.6.4 and 4.7.4). Visible lasers that have the capability of oscillating at several different wavelengths are generally tuned from line to line by replacing one laser mirror with a Littrow prism whose front face is set at Brewster's angle and whose back face is given a high-reflectance coating as shown in Figure 4.116a. As shown in Figure 4.116b, one can alternatively use a separate intracativity prism, designed so the intracavity beam passes through both its faces at Brewster's angle. Infrared gas lasers that can oscillate at several different wavelengths are tuned by replacing one resonator mirror with a gold-coated or solid metal diffraction grating mounted in Littrow, as shown in Figure 4.116c. The laser oscillation wavelength then satisfies

$$m\lambda = 2d\sin\theta \qquad (4.208)$$

where d is the spacing of the grating grooves, θ is the angle of incidence, and m is an integer.

A laser will oscillate only if its gain exceeds all the losses in the system, which include mirror transmission losses, absorption, and scattering at mirrors and windows, and any inherent losses of the amplifying medium itself—the last generally being significant only in solid-state, liquid, and semiconductor lasers. It is particularly important, therefore, to keep the mirrors and windows of a laser system clean and free of dust. Most commercial systems incorporate a flexible, sealed enclosure between Brewster windows and mirrors to exclude contaminants.

Figure 4.117 shows some design features of a simple lamp-pumped, solid-state laser oscillator. This design incorporates a linear pulsed or continuous lamp and a crystalline or glass laser rod with Brewster windows mounted inside a metal elliptical reflector, which serves to reflect pumping light efficiently into the laser rod.

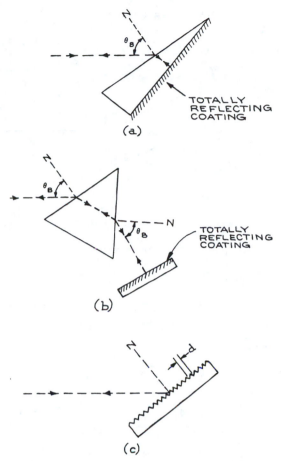

Figure 4.116 Methods for wavelength-selective reflection in laser systems: (a) Brewster's-angle Littrow prism; (b) intracavity Brewster's-angle prism; (c) reflective diffraction grating. N = surface normal.

Although such solid-state lasers are rarely used in experiments where extreme frequency stability and narrow linewidth operation are required, it is still good practice to incorporate features such as stable, kinematic resonator design and thermal isolation of the hot lamp from the resonator structure. Elliptical reflector housings for solid-state lasers (and flashlamp-pumped dye lasers) can be made in two halves by horizontally milling a rectangular aluminum block with the axis of rotation of the milling cutter set at an angle of arccos (a/b), where a and b are the semiminor and semimajor axes of the

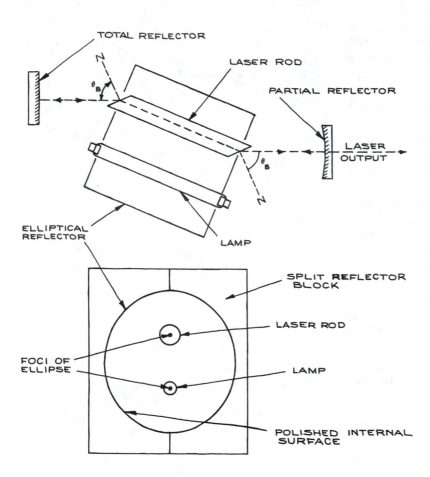

Figure 4.117 Schematic diagram of a simple solid-state laser system.

ellipse. This process is repeated for a second aluminum block. The two halves are then given a high polish, or plated, and joined together with locating dowel pins. For further details of solid-state laser design and particulars of other arrangements for optical pumping the reader should consult Röss[66] or Koechner.[67] Flashlamp-pumped dye lasers share some design features with solid-state lasers. The main difference is that the former require excitation of short pulse duration ($< \mu$s) with special flashlamps in low-inductance discharge circuitry.[68–71]

4.6.3 Specific Laser Systems

Gaseous Ion Lasers

Gaseous ion lasers, of which the argon, krypton, and helium-cadmium are the most important, generate narrow linewidth, highly coherent visible and ultraviolet radiation with powers in the range from milliwatts to a few tens of watts CW. Helium-cadmium lasers use low current-density (a few A cm^{-2}), air-cooled discharge structures, which in many respects are similar to those of helium-neon lasers. They are limited to a few tens of milliwatts in conveniently available output power. Argon and krypton ion lasers use very high current-density discharges (100–2000 A cm^{-2}) in special refractory discharge structures usually made of tungsten disks with ceramic spacers as shown in Figure 4.118. In operation, these disks cool themselves by radiating through an outer fused salica envelope surrounded by coolant water. Commercially available powers range up to a few tens of watts, but outputs up to 1 kW in the visible have been reported. Low power (< 100 mW) argon ion lasers can be forced air cooled. Comprehensive information about ion lasers can be found elsewhere.[72,73] Ion lasers are commercially available from American Laser, Coherent, Lexel Laser, and Spectra Physics. Argon and krypton ion lasers have been widely used for pumping CW dye lasers and Ti:sapphire lasers. They are rapidly being replaced in these applications, however, by frequency-doubled Nd lasers, which are more and more likely to be pumped themselves by semiconductor diode lasers. Such lasers can be built in the laboratory if long-term reliability and convenience are not a prime consideration. Construction of CW argon or krypton ion lasers is not recommended. Pulsed ion lasers operate at high current densities but have low duty cycles, so that ambient cooling is adequate. These lasers are not in widespread experimental use.

Helium-Neon Lasers

Of all types of laser, helium-neon lasers come closest to being the ideal classical monochromatic source. They can have very good amplitude (~ 0.1%) and frequency stability (1 part in 10^8 without servo frequency control). Servo frequency-controlled versions are already in use as secondary frequency standards, and have excellent temporal and spatial coherence.[74] Typical available power outputs range up to 50 mW. Single-frequency versions with power outputs of 1 mW, which are ideal for interferometry and optical heterodyne experiments, are available from Spectra-Physics and Aerotech. Although 632.8 nm is the routinely available wavelength from He-Ne lasers, other wavelengths such as 543 nm or 1.15 or 3.39 µm are also available. Helium-neon lasers of low power are very inexpensive, costing from about $100, and are ideal for alignment purposes.

Helium-Cadmium Lasers

These lasers are similar in many ways to helium-neon lasers. They operate at relatively low current densities and use discharge tube structures that do not require water cooling. They are quite important in applications requiring deep blue and ultraviolet light at powers of ten of miliwatts. They operate at 325 nm and 441.6 nm. These wavelengths, especially 325 nm, are useful in applications where a photochemically active source is required. They are widely used in photolithography. Helium-cadmium lasers are available from Kimmon Electric Co. and Melles Griot.

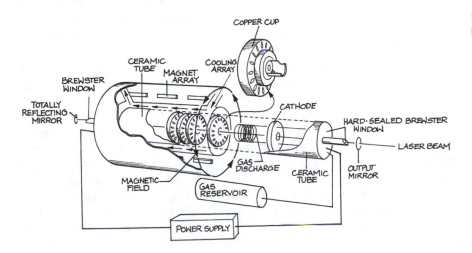

Figure 4.118   Construction of a high power, water-cooled ion laser.

CO$_2$ Lasers

Both pulsed and CW CO$_2$ lasers are easy to construct in the laboratory. A typical low-pressure CW version would incorporate most of the features shown in Figure 4.89. A water-cooled Pyrex discharge tube of internal diameter 10 to 15 mm, 1 to 2 m long, and operating at a current of about 50 mA, is capable of generating tens of watts of output at 10.6 μm. The best operating gas mixture is CO$_2$-N$_2$-He in the ratio 1:1:8 or 1:2:8 at a total pressure of about 25 mbar in a 10 mm diameter tube. ZnSe is the best material to use for the Brewster windows, as it is transparent to red light and permits easy mirror alignment. The optimum output mirror reflectance depends on the ratio of the tube length to diameter L/d roughly according to[75]

$$R \sim 1 - L/500d \qquad (4.209)$$

The output mirror is usually a dielectric-coated germanium, gallium arsenide, or zinc selenide substrate. Such mirrors are available from Janos, Laser Optics, Coherent, Infrared Industries, Laser Research Optics, Unique Optical, II-VI, and Rocky Mountain Instruments, among others. The total reflector can be gold-coated in lasers with outputs of a few tens of watts. For tunable operation, gratings are available from American Holographic, Gentec, Jobin-Yvon, Perkin-Elmer, Richardson Grating Laboratory, and Rochester Photonics. When a grating is used to tune a laser with Brewster windows, the grating should be mounted so that the E vector of the laser beam is orthogonal to the grating grooves. An excellent review of low pressure CW CO$_2$ gas lasers has been given by Tyte.[75]

CW CO$_2$ lasers can also be operated in a waveguide configuration, in which a much higher pressure gas mixture is excited in a small size discharge capillary.[76] These lasers provide substantially greater power output per volume than the low pressure variety. They can be excited by direct current, but radiofrequency excitation is now popular and has now solved the unreliability problems that afflicted early commercial versions. Current suppliers of these compact lasers are Edinburgh Instruments, DeMaria Electroptics Systems, Rofin-Sinar, Trumpf, Universal Laser Systems, and Synrad.

Transversely excited atmospheric-pressure (TEA) CO$_2$ lasers are widely used when high-energy, pulsed infrared energy near 10 μm is required. These lasers operate in discharge structures where the current flow is transverse to the resonator axis through a high-pressure ($\geq$ 0.5 bar) mixture of CO$_2$, N$_2$, and He, typically in the ratio 1:1:5. To achieve a uniform, pulsed glow discharge through a high-pressure gas, special electrode structures and preionization techniques are used to inhibit the formation of localized spark discharges. Many very high-energy types utilize electron-beam excitation. Discharge-excited TEA lasers are relatively easy to construct in the laboratory. A device with a discharge volume 60 cm long × 5 cm wide with a 3 cm electrode spacing excited with a 0.2 μF capacitor charges to 30 kV will generate several joules in a pulse about 150 ns long. There are very many different designs for CO$_2$ TEA lasers; particulars of various types can be found in the *IEEE Journal of Quantum Electronics* and the *Journal of Applied Physics*. A particularly good design that is easy to construct is derived from a vacuum ultra-violet photopreionized design described by Seguin and Tulip.[77] A diagram of the discharge structure is shown in Figure 4.119, together with its associated capacitor-discharge circuitry. The cathode is an aluminum Rogowski $2\pi/3$ profile;[78] the anode is a brass mesh, beneath which is a section of double-sided copper-clad printed circuit board machines into a matrix of separate copper sections on its top side. This printed circuit board provides a source of very many surface sparks for photopreionization of the main discharge. The laser is fitted with two separate capacitor banks—each circuit-board spark channel is energized by a 3600 pF Sprague or similar *doorknob* capacitor, while the main discharge is energized with a 0.2 μF low-inductance capacitor. Such capacitors are available from Hi Voltage Components, Maxwell Labs, and CSI Technologies among others. Both capacitor banks are switched with a single spark gap, which incorporates a Champion marine spark plug for triggering. The end of the spark plug, which has an annular gap, is ground down to its ceramic insulator and is triggered by an EG&G Model TM-11A Trigger generator. A good double Rogowski TEA laser design has been given by Sequin, Manes, and Tulip.[79] TEA lasers based on the multiple pin-discharge design first described by Beaulieu are also easy to construct, but they do not give

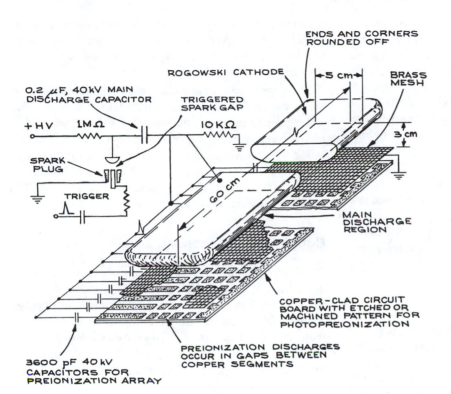

Figure 4.119 Cutaway view of vacuum-ultraviolet photopreionized Rogowski TEA laser structure

output energies comparable to those of volume-excited designs.[80] They do, however, lend themselves well to laser oscillation in many different gases—for example, HF formed by discharge excitation of H_2-F_2 mixtures.

CO Lasers

CW CO lasers are essentially similar in construction to CO_2 lasers except that the discharge tube must be maintained at low temperatures—certainly below 0°C. They operate in a complex mixture of He, CO, N_2, O_2, and Xe with, for example, 21 mbar He, 0.67 mbar N_2, 0.013 mbar O_2, and 0.4 mbar Xe.

Exciplex (Excimer) Lasers

Exciplex lasers operate in high-pressure mixtures such as Xe-F_2, Xe-Cl_2, Kr-Cl_2, Kr-F_2, and Ar-F_2. They are commonly also called *excimer* lasers, although this is not strictly correct scientific terminology except in the case of

the F_2^* laser. Both discharge TEA and E-beam-pumped configurations are used. Under electron bombardment, a series of reactions occur that lead to the formation of an excited complex (exciplex) such as XeF,[*] which is unstable in its ground state and therefore dissociates immediately on emitting light. Commercially available excimer lasers are sources of intense pulsed ultraviolet radiation. The radiation from these lasers is not inherently of narrow linewidth, because the laser transition takes place from a bound to a repulsive state of the exciplex, and therefore is not of well-defined energy. Excimer lasers are attractive sources for pumping dye lasers and whenever intense pulsed UV radiation is required in, for example, photolithography or photochemistry.

The beam quality from an excimer laser is not normally very high. These are *multimode* lasers, with often a rectangular output beam profile. These lasers are widely available commercially from companies such as Lambda Physik, Potomac Photonics, MPB Technologies, JP Sercel Associates, (JPSA), SOPRA, and Gam Laser.

Excimer lasers share technological design features in common with both CO_2 TEA lasers and nitrogen lasers. They use rapid-pulsed electrical discharges through the appropriate high pressure gas mixture. The discharge design must be of low inductance and the excitation is generally faster than is necessary to operate a CO_2 TEA laser. Constructional details can be found in journals such as the IEEE Journal of Quantum Electronics or the Journal of Applied Physics. Because these lasers use toxic gases such as fluorine and chlorine, appropriate gas handling safety precautions must be taken in their use. The gases used, which are usually diluted mixtures of a halogen in a noble gas, should be housed in a vertical gas cabinet and exhaust gases from the laser should be sent through a chemical *scrubber* or filter cartridge before venting into the atmosphere.

Nitrogen (N_2) Lasers

Molecular-nitrogen (N_2) lasers are sources of pulsed ultraviolet radiation at 337.1 nm. Commercially available low energy devices usually operate on a flowing nitrogen fill at pressures of several tens of mbar, and generate output powers up to about 1 MW in pulses up to about 10 ns long. These lasers are reliable and easy to operate, and are widely used for pumping low energy pulsed dye lasers. Nitrogen lasers, by virtue of their internal kinetics, only operate in a pulsed mode. They use very rapid transversely excited discharges, usually between two parallel, rounded electrodes energized with small, low-inductance capacitors charged to about 20 kV. The whole discharge arrangement must be constructed to have very low inductance. Special care must be taken with insulation in these devices; the very rapidly changing electric fields present can punch through an insulating material to ground under circumstances where the same applied dc voltage would be very adequately insulated. There are several good N_2 laser designs in the literature.[81–86] These designs fall into two general categories: Blumlein-type distributed capacitance discharge structures, and designs based on discrete capacitors. The latter, although they do not give such high-energy outputs from a given laser as the very rapid-discharge Blumlein-type structures, are much easier to construct in the laboratory and are likely to be much more reliable. The capacitors used in these lasers

need to be of low inductance and capable of withstanding the high voltage reversal and rapid discharge to which they will be subjected. Suitable capacitors for this application can be obtained from Murata or Sprague.

The output beam from a transversely excited N_2 laser is by no means Gaussian. It generally takes the form of a rectangular beam, typically ~ 0.5 cm × 3 cm, with poor spatial coherence. Such a beam can be focused into a line image with a cylindrical lens for pumping a dye cell in a pulsed dye laser. Nitrogen lasers are available commercialy from Laser Science, SOPRA, LTB Lasertechnik, and Opton Laser.

Copper Vapor Lasers

These are high-pulse repetition frequency (prf) gas lasers that operate using high temperature gas mixtures of copper vapor in a helium or neon buffer gas. With individual pulse energies up to several millijoules and prf up to 20 kHz, these lasers provide average powers up to several watts. They are used for pumping pulsed dye lasers. The principal operating wavelengths are 510.6 and 578.2 nm. Pulsed gold vapor lasers are also commercially available, which are similar in most respects. They operate at 627.8 nm.

CW Solid-State Lasers

The most important laser in this category is the Nd:YAG laser, where YAG (yttrium aluminum garnet $Y_3Al_sO_{12}$ is the host material for the actual lasing species-neodyminum ions, Nd^{3+}. Other host materials are also used: YLF (yttrium lithium fluoride $LiYF_4$), YVO_4 (yttrium vanadate), YALO (yttrium aluminum oxide $YAlO_3$) and GSGG (scandium-substituted gadolinium gallium garnety, $(Gd_3Sc_2Ga_3O_{12})$. The principal neodymium output wavelength is 1.06 μm, but other infrared wavelengths, especially 1.32 μm, are available. The 1.06 μm can be efficiently doubled to yield a CW source of green radiation. This source of green radiation has replaced the argon ion laser in many lasers, especially since the neodymium laser itself can be pumped so conveniently with appropriate semiconductor lasers.

Ti:sapphire is an attractive laser material for generating tunable radiation in the 700 to 1060 nm range. These lasers are generally pumped by argon-ion lasers on frequency-doubled neodymium lasers.

Fiber lasers, especially using erbium (Er)-doped fibers, have interesting properties. They can be operated in a *mode-locked* mode to generate very short pulses (< 1ps) in the infrared. The most important application of erbium-doped fiber, however, is in erbium-doped optical fiber amplifiers (EDFAs), which are widely used in fiber optic communication networks to amplify 1.55 μm semiconductor laser radiation. *F*-center lasers are doped-crystal lasers that are pumped with argon, krypton, or dye lasers.[87] They provide tunable operation in two regions—between 1.43 and 1.58 μm and between 2.2 and 3.3 μm—but are somewhat inconvenient because the crystal must be cooled with liquid nitrogen.

Pulsed Solid-State Lasers

Three lasers are most important in this category Nd:YAG, Nd:glass, and ruby. Nd:YAG and Nd:glass lasers oscillate at the same principal wavelength—1.06 μm. Nd:YAG is generally used in high-pulse repetition rate systems and/or where good beam quality is desired. High-energy Nd laser systems frequently incorporate an Nd:YAG oscillator and a series of Nd:glass amplifiers. Ruby lasers are still widely used in situations in which a high-energy visible output (694.3 nm) is required.

All these pulsed solid-state lasers use fairly long flash excitation (~1 ms). Unless the laser pulse output is controlled, they generate a very untidy optical output, consisting of several hundred microseconds of random optical pulses about 1 μs wide, spaced a few microseconds apart. This mode of operation, called *spiking*, is rarely used. The laser is usually operated in a *Q*-switched mode. In this mode, the laser cavity is blocked with an optical shutter such as an electrooptic modulator or Kerr cell, as illustrated in Figure 4.120. An appropriate time after flashlamp ignition, after the population inversion in the laser rod has had time to build to a high value, the shutter is opened. This operation, which changes the *Q* of the laser cavity from a low to a high value, causes the emission of a very large short pulse of laser energy. The *Q*-switched pulse can contain nearly as much energy as would be emitted in the *spiking* mode and is typically 10–20 ns long. Passive *Q*-switching can also be accomplished with a *bleachable* dye solution placed in an optical cell inside the laser cavity. The dye is opaque at low incident light intensities and inhibits the buildup of laser oscillation. When the laser acquires enough gain to overcome the intracavity dye absorption loss, however, laser oscillation begins, the dye bleaches, and a *Q*-switched pulse results. A bleachable dye cell with a short relaxation time, placed close to one of the laser mirrors, will frequently lead to mode-locked operation within the Q-switched pulse (see Section 4.6.4).[88-90]

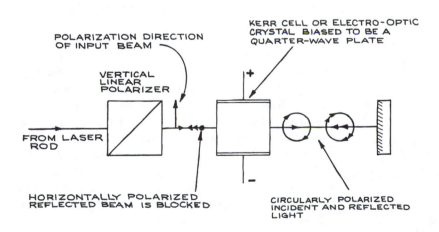

Figure 4.120　Schematic arrangement for Q-switching. Laser oscillation occurs when the voltage bias on the Kerr cell or electro-optic crystal is removed, thereby allowing the reflected beam to remain linearly polarized in the vertical direction.

Semiconductor Lasers

Three principal types of semiconductor lasers are commercially available—tunable diode lasers, which utilize lead-salt semiconductors such as PbSSe, PbSnSe, and PbSnTe and can provide tunable operation over limited regions anywhere between 3 and 30 μm; GaAs or GaAlAs lasers, which operate at fixed wavelengths in the region between 0.75 and 0.9 μm; and InGaAsP lasers,[*] which operate principally in the 1.3 and 1.55 μm regions. GaAlAs lasers are very inexpensive and reliable. They are widely used in communication and information-processing systems, in compact-disc players, or whenever a low-power source of fairly coherent radiation in the near infrared is required. InGaAsP lasers have been developed to operate at the optimum wavelengths for minimum dispersion and lowest loss in optical-fiber communication systems. New-infrared semiconductor lasers are described in detail in many specialized texts.[91–100]

Figures 4.121 through 4.124 give examples of some representative structures that are used in current semiconductor lasers. In all these structures, the current flow through the device is controlled so that a specific volumn of the active region is excited. This region is generally very small. The active length is not generally more than 1 to 2 mm. The lateral dimensions of the active region are generally in the 5 to 10 μm range, however, in single quantum well (SQW) and multiple quantum well (MQW) lasers the dimension of the active region in the direction of current flow can be as small as 10 to 100 nm.

Laser oscillation occurs in a resonant Fabry-Perot structure made up of the two end facets of the laser. The active region acts like a waveguide to confine the laser beam, which ideally emerges from the emitting facet as an elliptical Gaussian beam. This is a Gaussian beam whose spotsizes w_x, w_y are different in the two orthogonal (x,y) directions perpendicular to the z propagation direction of the laser beam. The smaller spot size is generally in the direction perpendicular to the junction. This asymmetry of the laser beam causes its beam divergence to be larger in the direction perpendicular to the junction than it is in the orthogonal direction, as shown schematically in Figure

---

[*] The phosphorous is often omitted, and the precise stoichiometry of (In,Ga) and (As,P) determines the laser wavelength.

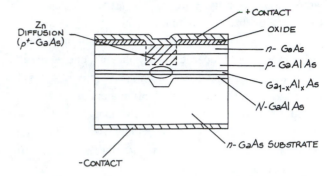

Figure 4.121 Double heterostructure laser diode design using a stripe excitation region confined by an oxide layer and deep zinc diffusion.

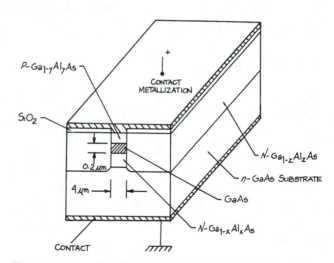

Figure 4.122 Double heterostructure laser of a buried stripe design.

4.125. Commercial diode lasers are often supplied with an *anamorphic* beam expander, which circularizes the laser beam and equalizes its orthogonal beam divergence angles; however, in general, the beam quality from any semiconductor laser is inferior to that from most gas lasers. In addition, the spectral properties of semiconductor lasers are inferior to helium-neon lasers. Their linewidths are typically several MHz, so their coherence lengths are short. This can make them less suitable in most interferometric applications than most gas lasers or diode-pumped solid-state lasers.

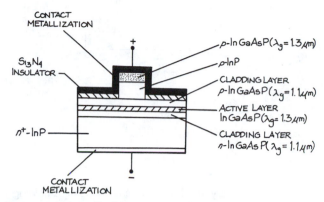

Figure 4.123   Cladded ridge double heterostructure semiconductor laser design.

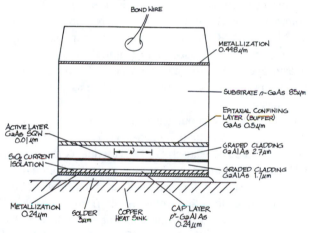

Figure 4.124   Graded Index Separate Confinement Heterostructure (GRINSCH) semiconductor laser design. Typical dimensions of the various layers are indicated. For a narrow stripe laser $w \sim 5$ μm, for a broad stripe laser $w \sim 60$ μm.

Despite their enormous commercial importance in telecommunications applications, near infrared diode lasers do not have many spectroscopic applications. They are not broadly tunable and do not provide many specific wavelengths below the red region of the spectrum. Blue-green diode lasers are now available but their power outputs are low (~ 1–10 mW).

High power diode lasers are available with powers of tens of watts, or higher. They are frequently used as pump

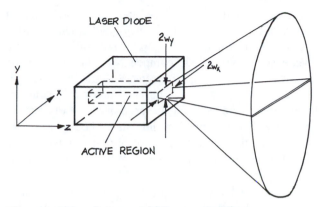

Figure 4.125   Schematic diagram of how a semiconductor laser emits a quasi-elliptical output beam because of asymmetrical beam confinement in two orthogonal directions inside the laser structure.

sources in diode-pumped solid-state lasers; however, as far as spectroscopy purity is concerned they are closer in character to a light emitting diode than they are to high coherence lasers such as helium-neon, helium-cadmium or diode-pumped Nd:YAG.

The tunable lead-salt diode lasers are now commonly used in high-resolution spectroscopy.[101] They generally emit several modes, which, because of the very short length of the laser cavity, are spaced far enough in wavelength for a single one to be isolated by a monochromator. The output laser linewidths that are obtainable in this way are about 10^{-4} cm^{-1} (3 MHz). Scanning is accomplished by changing the temperature of the cryogenically cooled semiconductor or the operating current. Complete tunable-semiconductor-laser spectroscopic systems are available from Spectra-Physics (LAD) or Quantum Associates.

Diode-Pumped CW Solid-State Lasers

Because of a fortuitous coincidence between the output of GaAlAs semiconductor lasers working near 809 nm and a strong absorption of the neodymium ions (Nd^{3+}) in YAG, YLF, or YVO$_4$, it is very efficient to optical pump Nd^{3+} lasers with diode lasers. These lasers have become widely available commercially. They can also be efficiently frequency-doubled to 530 nm, so have begun to replace ion lasers in applications where intense green light is required. Green output powers up to 70W are available.

Frequency-tripled operation at 355 nm is available at powers in excess of 3W—frequency-quadrupled operation at 266 nm provides powers up to 600 mW. There are several supplies of diode pumped Nd^{3+} lasers including Lightwave Electronics, Coherent, Spectra Physics, A-B Lasers, Quantronix, Lee Laser, Photonics Industries International and Spectra Laser Systems Ltd., among others.[27] Diode-pumped solid state lasers are also available at other wavelengths, notably 2.1 μm in holmium (Ho^{3+}), 1.6 μm and 2.8 μm in erbium (Er^{3+}) and 2.3 μm in thulium (Tm^{3+}).

The principal advantages of this type of laser are small size, narrow linewidth, and single-frequency operation.[102]

CW Dye Lasers

CW dye lasers usually take the form of a planar jet of dye solution continuously sprayed at Brewster's angle from a slit nozzle, collected, and recirculated. The jet stream is in a spherical mirror laser cavity, where it is pumped by the focused radiation of a CW ion laser or a frequently-doubled Nd:YAG laser, as shown in Figure 4.126. These lasers can generally be regarded as wavelength converters—-they have essentially the linewidth, frequency, and amplitude stability of the pumping source. For example, narrow-linewidth operation requires an etalon-controlled single-frequency ion laser. CW dye lasers are quite efficient at converting the wavelength of the pump laser—at pump powers far enough above threshold, the conversion efficiency ranges from 10 to 45 percent. Laser dyes are abailable from Exciton, Lambda, Physik, and Radiant Dyes Laser Accessories, among others. Dye lasers are rapidly being replaced wherever possible with tunable solid state devices. The dyes that are used photodegrade, and liquid spills are common. The laboratories of most dye-laser users are stained with the dyes that they use! For further details of CW dye-laser operation, consult Schafer.[68]

Pulsed Dye Lasers

CW dye lasers are not as frequently constructed in the laboratory as are pulsed dye lasers. In its simplest form, a pulsed dye laser consists of a small rectangular glass or quartz cell placed on the axis of an optical resonator and

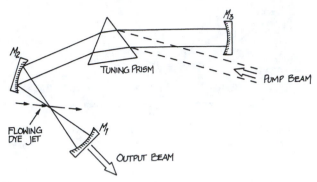

Figure 4.126   Layout of a CW dye laser using a flowing dye jet.

excited with a focused line image from a pump laser. Two simple geometrics are shown in Figure 4.127. The most commonly used pump lasers are N_2 and frequency-doubled Nd:YAG, although ruby, excimer, and copper-vapor lasers are also used. In its simplest form, this type of pulsed dye laser will generate a broad-band laser output (5–10 nm). Various additional components are added to the basic design to give tunable, narrow-linewidth operation. A very popular design that incorporates all the desirable features of a precision-tunable visible source has been described by Hänsch.[102] The essential components of a Hansch-type pulsed dye laser are shown in Figure 4.128. The pump laser is focused, with a cylindrical and/or a spherical lens, to a line image in the front of a dye cell. The line of excited dye solution (for example a 5×10^{-3} molar solution of Rhodamine 6G in ethanol) is the gain region of the laser. The laser beam from the gain region is expanded with a telescope in order to illuminate a large area of a high dispersion, Littrow-mounted echelle grating. Rotation of the grating tunes the center wavelength of the laser. Without the telescope, only a small portion of the grating is illuminated, high resolution is not achieved, and the laser linewidth will not be very narrow. With telescope and grating alone, the laser linewidth will be on the order of 0.005 nm at 500 nm (~0.2 cm^{-1}). Even narrower linewidth can be achieved by including a tilted Fabry-Perot etalon in the cavity. A suitable etalon will have a finesse of 20 and a free spectral range (see Section 4.7.4) below about 1 cm^{-1} in this case. Line widths as narrow as 4×10^{-4} nm can be achieved. A few experimental points are

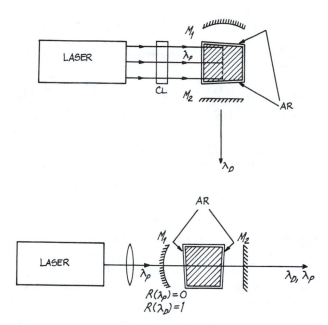

Figure 4.127   Two simple pulsed dye laser arrangements:
(a) transverse pumping; (b) longitudinal pumping.

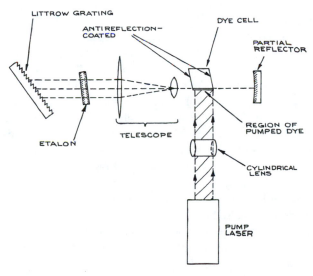

Figure 4.128   Hänsch-type pulsed dye laser.

worthy of note. The laser, for example, is tuned by adjusting grating and/or etalon. Alignment of the telescope to give a parallel beam at the grating is quite critical. The dye should be contained in a cell with slightly skewed faces to prevent spurious oscillation. The dye solution can remain static when low-power (10–100 kW) pump lasers are used, magnetically stirred when pump lasers up to about 1 MW are used, or continuously circulated from a reservoir when higher-energy or high-repetition-rate pump lasers are used. Suitable dye cells are available from Anderson Lasers, Esco, Hellma Cells, and NSG Precision Cells. Cells can be constructed from stainless steel with Brewster window faces when end-on pumping with a high-energy laser is used (as shown in Figure 4.129). The dye can be magnetically stirred by placing a small Teflon-

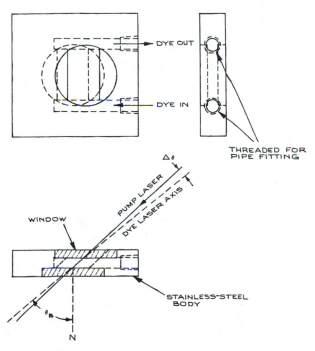

Figure 4.129   Flowing dye cell for end-pumped operation. N = normal to windows; θ_B = Brewster's angle, approximately the same for both pump and dye laser wavelengths; $\Delta\theta$ = a small angle. Dye solution flows laminarly in the channel between the windows.

coated stirring button in the bottom of the dye cell. Suitable pumps are available from Micropump Corporation for using continuous circulation of dye. Once a narrow linewidth, pulsed dye laser oscillator has been constructed, its output power can be boosted by passing the laser beam through one or more additional dye cells, which can be pumped with the same laser as the oscillator. In a practical oscillator-amplifier configuration, 10 percent of the pump power will be used to drive the oscillator and 90 percent will be used to drive the amplifier(s). Further details of Hänsch-type dye-laser construction are available in several publications.[102–105]

Two other approaches to achieving narrow-linewidth laser oscillation are worthy of note. The intracavity telescope beam expander can be replaced by a multiple-prism beam expander (see Section 4.3.4), which expands the beam onto the grating in one direction only. Such an arrangement does not need to be focused. An alternative, simple approach described by Littman is to eliminate the beam expander and instead use a Littrow-mounted holographic grating in grazing incidence as shown in Figure 4.130.[106,107] Linewidths below 0.08 cm^{-1} can be obtained without the additional use of an intracavity etalon.

4.6.4 Laser Radiation

Laser radiation is highly monochromatic in most cases, although flashlamp-pumped dye lasers in a worst case may have linewidths as large as 100 cm^{-1} (2.5 nm at 500 nm). The spectral characteristics of the radiation vary from laser to laser but will be specified by the manufacturer. Lasers are usually specified as single-mode or multimode.

A *single-mode* laser emits a single-frequency output. Continuous wave CO_2 lasers and other middle- and far-infrared lasers usually operate in this way, and helium-neon and argon-ion lasers can also be caused to do so. The single output frequency will itself fluctuate randomly, typically over a bandwidth of 100 kHz to 1 MHz, unless special precautions are taken to stabilize the laser—usually by stabilizing the spacing between the two mirrors that constitute its resonant cavity.

Multimode lasers emit several modes—called longitudinal modes—spaced in frequency by $c_0/2L$, where

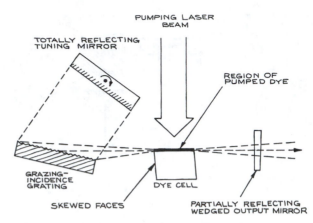

Figure 4.130 Littman-type dye laser.

c_0 is the velocity of light in vacuo and L is the optical path length between the resonator mirrors. In a medium of refractive index n and geometric length l, $L = nl$. There may, in fact, be several superposed combs of equally spaced modes in the output of a multimode laser if it is not operating in a single transverse mode. The transverse modes specify the different spatial distributions of intensity that are possible in the output beam. The most desirable such mode—which is standard in most good commercial lasers—is the fundamental TEM$_{00}$ mode, which has a Gaussian radial intensity distribution (see Section 4.2.4). A laser will generally operate on multiple longitudinal modes if the linewidth Δv of the amplifying transition is much greater than $c_0/2L$. In this case, the output frequencies will span a frequency range on the order of Δv. In short-pulse lasers, the radiation oscillating in the laser cavity may not be able to make many passes during the duration of the laser pulse, and (particularly if the linewidth of the amplifying transition is very large) the individual modes may not have time to become well characterized in frequency. This is the situation that prevails in pulsed dye and excimer lasers. Additional frequency-selective components must be included in such lasers, such as etalons and/or diffraction gratings, to obtain narrow output linewidths.

Laser radiation is highly collimated—beam divergence angles as small as 1 mrad are quite common. This makes a laser the ideal way to align an optical system, which is far superior to traditional methods using point sources and autocollimators. As mentioned previously, lasers are highly coherent. Single-mode lasers have the highest temporal coherence—with coherence lengths that can extend to hundreds of kilometers. Fundamental transverse-mode lasers have the highest spatial coherence.

In multimode lasers, the many output frequencies lie underneath the gain profile, as shown in Figure 4.114. Laser oscillation can be restricted to a single transverse mode by appropriate choice of mirror radii and spacing. The available output frequencies are then uniformly spaced longitudinal modes a frequency $c_0/2L$ apart, where L is the optical spacing of the laser mirrors. Laser oscillation can be restricted to a single one of these longitudinal modes by placing an etalon inside the laser cavity. The etalon should be of such a length l that its free spectral range $c/2l$ is greater than the width of the gain profile Δv. Its finesse should be high enough so that $c/2lF \leq c_0/2L$. The etalon acts as a filter that allows only one longitudinal mode to pass without incurring high loss.

Many lasers can be operated in a *mode-locked* fashion. Then the longitudinal modes of the laser cavity become locked together in phase and the output of the laser becomes a train of uniformly spaced, very narrow pulses. The spacing between these pulses corresponds to the round-trip time in the laser cavity $c_0/2L$. The temporal width of the pulses is inversely proportional to the gain bandwidth, therefore the broader the gain profile, the narrower the output pulses in mode-locked operation. An argon ion laser with a gain profile 5 GHz wide, for example, can yield mode-locked pulses 200 ps wide. A mode-locked dye laser with a linewidth of 100 cm^{-1} (3×10^{12} Hz) can, in principle, give mode-locked pulses as short as about 0.3 ps.

Even shorter pulses than this can be obtained with ring-dye lasers, with mode-locked Ti:sapphire lasers or with colliding pulse mode-locking techniques[13], which, with pulse compression techniques, can generate pulses a few tens of femtoseconds (10^{-15} s) long. Ultrafast laser systems are available from Clark-MXR, Coherent, Quantronix, Spectra Physics, and Thomson-CSF Laser. Pulse compressors are available from Femtochrome Research.

4.6.5 Coupling Light from a Source to an Aperture

A common problem in optical experiment design involves either the delivery of light from a source to a target, or collection of light from a source and delivery of the light to a collection aperture. The collection aperture might be the active area of a photodetector, the entrance slit of a spectrometer, or the facet of an optical fiber. A problem of specific importance is the focusing of a laser beam to a small spot. Sometimes the solution to one of these problems will involve an imaging system, but it might also involve a non-imaging light collector.

Collection of light from a point source and its delivery to a target aperture. This situation is illustrated in Figure 4.131. The solid angle $\Delta\Omega$ subtended by a circular aperture $S = \pi a^2$ at distance R from the source is

$$\Delta\Omega = 2\pi(1 - \cos\theta) \qquad (4.210)$$

which for $R >> a$ can be written as

$$\Delta\Omega = \frac{S}{R^2} \qquad (4.211)$$

If the radiant power of the source is P watts, then the power, collected by the aperture is

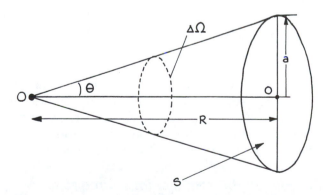

Figure 4.131   Geometry used to determine the collection efficiency of an aperture when it is illuminated by light from a point source.

$$P' = P\frac{\Delta\Omega}{4\pi} \tag{4.212}$$

To increase the light delivered to the aperture S, a lens system can be placed between O and O' as shown in Figure 4.13. To maximize light collection, the lens used should have a small f/number (f/#). The point source is placed close to the focal point of the lens, so that

$$\cos\theta \sim f/\sqrt{a^2 + f^2} \tag{4.213}$$

and

$$P' = \frac{P}{2}\left(1 - \frac{f}{\sqrt{a^2 + f^2}}\right) \tag{4.214}$$

Since the f/# is

$$f/\# = \frac{f}{2a} \tag{4.215}$$

Equation 4.124 becomes

$$P' = P\left(\frac{1}{2} - \frac{f/\#}{\sqrt{1 + 4f/\#}}\right) \tag{4.216}$$

with an f/1 lens

$$P' = P\left(\frac{1}{2} - \frac{1}{\sqrt{5}}\right) = 0.053P \tag{4.217}$$

In practice an aspheric lens should be used in this application, especially if the aperture S is of small size. If a spherical reflector is placed behind the source as shown in Figure 4.132b so that the source lies at its center of curvature, then the collection efficiency can be doubled.

Coupling light from a point source to an optical fiber.
Whether the source is actually a point source or is of small finite size relative to other distances in the geometry of Figure 4.132c is not very important.

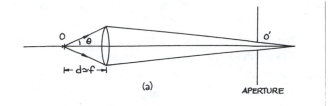

(a)

APERTURE

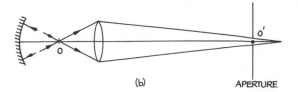

(b)

APERTURE

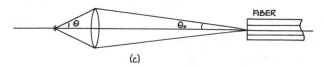

(c)

FIBER

Figure 4.132 Optical arrangements for optimizing coupling of a point source to an aperture: (a) lens coupling; (b) lens and spherical mirror coupling; (c) coupling of a point source to an optical fiber.

For coupling light to a fiber, the source must be imaged onto the front end of the cleaved fiber, but the light rays must remain within the numerical aperture of the fiber: $\sin\theta_0 \leq NA$. In principle, this can be accomplished by placing the source close to the focal point of the lens so that linear magnification (v/u) is large. In practice, the maximum magnification that can be used will be limited by the core diameter of the fiber and the finite size of the source.

Collection of light from an extended source and its delivery to a aperture.
This situation is shown in Figure 4.133. It is not possible, in this case, to image the extended source to a point—the best that can be done is to image as large an area of the source as possible onto the aperture S. If an axial point on the source is imaged to the center of the aperture S, then the area of the source that can be imaged onto the aperture S is

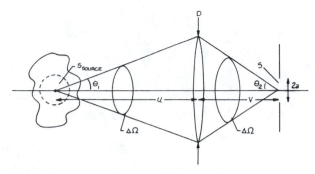

Figure 4.133 Geometry of an extended source and collection aperture used to describe the collection efficiency.

$$S_{\text{source}} = \frac{S}{m^2} \qquad (4.218)$$

since $m=(v/u)$ is the linear magnification of the system. A high-quality imaging system would need to be used in this situation, especially if the aperture S is small, otherwise aberrations will affect the size of the image.

If the brightness of the extended source in Figure 4.133 is $B_1(W/m^2/sr)$, then the power collected by the lens is approximately

$$P_1 = 2\pi B_1 S_{\text{source}}(1 - \cos\theta_1) \qquad (4.219)$$

This light is imaged onto the collection aperture of area S where it occupies a solid angle $\Delta\Omega_2$ determined by the angle θ_2, where

$$\Delta\Omega_2 = 2\pi(1 - \cos\theta_2) \qquad (4.220)$$

The brightness of the image is

$$B_2 = \frac{P_1}{2\pi S(1 - \cos\theta_2)} = \frac{B_1 S_{\text{source}}}{S}\frac{(1 - \cos\theta_1)}{(1 - \cos\theta_2)} \qquad (4.221)$$

which gives

$$B_2 = \frac{B_1 \sin^2(\theta_1/2)}{m^2 \sin^2(\theta_2/2)} \qquad (4.222)$$

For a paraxial system $\sin(\theta_1/2) \simeq (\theta_1/2)$, so

$$B_2 = \frac{B_1(\theta_1^2)}{m^2(\theta_2^2)} \qquad (4.223)$$

The ratio $\left(\dfrac{\theta_2}{\theta_1}\right)$ is the angular magnification, m', of the system. Therefore, since $mm' = 1$

$$B_2 = B_1 \qquad (4.224)$$

This is an example of the *brightness* theorem, which states that the brightness of an image can not be greater than the brightness of the object.[*]

4.6.6 Optical Modulators

In many optical experiments, particularly those involving the detection of weak light signals or weak electrical signals generated by some light-stimulated phenomenon, the signal-to-noise ratio can be considerably improved by the use of phase-sensitive detection. The principles underlying this technique are discussed in Section 6.8.3. To modulate the intensity of a weak light signal falling on a detector, the signal must be periodically interrupted. This is most easily done with a mechanical chopper. Weak electrical signals that result from optical stimulation of some phenomenon can also be modulated in this way by chopping the radiation from the stimulating source.

Modulation of narrow beams of light such as laser beams, or of extended sources that can be focused onto a small aperture with a lens, is easily accomplished with a tuning-fork chopper. Such devices are available from Boston Electronics, Electro-Optical Products, and Scitec Instruments. The region chopped can range up to several millimeters wide and a few centimeters long at frequencies from 5 Hz to 3 kHz. For chopping emission from extended sources, or over large apertures, rotating chopping wheels are very convenient. The chopping wheel can be made of

---

[*] In practice, in any real system, transmission losses, and aberrations will cause $B_2 < B_1$.

any suitable rigid, opaque material and should have radially cut apertures of one of the forms indicated in Figure 4.134. This form of aperture ensures that the mark-to-space ratio of the modulated intensity is independent of where the light passes through the wheels. By cutting very many slots in the wheel, very high modulation rates can be achieved—up to 100 kHz with a 20,000 rpm motor and a wheel with 300 slots, for example. The chopping wheel should usually be painted or anodized matt black. For chopping intense laser beams (in excess of perhaps 1 watt), however, it may be better to make the wheel reflective so that unwanted beam energy can be reflected into a beam dump. Reflective glass chopping wheels can be used at low speeds in experiments where a light beam must be periodically routed along two different paths.

To obtain an electrical reference signal that is synchronous with a mechanical chopping wheel, a small portion of the transmitted beam—if the latter is intense enough—can be reflected onto a photodiode or phototransistor. An auxiliary tungsten filament lamp and photodiode can also be mounted on opposite sides of the wheel. If these are mounted on a rotatable arm, the phase of the reference signal they provide can be adjusted mechanically. Chopping wheels with built-in speed control and electrical rotation reference signals are commercially available from DL Instrument, EG&G Princeton Applied Research, New Focus, Stanford Research Systems, and Thorlabs.

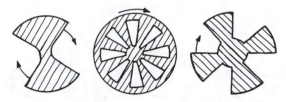

Figure 4.134   Some chopping-wheel designs.

Electro-optic modulators (Pockel's cells) can also be used in special circumstances, particularly for modulating laser beams or when mechanical vibration is undesirable. The operation of these devices is shown schematically in Figure 4.135. Light passing through the electro-optic crystal is plane-polarized before entry and then has its state of polarization altered by an amount that depends on the voltage applied to the crystal. The effect is to modulate the intensity of the light transmitted through a second linear polarizer placed behind the electro-optic crystal. By choosing the correct orientation of the input polarizer relative to the axes of the electro-optic crystal and omitting the output polarizer, the device becomes an optical phase modulator. For further details of these electro-optic amplitude and phase modulators the reader is referred to the books by Yariv,[12] Kaminow,[108] and Davis.[13] Electro-optic modulators are available from Cleveland Crystals, Conoptics, II-VI, JDS Uniphase, Meadowlark Optics, New Focus, and Quantum Technology, among others.

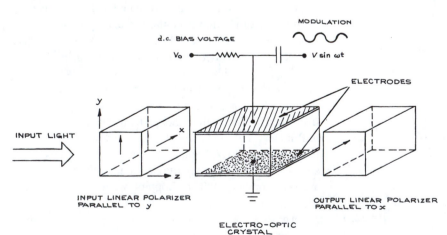

Figure 4.135   An electro-optic amplitude modulator using a transversely operated electro-optic crystal. The dc bias voltage can be used to adjust the operating point of the modulator so that, in the absence of modulation, the output intensity is a maximum, a minimum, or some intermediate value.

Acousto-optic modulators which operate by diffraction effects induced by sound wave-produced periodic density variations in a crystal are also available.[27] The most widely used materials for these devices are $LiNbO_3$, TeO_2, and fused quartz. In their operation, a radio frequency (rf) sound wave is driven through the material from a piezoelectric transducer, usually fabricated from $LiNbO_3$ or ZnO. A schematic diagram of how such an acousto-optic modulator works is given in Figure 4.136. Incident light that makes an appropriate angle θ_B, with the sound wavefronts, will be diffracted if it simultaneously satisfies the condition for constructive interference and reflection from the sound wavefronts. This condition is $\sin \theta_B = \lambda/\lambda_s$, where λ is the laser wavelength in the acousto-optic material, and λ_s is the sound wavelength. The device is used as an amplitude modulator by amplitude-modulating the input rf, which will be set to a specific optimum frequency for the device being used. Increase of drive power increases the diffracted power I_1 and reduces the power of the undeviated beam I_0 and vice versa. The device also functions as a frequency shifter. In Figure 4.99, the laser beam reflects off a moving sound wave and is Doppler shifted so that the beam I_1 is at frequency $\omega + \omega_s$, where ω_s is the frequency of the sound wave. Acoustooptic devices are available from many suppliers, including Anderson Lasers, Aurora Photonics, Crystal Technology, IntraAction, Isomet, and NEOS Technologies. Further details about these devices can be found in References 13, 109, and 110, located at the back of this chapter.

Very many liquids become optically active upon application of an electric field—that is, they rotate the plane of polarization of a linearly polarized beam passing through them. This phenomena is the basis of the Kerr cell, which can be used as a modulator but is more commonly used as an optical shutter (for example, in laser Q-switching applications as discussed in Section 4.6.3). A typical Kerr cell uses nitrobenzene placed between two plane electrodes across which a high voltage is applied. This voltage is typically several kilovolts and is sufficient to rotate the plane of polarization of the incident light passing between the plates by 45 or 90°. Kerr cells are rarely used these days and have been superseded by electro-optic modulators.

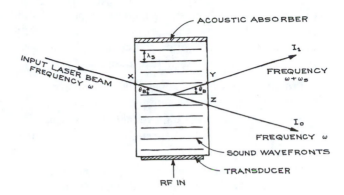

Figure 4.136 Schematic diagram showing the operation of an acousto-optic device in the Bragg regime. For simplicity, the refraction of the laser beam at X, Y, and Z is not shown.

Closely related to optical modulators are optical deflectors, which can be used for the spatial scanning or switching of light beams. Two principal types are commonly used—electromechanical devices that utilize small mirrors mounted on galvanometer suspensions (available from Cambridge Technology, GSI Lumonics, and NEOS Technologies), and beam deflectors that utilize the acousto-optic effect. In the latter, the deflection angle shown in Figure 4.136 is controlled by changing the rf drive frequency. These devices are available from Aurora Photonics, Crystal Technology, IntraAction, Isomet, and NEOS Technologies. Electro-optic scanners are also available, but are less widely used. These devices are generally designed for small deflections. They are available from Conoptics.

Spatial light modulators are multiple element devices, which generally use liquid crystals, work in transmission, and can modify the characteristics of an entire wavefront in a pixellated fashion. They are available from Display Tech and Meadowlark Optics.

4.6.7 How to Work Safely with Light Sources

Light sources, whether coherent or incoherent, can present several potential safety hazards in the laboratory. The primary hazard associated with the use of home-built optical sources is usually their power supply. Lasers and flashlamps, in particular, generally operate with potentially lethal high voltages, and the usual safety considerations for constructing and operating such power supplies should be followed in their design:

- Provide a good ground connection to the power supply and light-source housing.
- Screen all areas where high voltages are present.
- Install a clearly visible indicator that shows when the power supply is activated.
- HV power supplies, particularly those that operate with pulsed power, can remain dangerous even after the power is turned off, unless energy-storage capacitors are automatically shunted to ground. Always short the capacitors in such a unit to ground after the power supply is turned off before working on the unit.
- As a rule of thumb, keep a gap of about one inch for every 10 kV between high-voltage points and ground.
- To avoid excessive corona at voltages above about 20 kV, make sure that high-voltage components and connections have no sharp edges. Where such points must be exposed, corona can be reduced by installing a spherical metal corona cap on exposed items such as bolts or capacitor terminals. Resistor and diode stacks can be potted in epoxy or silicone rubber to prevent corona.
- Sources that are powered inductively or capacitively with rf or microwave power can burn fingers that come too close to the power source even without making contact.

Commercial optical sources are generally fairly safe electrically and are likely to be equipped with safety features, such as interlocks, which the experimentalist may not bother to incorporate into home-built equipment. However, *caveat emptor*: Remember the maxim "it's the volts that jolts, but it's the mills that kills."

Other general precautions with regard to electric shock include the following:

- Avoid wearing metallic objects such as watches, watchbands, and rings.
- If any operations must be performed on a line circuit, wear well-insulated shoes and, if possible, use only one hand.
- Keep hands dry—do not handle electrical equipment if you are sweating.

- Learn rescue and resuscitation procedures for victims of electric shock: Turn off the equipment, remove the victim by using insulated material; if the victim is not breathing, start mouth-to-mouth resuscitation; if there is no pulse, begin CPR procedures immediately; summon medical assistance; continue resuscitation procedures until relieved by a physician. If the victim is conscious but continues to show symptoms of shock, keep the person warm.

Other hazards in the use of light sources include the possibility of eye damage, direct burning (particularly by exposure to the beam from a high-average-power laser), and the production of toxic fumes. The last is most important in the use of high-power CW arc lamps, which can generate substantial amounts of ozone. The lamp housing should be suitably ventilated and the ozone discharged into a fume hood or into the open air. Some lasers operate using a supply of toxic gas, such as hydrogen fluoride and the halogens. Workers operating such systems must be sufficiently experienced to work safely with these materials. The *Matheson Gas Data Book*, published by Matheson Gas Products, is a comprehensive guide to laboratory gases, detailing the potential hazards and handling methods appropriate to each.

The potential eye hazard presented by most incoherent sources is not great. If a source appears very bright, one should not look at it, just as one should not look directly at the sun. Do not look at sources that emit substantial ultraviolet radiation; at the least, severe eye irritation will result—imagine having your eyes full of sand particles for several days! Long-term exposure to UV radiation should be kept below 0.5 μW/cm^2. Ordinary eyeglasses will protect the eyes from ultraviolet exposure to some extent, but plastic goggles that wrap around the sides are better. Colored plastic or glass provides better protection than clear material. The manufacturer's specification of ultraviolet transmission should be checked, since radiation below 320 nm must be excluded from the cornea.

The use of lasers in the laboratory presents an optical hazard of a different order. Because a laser beam is generally highly collimated and at least partially coherent, if the beam from a visible or near-infrared laser enters the pupil of the eye it will be focused to a very small spot on the retina (unless the observer is very near-sighted). If this focused spot happens to be on the optic nerve, total blindness may result. If it falls elsewhere on the retina, an extra blind spot may be generated. Although the constant motion of the human eye tends to prevent the focused spot from remaining at a particular point on the retina for very long, always obey the following universal rule: *Never look directly down any laser beam either directly or by specular reflection.* In practice, laser beams below 1 mW CW are probably not an eye hazard, but they should still be treated with respect. A beam must be below about 10 μW before most workers would regard it as really safe.

The safe exposure level for CW laser radiation depends on the exposure time. For pulsed lasers, however, it is the maximum energy that can enter the eye without causing damage that is the important parameter. In the spectral region between 380 and 1.5 μm, where the interior material of the eye is transparent, the maximum safe dose is on the order of 10^{-7} J/cm^2 for Q-switched lasers and about 10^{-6} J/cm^2 for non-Q-switched lasers. If any potential for eye exposure to a laser beam or its direct or diffuse reflection exists, it is advisable to carry out the experiment in a lighted laboratory—in a darkened laboratory, the fully dark-adapted human eye with a pupil area of about 0.5 cm^2 presents a much larger target for accidental exposure. The Laser Institute of America publishes a guide that shows the maximum permitted exposure in terms of power and direction for a range of laser wavelengths.[111]

Infrared laser beams beyond about 1.5 μm are not a retinal hazard, as they will not penetrate to the retina. Beams between 1.5 and 3 μm penetrate the interior of the eye to some degree. Because its energy is absorbed in a distributed fashion in the interior of the eye, 1.55 μm is one of the safest laser wavelengths. Lasers beyond 3 μm are a burn hazard, and eye exposure must be avoided for this reason. CO_2 lasers, for example, which operate in the 10 μm region, are less hazardous than equivalent-intensity argon-ion or neodymium lasers. Neodymium lasers represent a particularly severe hazard—they are widely used, emit substantial powers and energies, and operate at the invisible wavelength of 1.06 μm. This wavelength easily penetrates to the retina, and the careless worker can suffer severe eye damage without any warning. The use of safety goggles is strongly recommended when using such lasers. These goggles absorb or reflect 1.06 μm but allow normal transmission in at least part of the visible spectrum. Laser safety goggles are available

commercially from several sources, such as Bollé, Control Optics, Glendale/Dalloz Safety, Kentek, Lase-R Shield, Laser Peripherials, NOIR, and UVEX. The purchaser must usually specify the laser wavelength(s) for which the goggles are to be used.

Unfortunately, one problem with goggles prevents their universal use. Because the goggles prevent the laser wavelength from reaching the eyes, they prevent the alignment of a visible laser beam through an experiment by observation of diffuse reflection of the beam from components in the system. When such a procedure must be carried out, we recommend caution and the operation of the laser at its lowest practical power level during alignment. A weak, subsidiary alignment laser can frequently be used to check the potential path of a high-power beam before this is turned on—a strongly recommended procedure. For infrared laser beams thermal sensor cards are available that will reveal the location of an infrared laser beam through thermally activated fluorescence or fluoresence quenching. They are available from Applied Scintillation Technology, Laser S.O.S., Macken Instruments, and Newport Instruments. Some of these sensor cards or plates, especially for CO_2 laser beam location, require illumination with a UV source. Where an infrared laser beam strikes the illuminated surface, a dark spot appears. Used Polaroid film and thermally sensitive duplicating paper are also useful for infrared-beam tracking.

The beam from an ultraviolet laser can generally be tracked with white paper, which fluoresces where the beam strikes it. A potential burn and fire hazard exists for lasers with power levels above about 1 W/cm^2, although the fire hazard will depend on the target. Black paper burns most easily. Very intense beams can be safely dumped onto pieces of firebrick. Pulsed lasers will burn at output energies above about 1 J/cm^2.

In the United States, commercial lasers are assigned a rating by the FDA Center for Devices and Radiological Health, which identifies their type, power output range, and potential hazard. It must be stressed, however, that lasers are easy to use safely—accidents of any sort have been rare. Their increasing use in laboratories requires that workers be well-informed of the standard safety practices. For further discussion of this and all aspects of laser safety and related topics, we recommend both the book by Sliney and Wolbarsht[112] and a series of articles on laser safety in the *CRC Handbook of Laser Science and Technology*, Vol. 1.[113]

4.7 OPTICAL DISPERSING INSTRUMENTS

Optical dispersing instruments allow the spectral analysis of optical radiation or the extraction of radiation in a narrow spectral band from some broader spectral region. In this general category, we include interference filters, prism and grating monochromators, spectrographs and spectrophotometers, and interferometers. Interferometers have additional uses over and above direct spectral analysis—including studies of the phase variation over an optical wavefront, which allow the optical quality of optical components to be measured. Spectrometers, or *monochromators* as they are generally called, are optical filters of tunable center wavelength and bandwidth. The output narrow-band radiation from these devices is generally detected by a photon or thermal detector, which generates an electrical signal output. *Spectrographs*, on the other hand, record the entire spectral content of an extended bandwidth region either photographically, or more commonly these days with a linear detector array. *Spectrophotometers* are complete commercial instruments, which generally incorporate a source or sources, a dispersing system (which may involve interchangeable prisms and/or gratings), and a detector. They are designed for recording ultraviolet, visible, or infrared, absorption, and emission spectra.

Spectrofluometers are instruments to record fluorescence spectra. In such an application, two principal modes of spectral usage exist. The emission fluorescence spectrum is the spectral distribution of emission produced by a particular monochromatic excitation wavelength. The excitation spectrum is the total fluorescence emission recorded as the wavelength of the excitation source is scanned.

Before giving a discussion of important design considerations in the construction of various spectrometers and interferometers, it is worthwhile mentioning two important figures of merit that allow the evaluation and comparison of the performance of different types of optical dispersing instruments. The first is the resolving power, $\mathcal{R} = \lambda/\Delta\lambda$, which has already been discussed in connection with diffraction gratings (Section 4.3.5). The second is the luminosity, which is the flux collected by a detector at the output of a spectrometer when the source at the input has a radiance of unity. The interrelation between resolving power and luminosity for spectrometers using prisms, gratings, and Fabry-Perot etalons has been dealt with in detail by Jacquinot.[114]

Prism spectrometers have almost disappeared from use in recent years, however, because of their simple mode of operation—they serve as an archetype in discussing spectrometer performance.

Figure 4.137, which shows the essential components of a prism monochromator, will serve to illustrate the points made here. High resolution is clearly obtained in this arrangement by using narrow entrance and exit slits. The maximum light throughput of the spectrometer results when the respective angular widths W_1, W_2 of the input and output slits θ_1, θ_2 satisfy

$$W_1 = W_2 \tag{4.225}$$

where

$$W_1 = \frac{\theta_1}{(d\alpha/d\lambda)_\delta}, \; W_2 = \frac{\theta_2}{(d\delta/d\lambda)_\alpha} \tag{4.226}$$

The angles α, δ are the same as those used in Figure 4.43. The input and output dispersions $(d\alpha/d\lambda)^\delta$ and $(d\delta/d\lambda)^\alpha$ are only equal in a position of minimum deviation, or with a prism (or grating) used in a Littrow arrangement. If diffraction at the slits is negligible, the intensity distribution at the output slit when a monochromatic input is used is a triangular function as shown in Figure 4.138, where W is the spectral width of the slits given by

$$W = W_1 = W_2 \tag{4.227}$$

The limit of resolution of the monochromator is

$$\delta\lambda = W = \frac{\theta_2}{(d\delta/d\lambda)_\alpha} \tag{4.228}$$

The maximum flux passing through the output slit is

$$\Phi = TE_e(\lambda)S1\theta_2/f_2 \tag{4.229}$$

where S is the normal area of the output beam, T is the transmittance of the prism (or efficiency of the grating in

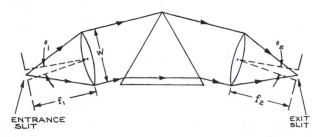

Figure 4.137 Basic elements of a prism monochromator.

the order used) at the wavelength being considered, l is the height of the entrance and exit slits (equal), and $E_e(\lambda)$ is the radiance of the monochromatic source of wavelength λ illuminating the entrance slit. $l\theta_2/f_2$ is the solid angle that the exit slit subtends at the output focusing lens. Equation 4.229 can be written as

$$\Phi = \frac{TE_e(\lambda)Sl\lambda(d\delta/d\lambda)_\alpha}{f_2 \Re} \qquad (4.230)$$

which clearly shows that the output flux is inversely proportional to the resolving power (for a continuous source at the entrance slit, Equation 4.230 is multiplied by an additional factor $\lambda/\Re$). The efficiency E of the monochromator is defined by the equation

$$E = \frac{TSl}{f}\left(\frac{d\delta}{d\lambda}\right)_\alpha \qquad (4.231)$$

where we have assumed the usual case in which $f_1 = f_2 = f$.

To collect all the light from the collimating optics in the arrangement of Figure 4.137, the prism or grating should have a normal area at least as large as the aperture of the collimating optics. The height of the prism (or diffraction grating) should therefore be $h \geq W$. The ratio f/W, which is a measure of the light gathering power of the monochromator, defines its $f/\#$. The most efficient way to use a spectrometer of any kind is to send the input radiation into the entrance slit within the cone of angles defined by the $f/\#$. The maximum resolution is obtained if the input $f/\#$ matches the spectrometer $f/\#$. If radiation enters the spectrometer with an $f/\#$ that is too small, this radiation overfills the dispersing element and some is

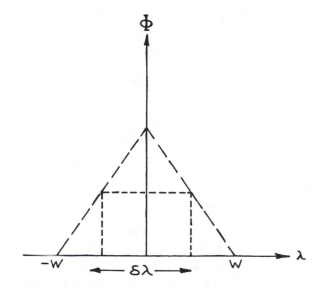

Figure 4.138 Intensity distribution at the output slit of a monochromator whose entrance and exit slits have the same spectral width W.

wasted. This can be illustrated with the aid of Figure 4.139, which shows the use of a point source directly at the entrance slit. In Figure 4.139b, a point source is imaged on the entrance slit using a lens that matches the input radiation to the $f/\#$ of the spectrometer. In Figure 4.139a, the fractional useful light collection from the source is

$$\phi_1 \cong \frac{\pi W^2}{4f^2} = \frac{\pi}{4F^2} \qquad (4.232)$$

where F is the $f/\#$ of the spectrometer. In contrast, in Figure 4.139b it is approximately

$$\phi_2 \cong \frac{\pi d^2}{4u^2} \qquad (4.233)$$

which can be written as

$$\phi_2 = \frac{\pi d^2(v-f)^2}{4f^2 v^2} \qquad (4.234)$$

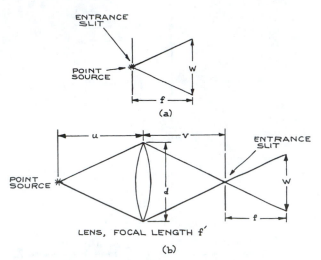

Figure 4.139 Schematic diagrams illustrating geometrical factors involved in optimizing light collection efficiency by a spectrometer of aperture W: (a) point source directly at entrance slit; (b) remote point source imaged on entrance slit with a lens.

and substituting $v/d = F$, $f/d = F'$ (the $f/\#$ of the lens), we have

$$\phi_2 = \frac{\pi(F/F' - 1)^2}{4F^2} / F' \le F \qquad (4.235)$$

Thus, equally efficient light gathering to that which would be obtained with the point source at the entrance slit results if

$$F' = F/2 \qquad (4.236)$$

More efficient light gathering results if F' is less than this value, but, if the lens and point source are incorrectly positioned, so that $v/d < F$, then not all the light collected by the lens reaches the dispersing element of the spectrometer.

It is quite common for spectrometers to be used with a fiber optic light delivery system. If the NA of the fiber is too large and the fiber optic is placed at the entrance of the spectrometer, then overfilling of the dispersing element will occur. In this case, correction optics must be included to match the fiber to the $f/\#$ of the spectrometer.

4.7.1 Comparison of Prism and Grating Spectrometers

Following Jacquinot, we use Equation 4.230 to compare prism and grating instruments.[114] Assume identical values of l and f_2 since, for a given degree of aberration of the system, their values are identical for prisms and gratings. Blazed-grating efficiencies can be high, so T is assumed to be similar for prism and grating. The comparison of luminosity under given operating conditions of wavelength and resolution therefore depends solely on the quantity $S(d\delta/d\lambda)^\alpha$ for a prism and $S(d\beta/d\lambda)^\alpha$ for a grating.

For the prism,

$$\frac{d\delta}{d\lambda} = \frac{t}{W}\frac{dn}{d\lambda} \qquad (4.237)$$

where for maximum dispersion the whole prism face is illuminated, so that t is the width of the prism base. $S = hW$ and $th = A$, the area of the prism base, and so

$$S\frac{d\delta}{d\lambda} = A\frac{dn}{d\lambda} \qquad (4.238)$$

For the grating,

$$\frac{d\beta}{d\lambda} = \frac{m}{d\cos\beta} \qquad (4.239)$$

$S = A\cos\beta$, where A is the area of the grating. If it is assumed that the grating is used in Littrow, then $m\lambda = 2d\sin\beta$, and therefore

$$S\frac{d\beta}{d\lambda} = \frac{2A\sin\beta}{\lambda} \qquad (4.240)$$

The ratio of luminosities for a prism and grating of comparable size is therefore

$$\rho = \frac{\Phi(\text{prism})}{\Phi(\text{grating})} = \frac{\lambda dn/d\lambda}{2\sin\beta} \qquad (4.241)$$

This ratio can be improved by a factor of two if the prism is also used in Littrow. For a typical value of 30° for β, Equation 4.241 predicts that a grating is always superior to a prism instrument. Except for a few materials in restricted regions of wavelength where $dn/d\lambda$ becomes large and ρ may reach 0.2 to 0.3, a grating is more luminous than a prism by a factor of 10 or more. The only potential advantage of a prism instrument over a grating is the absence of overlapping orders. A grating instrument illuminated simultaneously with 300 nm and 600 nm, for example, will transmit both at the same angular position—a prism instrument will not. This minor problem is easily solved with an appropriate order-sorting color or interference filter.

Jacquinot has demonstrated that a Fabry-Perot etalon or interferometer (see Section 4.7.4) used at a high or moderate resolving power is superior in luminosity to a grating instrument by a factor that can range from 30 to 400 or more.[114] An etalon cannot generally be used alone because of its many potential overlapping orders. It is usually coupled with a grating or prism monochromator in high-resolution spectroscopic applications, except in those cases where the source is already highly monochromatic.

Table 4.13 gives a comparison of the performance characteristics of prism and grating monochromators and Fabry-Perot interferometers. From the vacuum ultraviolet (prism instruments cannot be used below about 120 nm) to the far infrared, grating instruments are much superior to comparable-size prism instruments in both resolving power and luminosity. In recent years, the advantages of grating instruments over prism instruments has virtually eliminated the latter. This has primarily resulted because of improvements in grating fabrication. Holographic plane gratings have fewer and weaker ghosts than metal gratings, and holographic curved gratings have allowed aberration reduction in spectrometers. Apart from the minor inconvenience of potential overlapping orders, grating instruments are almost always to preferred over prism instruments. There is one exception worth noting. When a monochromator is being used to observe some weak optical emission in the simultaneous presence of a strong laser beam, unless the optical emission is too close in wavelength to the laser for prism resolution to be adequate. In this application, a prism avoids the problems of ghost emission or grating scattering. A good prism has

very low scattered light. It is common to predisperse a light beam with a prism before sending the beam into the entrance slit of a monochromator in order to avoid both the overlapping-order problem and spurious signals from a very strong light signal at another wavelength. Fabry-Perot interferometers have very high resolution and luminosity, but require careful and regular adjustment to maintain high performance. They are used only where their very high resolution is essential, frequently in conjunction with a grating monochromator for preisolation of a narrow spectral region.

A final point to note with regard to the use of grating spectrometers—these instruments are polarization sensitive. If unpolarized light strikes a grating, the diffracted light will be partially polarized because the grating efficiency is wavelength, and S and P polarization dependent.

4.7.2 Design of Spectrometers and Spectrographs

The construction of monochromators and spectrographs is a specialized undertaking. The availability of good commercial instruments generally makes it unnecessary for anyone but the enthusiast to contemplate their construction in the laboratory. On the other hand, the principles that underlie good design in a dispersing instrument are not complex. Some essential features of such instruments can be illustrated with reference to particular designs that are used both in laboratory and commercial instruments. Attention will be restricted to monochromators—spectrographs are similar, apart from having an array detector or photographic plate instead of an exit slit, as shown in Figure 4.140. Spectrographs are also simpler in construction because wavelength scanning is not required and all optical components of the system remain fixed. The discussion here should also allow intelligent evaluation and comparison of commercial instruments. For further details of the many different instrument designs that have been used for the construction of monochromators, spectrographs, and spectrophotometers, see References 115 and 116, The Photonics Design and Applications Handbook, and the catalogs of instrument manufacturers.[117]

TABLE 4.13 MAIN CHARACTERISTICS OF OPTICAL DISPERSING ELEMENTS

Element	Advantages	Disadvantages
Multilayer dielectric interference filter	High throughput at center frequency, ~ 50%. Tunable to some extent by tilting.	Low resolution—minimum bandwidth ~1 nm. None available at short wavelengths ≤ 200 nm.
Prism spectrometer	Dispersion and resolution increase near absorption edge of prism (however, transmission falls). No ghosts. Scattered light can be lower than in grating spectrometer. No order sorting necessary.	Cannot be used below 120 nm (with LiF prism). Resolution not as good as grating spectrometer, best ~ 0.01 nm. Resolution particularly poor in infrared.
Grating spectrometer	Resolution can be high (≤ 0.001 nm). Can be used from vacuum UV to far IR. Dispersion independent of wavelength. Luminosity depends on blaze angle, but generally much higher than for prism spectrometer.	Ghosts can be a problem. Expensive if really low scattered light required—may need double- or triple-grating instrument. Need to remove unwanted orders (not necessarily a serious problem).
Fabry-Perot interferometer	Very high light throughput—much higher than for grating spectrometer. Very high resolution—tens of MHz at optical frequencies ($\geq 10^{-5}$ nm).	Many orders generally transmitted simultaneously. Cannot be used at short wavelengths (≤ 200 nm). Transmission maximum can be scanned only over narrow range. Most difficult dispersing element to use.

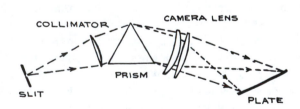

Figure 4.140   Schematic diagram of a simple prism spectrograph. (From R. J. Meltzer, "Spectrographs and Monochromators," in *Applied Optics and Optical Engineering*, Vol. 5, R. Kingslake, ed., Academic Press, New York, 1969; by permission of Academic Press.)

Figure 4.141 shows a Littrow prism monochromator. The light from the input slit is collimated with an off-axis paraboloidal mirror. The output wavelength is changed by rotating the Littrow mirror. If the output wavelength must change uniformly with rotation of a drive shaft, then the rotation of the mirror requires a cam. Figure 4.142 shows a Czerny-Turner prism configuration. The collimating mirrors are spherical. In the geometry of Figure 4.142a, the coma of one mirror compensates that of the other, while in Figure 4.142b these comas are additive. The wavelength is changed by rotating the prism.

In any of these spectroscopic instruments, the optical arrangement should be enclosed in a light-tight enclosure painted on its interior with black, nonreflective paint. Table 4.14 lists the various *black* coatings that are available for this purpose, and in any application where broad reflection over a range of angles from a surface must be minimized. This helps to minimize scattered light. It is essential that light passing through the entrance slit not be able to reach the exit slit without going through the illuminating optics and prism. To this end, it is common practice to incorporate internal black-painted baffles in the monochromator enclosure. The most likely source of scattered light is overillumination of the input

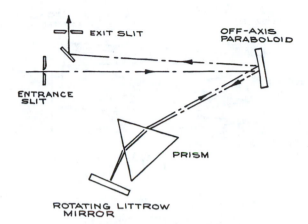

Figure 4.141   Littrow monocromator. (From R. J. Meltzer, "Spectrographs and Monochromators" by permission of Academic Press.)

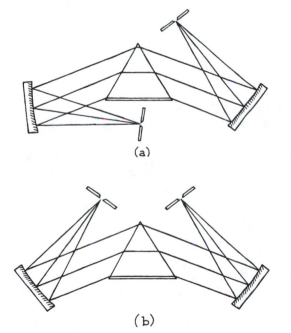

(a)

(b)

Figure 4.142   Czerny-Turner monochromator arrangements; (a) coma of one mirror compensates that of the other; (b) comas are additive.

TABLE 4.14 BLACK MATERIALS AND COATINGS

Material	Absorptivity
Anodized Black	0.88
Carbon Lampblack	0.84
Chemglaze Black Paint Z306	0.91
Delrin Black Plastic	0.87
Electro Optical Industries Mid-Temperature Black Coating (up to 200°C)	0.965
Electro-Optical Industries High-Temperature Black Coating (up to 1400°C)	0.93
Martin Black Velvet Paint	0.94
3M Black Velvet Paint	0.91
Parsons Black Paint	0.91
Polyethylene Black Plastic	0.92
Anodized Aluminum	0.84

collimator—input radiation should not enter with an *f/#* smaller than that of the monochromator. To check for stray light, visual observation through the exit slit in a darkened room when the entrance slit is illuminated with an intense source will reveal where additional internal baffles would be helpful. Stray light is more likely to be a problem in a Littrow arrangement than in a straight-through type such as the Czerny-Turner.

In the visible and near infrared, lens collimating optics can be used, as in the design shown in Figure 4.140, for example. Good lenses, such as camera lenses, should be used. Mirror collimating optics are useful over a much broader wavelength region. For ultraviolet use they should be aluminum overcoated with MgF_2, and they are frequently of this type even in instruments designed for longer wavelengths.

Slit design is important in achieving high resolution. Entrance and exit slits of equal width are generally used. If the source is small, having a height much less than the height of the prism or grating, parallel-sided slits are adequate. Adjustable versions are available from Ealing and Melles Griot. Fixed slits can be easily made using razor blades. If the source is high compared to the height of the dispersing element, a straight entrance slit produces a curved image at the exit slit. In the case of a prism, this happens because light passing through the prism at an angle inclined to the meridian plane suffers slightly more deviation than light in the meridian plane. Light striking a grating in a plane that is not perpendicular to the grating grooves sees an apparently smaller groove spacing, and is consequently deviated more than light in the plane perpendicular to the grating. To maintain high resolution in the presence of image curvature, curved entrance and exit slits should be used, each with a radius equal to the distance of the slit from the axis of the system. This slit arrangement is shown in Figure 4.143, which also shows a very good, and widely used, grating monochromator design due to Fastie[118] (based on a design originally described by Ebert[119] for use with a prism). It is also quite common for the single spherical mirror in Figure 4.143 to be replaced by a pair, in which case the design becomes the Czerny-Turner one. Such an arrangement is shown in Figure 4.144. In either case, the wavelength is tuned by rotating the grating. The wavelength variation at the exit slit is much more nearly linear with grating rotation than with prism rotation. If linearity is required, a cam, sine-drive, or

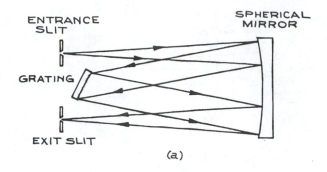

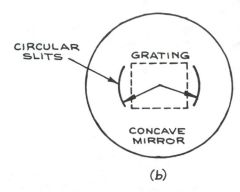

Figure 4.143   (a) Ebert-Fastie monochromator; (b) curbed entrance and exit slit configuration for use with the above.

other such arrangement is used for rotating the grating or prism.[116] In a Czerny-Turner grating instrument, the entrance and exit slits are generally arranged asymmetrically with respect to the grating as this allows coma to be corrected in the working wavelength range.

Grating monochromators for use in the vacuum ultraviolet generally have a concave grating. Such a grating, which has equally spaced grooves on the chord of a spherical mirror, combines both dispersion and focusing in one element. For maximum resolution both the object and image formed by a concave grating must be on a circle called the *Rowland circle*, whose diameter is the principal radius of the spherical surface on which the grating is ruled, as shown in Figure 4.145. The use of a concave-grating monochromator (or spectrograph) entails mounting grating and slits (or grating, slit, and photographic plate) on a Rowland circle. Adjustment of the instrument will involve moving the grating, slits, or plate on

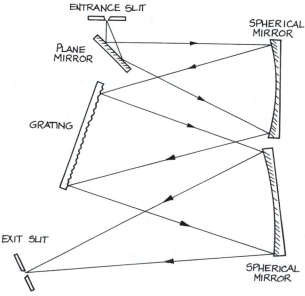

Figure 4.144   Arrangement of a Czerny-Turner grating monochromator.

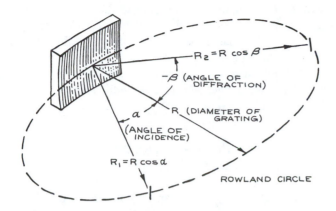

Figure 4.145   Rowland circle of a concave diffraction grating. For maximum resolution, both object and image must be on the Rowland circle. R = object (entrance slit) distance: R_2 = image (exit slit) distance. (From R. J. Meltzer, "Spectrographs and Monochromators" by permission of Academic Press.)

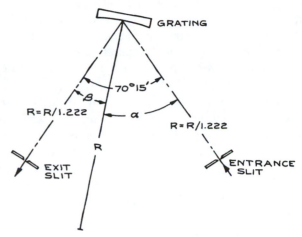

Figure 4.146   Layout of the Seya-Namioka monochromator.

the circle. Various mounting configurations such as the Rowland, Abney, Paschen, Eagle, Wadsworth, and grazing-incidence mounting have been used to accomplish this.

A convenient concave-grating design widely used in vacuum-ultraviolet monochromators is the Seya-Namioka shown in Figure 4.146.[120,121] In this arrangement, by judicious choice of slit distances from the grating and slit angular separation, the only adjustment needed for wavelength adjustment is rotation of the grating. In comparison with an ideal Rowland-circle mounting, only a small resolution loss results over a wide wavelength range. Vacuum-ultraviolet instruments possess the added complication that all interior adjustments of the instrument must be transmitted through the walls of the vacuum chamber that houses the optical arrangement. Various kinds of rotary-shaft and bellows seals and magnetically coupled drives exist for this purpose.[122]

Multiple monochromators use multiple dispersing elements in series to obtain greater dispersion and/or lower scattered light. In a double-grating monochromator, for example, two diffraction gratings move simultaneously—the desired diffracted wavelength from the first grating is diffracted once again from the second grating before being focused on the exit slit. The considerable reduction in

scattered light that can be obtained with multiple monochromators makes them particularly useful for analyzing weak emissions close in wavelength to a strong emission, as in Raman spectroscopy. Instruments with as many as four gratings in series are commercially available. Grating monochromators, spectrographs, and spectrophotometers are available from many manufacturers[27] including Acton Research, American Holographic, CVI Spectral Products, Jobin-Yvon/Spex (also often listed as Instruments SA), Minuteman Laboratories, Oriel, Thermo Jarrell Ash (TJA) (which incorporates the old companies Jarrell-Ash, Aminco-Bowman, Unicam, and Varian Associates). Fiber-coupled spectrometers are manufactured by Stellar Net and Ocean Optics, which offers a miniature spectrometer as a plug-in card for a PC. Acton Research, McPherson, and Minuteman Laboratories are noted for their vacuum instruments, Jobin-Yvon and Oriel for their multiple grating spectrometers, and TJA for their large high-resolution spectrometers and spectrographs.

Imaging Spectrographs (sometimes called optical multichannel analysers [OMAs]) image an entire spectrum onto either a linear photodiode area (PDA) or onto a linear CCD. These instruments often have the capability for recording complete spectra from several sources at once. Manufacturers include Jobin-Yvon, Olis, Oriel, Roper Scientific, Sensor Physics and TJA. Note that CCD arrays are approximately 100 times more sensitive than PDA so are best used for the lowest light level applications. CCDs must be used with appropriate shutters or light modulation otherwise image *smear* will occur. This happens if new light falls on the array while the previous deposited charge distribution is being read out.

Wavemeters are optical devices that are designed to provide direct electronic readout of the wavelength of monochromatic radiation that is directed into them. They are most frequently used in conjunction with a tunable laser system. These convenient instruments incorporate a Fabry-Perot or Fizeau interferometer.[7] Manufacturers include Burleigh and Lasertechnics.

4.7.3 Calibration of Spectrometers and Spectrographs

To calibrate the wavelength scale of a spectrometer or spectrograph, the entrance slit is illuminated with a reference source that has characteristic spectral features of accurately known wavelength. It is quite common to mount a small reference lamp—mercury, neon, and hollow-cathode iron lamps are the most popular—near the entrance slit and use a small removable mirror or beamsplitter to superimpose the spectrum of the reference on the unknown spectrum under study. To calibrate a grating instrument through its various orders, a helium-neon laser is ideal. If the slits are set to their narrowest position and a low-power (< 1 mW) He-Ne laser is directed at the entrance slit, the various grating settings that transmit a signal (which can be safely observed by eye at the exit slit) correspond to the wavelengths 0, 632.8 nm, 632.8 × 2 nm, 632.8 × 3 nm, and so on. By interpolation between the observed values, a reliable calibration curve can be obtained even for instruments operating far into the infrared.

The efficiency of the instrument at a given wavelength can be determined with any suitable detector. A laser or other narrow-band source at the desired wavelength illuminates the entrance slit, or a duplicate of it, and the transmitted power is measured with the detector placed directly behind the slit. A similar measurement is then made at the exit slit of the whole instrument. The relative spectral efficiency can be determined by illuminating the entrance slit with a blackbody source. A lens should not be used for focusing light on the entrance slit, as its collection and focusing efficiency will vary with wavelength. The signal at the output slit is then measured as a function of wavelength with a spectrally flat detector, such as a pyroelectric detector, thermopile, or Golay cell (see Section 4.8.5), and normalized with respect to the blackbody distribution.

4.7.4 Fabry-Perot Interferometers and Etalons

Fabry-Perot interferometers and etalons are optical filters that operate by multiple-beam interference of light reflected and transmitted by a pair of parallel flat or coaxial spherical reflecting interfaces. In its simplest form, an *etalon* consists of a flat, parallel-sided slab of transparent material that may or may not have semireflecting coatings on each face. An etalon may also consist of a pair of parallel, air-spaced flat mirrors of fixed spacing. If the reflective surfaces have adjustable spacing, or if their effective optical spacing can be adjusted by changing the gas pressure between the surfaces, the device is called a *Fabry-Perot interferometer*. These devices can be analyzed by the impedance concepts discussed in Section 4.2.3, but it is instructive to discuss their operation from the standpoint of multiple-beam interference.

Figure 4.147 shows the successive reflected and transmitted field amplitudes of a plane electromagnetic wave striking a plane-parallel Fabry-Perot etalon or interferometer at angle of incidence θ'. Although the reflection coefficients at the two reflecting interfaces may be different, only the case where they are equal will be analyzed. Almost all practical devices are built this way. The optical-path difference between successive transmitted waves is $2nl\cos\theta$, where l is the interface spacing and n is the refractive index of the medium between these interfaces (air, glass, quartz, or sapphire, for example). If refraction occurs at the reflecting interfaces, θ and θ' will be related by Snell's law. The phase difference between successive transmitted waves is

$$\delta = 2kl\cos\theta + 2\varepsilon \tag{4.242}$$

where

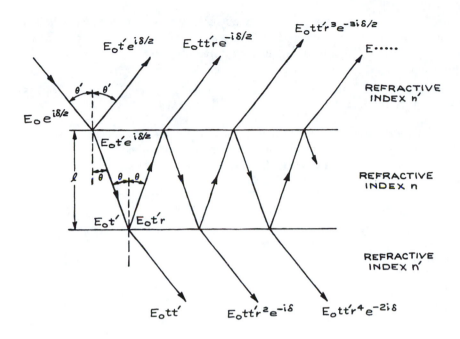

$$k = \frac{2\pi}{\lambda} = \frac{2\pi n}{\lambda_0} = \frac{2\pi n v}{c_0} \qquad (4.243)$$

and ε is the phase change (if any) on reflection.

The total resultant transmitted complex amplitude is

$$E_T = E_0 tt' + E_0 tt'r^2 e^{-i\delta} + E_0 tt'r^4 e^{-2i\delta} + \dots \qquad (4.244)$$

$$= \frac{E_0 tt'}{1 - r^2 e^{-i\delta}}$$

Here, r and t are the reflection and transmission coefficients of the waves passing from the medium of refractive index n to that of refractive index n' at each interface, r' and t' are the corresponding coefficients for passage from n' to n, and E_0 is the amplitude of the incident electric field.

The total transmitted intensity is

$$I_T = \frac{|E_T|^2}{2Z} = \frac{I_0 |tt'|^2}{|1 - r^2 e^{-i\delta}|^2} \qquad (4.245)$$

Here $|tt'| = T$ and $|r|^2 = |r'|^2 = R$, where T and R are the transmittance and reflectance of each interface, respectively.

If there is no energy lost in the reflection process,

$$T = 1 - R \qquad (4.246)$$

If the etalon had slightly absorbing reflective layers, this would no longer be true—if each layer absorbed a fraction A of the incident energy, then

$$T = 1 - R - A \qquad (4.247)$$

If $A = 0$, Equation 4.245 gives

$$\frac{I_T}{I_0} = \frac{1}{1 + \dfrac{4R}{(1-R)^2} \sin^2 \dfrac{\delta}{2}} \qquad (4.248)$$

This variation of transmitted intensity with δ is called an Airy function, and is illustrated in Figure 4.148 for several values of R. The variation of reflected intensity with δ is just the inverse of Figure 4.148, since

$$\frac{I_R}{I_0} = 1 - \frac{I_T}{I_0} \qquad (4.249)$$

If $A \neq 0$, Equation 4.248 becomes

$$\frac{I_T}{I_0} = \frac{\left(\frac{T}{1-R}\right)^2}{1 + \frac{4R}{(1-R)^2}\sin^2\frac{\delta}{2}} \qquad (4.250)$$

In either case, maximum transmitted intensity results when

$$\delta = 2m\pi \qquad (4.251)$$

where m is an integer. If the phase change on reflection is neglected, this reduces to

$$2l\cos\theta = m\lambda \qquad (4.252)$$

where λ is the wavelength of the light in the medium between the two reflecting surfaces (plates). In normal incidence, transmission maxima result when

$$1 = m\lambda/2 \qquad (4.253)$$

Even if the phase change on reflection is included, the necessary adjustment in spacing to go from one transmission maximum to the next is one half wavelength. The transmitted intensity as a function of plate separation when a Fabry-Perot interferometer is illuminated normally with plane monochromatic light, for example, is therefore also shown in Figure 4.148. In an ideal Fabry-Perot device, the overall transmittance at a transmission-intensity maximum is unity, whereas in a practical device, which will invariably have some losses, the maximum overall transmittance is reduced by a factor $[T/(1-R)]^2$. As is clear from Figure 4.148, the transmission peaks of the device become very sharp as R approaches unity.

Free Spectral Range, Finesse, and Resolving Power

If a particular transmission maximum for fixed l occurs for normally incident light of frequency v_0, then

$$v_0 = \frac{mc}{2l} \qquad (4.254)$$

where c is the velocity of light in the material between the plates. Of course, the device will also show transmission

maxima at all frequencies $v_0 \pm pc/2l$ as well, where p is an integer. The frequency between successive transmission maxima is called the *free spectral range*:

$$\Delta v = \frac{c}{2l} \qquad (4.255)$$

When R is close to unity, all phase angles δ within a transmission maximum differ from the value $2m\pi$ by only a small angle. We can therefore write

$$\delta = \frac{4\pi vl}{c} = \frac{4\pi v_0 l}{c} + \frac{4\pi(v-v_0)l}{c} \qquad (4.256)$$

which can be written in the form

$$\delta = 2m\pi + \frac{2\pi(v-v_0)}{\Delta v} \qquad (4.257)$$

Equation 4.248 becomes

$$\frac{I_T}{I_0} = \frac{1}{1 + \frac{4R\pi^2}{(1-R)^2}\left(\frac{v-v_0}{\Delta v}\right)^2} \qquad (4.258)$$

Writing $\pi\sqrt{R}/(1-R) = F$, where F is called the *finesse*, the shape of a narrow transmission maximum can be written as

$$\frac{I_T}{I_0} = \frac{1}{1 + [2(v-v_0)/\Delta v_{1/2}]^2} \qquad (4.259)$$

Here $\Delta v_{1/2}$—the full width at half maximum transmission of the transmission peak—is given by

$$\Delta v_{1/2} = \Delta v/F \qquad (4.260)$$

The *resolving power* of a Fabry-Perot device is a measure of its ability to distinguish between two closely spaced monochromatic signals. A good criterion for determining this is the *Rayleigh criterion*, which recognizes two closely spaced lines as distinguishable if their half-intensity points on opposite sides of the two line

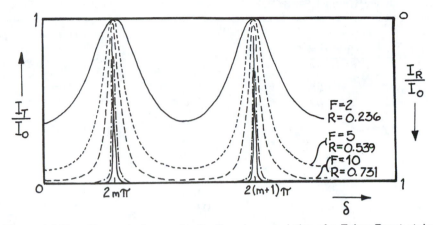

Figure 4.148   Transmission and reflection characteristics of a Fabry-Perot etalon.

shapes are coincident, as shown in Figure 4.149. For a Fabry-Perot device, therefore, the resolving power is

$$\mathcal{R} = \frac{v}{\Delta v_{1/2}} = \frac{Fv}{\Delta v} \qquad (4.261)$$

It would appear that very high resolving powers can be obtained by the use of high-reflectance plates (and consequent high finesse values, as illustrated in Table 4.15) and/or large plate spacings (and consequent small free spectral ranges). Additional practical considerations, however, limit how far these approaches to high resolution can actually be used.

Practical Operating Configurations and Performance Limitations of Fabry-Perot Systems

The use of very high reflectance values to obtain improved resolving power is limited by the ability of the optical polisher to achieve ultraflat optical surfaces. Very good-quality plates for use in the visible can be obtained with flatnesses of $\lambda/200$ over regions a few centimeters in diameter. Some claims of flatness in excess of $\lambda/200$ are also seen. These should be viewed with skepticism—such a degree of flatness is extremely difficult to verify. If the plates have surface roughness Δs, the average error in plate

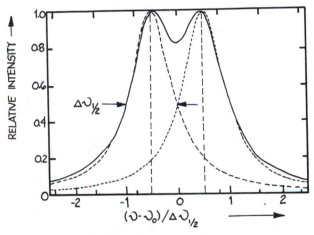

Figure 4.149   Two monochromatic spectral lines just resolved according to the Rayleigh criterion.

spacing is $\sqrt{2}\Delta s$. This gives rise to a spread in the frequency of transmission maxima equal to

$$\Delta v_s \cong \frac{mc\Delta s}{\sqrt{2}l^2} \cong \sqrt{2}v_0\frac{\Delta s}{l} \qquad (4.262)$$

The flatness-limited finesse is therefore

TABLE 4.15 PROPERTIES OF FABRY-PEROT ETALONS

Plate Separation l (cm)	Reflectance R (%)	Free Spectral Range $\Delta v = c/2Z(Hz)$	Finesse $F = \dfrac{\pi\sqrt{R}}{1-R}$	Resolving Power $\mathfrak{R} = Fv/\Delta v$
0.1	80	1.5×10^{11}	14	5.6×10^{4}
0.1	90	1.5×10^{11}	30	1.2×10^{5}
0.1	95	1.5×10^{11}	61	2.44×10^{5}
0.1	99	1.5×10^{11}	313	1.25×10^{6}
1.0	80	1.5×10^{10}	14	5.6×10^{5}
1.0	90	1.5×10^{10}	30	1.2×10^{6}
1.0	95	1.5×10^{10}	61	2.44×10^{6}
1.0	99	1.5×10^{10}	313	1.25×10^{7}
10.0	80	1.5×10^{9}	14	5.6×10^{6}
10.0	90	1.5×10^{9}	30	1.2×10^{7}
10.0	95	1.5×10^{9}	61	2.44×10^{7}
10.0	99	1.5×10^{9}	313	1.25×10^{8}

Note: Properties shown are at 500 nm (600 THz).

$$F_s = \frac{\Delta v}{\Delta v_s} \cong \frac{1}{\sqrt{2}}\left(\frac{\Delta v}{v_0}\right)\frac{l}{\Delta s} \cong \frac{\lambda}{2\sqrt{2}\Delta s} \qquad (4.263)$$

Two plates with surface roughness of $\lambda/200$, for example, would have a flatness-limited finesse of about 71. In practice, Fabry-Perot plates may not be randomly rough, and their stated flatness figure may represent the average deviation from parallelism of the two plates. In this case, the parallelism-limited finesse is

$$F_p = \frac{\lambda}{2\Delta s} \qquad (4.264)$$

where Δs is the deviation from parallelism over the used aperture of the system. With a 1 cm aperture, for example, a 0.01" deviation from parallelism would imply $\Delta s = 4.84$ nm and a parallelism-limited finesse of only 56 at 546 nm. Consequently, Fabry-Perot plates must be held very nearly parallel to achieve high resolution. Their separation should also not be allowed to drift. A random variation of Δs will limit the finesse as predicted by Equation 4.263. A fractional length change of 10^{-6} in a 1 cm spacing Fabry-

Perot device will limit the finesse to only about 27. It must, therefore, be isolated from vibrational interference to achieve high finesse. If the frequencies of the transmission maxima are not to drift, the change in plate spacing due to thermal expansion must also be minimized by constructing the device of low-expansion materials such as Invar,[123] Super-Invar,[124] fused silica,[125] or special ceramics.[126]

The resolving power of a Fabry-Perot system is also limited by the range of angles that can be transmitted through the instrument. One popular configuration is shown in Figure 4.150. A small source, or light from an extended source that passes through a small aperture, is collimated by a lens, passes through the interferometer, and is focused onto a second small aperture in front of a detector. The maximum angular spread of rays passing through the system will be governed by the aperture sizes W_1 and W_2 and the two focal lengths f_1 and f_2. In the paraxial-ray approximation, the angular width $2\Delta\theta$ of the paraxial-ray bundle that comes from, or is delivered to, an aperture of diameter W is equal to the angular width $2\Delta\theta'$ that traverses the system, as illustrated in Figure 4.151. In this case,

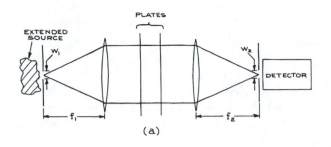

(a)

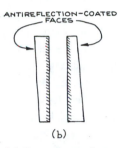

ANTIREFLECTION—COATED FACES

(b)

Figure 4.150   (a) Operating scheme for a Fabry-Perot interferometer, using collimated light; (b) detail of typical plate configuration (the wedge angle of the reflecting plates is exaggerated).

$$\Delta\theta \cong W/2f \qquad (4.265)$$

In Figure 4.148, the resolving power will be limited by W_1/f_1 or W_2/f_2—whichever is the larger. The transmitted frequency spread associated with an angular spread $\Delta\theta$ around the normal direction is, from Equations 4.252 and 4.254,

$$\Delta v_{1/2} = v_0(\Delta\theta)^2 / 2 \qquad (4.266)$$

The corresponding aperture-limited finesse is

$$F_A = \frac{\lambda}{l(\Delta\theta)^2} = \frac{4\lambda f^2}{lW^2} \qquad (4.267)$$

Using, for example, a 1 mm diameter aperture, a 10 cm focal-length lens, and a 1 cm air spaced Fabry-Perot device at 546.1 nm, the aperture-limited finesse is 218. The operating conditions of a Fabry-Perot system should be arranged so that F_A is at least three times larger than the

desired operating finesse, which usually is ultimately limited by the flatness-limited finesse. The apertures in Figure 4.112 should be circular and accurately coaxial with both lenses, or the aperture finesse will be further reduced. If the limiting aperture (either a lens or the Fabry-Perot device itself) is D, there is also a diffraction limit to the finesse, which on axis is

$$F_D = \frac{2D^2}{\lambda nl} \qquad (4.268)$$

At the pth fringe off axis,

$$F_D = \frac{D^2}{2p\lambda nl} \qquad (4.269)$$

where n is the refractive index of the material between the plates. The overall finesse of a Fabry-Perot system, F_1, is related to the individual contributions to the finesse, F_i, by

$$F_1^{-2} = \sum_i F_i^{-2} \qquad (4.270)$$

In the operating configuration of Figure 4.150, if the plate spacing is fixed, then with a broad-band source all frequencies that satisfy Equation 4.252 will be transmitted. These frequencies become more closely spaced as the plate separation l is increased. There is a concommitant increase in resolution, but this may not prove useful if any of the spectral features under study are broader than the free spectral range. In this case, simultaneous transmission of two broadened spectral features always occurs. For two lines of wavelength λ_1 and λ_2 and of FWHM $\Delta\lambda$, for example, it is usually possible to find two integers m_1 and m_2 such that

$$m_1\lambda'_1 = 2l, \left|\lambda'_1 - \lambda_1\right| \leq \Delta\lambda \qquad (4.271)$$

$$m_2\lambda'_2 = 2l, \left|\lambda'_2 - \lambda_2\right| \leq \Delta\lambda \qquad (4.272)$$

For isolated, high-resolution studies of the spectral feature at λ_1, all potential interference features such as λ_2 must be filtered out. This is done with a color or

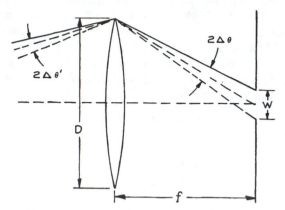

Figure 4.151 Angular factors involved in the collection of light transmitted through a Fabry-Perot interferometer by a circular aperture. In the paraxial ray approximation, $\Delta\theta' = \Delta\theta$.

interference filter—or a prism or grating monochromator—before the signal is sent to the interferometer. The choice of prefiltering element will depend on the relative wavelength spacing of λ_1 and λ_2. The optical throughput of filters can be very high—50 percent or more—but will not allow prefiltering of lines closer than a few nanometers. When very high throughput coupled with high resolution is essential, one or more additional Fabry-Perot devices of appropriate free spectral range can be used. An etalon of spacing 0.01 mm, for example, has a free spectral range of about 15 nm. If it has a finesse of 50, its useful transmission bandwidth is 0.3 mn. Such an etalon could be coupled with etalons of spacing 0.1 mm and 1 mm with similar finesse to isolate a band only 0.003 nm wide from an original relatively broad frequency band transmitted through a filter. The final transmitted wavelength of such a combination of etalons can be tuned by tilting.

The wavelength of maximum transmission of a single etalon varies as

$$\lambda = \lambda_0 \cos\theta \qquad (4.273)$$

where λ_0 is the wavelength for maximum transmission in normal incidence. In the interferometer arrangement shown in Figure 4.150, wavelength scanning can be accomplished in two ways—by pressure scanning, or by adjusting the

axial position of one of the plates with a piezoelectric transducer.[127] To accomplish pressure scanning, the interferometer is mounted in a vacuum- or pressure-tight housing into which gas can be admitted at a slow steady rate. The resultant change of refractive index of the gas between the plates, which is approximately linear with pressure, causes a continuous, unidirectional scanning of the center transmitted frequency. The interferometer should be mounted freely within the housing so that the pressure differential between the interior and exterior of the housing causes no deformation of the interferometer. Pressure scanning is not convenient where rapid or bidirectional scanning is required. Because no geometrical dimensions of the interferometer change during the scan, no change in finesse should occur during scanning.

Piezoelectric scanning is very convenient, but the transducer material must be specially selected to give uniform translation of the moving plate without any tilting—which would cause finesse reduction during scanning. If separate piezoelectric transducers are mounted on the kinematic-mounted alignment drives of the plate holders, it is possible to trim the finesse electronically after good mechanical alignment has been achieved by micrometer adjustment.[103] The alignment of the plates during scanning can be maintained with a servo system.[128,129]

A Fabry-Perot etalon can be used in the configuration shown in Figure 4.152 to give a high-resolution display of a narrow-band spectral feature that cannot be displayed with a scanning interferometer because, for example, the optical signal is a short pulse. In the arrangement shown in Figure 4.152, all rays from a monochromatic source that are incident at angle θ_m on the plates will be transmitted and brought to a focus in a bright ring on a screen if

$$\cos\theta_m = \frac{m\lambda}{2l} \qquad (4.274)$$

In the paraxial approximation, θ_m is a small angle and the radius ρ_m of a ring on the screen is

$$\rho_m = f\theta_m \qquad (4.275)$$

which can be written as

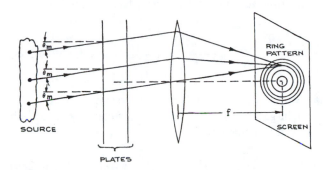

Figure 4.152   Arrangement for observing a ring interference pattern with a Fabry-Perot etalon.

$$\rho_m = f\sqrt{2 - \frac{m\lambda}{l}} \qquad (4.276)$$

The radius of the smallest ring corresponds to the largest integer m, for which $m\lambda/2l \leq 1$. Successive rings going out from the center of the pattern correspond to integers $m-1$, $m-2$, and so on.

If there is a bright spot at the center of the ring pattern, the radius of the pth ring from the center is

$$\rho_p = f\sqrt{\frac{p\lambda}{l}} \qquad (4.277)$$

This ring will be diffuse because of the finite finesse of the etalon, even if the source is highly monochromatic. The FWHM at half maximum intensity of the pth ring in this case is

$$\Delta\rho_p = \frac{\rho_p}{2pF} \qquad (4.278)$$

which can be verified by expanding Equation 4.248 in the angle θ. This ring width is the same as would be produced by a source of FWHM $\Delta\lambda = \lambda^2/2lF$.

Equation 4.278 predicts that the total flux through each ring of the pattern will be identical, since the area of each ring is

$$\Delta A = 2\pi\rho_p\Delta\rho_p = \pi f^2/2lF \qquad (4.279)$$

The ratio of ring width to ring spacing for the pth ring in the pattern is

$$\frac{\Delta\rho_p}{\rho_{p+1} - \rho_p} = \left[2Fp\left(1 - \sqrt{1 + \frac{1}{p}}\right)\right]^{-1} \qquad (4.280)$$

which is close to $1/F$ except for the first few rings. Observation of the ring pattern from a quasimonochromatic source will therefore yield its approximate bandwidth $\Delta\lambda$ as

$$\Delta\lambda \sim \frac{\lambda}{2l} \frac{\Delta\rho}{\rho_{p+1} - \rho_p} \qquad (4.281)$$

A convenient way of making a qualitative visual observation of this kind is to use an etalon in conjunction with a small telescope as shown in Figure 4.153.

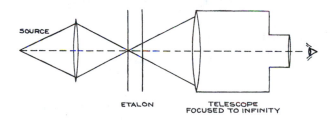

Figure 4.153   Arrangement for visual observation of Fabry-Perot fringes.

Luminosity, Throughput, Contrast Ratio, and Étendue

The luminosity of a Fabry-Perot system is, as already mentioned, much higher than that of any prism or grating monochromator. In the arrangement of Figure 4.150, the luminosity is

$$\Phi = \left(\frac{T}{1-R}\right)^2 E_e(\lambda) S \frac{\pi W_2^2}{4f_2} \qquad (4.282)$$

where $\pi W_2^2/4f_2$ is the solid angle subtended by the output circular aperture at the output focusing lens, and S is the illuminated area of the plates. If the system has an aperture-limited finesse, then from Equation 4.267 the luminosity becomes

$$\Phi = \left(\frac{T}{1-R}\right)^2 E_e(\lambda) \frac{\pi \lambda S}{\rho F_A} \qquad (4.283)$$

which can be written in terms of the resolving power $\mathscr{R}$ as

$$\Phi = \left(\frac{T}{1-R}\right)^2 E_e(\lambda) \frac{2\pi S}{\mathscr{R}} \qquad (4.284)$$

The factor $[T/(1-R)]^2$ is called the throughput of the system. To compare the luminosity predicted by Equation 4.284 with that of a grating, we shall assume that the throughput of the Fabry-Perot system is equal to the corresponding efficiency factor T of the grating in Equation 4.230. In this case—for identical resolving powers $\mathscr{R}$—from Equations 4.284 and 4.230 we have

$$(4.285)$$

$$\frac{\Phi(F.P.)}{\Phi(grating)} = \frac{\pi S}{(A\sin\beta)(l/f_2)}$$

where l/f_2 is the angular slit height at the output-focusing element of the grating monochromator. For a blaze angle of 30° and equal Fabry-Perot aperture and grating area

$$\frac{\Phi(F.P.)}{\Phi(grating)} = \frac{\pi}{l/f_2} \qquad (4.286)$$

For a 0.3 m focal-length grating instrument and a slit height of 3 cm (which is a typical maximum value for a small instrument of this size), the luminosity ratio is 71.4. This probably represents an optimistic comparison as far as the grating instrument is concerned, since high-resolution grating monochromators typically have focal lengths in excess of 0.75 m.

The *contrast ratio* of a Fabry-Perot system is defined as

$$C = \frac{(I_t/I_0)_{max}}{(I_t/I_0)_{min}} \qquad (4.287)$$

which from Equation 4.248 clearly has the value

$$C = \left(\frac{1+R}{1-R}\right)^2 = 1 + \frac{4F^2}{\pi^2} \qquad (4.288)$$

The *étendue* U of a Fabry-Perot system is a measure of its light-gathering power for a given frequency bandwidth $\Delta V_{1/2}$. It is defined by

$$U = \Omega S \qquad (4.289)$$

where A is the area of the plates and Ω is the solid angle within which incident rays can travel and be transmitted with a specified frequency bandwidth. From Equation 4.266 (since $\Omega = \pi \Delta \theta^2$),

$$\Delta v_{1/2} = \frac{v_0 \Omega}{2\pi} \qquad (4.290)$$

Therefore,

$$U = \Omega S = \frac{2\pi \Delta v_{1/2}}{v_0} \frac{\pi D^2}{4} \qquad (4.291)$$

which can be written in the form

$$U = \Omega S = \frac{\pi^2 D^2 \lambda}{4lF} \qquad (4.292)$$

Confocal Fabry-Perot Interferometers

Fabry-Perot interferometers with spherical concave mirrors are generally used in a confocal arrangement, as shown in Figure 4.154. In this configuration, the various characteristics of the interferometer can be summarized as follows:

$$\left(\frac{I_T}{I_0}\right)_{max} = \frac{T^2}{2(1-R)^2} \qquad (4.293)$$

$$\Delta v(\text{FSR}) = c/4l \qquad (4.294)$$

$$\text{reflectivity-limited finesse} = \frac{\pi\sqrt{R}}{(1-R)} \qquad (4.295)$$

The approximate optimum radius of the apertures used in an arrangement similar to Figure 4.112 is $1.15\,(\lambda\,r_3/F_1)^{1/4}$, where r is the radius of the spherical mirrors.

The characteristics of all the possible combinations of two spherical mirrors that can be used to form a Fabry-Perot interferometer have been very extensively studied in connection with the use of such mirror arrangements to form the resonant cavities of laser oscillators.[12-14] In that application, these spherical-mirror resonators support Gaussian beam modes, and it is with such Gaussian beams that highest performance can be achieved with a spherical-mirror Fabry-Perot interferometer. A confocal Fabry-Perot interferometer is therefore ideal for examining the spectral content of a Gaussian laser beam. The beam to be studied should be focused into the interferometer with a lens that matches the phase-front curvature and spot size of the laser beam to the radius of the Fabry-Perot mirror and the spot size of the resonant modes. (See Section 4.2.5.)

4.7.5 Design Considerations for Fabry-Perot Systems

Fabry-Perot interferometers and etalons are extremely sensitive to angular misalignment and fluctuations in plate spacing. Consequently, their design and construction are more demanding than for other laboratory optical instruments. The plates must be extremely flat, as already mentioned, and must be held without any distortion in high-precision alignment holders. The plate spacing must be held constant to about 1 nm in a typical high-finesse device for use in the visible, and the angular orientation must be maintained to better than 0.01 second of arc. Small-plate-spacing etalons are generally made commercially by optically contacting two plates to a precision spacer, or spacers, made of quartz. Such etalons are available from IC Optical Systems Ltd., Burleigh, Virgo, CVI, and Klinger, among others. Etalons of thickness in excess of 1 or 2 mm for use in the visible can be made from solid plane-parallel pieces of quartz or sapphire with dielectric coatings on their faces, or from materials such as germanium or zinc selenide for infrared

use. Fabry-Perot interferometer plates are almost always slightly wedge-shaped and antireflection-coates on their outside faces to eliminate unwanted reflections, as shown in Figure 4.150b. To prevent distortion of the plates they should be kinematically mounted in their holders. A good way to do this is to use three quartz or sapphire balls cemented to the circumferences of the plates and held in ball, slot, and plane locations with small retaining clips.

For the construction of Fabry-Perot interferometers, low-expansion materials should be used. The plate holders can be held at a fixed spacing by having them spring-loaded against three quartz or ceramic spacer rods or by the use of an Invar or Super-Invar spacer structure.

After a Fabry-Perot interferometer is aligned mechanically, it will creep slowly out of alignment, because of the easing of micrometer threads, for example. The alignment can be returned by having a piezoelectric element incorporated in each plate-orientation adjustment screw.[127] Fabry-Perot interferometers are made commercially by Burleigh, Hovemere, and Zygo.

4.7.6 Double-Beam Interferometers

In a double-beam interferometer, waves from a source are divided into two parts at a beamsplitter and then recombined after traveling along different optical paths. The most practically useful of these interferometers are the Michelson and the Mach-Zehnder.

The Michelson Interferometer

The operation of a Michelson interferometer can be illustrated with reference to Figure 4.155. Light from an extended source is divided at a beamsplitter and sent to two plane mirrors M_1 and M_2. If observations of a broad-band (temporally incoherent) source are being made, a compensating plate of the same material and thickness as the beamsplitter is included in arm 2 of the interferometer. This plate compensates for the two additional traverses of the beamsplitter substrate made by the beam in arm 1. Because of the dispersion of the beamsplitter material, it produces a phase difference that is wavelength-dependent, and without the compensating plate, interference fringes would not be observable with broad-band illumination (such as white light). A maximum of illumination results at angle θ in Figure 4.155 if the phase difference between the two beams coming from M_1 and M_2 is an integral multiple of 2. If M_2' represents the location of the reflection of M_2 in the beamsplitter, the condition for a maximum at angle θ is

$$2l\cos\theta = m\lambda \qquad (4.296)$$

The positioning of M_1 at M_2 corresponds to $l = 0$, $m = 0$. If the output radiation is focused with a lens, a ring pattern is produced. A clear ring pattern is only visible if

$$2l\cos\theta \leq l_c \qquad (4.297)$$

where l_c, is the coherence length of the radiation from the source. The clearness of the rings is usually described in terms of their *visibility* V, defined by the relation

$$V_{\text{max}} = \frac{I_{\text{max}} - I_{\text{min}}}{I_{\text{max}} + I_{\text{min}}} \qquad (4.298)$$

To obtain a ring-shaped interference pattern, M_1 and M_2 must be very nearly parallel, otherwise a pattern of curved dark and light bands will result. To study the fringes, mirror M_1 must be scanned in a direction perpendicular to its surface. This can be done with a precision micrometer or a piezoelectric transducer if only small motion is desired. Because interference fringes are only observed with a broadband source illuminating a Michelson (or other double-beam) interferometer when the two interferometer arms are of almost equal length, mode of operation is therefore extraordinarily sensitive. Such *white-light interferometry* forms the basis of a number of sensor systems for phenomena such as vibration, temperature, and pressure.

In practice, Michelson interferometers are rarely used in the configuration shown in Figure 4.155, but are used instead with collimated illumination (a laser beam, or parallel light obtained by placing a point source at the focal point of a converging lens) as shown in Figure 4.156. In this case, the output illumination is (almost) perfectly uniform, being a maximum if

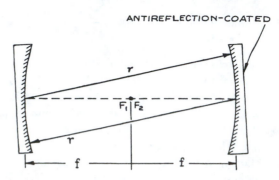

ANTIREFLECTION-COATED

Figure 4.154 Confocal Fabry-Perot interferometer.

$$2l = m\lambda \qquad (4.299)$$

where $2l$ is the path difference between the two arms. If monochromatic illumination of angular frequency ω is used, we can write the fields of the waves from arms 1 and 2 as

$$V_1 = V_0 e^{i(\omega t - \phi_1)} \qquad (4.300)$$
$$V_2 = V_0 e^{i(\omega t - \phi_2)}$$

where it is assumed that the beamsplitter divides the incident beam into two equal parts. ϕ_1 and ϕ_2 are the optical phases of these two beams, which are dependent on the optical path lengths traveled in arms 1 and 2.

The output intensity is

$$I \propto (V_1 + V_2)^*(V_1 + V_2) \qquad (4.301)$$

which gives

$$I = I_0[1 + \cos(\phi_2 - \phi_1)] = I_0\left[1 + \cos\left(\frac{4\pi v l}{c}\right)\right] \qquad (4.302)$$

in which I_0 is the intensity of the light from one arm alone. Suppose an optical component such as a prism or lens is placed in one arm of the interferometer, as shown in Figure 4.157, with the plane mirror replaced by an appropriate radius spherical mirror in the case of the lens, with quasi-monochromatic illumination used. The output fringe pattern will then reveal inhomogeneities and defects in the inserted component. A Michelson interferometer

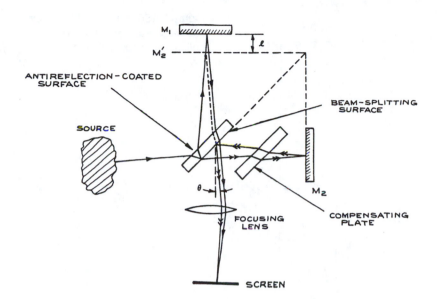

Figure 4.155   Michelson interferometer.

used in this way for testing optical components is called a *Twyman-Green interferometer*.[9]

Equation 4.302 illustrates why Michelson (and all other double-beam) interferometers are not directly suitable for spectroscopy. The distance the movable mirror must be adjusted to go from one maximum to the next is

$$\Delta l = \lambda / 2 \tag{4.303}$$

The distance required to go from a maximum- to a half maximum-intensity point is

$$\Delta l = \lambda / 8 \tag{4.304}$$

The effective finesse is therefore only 4—sharp spectral peaks with monochromatic illumination are not obtained as they are in multiple-beam interferometers.

Fourier Transform Spectroscopy

In an arrangement such as Figure 4.156, if the movable mirror is scanned at a steady velocity v, the resultant intensity as a function of position can be used to obtain considerable information about the spectrum of the source. If the source has intensity $I(\nu)$ at ν, the contribution to the intensity from the small band $d\nu$ at ν is

$$I_\nu(l) = I(\nu)\left(1 + \cos\frac{4\pi\nu l}{c}\right)d\nu \tag{4.305}$$

where $l = vt$. The total observed output from the full spectrum of the source is

$$I(l) = \int_{\nu=0}^{\infty} I\nu 1 + \cos\frac{4\pi\nu l}{c}d\nu \tag{4.306}$$

$$= \frac{I_0}{2} + \int_{\nu=0}^{\infty} I\nu\cos\frac{4\pi\nu l}{c}d\nu$$

The second term on the second line is essentially the cosine transform of $I(\nu)$. $I(\nu)$ can therefore be found as the cosine transform of $I(l)$–$(I_0/2)$:

$$I(\nu) = \frac{2}{c} \int_{l=0}^{\infty} Il - \frac{I_0}{2}\cos\frac{4\pi l\nu}{c}dl \tag{4.307}$$

This method for obtaining the spectrum of a source as the Fourier transform of an intensity recorded as a function of scanning mirror position is called *Fourier Transform Spectroscopy*. Such Fourier Transform spectrometers have high luminosity and collect spectral information very

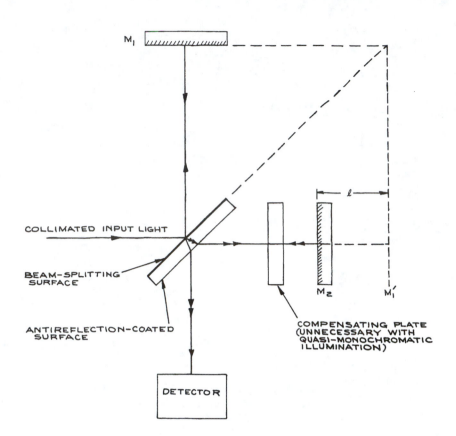

Figure 4.156   Simple Michelson interferometer for use with collimated input radiation.

efficiently—the whole emission spectrum is in reality sampled all at once, rather than one feature at a time as in a conventional dispersive grating spectrometer. This is often called the Fellgett or Jacquinot advantage.

Commercial Fourier Transform spectrometers have become widely available and easy to use. Suppliers include Bio-Rad Laboratories, Bomen, Bruker, Melles Griot, MIDAC, Nicolet Instruments, Oriel, and Perkin-Elmer. These instruments have replaced grating instruments in many analytical applications.

The Mach-Zehnder Interferometer

The Mach-Zehnder interferometer, shown in Figure 4.158, is widely used for studying refractive-index distributions in gases, liquids, and solids. It is particularly widely used for studying the density variations in compressible gas flows,[130] the thermally induced density (and consequent refractive-index) changes behind shock fronts, and thermally induced density changes produced by laser beams propagating through transparent media.[131,132] If the light entering the interferometer is perfectly collimated, uniform output illumination results, which is a maximum if the path difference between the two arms is an integral number of wavelengths. If the image of M_1 in beamsplitter B_2 is not parallel to M_2, the output fringe pattern is essentially the interference pattern observed from a wedge—a series of parallel bright and dark bands. If the medium in arm 2 is perturbed in some way so that a spatial variation in refractive index results, the output fringe pattern changes. Analysis of the new fringe pattern reveals the spatial distribution of refractive-index variation along the beam path in arm 2. Various design variations on the Mach-Zehnder interferometer exist, notably the Jamin and the Sirks-Pringsheim.[9]

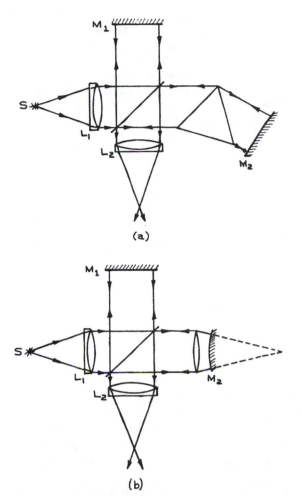

(a)

(b)

Figure 4.157   Twyman-Green inteferometer: (a) arrangement for testing a prism; (b) arrangement for testing a lens. S is a monochromatic source L_1 is a collimating lens, and L_2 is a focusing lens. (Adapted with permission from M. Born and E. Wolf, Principles of Optics, 3rd ed., Pergamon Press, Oxford, © 1965.)

Design Considerations for Double-Beam Interferometers

Michelson and Mach-Zehnder interferometers are much easier to build in the laboratory than multiple-beam interferometers such as the Fabry-Perot. Because the beams only reflect off the interferometer mirrors once, mirrors of flatness $\lambda/10$ or $\lambda/20$ are quite adequate for

building a good instrument. The usual precautions should be taken in construction—stable rigid mounts for the mirrors and beamsplitters and, if long-term thermal stability is required, low-expansion materials. Piezoelectric control of mirror alignment is not generally necessary, as these instruments are less sensitive to small angular misalignments than Fabry-Perot interferometers. To achieve freedom from alignment disturbances along one or two orthogonal axes, the plane mirrors of a Michelson interferometer can be replaced with right-angle prisms or corner-cube retroreflectors, respectively. Some care should be taken to avoid source-polarization effects in these instruments—the beamsplitters are generally designed to reflect and transmit equal amounts only for a specific input polarization. If polarizing materials are placed in either arm of the interferometer, no fringes will be seen if, by any chance, the beams from the two arms emerge orthogonally polarized.

4.8 DETECTORS

The development of optical detectors has occurred in common with other branches of electronics, by a series of advances through the use of gas-filled tubes and vacuum tubes to various semiconductor devices. Gas-filled and vacuum tube devices still retain some niche electronics applications, such as high voltage switching, microwave power amplification, and specialized audio. Gas-filled photodetectors, which used to offer some degree of signal amplification in low-cost consumer applications have essentially disappeared. Vacuum tube photodetectors, however—especially photomultiplier tubes—remain in widespread use for detection of radiation below about 1 μm. This is especially true for low light level signal detection (photon counting), and in applications where a large detector sensitive surface area is required (in scintillation counters). The designer of an optical system is therefore confronted with a very broad range of detector types—vacuum-tube, semiconductor, and thermal. The last category includes thermopiles, pyroelectric detectors, Golay cells, and bolometers, each with its own advantages and disadvantages.

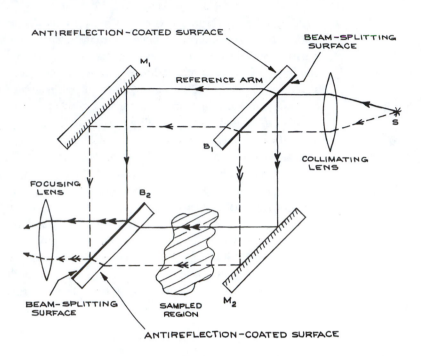

Figure 4.158 Mach-Zehnder interferometer. M_1, M_2 are totally reflecting mirrors; B_1, B_2 are beamsplitters (usually equally reflective and trasnmissive). The distribution of output illumination reflects the refractive index distribution in the sampled region.

The choice of detector for a given application will almost invariably involve a choice from among the numerous commercially available devices in each of these categories. The construction of optical detectors is rarely undertaken in the laboratory except by the specialist. The choice of detector will be governed by factors such as the following:

1. The wavelength region of the radiation to be detected

2. The intensity of the radiation to be detected

3. The time response required to resolve high-speed events

4. The detector surface area required

5. The environmental conditions under which the detector is to be used

6. The cost

The choice of detector should not, however, be an exercise performed in isolation. The other components of the optical system may influence it, and vice versa. This is particularly true if very weak radiation is to be detected. In such a situation, action should be taken to maximize the light signal actually reaching the detector. This will involve a careful choice of optical components to be placed between source and detector—suitable light collection and focusing optics, for example. If the light from the source must pass through a dispersing element (monochromator, etalon, etc.) before reaching the detector, then the choices of dispersing element and detector are likely to be interrelated. Some dispersing elements have high resolution (see Section 4.6) but low light throughput; others the reverse. Still others have high throughput and resolution but can only be scanned in a restricted way, if at all. In a given experimental situation, the various requirements of resolution, light throughput, scanability, and available detector sensitivities must be offset one against another in arriving at a sensible compromise.

Detectors can be classified as photon detectors or thermal detectors. In *photon detectors*, individual incident photons interact with electrons within the detector material. This leads to detectors based on *photoemission*, where the absorption of a photon frees an electron from a material; *photoconductivity*, where absorption of photons increases the number of charge carriers in a material or changes their mobility; the *photovoltaic effect*, where absorption of a photon leads to the generation of a potential difference across a junction between two materials; the *photon-drag effect*, where absorbed photons transfer their momentum to free carriers in a heavily doped semiconductor; and photochemical detectors such as the photographic plate.

In *thermal detectors*, the absorption of photons leads to a temperature change of the detector material, which may be manifest as a change in resistance of the material, as in the *bolometer*; the development of a potential difference across a junction between two different conductors, as in the *thermopile*; or a change of internal dipole moment in a temperature-sensitive ferroelectric crystal, as in a *pyroelectric* detector. With the exception of the last, thermal detectors have slow time response when compared to photon detectors. They do, however, respond uniformly at all wavelengths (in principle).

Photon detectors can be further classified by the spectral region to which they are sensitive. Most photoemissive detectors operate from the vacuum ultraviolet through to about 1.2 μm and rapidly become insensitive beyond 1 μm. New semiconductor photocathodes are becoming available, however, that are sensitive out to 1.7 μm. Photoconductive and photovoltaic detectors based on different materials are used right from the visible into the very far infrared to at least 1000 μm. Photographic films, which are also photon detectors, generally respond best in the visible, although films marginally sensitive out to about 1.2 μm are available.

This brief survey is not intended to be exhaustive. Many other detection mechanisms have been described, but not all are commonly used or are commercially available. For more information and details about detectors in general than will be found here, the reader should consult references 13, 50, 51, 93, and 133 through 135 and the catalogs of manufacturers listed at the end of this chapter.

4.8.1 Figures of Merit for Detectors

Before discussing the characteristics of commonly used and commercially available detectors individually, the various figures of merit used by manufacturers to specify how their products will be described. These are the noise-equivalent power (NEP), detectivity (D^*), responsivity ($\mathscr{R}$), and time constant (τ), which allow questions to be answered such as: What is the minimum light intensity falling on the detector that will give rise to a signal voltage equal to the noise voltage from the detector? What signal will be obtained for unit irradiance? How does the electrical signal from the detector vary with the wavelength of the light falling on it? What is the modulation frequency response of the detector or its ability to respond to short light pulses?

Noise-Equivalent Power

The *noise-equivalent power* (NEP) is the rms value of a sinusoidally modulated radiant power falling upon a detector that gives an rms signal voltage equal to the rms noise voltage from the detector. The NEP is usually specified in terms of a blackbody source at 500K, the reference band width for the detection of signal and noise (usually 1 or 5 Hz), and the modulation frequency of the radiation (usually 90, 400, 800, or 900 Hz). A noise-equivalent power written NEP (500 K, 900, 1), for example, implies a blackbody source at 500 K, a 900 Hz modulation frequency, and a 1 Hz detection bandwidth. If a radiant intensity I (W m^{-2}) falls on a detector of sensitive area A (m^{-2}), and if signal and noise voltages V_s and V_n are measured with bandwidth Δf (Hz) (which is small enough so that the noise voltage frequency spectrum is flat within it), then the NEP measured in W Hz$^{-1/2}$ is

$$\text{NEP} = \frac{\text{IA}}{\sqrt{\Delta f}} \frac{V_\text{N}}{V_\text{S}} \tag{4.308}$$

Detectivity

At one time, detectivity was defined as the reciprocal of the NEP. Most detectors, however, exhibit an NEP that is proportional to the square root of the detector area so a detector area-independent *detectivity* D^* is now used, specified by

$$D^* = \frac{\sqrt{A}}{\text{NEP}} \tag{4.309}$$

D^* is specified in the same way as NEP: for example, $D^*(500 \text{ K}, 900, 1)$. To specify the variations in response of a detector with wavelength, the spectral detectivity is used. The symbol $D^*(\lambda, 900, 1)$ would therefore specify the response of the detector to radiation of wavelength λ, modulated at 900 Hz and detected with a 1 Hz bandwidth. The units of D^* are cm $\text{Hz}^{1/2}\text{W}^{-1}$. Curves of $D^*(\lambda)$ for various detectors are shown in Figure 4.160.

Responsivity

The responsivity $\mathcal{R}$ of a detector specifies its response to unit irradiance:

$$\mathcal{R} = V_S/\text{IA} \qquad (4.310)$$

A similar figure of merit, usually used to characterize photoemissive detectors, is the *radiant sensitivity S*, which is the current per unit area of the photoemissive surface produced by unit irradiance:

$$S = \frac{i_S}{P} \qquad (4.311)$$

where i_S is the total current from the detector and P the total radiant power falling on it. Some curves showing the spectral variation of radiant sensitivity for different photoemissive surfaces are given in Figure 4.159.

Frequency Response and Time Constant

The frequency response of a detector is the variation of responsivity or radiant sensitivity as a function of the modulation frequency of the incident radiation. The signal from the detector should be ac-coupled, otherwise the generated dc signal that is produced as the detector begins to fail to respond to the modulation will also be detected. The frequency variation of $\mathcal{R}$ and the time constant τ of the detector are generally related according to

$$\mathcal{R}(f) = \frac{\mathcal{R}(0)}{(1 + 4\pi^2 f^2 \tau^2)^{1/2}} \qquad (4.312)$$

as shown in Figure 4.161.

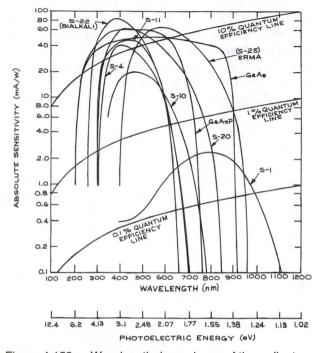

Figure 4.159   Wavelength dependence of the radiant sensitivity of several commercially available photocathode materials. (From *Handbook of Lasers*, R. J. Pressley, ed., CRC Press, Cleveland, 1971; by permission of the CRC Press.)

The responsivity has therefore decayed to a value of $\mathcal{R}(0)/\sqrt{2}$ at a frequency $1/2\pi\tau$. τ is in reality, therefore, a simple measure of the ability of the detector to respond to a sharply rising or falling optical signal.

A very important parameter that determines the frequency response of a detector is its capacitance. For driving a 50 ohm load, for example, a capacitance of 1 pF will limit the time constant to 50 ps. With the exception of photomultiplier tubes (PMTs) there is a direct relationship between detector size and time constant. PMTs with a sensitive area of 2000 mm^2 can respond as quickly as 1 to 2 ns. For semiconductor detectors, however, there is no escaping the size/time-constant relationship. A semiconductor with a 1 mm sensitive area, for example, will have a limiting time response of about 1 ns and a bandwidth of 160 MHz. On the other hand, a small size

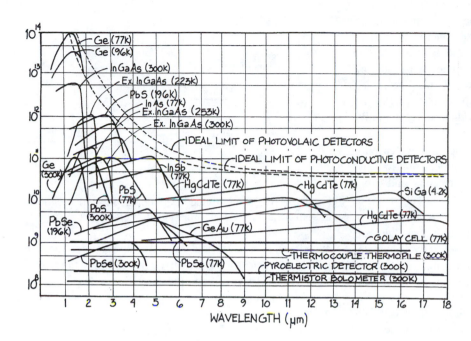

Figure 4.160 Variation of the detectivity (D^*) as a function of wavelength for a comprehensive set of different detectors.

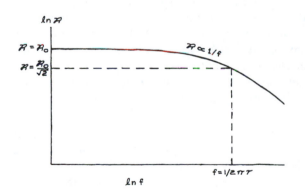

Figure 4.161 Typical dependence of detector responsivity $\mathscr{R}$.

direct coupling to optical fiber will have a bandwidth out to 20 GHz.

Noise

The random fluctuations in the output voltage or current from a detector set a lower limit to the radiant power that can be detected, given the detector temperature and operating conditions, source modulation frequency, and electronic-detection-system bandwidth. These fluctuations are from a variety of sources, some of which can be reduced by operating the detector in an appropriate way. The most important forms of noise are thermal noise, which results from the random motion of charge carriers in the detector—shot noise, which is primarily important in vacuum and gas-filled tubes and arises from the random, discrete nature of the photoemission process; current noise, which is important in semiconductor detectors and is associated with random conductance fluctuations; and *generation-recombination* (gr) noise, also important in semiconductors, which results from the random recombination of electrons and holes in the material. Thermal noise and shot noise have so-called white spectra—they contribute equal noise signals in a given detection bandwidth at any frequency. Current noise generally shows a $1/f$ frequency dependence (it is frequently referred to as $1/f$ noise) and is therefore the dominant source of noise to be dealt with at low frequencies. As shown in Figure 4.162, gr noise contributes at frequencies up to a value of the order of the reciprocal of the carrier lifetime, and then falls off rapidly with increasing frequency.

In a practical detection system, the detector is generally coupled to various forms of electronic processing systems, preamplifiers, amplifiers, filters, lock-in detectors, and so on. This electronic system necessarily creates additional noise; in a well-designed system this is kept to a minimum.

It will be seen in the following sections that there are various ways to reduce, and even essentially eliminate, detector noise. There is a fundamental fluctuation component in the signal from the detector, however, that cannot be removed. This results from fluctuations in the arrival rate of photons from the source that is being observed. The ultimate performance of any detector is therefore *photon noise-limited*. The photon-noise limit is different for thermal detectors, which are sensitive to total absorbed radiation, and photon detectors, which respond, at least in a microscopic way, to the absorption of individual quanta.

For a thermal detector of absorptivity and emissivity ε, at absolute temperature T_1, completely enclosed by an environment at temperature T_2, the photon-noise-limited detectivity can be shown to be

$$D^* = \frac{4.0 \times 10^{16} \varepsilon^{1/2}}{(T_1^5 + T_2^5)^{1/2}} \mathrm{cmHz}^{1/2}\mathrm{W}^{-1} \qquad (4.313)$$

For a photon detector, the limiting spectral detectivity is

$$D^*(\lambda) = \frac{c_0 \eta(\nu)}{2h\nu\pi^{1/2}\left[\int_{\nu_0}^{\infty} \frac{\eta(\nu')\nu'^2 e^{h\nu'/kT_2}}{(e^{h\nu'/kT_2}-1)^2}d\nu'\right]^{1/2}} \qquad (4.314)$$

where $\eta(\nu)$ is the *quantum efficiency* (the average number of charge carriers produced for each incident photon

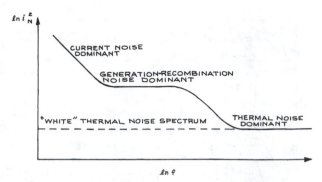

Figure 4.162 Schematic frequency dependence of noise in semiconductor detectors; i_N is the noise current.

absorbed), $\lambda = c_0/\nu$, and ν_0 is the lowest frequency to which the detector is sensitive. Some evaluations from Equation 4.314 have been given by Kruse, McGlauchlin, and McQuistan, who also give a detailed discussion of noise in optical detectors.[50]

The signal-to-noise (S/N) performance of a detector in a given situation can usually be calculated from the detector characteristics specified by the manufacture and the chateracteristics of the electronic system that the detector will drive. Specifically, the dark noise level, the noise figure (especially for devices with internal gain such as avalanche photodiodes (APDs)), and the quantum efficiency (or responsivity) of the detector and the input impedance of the following electronics and its noise characteristics are important. Davis has provided several examples of how to calculate detector noise performance in various situations.[13]

If the detector manufacturer specifies D^* or NEP—and not all do—this calculation can be simplified. The dark noise current will frequently be specified, often in units of nA/$\sqrt{\mathrm{Hz}}$ or pA/$\sqrt{\mathrm{Hz}}$, or sometimes just in nA or pA. Noise generally increases with the square root of the bandwidth used. A detector with a bandwidth of 100 MHz and a dark current of 5 nA, for example, has a dark noise of 0.5 pA/$\sqrt{\mathrm{Hz}}$.

4.8.2 Photoemissive Detectors

Photoemissive detectors include vacuum photodiodes, photomultiplier tubes, microchannel plates and photo-channeltrons. These are all photon detectors that utilize the photoelectric effect. When radiation of frequency v falls upon a metal surface, electrons are emitted, provided the photon energy hv is greater than a minimum critical value ϕ—called the *work function*—which is characteristic of the material being irradiated. A simplified energy-level diagram illustrating this effect for a metal-vacuum interface is shown in Figure 4.163a. For most metals, ϕ is in a range from 4 to 5 eV (1 eV $\leftrightarrow$ 1.24 μm), although for the alkali metals it is lower—for example, 2.4 eV for sodium and 1.8 eV for cesium. Pure metals or alloys, particularly beryllium-copper, are used as photoemissive surfaces in ultraviolet and vacuum-ultraviolet detectors. Lower work functions—and consequently sensitivities that can be extended into the infrared—can be obtained with special semiconductor materials. Figure 4.163b shows a schematic energy-level diagram of a semiconductor-vacuum interface. In this case, the work function is defined as $\phi = E_{vac} - E_F$, where E_F is the energy of the Fermi level. In a pure semiconductor, the Fermi level is in the middle of the band gap, as illustrated in Figure 4.163b. In a *p*-type doped semiconductor, E_F moves down toward the top of the valence band, while in *n*-type material it moves up toward the bottom of the conduction band. ϕ is therefore not as useful a measure of the minimum photon energy for photoemission as it is for a metal. The electron affinity χ is a more useful measure of this minimum energy for a semiconductor. Except at absolute zero, photons with energy $hv > \chi$ cause photoemission. Photons with energy $hv > E_g$ lead to the production of carriers in the conduction band. This leads to intrinsic photoconductivity, which is the operative detection mechanism in various infrared detectors, such as InSb.

Vacuum Photodiodes

Once electrons are liberated from a photoemissive surface (a photocathode), they can be accelerated to an electrode positively charged with respect to the cathode—the anode—and generate a signal current. If the acceleration of photoelectrons is directly from cathode to anode through a vacuum, the device is a vacuum photodiode. Because the electrons in such a device take a very direct path from anode to cathode and can be accelerated by high voltages—up to several kilovolts in a small device—vacuum photodiodes have the fastest response of all

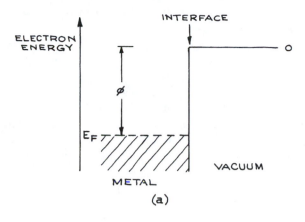

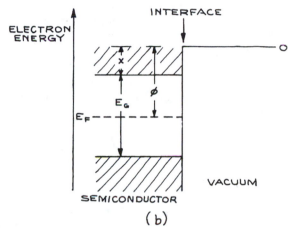

Figure 4.163   Band structure of (a) a metal vacuum interface and (b) a pure semiconductor-vacuum interface (ϕ = work function; X = electron affinity; E_g = band-gap energy; E_F = Fermi level).

photoemissive detectors. Risetimes of 100 ps or less can be achieved. External connections and electronics are generally the limiting factors in obtaining short risetimes from such devices. Vacuum photodiodes, however, are not very sensitive since one electron at most can be obtained for each photon absorbed at the photocathode. In principle, of course, the limiting sensitivity of the device is set by the *quantum efficiency* η of the photoemissive surface, defined as

$$\eta = \frac{\text{electrons emitted}}{\text{photons absorbed}} \qquad (4.315)$$

Practical quantum efficiencies for photoemissive materials range up to about 0.4.

If the space between photocathode and anode is fired with a noble gas, photoelectrons will collide with gas atoms and ionize them, yielding secondary electrons. An electron multiplication effect therefore occurs. Because the mobility of the electrons moving from cathode to anode through the gas is slow, however, these devices have a long response time—typically about 1 ms. Gas-filled photocells are no longer competitive with solid-state detectors and have disappeared from use.

Photomultipliers

If photoelectrons are accelerated *in vacuo* from the photocathode and allowed to strike a series of secondary electron emitting surfaces—called *dynodes*—held at progressively more positive voltages, a considerable electron multiplication can be achieved and a substantial current can be collected at the anode. Such devices are called *photomultipliers*. Practical gains of 10^9 (anode electrons per photoelectron) can be achieved from these devices for short light pulses. Continuous gains must be held lower—to perhaps 10^7—a limit imposed primarily by the ability of the final dynodes in the chain to withstand the thermal loading produced by continuous electron impact. Because of their very high gain, photomultipliers can generate substantial signals when only a single photon is detected. An anode pulse of 2 ns duration containing 10^9 secondary photoelectrons produced from a single photoelectron, for example, will generate a voltage pulse of 4 V across 50 ohms. This, coupled with their low noise, makes photomultipliers very effective single-photon detectors. Photomultiplier D^* values can range up to 10^{16} cm $Hz^{1/2}W^{-1}$. Only the dark-adapted human eye, which can detect bursts of about 10 photons in the blue, comes close to this sensitivity.

The time response characteristics of photomultiplier tubes depend to a considerable degree on their internal dynode arrangement. The response of a given device can be specified in terms of the output signal at the anode that results from a single photoelectron emission at the photocathode. This is illustrated in Figure 4.164. Because electrons passing through the dynode structure can generally take slightly different paths, secondary electrons

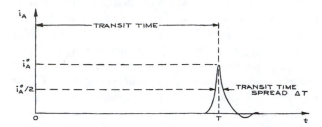

Figure 4.164 Typical anode pulse produced by a single photoelectron emission at the cathode of a photomultiplier tube. *t* is the time following photoemission. The transit-time *T*, the transit-time speed ΔT, and the peak anode current i_A^0 all fluctuate from pulse to pulse.

arrive at the anode at different times. The anode pulse has a characteristic width called the *transit-time spread*, which usually ranges from 0.1 to 20 ns. The time interval between photoemission at the cathode and the appearance of an anode pulse is called the *transit time*, and is usually a few tens of nanoseconds. The transit time and transit-time spread also fluctuate slightly from one single photoelectron-produced anode pulse to the next. Although this is a fine point, it may be something to worry about in precision-timing experiments. The principle suppliers of photomultiplier tubes are Hamamatsu, Burle Industries,[136] Electron Tubes, Phillips (Amperex), and EMR.

There are seven main types of dynode structure in common usage in photomultiplier tubes—these are illustrated in Figure 4.165. The circular cage structure (used, for example, in the Hamamatsu 1P28 photomultiplier tube) is compact and can be designed for good electron-collection efficiency and small transit-time spread. This dynode structure works well with opaque photocathodes, but is not very suitable for high-amplification requirements where a larger number of dynodes is required. The box-and-grid and Venetian-blind structures (used, for example, in the Hamamatsu Models 464 and R1513, respectively) offer very good electron-collection efficiency. Because they collect multiplied electrons independently of their path through the dynode structure, a wide range of secondary-electron trajectories is possible, leading to a large transit-time spread and slow response. Typical response times of these types of tube are 10 to 20 ns. Venetian-blind tubes can easily be extended to many dynode stages and have a very stable gain in the presence of small power supply fluctuations. These tubes also have an optically opaque dynode structure, which provides very low dark-current noise when operated under appropriate conditions.

The focused dynode structure (used, for example, in the Phillips 56 AVP and other 56-series tubes[137]) is designed so that electrons follow paths of similar length through the dynode structure. To accomplish this, electrons that deviate too much from a specified range of trajectories are not collected at the next dynode. These types of tube offer short response times, typically 1 to 2 ns. Some recently developed tubes are even faster.

The mesh type of dynode structure has a series of crossed fine mesh dynodes arranged in close proximity. The structure allows compact construction, high immunity to magnetic fields, as well as uniformity of response and good pulse linearity. Versions of this type with special anodes provide position sensitive performance. In this type of device, the output signal from a series of anodes provides direct information on the location of illumination on the photocathode surface.

The metal channel dynode structure provides a very compact dynode structure and fast response becomes of the closeness between each stage of dynodes. This structure also allows position-sensitive detection.

Closely related to these fundamental photomultiplier dynode structures is the microchannel plate (MCP). Instead of traditional dynodes, photoelectrons enter an array of millions of microglass tubes fused together in an array. Each tube has a resistive coating on its inner surface that acts as a continuous electron multiplier. Each tube individually can be referred to as a channel electron multiplier (or channeltron). In these tubes, electrons make multiple reflections from the tube walls generating secondary electrons. MCPs have fast time response, good immunity from external magnetic fields, and positive-sensitive capability if provided with multiple anodes. These devices are commercially available from Galileo, Hamamatsu, and Nano Sciences,

Other types of multiplier structures, not shown in Figure 4.165, are also available. Traveling-wave phototubes and various types of crossed-field photomultiplier have very short response times, down to 0.1 ns. Some designs of crossed-field photomultiplier used to be available commercially.[138] These tubes use a combination of static electric and magnetic fields to direct high-energy electrons along long paths and low-energy electrons along appropriately shorter paths between dynodes. This minimizes transit-time variations between secondary electrons of different energy.

Photocathode and Dynode Materials

The performance of a photomultiplier depends not only on its internal structure, but also on the photoemissive material of its photocathode and the secondary electron-emitting material of its dynodes. The selection of a photodiode or photomultiplier for a particular spectral response depends on an appropriate selection of photocathode material. The wavelength dependence of various commercially available photocathode materials is shown in Figure 4.160; some radiant sensitivities and quantum efficiencies are given in Table 4.16. The short wavelength cutoff of a given material depends on its work function. This cutoff is not sharp because, except at absolute zero, there are always a few electrons high up in the conduction band available for photoexcitation by low-energy photons. A few of these electrons will be emitted without any photostimulation because of their thermal excitation. This contributes the major part of the dark current observed from the photocathode. Materials that have low work functions, and are consequently more red- and infrared-sensitive, have higher—often much higher—dark currents than materials that are optimized for visible and ultraviolet sensitivity. Unless a phototube is to be used primarily for detecting long-wavelength radiation, we recommend selecting a photocathode with the shortest possible wavelength response.

Photocathodes are available in both opaque and semitransparent forms, depending on the model of phototube. In the semitransparent form, the photoelectrons are ejected from the thin photoemissive layer on the opposite side from the incident light. In both types, the photocathode has to perform two important functions: it must absorb incident photons, and allow the emitted photoelectron to escape. The latter event is inhibited if photons are absorbed too deep in a thick photoemissive layer, or if the photoelectron suffers energy loss from scattering in the layer. If the emitted photoelectron has too

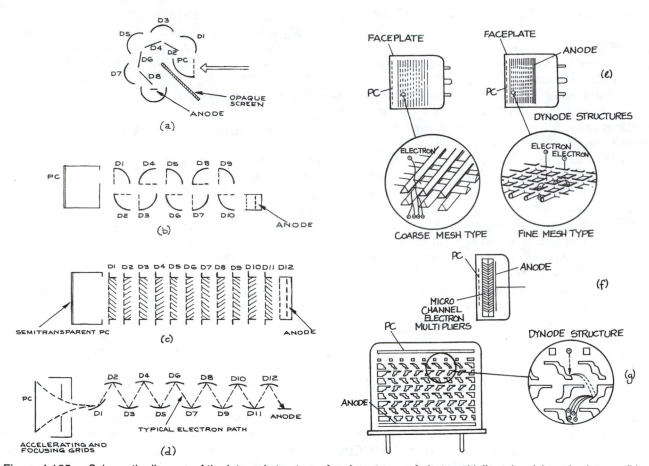

Figure 4.165 Schematic diagram of the internal structure of various types of photomultiplier tube; (a) squirrel cage; (b) box and grid; (c) venetian blind; (d) focused dynode; (e) mesh type; (f) microchannel electron multiplier; (g) metal channel dynode structure. D = dynode; PC = photocathode.

much energy, it can excite a further electron across the band gap. This pair-production phenomenon inhibits the release of photoelectrons from the photoemissive layer, and accounts for the ultraviolet cutoff characteristics of the different materials shown in Figure 4.160. With reference to Figure 4.163b, it can be shown that for pair production to occur the incident photon energy must be greater than $2E_g$. Photoelectrons have the best chance of escaping, and the photocathode its highest quantum efficiency for materials where $\chi < E_g$.

Because semi-transparent photocathodes allow photons to penetrate sufficiently for photoelectrons to escape on the far side of the layer, they are deposited as semiconductor layers on the insulating glass substrate of the entrance window. These photocathodes are less tolerant of relatively high light levels than are opaque photocathodes, which can be deposited on a conducture sulstrate. A conductive substrate preserves field distribution within the PMT, even at relatively large cathode currents. When an opaque photocathode is used, photons pass through the entrance window of the PMT and strike the photocathode, which is mounted separately as shown schematically in Figure 4.165a. Photoelectrons are emitted on the same side of the photocathode where photons are absorbed.

TABLE 4.16 CHARACTERISTICS OF PHOTOEMISSIVE SURFACES

Cathode	Radiant Sensitivity (mA/W)			Peak Quantum Efficiency (%)	λ_{peak} (μm)	Dark Current[a] (A/cm^2)
	515 nm	694 nm	1.06 μm			
S-1 Cs-O-Ag	0.6	2	0.4	0.08	800	9×10^{-13}
S-10 Cs-O-Ag-Bi[b]	20	2.7	0	5	470	4×10^{-16}
S-11 Cs_3Sb on MnO[c]	39	0.2	0	13	440	10^{-16}
S-20 (Cs)Na_2KSb (tri-alkali)	53	20	0	18	470	3×10^{-16}
S-22 (bi-alkali)	42	0	0	26	390	$1-6 \times 10^{-18}$
GaAs[d–g]	48	28	0	14	560[h]	10^{-16}
GaAsP[d]	60	30	0	19	400[h]	3×10^{-15}
InGaAs[d, e]	—	—	4.3	—	—[h]	3×10^{-14}
InGaAsP[d,f]	—	—	—	47	300[h]	2.5×10^{-13}
S-25 (ERMA)[i]	53	26	0	25	430	1×10^{-15}
Cs-Te (solar blind)	—	—	—	15	254	2.5×10^{-17}

Note: Table shows typical values, but these can vary greatly from one manufacturer to another.

[a] At room temperature.
[b] Cathode designated S-3 is similar.
[c] Several types of CsSb photocathodes exist where the CsSb is deposited on different opaque and semitransparent substrates and various window materials are used. These photocathodes have the designations S-4, 5, 13, 17, and 19 as well as S-11.
[d] NEA photoemitters.
[e] Available from Burle Industries.
[f] Available from ITT, Fort Wayne, Indiana.
[g] Available from Hamamatsu Corporation, Bridgewater, New Jersey.
[h] May show no wavelength of maximum quantum efficiency; quantum efficiency falls with increasing wavelength. However, exact spectral characteristics will depend on thickness of photoemitter and whether it is used in transmission or not.
[i] Extended-red S-20.

Practical photoemissive materials fall into two main categories: classical photoemitters and negative-electron-affinity (NEA) materials. Classical photoemitters generally involve an alkali metal or metals, a group-V element such as phosphorus, arsenic, antimony, or bismuth, and sometimes silver and/or oxygen. Examples are the Ag-O-Cs (S1) photoemitter, which has the highest quantum efficiency beyond about 800 nm of any classical photoemitter, and Na_2KSbCs—the so-called tri-alkali (S-20) cathode.

NEA photoemitters have been developed only within the last ten years. These materials utilize a photoconductive single-crystal semiconductor substrate with a very thin surface coating of cesium and usually a small amount of oxygen. The cesium (oxide) layer lowers the electron affinity below the value it would have in the pure semiconductor, achieving an effectively negative value. Examples of such NEA photoemitters are GaAs (CsO) and InP (CsO). NEA emitters can offer very high quantum efficiency and extended infrared response. GaAs (CsO), for example, has higher quantum efficiency in the near infrared then an S-1 photocathode. Commercial photomultipliers with NEA photocathodes are available from Burle Industries and Hamamatsu, who offer response out to 1.7 μm in a tube cooled to $-100°$C. Development of NEA photoemitters is continuing, and the experimentalist seeking long-wavelength response would be well advised to follow the literature and check for the introduction of new commercial tubes. For further details about NEA materials, the interested reader should consult the article by Zwicker.[139]

The performance of the dynode material in photomultiplier tubes is specified in terms of the secondary-emission ratio δ as a function of energy. For a phototube with n dynodes, the gain is δ^n, In the past, the commonest dynode materials were CsSb, AgMgO, and BeCuO. The last is also used as the primary photoemitter in windowless photomultipliers that are operated *in vacuo* for the detection of vacuum-ultraviolet radiation. BeCuO can be reactivated after exposure to air. The above materials have δ values of 3 to 4. Newer NEA dynode materials have much higher δ values—in particular, that of GaP can range up to 40 for an incident-electron input energy of 800 eV. With such high δ values, a photomultiplier tube needs fewer dynodes for a given gain, which means a more compact and faster-response tube can be built. In many commercial photomultipliers, the first dynode at least is now frequently made of GaP. This offers improved characterization of the single photo-electron response of the tube, which is important in designing a system for optimum signal-to-noise ratio.

Practical Operating Considerations for Photomultiplier Tubes

Detailed advice is given here on the use of photomultipliers in experimental apparatus.

(i) Dynode chains.
The accelerating voltages are supplied to the dynodes of a photomultiplier by a resistive voltage divider called a *dynode chain*. The relative resistance values in the chain determine the distribution of voltages applied to the dynodes. The total chain resistance R determines the chain current at a given total photocathode-anode applied voltage. Some phototubes are supplied with an integral dynode chain, but most are not. Adequate dynode-chain designs are generally supplied by the manufacturer with each tube. Alternative designs intended for certain types of response, however, such as maximum gain for short pulses or highest linearity, can frequently be found by consulting the literature. Two examples of such dynode-chain designs for use with Phillips 5-series tubes are given in Figure 4.166. The total

dynode-chain resistance is chosen so that the chain current is at least 100 times greater than the average dc anode current to be drawn from the tube. This current is specified for each tube and may range as high as 1 mA for high-gain tubes. The operation of photomultiplier tubes at such high average currents, however, is not recommended for very long. Long-term dc anode currents should be kept between 1 and 10 µA for most tubes.

There are several interesting features of the dynode chains shown in Figure 4.166. The photocathode is generally operated at negative potential with the anode near ground. This makes it easy to couple the output signal to other electronics. The tube can be operated with the cathode grounded, but in this case the signal from the anode, which is now at high positive potential, must be coupled through high-voltage capacitors or an insulated transformer as shown in Figure 4.167. The photocathode is also often connected to a conducting shield, which is either a painted coat of colloidal graphite (Aquadag) on the outside of the tube envelope, or a metal foil wrapped around the tube envelope. If the photocathode is at negative potential, appropriate insulation must be used around the tube if the tube is in close proximity to any grounded metal parts inside its housing.

Many photomultiplier tubes have one or more focusing electrodes between the photocathode and the first dynode. The voltage on these electrodes can be adjusted to optimize the collection of photoelectrons from the photocathode. The last three or four dynodes are decoupled with high-voltage, high-frequency capacitors (disc ceramics work well in this application). These capacitors prevent depression of the dynode-chain voltage on the last few dynodes when a large pulse of electrons passes through the tube (the secondary-electron current drawn from each dynode must be supplied by the dynode chain). For high-frequency applications, the inclusion of small damping resistors (typically 50 Ω to 1 kΩ) between the decoupling capacitors and the dynodes is recommended. The resistor R_A—the anode resistor—determines the actual voltage that will be produced by a given anode current, since the photomultiplier serves as a current source. R_A should not be so large that the voltage developed across it is significant compared with the voltage between anode and last dynode. A value of R_A between 1 and 10 MΩ is usual. Lower values than this can

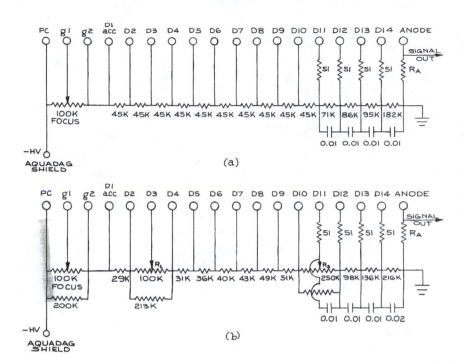

Figure 4.166   Dynode chains for Phillips 56-series photomultiplier tubes: (a) chain for high gain and fast time response; (b) chain for high linearity (R_1 and R_2 adjusted for optimum performance). PC = photocathode; g1, g2 = grids; acc = accelerator grid; D1 to D14 = dynodes. Resistance values are in ohms, capacitance values in μF. For very low-level light detection, all resistances can be scaled upward to reduce overall chain current.

be obtained by terminating the signal lead from the anode with a second resistor. In many cases, the effective value of the anode load is set by the input impedance of the electronics to which the tube is connected. In high-frequency applications, such as single-photon counting, the signal from R_A should be coupled out through a 50 Ω coaxial cable terminated in 50 Ω. In such applications, the cable connection to the anode should be shielded as close to the anode as possible, even going so far as to insert R_A under the grounded screen of the coaxial cable. Some high-frequency tubes are supplied with an integral coaxial connector on the anode. In any case, the aim is to reduce parasitic inductance and capacitance associated with the anode connection. Even with a 50 Ω load, for example, a parasitic capacitance greater than 20 pF will limit the response time of the tube to 1 ns.

The actual response-time behavior of the photomultiplier can be determined by observing its single photoelectron response. This is done by looking at the anode pulses with a fast oscilloscope. The photocathode should not need to be illuminated for this to be done; sufficient noise pulses will usually be observed. The pulses should appear as in Figure 4.164. If they have too much of an exponential tail, they are probably limited by the anode resistor and parasitic capacitance. These anode pulses reflect the time distribution and number of secondary electrons reaching the anode following single (or multiple) photoelectron emissions from the photocathode. If the height distribution of anode pulses is measured, a distribution such as is shown in Figure 4.168 will probably be seen.

(ii) Mounting photomultiplier tubes.

Photomultiplier tubes are generally mounted by clamping their plastic socket inside a cylindrical tube housing. The housing should be light-tight—a photomultiplier should never be exposed to ambient lighting when its high-voltage dynode chain is on because the resultant large secondary-electron current will disintegrate the last dynodes and may strip the photocathode layer itself. If the total power dissipation of the dynode chain is low, the whole dynode chain may be mounted directly on the tube socket. If space is not at a premium, however, the dynode chain can be remote from the tube base and connected to it

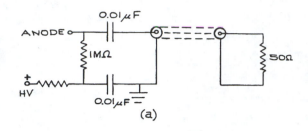

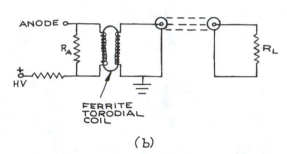

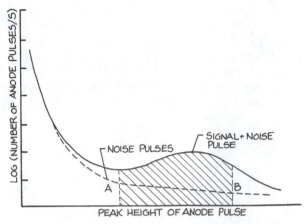

Figure 4.168 Schematic photomultiplier anode pulse-height distribution likely to be observed in practice. The best signal/noise ratio would be obtained in a photon-counting experiment by collecting only anode pulses in a height range roughly indicated by the shaded region *AB*.

Figure 4.167 Anode signal coupling methods for a photomultiplier operated with the anode at high positive potential: (a) capacitor coupling for fast risetime, short pulse operation (the component sizes are typical; the capacitors should be high frequency, high voltage types); (b) transformer coupling for lower-frequency, modulated operation (the high-voltage winding on the transformer should be sufficiently well insulated for isolation from the core; the size of R_A will depend on various factors, such as transformer primary impedance and operating frequency).

with high voltage-insulated wire. The damping resistors, decoupling capacitors, and anode resistor should, however, be kept in close proximity to the tube. The photomultiplier should be mounted so there is no optical path from dynode chain to photocathode.

Photomultipliers are fragile and should be mounted so they are not subject to stress. This is particularly important if a tube is to be cooled. Rugged (and expensive) photomultipliers that can withstand severe vibrations and accelerations are available from ADIT, Hamamatsu, and EMR.

Noise in Photomultiplier Tubes

Noise in photomultipliers comes from several sources:

1. Thermionic emission from the photocathode

2. Thermionic emission from dynodes

3. Field emission from dynodes (and photocathode) at high interdynode voltages

4. Radioactive materials in the tube envelope (for example, ^{40}K in glass)

5. Electrons striking the tube envelope and causing fluorescence

6. Electrons striking the dynodes and causing fluorescence

7. Electrons colliding with residual atoms of vapor in the tube (cesium for example) and causing fluorescence

8. Cosmic rays

Noise from photomultipliers is always greater after they have been exposed to light (without high voltage applied). They gradually become quieter after operation under dark conditions for an extended period.

Noise from thermionic emission can be reduced considerably by cooling the tube (particularly for S-1 tubes). A factor of 10 to 1000 reduction in noise can be obtained by cooling to about −20°C. S-1 tubes need to be cooled to low temperatures, but temperatures below about –70°C are not recommended. Too much cooling has a marginal effect on reducing thermionic emission—which will already be negligible for most tubes at –20°C—and may have deleterious effects. At low temperatures the conductivity of the photocathode layer falls, which causes it to no longer be an equipotential. The resultant distorted field distribution between cathode and dynodes will increase the possibility of noise from sources such as item 5 above. Two convenient designs of photomultiplier tube-coolers are shown in Figure 4.169. In the design shown in Figure 4.169a, a copper or aluminum cylinder encloses the tube (without touching it). This tube is in thermal contact with a chamber containing dry ice (liquid nitrogen). Condensation on the face of the photomultiplier tube is minimized, since moisture preferentially condenses and freezes on the much colder metal tube. In the second design, shown in Figure 4.169b, cold nitrogen gas boiled off from a liquid-nitrogen Dewar flows through a copper tube surrounding the photomultiplier, and finally sprays over its photocathode surface to minimize water-vapor condensation. The rate of supply of cold nitrogen gas can be adjusted to provide a range of temperatures.

Photomultiplier tube coolers are available from Amherst Scientific, Electron Tubes, Hamamatsu, and Products for Research.

Noise from field emission may be a problem with certain high-voltage, high-gain tubes—one solution is to reduce the voltage. In single photon-counting experiments, larger than normal anode pulses may result from field emission and can be rejected with an appropriate window discriminator.[140]

Noise from radioactive envelope materials should generally be negligible. Noise from electrons striking the tube envelope can be reduced by a shield around the tube held at photocathode potential. Noise from electrons striking the dynodes and causing fluorescence causes the fewest problems in Venetian blind tubes, which have little or no direct optical path from dynodes to photocathode. Noise from gas in the tube can be a problem in old tubes, which frequently become gassy; solution: buy a new tube. Noise from cosmic rays is usually unimportant; it can be reduced by shielding.

Of all these sources of noise, thermionic emission from the photocathode is generally by far the most important. For a given photocathode material, the total noise will

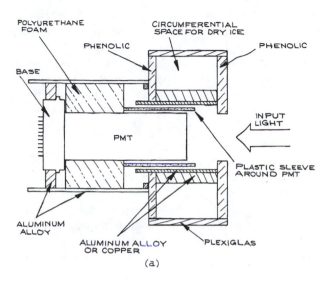

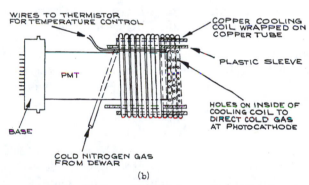

Figure 4.169 Photomultiplier tube coolers: (a) using dry ice reservoir in thermal contact with a metal tube close to, but electrically insulated from, the photocathode; (b) using cold nitrogen gas circulated around tube and sprayed onto photocathode surface (a thermistor monitors temperature and allows control of the rate of supply of cold gas).

depend on the photocathode area. Under comparable conditions, therefore, tubes with small photocathode area like the S-20 ITT FW-130—with a photocathode diameter of 0.25 cm—are less noisy than tubes with large photocathodes like the 5 cm photocathode of the S-20 Phillips 56 TVP. In many experiments, a large photocathode area is unnecessary—light coming through a monochromator slit illuminates a very small area. The effective photocathode area, and consequently its thermionic noise, can be reduced by wrapping a magnetic coil around the photocathode, which prevents electrons other than those from the center of the photocathode from reaching the first dynode. Under circumstances where photomultipliers must be operated in close proximity to magnetic fields, magnetic shields to enclose the tube are available from Ad-Vance Magnetics, Amuneal, Magnetic Shield Corp. (Perfection Mica), Mμ Shield, and Products for Research

In the anode pulse-height distributions shown in Figure 4.168, small anode pulses are much more likely to correspond to noise than are pulses in the middle of the distribution. In photon-counting experiments, these small pulses should be rejected with a discriminator. A window discriminator provides simultaneous rejection of these pulses and those which are much larger than average.[140]

Ultraviolet and Vacuum-Ultraviolet Detection with Photodiodes and Photomultipliers

For detection of radiation down to about 105 nm, phototubes with windows of LiF are available, as are tubes with windows of MgF_2 or sapphire. For detection of even shorter wavelengths, windowless phototubes or channeltrons are used. For detection of radiation down to 58 nm or beyond, conventional photomultiplier tubes sensitive to visible light (S-11) can be used if the outer surface is coated with a layer of sodium salicylate—or if a sodium salicylate-coated disc is placed close to the photocathode. Sodium salicylate fluoresces in the visible with almost unity quantum efficiency and converts incident vacuum-ultraviolet radiation to a wavelength where a conventional tube can detect it. The response of the sodium salicylate is also fast so that 1 ns ultraviolet detection is possible. To prepare a sodium-salicylate-coated window, a solution of sodium salicylate in reagent-grade methanol (80 g/liter) is atomized and sprayed onto the window, which is placed about 10 cm above the atomizer. Commercial nasal sprays pressurized with dry nitrogen are suitable for this purpose. The atomizer should be far enough from the window so that no large drops can be deposited. The method works better if the window is kept warm (50–70°C). Spraying should be continued until the density of the sodium salicylate layer reaches about 20 g m^{-2}, which can be determined by weighing.

4.8.3 Photoconductive Detectors

Photoconductive detectors can operate through either intrinsic or extrinsic photoconductivity. The physics of intrinsic photoconductivity is illustrated in Figure 4.170a. Photons with energy $h\nu > E_g$ excite electrons across the band gap. The electron-hole pair that is thereby created for each photon absorbed leads to an increase in conductivity—which comes mostly from the electrons. Semiconductors with small band gaps respond to long-wavelength infrared radiation but must be cooled accordingly, otherwise thermally excited electrons swamp any small photoconudctivity effects. Table 4.17 lists commonly available intrinsic photoconductive detectors together with their usual operating temperature and the limit of their long-wavelength response λ_0, together with some representative figures for detectivities and time constants. Note that silicon and germanium are also operated in both photovoltaic and avalanche modes (see Section 4.8.4).

If a semiconductor is doped with an appropriate material, impurity levels are produced between the valence and conduction bands as shown in Figure 4.170b. Impurity levels that are able to accept an electron excited from the conduction band are called *acceptor levels*, whereas impurity levels that can have an electron excited from them into the conduction band are called *donor levels*. In Figure 4.170, photons with energy $h\nu > E_A$ excite an electron to the impurity level, thus leaving a hole in the valence band and thereby giving rise to p-type extrinsic photoconductivity. Photons with energy $h\nu > E_D$ will excite an electron into the conduction band, giving n-type extrinsic photoconductivity. Gold-doped germanium, for example, has an acceptor level 0.15 eV above the valence band and is an extrinsic p-type photoconductor, as is copper-doped germanium, which has an acceptor level 0.041 eV above the valence band. These are two commonly used extrinsic photoconductive detectors, responding out to about 9 µm and 30 µm, respectively. Curves showing the variation of their D^* with wavelength are given in Figure 4.171. Table 4.18 lists the operating characteristics of these and some other commercially available extrinsic photoconductive detectors. Detectors listed with operating temperatures below 28 K will, in practice, frequently be operated at 4 K, since liquid helium

TABLE 4.17 INTRINSIC PHOTOCONDUCTIVE DETECTORS

Semiconductor	T (K)	E_g (eV)	λ_0 (μm)	$D^*(max)$ ($cm\,Hz^{1/2}\,W^1$)	τ
CdS	295	2.4	0.52	3.5×10^{14}	~ 50 ms
CdSe	295	1.8	0.69	2.1×10^{11}	~ 10 ms
Si	295[a]	1.12	1.1	$\leq 2 \times 10^{12}$	—[b]
Ge	295[a]	0.67	1.8	10^{11}	10 ns[c]
PbS	295	0.42	2.5	$\leq 2 \times 10^{11}$	—[d]
	195	0.35	3.0	$\leq 5 \times 10^{11}$	—[d]
	77	0.32	3.3	$\leq 8 \times 10^{11}$	—[d]
PbSe	295	0.25	4.2	$1 \times 10^9 – 5 \times 10^9$	1 μs
	195	0.23	5.4	$1.5 – 4 \times 10^{10}$	30–50 μs
	77	0.21	5.8	$\leq 3 \times 10^{10}$	50 μs
InSb[e]	77	$\cong 0.23$	5.5–7.0	$\leq 3 \times 10^{10}$	0.1–1 μs
$Hg_{0.8}Cd_{0.2}Te$	77	≤ 0.1	12–25	$10^9 – 10^{11}$	> 1 ns

[a] Increased sensitivity can be obtained by cooling.
[b] Detectors with time constants from 50 ps upward and various detectivities are available.
[c] Detectors operated in a photovoltaic mode with time constants from 120 ps upward and various detectivities are available.
[d] Detectors with time constants ranging from about 100 μs to 10 ms and varying detectivities are available.
[e] More commonly operated in a photovoltaic mode.

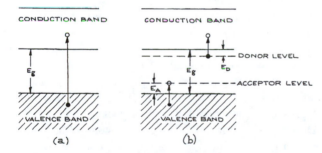

Figure 4.170 Mechanism for (a) intrinsic photoconductivity; (b) extrinsic photoconductivity.

is a safe, available coolant, although its use involves more complicated technology than the use of liquid nitrogen.

All photoconductive detectors, whether intrinsic or extrinsic, are operated in essentially the same way, although there are wide differences in packaging geometry. These differences arise from differing operating temperatures and speed-of-response considerations. A schematic diagram which shows the main construction features of a liquid nitrogen-cooled photoconductive or photovoltaic detector is given in Figure 4.172. Uncooled detectors can be of much simpler construction—for example, in a transistor or flat solar-cell package.

One feature of the cooled detector design shown in Figure 4.172 is worthy of note. The field of view of the detector is generally restricted by an aperture, which is kept at the temperature of the detection element. This shields the detector from ambient infrared radiation, which peaks at 9.6 μm. For detection of low-level narrow-band infrared radiation, the influence of background radiation can be further reduced by incorporating a cooled narrow-band filter in front of the detector element. The filter will only radiate beyond the cutoff wavelength of the detector, and it restricts transmitted ambient radiation to a narrow band. Liquid helium-cooled detectors generally have a double Dewar—an outer one for liquid nitrogen surrounds the inner for liquid helium. Liquid nitrogen-cooled detectors—such as HgCdTe—are recommended in preference to liquid helium-cooled detectors—such as Ge:Cu—whenever there is a choice. Some miniature-

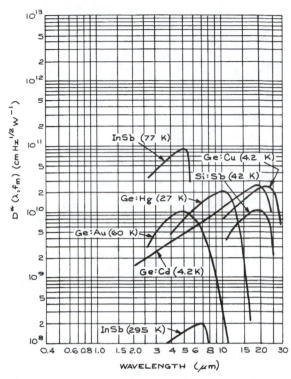

Figure 4.171 $D^*(\lambda)$ as a function of wavelength for various photoconductive detectors. (Courtesy of Hughes Aircraft Company.)

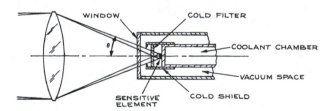

Figure 4.172 Radiation-sheilded liquid nitrogen-cooled photoconductive or photovoltaic infrared detector assembly.

package commercial detectors can be cooled with Joule-Thompson refrigeration units driven with compressed gas, which eliminates the need for externally supplied liquified-gas coolant. Thermoelectrically cooled (Peltier effect) infrared detector packages are available from New England Photoconductor Hamamatsu, Textron, Cal

Sensors, and Oriel for the operation of PbS or PbSe detectors down to temperatures of 193 K. Further details of the operating principles behind these and other refrigeration techniques used with infrared detectors are given by Hudson.[51]

Figure 4.173 shows a basic biasing circuit commonly used for operating photoconductive detectors. R_d is the detector dark resistance. It is easy to see that the change in voltage, ΔV, that appears across the load resistor R_L for a small change ΔR in the resistance of the detector is

$$\Delta V = \frac{-V_0 R_L \Delta R}{(R_d + R_L)^2} \qquad (4.316)$$

This is at a maximum when $R_L = R_d$. It is therefore common practice to bias the detector with a load resistance equal to the detector's dark resistance. The bias voltage is selected to give a bias current through the detector that gives optimum detectivity. This bias current will generally be specified by the manufacturer. Figure 4.174 shows a simple op-amp circuit for operating a photoconductive detector.

To obtain fast response from a photoconductive detector great care must be taken to minimize the stray capacitance C_s in the input circuit to the preamplifier. Otherwise, the time constant of the detector will be limited by R_S. The ultimate limits to the speed of an actual detector are set by its internal capacitance C_d and the majority-carrier lifetime. Many commercial detectors are manufactured so that the stray capacitance of the detector and its connection leads is very small. This is generally true of high-speed commercial detectors packaged in all-metal Dewars with integral coaxial bias connections. Metal Dewar packages are preferable to glass ones: although the latter are cheaper, they are very fragile.

If a detector is not specifically packaged to minimize stray inductance, it is possible to reduce the stray capacitance of the detector package and improve its speed of response by a technique called *bootstrapping*, which is illustrated in Figure 4.175. All bias leads to the detector and the leads to the preamplifier are double-shielded. The preamplifier should ideally have a high input impedance, a low input capacitance, a wide bandwidth, and 50 ohm output impedance. Such preamplifiers can be constructed, or bought, from detector suppliers or Perry Amplifier. The

TABLE 4.18 EXTRINSIC PHOTOCONDUCTIVE DETECTORS

Semiconductor	Impurity	T (K)	λ (μm)	D*	τ	Suppliers[a]
Ge:Au[b]	p-type	77	8.3	3×10^9–10^{10}	30 ns	B, J
Ge:Hg[c]	p-type	< 28	14	1–2×10^{10}	> 0.3 ns	J, M
Ge:Cd	p-type	< 21	21	2–3×10^{10}	10 ns	J, M
Ge:Cu[c]	p-type	< 15	30	1–3×10^{10}	> 0.4 ns	
Ge:Zn[b, c]	p-type	< 12	38	1–2×10^{10}	10 ns	J
Ge-Be	p-type	< 3	115	2×10^{10}	> 1 μs	I
Ge:In	p-type	4	111	—	< 1 μs	
Si:Ga	p-type	4	17	10^9–10^{10}	> 1 μs	I
Si:As	n-type	< 20	23	1–3.5×10^{10}	0.1 μs	

[a] B:EDO/Barnes Engineering; I: Infrared Labs; J: Judson; M: Mullard.
[b] Sometimes also contain silicon.
[c] Sometimes also contain antimony.

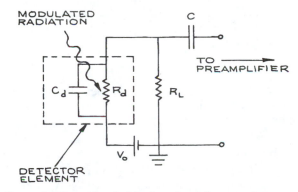

Figure 4.173   Simple biasing circuit for operating a photoconductive detector with modulated radiation.

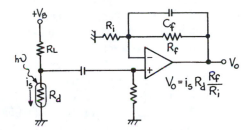

Figure 4.174   Simple op-amp circuit for operating a photoconductive detector.

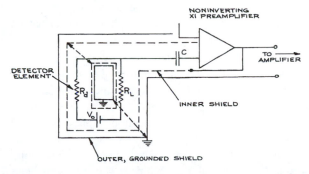

Figure 4.175   Arrangement for *bootstrapping* an infrared detector to minimize effects of parasitic capacitance.

inner shield is not grounded, but is connected directly to the low-impedance output of the unit-gain preamplifier. The inner shield is therefore *bootstrapped* to the signal voltage and stray capacitance is effectively eliminated. By this means, the speed of response of a detector can be improved substantially—from 1 μs down to 50 ns. An alternative way to improve the speed of response, at the expense of signal (but not detectivity), is to reduce the value of the load resistor.

A few photoconductive detectors are worthy of brief extra comment. Silicon and germanium are much more commonly used for photodiodes, frequently in an avalanche mode. These devices are discussed in Section 4.7.4. Lead sulfide detectors have high impedance, 0.5 to 100 MΩ, and slow response, but are the most sensitive detectors in the spectral region between 2, and 3 μm and can be used uncooled. $D^*(\lambda)$ curves for these detectors are shown in Figure 4.176. They are available from Cal Sensors, Hamamatsu, New England Photoconductor, and Textron. Lead selenide is sensitive to longer wavelengths than lead sulfide, but InAs operated in a photovoltaic mode is to be preferred in this spectral region (3–4 μm). Gold-doped germanium is a simple detector to use in the spectral region between 1 and 9 μm—it is superior to cooled PbSe in this region. InSb operated in a photovoltaic mode is probably the detector of choice between 4 and 5.3 μm. Beyond about 5 μm, HgCdTe is the detector of choice and is to be preferred to detectors such as Ge:Hg and Ge:Cu, which must be operated below liquid-nitrogen temperature. HgCdTe comes in different stochiometries, generally represented as $Hg_xCd_{1-x}Te$. Different stoichiometries have different spectral responses. By varying its composition, HgCdTe exceeds all other 77°K detectors in performance between 6 and 12 μm. Sometimes these detectors also contain zinc and are designated HgCdZnTe. Gold-doped germanium is still sufficiently responsive at 10.6 μm to be used as a CO_2 laser detector. Ge:Au detectors are available from EDO Barnes and EG&G Optoelectronics (Judson). HgCdTe detectors are available from EG&G Optoelectronics, Fermionics, Hamamatsu, Infrared Associates, Kolmar Technologies, and Oriel. For further details on the above detectors, the reader should consult references 50, 51, and 141 at the end of this chapter. Suppliers of other detectors are listed in reference 27.

The use of extrinsic photoconductivity for the detection of far-infrared radiation requires the introduction of appropriate doping material into a semiconductor in order to generate an acceptor or donor impurity level extremely close to the valence or conduction bands, respectively. Well-characterized impurity levels can be generated in germanium in this way using gallium,[142] indium,[143] boron,[144] or beryllium doping. The long-wavelength sensitivity limit is restricted to about 120 μm but can be extended to 200 μm in Ge:Ga by stressing the crystal. Longer wavelength-sensitive, extrinsic photoconductivity can be observed in appropriately doped InSb in a magnetic

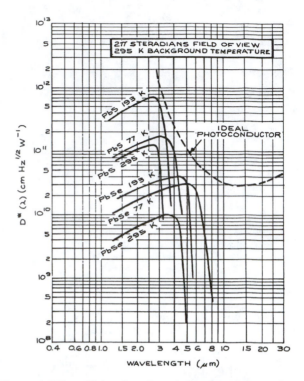

Figure 4.176 D^* is a function of wavelength for various lead-salt photoconductive detectors. (Courtesy of Hughes Aircraft Company.)

field. Bulk InSb can be used more efficiently for infrared detection in a different photoconductive mode entirely.[145] Even at the low temperatures, at which far-infrared photoconductive detectors operate ($\leq$ 4 K), there are carriers in the conduction band. These free electrons can absorb far-infrared radiation efficiently and move into higher-energy states within the conduction band. This change in energy results in a change of mobility of these free electrons, which can be detected as a change in conductivity using a circuit such as the one shown previously in Figure 4.174. Some typical wavelength response curves are shown in Figure 4.177. These *hot-carrier-effect* photoconductive detectors or hot-electron *bolometers* can be used successfully over a wavelength range extending from 50 to 10^4 μm; they have detectivities up to 2×10^{12} cm $Hz^{1/2}W^{-1}$ and response times down to 10 ns or less. They are frequently operated in a large

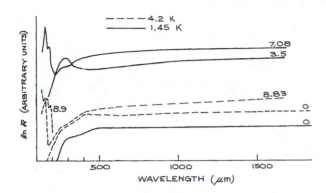

Figure 4.177   Responsivity as a function of wavelength of a hot carrier InSb photoconductive detector operated at two different temperature and various magnetic field strengths. The number on each curve is the magnetic field strength in untis of 10^5 A/m.

magnetic field (several hundred kA/m or more). These detectors are available from Infrared Laboratories and QMC Instruments. Magnetically tuned detectors are available from QMC Instruments. These detectors can also be tuned in wavelength response by operating them in a magnetic field.[145,146]

4.8.4 Photovoltaic Detectors (Photodiodes)

In a photovoltaic detector, photoexcitation of electron-hole pairs occurs near a junction when radiation of energy greater than the band gap is incident on the junction region. Extrinsic photoexcitation is rarely used in photovoltaic photodetectors. The internal energy barrier of the junction causes the electron and hole to separate, creating a potential difference across the junction. This effect is illustrated for a p–n junction in Figure 4.178. Other types of structure are also used, such as p–i–n, Schottky-barrier (a metal deposited onto a semiconductor surface) and heterojunction (a junction between two different semiconductors). The p-n and p-i-n structures are the most commonly used. All these devices are commonly called photodiodes. The characteristics of some important photodiodes are listed in Table 4.19. Important photodiodes include silicon for detection of radiation between 0.1 and 1.1 μm, germanium

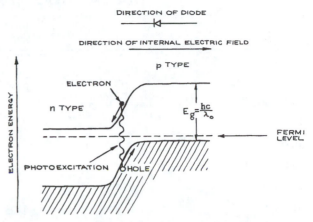

Figure 4.178   Photoexcitation at a p–n junction.

for use between 0.4 and 1.8 μm and indium (gallium) arsenide between 1 and 3.8 μm. Responsivity variations with wavelength for photodiodes that are most important for near infrared (communications) applications are shown in Figure 4.179. Other important photodiodes include indium antimonide between 1 and 7 μm, lead-tin telluride between 2 and 18 μm, and mercury-cadmium telluride between 1 and 12 μm. Some typical curves of $D^*(\lambda)$ are shown in Figure 4.180. These spectral response regions are not all necessarily covered by a detector operating at the same temperature; for example, InSb responds to 7 μm at 300 K but to wavelengths no longer than 5.6 μm at 77 K. The wavelength response of PbSnTe and HgCdTe depends also on the stoichiometric composition of the crystal. All these photodiodes have very high quantum efficiency, defined in this case as the ratio of photons absorbed to mobile electron-hole pairs produced in the junction region. Values in excess of 90 percent have been observed in the case of silicon.

When a photodiode detector is illuminated with radiation of energy greater than the band gap, it will generate a voltage and can be operated in the very simple circuit illustrated in Figure 4.181a. It is much better to generate a photodiode detector in a reverse-biased mode, however, as shown in Figure 4.181b, where positive voltage is applied to the n-type side of the junction and negative to the p-type. In this case, the observed photosignal is seen as a change in current through the load resistor. The difference between the two modes of

TABLE 4.19 PHOTOVOLTAIC DETECTORS (PHOTODIODES)

Semiconductor	$T (K)$	Wavelength Range (μm)	$D^*(max)$	τ	Suppliers[a]
Si	300	0.2–1.1	$\leq 2 \times 10^{13}$	—	—[b]
Ge	300	0.4–1.8	10^{11}	0.3 ns	(1)
InAs	300	1–3.8	$< 4 \times 10^9$	5 ns–1μs	(2)
InAs	77	1–3.2	4×10^{11}	0.7μs	(2)
InSb	300	1–7	1.5×10^8	0.1 μs	(3)
InSb	77	1–5.6	$< 2 \times 10^{11}$	>25 ns	(4)
PbSnTe	77	2–18[c]	$< 10^{11}$	20 ns–1μs	(5)
HgCdTe	77	1–25[c]	10^9–10^{11}	>1.6 ns	(6)

[a] (1) Available from Perkin-Elmer Optoelectronics (formerly EG&G Optoelectronics and Judson), Optoelectronics, and Oriel; (2) Electro-Optical Systems, Hamamatsu, Perkin-Elmer Optoelectronics (formerly EG&G Optoelectronics and Judson); (3) Electro-Optical Systems, Hamamatsu, and Kolmar Technologies; (4) Electro-Optical Systems, Kolmar Technologies, Perkin-Elmer Optoelectronics; (5) Kolmar Technologies; (6) Boston Electronics, Fermionics, Kolmar Technologies, Hamamatsu, II-VI, Infrared Associates, Kolmar Technologies, Oriel, and Perkin-Elmer Optoelectronics

[b] A very wide range of silicon photoconductive and photovoltaic detectors are available with NEP figures down to 10^{-16} WHz$^{-1/2}$ and time constants down to 6 ps. References 15 and PBG contains a list of suppliers.

[c] Range varies from one supplier to another.

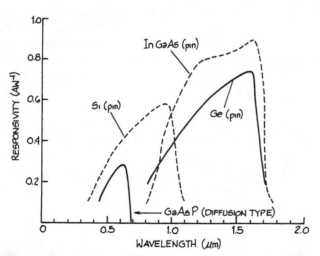

Figure 4.179   Responsivity of important near-infrared photodiodes.

operation can be easily seen from Figure 4.182, which shows the I-V characteristic of a photodiode in the dark and in the presence of increasing levels of illumination. At a given level of illumination, the photodiode can generate either an open-circuit voltage V_{oc} or a short-circuit current I_{Sc}. A photodiode responds much more linearly to changes in light intensity and has greater detectivity when operated in the reverse-biased mode. Ideal operation is obtained when the diode is operated in the current mode with an operational amplifier that effectively holds the photodiode voltage at zero—its optimum bias point. Two simple practical circuits which can be used to operate a photodiode in this way are shown in Figures 4.183 and 4.184. In Figure 4.184, the bias voltage V_B is not necessary, but for many photodiodes will improve the speed of response, albeit at the expense of an increase in noise. Integrated packages incorporating a photodiode and operational amplifier are available from EG&G, UDT, RCA, Centronic, and Silicon Detector Corporation. The p–i–n structure is most commonly used in these devices because its performance, in terms of quantum efficiency (number of useful carriers generated per photon absorbed) and frequency response, can be readily optimized. These devices have very low noise and fast response. In practice, the limiting sensitivity that can be obtained with them will be determined by the noise of the associated amplifier circuitry.

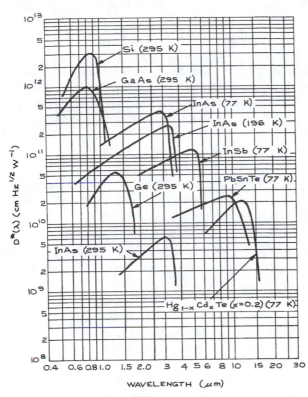

Figure 4.180 D^* as a function of wavelength for various photovoltaic detectors. (Courtesy of Hughes Aircraft Company.)

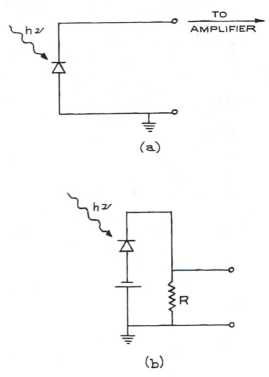

Figure 4.181   Photovoltaic detector operated in (a) open-circuit mode; (b) reverse-biased mode.

If the reverse bias voltage on a photodiode is increased, photoinduced charge carriers can acquire sufficient energy transversing the junction region to produce additional electron-hole pairs. Such a photodiode exhibits current gain and is called an *avalanche photodiode* (APD). It is in some respects the solid-state analog of the photomultiplier. Avalanche photodiodes are noisier than *p–i–n* photodiodes, but because they have internal gain, the practical sensitivity that can be achieved with them is greater. For detection of weak light signals, particularly short pulses, the signal-to-noise ratio will not be dominated by the associated electronics. A good discussion of the design, fabrication, and operating characteristics of various photodiode detectors has been given by Gowar.[92] If such an APD with a very low dark current is operated in the *Geiger* mode, where the bias

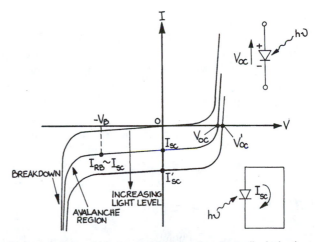

Figure 4.182   I-V characteristics of a photodiode in the dark and with increasing levels of illumination.

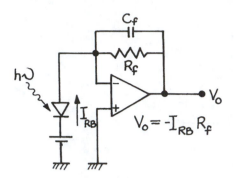

Figure 4.183 Simple op-amp circuit for reverse-bias operation of a photodiode. If the dc bias is not included, then the detector is held at zero voltage and is operated in a *current mode*.

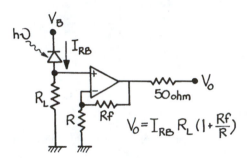

Figure 4.184 Op-amp circuit for reverse-bias operation of a photodiode.

voltage is set greater than the diode breakdown voltage, a single photon can trigger a substantial circuit pulse. To prevent the diode from permanent damage it must be operated with a *quenching* current, which reduces the bias voltage to stop the avalanche process. APDs operated in this way are true *single photon* detectors, and can replace PMTs in many photon-counting applications. Complete Geiger-mode APD modules are available from Perkin Elmer Optoelectronics and EG&G Optoelectronics.

4.8.5 Detector Arrays

Arrays of individual photodetector elements, based on photodiode, photoconductive, pyroelectric, CMOS, and CCD elements, are widely available for use in both the visible and the infrared. Such arrays are useful in a variety of applications, such as spectral analysis, image analysis, and position sensing. Three specific types are worthy of mention. Multiple element linear arrays, often called *reticons*, are widely used in spectrographs. Light from the diffraction grating in such an instrument falls on a such an array instead of on an exit slit. Arrays based on silicon. InGaAs, HgCdTe, PbS, PbSe are available from various manufacturers including Cal-Sensors, Kodak, EG&G Optoelectronics, Fermionics, Hamamatsu, Kolmar, Technologies, Photonic Detectors, Sens Array, Textron-Systems, and UDT Sensors. The number of pixels in a linear detector can be as large as 10,2000, although 1024 or 2048 element arrays are most common. Linear arrays can also be used for position sensing.

Quadrant detectors are particularly useful in laser-beam tracking and similar operations. These detectors have four contiguous active elements with very little inactive area between them, as shown in Figure 4.185. If a light beam is centered on the detector, all four elements provide the same signal. By the use of a circuit such as the one shown in Figure 4.185, simultaneous display of the X and Y displacements of the beam from the center of the array can be obtained. Quadrant detectors are available from Advanced Photonix, EG&G Optoelectronics, Photonic Detector, Silicon Sensors, and UDT Sensors. Two-dimensional arrays of photodetectors generally show up in imaging systems. CCD or CMOS arrays based on silicon are common in visible and near-infrared digital camera systems InGaAs arrays provide performance between 0.9 μm and 2.2 μm—a comprehensive list of suppliers of array detectors in different spectra regions is available.[27]

4.8.6 Thermal Detectors

Thermal detectors, in principle, have a detectivity that is independent of wavelength from the vacuum ultraviolet upward, as shown in Figure 4.186. The absorbing properties of the *black* surface of the detector will, however, generally show some wavelength dependence, and the necessity for a protective window on some detector elements may limit the useful spectral bandwidth of the device. Most commonly available thermal detectors, although by no means as sensitive as various types of photon detectors, achieve spectral response very far into

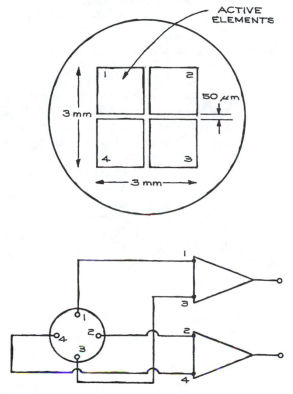

Figure 4.185 Schematic diagram showing the construction of a quadrant detector with a circuit suitable for providing X- and Y-beam deflection readouts in a beam-centering or beam-tracking application.

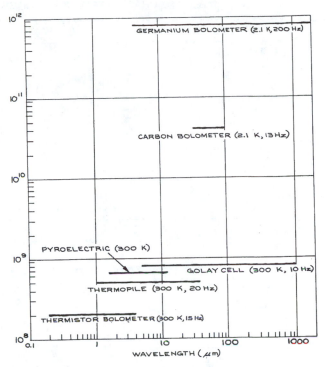

Figure 4.186 Typical $D^*(\lambda)$ curves for various thermal detectors assuming total adsorption of incident radiation. The operating temperature, modulation frequency, and typical useful wavelength range are shown for each detector. In each case, the detector is assumed to a view a hemispherical surround at a temperature of 300 K.

the infrared—to the microwave region, in fact—while conveniently operating at room temperature. The operating characteristics of some commercially available thermal detectors are given in Table 4.20. Each of these detectors is discussed briefly below. Putley gives a more detailed discussion.[47]

Thermopile

Thermopiles, although they are one of the earliest forms of infrared detector, are still widely used. Their operation is based on the Seebeck effect, where heating the junction between two dissimilar conductors generates a potential difference across the junction. An ideal device should have a large Seebeck coefficient, low resistance (to minimize ohmic heating), and a low thermal conductivity (to minimize heat loss between the hot and cold junctions of the thermopile). These devices are usually operated with an equal number of hot (irradiated) and cold (dark) junctions, the latter serving as a reference to compensate for drifts in ambient temperature. Both metal (copper-constantan, bismuth-silver, antimony-bismuth) and semiconductor junctions are used as the active elements. The junctions can take the form of evaporated films, which improves the robustness of the devices and reduces their time constant, although this is still slow (0.1 ms at best). Because a thermopile has very low output impedance, it must be used with a specially designed low-noise amplifier, or with a step-up transformer as shown in Figure 4.187. Such transformers can be conveniently built in the

TABLE 4.20 CHARACTERISTICS OF COMMERCIALLY AVAILABLE THERMAL DETECTORS

Device	$D^*(cm\,Hz^{1/2}W^{1})$	Time Constant	Suppliers
Thermopile	$(1–4)\times10^{8}$	20 μs–60 ms	EDO/Barnes, Eppley, International Light, Oriel, Molectron, Newport, Gentec, Scientech
Pyroelectric	$10^{6}–10^{9}$	> 100 ps	Molectron, EDO/Barnes, Oriel, Laser Probe, Sens Array, Spiricon, Eltec
Bolometer	$2.5–10^{8}$	1 ms	EDO/Barnes, Cambridge Research and Instrumentation (CRI), Infrared Laboratories, QMC Instruments, Sens Array
Golay Cell	NEP $(W\,Hz^{1/2}) < 10^{-10}$	$\cong$10 ms	QMC Instruments

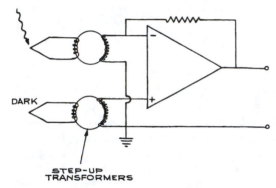

Figure 4.187 Operating circuit for a thermopile using a dark junction (or junctions) and a differential amplifier for compensation.

laboratory by winding the primary and secondary coils on a small ferrite torus.

Although thermopiles, along with other thermal detectors, do not have absolutely flat spectral response over unlimited wavelength regions, they can be remarkably flat within restricted regions—in the visible or from 1 to 10 μm, for example. In addition, such detectors with calibrations traceable to the national Bureau of Standards are available. They are therefore invaluable in the absolute calibration (both for radiant sensitivity and for spectral response) of light sources, detectors, and spectrometers. Thermopiles are available from EDO

Barnes, The Eppley Laboratory, Molectron, Scientech, and Sensor Physics.

Pyroelectric Detectors

These are detectors that utilize the change in surface charge that results when certain asymmetric crystals (ones that can possess an internal electric dipole moment) are heated. The crystalline material is fabricated as the dielectric in a small capacitor, and the change in charge is measured when the element is irradiated. These devices, therefore, are inherently ac detectors. If the chopping frequency of the input radiation is slow compared to the thermal relaxation time of the crystal, the crystal remains close to thermal equilibrium and the current response is small. When the chopping period becomes shorter than the thermal relaxation time, much greater heating and current response results. The responsivity of the detector in this case can be written as

$$R = \frac{p(T)}{\rho C_{p}d} \quad (AW^{-1}) \qquad (4.317)$$

where $p(T)$ is the pyroelectric coefficient at temperature T, d is the spacing of the capacitor electrodes, and ρ and C_{p} are the density and specific heat of the crystal, respectively.

The equivalent circuit of a pyroelectric detector is a current source in parallel with a capacitance, which can range from a few to several hundred picofarads. For optimum performance, the resultant high impedance must be matched to a high-input-impedance, low-output-impedance amplifier. Two examples of such circuits are given in Figure 4.188. Pyroelectric detectors are available from EDO Barnes, Eltec, Sentec, Molectron, Newport, Oriel, SensArray, and Thermometrics. These packages frequently contain two pyroelectric elements connected with reverse polarity so that the detector only responds to the difference in illumination between the two elements. This is a useful way to reduce the effect of background radiation. For faster response, the capacitance of the detector must be shunted with a small resistor (though this reduces the detectivity). Response times as short as 2 μs have been seen in detectors operated in this way without amplification when illuminated with high-intensity mode-locked laser pulses.

Pyroelectric detectors are robust and have frequency responses extending from a few hertz to 100 GHz or so. Their detectivities are comparable with those of thermopiles, and they also have flat spectral response. They can consequently replace the thermopile in many applications as a convenient, room temperature, wide-spectral-sensitivity detector of infrared and visible light. Pyroelectric detector arrays are available from EDO Barnes and SensArray. Pyroelectric detectors are widely used for the detection of very short-duration infrared laser pulses. They are much more sensitive than photon-drag detectors for this purpose.[148] High-energy pulses, however, generate acoustic waves in the detector crystal, which give rise to spurious signals. These acoustic signals are more of a problem in the observation of long laser pulses (> 100 ns) than they are with short pulses.

Bolometer

The resistance of a solid changes with temperature according to a relation of the form

$$R(T) = R_0[1 + \gamma(T - T_0)] \tag{4.318}$$

where γ is the temperature coefficient of resistance, typically about 0.05 K^{-1} for a metal, and R_0 is the resistance at temperature T_0.

A *bolometer* is constructed from a material with a large temperature coefficient of resistance. Absorbed radiation heats the bolometer element and changes its resistance. Bolometers utilize metal, semiconductor, or almost superconducting elements. Metal bolometers utilize fine wires (platinum or nickel) or metal films. The mass of the element must be kept small in order to maximize its temperature rise. Even so, the response time is fairly long ($\geq$ 1ms). Semiconductor bolometer elements (thermistors) have larger absolute values of γ and have largely replaced metals except where very long-term stability is required.

Bolometer elements can be operated in several ways—a simple bias circuit for a single element is shown in Figure 4.189a. It is usual, however, to operate the elements in pairs in a bridge circuit, as shown in Figure 4.189b. One

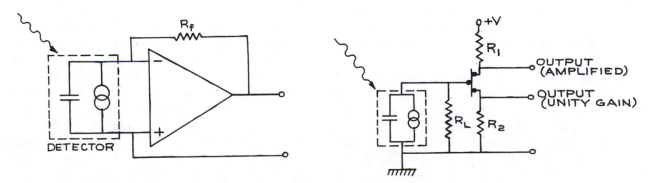

Figure 4.188   Two examples of operating circuits for pyroelectric detectors.

element is irradiated, while the second serves as a reference and compensates for changes in ambient temperature. Thermistors have a negative I-V characteristic above a certain current and will exhibit destructive thermal runaway unless operated with a bias resistor. It is, therefore, usually best to operate the thermistor at currents below the negative-resistance part of its I-V characteristic. Bolometers can operate to wavelengths up to 1000 μm, ususally limited by the transmission of their entrance window. These devices are available from EDO Barnes, Infrared Laboratories, SensArray, and Thermometrics.

The Golay Cell

In a Golay cell (named for its Inventor, M.J.E. Golay),[149] radiation is absorbed by a metal film that forms one side of a small sealed chamber containing xenon (used because of its low thermal conductivity). Another wall of the chamber is a flexible membrane, which moves as the xenon is heated. The motion of the membrane is used to change the amount of light reflected to a photodetector. The operating principle and essential design features of a modern Golay cell are shown in Figure 4.190. Although these detectors are fragile, they are quite sensitive and are still widely used for far-infrared spectroscopy. Golay cells are available from QMC Instruments.

4.8.7 Detector Calibration

In certain cases, the data supplied by a detector manufacturer on parameters such as $D^*(\lambda)$ or τ will only indicate a range of values within which the detector's characteristics will fall, or the data may not be sufficiently reliable. If necessary, more accurate detector calibration can be carried out.

To determine $D^*(\lambda)$ for a detector, its response must be measured with a source giving a known irradiance $E_e(\lambda)$ at the detector. This can be done either with a calibrated source such as a blackbody, or by comparison with another detector such as a thermopile, which has a calibrated response. To determine the temporal response function of a detector operated in a given configuration, all that is necessary is to irradiate the detector with a very short light

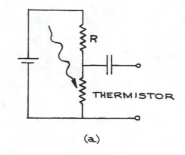

(a)

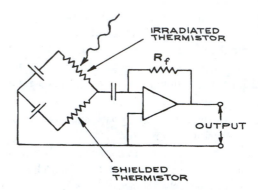

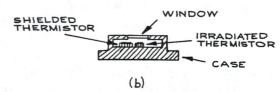

(b)

Figure 4.189   Operating circuits for thermistor bolometers: (a) simple bias circuit; (b) bridge circuit using compensating shielded thermistor, with device construction shown.

pulse and record the output pulse shape on a fast oscilloscope. For response functions below about 1 ns, a sampling oscilloscope should be used in conjunction with repetitive short-light-pulse irradiation of the detector. For detectors with response faster than about 1 ns, sufficiently short light pulses can be obtained from a mode-locked Nd:YAG, Nd:glass, ruby, or dye laser. Pulses from mode-locked argon or CO_2 lasers may not be short enough to calibrate a very fast detector. To obtain short light pulses outside conveniently available wavelength regions, nonlinear harmonic generation or mixing schemes can be

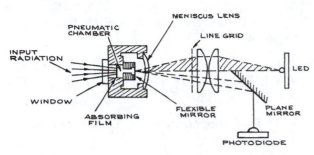

Figure 4.190   Schematic design of the Golay detector. The top half of the line grid is illuminated by the LED and imaged back on the lower half of the grid by the flexible mirror and meniscus lens. Any radiation-induced deformation of the flexible mirror moves the image of the line grid and changes the illunination reaching the photodiode.

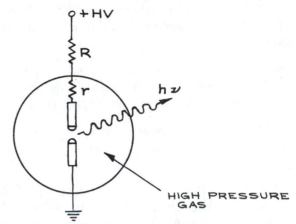

Figure 4.191   Simple design for high-pressure, nanosecond-duration pulsed light source. Generally $R \gg r$, where r ranges from 1 kΩ to 1 MΩ.

used, but the difficulties of such techniques can be severe. To determine the time response of a detector that responds slower than 1 to 2 ns, any pulsed light source of much shorter duration than the detector response can be used. Short-duration (~10 ns) flashlamps are available from EG&G Optoelectronics and Xenon Corporation.

Very short-duration, low-energy, pulsed light sources are easily made in the laboratory. A very simple design is shown in Figure 4.191. A high-voltage discharge between two electrodes spaced by .a few millimeters or less in a high-pressure gas is used. Hydrogen works very well, and even air at atmospheric pressure is satisfactory. To obtain a short-duration flash, only the self-capacitance C_s of the electrodes must be allowed to discharge. To accomplish this, a small charging resistor must be placed very close to the high-voltage electrode—a larger series charging resistor can be placed farther from the electrode. The tips from ballpoint pens make excellent electrodes in this application. The flash duration is proportional to L/p, where L is the electrode spacing and p the gas pressure. The breakdown voltage is proportional to L and also approximately to p—the capacitance of the electrode gap is proportional to L, so the flash energy $\frac{1}{2}C_s V^2$ is proportional to $L^3 p^2$. For short-duration, high-energy flashes, therefore, p should be high. The repetition frequency of the lamp, which is most easily operated in a free-running mode, will depend on the charging time constant.

CITED REFERENCES

1. *Proceeding of the Symposium on Quasi-Optics,* New York, June 8–10, 1964, J. Fox, ed., Polytechnic Press of the Polytechnic Institute of Brooklyn, N.Y., 1964.

2. *Proceeding of the Symposium on Submillimeter Waves*, New York, March 31–April 2, 1970, J. Fox, ed., Polytechnic Press of the Polytechnic Institute of Brooklyn, N.Y., 1964.

3. K. M. Baird, D. S. Smith, and B. G. Whitford, "Confirmation of the Currently Accepted Value 299792458 Metres per Second for the Speed of Light," Opt. Commun., **31**, 367–368 (1979).

4. E. R. Cohen and B. N. Taylor, Committee on Data for Science and Technology, CODATA Bulletin, No. 63, 1986.

5. G. W. C Kaye and T. H. Laby, *Tables of Physical and Chemical Constants*, 15th ed. Longman, London, 1986.

6. C. D. Coleman, W. R. Bozman, and W. F. Meggers, Table of Wavenumbers, U.S. Nat'l. Bur. Std. Monograph 3, Vols. 1–2, 1960.

7. M. W. Urban, *Attenuated Total Reflectance Spectroscopy of Polymers: Theory and Practice*, American Chemical Society, Washington, D.C., 1996.

8. R. Wiesendanger, *Scanning Probe Microscopy and Spectroscopy*, Cambridge University Press, Cambridge, 1994.

9. M. Born and E. Wolf, *Principles of Optics*, 7th ed. Cambridge University Press, Cambridge, 1999.

10. E. E. Wahlstrom, *Optical Crystallography*, 5th ed. Wiley, New York, 1979.

11. W. A. Shurcliff, *Polarized Light: Production and Use*, Harvard University Press, Cambridge, Mass., 1962.

12. A. Yariv, *Optical Electronics in Modern Communications*, 5th ed. Oxford University Press, New York, 1997.

13. C. C. Davis, *Lasers and Electro-Optics*, Cambridge University Press, Cambridge, 1996.

14. H. Kogelnik and T. Li, "Laser Beams and Resonators," *Proc. IEEE*, **54**, 1312–1329, 1966.

15. R. Kingslake, *Optical System Design*, Academic Press, Orlando, 1983.

16. Code V. Available from Optical Research Associates, 3280 East Foothill Boulevard, Pasadena, CA 91107. Tel: (626) 795–9101, Fax: (626) 795–9102

17. Zemax. Available from Focus Software, Inc., P. O. Box 18228, Tucson, Arizona 85731. Tel: (520) 733–0130, Fax: (520) 733–0135

18. Oslo. Developed by Sinclair Optics, Inc., 6780 Palmyra Rd., Fairport NY 14450, Tel: (716) 425–4380 Fax: (716) 425–4382. Distributed by Lambda Research Corporation, 80 Taylor St., P.O. Box 1400, Littleton, MA 01460–4400, Tel: (978) 486–0766, Fax: (978) 486–0765.

19. Solstis. Available from Optis, BP 275, 83078 Toulon Cedex 9, France, Tel: 33-4-94-08-66-99, Fax: 33-4-94-08-66-94.

20. Optikwerk. Available from Optikwerk, Inc., P.O. Box 92607, Rochester, NY 14692, Tel: (716) 321–1821, Fax: (716) 321-1809.

21. R. W. Ditchburn, *Light*, 3rd ed. Academic Press, New York, 1976.

22. L. Levi, *Applied Optics*, Vol. 1, Wiley, New York, 1968.

23. W. J. Smith, *Modern Optical Engineering*, 2nd ed. McGraw-Hill, New York, 1990.

24. R. R. Shannon, *The Art and Science of Optical Design*, Cambridge University Press, Cambridge, 1997.

25. S. Ramo, J. R. Whinnery, and T. Van Duzer, *Fields and Waves in Communication Electronics*, 3rd ed. Wiley, New York, 1994.

26. Diffraction-limited spherical lenses are available from Melles Griot, Optics for Research, Oriel, Special Optics, and J. L. Wood Optical Systems, among others (see following reference).

27. Laser Focus Buyers Guide, published annually by Pennwell Publishing Co., 1421 South Sheridan, Tulsa, OK 74112 01460, lists a large number of suppliers of a wide range of optical components and systems as does *The Photonics Buyers' Guide*, published annually by Photonics Spectra, Laurin Publishing Co., Berkshire Common, P.O. Box 4949, Pittsfield, MA 01202-4949. Additional valuable listings of this sort are: *Lasers and Optronics Buying Guide*, published annually by Cahners, 301 Gibraltar Drive, Box 650, Morris Plains, NJ 07950-0650; and *Lightwave*, 98 Spit Brook Rd., Ste. 100, Nashua, NH 03062-5737.

28. The "float" method for manufacturing plate glass involves the drawing of the molten glass from a furnace, where the molten glass floats on the surface of liquid tin. The naturally flat surface of the liquid metal ensures the production of much larger sheets of better-flatness glass than was possible by earlier techniques.

29. A. B. Meinel, "Astronomical Telescopes," in *Applied Optics and Optical Engineering*, Vol. 5, R. Kingslake, ed., Academic Press, New York, 1969.

30. W. Brouwer and A. Walther, "Design of Optical Instruments," in *Advanced Optical Techniques*, A. C. S. Van Heel, ed., North-Bouand, Amsterdam, 1967.

31. Celestron Telescopes are available from Celestron International, 2835 Columbia St., P.O. Box 3578, Torrance, CA 90503; (310) 328-9560.

32. Meade Telescopes are available from Meade Instruments Corporation, 6001 Oak Canyon, Irvine, CA 92618-5200; (949) 451-1450

33. W. T. Wellford and R. Winston, *High Collection Nonimaging Optics*, Academic Press, San Diego, 1989.

34. A. Girard and P. Jacquinot, "Principles of Instrumental Methods in Spectroscopy," in *Advanced Optical Techniques*, A. C. S. Van Heel, ed., North-Bouand, Amsterdam, 1967.

35. J. D. Strong, *Procedures in Experimental Physics*, Prentice-Hall, Englewood Cliffs, NJ, 1938.

36. *Handbook of Chemistry and Physics*, 81st ed. David R. Lide, ed., CRC Press, Boca Raton, Fla., 2000.

37. *Lightwave*, 98 Spit Brook Rd., Ste. 100, Nashua, NH 03062-5737.

38. U. Hochuli and P. Haldemann, "Indium sealing techniques," Rev. Sci. Instr. **43**, 1088–1089 (1972).

39. Epoxy-removing solvents are available from Electron Microscopy Services, Oakite, or Hydroclean.

40. T. C. Poulter, "A Glass Window Mounting for Withstanding Pressure of 30,000 Atmospheres," *Phys. Rev.* **35**, 297 (1930).

41. W. Paul, W. W. Meis, and J. M. Besson, "Windows for Optical Measurements at High Pressures and Long Infrared Wavelengths," *Rev. Sci. Instr.*, **39**, 928–930 (1968).

42. *High Pressure Technology*, I. L. Spain and J. Paawe, eds., Dekker, New York, 1967.

43. *Compilation of ASTM Standard Definitions*, 3rd ed. American Society for Testing and Materials, Philadelphia, 1976.

44. Glass, quartz, and sapphire vacuum window assemblies are available from Adolf Meller, Ceramaseal, Varian, and Vacuum Generators.

45. *Spectroscopic Techniques for Far Infa-Red, Submillimetre and Millimetre Waves*, D. H. Martin, ed., North Holland, Amsterdam, 1967.

46. K. D. Moller and W. G. Rothschild, *Far-Infrared Spectroscopy*, Wiley-Interscience, 1971.

47. A. Hadni, *Essentials of Modern Physics Applied to the Study of the Infrared*, Pergamon Press, Oxford, 1967.

48. G. A. Fry, "The Eye and Vision," in *Applied Optics and Optical Engineering*, Vol. 2, R. Kingslake, ed., Academic Press, 1965.

49. C. C. Davis and R. A. McFarlane, "Lineshape Effects in Atomic Absorption Spectroscopy," *J. Quant. Spect. Rad. Trans.*, **18**, 151–170 (1977).

50. P. W. Kruse, L. D. McGlauchlin, and R. B. McQuistan, *Elements of Infrared Technology: Generation, Tranmission and Detection*, Wiley, New York, 1969.

51. R. D. Hudson, Jr., *Infrared System Engineering*, Wiley-Interscience, New York, 1969.

52. J. A. R. Samson, *Techniques of Vacuum Ultraviolet Spectroscopy*, Wiley, New York, 1967.

53. Available from EG&G Optoelectronics, ILC, Verre & Quartz, and Xenon Corporation.

54. Available from EG&G Optoelectronics, Verre & Quartz, Resonance, and Xenon Corporation.

55. J. P. Markiewiez and J. L. Emmett, "Design of Flashlamp Driving Circuits," *IEEE J. Quant. Electron.* **QE-2**, 707–711 (1966).

56. J. G. Edwards, "Some Factors Affecting the Pumping Efficiency of Optically Pumped Lasers," *Appl. Opt.*, **6**, 837–843 (1967).

57. H. J. Baker and T. A. King, "Optimization of Pulsed UV Radiation from Linear Flashtubes," *J. Phys. E.: Sci. Instr.,* **8**, 219–223 (1975).

58. W. T. Silfvast, *Laser Fundamentals*, Cambridge University Press, 1996.

59. B. E. A. Saleh and M. C. Teich, *Fundamentals of Photonics*, Wiley, New York, 1991.

60. A. E. Siegman, *Lasers*, University Science Books, Mill Valley, Calif., 1986.

61. *Quantum Electronics*, H. Rabin and C. L. Tang, eds., Vols. 1A and 2A, Nonlinear Optics, Academic Press, New York, 1975.

62. *Nonlinear Optics*, Proceedings of the Sixteenth Scottish Universities Summer School in Physics, 1975, P. G. Harper and B. S. Wherrett, eds., Academic Press, London, 1977.

63. N. Bloembergen, *Nonlinear Optics*, Benjamin, New York, 1965.

64. F. Zernike and J. E. Midwinter, *Applied Nonlinear Optics*, Wiley, New York, 1973.

65. Robert W. Boyd, *Nonlinear Optics*, Academic Press, Boston, 1992.

66. *Lasers, Light Amplifiers and Oscillators*, D. Ross, Academic Press, New York, 1969.

67. W. Koechner, *Solid State Laser Engineering*, 5th revised and updated ed., Springer-Verlag, Berlin, 1999.

68. *Dye Lasers*, F. P. Schafer, ed., Topics in Applied Phys., Vol.1, 3rd revised and enlarged ed., Springer-Verlag, Berlin, 1990.

69. H. W. Furumoto and H. L. Ceccon, "Optical Pumps for Organic Dye Lasers," *Appl. Opt.*, **8**, 1613–1623 (1969).

70. J. F. Holzrichter and A. L Schawlow, "Design and Analysis of Flashlamp Systems for Pumping Organic Dye Lasers," *Ann. N.Y. Acad. Sci.* **168**, 703–714 (1970).

71. T. B. Lucatorto, T. J. McIlrath, S. Mayo, and H. W. Furumoto, "High-Stability Coaxial Flashlamp-Pumped Dye Laser," *Appl. Opt.*, **19**, 3178–3180 (1980).

72. C. C. Davis and T. A. King, "Gaseous Ion Lasers," in *Advances in Quantum Electronics*, Vol. 3, D. W. Goodwin, ed., Academic Press, London, 1975, pp. 169–454.

73. W. B. Bridges, "Ion Lasers," in *Handbook of Laser Science and Technology*, Vol. 1; Lasers in All Media, M. J. Weber, ed., CRC Press, Boca Raton, Fla., 1982.

74. C. C. Davis, "Neutral Gas Lasers," in *Handbook of Lasers Science and Technology*, Vol. 1: Lasers in All Media, M. J. Weber, ed., CRC Press, Boca Raton, Fla., 1982.

75. D. C. Tyte, "Carbon Dioxide Lasers," in *Advances in Quantum Electronics*, Vol. 1, D. W. Goodwin, ed., Academic Press, London, 1970.

76. J. J. Degnan, "The Waveguide Laser: A Review," *Appl. Phys.*, **11**, 1–33 (1976).

77. H. Seguin and J. Tulip, "Photoinitiated and Photosustained Laser," *Appl. Phys. Lett.*, **21**, 414–415 (1972).

78. J. D. Cobine, *Gaseous Conductors*, Dover, N.Y., 1958.

79. H. J. Seguin, K. Manes, and J. Tulip, "Simple Inexpensive Laboratory-Quality Rogowski TEA Laser," *Rev. Sci. Instr.*, **43**, 1134–1139 (1972).

80. A. J. Beaulieu, "Transversely Excited Atmospheric Pressure CO_2 Laser," *Appl. Phys. Lett.*, **16**, 504–505 (1970).

81. D. Basting, F. P. Schafer, and B. Steyer, "A Simple, High Power Nitrogen Laser," *Opto-electron.*, **4**, 43–49 (1972).

82. P. Schenck and H. Metcalf, "Low Cost Nitrogen Laser Design for Dye Laser Pumping," *Appl. Opt.*, **12**, 183–186 (1973).

83. C. P. Wang, "Simple Fast-Discharge Device for High-Power Pulsed Lasers," *Rev. Sci. Instr.*, **47**, 92–95 (1976).

84. A. J. Schwab and F. W. Bouinger, "Compact High-Power N2 Laser: Circuit Theory and Design," *IEEE J. Quant. Electron.*, **QE-12**, 183–188 (1976).

85. C. L. Sam, "Small-Size Discrete-Capacitor N_2 Laser," *Appl. Phys. Lett.*, **29**, 505–506 (1976).

86. M. Feldman, P. Lebow, F. Raab, and H. Metcalf, "Improvements to a Home-Built Nitrogen Laser," *Appl. Opt.*, **17**, 774–777 (1978).

87. L. F. Mollenauer, "Dyelike Lasers for the 0.9 μm Region Using F_2^+ Centers in Alkali Halides," *Opt. Lett.*, **1**, 164 (1977). See also *Opt. Lett.* **3**, 48–50 (1978); **4**, 247–299 (1979); **5**, 188–190 (1980).

88. R. C. Greenhow and A. J. Schmidt, "Picosecond Light Pulses," in *Advances in Quantum Electronics*, Vol. 2, D. W. Goodwin, ed., Academic Press, London, 1973.

89. S. L. Shapiro, ed., *Ultrashort Light Pulses, Picosecond Techniques and Applications*, Topics in Applied Physics, Vol. 18, Springer, Berlin, 1977.

90. *Picosecond Optoelectronic Devices*, C. H. Lee, ed., Academic Press, Orlando, Fla., 1984.

91. P. Bhattacharya, *Semiconductor Optoelectronic Devices*, 2nd ed., Prentice-Hall, Upper Saddle River, N.J., 1997.

92. J. Gowar, *Optical Communication Systems*, 2nd ed., Prentice Hall, Englewood Cliffs, NJ, 1993.

93. D. Wood, *Optoelectronic Semiconductor Devices*, Prentice-Hall, New York, 1994.

94. G. P. Agrawal and N. K. Dutta, *Long-Wavelength Semiconductor Lasers*, Van Nostrand Reinhold, New York, 1986.

95. G. H. B. Thompson, *Physics of Semiconductor Laser Devices*, Wiley, Chichester, 1980.

96. S. L. Chuang, *Physics of Optoelectronic Devices*, Wiley, New York, 1995.

97. J. Singh, *Semiconductor Optoelectronics*, McGraw-Hill, New York, 1995.

98. *Semiconductor Devices for Optical Communication*, Topics in Applied Physics, Vol. 39, H. Kressel, ed., 2nd updated ed., Springer-Verlag, Berlin, 1982.

99. *Semiconductor Lasers*, E. Kapon, ed., Vols. 1 and 2, Academic Press, San Diego, 1999.

100. E. D. Hinkley, K. W. Nill, and F. A. Blum. "Infrared Spectroscopy with Tunable Lasers," in *Topics in Applied Physics*, Vol. 2, H. Walther, ed., Springer, Berlin, 1976, pp. 125–196.

101. T. J. Lane and R. L. Byer, "Monolithic, Unidirectional Single-Mode Nd: YAG Ring Laser," *Optics Lett.*, **10**, 65–67 (1985).

102. T. W. Hansch, "Repetitively Pulsed Tunable Dye Laser for High Resolution Spectroscopy," *Appl. Opt.*, **11**, 895–898 (1972).

103. R. Wallenstein and T. W. Hansch, "Linear Pressure Tuning of a Multielement Dye Laser Spectrometer," *Appl. Opt.*, **13**, 1625–1628 (1974).

104. R. Wallenstein and T. W. Hansch, "Powerful Dye Laser Oscillator-Amplifier System for High-Resolution Spectroscopy," *Opt. Commun.*, **14**, 353–357 (1975).

105. G. L. Eesley and M. D. Levenson, "Dye-Laser Cavity Employing a Reflective Beam Expander," *IEEE J. Quant. Electron*, **QE-12**, 440–442 (1976).

106. M. G. Littman and H. J. Metcalf, "Spectrally Narrow Pulsed Dye Laser without Beam Expander," *Appl. Opt.*, **17**, 2224–2227 (1978).

107. M. Littman and J. Montgomery, "Grazing-Incidence Designs Improve Pulsed Dye Lasers," *Laser Focus/ Electro-optics*, **24**, 70–86 (1988).

108. I. P. Kaminow, *An Introduction to Electro-optic Devices*, Academic Press, New York, 1974.

109. A. Korpel, *Acousto-Optics*, Marcel Dekker, New York, 1988, seee also "Acousto-optics: A Review of Fundamentals," *Proc. IEEE*. **69**, 48–53 (1981).

110. I. C. Chang, "Acousto-optic Devices and Applications," *IEEE Trans. Sonics and Ultrasonics*, **SU-23**, 2–22 (1976).

111. American National Standard for Safe Use of Lasers, ANSI Z136-2-2000, Published by the Laser Institute of America, Ste. 128, 13501 Ingenuity Drive, Orlando, FL 32826.

112. D. Sliney and M. Wolbarsht, *Safety with Lasers and Other Optical Sources: A Comprehensive Handbook*, Plenum, New York, 1980.

113. M. J. Weber, ed., *CRC Handbook of Laser Science and Technology*, Vol. 1, Lasers and Masers, CRC Press, Boca Raton, Fla., 1982.

114. P. Jacquinot, "The Luminosity of Spectrometers with Prisms, Gratings, or Fabry-Perot Etalons," *J. Opt. Soc. Am.*, **44**, 761–765 (1954).

115. *Applied Optics and Optical Engineering*, Vol. 5: Optical Instruments Part II, R. Kingslake, ed., Academic Press, New York, 1969.

116. J. F. James and R. S. Sternberg, *The Design of Optical Spectrometers*, Chapman and Hall, London, 1969.

117. The Photonics Design and Applications Handbook, published annually by Laurin Publishing Co. Inc., P.O. Box 4949, Pittsfield, MA 01202-9985. (413) 499-0514.

118. W. G. Fastie, "A Small Plane Grating Monochromator," *J. Opt. Soc. Am.*, **42**, 641–647 (1952).

119. H. Ebert, "Zwei Formen von Spectrograhen," *Annalen der Physik und Chemie*, **38**, 489–493 (1889).

120. M. Seya, "A New Mounting of Concave Grating Suitable for a Spectrometer," *Sci. Light (Tokyo)*, **2**, 8–17 (1952).

121. T. Namioka, "Constitution of a Grating Spectrometer," *Sci. Light (Tokyo)*, **3**, 15–24 (1952).

122. Available from numerous suppliers of vacuum equipment—for example, Ceram Tec (Ceramaseal Division), Edwards, Ferrofluidics Corp., Perkin Elmer, Vacuum Generators (VG), Varian, and Veeco.

123. Available from Carpenter Technology, P.O. Pox 14662, Reading, PA 19612-4662; Tel: (800) 338-4592 and (610) 208-2000, FAX: (610) 208-2361.

124. Available from Burleigh.

125. Available from Heraeus-Amersil, Dynasil, Esco, and Quartz Scientific., among others.

126. Available from Corning Glass Works, Optical Products Department.

127. Piczoelectric transducers are available from American Piezo Ceramics, Burleigh, EDO, Polytec, Queensgate Instruments, and Xinetics, among others.

128. C. F. Bruce, "On Automatic Parallelism Control in a Scanning Fabry-Perot Interferometer," *Appl. Opt.*, **5**, 1447–1452 (1966).

129. Commercial Fabry-Perot Interferometers that can incorporate this feature are available from Burleigh.

130. R. Landenburg and D. Bershader, "Interferometry," in *Physical Measurements in Gas Dynamics and Combustion*, R. Ladenburg, ed., Vol. 9 of *High Speed Aerodynamics and Jet Propulsion*, Princeton University Press, Princeton, N.J., 1954.

131. P. R. Longaker and M. M. Litvak, "Perturbation of the Refractive Index of Absorbing Media by a Pulsed Laser Beam."

132. D. C. Smith, "Thermal Defocusing of CO_2 Laser Radiation in Gases," *IEEE J. Quant. Electron.*, **QE-5**, 600–607 (1969).

133. "Optical and Infrared Detectors," in *Topics in Applied Physics*, R.J. Keyes, ed., Springer, Berlin, 1977.

134. *Infrared Detectors*, R. D. Hudson, Jr., and J. W. Hudson, eds., Benchmark Papers in Optics, Vol. 2, Dowden, Hutchinson and Ross, Stroudsburg, Pa. 1975.

135. R. H. Kingston, *Detection of Optical and Infrared Radiation*, Springer Series in Optical Sciences, Vol. 10, Springer, Berlin, 1978.

136. RCA photomultiplier tubes are supplied by Burle Industries,1000 New Holland Ave., Lancaster, Pa. 17601-5688.

137. Phillips photomultiplier tubes were formerly available from Amperex, but these brand names are now manufactured by Photonis, Avenue Roger Roncier, Z.I. Beauregard, B.P. 520, 19106 BRIVE Cedex, France; Phone: +33 (0) 555 86 37 00, FAX: +33 (0) 555 86 37 73. These tubes are also available from Richardson Electronics Ltd., 40W267 Keslinger Rd., PO Box 393, LaFox, IL, 60147-0393; Phone: (630) 208-2200, FAX: (630) 208-2450.

138. Static crossed-field photomultipliers used to be available from ITT. See also V. J. Koester, "Improved timing resolution in time-correlated photon-counting with a static crossed-field photomultiplier," *Anal. Chem.* **51**, 458–459 (1979).

139. H. R. Zwicker, "Photoemissive Detectors," in *Optical and Infrared Detectors*, R. S. Keyes, ed., *Topics in Applied Physics*, Vol. 19, pp. 149–196, Springer, Berlin, 1977.

140. "Window" discriminators are available from EG&G amd LeCroy.

141. D. Long, "Photovoltaic and Photoconductive Infrared Detectors," in *Optical and Infrared Detectors*, R. J. Keyes, ed., *Topics in Applied Physics*, Vol. 19, pp. 101–147, Springer, Berlin, 1977.

142. W. J. Moore and H. Shenker, "A High-Detectivity Gallium-Doped Germanium Detector for the 40–120 Region," *Infrared Phys.*, **5**, 99–106 (1965).

143. F. J. Low, "Low Temperature Germanium Bolometer," *J. Opt. Soc. Am.*, **51**, 1300–1304 (1961). See also B. T. Draine and A. J. Sievers, "High responsivity, low-noise germanium bolometer for fard infrared," *Opt. Commun.*, **16**, 425–429 (1976).

144. H. Shenker, W. J. Moore, and E. M. Swiggard, Infrared Photoconductive Characteristics of Boron-Doped Germanium," *J. Appl. Phys.*, **35**, 2965–2970 (1964).

145. E. H. Putley, "Indium Antimonide Submillimeter Photoconductive Detectors," *Appl. Opt.*, **4**, 649–656 (1965).

146. M. A. C. S. Brown and M. F. Kimmitt, "Far-Infrared Resonant Photoconductivity in Indium Antimonide," *Infrared Phys.*, **5**, 93–97 (1965).

147. E. H. Putley, "Thermal Detectors," in *Optical and Infrared Detectors*, R. J. Keyes, ed., *Topics in Applied Physics*, Vol. 19, Springer, Berlin, 1977.

148. A. F. Gibson, M. F. Kimmitt, and A. C. Walker, "Photon Drag in Germanium," *Appl. Phys. Lett.*, **17**, 75–77 (1970). See also R. Kesselring, A. W. Kalin, and F. K. Kneubuhl, "Fast midinfrared dectectors," *Infrared Physics*, **33**, 423–436 (1992).

149. M. J. E. Golay, "A Pneumatic Infra-Red Detector," *Rev. Sci. Instrum.*, **18**, 357–362 (1947).

GENERAL REFERENCES

Comprehensive (General) Optics Texts

M. Born and E. Wolf, *Principles of Optics*, 7th ed., Cambridge University Press, Cambridge, 1999.

E. Hecht, *Optics*, 3rd ed., Addison-Wesley, Reading, Massachusetts, 1998.

R. W. Ditchburn, *Light*, 3rd ed, Academic Press, New York, 1976.

F. A. Jenkins and H. E. White, *Fundamentals of Optics*, 4th ed, McGraw-Hill, New York, 1976.

M. V. Klein and T. E. Furtak, *Optics*, 2nd edition, Wiley, New York, 1986.

R. S. Longhurst, *Geometrical and Physical Optics*, 3rd ed., Longman, London, 1973.

F. G. Smith and J. H. Thomson, *Optics*, 2nd ed., Wiley, Chichester, 1988.

Applied Optics

W. J. Smith, *Modern Optical Engineering*, 2nd ed., McGraw-Hill, New York, 1990.

Applied Optics and Optical Engineering, R. Kingslake, ed., Academic, New York, Vol. 1, 1965; Vol. 2, 1965; Vol. 3, 1965; Vol. 4, 1967; Vol. 5, 1969.

L. Levi, *Applied Optics*, Wiley, New York, Vol. 1, 1968; Vol. 2, 1980.

Lens Design

W. J. Smith, *Modern Lens Design: A Resource Manual*, McGraw-Hill, New York, 1992.

R. R. Shannon, *The Art of Science and Optical Design*, Cambridge University Press, Cambridge, 1997.

M. Laikin, *Lens Design*, Marcel Dekker, New York, 1991.

Electro-Optic Devices

M. A. Karim, *Electro-Optical Devices and Systems*, PWS-Kent Publishing, Boston, 1990.

I. P. Kaminow, *An Introduction to Electro-optics*, Academic Press, New York, 1974.

A. Yariv, *Optical Electronics in Modern Communications*, 5th ed., Oxford University Press, New York, 1997.

Far-Infrared Techniques

M. F. Kimmitt, *Far-infrared Techniques*, Pion, London, 1970.

A. Hadni, *Essentials of Modern Physics Applied to the Study of the Infrared*, Pergamon Press, Oxford, 1967.

K. D. Moller and W. G. Rothschild, *Far-Infrared Spectroscopy*, Wiley-Interscience, New York, 1971.

L. C. Robinson, *Physical Principles of Far-Infrared Radiation, Methods in Experimental Physics*, Vol. 10, L. Marton, ed., Academic Press, New York, 1973.

Fiber Optics

J. Hecht, *Understanding Fiber Optics*, 3rd ed., Prentice-Hall, Upper Saddle River, N.J., 1999.

A. Ghatak and K. Thyagarajan, *Introduction to Fiber Optics*, Cambridge University Press, Cambridge, 1998.

A. Ghatak, A. Sharma, and R. Tewari, *Understanding Fiber Optics on a PC*, Viva Books Private Ltd., New Delhi, 1994.

P. K. Cheo, *Fiber Optics and Optoelectronics*, 2nd ed., Prentice-Hall, Englewood Cliffs, N.J., 1990.

Leonid Kazovsky, Serio Benedetto, and Alan Willner, *Optical Fiber Communication Systems*, Artech House, Boston, 1996.

J. Gowar, *Optical Communication Systems*, 2nd ed., Prentice-Hall, Englewood Cliffs, N.J., 1993.

Introduction to Integrated Optics, M. K. Barnoski, ed., Plenum, New York, 1974.

D. Marcuse, *Principles of Optical Fiber Measurements*, Academic Press, San Diego, 1981.

T. Okoshi, *Optical Fibers*, Academic Press, New York, 1982.

Optical Fiber Technology, D. Gloge, ed., IEEE, New York, 1976.

J. C. Palais, *Fiber Optic Communications*, 4th ed., Prentice-Hall, Upper Saddle River, N.J., 1998.

A. D. Snyder and J. D. Love, *Optical Waveguide Theory*, Chapman and Hall, London and New York, 1983.

Filters

Handbook of Chemistry and Physics, 81st ed., David R. Lide, ed., CRC Press, Boca Raton, Fla., 2000.

Handbook of Lasers, R. J. Pressley, ed., CRC Press, Cleveland, 1971.

L. Levi, *Applied Optics*, Vol. 2, Wiley, New York, 1980.

H. A. Macleod, *Thin-Film Optical Filters*, American Elsevier, New York, 1969.

E. J. Bowen, *Chemical Aspects of Light*, 2nd revised ed., The Clarendon Press, Oxford, 1946.

Incoherent Light Sources

Advanced Optical Techniques, A. C. S. Van Heel, ed., North Bouand, Amsterdam, 1967.

Applied Optics and Optical Engineering, R. Kingslake, ed., Vol. 1, Academic Press, New York, 1965.

Handbook of Lasers, R. J. Pressley, ed., CRC Press, West Palm Beach, Fla., 1971.

L. Levi, *Applied Optics*, Vol. 1, Wiley, New York, 1968.

Infrared Technology

The Infrared Handbook, William L. Wolfe and George J. Zissis, eds., Office of Naval Research, Washington, D.C., 1985.

A. R. Jha, *Infrared Technology*, Wiley, New York, 2000.

A. Hadni, *Essentials of Modern Physics Applied to the Study of the Infrared*, Pergamon Press, Oxford, 1967.

R. D. Hudson, *Infrared System Engineering*, Wiley-Interscience, New York, 1969.

P. W. Kruse, L. D. McGlauchlin, and R. B. McQuistan, *Elements of Infrared Technology*, Wiley, New York, 1962.

Interferometers and Interferometry

P. Hariharan, *Optical Interferometry*, Academic Press, Orlando, 1985.

G. Hernandez, *Fabry-Perot Interferometers*, Cambridge University Press, Cambridge, 1986.

R. Jones and C. Wykes, *Holographic and Speckle Interferometry*, 2nd ed., Cambridge University Press, Cambridge, 1989.

Laser Speckle and Related Phenomena, J. C. Dainty, ed., Topics in Applied Physics Vol. 9, Springer-Verlag, Berlin; New York, 1975.

M. Francon, *Laser Speckle and Applications in Optics*, Academic Press, New York, 1979.

M. Born and E. Wolf, *Principles of Optics*, 7th ed., Cambridge University Press, Cambridge, 1999.

M. Francon, *Optical Interferometry*, Academic Press, New York, 1966.

W. H. Steel, *Interferometry*, 2nd ed., Cambridge University Press, Cambridge, 1983.

S. Tolansky, *An Introduction to Interferometry*, Longman, London, 1955.

Fourier Optics and Holography

J. Goodman, *Introduction to Fourier Optics*, 2nd ed., McGraw-Hill, New York, 1996.

P. Hariharan, *Optical Holography*, Cambridge University Press, Cambridge, 1984.

Lasers

A. Yaiv, *Optical Electronics in Modern Communications*, 5th ed., Oxford University Press, New York, 1997.

C. C. Davis, *Lasers and Electro-Optics*, Cambridge University Press, Cambridge, 1996.

Clifford R. Pollack, *Fundamentals of Optoelectronics*, Irwin, Chicago, 1995.

William T. Silfvast, *Laser Fundamentals*, Cambridge University Press, Cambridge, 1996.

B. E. A. Saleh, and M. C. Teich, *Fundamentals of Photonics*, Wiley, New York, 1991.

Handbook of Laser Science and Technology, Vol. 1: *Lasers and Masters*; Vol. 2: *Gas Lasers*, M. Weber, ed., CRC Press, Boca Raton, Fla., 1982.

A. Maitland and M. H. Dunn, *Laser Physics*, North-Holland, Amsterdam, 1969.

D. C. O'Shea, W. R. Callen, and W. T. Rhodes, *Introduction to Lasers and Their Applications*, Addison-Wesley, Reading, Mass., 1977.

J. T. Verdeyen, *Laser Electronics*, 3rd ed., Prentice-Hall, Englewood Cliffs, N.J., 1995.

A. Yariv, *Quantum Electronics*, 3rd ed., Wiley, New York, 1989.

A. E. Siegman, *Lasers*, University Science Books, Mill Valley, Calif., 1986.

Nonlinear Optics

Robert W. Boyd, *Nonlinear Optics*, Academic Press, Boston, 1992.

N. Bloembergen, *Nonlinear Optics*, Benjamin, New York, 1965.

Nonlinear Optics, P. G. Harper and B. S. Wherrett, eds., Academic Press, New York, 1977.

Quantum Electronics, Vols. 1A and 2A, Nonlinear Optics, C. L. Tang and H. Rabin, eds., Academic Press, New York, 1975.

F. Zernike and J. E. Midwinter, *Applied Non-linear Optics*, Wiley, New York, 1973.

Y. R. Shen, *The Principles of Nonlinear Optics*, Wiley, New York, 1984.

Optical Component and Instrument Design

Advanced Optical Techniques, A. C. S. Van Heel, ed., North Holland, Amsterdam, 1967.

Applied Optics and Optical Engineering, R. Kingslake, ed., Academic Press, New York, Vol. 3, 1965; Vol. 4, 1967; Vol. 5, 1969.

Optical Detectors

A. Rogalksi, *Infrared Detectors*, Gordon and Breadh, Amsterdam, 2000.

J. Gowar, *Optical Communications Systems*, 2nd ed., Prentice-Hall, Englewood Cliffs, N.J., 1993.

E. L. Dereniak and G. D. Boreman, *Infrared Detectors and Systems*, Wiley, New York, 1996.

E. L. Dereniak and Devon G. Crowe, *Optical Radiation Detectors*, Wiley, New York, 1984.

J. Graeme, *Photodiode Amplifiers*, McGraw-Hill, New York, 1996.

J. B. Dance, *Photoelectric Devices*, Iliffe, London, 1969.

P. W. Kruse, L. D. McGlauchlin, and R. B. McQuistan, *Elements of Infrared Technology*, Wiley, New York, 1962.

L. Levi, *Applied Optics*, Vol. 2, Wiley, New York, 1980.

Optical and Infrared Detectors, R. S. Keyes, ed., Topics in Applied Physics, Vol. 19, Springer, Berlin, 1977.

Optical Materials

Applied Optics and Optical Engineering, Vol. 1, R. Kingslake, ed., Academic Press, New York, 1965.

Handbook of Laser Science and Technology, Vols. III–V: *Optical Materials*, M. Weber, ed., CRC Press, Boca Raton, Fla., 1986–87.

P. W. Kruse, L. D. McGlauchlin, and R. B. McQuistan, *Elements of Infrared Technology*, Wiley, New York, 1962.

A. J. Moses, *Optical Material Properties*, IFI/Plenum, New York, 1971.

Optical Safety

Handbook of Laser Science and Technology, Vol. 1: *Lasers and Masers*, M. J. Weber, ed., CRC Press, Boca Raton, Fla., 1982.

D. Sliney and M. Wolbarsht, *Safety with Lasers and Other Optical Sources: A Comprehensive Handbook*, Plenum, New York, 1980.

Polarized Light and Crystal Optics

R. M. A. Azzam and N. M. Bashara, *Ellipsometry and Polarized Light*, North-Bouand, Amsterdam, 1977.

W. A. Shurcliff, *Polarized Light*, Harvard University Press, Cambridge, Mass., 1962.

E. Wahlstrom, *Optical Crystallography*, 5th ed., Wiley, New York, 1979.

A. Yariv and P. Yeh, *Optical Waves in Crystals*, Wiley, New York, 1983.

Spectrometers

J. F. James and R. S. Sternberg, *The Design of Optical Spectrometers*, Chapman and Hall, London, 1969.

H. S. Strobel, and W. R. Heinemanm *Chemical Instrumentation*, 3rd ed., Wiley, New York, 1989.

Spectroscopy

Atomic Absorption Spectrometry, M. Pinta, ed., Wiley, New York, 1975.

J. R. Edisbury, *Practical Hints on Absorption Spectrometry*, Hilger and Watts, London, 1966.

R. J. Reynolds and K. Aldous, *Atomic Absorption Spectroscopy*, Barnes and Noble, New York, 1970.

R. A. Sawyer, *Experimental Spectroscopy*, Prentice-Hall, Englewood Cliffs, N.J., 1951.

Spectroscopy, D. Williams, ed., *Methods of Experimental Physics*, Vol. 13, Parts A and B, Academic Press, New York, 1968

S. Walker and H. Straw, *Spectroscopy*, Vol. I: *Microwave and Radio Frequency Spectroscopy*; Vol. II: *Ultraviolet, Visible, Infrared and Raman Spectroscopy*, Macmillan, New York, 1962.

Submillimeter Wave Techniques

Instrumentation for Submillimeter Spectroscopy, E. Kollberg, ed., SPIE Proceedings, Vol. 598, SPIE, Bellingham, Washington, 1986.

G. W. Chantry, *Submillimeter Spectroscopy*, Academic Press, New York, 1971.

Spectroscopic Techniques, D. H. Martin, ed., North-Holland, Amsterdam, 1967.

Tables of Physical and Chemical Constants

G. W. C. Kaye and T. H. Laby, *Tables of Physical and Chemical Constants*, 15th ed., Longman, London, 1986.

Tables of Spectral and Laser Lines

M. J. Weber, *Handbook of Lasers*, CRC Press, Boca Raton, Fla., 2001.

M. J. Weber, *Handbook of Laser Wavelengths*, CRC Press, Boca Raton, Fla., 1999.

G. R. Harrison, MIT Wavelength Tables, *MIT Press*, Cambridge, Mass., 1969.

A. R. Striganov and N. S. Sventitskii, *Tables of Spectral Lines of Neutral and Ionized Atoms*, IFI/Plenum, New York, 1968.

A. N. Zaidel', V. K. Prokof'ev, S. M. Raiskii, V. A. Slavnyi, and E. Ya. Shreider, *Tables of Spectral Lines*, IFI/Plenum, New York, 1970.

Ultraviolet and Vacuum-Ultraviolet Technology

The Middle Ultraviolet: Its Science and Technology, A. E. S. Green, ed., Wiley, New York, 1966. J. A. R. Samson, *Techniques of Vacuum Ultraviolet Spectroscopy*, Wiley, New York, 1967.

MANUFACTURERS AND SUPPLIERS

A-B Lasers
336 Baker Ave.
Concord, MA 01742
(978) 635-9100
FAX: (978) 635-9199

Acton Research Corp.
525 Main St.
P.O. Box 2215
Acton, MA 01720-6215

Ad-Vance Magnetics
625 Monroe St.
P.O. Box 69
Rochester, IN 46975
(219) 223-3158
FAX: (219) 223-2524

Aerotech, Inc.
101 Zeta Dr.
Pittsburgh, PA 15238-2897
(412) 963-7470
FAX: (412) 963-7459

Alcatel Optronics
Subsidiary of Alacatel Alstom
Route de Villejust
F-19625 Nozay
France
+33-1-64-49-49-10
FAX: +33-1-64-49-49-61

American Holographic
601 River St.
Fitchburg, MA 01420
(978) 343- 0096
FAX: (978) 348-1864

American Laser Corp.
1832 S. 3850 W
Salt Lake City, UT 84104-4911
(801) 972-1311
FAX: (801) 972-5251

American Acoustical Products
6 October Hill Rd.
Holliston, MA 01746
(508) 429-1165
FAX: (508) 429-8543

Amherst Systems, Inc.
30 Wilson Rd.
Buffalo, NY 14221
(716) 631-0610
FAX: (716) 631-0629

Amorphous Materials, Inc.
3130 Benton St.
Garland, TX 75042-7410
(972) 494-5624
FAX: (972) 272-7971

Amuneal
4737 Darrah St.

Philadelphia, PA 19124-2705
(215) 535-3000
FAX: (215) 743-1715

Analog Modules
126 Baywood Ave.
Longwood, FL 32750-3426
(407) 339-4355
FAX: (407) 834-3806

Anderson Lasers
2883 W. Royalton Rd.
Broadview Heights, OH 44147
(440) 237-6629
FAX: (440) 237-0656

AOET, Inc.
14235 Ambaum Blvd. SW
P.O. Box 48187
Seattle, WA 98166
(800) 745-5715
FAX: (206) 244-5829

Apex Microtechnology
5980 N. Shannon Rd.
Tucson, AZ 85741-5103
(520) 690-8600
FAX: (520) 888-3329

Applied Scintillation Technologies, Inc.
11 President Point Dr., Building 11, Ste. A3
Annapolis, MD 21403
(410) 263-6005
FAX: (410) 263-4495

Argyle International, Inc.
207 Laurel Circle
Princeton, NJ 08540
(609) 393-5660
FAX: (609) 924-2679

Atochem North America (Atofina)
2000 Market St.
Philadelphia, PA 19103-3222
(215) 419-7000
FAX: (800) 446-2800

Aurora Photonics, Inc.
3350 Scott Blvd., Building 20
Santa Clara, CA 95054
(408) 748-2960
FAX: (408) 748-2962

B & W Tek, Inc.
P.O. Box 451
Newark, DE 19715-0451
(301) 368-7824
FAX: (302) 368-7830

Bicron
Bicron Crystal Products
750 S. 32nd St.
Washougal, WA 98671
(360) 835-8566
FAX: (360) 835-9848

Biorad
3316 Spring Garden St.
Philadelphia, PA 19104
(215) 381-7800
FAX: (215) 662-0585

Bolle: Products are available from many suppliers of protective eyewear.

Bomem USA
Drummond Plaza Office Park
Building 1, Ste. 1105
Newark, DE 19711

Bond Optics, Inc.
Etna Rd.
P.O. Box 422
Lebanon, NH 03766
(603) 448-2300
FAX: (603) 448-5489

Boston Electronics
72 Kent St.
Brookline, MA 02445-7360
(617) 566-3821
FAX: (617) 731-0935

Bruker Instruments
19 Fortune Dr.
Manning Park
Billerica, MA 01821
(978) 667-9580
FAX: (978) 663-9177

Bulbtronics
45 Banfi Plaza
Farmington, NY 11735
(516) 249-2272
FAX: (516) 249-6066

Burle Electron Tubes
1000 New Holland Ave.
Lancaster, PA 17601-5688
(800) 366-2875 or (717) 295-6888
(717) 295-6773 (Int. Calls)
FAX: (717) 295-6096

Burleigh Instruments, Inc.
P.O. Box E
Burleigh Park
Fishers, NY 14453
(716) 924-9355
FAX: (719) 924-9072

Calmar Optcom
958 San Leandro Ave., #700
Mountain View, CA 94043
(650) 938-7299
FAX: (650) 938-6810

Cal-Sensors
5460 Skyline Blvd.
Santa Rosa, CA 95403
(707) 545-4181
FAX: (707) 545-5113

Cambridge Research and Instrumentation
35-B Cabot Rd.
Woburn, MA 01801
(781) 935-9099
FAX: (781) 935-3388

Cambridge Technology
109 Smith Pl.

Cambridge, MA 02138
(617) 441-0600
FAX: (617) 497-8800

Canon: For the nearest dealer call 1-800-OK-CANON

Cathodeon
Nuffield Rd.
Cambridge CB4 1TW, England
44 1223 424100
FAX: 44 1223 426338

Chance-Pilkington: see Pilkington Special Glass

Clark-MXR, Inc.
7300 Huron River Dr.
Dexter, MI 48130-1065
(734) 426-2803
FAX: (734) 426-6288

Cleveland Crystals, Inc.
19306 Redwood Ave.
Cleveland, OH 44110-2738
(216) 486-6100
FAX: (216) 486-6103

Coherent General
1 Picker Rd.
Sturbridge, MA 01566

Coherent, Inc.
Components Group
2301 Lindbergh St.
Auburn, CA 95603
(800) 343-4912
FAX: (530) 889-5366

Coherent, Inc.
Laser Products Division
3210 Porter Dr.
P.O. Box 10321
Palo Alto, CA 94304

Conoptics, Inc.
19 Eagle Rd.
Danbury, CT 06810
(800) 748-3349
FAX: (203) 790-6145

Continental Optical Corp.
15 Power Dr.
Hauppauge, NY 11788
(516) 582-3388
FAX: (516) 582-1054

Continuum
3150 Central Expy
Santa Clara, CA 95051
(408) 727-3240
FAX: (408) 727-3550

Control Optics
13111 Brooks Dr., Ste. J
Baldwin Park, CA 91706
(626) 813-1991
FAX: (626) 813-1993

Corning Glass Works
Optical Products Depart.
Corning, NY 14831

Crismatec products are obtained from Bicron

Crystal Technology, Inc.
1040 E. Meadow Circle
Palo Alto, CA 94303
(650) 856-7911
FAX: (650) 354-0173

Crystran Ltd.
27 Factory Rd.
Poole
Dorset BH16 5SL
UK
+44-1202-632833
FAX: +44-1202-632832

CSI Technologies
Capacitor Division
810 Rancheros Dr.
San Marcos, CA 92069-3009
(760) 747-4000
FAX: (760) 743-5094

CTS Piezoelectric Products
4800 Alameda Blvd., NE

Albuquerque, NM 87113
(505) 348-4361

CVI Laser Corp.
200 Dorado Pl., SE
Albuquerque, NM 87192
(505) 296-9541
FAX: (505) 298-9908

DeMaria ElectroOptics Systems (DEOS)
1280 Blue Hills Ave.
Bloomfield, CT 06002
(860) 243-9557
FAX: (860) 243-9577

Digital Optics
5900-J Northwoods Pkwy.
Charlotte, NC 28269
(704) 599-9191
FAX: (704) 599-4997

DL Instruments
725 West Clinton St.
Ithaca, NY 14850
(607) 277-8498
FAX: (607) 277-8499

Diffraction Products, Inc.
P.O. Box 1030
Woodstock, IL 60098
(815) 338-6768
FAX: (815) 338-7167

Digilab—see Biorad

Display Tech, Inc.
2602 Clover Basin Dr.
Longmont, CO 80503
(303) 772-2191
FAX: (303) 772-2193

Dynasil Corp. of America
385 Cooper Rd.
W. Berlin, NJ 08091-9145
(856) 767-4600
FAX: (856) 767-6813

Eastman Kodak Co.
1447 St. Paul St.
Rochester, NY 14650

Edinburgh Instruments Ltd.
2 Bain Square
Kirkton Campus
Livingston, EH5 47DQ, UK
FAX: 01506-425320

Edmund Industrial Optics
101 E. Gloucester Pike
Barrington, NJ 08007
(609) 573-6250
FAX: (609) 573-6295

Edmund Scientific
101 E. Gloucester Pike
Barrington, NJ 08007
(609) 573-6250
FAX: (609) 573-6295

EDO Corp.
Barnes Engineering Division
88 Long Hill Cross Rd.
Shelton, CT 06484-0867
(609) 926-1777

EDO Electro-Ceramic Products
2645 S 300 W
Salt Lake City, UT 84115
(801) 486-7841
FAX: (801) 484-3301

EEV Inc.
4 Westchester Plaza
Elmsford, NY 10523
(914) 592-6050
FAX: (914) 682-8922

EG&G Inc.
Electro-Optics Div.
35 Congress St.
Salem, MA 01970

EG&G Princeton Applied Research
P.O. Box 2565
Princeton, NJ 08543-2565

EG&G Reticon Corp.
345 Potrero Ave.
Sunnyvale, CA 94086-4197

Eksma
Mokslininku St. 11
2600 Vilnius
Lithuania
370-2-729900
FAX: 370-2-729715

Electro-optical Products Corp.
P.O. Box 650441
Freshmeadows, NY 11365
(718) 776-4960
FAX: (718) 776-4960

Electro-Optical Systems Inc.
1039 W. Bridge St.
Phoenixville, PA 19460
(610) 935-5838
FAX: (610) 935-8548

Electro-Optical Technology, Inc.
1030 Hastings St.
Traverse City, MI 49686

Electron Tubes
Bury St.
Ruislip Middlesex HA4 7TA, England
44-1895-630771
FAX: 44-1895-635953

Electron Tubes
100 Forge Way, Unit F
Rockaway, NJ 07866
(973) 586-9594
FAX: (973) 586-9771

Eltec
P.O. Box 9610
Daytona Beach, FL 32120

(904) 253-5328
FAX: (904) 258-3791

EMI Gencom—see Thorn EMI Gencom

EMR Photoelectric
Division of Schlumberger
153 East 53rd St., 57th Floor
New York, NY 10022-4624
(609) 799-1000
FAX: (609) 799-2247

Eppley Laboratory Inc.
12 Sheffield Ave.
Newport, RI 02840
(401) 847-1020
FAX: (401) 847-1031

Equilasers
536 Weddell Dr., Ste. 6
Sunnyvale, CA 94089
(408) 734-2700
FAX: (408) 734-2929

Esco
171 Oak Ridge Rd.
Oak Ridge, NJ 07438
(973) 697-3700
FAX: (973) 697-3011

Exciton
P.O. Box 31126
Dayton, OH 45437-0126
(937) 252-2989
FAX: (937) 258-3937

Femtochrome Research
2123 4th St.
Berkeley, CA 94710-2240
(510) 644-1869
FAX: (510) 644-0118

Fermionics
4555 Runway St.
Simi Valley, CA 93063
(805) 582-0155
FAX: (805) 582-1623

Fiberall
250 Jericho Turnpike, Ste. 601
Floral Park, NY 11001
(516) 354-5600
FAX: (516) 354-5645

Fiberguide Industries
1 Bay St.
Stirling, NJ 07980
(908) 647-6601
FAX: (908) 647-8464

Flashlamps (Verre et Quartz)
30, Route d'Aulnay
93140 Bondy
France
01-48-49-74-21
FAX: 01-48-48-44-22

Force, Inc.
825 Park St.
P.O. Box 2045
Christiansburg, VA 24073
(540) 382-0462
FAX: (540) 381-0392

Fotono
Stegne 7
SLO-1210 Ljubljana, Slovenia
386-61-15-11-215
FAX: 386-61-15-11-145

Fresnel Technologies, Inc.
101 W. Morningside Dr.
Fort Worth, TX 76110
(817) 926-7474
FAX: (817) 926-7146

Galileo
P.O. Box 550
Sturbridge, MA 01566
(508) 347-9191
FAX: (508) 347-3849

GAM Laser, Inc.
6901 TPC Dr., #450
Orlando, FL 32822

(407) 851-8999
FAX: (407) 850-0700

Gamma Scientific
8581 Aero Dr.
San Diego, CA 92123
(619) 279-8034
FAX: (619) 576-0928

GBC Scientific Equipment
3930 Ventura Dr., Ste. 350
Arlington Heights, IL 60004
(847) 506-1900
FAX: (847) 506-1901

Gentec
2625 Dalton St.
Sainte-Foy, P.Q. G1P 3S9, Canada
(418) 651-8003
FAX: (418) 651-6695

Glendale/Dalloz Safety—see GPT Glendale

Gould Precision Optics
151 Court St., Ste. 106
Binghamton, NY 13901
(607) 722-7606
FAX: (607) 722-7939

GPT Glendale
5300 Region Ct.
Lakewood, FL 33815
(863) 687-7266
FAX: (863) 687-0431

Graseby
97 Cray Ave., St. Mary Cray
River House
Orpington, Kent BR5 4HE
44 1689 873134
FAX: 44 1689 878527

GSI Lumonics
105 Schneider Rd.
Kanata, Ontario
Canada K2K 1Y3

(613) 592-1460
FAX: (613) 592-5706

Hamamatsu Corp.
360 Foothill Rd.
P.O. Box 6910
Bridgewater, NJ 08807
(908) 231-0960
FAX: (908) 231-1218

Hampton Controls
Wendel Rd.
P.O. Box 187
Wendel, PA 15691-0187
(724) 861-0150 or (724) 861-0160

Harrick Scientific Corp.
Box 1288, 88 Broadway
Ossining, NY 10562
(914) 762-0020 or (800) 248-3847
FAX: (914) 762-0914

Hellma
Box 544
Borough Hall Station
Jamaica, NY 11424-9989
(718) 544-9166
FAX: (718) 263-6910

Heraeus-Amersil Inc.
650 Jernees Mill Rd.
Sayreville, NJ 08872

Hewlett Packard
5601 Lindero Canyon Rd.
Westlake Village, CA 91362
(818) 879-6200

Hi Voltage Components
699 4th St. NW Ste. A
Largo, FL 33770-2408
(813) 585-3884

Hovemere
56 Hayes St.
Hayes, Bromley
Kent BR2 7LD, UK

+44-208-462-4921
FAX: +44-208-462-7841

Howden Laser
Unit T, Charles Bowman Ave.
Claverhouse Ind'l Park
Dundee DD4 9UB, Scotland
+44-1382-518200
FAX: +44-1382-518228

Hoya Optics, Inc.
960 Rincon Circle
San Jose, CA 95131
(408) 321-7705
FAX: (408) 321-9216

IC Optical Systems
190-192 Ravenscroft Rd.
Beckenham
Kent BR3 4TW, UK
+44-208-778-5094
FAX: +44-208-676-9816

II-VI, Inc.
375 Saxonburg Blvd.
Saxonburg, PA 16056-9499
(724) 352-1504
FAX: (724) 352-4980

ILC Technology
399 W. Java Dr.
Sunnyvale, CA 94089
(408) 745-7900
FAX: (408) 744-0829

Infrared Associates Inc.
2851 SE Monroe St.
Stuart, FL 34997
(561) 223-6670
FAX: (561) 223-6671

Infrared Industries Inc., Detector Div.
12151 Research Pkwy.
Orlando, FL 32826

Infrared Laboratories, Inc.
1808 E. 17th St.

Tucson, AZ 85719
(520) 622-7074
FAX: (520) 623-0765

Infrared Optical Products
120 Secatouge Ave.
P.O. Box 292
Farmingdale, NY 11735
(516) 694-6035
FAX: (516) 694-6049

Inrad
181 Legrand Ave.
Northvale, NJ 07647
(201) 767-1910
FAX: (201) 767-9644

Instruments SA (see John Yvin)

Intelit, Inc.
7740 W Manchester, Str. 110
Playa Del Rey, CA 90293
(310) 301-6761
FAX: (310) 301-6763

International Light
17 Graf Rd.
Newburyport, MA 01950-4092
(978) 465-5923
FAX: (978) 462-0759

International Scientific Products
1 Bridge St.
Irvington, NY 10533-1543
(914) 591-3070
FAX: (914) 591-3715

IntraAction Corp.
3719 Warren Ave.
Bellwood, IL 60104
(708) 547-6644
FAX: (708) 547-0687

Isomet Corp.
5263 Port Royal Rd.
Springfield, VA 22151

(703) 321-8301
FAX: (703) 321-8546

ISS
2604 N Mattis Ave.
Champaign, IL 61822
(217) 359-8681
FAX: (217) 359-7879

ITT
1919 W Cook Rd.
P.O. Box 3700
Fort Wayne, IN 46801-3700
(219) 451-6000
FAX: (219) 487-6126

Janos Technology Inc.
Rt. 35, 1068 Grafton Rd.
Townshend, VT 05353-7702
(802) 365-7714
FAX: (802) 365-4596

Jarrell-Ash—see Thermo-Jarrel-Ash

JDS Uniphase
Fiberoptics Products Division
570 W. Hunt Club Rd.
Nepean, Ontario
Canada K2G 5WB
(613) 727-1303
FAX: (613) 727-8284

JDS Uniphase
Commercial Lasers Division
163 Baypointe Pkwy
San Jose, CA 95134
(408) 434-1800
FAX: (408) 894-06237

Jobin-Yvon, Inc./Horiba
380 Park Ave.
Edison, NJ 08820
(732) 549-5125
FAX: (732) 494-8660

JPSA (JP Sercel Associates)
17D Clinton Dr.

Hollis, NH 03049-6474
(603) 595-7048
FAX: (603) 598-3835

Judson (now part of Perkin Elmer Optoelectronics)
111 Park Dr.
Montgomeryville, PA 18936
(800) 995-0602
(215) 368-0700
FAX: (215) 368-4790

Kaiser Optical Systems
P.O. Box 983
Ann Arbor, MI 48106-0983

Karl Lambrecht Corp. (KLC)
4204 N. Lincoln Ave.
Chicago, IL 60618
(773) 472-5442
FAX: (773) 472-2724

Kentek Corp.
19 Depot St.
Pittsfield, NH 03263
(603) 435-7201
FAX: (603) 435-7441

Kigre, Inc.
100 Marshland Rd.
Hilton Head, SC 29928
(843) 681-5800
FAX: (843) 681-4559

Kimmon Electric Co., Ltd.
7002 S. Revere Pkwy., Ste. 85
Englewood, CO 80112
(303) 754-0401
FAX: (303) 754-0324

Klinger, now Micro-Controle S.A.
subsidiary of Newport Corporation
3 bis rue Jean Mermoz
BP 189, F-91006 Evry, Cedex
France
+33-1-60-91-68-68
+33-1-60-91-68-69

Kodak (Eastman Kodak Company)
Kodak Park Site
Rochester, NY 14652
Kodak Information Center
(800) 242-2424

Kolmar Technologies, Inc.
1400 General Arts Rd., Unite E
Conyers, GA 30012
(770) 918-0334
FAX: (770) 918-0232

Kopp Glass, Inc.
2108 Palmer St.
Pittsburgh, PA 15218-2516
(412) 271-0190
FAX: (412) 271-4103

Lambda Physik
3201 W. Commercial Blvd., #110
Fort Lauderdale, FL 33309
(954) 486-1500
FAX: (954) 486-1501

Lambda Research Optics
17605 Fabrica Way, Ste. A-C
Cerritos, CA 90703-7021
(714) 228-1192
FAX: (714) 228-1193

Lasag
601 Campus Dr., Ste. B5
Arlington Heights, IL 60005
(847) 593-3021
FAX: (847) 593-5062

Lase-R Shield
A Bacou USA Co.
7011 Prospect Place NE
Albuquerque, NM 87110-4311
(505) 872-3400
FAX: (505) 872-3500

Laser Inc.—see Coherent General

Laser Components
5460 Skylane Blvd.

Santa Rosa, CA 95403
(707) 568-1642
FAX: (707) 568-1652

Laser Consultants Inc.
344 W. Hills Rd.
Huntington, NY 11743
(516) 423-4905
FAX: (516) 424-0422

Laser Diode, Inc.
4 Olsen Ave.
Edison, NJ 08820
(732) 549-9001
FAX: (732) 906-1559

Laser Optics Inc.
111 Wooster St.
Bethel, CT 06801
(203) 744-4160
FAX: (203) 798-7941

Laser Peripherals, Inc.
5484 Feltl Rd.
Minnetonka, MN 55343
(800) 966-5273
(612) 930-9065
FAX: (612) 930-9069

Laser Physics
8050 South 1300 W
West Jordan, UT 84088
(801) 256-0962
FAX: (801) 256-3879

Laser Probe, Inc.
23 Wells Ave.
Utica, NY 13502-2521
(315) 797-4492
FAX: (315) 797-0696

Laser S.O.S.
Unit, Burrel Rd.
St. Ives Ind. Estate
St. Ives, Cambs PE17 4LE
England

44-1480-460990
44-1480-469978

Laser System Devices, Inc.
5645 E. General Washington Dr.
Alexandria, VA 22312
(703) 642-5758
FAX: (702) 642-5838

Lasertechnics, Inc.
5500 Wilshire Ave. NE
Albuquerque, NM 87113
(505) 822-1123
FAX: (505) 821-2213

LeCroy Corp.
700 Chestnut Ridge Rd.
Chestnut Ridge, NY 10977-6499
(914) 578-6020
FAX: (914) 821-2213

Lee Laser, Inc.
7605 Presidents Dr.
Orlando, FL 32809
(407) 812-4611
FAX: (407) 850-2422

Leica Microsystems
2345 Waukegan Rd.
Bannockburn, IL 60015
(800) 248-0123
FAX: (847) 405-0030

Leitz, now Leica

Lexel Laser, Inc.
48503 Milmont Dr.
Fremont, CA 94530
(510) 770-0800
FAX: (510) 651-6598

Lexitek, Inc.
14 Mica Lane, Ste. 6
Wellesley, MA 02481-1708
(781) 431-9604
FAX: (781) 431-9605

Light Age, Inc.
2 Riverview Dr.
Somerset, NJ 08873
(732) 563-0060
FAX: (732) 563-1571

Lightwave Electronics
2400 Charleston Rd.
Mountain View, CA 94043
(650) 962-0755
FAX: (650) 962-1661

Linos Photonics

Litton Poly-Scientific
1213 N. Main St.
Blacksburg, VA 24060-3100
(540) 953-4751
FAX: (540) 953-1841

LTB Lasertechnik Berlin GmbH
Rudower Chaussee 5
D-12489 Berlin, Germany
49-30-6392-6190
FAX: 49-30-6392-6199

3M
Decorative Products Div.
3M Center
St. Paul, MN 55101

3M Speciality Optical Fibers

Macken Instruments
3644 Airway Dr.
Santa Rosa, CA 95403
(707) 566-2110
FAX: (707) 566-2119

Macrooptica Lts.
Anohina 13-112
Moscow 117602
Russia
095-956-7624
FAX: 095-956-6137

Magnetic Shield Corp.
(Perfection Mica)

740 N. Thomas Dr.
Bensenville, IL 60106-1643
(630) 766-7800
FAX: (630) 766-2813

Matheson Electronic Products Group
625 Wool Creek Dr., Ste. A
San Jose, CA 95112
(408) 971-6500
FAX: (408) 275-8643

Maxwell Energy Products
4949 Greencraig Lane
San Diego, CA 92123
(619) 576-7545
FAX: (619) 576-7672

McPherson
7A Stuart Rd.
Chelmsford, MA 01824
(978) 256-4512
FAX: (978) 250-8625

Meadowlark Optics
5964 Iris Parkway
P.O. Box 1000
Frederick, CO 80530
(303) 833-4333
FAX: (303) 833-4335

Adolf Meller Co., now Meller Optics
120 Corliss St.
Providence, RI 02940
(401) 331-3717
FAX: (401) 331-0519

Melles Griot
16542 Millikan Ave.
Irvine, CA 92714
(949) 261-5600
FAX: (949) 261-7790

Mesh Tel
7740 W. Manchester, Ste. 110
Playa del Rey, CA 90293
(310) 394-3694
FAX: (310) 393-8345

Metrotek Industries
33 W. Main St.
Elmsford, NY 10523-2413
(914) 347-4112
FAX: (914) 347-4203

Micron Techniques Ltd.
22 Ashley Walk
Mill Hill, London, NW7 1DU
United Kingdom
+44-20-8343-4836
FAX: +44-20-8343-4286

Mikron Instrument Company, Inc.
16 Thornton Rd.
Oakland, NJ 07436
(201) 405-0900
(800) 631-0176
FAX: (201) 405-0090

Midac Corp.
17911 Fitch Ave.
Irvine, CA 92614
(949) 660-8558
FAX: (949) 660-9334

Minolta Corp.
101 Williams Dr.
Ramsey, NJ 07446
(201) 818-3517
FAX: (201) 825-4374

Minuteman Laboratories Inc.
7A Stuart Rd.
Chelmsford, MA 01824

Mm Shield Co.
5 Springfield Rd.
P.O. Box 439
Goffstown, NJ 03045-0439

Modern Optics
(no longer in business, but tables can
be found at used equipment vendors)

Molecular Technology GmbH
Rudower Chaussee 29-31 (OWZ)

12489 Berlin
Germany
+4930-6392-6620
FAX: +4930-6392-6622

Molectron Detector Inc.
7470 SW Bridgeport Rd.
Portland, OR 97224
(503) 620-9069
FAX: (503) 620-8964

MPB Technology, Inc.
151 Hymus Blvd.
Pointe Claire, P.Q. H9R 1E9
Canada
(514) 694-8751
FAX: (514) 694-0776

NanoSciences
83 Prokop Rd.
Oxford, CT 06478
(203) 881-2827
FAX: (203) 881-2855

NEOS
4300-C Fortune Place
West Melbourne, FL 32904
(407) 676-9020
FAX: (407) 722-4499

New England Photoconductor
253 Mansfield Ave.
P.O. Box M
Norton, MA 02766
(508) 285-5561
FAX: (508) 285-6480

National Photocolor
428 Waverly Ave.
P.O. Box 586
Mamaroneck, NY 10543
(914) 687-8111
FAX: (914) 698-3629

New Focus, Inc.
2630 Walsh Ave.
Santa Clara, CA 95051-0905

(408) 980-8088
FAX: (408) 980-8883

Newport Corporation
1791 Deere Ave.
P.O. Box 19607
Irvine CA 90623
(949) 863-3144
FAX: (949) 253-1680

New Wave Research
47613 Warm Springs Blvd.
Fremont, CA 94539
(510) 249-1550
FAX: (510) 249-1551

Nicolet Analytical Instruments
5225 Verona Rd.
Madison, WI 53711
(608) 276-6100
FAX: (608) 273-5046

Nikon Inc.
1300 Walt Whitman Rd.
Melville, NY 11747-3064
(516) 547-4200
FAX: (516) 547-0308

NoIR Laser Co. LLC
6155 Pontiac Tr.
P.O. Box 159
South Lyon, MI 48178
(800) 521-9746
FAX: (734) 769-1708

J.A. Noll Co.
P.O. Box 312
Monroeville, PA 15146-0312
(412) 856-7566
FAX: (412) 372-1752

NSG Precision Cells, Inc.
195G Central Ave.
Farmingdale, NY 11735
(516) 249-7474
FAX: (516) 249-8575

Oakite, now Chemetall Oakite
50 Valley Rd.
Berkeley Heights, NJ 07922
(800) 526-4473 or (908) 464-6900
FAX: (908) 464-4658

OCLI Optical Coating Laboratory GimbH
Tilsiter Str. 12
D-64354 Reinheim, Germany
49-6162-93210
49-6162-932119

O'Hara
50 Columbia Rd.
Branchburg
Somerville, NJ 08876-3519
(908) 218-0100
FAX: (908) 218-1685

Omega Optical
3 Grove St.
P.O. Box 573
Brattleboro, VT 05302-0573
(802) 254-2690
FAX: (802) 254-3937

Opthos Instrument Co.
17805 Caddy Dr.
Rockville, MD 20855
(301) 926-0589
FAX: (301) 963-3060

Optical Engineering Inc.
2495 Bluebell Dr.
Santa Rosa, CA 95402
(707) 528-1080
FAX: (707) 528-1081

Optics Plus Inc.
1369 E. Edinger Ave.
Santa Ana, CA 92705
(714) 972-1948
FAX: (714) 835-6510

Optics for Research Inc.
P.O. Box 82
Caldwell, NJ 07006

(973) 228-4480
FAX: (973) 228-0915

Opti-Forms, Inc.
42310 Winchester Rd.
Temecula, CA 92590
(909) 676-4724
FAX: (909) 676-1178

Optometrics
Stony Brook Industrial Park
Nemco Way
Ayer, MA 10432
(978) 772-1700
FAX: (978) 772-0017

Opton Laser International
Parc Club d'Orsay Univ., Bat 0
29 rue J. Rostand
F-91893 Orsay, Cedex, France
33-1-69-41-04-05
FAX: 33-1-69-41-32-90

OptonSigma
2001 Deere Ave.
Santa Ana, CA 92705
(949) 851-5881
FAX: (949) 851-5058

Optovac Inc.
EM Industries Div.
24 E. Brookfield Rd.
P.O. Box 248
North Brookfield, MA 01535
(508) 867-6444
FAX: (508) 867-8349

ORC Lighting Products
1300 Optical Dr.
Azusa, CA 91703-3295
(626) 815-3100
FAX: (626) 815-3074

Oriel Corp.
250 Long Beach Blvd.
Stratford, CT 06497

(203) 377-8282
FAX: (203) 375-0851

Oxford Lasers Ltd.
Abingdon Science Park, Barton Lane
Abingdon, Oxford OXF 3YR U.K.
44-0-1235-554211
FAX: 44-0-1235-554311

Pacific Optical
(subsidiary of Bourns, Inc.)
1190 Columbia Ave.
Riverside, CA 92507
(909) 784-2139
FAX: (909) 784-3100

Parallax Technology, Inc.
30 Spinelli Pl.
Cambridge, MA 02138
(617) 499-0050
FAX: (617) 499-0051

Pennwalt—see Atochem North America

The Perkin-Elmer Corp.
761 Main Ave.
Norwalk, CT 06859-0012
(203) 762-4000
FAX: (203) 762-4228

Perry Amplifier
138 Fuller St.
Brookline, MA 02146
(617) 734-0844
FAX: (617) 734-0844

Piezoelectric Products—see CTS Piezolelectric Products

Piezo Systems
186 Massachusetts Ave.
Cambridge, MA 02139
(617) 547-1777
FAX: (617) 354-2200

Photonic Detectors
90-A W. Cochran St.
Simi Valley, CA 93065

(805) 527-3900
FAX: (805) 527-3931

Photonics Industries International
23 E. Loop Rd.
Stoney Brook, NY 11790-3350
(516) 444-8822
FAX: (516) 444-8821

Physik Instruments
Polytec-Platz 1-7
P.O. Box 162
D-76337 Waldbronn, Germany
49-7243-604-100
FAX: 49-7243-604-145

PLX
40 W. Jeffryn Blvd.
Deer Park, NY 11729
(516) 586-4190
FAX: (516) 586-4196

Pilkington Special Glass
Glascoed Rd.
St. Asaph, Denbighshire
Wales
44-1745-583301
FAX: 44-1745-584358

Polytec
1342 Bell Ave.
Suite 3A
Tustin, CA 92780
(714) 850-1835
FAX: (714) 850-1831

Polytec Optronics
3001 Redhill Ave.
Costa Mesa, CA 92626
(714) 850-1835
FAX: (714) 850-1831

Potomac Photonics
4445 Nicole Dr.
Lanham, MD 20706
(301) 459-3031
FAX: (301) 459-3034

Precision Optical
869 W. 17th St.
Costa Mesa, CA 92627
(949) 631-6800
FAX: (949) 642-7501

QMC Instruments
Queen Mary and Westfield College
327, Mile End Rd.
London E1 4NS England
44-81-980-1288
44-81-981-8337

Quantel
BP 23-Ave. de l'Atlantique
Les Ulis, France
33-1-6929-1700
33-1-6929-1729

Quantronix
41 Research Way
East Setauket, NY 11733-3454
(516) 273-6900
FAX: (516) 273-6958

Quantum Technology Inc.
108 Commerce St.
Lake Mary, FL 32746-6212
(407) 333-9348
FAX: (407) 333-9352

Queensgate Instruments Ltd.
Queensgate House
Waterside Park
Brackmell, Berks RG12 1RB, England
44-1344-484111
FAX: 44-1344-484115

Radiant Dyes Laser Accessories
P.O. Box 1462
D-42907 Wermelskirchen, Germany
49-2196-80161
FAX: 49-2196-3422

Reflex Analytical
P.O. Box 119
Ridgewood, NJ 07451

(201) 444-8958
FAX: (201) 670-6737

Research Electro-Optics, Inc.
1855 S. 57th Ct.
Boulder, CO 80301
(303) 938-1960
FAX: (303) 447-3297

Reynard Corporation
1020 Calle Sombra
San Clementa, CA 92673-6227
(949) 366-8866
FAX: (949) 498-9528

Richardson Grating Laboratory
705 St. Paul St.
Rochester, NY 14605
(716) 262-1331
FAX: (716) 454-1568

Rochester Photonics
330 Clay Rd.
Rochester NY 14623-3227
(716) 272-3010
FAX: (716) 454-1568

Rocky Mountain Instruments
106 Laser Dr.
Lafayette, CO 80026
(303) 664-5000
FAX: (303) 664-5001

Rofin Sinar
45701 Mast St.
Plymouth, MI 48170
(734) 455-5400
FAX: (734) 455-2741

Rolyn Optics Co.
706 Arrowgrand Cir.
Covina, CA 91722
(626) 915-5707
FAX: (626) 915-1379

Schott Glass Technologies Inc.
400 York Ave.

Duryea, PA 18642
(570) 457-7458
FAX: (570) 457-6960

Scitec Instruments Ltd.
Bartles Industrial Estate
North St. Redruth
Cornwall TR151HR England
44-1209-314608
44-1209-314609

SLM-Aminco—see Thermo-Spectronics

SensArray
3 Ray Ave.
Burlington, MA 01803
(781) 273-7373
FAX: (781) 273-2552

SensorPhysics
105 Kelleys Trail
Oldsmar, FL 34677
(727) 781-4240
FAX: (727) 781-7942

Sensor Technology Ltd.
P.O. Box 97
Collingwood
Ontario, Canada L9Y 3Z4
(705) 444-1440
FAX: (705) 222-6787

Solar Laser Systems
77 Partisansky Ave.
Minsk 220107, Belarus
375(17) 285-46-62, 285-46-63, 285-46-61
FAX: 375(17) 246-48-33

Sopra
68 Rue Pierre Joigneaux
Bois Columbes, France 92270
33-1-47-81-09-49
FAX: 33-1-42-42-29-34

Space Optics Research Labs (SORL)
7 Stuart Rd.
Chelmsford, MA 01824

(978) 256-8640
FAX: (978) 256-5605

Spaw Industries, Inc.
2051 Spawr Cir.
Lake Havasu City, AZ 86403
(520) 453-8800
FAX: (520) 453-8811

Specac, Inc.
500 Technology Ct.
Smyrma, GA 30082
(770) 803-1106
FAX: (770) 319-2488

Spectra Physics (LAD)—see Spectra Physics

Spectral Energy
67 Woodland Ave.
Westwood, NJ 07675
(201) 664-0876
FAX: (201) 664-1214

Spectra-Physics Inc.
Laser Products Div.
1335 Terra Bella Ave.
Mountain View, CA 94042
(650) 961-2550
FAX: (650) 964-3584

Spectron Laser Systems
8 Consul Rd.
Rugby, Warwickshire CV21 1PB
44-1788-544694
FAX: 44-1788-575379

Spectrogon
24B Hil Rd.
Parsippany, NJ 07054-1001
(973) 331-1191
FAX: (973) 331-1373

Spectrum Associates
4 Harvard St.
Conway, NH 07885
(978) 531-0939
FAX: (978) 531-0939

Special Optics
315 Richard Mine Rd.
Wharton, NJ 03818
(973) 366-7289
FAX: (973) 366-7407

Spindler and Hoyer
459 Fortune Blvd.
Milford, MA 01757-1745

Spiricon, Inc.
2600 N. Main
Logan, UT 84321
(435) 753-3729
FAX: (435) 753-5231

Stanford Research Systems (SRS)
1290-D Reamwood Ave.
Sunnyvale, CA 94089
(408) 744-9040
FAX: (408) 744-9049

Staveley Sensors Inc.
East Hartford, CT
(509) 736-2751

Surface Finishes
42 Official Rd.
Addison, IL 60101-4592
(630) 543-6682
FAX: (630) 543-4013

Synrad
6500 Harbour Heights Pkwy.
Mukilteo, WA 98275
(425) 349-3500
FAX: (425) 485-4882

Textron
1309 Dynamic St.
Petaluma, CA 94954
(707) 763-5900
FAX: (707) 762-7383

Thermo Jarrel Ash
Division of Thermo Elemental (USA)
27 Forge Pkwy.

Franklin, MA 02038
(508) 520-1880
FAX: (508) 520-1732

Thermometrics
808 US Hwy 1
Edison, NJ 08817
(732) 287-2870
FAX: (732) 287-8847

Thermo-Spectronic
Mercers Row,
Cambridge CB5 8HY, U.K.
44-1223-446655
FAX: 44-1223-446644

Thomson-CSF
Route Departementale 128
PB 46
F91401 Orsay, Cedex, France
33-1-69-33-00-00
FAX: 33-1-69-33-03-21

ThorLabs
435 Route 206
Newton, NJ 07860-0366
(973) 579-7227
FAX: (973) 383-8406

TMC – Technical Manufacturing Corp.
15 Centennial Dr.
Peabody, MA 01960
(978) 532-6330 or (800) 542-9725
FAX: (978) 531-8682

Trumpf
Farmington Industrial Pk.
Hyde Rd.
Farmington, CT 06032
(860) 677-9741
FAX: (860) 678-1704

TVC
2527 Foresight Cir.
Grand Junction, CO 81505
(970) 241-3308
FAX: (970) 241-6618

Universal Laser Systems
16008 N. 81st St.
Scottsdale, AZ 85260-1803
(602) 483-1214
FAX: (602) 483-5620

Ushio
10550 Camden Dr.
Cypress, CA 90630
(714) 236-1200
FAX: (714) 229-3180

Uvex Safety
10 Thurber Blvd.
Smithfield, RI 02917
(401) 232-1200
FAX: (401) 232-3180

V-A Optical Labs, Inc.
60 Red Hill Ave.
San Anselmo, CA 94960
(415) 459-1919
FAX: (415) 459-7216

Varian Associates, Instruments Div.
611 Hansen Way
Palo Alto, CA 94303
(650) 213-8000
FAX: (650) 213-2800

Veeco Metrology Group
sub of Veeco Instruments, Inc.
2650 E. Elvira Rd.
Tucson, AZ 85706
(520) 741-1297
FAX: (520) 294-1799

Velmex Inc.
7550 State Rt. 5 and 20
Bloomfield, NY 14469
(716) 657-6151
FAX: (716) 657-6153

Verre et Quartz—see Flashlamps (Verre et Quartz)

Virgo – now VLOC
subsidiary of II-VI

7826 Photonics Dr.
New Port Richey, FL 34655
(727) 375-8562
FAX: (727) 375-5300

Vitro Technology
955 Connecticut Ave., Ste 1113
Bridgeport, CT 06607-1222
(203) 366-0660
FAX: (203) 366-8820

Wavelength Optics
3353 Bradshaw Rd., Ste. 209
Sacramento, CA 95827
(916) 368-9283
FAX: (916) 368-4553

Xenon Corporation
20 Commerce Way
Woburn, MA 01801-1006
(781) 938-3594
FAX: (781) 933-8084

Zeiss
1 Zeiss Dr.
Thornwood, NY 10594
(914) 747-1800
FAX: (914) 681-7446

Zygo Corp.
Laurel Brook Rd.
Middlefield, CT 06455
(860) 347-8506
FAX: (860) 347-8372

Suppliers of Optical Windows

Several of the suppliers listed will supply lenses, prisms, and other components fabricated from these materials:

AMTIR (GeAsSe Glass): Harrick Scientific, Janos, REFLEX Analytical

Arsenic trisulphide: Infrared Optical Products, Harrick Scientific, REFLEX Analytical, Sterling Precision Optics

Barium fluoride: Crystran, Infrared Optical Products, International Scientific Products, Janos, Macrooptica, Molecular Technology

Cadmium sulfide: Cleveland Crystals

Cadmium sclenide: Cleveland Crystals

Cadmium telluride (Irtran 6): Cleveland Crystals, International Scientific Products, Janos, Laser Research Optics, Molecular Technology

Calcium carbonate (calcite): Karl Lambrecht Corp (KLC), Thin Film Lab

Calcium fluoride (Irtran 3): Crystran, Infrared Optical Products, International Scientific Products, KLC, Meller Optics, Molecular Technology, Newport, Opto-Sigma, REFLEX Analytical, Rocky Mountain Instruments, Sterling Precision Optics, Thorlabs

Cesium bromide: Argus International, Crystran, Harrick Scientific, Janos

Cesium iodide: Crystran, Molecular Technology, REFLEX Analytical

Diamond: Argus International, Coherent Photonics Group, II-VI, International Scietific Products, Laser Power Optics, REFLEX Analytical, Sterling Precision Optics, Optics for Research, Oriel

Gallium arsenide: Infrared Optical Products, II-VI, International Scientific Products, Laser Power Optics, Laser Research Optics, Meller Optics, REFLEX Analytical, Rocky Mountain Instruments, Sterling Precision Optics

Glasses: Coherent, Ealing, Edmund Scientific. Newport, Opto-Sigma, Rocky Mountain Instruments, Rolyn, Schott, Sterling Precision Optics, Thorlabs

Lithium fluoride: Crismatec, Crystran, Infrared Optical Products, International Scientific Products, Janos, Macrooptica, Molecular Technology, REFLEX Analytical, Sterling Precision Optics

Magnesium fluoride: (Irtran 1): Crystran, Infrared Optical Products, International Scientific Products, Janos,

KLC, Macrooptica, Meller Optics, Molecular Technology, Newport, REFLEX Analytical, Rocky Mountain Instruments, Sterling Precision Optics

Magnesium oxide (Irtran 5): Crystran, Harrick Scientific

Potassium bromide: Crystran, International Scientific Products, Janos, Macrooptica, Molecular Technology

Potassium chloride: Crystran, Janos, Macrooptica, Molecular Technology, Optovac

Potassium iodide: Crystran

Quartz (crystalline): Crismatec, Crystran, Esco, Infrared Optical Products, International Scientific Products, Janos, Meller Optics, Molecular Technology, Newport, Opto-Sigma, REFLEX Analytical, Sterling Precision Optics, Continental Optical, Optics for Research, Oriel, Adolf Meller

Sapphire: Infrared Optical Products, International Scientific Products, Laser Power Optics, Meller Optics, Newport, Opto-Sigma, REFLEX Analytical, Rocky Mountain Instruments, Rolyn, Sterling Precision Optics

Silica (fused): Esco, International Scientific Products, Janos, Macrooptica, Meller Optics, Newport, Opto-Sigma, REFLEX Analytical, Rolyn, Schott, Sterling Precision Optics, Thorlabs, Continental Optical, Optical Coating Lab (OCLI)

Silicon: Crismatec, Crystran, Infrared Optical Products, International Scientific Products, Janos, Laser Power Optics, Macrooptica, Meller Optics, Molecular Technology, REFLEX Analytical, Rocky Mountain Instruments, Sterling Precision Optics

Silver bromide: Crystran, Harrick Scientific, REFLEX Analytical

Silver chloride: Crystran, Harrick Scientific, REFLEX Analytical

Sodium chloride: Crystran, International Scientific Products, Macrooptica, Molecular Technology

Sodium fluoride: Crystran

Strontium fluoride: Crystran, Harrick Scientific

Strontium titanate: Commercial Crystal Labs, Harrick Scientific, Hibshman-Pacific Optical

Tellurium: Molecular Technolgy

Thallium bromide: Crystran

Thallium bromoidide (KRS-5): Crystran, Infrared Optical Products, International Scientific Products, Janos, Macrooptica, Molecular Technology, Sterling Precision Optics

Thallium chlorobromide (KRS-6): Crystran, Macrooptica, Molecular Technology

Titanium dioxide (rutile): Commercial Crystal Labs, Harrick Scientific, International Scientific Products, Janos, Laser Power Optics, Laser Research Optics, Meller Optics, Molecular Technology, REFLEX Analytical, Rocky Mountain Instruments, Sterling Precision Optics

Zinc sulphide (Irtran 2): Crystran, Infrared Optical Products, Laser Power Optics, Laser Research Optics, REFLEX Analytical, Rocky Mountain Instruments, Sterling Precision Optics

Zirconium dioxide: Harrick Scientific

CHAPTER 5   CHARGED-PARTICLE OPTICS[*]

Fifty years ago, devices employing charged-particle beams were confined to the purview of a small group of physicists studying elementary processes. Beams of ions or electrons are used today by chemists, biologists, and engineers to probe various materials and investigate discrete processes. Physicists are constructing beam machines to control the momentum of interacting particles with energies from a few tenths of an electron volt to trillions of electron volts. Chemists routinely use mass spectrometers as analytical tools and various electron spectrometers to probe molecular structures. The electron microscope is one of the primary tools of the modern biologist. Charged-particle beam technology has also spread to industry, where electron-beam machines are used for cleaning surfaces and welding, and ion-beam devices are used in the preparation of semiconductors.

The properties of charged-particle beams are analogous in many respects to those of photon beams—hence the appellation *charged-particle optics*. In the following sections, the laws of geometrical optics will be covered insofar as they apply to charged-particle beams. The consequences of the coulombic interaction of charged particles will be considered. In addition, we shall discuss the design of electron and ion sources as well as the design of electrodes that constitute optical elements for manipulating beams of charged particles. We shall consider primarily electrostatic focusing by elements of cylindrical symmetry, and restrict discussion to particles of sufficiently low kinetic energies that relativistic effects can be ignored.

A number of excellent books and review articles are available to the reader who wishes to pursue the topic of electron and ion optics in greater detail.[1-6]

5.1 BASIC CONCEPTS OF CHARGED-PARTICLE OPTICS

5.1.1 Brightness

The *brightness* β of a point on a luminous object is determined by the differential of current dI, that passes through an increment of area dA about the point and shines into a solid angle $d\Omega$:

$$\beta = \frac{dI}{dA\,d\Omega} \qquad (\mu\text{A cm}^{-2}\text{sr}^{-1})$$

In electron or ion optics, a luminous object is usually defined by an aperture called a *window* that is uniformly illuminated from one side by a stream of charged particles. The angular spread of particles emanating from the window is limited by a second aperture called a *pupil*. This situation, in the xz plane, is illustrated in Figure 5.1. It will be assumed that the system is cylindrically symmetric about the z-axis and the distance between the window and

---

* The science of charged-particle optics owes a great deal to the principles of design formulated by C. E. Kuyatt (1930–1998). The authors are especially indebted to Dr. Kuyatt for his advice and for permission to present parts of the contents of his "Electron Optics Lectures" in the beginning of this chapter.

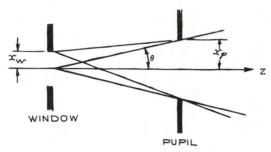

Figure 5.1   A pupil determines the half angle θ of the bundle of rays emanating from each point on an object defined by a window.

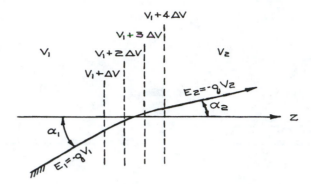

Figure 5.2   A charged-particle trajectory exhibits refraction at a potential gradient. It is assumed that the particle originated at ground potential with no kinetic energy.

pupil is sufficiently great that the angular spread of rays emanating from each point on the object is the same. Then the integrated brightness of the object outlined by the window is

$$\beta = \frac{I}{\pi^2 x_w \theta^2}$$

where x_w is the radius of the window and θ is the half angle defined by the pupil relative to a point at the window.

5.1.2 Snell's Law

When a beam of charged particles enters an electric field, the particles will be accelerated or decelerated and the trajectory will depend on the angle of incidence with respect to the equipotential surfaces of the field. This effect is analogous to the situation in optics when a light ray passes through a medium in which there is a change of refractive index. Figure 5.2 illustrates the behavior of a charged-particle beam as it passes from a region of uniform potential V_1 to a region of uniform potential V_2. The initial and final kinetic energies of a particle of charge q that originates *at ground potential* with no kinetic energy are $E_1 = -qV_1$ and $E_2 = -qV_2$, respectively (i.e., for V in volts and E in eV, the charge q on an electron is -1, and the charge of a singly charged positive ion is $+1$). α_1 and α_2 are the angles of incidence and refraction with respect to the normals to the equipotential surfaces that separate the field-free regions.

The quantity in charged-particle optics that corresponds to the index of refraction is the particle velocity,

proportional to the square root of the particle energy. Thus, the charged-particle analog of Snell's law is

$$\sqrt{E_1}\sin\alpha_1 = \sqrt{E_2}\sin\alpha_2$$

This property can clearly be exploited in charged-particle optics, as in light optics, to make lenses by shaping the equipotential surfaces. This is equivalent to varying the refractive index and shape of lenses for light.

5.1.3 The Helmholtz-Lagrange Law

Consider the situation illustrated in Figure 5.3, in which an object defined by a window at z_1 is imaged at z_2. The rays emanating from a point (x_1, z_1) at the edge of the object are limited by a pupil to fall within a cone defined by a half angle θ_1, which is called the *pencil angle*. The angle of incidence α_1 of the central ray from $(x_1, z_1$) on the plane of the pupil is referred to as the *beam angle*. The rays emanating from each point on the image appear to be limited by an aperture that corresponds to the image of the object pupil. An image pencil angle θ_2 and beam angle α_2 from a point (x_2, z_2) at the edge of the image are defined with respect to this image pupil.

For most practical electron optical systems, the pencil and beam angles are small and the approximation $\sin\theta = \theta$ is valid. The treatment of optics based on this approximation is called *Gaussian* or *paraxial* optics.

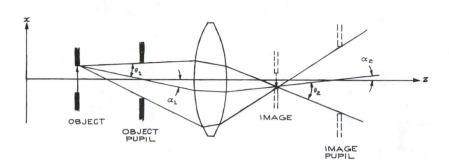

Figure 5.3 Relation of image pencil angle θ_2 and beam angle α_2 to object pencil angle θ_1 and beam angle α_1.

The image pencil angle is determined by the object pencil angle. This relation is given to first order by the Helmholtz-Lagrange law:

$$x_1 \theta_1 \sqrt{E_1} = x_2 \theta_2 \sqrt{E_2}$$

$$\sqrt{\frac{E_1}{E_2}} = Mm \qquad M = \frac{x_2}{x_1} \qquad m = \frac{\theta_2}{\theta_1}$$

where M and m are the linear and angular magnifications, respectively.

The current I_1 through the object window that falls within the object pupil is the same as the current I_2 through the image window and pupil. Setting $I_1 = I_2 = I$, it follows that

$$\frac{I}{E_1 \theta_1^2 x_1^2} = \frac{I}{E_2 \theta_2^2 x_2^2}$$

Furthermore, if β_1 and β_2 are the brightnesses of the object and image, respectively, then this equality can be written:

$$\frac{\beta_1}{E_1} = \frac{\beta_2}{E_2}$$

demonstrating that *the ratio of brightness to energy is conserved from object to image.*

5.1.4 Vignetting

The foregoing discussion assumes that the object radiates uniformly. To the extent that the pencil angles are the same for all points on the object, it follows that the image is uniformly illuminated. If, however, a second pupil is added to the system as illustrated in Figure 5.4, the illumination will vary across the image. Such a situation is called *vignetting*. The one case in which an aperture, in addition to the object window and pupil, does not produce vignetting is

when an aperture is placed at the location of the image of the original window or pupil. Such an aperture is sometimes employed to skim off stray current resulting from scattering from slit edges and from aberrations. An aperture is also placed at the exit window of an energy or momentum analyzer to define its resolution. Except for *spatter apertures* or resolving apertures, an electron-optical system should have only two apertures.

5.2 ELECTROSTATIC LENSES

It is a simple matter to produce axially symmetric electrodes that, when electrically biased, will produce equipotential surfaces with shapes similar to those of optical lenses. A charged particle passing across these surfaces will be accelerated or decelerated, and its path will be curved so as to produce a focusing effect. The chief difference between such a charged-particle lens and an optical lens is that the quantity analogous to the refractive index—namely the particle velocity—varies continuously across an electrostatic lens, whereas a discontinuous change of refractive index occurs at the surfaces of an optical lens. Charged-particle lenses are *thick* lenses, meaning that their axial dimensions are comparable to their focal lengths.

5.2.1 Geometrical Optics of Thick Lenses

In the case of a thick lens, it is not correct to measure focal distances from a plane perpendicular to the axis through the center of the lens. The focal points of a thick lens are

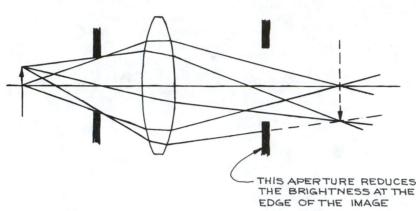

Figure 5.4   Vignetting.

THIS APERTURE REDUCES
THE BRIGHTNESS AT THE
EDGE OF THE IMAGE

located by *focal lengths* f_1 and f_2, measured from *principal planes* H_1 and H_2, respectively. As shown in Figure 5.5, the principal planes of a charged-particle lens are always crossed, and they are both located on the low-voltage side of the central plane M of the lens. The locations of the focal points with respect to the central plane are given by F_1 and F_2, and hence the distances from the central plane to the principal planes are $F_1 - f_1$ and $F_2 - f_2$, respectively. The object and image distances with respect to the principal planes are p and q, and, with respect to the central plane, are P and Q, respectively. All distances are positive in the direction indicated by the arrows in Figure 5.5.

As in light optics, it is possible to graphically construct the image produced by a lens if the cardinal points of the lens are known. This procedure is illustrated in Figure 5.6. It is only necessary to trace two principal rays. From a point on the object, draw a ray through the first focal point and thence to the first principal plane. From the point of intersection with this plane, draw a ray parallel to the axis. Draw a second ray through the object point and parallel to the axis. From the point of intersection of this ray with the second principal plane, draw a ray through the second focal point. The intersection of the first and second principal rays gives the location of the image point corresponding to the object point.

A number of important relationships can be derived geometrically from Figure 5.6. The linear and angular magnifications are given by

$$M = \frac{f_2 - q}{f_2} = \frac{f_1}{f_1 - p}$$

and

$$M = \frac{f_1 - p}{f_2} = \frac{f_1}{f_2 - q}$$

where negative magnification implies an inverted image.

The object and image distances from the central plane are

$$P = F_1 - \frac{f_1}{M}$$

and

$$Q = F_2 = Mf_2$$

The object and image distances from the principal planes are related by Newton's law,

$$(p - f_1)(q - f_2) = (P - F_1)(Q - F_2) = f_1 f_2$$

and Newton's formula,

$$\frac{f_1}{p} + \frac{f_2}{q} = 1$$

For lenses that are not very strong, the principal planes are close to the central plane. Then

$$p \to P$$

$$q \to Q$$

and

$$f_1 \cong f_2 \cong f$$

to give an approximate form of Newton's formula that is quite useful in the initial stages of design work:

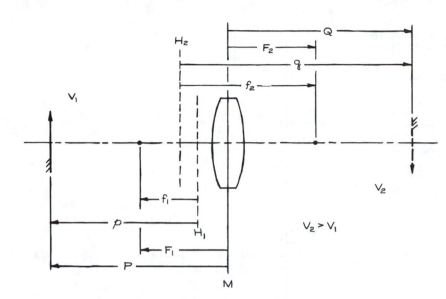

Figure 5.5 Lens parameters. H_1 and H_2 are the principal planes.

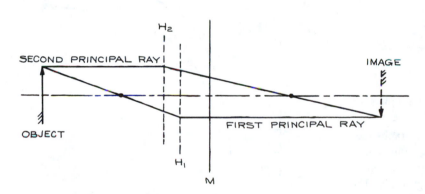

Figure 5.6 Graphical construction of an image.

$$\frac{1}{P} + \frac{1}{Q} = \frac{1}{f}$$

Furthermore, Spangenberg has observed that for the lenses discussed in succeeding sections,[1]

$$M \cong -0.8\frac{Q}{P}$$

5.2.2 Cylinder Lenses

The most widely used lens for focusing charged particles with energies of a few eV to several keV is that produced by two cylindrical coaxial electrodes biased at voltages corresponding to the desired initial and final particle energies. Figure 5.7 illustrates the equipotential surfaces associated with the field produced by two coaxial cylinders biased at voltage V_1 and V_2, respectively.

The focal properties of a *two-cylinder lens* depend upon the diameters of the cylinders, the spacing between them, and the ratio E_2/E_1 of the final to initial kinetic energies of the transmitted particles. The lens properties scale with the diameter, so all dimensions are taken in units of the diameter D of the larger cylinder. If the gap between cylinders is large, the focal properties become sensitive to the cylinder wall thickness and external fields are liable to penetrate to the lens. Most lenses are constructed with a gap $g = 0.1D$. The particle energies depend upon the electrical potentials (bias voltages) applied to the

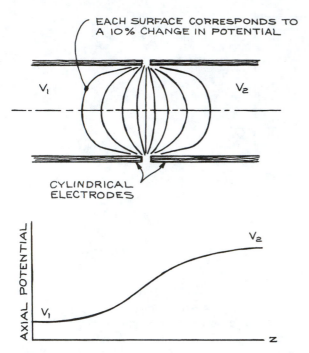

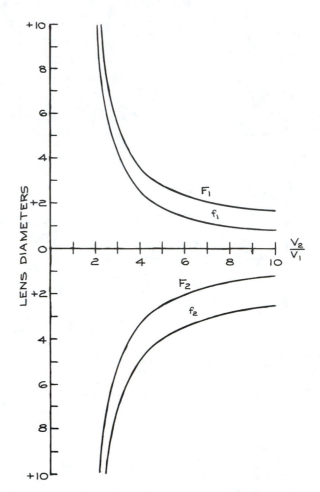

Figure 5.7 Potential distribution in a two-cylinder lens.

Figure 5.8 Focal properties of a two-cylinder lens with coaxial cylinders of the same diameter D and a gap between of 0.1D. (From E. Hartung and F. H. Read, *Electrostatic Lenses*, Elsevier, New York, 1976; by permission of Elsevier Publishing Company.)

elements. It is assumed that the bias voltage supplies are referenced to the particle source, so that $V_2/V_1 = E_2/E_1$.

Focal properties are usually presented as graphs (see Figure 5.8) or tables of F_1, f_1, F_2, and f_2 as a function of V_2/V_1. For simple lenses it is also convenient to plot P *versus* Q for different V_2/V_1 ratios as in Figure 5.9. The chief sources of information on the focal properties of electrostatic lenses are the book *Electrostatic Lenses* by Harting and Read[7] and various journal articles by Read and his coworkers.[8] These data are determined from numerical solutions of the equation of motion of an electron in a lens field and are thought to be accurate to 1 to 3 percent. Unfortunately, these sources contain no information applicable to systems with virtual objects or images. For two-cylinder lenses, Natali, DiChio, Uva, and Kuyatt have computed P-Q curves that extend to negative values of the object and image distances.[9]

To obtain desired focal properties with a given acceleration ratio, two or more two-cylinder lenses can be used in series. The lens field extends about one diameter D on either side of the midplane gap. When two gaps are

sufficiently close that their lens fields overlap, it is convenient to treat the combination as a single lens. Such a *three-cylinder lens* has the advantage that it can produce an image of a fixed object at a fixed image plane for a range of final-to-initial energy ratios. Alternatively, it can produce a variable image location with a fixed energy ratio.

A three-cylinder lens for which the initial and final energies differ is called an *asymmetric lens*. It can be used

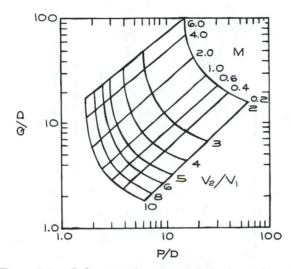

Figure 5.9   P-Q curves for a two-cylinder lens with equal diameter cylinders and a gap of $0.1D$. (From E. Hartung and F. H. Read, *Electrostatic Lenses*, Elsevier, New York, 1976; by permission of Elsevier Publishing Company.)

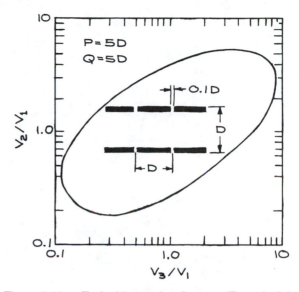

Figure 5.10   Typical "zoom-lens" curve. (From A. Adams and F. H. Read, "Electrostatic Cylinder Lenses III," *J. Phys.*, **E5**, 156; copyright 1972 by the Institute of Physics.)

with an electron or ion source to produce a variable-energy beam. Such lenses are also used to focus an image of the exit slit of a monochromator on a target while allowing for a variation of the particle acceleration between monochromator and target.

The focal properties of three-cylinder lenses are not conveniently presented graphically, since each focal length is a function of two independent variables. These variables are usually taken to be V_2/V_1 and V_3/V_1, where V_1, V_2, and V_3 are the voltages on the first, second, and third lens elements. Harting and Read and Heddle present data on three-element lenses for V_3/V_1 values between 1.0 and 30.0 in the form of tables of lens parameters (f_1, f_2, F_1, F_2) as a function of V_2/V_1.[7,10] They also plot *zoom-lens curves* as graphs of V_2/V_1 *versus* V_3/V_1 for selected values of P and Q. An example is given in Figure 5.10.

Read has defined a length f that approximately satisfies the relation

$$\frac{1}{f} = \frac{1}{P} + \frac{1}{Q}$$

for three-element lenses. For use in initial design work, V_2/V_1 is graphed as a function of f/D for various values of V_3/V_1.[3]

A three-element lens for which $V_3/V_1 = 1$ is called an *einzel lens* or *unipotential lens*. Such a lens produces focusing without an overall change in the energy of the transmitted particle. The focal properties are symmetrical: $f_1 = f_2 = f$ and $F_1 = F_2 = F$. Figure 5.11 illustrates the focal properties of a typical einzel lens. For any desired object and image distance there are two focusing conditions—a decelerating mode with $V_2/V_1 < 1$ and an accelerating mode with $V_2/V_1 > 1$. The accelerating mode is preferred, because deceleration in the middle element causes expansion of the transmitted beam resulting in aberrations and unwanted interactions with the lens surfaces.

5.2.3 Aperture Lenses

An aperture in a plane electrode separating two uniform-field regions will have a focusing effect if the field on one side is greater than on the other. As illustrated in Figure 5.12, this lens results from a bulge in the equipotential surfaces caused by field penetration through the aperture. A lens of this type is called a *Calbick lens*. Single-aperture focusing is important in electron-gun and ion-source

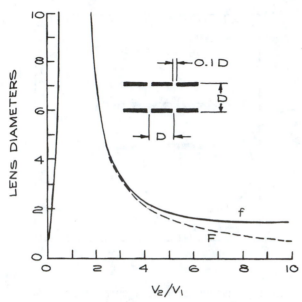

Figure 5.11   Focal properties of an einzel lens with the intermediate cylinder of length $A = D$ and lens gap $g = 0.1D$. (From E. Hartung and F. H. Read, *Electrostatic Lenses*, Elsevier, New York, 1976; by permission of Elsevier Publishing Company.)

design because these devices usually have an aperture at their output.

The Calbick lens is the analog of a thin lens. The electric fields are established by the potentials V_1, V_2, and V_3 applied to the three planar elements as shown in Figure 5.12. In this case it is assumed that electrons or ions originate at potential V_1 with essentially no kinetic energy. For a circular aperture,

$$f_1 = f_2 = F_1 = F_2 = \frac{4V_2}{\mathcal{E}_B - \mathcal{E}_A}$$

where the electrical field intensities in the uniform-field regions are

$$\mathcal{E}_A = \frac{V_2 - V_1}{a} \qquad \mathcal{E}_B = \frac{V_3 - V_2}{b}$$

The ratio of aperture potential to radius must be larger than $\mathcal{E}_A$ or $\mathcal{E}_B$. For a long slot in the xy plane,

$$f_x = \frac{2V_2}{\mathcal{E}_B - \mathcal{E}_A}$$

and

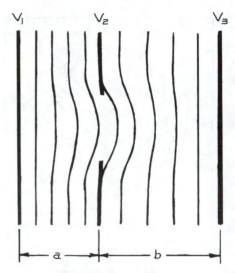

Figure 5.12   Protrusion of equipotential surfaces through an aperture seperating regions of different field strength.

$$f_y \cong \infty$$

when the short dimension of the slot is in the x-direction. The focal properties of a slot are important in the design of electron guns that employ a ribbon filament.[11]

Lenses can also be constructed using two or more apertures. Read has computed the properties of two-aperture and three-aperture lenses.[7,8]

In practice, concentric cylindrical mounts support the circular-aperture electrodes of an aperture lens. The focusing effects of these cylindrical elements can be ignored if the cylinder diameter is at least three times greater than the aperture diameter.

5.2.4 Matrix Methods

A useful description of the charged-particle trajectories through a focusing system can be formulated using the matrix methods introduced in Section 4.2.2. A particle trajectory through a system that is cylindrically symmetrical about the z-axis is determined by (r, $dr/dz = r'$, z), where r is the radial distance of the trajectory from the axis at z. A trajectory through a plane perpendicular to the central axis at z is represented by a vector

$$\begin{pmatrix} r \\ r' \end{pmatrix}$$

The effect of a displacement in a field-free region from z_1 to z_2 is given by

$$\begin{pmatrix} r_2 \\ r'_2 \end{pmatrix} = \mathbf{M}(z_1 \to z_2)\begin{pmatrix} r'_1 \\ r'_1 \end{pmatrix}$$

where the *ray transfer matrix* is

$$\mathbf{M}(z_1 \to z_2) = \begin{pmatrix} 1 & \Delta z \\ 0 & 1 \end{pmatrix} \qquad \Delta z = z_2 - z_1$$

A lens can be represented as if its effect were confined to the region between the principal planes. The matrix operating between the principal planes is

$$\mathbf{M}(H_1 \to H_2) = \begin{pmatrix} 1 & 0 \\ -\dfrac{1}{f_2} & \dfrac{f_1}{f_2} \end{pmatrix}$$

Consider, for example, a ray originating at a point on an object at z_1 and passing through a point on an image at z_2. The trajectory at z_2 can be determined from the trajectory at z_1 by

$$\begin{pmatrix} r_2 \\ r'_2 \end{pmatrix} = \mathbf{M}(H_2 \to z_2)\mathbf{M}(H_1 \to H_2)\mathbf{M}(z_1 \to H_1)\begin{pmatrix} r_1 \\ r'_1 \end{pmatrix}$$

In terms of the parameters defined in Figure 5.5, the displacement between the object and the first principal plane is $P - F_1 + f_1$; thus

$$\mathbf{M}(z_1 \to H_1) = \begin{pmatrix} 1 & P - F_1 + f_1 \\ 0 & 1 \end{pmatrix}$$

and similarly

$$\mathbf{M}(H_2 \to z_2) = \begin{pmatrix} 1 & Q - F_2 + f_2 \\ 0 & 1 \end{pmatrix}$$

Substitution yields

$$\begin{pmatrix} r_2 \\ r'_2 \end{pmatrix} = \begin{pmatrix} \dfrac{F_2 - Q}{f_2} & \dfrac{f_1 f_2 - (Q - F_2)(P - F_1)}{f_2} \\ -\dfrac{1}{f_2} & \dfrac{F_1 - P}{f_2} \end{pmatrix}\begin{pmatrix} r_1 \\ r'_1 \end{pmatrix}$$

This transformation matrix is generally applicable to the problem of tracing a trajectory through the field of a lens. For the two-cylinder lens, DiChio et al. have derived simple analytical expressions for the elements of the matrix.[12] When r_1 and r_2 refer to points in the object and image planes, respectively, the diagonal elements of the transformation matrix are the linear and angular

magnification. Furthermore, the focusing condition given by Newton's formula requires the numerator of the upper right element to be zero. Thus, the transformation matrix between object and image is

$$\mathbf{M}(\text{object} \to \text{image}) = \begin{pmatrix} M & 0 \\ -\dfrac{1}{f_2} & m \end{pmatrix}$$

The matrix formulation is a desirable alternative to graphical ray tracing for a system of lenses that are so close to one another that an image does not appear between each lens and the next. Consider, for example, the system of two lenses in Figure 5.13. In this case the transformation matrix is

$$\mathbf{M}(z_1 \to z_2) = \mathbf{M}(H'_2 \to z_2)\mathbf{M}(H'_1 \to H'_2) \times$$
$$\mathbf{M}(H_2 \to H'_1)\,\mathbf{M}(H_1 \to H_2)\mathbf{M}(z_1 \to H_1)$$

In general, the image distance L_3 will be unknown initially. It can, however, be determined from the focusing condition that requires the upper right element of $\mathbf{M}(z_1 \to z_2)$ to be zero.

It is occasionally necessary to determine a particle trajectory in a uniform field region. The transformation matrix from z_1 to z_2 in a uniform field in the z-direction is

$$\mathbf{M}(\text{uniform field}) = \begin{pmatrix} 1 & \dfrac{2\Delta z V_1}{V_2 - V_1}\sqrt{\dfrac{V_2}{V_1}} - 1 \\ 0 & \sqrt{\dfrac{V_1}{V_2}} \end{pmatrix}$$

where V_1 and V_2 are the potentials of the planar equipotential surfaces perpendicular to the z-axis at z_1 and z_2.

5.2.5 Aberrations

There are three types of aberrations that must be considered when designing charged-particle optics:

- *Geometrical aberrations* due to deviations from the paraxial assumption that the angle θ between a ray and the central axis is sufficiently small that $\theta \approx \sin\theta \approx \tan\theta$.
- *Chromatic aberrations* caused by variations in the kinetic energies of transmitted particles.
- *Space charge* due to electron-electron interactions.

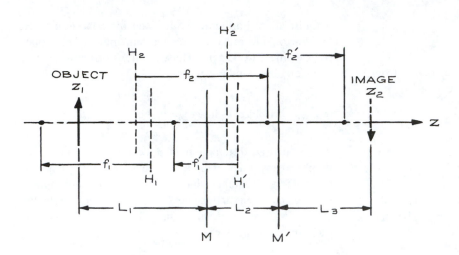

Figure 5.13　A compound lens system.

The *geometrical aberration* associated with a lens is defined in terms of the radius Δr_2 of a spot in the image plane formed by rays within the pencil angle θ_1 that emanate from an object point on the axis, as shown in Figure 5.14. The important geometrical errors can be expressed in terms of the coefficients of a power series in θ_1:

$$\Delta r_2 = b\theta_1^2 + c\theta_1^3 K$$

An imaging system is said to be *second-order focusing* if $b = 0$, and *third-order focusing* if $c = 0$. An axially symmetric system is always at least second-order-focusing, since only terms of odd order appear in the expansion of Δr_2. The coefficient c is referred to as the *third-order aberration coefficient*. Higher-order coefficients can usually be ignored. For an object point on a lens axis, the error represented by c is the spherical aberration. Harting and Read have computed spherical-aberration coefficients C_s defined by

$$c = MC_S$$

for all of the lenses for which they give focal properties.[7] For object points off axis, other types of aberrations must be considered (coma, field curvature, distortion, and astigmatism). Most off-axis aberrations can be related to the magnitude of C_s. When the object is small compared to the lens diameter, the spherical-aberration coefficient gives a reasonable measure of the total geometrical aberration.

Figure 5.14　Geometrical aberration of the image of a point.

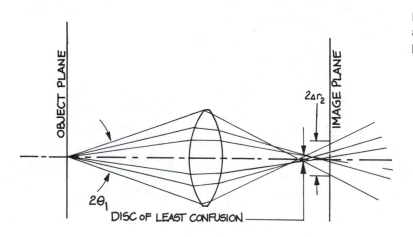

As can be seen from Figure 5.14, the pencil of rays emanating from an object point achieves a minimum diameter at a point slightly in front of the image plane. This minimum diameter is the *circle of least confusion*. To achieve the sharpest focus it is often possible to vary the lens voltages so as to place the circle of least confusion on the desired image plane. The radius of this circle is

$$\frac{1}{4}\Delta r_2$$

and thus C_s can be reduced by a factor of four in practice.

The diameter of the bundle of rays emanating from an object is usually determined by a pupil that limits the portion of the lens that is illuminated by rays from the object. The extent of illumination is specified by a *filling factor*, which is the ratio of the diameter of the bundle of rays near the lens gap to the lens diameter. Because the spherical aberration depends upon the third power of the angle between the limiting ray and the axis, it is obvious that aberration increases rapidly with increasing filling factor. In practice, a lens system should not be designed with a filling factor in excess of about 50 percent.

For an object of finite extent, the worst aberration occurs for off-axis image points. The magnitude of the aberration depends upon the maximum value of the angle of a ray from the edge of the object relative to the central axis of the lens system. From Figure 5.3 it can be seen that this angle is the sum of the object pencil angle and beam angle. In many lens systems, the image produced by one

lens serves as an object for succeeding lenses. The pencil angle associated with this intermediate image is determined by the Helmholtz-Lagrange law from the pencil angle associated with the initial object. There are, however, no such physical restrictions on the beam angle from the intermediate image. In fact, as shown in Figure 5.15, if an object pupil is placed at the first focal point of a lens, then the corresponding image pupil is at infinity and the image beam angle is zero. Such an arrangement obviously reduces aberrations produced by lenses that treat this image as an object.

Chromatic aberration depends upon the relative spread in particle energies, $\Delta E/E$, passing through a lens. Variations in particle energies can occur because of conditions in the particle source or within the lens. An energy spread of 0.2 to 0.4 eV is characteristic of electron sources. Some ion sources yield ions in an energy bandwidth as great as 30 eV. Energy variations may also be caused by fluctuations in the voltage from the power supplies that establish both source and lens-element potentials.

For a given lens, the extent of chromatic aberration can best be determined from a calculation of the image distances for particles with energies at the extremes of the anticipated energy distribution. Consider, for example, the case of a source at ground potential that produces particles with energies between 0 and ΔE. For a lens of voltage ratio V_2/V_1 a distance P from the source, determine the corresponding image position, $Q = f(P, V_2/V_1)$. Determine also

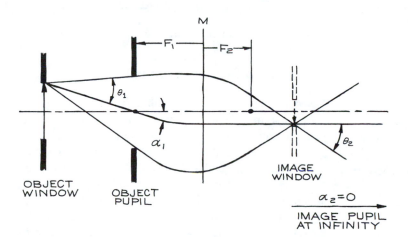

Figure 5.15   An object pupil at the first focal point moves the image pupil to infinity and reduces the image beam angle α_2 to zero.

$$Q + \Delta Q = f\left(P, \frac{V_2 - \Delta E/q}{V_1 - \Delta E/q} \right)$$

Then a point on axis at the source yields an image at Q of radius

$$\Delta r_2 = \theta_2 \Delta Q$$

or, by the Helmholtz-Lagrange law,

$$\Delta r_2 = \frac{\theta_1 \Delta Q}{M} \sqrt{\frac{V_2}{V_1}}$$

where θ_1 and θ_2 are the pencil angles at the source and the image, respectively.

In electron-beam devices, the main cause of chromatic aberration is the spread of electron kinetic energies emitted from a thermionic cathode as well as the potential drop across the emitting portion of the cathode. The energy of electrons at the maximum of the current distribution from a cathode is

$$E_{max} = 8.6 \times 10^{-5} T \text{ eV}$$

The width of the distribution is $\Delta E \approx E_{max}$.

For a bare tungsten filament, a temperature of about 3000 K is required to produce significant emission at this temperature, $\Delta E \cong 0.25 \text{eV}$. An oxide-coated cathode (available from Electron Technology) can be operated at temperatures as low as 1200 K, where $\Delta E \cong 0.1 \text{eV}$. The potential drop across the emitting portion of the cathode may add several tenths of an eV to the energy spread.

The *space-charge* effect arises from the mutual repulsion of particles of like charge. The effect increases with current density. As a result, even a focused beam will give a diffuse image. Furthermore, the beam will diverge away from the image more rapidly than predicted by geometrical optics.

Space charge places a limit on the current in a charged-particle beam. As an example of practical value, consider the problem of transmitting maximum current through a cylindrical element of length L and diameter D. As shown in Figure 5.16, the maximum current is achieved by focusing the beam on a point at the center of the tube with a pencil angle $\theta = D/L$. It can be shown that the space charged-limited current of electrons of energy $E = -qV$ is

$$I_{max}(\text{electrons}) = 38.5 V^{3/2} \frac{D^2}{L} \quad \mu A$$

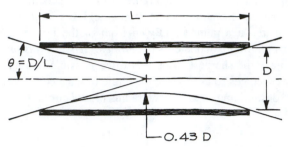

Figure 5.16   A beam focused to give maximum current through a tube.

and the maximum ion current is

$$J_{max}(\text{ions}) = 0.90 \frac{M^{1/2}}{n} V^{3/2} \frac{D^2}{L} \quad \mu A$$

where V is in volts, M is in amu, and n is the charge state.[2] The minimum beam diameter at the space-charge limit is $0.43D$.

Space charge also limits the current available from a surface that emits electrons or ions. The typical source geometry is the plane diode consisting of a planar anode at potential V a distance d from the cathode. Charged particles produced by thermionic emission from the cathode are accelerated toward the anode. The maximum electron current density at the anode is

$$J_{max}(\text{electrons}) = 2.34 \frac{V^{3/2}}{d^2} \quad \mu A \text{ cm}^{-2}$$

while for an ion-emitting cathode,

$$J_{max}(\text{ions}) = 0.054 \frac{n^{1/2}}{M} \frac{|V|^{3/2}}{d^2} \quad \mu A \text{ cm}^{-2}$$

with V in volts, d in cm, M in amu, and n the charge state.

5.2.6 Lens Design Example

Consider the problem of producing the image of the anode aperture of an electron source on the entrance plane of an energy analyzer as illustrated in Figure 5.17. The anode potential is $V_1 = 100$ V, and the analyzer entrance-plane potential is $V_2 = 10$ V. The anode aperture radius is $r_1 = 1.0$ mm, and the distance from anode to analyzer is $l = 10$ cm. It is desired that the image of the anode aperture serve as the entrance aperture for the analyzer with radius $r_2 = 0.5$ mm. So as not to overfill

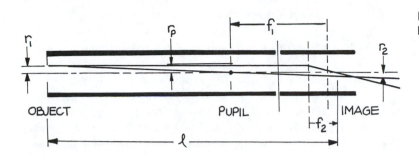

Figure 5.17   A simple lens design.

the analyzer, one must also have the image pencil angle $\theta_2 = 0.14$ and the image beam angle $\alpha_2 = 0$.

The lens to be designed is a decelerating lens where $V_2/V_1 = 0.1$ and magnification $M = 0.5$. Only the focal properties of accelerating lenses are customarily tabulated. The focal properties for the desired lens are just the reverse of those of the accelerating lens with $V_2/V_1 = 10$. From Harting and Read,[7] the focal properties of the $V_2/V_1 = 10$ lens (identified by a prime) are

$$f_i' = 0.80D \qquad F_i' = 1.62D$$
$$f_2' = 2.54D \qquad F_2' = 1.19D$$

The corresponding parameters for the $V_2/V_1 = 0.1$ lens are

$$f_i = 2.54D \qquad F_1 = 1.19D$$
$$f_2 = 0.80D \qquad F_2 = 1.62D$$

From the P-Q curve of Figure 5.9 for $V_2/V_1 = 10$ and $M' = 1/M = 2$, we estimate

$$P = Q' = 6.20D$$
$$Q = P' = 2.05D$$

Taking $P = 6.20D$, Newton's law gives

$$Q = \frac{f_1 f_2}{P - F_1} + F_2 = 2.03D$$

The overall length of the lens is

$$l = P + Q = 6.20D + 2.03D = 10 \text{ cm}$$

and thus

$$D = 1.22 \text{ cm}$$

In order for the image beam angle to be zero, the object pupil should be placed at the first focal point of the lens so that the image pupil is at infinity. The object pupil must define a pencil angle θ_1 that is consistent with the required image pencil angle. By the Helmholtz-Lagrange law,

$$\theta_1 = M \theta_2 \sqrt{\frac{V_2}{V_1}} = 0.022$$

This angle is determined by the radius r_p of an aperture at the first focal point:

$$r_p = \theta_1(P - F_1) = 0.13 \text{ cm}$$

To estimate the geometrical aberration of this lens, we calculate the spherical aberration for the lens used in reverse. This is necessary because spherical-aberration coefficients are available only for lenses with $V_2/V_1 > 1$. For $V_2/V_1 = 10$, Harting and Read give

$$\frac{M' C_S}{D} = 20$$

Thus for the lens used in reverse, the image of a point is a spot of diameter

$$\Delta r' = M' C_S \theta_2^3 = 0.067 \text{ cm}$$

For the lens used as designed, the size of the image of an object point is

$$\Delta r = M \Delta r' = 0.033 \text{ cm}$$

Thus spherical aberration will enlarge the image by 67 percent; however, if the lens voltage is adjusted slightly to bring the circle of least confusion onto the image plane, the spherical aberration can be reduced to 17 percent.

For a typical electron source, $\Delta E = 0.3$ eV. This energy spread will result in chromatic aberration at the image. To estimate this effect, calculate the image position $Q + \Delta Q$

for electrons at the extreme of the energy distribution where the deceleration ratio of the lens is

$$(V_2 - \Delta E_4)/(V_1 - \Delta E_4) = 10.3/100.3 = 0.103$$

As before, the focal properties can be determined from those of the corresponding accelerating lens.

For the lens with $V_2/V_1 = 0.103$,

$$f_1 = 2.58D \qquad F_1 = 1.22D$$
$$f_2 = 0.82D \qquad F_2 = 1.65D$$

The image position by Newton's law in this case is

$$Q + \Delta Q = \frac{f_1 f_2}{P - F_1} + F_2 = 2.07D$$

so

$$\Delta Q = 0.04D = 0.05 \text{ cm}$$

This displacement of the image plane will cause the image of a point object to appear as a disc of radius

$$\Delta r = \Delta Q \theta_2 = 0.007 \text{ cm}$$

in the original image plane. Chromatic aberration enlarges the image by about 14 percent.

The overall aberration of the lens as designed is about 30 percent. The filling factor is

$$\frac{2(\theta_1 + \alpha_1)P}{D} = 54\%$$

From the aberration calculations it can be seen that the aberrations would become serious if the filling factor were to exceed this value.

5.2.7 Computer Simulations

There are a variety of programs available for the computation of charged-particle trajectories in the field of an array of electrodes. These employ Green's function, finite-element, and surface-charge methods to compute the electrical field produced by the array. All provide for the possibility of an applied magnetic field and some solve for the field produced by an array of magnets. A number of the currently available programs and their sources are listed at the end of this chapter.

It must be emphasized that computer simulation programs cannot substitute for an understanding of the basic principles of electron optics described in this chapter. It is impossible to obtain the desired performance in the design of an electron optical system by arbitrarily choosing an array of electrodes and *fiddling* to obtain the desired beam properties. The positions of lenses, windows, and especially pupils should be determined from considerations of basic principles before turning to the computer to optimize a design or test its performance. A particular virtue in computer simulation programs is that they can be used to predict the properties of compound lenses that are more complex than the two- and three-element lenses that are described above and for which there are tabulated values of focal lengths and focusing properties.

Figure 5.18a shows a computer simulation of the simple cylindrical lens discussed in Section 5.2.6 and illustrated in Figure 5.17. The computed electric field at the lens gap is shown as equipotential surfaces. Electron trajectories are traced from points at the center and the upper edge of the object window. One of the three trajectories from each point passes through the center of the pupil; the other two are the extreme rays that can pass through the pupil. Figure 5.18b—a blowup of a portion of 5.18a—shows the location of the image formed. It is worth noting that the size of the image is half that of the object, as predicted in Section 5.2.6. The image distance, $Q = 1.93D$, is somewhat short of that predicted. The bundle of rays from both the center and the edge of the object emerge from the image with their central axes parallel to the axis of the lens. This implies that the beam angle $\alpha_2 = 0$ (see Figure 5.15). Figure 5.18c shows a lens system identical to that in 5.18a, except the pupil has been moved closer to the object and away from the first focal point of the lens. The result is a significant increase in the image beam angle.

5.3 CHARGED-PARTICLE SOURCES

An electron or ion gun consists of a source of charged particles, such as a hot metal filament or plasma, and an electrode structure that gathers particles from the source and accelerates them in a particular direction to form a beam. There is a wide range of practical electron and ion guns. Some simple devices and their principles of

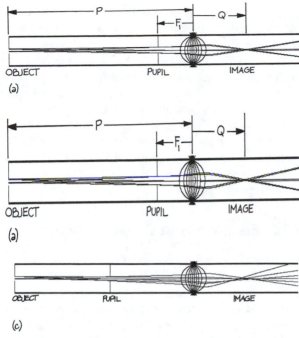

Figure 5.18   (a) MacSimion simulation of the cylindrical lens shown in Figure 5.17 and discussed in section 5.2.6. The voltage ratio $V_2/V_1 = 0.10$; the object is defined by an aperture window and the object distance from the lens gap is $P = 6.20D$. (D is the diameter); a pupil is placed at the first focal point of the lens. The electric field at the lens gap is shown by equipotential surfaces computed by the program. The trajectories of 100 eV electrons originating at the center and edge of the window have been traced by the computer program. (b) A blowup showing the image formed. The image distance from the lens gap is $Q = 1.93D$. (c) The same lens with the pupil moved closer to the object and away from the first focal point of the lens. The result is a significant increase in the beam angle at the image.

operation will be described in this section. In addition to simple devices that can be conveniently constructed for laboratory use, there are a number of very sophisticated guns that have been developed commercially. These include electron guns for cathode-ray tubes and electron-beam welders, and ion guns for sputter-cleaning apparatus and for ion-implantation doping of semiconductors.

5.3.1 Electron Guns

Electrons are produced by thermionic emission, field emission, photoelectric emission, and electron-impact ionization. Thermionic sources are most common.[13] These sources typically consist of a wire filament of some refractory metal, such as tungsten or tantalum, that is heated by an electrical current passing through it. In some cases, thermionic sources consist of a metal cup or button that is indirectly heated by an electrical heater or by electron bombardment of the back surface. The electrons from a hot filament possess energies of a few tenths of an electron volt. Mutual repulsion will cause the electrons to diffuse away from the source if they are not immediately accelerated into a beam. Many accelerator structures are used. Most use either diode (two-electrode) or triode (three-electrode) geometries, although some TV tubes employ a pentode geometry. The Pierce diode geometry is most common in laboratory devices.

Figure 5.19a illustrates a simple diode electron gun consisting of a plane emissive surface and a parallel plane anode. Electrons leave the cathode with a nominal energy E_k. The anode is biased at a positive potential V_a relative to the cathode, so that electrons from a spot on the cathode will appear at a spot on the anode with energies of approximately $E_a = -qV_a$. To admit the accelerated electrons to the system beyond, a hole is made in the anode. If the cathode and anode were infinite in extent, the space charge-limited current density given in Section 5.2.5 could be achieved at the anode, and the electron beam emerging from the anode hole would be characterized by a beam angle

$$\alpha = \frac{r_a}{3d}$$

where r_a is the radius of the anode hole and d the cathode-to-anode spacing. Since the most divergent electron arriving at the anode would be one emitted parallel to the cathode with energy E_k, it can be seen that the pencil angle characterizing the beam from the anode aperture would be

$$\theta = \sqrt{\frac{E_k}{E_a}}$$

For a cathode of finite extent, the space-charge interaction causes the beam to spread laterally within the gap between

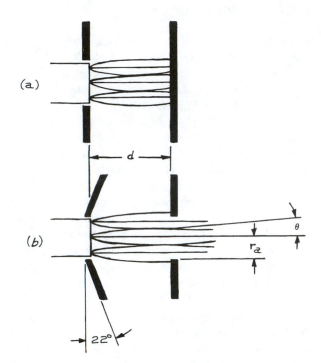

(a)

(b)

$22°$

Figure 5.19 (a) The plane diode; (b) the Pierce geometry that simulates the field of the plane diode.

cathode and anode. Pierce has shown that the electric field in an infinite space charge-limited diode can be reproduced in the region of a finite cathode by means of the conical cathode structure illustrated in Figure 5.19b.[2] The maximum current of electrons is

$$I_{max} = \pi r_a^2 J_{max} = 7.35 \frac{r_a^2}{d} V_a^{3/2} \quad \mu A$$

The corresponding ratio of brightness to energy (which is conserved) is

$$\frac{\beta}{E_a} = \frac{I_{max}}{E_a \pi^2 r_a^2 \theta^2}$$

$$= 0.74 \frac{E_a^{3/2}}{E_k d^2} \quad \mu A \, cm^{-2} \, sr^{-1} \, eV^{-1}$$

The design of multielectrode guns is a complicated process involving time-consuming experimentation. A number of special guns have been developed for purposes such as producing low-energy electron beams or very narrow beams. Klemperer and Barnett describe many of these,[3] and most can be constructed easily if they are really needed. In general, however, it is best to use the simplest device that will fulfill one's requirements.

The guns developed for TV tubes are the product of much industrial development. These guns will produce a beam of electrons of several hundred eV to several tens of thousands eV that is sharply focused to a spot at a distance of 10 to 20 cm. Cathode-ray tube guns can be obtained quite inexpensively from commercial suppliers such as Cliftronics. Specialized guns of similar design are available from Kimble Physics and other manufacturers of electron spectrometers.

5.3.2 Electron-Gun Design Example

Consider the problem of constructing a gun consisting of a Pierce diode and a lens system that can inject a beam of 20 to 200 eV electrons into a gas cell, as illustrated in Figure 5.20. It is desired that the beam current approach the space-charge limit at 20 eV. The overall length of the gun is to be about 10 cm, and the cell is to be located 4 cm in front of the gun. The cell is 2 cm long, with an input aperture of radius $r_2 = 0.1$ cm and an exit aperture of radius 0.15 cm. Treat the exit aperture as though its radius r_4 were 0.1 cm, so as to make the beam tight enough to minimize backscattering of electrons from the edges of this aperture.

Note that the electron gun system in Figure 5.20 is wired in such a way that the energies of the electrons through the system, E_1, E_2, and E_3, in eV, are numerically equal to the power supply voltages V_1, V_2, and V_3. This is, in general, a great convenience since electron energies can be read directly from the power supplies. In the following discussion, E_i and V_i can be used interchangeably.

The maximum current through the cell can be taken as the space charge-limited current through a cylinder with a diameter D equal to that of the gas-cell apertures and length L equal to the length of the gas cell. From Section 5.2.5, the maximum current of electrons with energy $E_3 = 20$ eV is in this case

$$I_{max} = 38.5 V_3^{3/2} \frac{D^2}{L}$$

$$= 34.4 \ \mu A$$

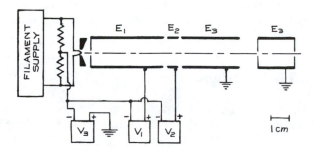

Figure 5.20 An electron gun system to inject a variable-energy beam into a gas cell.

To approach this limit, focus an image of the anode aperture on the center of the cell. Make the image radius $r_3 = 0.43r_2$, the minimum size of the space charge-limited beam.[14] The image pencil angle will be $\theta_3 = D/L = 0.1$ and the ratio of brightness to energy at this image is

$$\frac{\beta_3}{E_3} = \frac{I_{max}/\pi^2 r_3^2 \theta_3^2}{E_3}$$

$$= 9500 \; \mu A \; cm^{-2} \; sr^{-1} \; eV^{-1}$$

At the anode of the Pierce diode (Section 5.3.1), the ratio of brightness to energy is given by

$$\frac{\beta_1}{E_1} = 0.74 \frac{E_1^{3/2}}{E_k d^2}$$

where E_k is the mean kinetic energy of electrons emitted by the cathode, and E_1 is the energy of electrons at the anode that is held at the potential V_1 relative to the cathode. Now the ratio of brightness to energy is a conserved quantity, so

$$\frac{\beta_1}{E_1} = \frac{\beta_3}{E_3}$$

Substituting from the previous two equations yields

$$E_1 = \frac{\beta_3}{E_3} \frac{E_k d^{2/3}}{0.74}$$

For $E_k = 0.25$ eV (see Section 5.2.5), a cathode-to-anode spacing $d = 0.5$ cm, and a final electron energy of 20 eV, the kinetic energy of electrons at the anode, $E_1 = 86$ eV, and the potential to be applied to the anode is

$$V_1(E_3 = 20 \; eV) = 86 \; V$$

A similar calculation yields

$$V_1(E_3 = 200 \; eV) = 185 \; V$$

The current from the diode depends upon the radius of the anode aperture r_1 and the current density J at the anode:

$$I = J\pi r_1^2$$

With the lens system tuned to inject 20 eV electrons into the target cell ($E_3 = 20$ eV), the maximum current density at the anode is

$$J_{max} = \frac{2.34}{d^2} V_1^{3/2} \qquad V_1 = V_1(E_3 = 20 \; eV)$$

$$= 7500 \; \mu A \; cm^{-2}$$

To achieve maximum current in the cell at 20 eV, we choose the anode aperture to be

$$r_1 = \left(\frac{I_{max}}{\pi J_{max}} \right)^{1/2} = 0.04 \; cm$$

A similar calculation for $E_3 = 200$ eV would suggest using a larger anode aperture. This would produce an excess of current when the system is tuned to produce a low-energy beam, however, resulting in a large number of stray electrons.

Note that when the system is designed to approach the space-charge limit at 20 eV, the pencil angle at the anode is consistent with the pencil angle defined by the cell apertures. The pencil angle characteristic of the Pierce diode at the low-energy limit is

$$\theta_1 = \sqrt{\frac{E_k}{E_1}} = \sqrt{\frac{0.25}{86}} = 0.05$$

and the anode pencil angle that gives the space charge-limited pencil angle in the cell is, according to the Helmholtz-Lagrange law,

$$\theta_1 = \sqrt{\frac{E_k}{E_1}} \left(\frac{r_3}{r_1} \right) \theta_3 = 0.05$$

For final energies greater than 20 eV, these pencil angles will be less than the angle defined by the cell apertures, and scattering from the edges of the apertures will be minimized.

All that remains is to design a lens system that will image the anode aperture on the center of the gas cell. A variable-ratio lens is required, and in this case a three-cylinder asymmetric lens seems appropriate. The voltage ratios at the extremes of the desired operating conditions are

$$\frac{V_3}{V_1}(E_3 = 20 \ \ eV) = 0.23$$

and

$$\frac{V_3}{V_1}(E_3 = 200 \ \ eV) = 1.1$$

The desired magnification is

$$M = \frac{r_3}{r_1} \cong 1$$

The lens diameter must be chosen to give an acceptable filling factor. As a starting point, recall that

$$M \cong 0.8 \frac{Q}{P}$$

The lens is to focus the anode aperture on the center of the gas cell, so $P + Q = 10$ cm $+ 4$ cm $+ 1$ cm $= 15$ cm. The distance from the anode to the middle of the lens will be

$$P = \left(\frac{P}{P + Q}\right) \times 15 \ \ cm$$

$$= \left(\frac{P}{P + 1.25MP}\right) \times 15 \ \ cm$$

$$= 6.7 \ \ cm$$

For a final energy of 20 eV, the maximum diameter of the beam through the lens system will be approximately

$$2\theta_1 P = 0.67 \ \ cm; \qquad \theta_1(E_3 = 20 \ \ eV)$$

To achieve a filling factor less than 30 percent, choose $D = 2.0$ cm. Then

$$P = 3.35D$$

$$Q = 4.15D$$

and the nominal focal length of the desired lens (see Section 5.2.2) is

$$f = \left(\frac{1}{P} + \frac{1}{Q}\right)^{-1} = 1.85D$$

The calculations of Adams and Read or Harting and Read can then be used to determine the lens voltages that correspond to this nominal focal length.[7,8] The lens system can also be modeled by computer simulation and the potential to be applied to the central element of the three-cylinder lens can be determined empirically. Figure 5.21 shows the results of simulations for the projection of 20 and 200 eV electrons into the target cell. Figure 5.22 gives the central lens element voltage V_2 obtained from the simulation.

5.3.3 Ion Sources

An extensive range of parameters must be considered in choosing or designing an ion source. Foremost is the desired type of ionic species and charge state. The physical and chemical properties of the corresponding parent material determine to a large extent the means of ion production. Other important parameters are the desired current, brightness, and energy distribution. In some cases, it is also necessary to consider the efficiency of utilization of the parent material. Considering the number of variables involved, it is not surprising that many different types of ion sources have been developed to meet both scientific and industrial demands.[15]

Ion sources can be classified according to the ion production mechanism employed. The two most common means are surface ionization and electron impact. The electron-impact sources include both electron-bombardment sources and plasma sources. More exotic ion sources employ ion impact, charge exchange, field ionization, and photoionization. Three typical sources, illustrating different ion production mechanisms, will be discussed below.

In practice, the simplest means of producing ions is by thermal excitation of neutral species. This process, known as *surface ionization,* occurs efficiently when an atom or molecule is brought in contact with a heated surface whose work function exceeds the ionization potential of the atom or molecule. This requirement is fulfilled for alkali atoms (Li, Na, K, Rb, Cs) on surfaces of tungsten, iridium, or platinum. In order for ionization to compete with vaporization it is necessary that the surface be sufficiently hot that the substrate surface is only partially covered with neutrals, so that the work function of the surface is characteristic of the substrate rather than the material to be ionized.

In a surface-ionization ion gun, the source of neutral atoms can either be remote from or integral with the ionizing surface. In the remote type, the alkali metal is contained in an oven. A stream of metal vapor from the oven is directed toward the ionizer, which is a heated metal

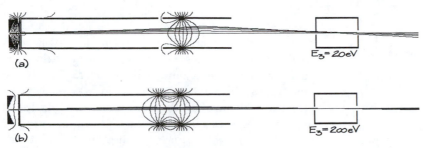

Figure 5.21   A computer simulation of the electric field and electron trajectories in the electron gun system shown schematically in Figure 5.20. The cathode-to-anode distance in the Pierce diode is $d = 0.5$ cm and the radius of the aperture in the anode is $r_1 = 0.04$ cm. The radii of the apertures in the target cell are $r_3 = 0.1$ cm and $r_4 = 0.15$ cm. The diameter of the lens $D = 2$ cm, the length of the central element $A = 0.5D$, and the gap between the lens elements is $g = 0.1D$. (a) Trajectories of electrons that pass through the cell with energy $E_3 = 20$ eV; $V_1 = 86$ V, $V_2 = 105$ V, and $V_3 = 20$ V. (b) Trajectories of electrons that pass through the cell with energy $E_3 = 200$ eV; $V_1 = 185$ V, $V_2 = 960$ V, and $V_3 = 200$ V.

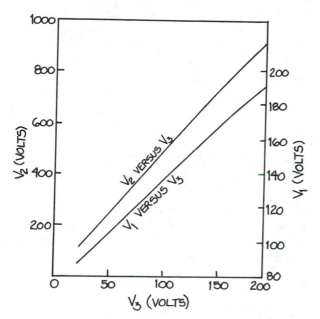

Figure 5.22   Lens voltages for the lens system of Figure 5.20.

heated, alkali atoms are generated, the atoms diffuse to the surface of the disc, and they are ionized as they leave the surface. These emitters with integral heaters are commercially available (e.g. from Spectra-Mat). As shown in Figure 5.23, a source is constructed by inserting the emitter into the cathode of a Pierce diode or into some other extractor electrode structure.

The majority of laboratory ion sources employ *electron-impact* ionization. The simplest configuration is illustrated in Figure 5.24. In this source, an electron beam from a simple diode gun is injected into an ionization chamber containing an appropriate parent gas. A transverse electric field of a few V/cm causes ions produced by electron impact to drift across the chamber and out through a skit in the side of the chamber. An extractor electrode system accelerates the ions into a beam. The ionic species produced will depend upon the parent gas, the gas pressure, and the electron energy. Singly charged ions are produced with maximum efficiency at electron energies of about 70 eV. Higher electron energies favor multiply charged ions. Lower electron energies yield only singly charged ions and reduce the extent of fragmentation in the event that the parent gas is a molecular species. Gas pressures are typically 1 to 10 mtorr. Higher pressures place an unreasonable gas load on the vacuum system and impede the flow of ions from the ionization chamber.

The efficiency of the electron-impact ionization source can be improved by imposing a magnetic field of a few hundred gauss coaxial with the electron beam. The field

cathode located within an electrode structure similar to that of an electron gun. The *anode* is negatively biased to accelerate positive ions into a beam. The most convenient alkali-ion emitter consists of a porous tungsten disc that has been infused with an alkali-containing mineral and mounted on an electrical heater. When the mineral is

Relatively high ion densities can be achieved in an electrical discharge through a gas. The ion density in a discharge can be further increased if the discharge is confined and compressed by a magnetic field. The duoplasmatron source illustrated in Figure 5.25 is the prototypical discharge-type ion source.[16] Applying a potential of 300 to 500 V between the heated cathode and the anode produces a discharge through a gas at about 100 mtorr. After the arc is struck, the discharge is maintained by passing a current 0.5 to 2 A through the ionized gas. In the duoplasmatron, the intermediate element known as the *zwischen* is one pole of a magnet. The axial magnetic field at the tip of the zwischen confines the discharge to a dense plasma bubble at the anode. Plasma leaks through a hole in the anode to fill a cup on the front face of the anode. An extractor electrode, biased at about 10 kV relative to the anode, withdraws ions from the surface of the plasma. Ion currents of 1 to 10 mA are easily obtained. In operation, several hundred watts of electrical power are dissipated in the arc and the electromagnet. In its original form, this source was liquid cooled. The author, however, has found air cooling to be acceptable for slightly smaller versions. The duoplasmatron is the brightest of ion sources. Copious quantities of singly charged atomic and diatomic ions can be obtained for these species for which an appropriate gaseous parent can be found. It is also possible to obtain a few microamperes of doubly charged ions. Negative ions can be obtained by reversing the extraction potential.[17] The chief disadvantages of plasma sources are their complicated construction and the fact that stable operation is only possible for a few days before cleaning and filament replacement are necessary.

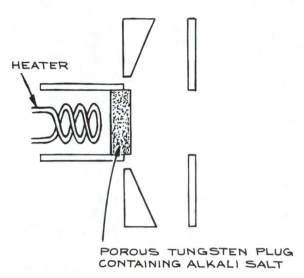

Figure 5.23   An alkali-ion source consisting of an indirectly heated emitter in a Pierce diode.

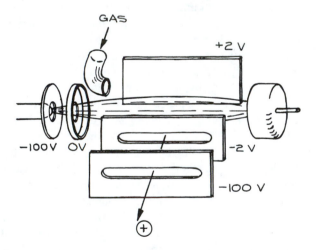

Figure 5.24   An electron-impact ion source.

confines the electron beam to a spiral around the beam axis, thus increasing the path length and maximizing the probability of collision with the gas molecules. Electron-bombardment sources of this type yield currents of only a few tens of microamperes. The chief advantages are simplicity of construction and an energy spread in the product ions of only a few eV.

5.4 ENERGY ANALYZERS

There are three basic means of measuring the energy of charged particles in a beam. These involve measuring the time of flight over a known distance, the retarding potential required to stop the particles, or the extent of deflection in an electric or magnetic field.

Because of the great velocities involved, the flight time of a charged particle over any reasonable distance is very short. Determination of the kinetic energy of a particle

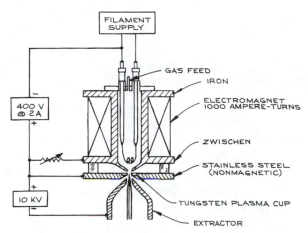

Figure 5.25   A duoplasmatron ion source.

from the time required to cover a distance of a few centimeters typically requires electronics with a response time of a few nanoseconds. Time-of-flight analyzers are used for energy analysis of electrons with energies less than 10 eV and ions below 1 keV.

An energy analysis of the particles in a beam can be performed by placing a grid or aperture in front of a particle collector and varying the potential on this element.[18] The current at the collector is the integrated current of particles whose energy exceeds the potential established by the grid. As the grid potential is reduced from that at which all current is cut off, the collector current increases. To obtain the energy distribution, the integrated current as a function of retarding potential must be differentiated. One drawback to this method is that only the component of velocity normal to the retarding grid is selected. There are also a number of practical difficulties. The ratio of the initial energy to the energy at the potential barrier varies rapidly for particles near the threshold for penetrating the retarding barrier. This gives rise to rapidly varying focusing effects near the threshold, so particles approaching slightly off axis are often deflected away from the collector. The result is that the transmission of the analyzer is unpredictable near the threshold. Another problem with retarding potential analyzers is that low-energy particles near the retarding grid are seriously affected by space charge and by stray electric and magnetic fields. Retarding potential analyzers are very

easily constructed and very compact, but the vagaries in their performance suggest that their use for high-resolution energy analysis be avoided if possible.

The energies of particles in a beam can be determined by passing the beam through an electric or magnetic field so that the deflection of the particle paths is a function of the particle energy per unit charge or momentum per unit charge. In these dispersive-type analyzers, the shape of the deflection field must be carefully controlled. Since it is generally easier to produce a shaped electric field than a shaped magnetic field, dispersive analyzers for particle energies up to several keV are usually of the electrostatic variety. For very high-energy particles, magnetic analyzers are preferred since production of the required electric fields would demand inconveniently large electrical potentials.

The energy pass band ΔE of an analyzer may be defined as the full width at half maximum (FWHM) of the peak that appears in a plot of transmitted current versus energy. Ideally, the transmission function is triangular. For a real analyzer, it resembles a Gaussian. We shall take ΔE to be half the full width of the transmission function. Such an approximation overestimates the pass band of a real analyzer.[19] To first order, the pass band is established by entrance and exit slits or apertures, usually of equal width w. The transmission function also depends upon the maximum angular extent to which particles can deviate from the central path leading from the entrance to the exit slit. This angular deviation is defined by angles $\Delta\alpha$ in the plane of deflection and $\Delta\beta$ in the perpendicular plane. If E is the central energy of particles transmitted through an analyzer, then the *resolution is*

$$\frac{\Delta E}{E} = aw + b(\Delta\alpha)^2 + c(\Delta\beta)^2$$

where a, b, and c are constants characteristic of the particular analyzer.

5.4.1 Parallel-Plate Analyzers

The simplest electrostatic analyzer employs a uniform field created by placing a potential difference across a pair of plane parallel plates as shown in Figure 5.26. With the entrance and exit slits in one of the plates, first-order

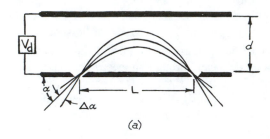

(a)

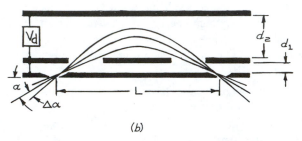

(b)

Figure 5.26   Parallel-plate analyzers with (a) the energy-resolving slits in one of the plates and (b) the slits in a field-free region.

focusing in the deflection plane is obtained when the angle of incidence of entering particles is about $\alpha = 45°$.

The deflection potential V_d (V), in relation to the incident energy E(eV), the plate spacing d, and the slit separation L, is given by

$$V_d = (E/q)\frac{2d}{L}$$

with the proviso that $d > L/2$ to prevent particles from striking the back plate. The resolution is

$$\frac{\Delta E}{E} = \frac{w}{L} + (\Delta\alpha)^2 + \frac{1}{2}(\Delta\beta)^2$$

A point at the entrance aperture is focused into a line of length $2\sqrt{2}\Delta\beta$, and the length of the exit slit is correspondingly greater than that of the entrance slit.

When the slits are placed in a field-free region[20] as in Figure 5.26b, optimum performance is obtained with the angle of incidence $\alpha = 30°$. The distance from the slits to the entrance plate of the analyzer

$$V_d = 2.6(E/q)\frac{d_2}{L}$$

where d_2 is the plate spacing. The deflection potential is

$$V_d = 2.6(E/q)\frac{d_2}{L}$$

This arrangement gives second-order focusing in the plane of deflection. The resolution is

$$\frac{\Delta E}{E} = 1.5\frac{w}{L} + 4.6(\Delta\alpha)^2 + 0.75(\Delta\beta)^2$$

and the image of a point on the entrance slit is a line of length $4L\Delta\beta$ at the exit slit.

Although the parallel plate is an attractive design because of its simple geometry, there are several problems. The entrance apertures or slits in the front plate are at the boundary of a strong field and act as lenses to produce unwanted aberrations. For the design with the energy-resolving slits in a field-free region, this problem can be alleviated by placing a fine wire mesh over these entrance apertures to mend the field. A large electrical potential must be applied to the back plate, creating a strong electrostatic field outside the analyzer. The apparatus in which the analyzer is installed must often be shielded from this field. In addition, the fringing field at the edges of the plates can penetrate into the deflection region since the gap between the plates is large. This problem can be solved by extending the edges of the plate well beyond the deflection region, or by placing compensating electrodes at the edges of the gap, as in Figure 5.27.

5.4.2 Cylindrical Analyzers

There are two well-known electrostatic analyzers that employ cylindrical electrodes. These are the radial cylindrical analyzer and the cylindrical mirror analyzer.

The *radial cylindrical analyzer* is shown in Figure 5.28. The radial electric field is produced by an electrical potential placed across concentric cylindrical electrodes. Particles are injected midway between the electrodes in a direction approximately tangent to the circular arc of radius R_0. First-order focusing is obtained if the cylindrical electrodes subtend an angle of $\pi/\sqrt{2} = 127°$. Assuming a charged particle that originates at ground potential with essentially no kinetic energy, and a mean pass energy $E = -qV$ (that is, the energy of the particle that travels along the central path of radius R_0), the potentials

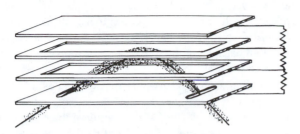

Figure 5.27 Guard electrodes at the edges of a parallel-plate analyzer to offset the distortion of the field caused by fringing.

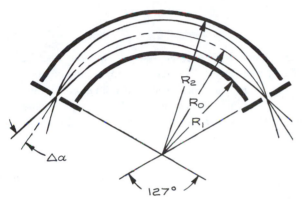

Figure 5.28 The radial cylindrical analyzer.

to be applied to the outer and inner cylindrical elements are

$$V_{\text{outer}} = V\left(1 - 2\ln\frac{R_2}{R_0}\right)$$

and

$$V_{\text{inner}} = V\left(1 - 2\ln\frac{R_1}{R_0}\right)$$

where R_2 and R_1 are the radii of the outer and inner cylinders, respectively. To be more general, the potentials on the electrodes must be set so that a charged particle that enters at the midradius R_0 with an energy equal to the pass energy E will *lose* an amount of energy equal to

$$2E\ln\frac{R_2}{R_0}$$

were it to travel to the outer electrode, and would *gain* an amount of energy equal to

$$2E\ln\frac{R_0}{R_1}$$

were it to travel to the inner electrode.

The resolution of this analyzer is

$$\frac{\Delta E}{E} = \frac{w}{R_0} + \frac{2}{3}(\Delta\alpha)^2 + \frac{1}{2}(\Delta\beta)^2$$

It is good practice to limit the angle of divergence in the plane of deflection so that

$$\Delta\alpha < \frac{2\sqrt{2}}{\pi}\frac{R_2 - R_1}{R_0}$$

in order to keep the filling factor below 50 percent. The radial-field analyzer gives focusing only in the plane of deflection. A point at the entrance slit is imaged as a line of length $\sqrt{2}\pi R_0 \Delta\beta$ at the exit slit.

The *cylindrical mirror analyzer* is similar to the parallel-plate analyzer except the deflection plates are coaxial cylinders. In fact, the parallel-plate analyzer can be considered a special case of the cylindrical mirror. In the usual geometry, shown in Figure 5.29, the source is located on the axis, and particles emitted into a cone defined by polar angle α pass through an annular slot in the inner cylinder. Particles of energy E are deflected so that they pass through an exit slot and are focused to an image on the axis. The cylindrical mirror is double-focusing—that is, focusing occurs in both the deflection plane and the perpendicular plane so that the image of a point at the source appears as a point at the detector. An obvious advantage of this analyzer is that particles at any azimuthal angle can be collected.

For optimum performance, the entry angle $a = 42.3°$, in which case the distance from source to detector is $L = 6.12R_1$.[21] The inner cylindrical plate is at the same potential as the source, and the potential on the outer cylinder, relative to the inner cylinder, is

$$V_{\text{outer}} = 0.763(E/q)\ln\frac{R_2}{R_1}$$

It is wise to choose $R_2 > 2.5R_1$ so that particles are not scattered from the outer electrode. The resolution of the axial-focusing cylindrical-mirror analyzer is approximately

$$\frac{\Delta E}{E} = 1.09\frac{w}{L}$$

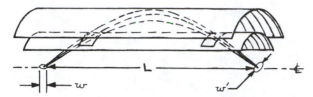

Figure 5.29 Axial-focusing cylindrical-mirror analyzer.

for a source of axial extent w and an energy-resolving aperture of diameter $w' = w \sin \alpha$ perpendicular to the axis (as shown in Figure 5.29).

If the source is not small and well defined, the entry and exit slots in the inner cylinder can be used to define the resolution of the cylindrical-mirror analyzer, as in Figure 5.30. Although the axial-focusing geometry gives focusing to second order, this arrangement is only first-order focusing.

5.4.3 Spherical Analyzers

For many applications, the most desirable analyzer geometry employs an inverse-square law field that is created by placing a potential across a pair of concentric spherical electrodes. Focusing in both the deflection plane and the perpendicular plane can be obtained using any sector portion of a sphere. As shown in Figure 5.31, the

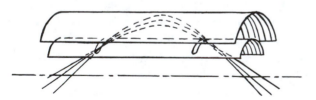

Figure 5.30 Cylindrical-mirror analyzer with energy-resolving slits in the inner electrode.

object and image lie on lines that are perpendicular to the entrance and exit planes, respectively, and tangent to the circle described by the midradius R_0. Furthermore, by Barber's rule, the object, the center of curvature of the spheres, and the image lie on a common line. The 180° spherical sector (see Figure 5.32) is often used because of

the compact geometry that results from folding the beam back onto a line parallel to its original path. As with the cylindrical mirror, the spherical analyzer can be used to collect all particles emitted from a point source at or near a particular polar angle.

Assuming a charged particle that originates at ground potential with essentially no kinetic energy, and a mean pass energy $E = -qV$ (that is, the energy of the particle that travels along the central path of radius R_0), the potentials to be applied to the outer and inner spherical elements are

$$V_{\text{outer}} = V \left[2\frac{R_0}{R_2} - 1 \right]$$

and

$$V_{\text{inner}} = V \left[2\frac{R_0}{R_1} - 1 \right]$$

where R_2 and R_1 are the radii of the outer and inner spheres, respectively. To be more general, the potentials on the electrodes must be set so that a charged particle that enters at the midradius R_0 with an energy equal to the pass energy E will *lose* an amount of energy equal to

$$2E \left[1 - \frac{R_0}{R_2} \right]$$

were it to travel to the outer electrode, and would *gain* an amount of energy equal to

$$2E \left[\frac{R_0}{R_1} - 1 \right]$$

were it to travel to the inner electrode.

The resolution of the 180° spherical sector (see Figure 5.32) is

$$\frac{\Delta E}{E} = \frac{w}{2R_0} + \frac{1}{2}(\Delta \alpha)^2$$

and the maximum deviation, w_{m}, of a trajectory from the central path within the analyzer is given by

$$\frac{w_{\text{m}}}{R_0} = \frac{\Delta E}{E} + \Delta \alpha + \frac{1}{2\Delta \alpha}\frac{w}{2R_0} + \frac{\Delta E^2}{E}$$

For the truncated spherical-sector shown in Figure 5.31, the resolution is approximately

$$\frac{\Delta E}{E} = \frac{w}{R_0(1 - \cos \phi) + l \sin \phi}$$

where ϕ is the angle subtended by the analyzer sector and l is the distance from the exit boundary of the analyzer to the exit aperture. Although it would appear that the

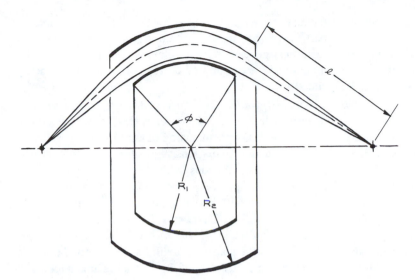

Figure 5.31   Focusing
of a spherical analyzer.

resolution can be increased arbitrarily by increasing l, the aberrations increase rapidly as the system becomes asymmetric. It is best to employ the symmetric geometry with ϕ in the range 60 to 180°.

In comparison with the parallel plate or cylindrical-mirror analyzers, the spherical analyzer has the advantage of requiring relatively low electrical potentials on the electrodes. Because the electrodes are closely spaced in the spherical analyzer, fringing fields are less of a problem and more easily controlled. The chief drawback of this type of analyzer is the difficulty of fabrication and mounting.

5.4.4 Preretardation

In all electrostatic-deflection analyzers, the pass band ΔE is a linear function of the transmitted energy. The absolute energy resolution of these devices can therefore be improved by retarding the incident particles prior to their entering the analyzer. In principle, the pass band of an analyzer can be reduced arbitrarily by preretardation; in practice, there are limitations on this technique. Charged particles incident on the entrance aperture of an analyzer can be slowed by a decelerating lens as in the example given in Section 5.2.6. It would be very difficult, however, to design a lens system to be used with analyzers that have an annular entrance slit. Space charge and stray electric and

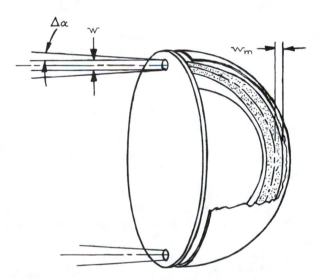

Figure 5.32   The hemispherical analyzer: w is the width of the entrance and exit apertures, $\Delta\alpha$ the maximum angular deviation of an incident trajectory, and w_m the maximum deviation of a trajectory from the central path through the analyzer.

magnetic fields must be considered. Because the flight path through an analyzer is long, it is usually not possible to reduce the energy of transmitted particles below about 2 eV before these effects result in severe aberrations. It is important to recall that if all particles of a particular energy in a beam are to be transmitted through an analyzer, the ratio of brightness to energy must be conserved (see Section 5.1.3). As the energy of particles incident on an analyzer is decreased, the current is ultimately reduced because the pencil angle of the beam exceeds the acceptance angle of the analyzer.

When using a decelerating lens with an analyzer, it is good practice to place an aperture on the high-energy side of the lens and design the lens so that the image of this aperture appears at the entrance plane of the analyzer. The need for a real entrance aperture is thereby eliminated, and there are no metal surfaces in the vicinity of the low-energy beam. Since space charge will cause the low-energy beam to expand at the entrance plane of an analyzer, current would be lost if a real entrance aperture were used. As demonstrated by Kuyatt and Simpson, space charge expansion at a virtual aperture is compensated by the spreading of the beam, with the result that the beam appears to originate at an aperture of about the same size as that in the absence of space charge (see Figure 5.16).[22]

5.4.5 The Energy-Add Lens

An electrostatic energy analyzer can be used as a monochromator to select the particles in a beam whose energies fall within a narrow range, or as an analyzer to determine the energy distribution of particles originating from a process under study. In the latter application, the pass band of the analyzer must be scanned over the energy range of interest. Scanning can be accomplished by varying the potential difference between the analyzer electrodes. Neither the transmission nor the energy resolution, however, will be constant. A more satisfactory method for determining the energy distribution of particles in a beam is to fix the analyzer potential at the voltage that allows transmission of particles of the highest desired energy and preaccelerate the incident particles by a variable amount. The energy that must be added to the particles so that they pass through the analyzer is then a measure of the difference between their initial energy and the energy necessary to transmit them with no preacceleration. In a scattering experiment, for example,

the analyzer might be set to transmit elastically scattered particles, and the distribution of energy lost by inelastically scattered particles would be scanned by monitoring the transmitted current as a function of energy added.

An *energy-add lens* must add back energy without disturbing the optics of the analyzer. Kuyatt has shown that the electron-optical analog of the *field lens* is suited to this application.[23] This lens is positioned so that its first principal plane coincides with the particle source to be examined. The source is then imaged onto the second principal plane with unit magnification. For a reasonably strong lens ($V_2/V_1 > 3$), the position of the principal planes is nearly independent of the voltage ratio and the separation of the planes is small. Particles from the source can therefore be accelerated by a variable amount without changing the apparent position of the source. An energy-add lens used in conjunction with a fixed-ratio decelerating lens for preretardation at the input of an electron energy analyzer is schematically illustrated in Figure 5.33. For transmission of electrons with energy A (eV) less than the maximum energy to be transmitted, the potential of the second element of the energy-add lens and *every electrode thereafter* is increased by A (V). For a positive ion analyzer, the potential would be decreased by A/n (V), where n is the charge state of the ions.

A problem with the energy-add lens is that the object pupil moves as the lens voltages change. While the apparent source remains stationary, the beam angle may vary over a considerable range as the lens is tuned. As a result, the lenses following the energy-add lens may be overfilled. This can be avoided by placing a second field lens downstream from the first at an intermediate image of the apparent source. The second field lens is then tuned to give a zero beam angle. So that there is no net change of energy as the beam angle is manipulated, an einzel lens (Section 5.2.2) can be used.

5.4.6 Fringing-Field Correction

For the radial-field cylindrical analyzer or the spherical analyzer, particles must pass through the fringing field at the edge of the electrode gap as they enter the deflection region. The electric field at the edge bulges out of the gap,

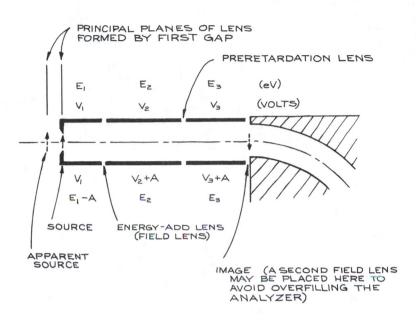

PRINCIPAL PLANES OF LENS
FORMED BY FIRST GAP

PRERETARDATION LENS

| E_1 | E_2 | E_3 | (eV) |
| V_1 | V_2 | V_3 | (VOLTS) |

| V_1 | V_2+A | V_3+A |
| E_1-A | E_2 | E_3 |

SOURCE   ENERGY-ADD LENS
(FIELD LENS)

APPARENT
SOURCE

IMAGE (A SECOND FIELD LENS
MAY BE PLACED HERE TO
AVOID OVERFILLING THE
ANALYZER)

Figure 5.33 An energy-add lens and fixed-ratio retarding lens at the input of an analyzer. The energies and corresponding lens voltages for electrons transmitted with no added energy are given above the lens. Energies and voltages for transmission of electrons that have lost energy *A* are shown below. The *source* is defined by an aperture that is finally imaged at the entrance plane of the analyzer.

and this curvature produces a focusing effect that causes undesirable aberrations. Herzog and Wollnik and Ewald have shown that this effect can be largely eliminated by placing a diaphragm at the entrance of the condenser gap.[24,25] As illustrated in Figure 5.34, the diaphragm, when properly located, produces a field that has the same effect as a field abruptly terminated at a distance d^* in front of the condenser gap. The appropriate location of the diaphragm (as a function of the dimensions given in Figure 5.34) can be determined for either a thick or thin diaphragm from the graphs in Figure 5.35.

5.4.7 Magnetic Energy Analyzers

Magnetic fields are employed in several charged particle energy analyzers and filters. The *trochoidal analyzer* (Figure 5.36a) has proven quite useful for the dispersion of very low-energy (0 to 10 eV) electrons and ions. This device employs a magnetic field aligned with the direction of the charged-particle beam, and an electric field perpendicular to this direction.[26] The trajectory of a particle injected into this analyzer describes a spiral and the guiding center of the spiral drifts in the remaining perpendicular direction. The drift rate depends upon the particle energy so that a beam entering the device is

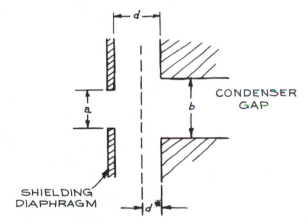

CONDENSER
GAP

SHIELDING
DIAPHRAGM

Figure 5.34 A shielding diaphragm to correct for focusing at the fringing field at the edge of an electrostatic condenser. The field appears as if it were terminated abruptly at a distance d^* from the gap.

dispersed in energy at the exit. The projection of the trajectory on a plane perpendicular to the electric field direction is a trochoid, hence the name. The device requires a very uniform magnetic field of about 100 gauss. The field is usually produced by a relatively large Helmholtz pair (see Section 5.6.5) that is mounted outside

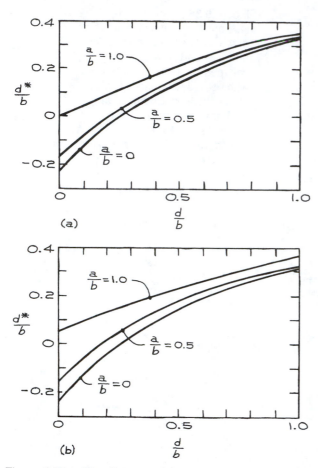

(a)

(b)

Figure 5.35   The distance of the apparent field boundary from the edge of the condenser electrodes as a function of the dimensions given in Figure 5.34 for (a) a thick diaphragm and (b) a thin diaphragm. (From H. Wollnik and H. Ewald, "The Influence of Magnetic and Electric Fringing Fields on the Trajectories of Charged Particles," *Nucl. Instr. Meth.*, **36**, 93 (1965); by permission of North-Holland Publishing Company.)

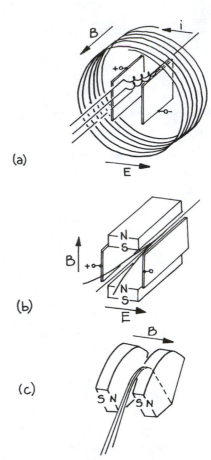

Figure 5.36   Charged-particle energy analyzers with magnetic fields: (a) the trochoidal analyzer with an electromagnet; (b) the Wien filter; (c) the sector magnet analyzer. Trajectories for electrons of different energies are shown. Magnet polarities are for electrons.

the vacuum system containing the analyzer. The result is that the entire experiment is immersed in the field. This may be a drawback in some cases, but on the other hand, for low-energy electrons or ions the field helps contain particles that otherwise would be lost from coulombic repulsion (i.e., space charge).

The Wien filter (Figure 5.36b) similarly employs crossed electric and magnetic fields; however, the fields are perpendicular to one another and both are perpendicular to the injected beam direction.[27] The Coulomb force induced by the electric field E deflects charged particles in one direction and the Lorentz force associated with the magnetic field B tends to deflect them in the opposite direction. The forces balance for one velocity, $v = |E|/|B|$, and particles of the corresponding energy are transmitted straight through to the exit aperture.

A magnetic field alone, perpendicular to the direction of a charged-particle beam, will provide energy dispersion provided that the particles are all of the same mass and charge (Figure 5.36c). Sector magnets such as those used in mass spectrometers are used in electron spectrometers for very high-energy electrons and ions—the advantage over electrostatic deflectors being that large electrical potentials are not required. Another advantage is that the deflection is in the direction parallel to the magnet pole face, making it possible to view the entire dispersed spectrum at one time. By contrast, energy dispersion in an electrostatic device results in a significant portion of the dispersed particles striking one or the other electrode.

5.5 MASS ANALYZERS

Mass analysis is more complicated than energy analysis of an isotopically pure beam since energy and momentum are independent variables in a mixed beam. A detailed discussion of the many types of mass analyzers that have been developed is beyond the scope of this book. The principles of operation of a few representative analyzers will be described.

5.5.1 Magnetic Sector Mass Analyzers

The deflection of ions in a perpendicular magnetic field is proportional to the particle momentum per unit charge. If all particles entering a magnetic field have the same energy per unit charge, then the field will separate particles according to their masses. In the usual configuration, a magnetic field is produced between the two parallel, sector-shaped polefaces of an electromagnet as illustrated in Figure 5.37. The focusing properties of this field are the same as those of the spherical electrostatic analyzer,[28] and the locations of the entrance and exit slits are given by Barber's rule (Section 5.4.3). The radius of curvature of ions of energy E in a magnetic field of intensity B is

$$R = \frac{144}{Bn}\sqrt{mE} \quad \text{cm}$$

with B in gauss, E in eV, m in amu, and n the charge state. In terms of the parameters specified in Figure 5.37, the resolution is

$$\frac{\Delta E}{E} = \frac{w}{R_0(1 - \cos\phi) + l\sin\phi}$$

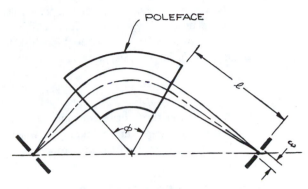

Figure 5.37   Focusing of the magnetic sector.

where R_0 is the radius of curvature of the central trajectory. To first order, there is no focusing in the plane perpendicular to the plane of deflection. When the curvature of the fringing field is taken into consideration, however, some focusing in the perpendicular plane results. Incident ion trajectories should be normal to the entrance plane to avoid focusing effects.

5.5.2 Wien Filter

The Wien filter, employing mutually perpendicular electric and magnetic fields normal to the ion trajectory (see Figure 5.36b), can serve to disperse charge particles according to either mass or energy.[27] Ions of a unique velocity v will experience equal and opposite forces through interaction with the two fields (the Coulomb force and the Lorentz force) and be transmitted in a straight line. If all of the ions injected into the fields of the Wien filter have the same kinetic energy, ions of a unique mass are transmitted straight through. If the injected ions have all been accelerated through the same potential (relative to their source), ions of a unique mass-to-charge ratio are transmitted. In most Wien filters, the magnetic field is supplied by a pair of permanent magnets, and the electric field is produced by placing an electrical potential V_d between a pair of parallel electrodes separated by a distance d. The velocity of ions transmitted straight through the Wien filter is

$$v = 10^8 \frac{V_d}{dB} \quad \text{cm s}^{-1}$$

with V_d in volts, d in cm, and B in gauss. The mass spectrum of ions is scanned by varying V_d. If slits of width w are placed at the entrance and exit of a Wien filter of length L, the mass resolution for ions of energy E is

$$\frac{\Delta m}{m} = \frac{2dEw}{qV_dL^2}$$

assuming the incident ion trajectories are normal to the entrance plane. Boersch, Geiger, and Stickel have made a detailed analysis of the Wien filter focusing properties.[29]

5.5.3 Dynamic Mass Spectrometers

A class of analyzers known as *dynamic mass spectrometers* use time-varying electric or magnetic fields or timing circuits to disperse ions according to their masses.[30] The quadrupole mass analyzer and the linear time-of-flight mass spectrometer are the two most successful designs of this type.

The *quadrupole mass analyzer* illustrated in Figure 5.38 employs a time-varying electric quadrupole field. For a particular field intensity and frequency, only ions of a unique mass-to-charge ratio will follow a stable path and pass through the field. As shown in the figure, the quadrupole field is approximated by a square array of four cylindrical electrodes parallel to the axis along which ions are injected. The potential applied to the vertical pair of electrodes is

$$V_v = U + V\cos 2\pi ft$$

while for the horizontal pair it is

$$V_h = -U + V\cos(2\pi ft + \pi)$$

where f refers to an rf frequency. For optimum performance, the ratio of the dc field to the rf field is adjusted so that $U/V \cong 0.17$. For singly charged ions, the mass of the ions that are transmitted is

$$m = 0.14\frac{V}{r^2 f^2} \text{ amu}$$

with V in volts, r in cm, and f in MHz. Mass scanning is usually accomplished by varying the rf amplitude V, while

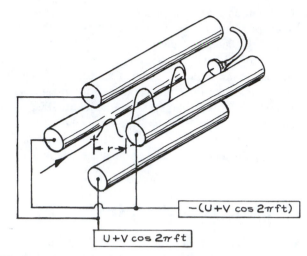

Figure 5.38 The quadrupole mass filter.

fixing the ratio of dc to rf voltage (U/V) and the frequency f. Quadrupole analyzers are compact and offer the advantages of high transmission, fast scanning, and insensitivity to the initial ion energy. These instruments are available commercially as residual-gas analyzers for high-vacuum systems and are suitable for use in many scattering experiments.

The *time-of-flight mass spectrometer* depends upon the fact that the velocity of ions of the same kinetic energy is a function of mass. In this spectrometer, ions that have been accelerated to an energy of about 1 keV are admitted to a long drift tube in short bursts by a negative electrical pulse applied to a grid at the entrance of the drift region. The lightest ions travel most rapidly and are the first to arrive at the detector, located at the end of the drift tube. Heavier ions arrive in the order of their masses. The obvious advantage of this design is that the entire mass spectrum is scanned in a few microseconds. The spectrum can also be scanned thousands of times each second, making this system ideal for studying rapidly varying processes. Its chief drawback is its size, since the length of the flight tube required to achieve good resolution may be greater than a meter.

5.6 ELECTRON- AND ION-BEAM DEVICES: CONSTRUCTION

To realize the design of a charged-particle optical system, it is necessary to create an environment where a stream of particles can travel without loss of momentum and to make electrodes and pole pieces that faithfully produce the electrical and magnetic fields necessary to deflect the particles.

5.6.1 Vacuum Requirements

Charged particles lose energy through interactions with gas molecules, therefore ions or electrons can only be transported in a vacuum. The required pressure depends upon the distance they must travel. As noted in Section 3.1.2, the mean free path at a pressure of 1 mtorr is a few centimeters, while at 10^{-5} torr this increases to a few meters. It follows that high vacuum is required for the operation of a charged-particle optical system.

Conducting surfaces must be kept clean, since contamination with an insulating material will result in the buildup of surface charges that causes an unpredictable deflection of charged particles passing nearby. Clean electrode surfaces are particularly important for particles of energy less than 100 eV and when high spatial resolution is required. All electrodes must be clean and free of hydrocarbon contamination before installation in the vacuum system (see Section 3.6.3). An oil-free vacuum system with turbomolecular pumps, ion pumps, or sorption pumps (see Section 3.4.3) is most desirable, although a system evacuated with a properly trapped oil diffusion pump (see Sections 3.5.5 and 3.4.2) is adequate in many cases. Polyphenyl ether diffusion-pump fluids such as Convalex-10 (Consolidated Vacuum) or Neovac Sy (Varian) have been found to be the least offensive diffusion-pump oils for electron-optical systems. In low-energy beam devices, a daily bake to 200 to 300°C will keep electrode surfaces clean and ensure stable operation. Hydrocarbons on aperture surfaces exposed to charged particles create a particularly obnoxious problem. Bombardment of the adsorbed hydrocarbons produces an insulating carbon polymer that adheres tenaciously to the underlying metal. This carbon material can only be removed by high-temperature baking (400°C), by vigorous application of an abrasive, or by etching with a strong NaOH solution.

When baking is impractical, electrode surface quality and stability are often improved by a coating of carbon black. A surface can be blacked by brushing with a sooty acetylene flame, but the coating produced in this manner does not adhere well. A superior coating can be produced by spraying or brushing the surface with a thin water or ethanol slurry of colloidal graphite (such as Aquadag, made by Acheson Colloids Co.). The tenacity of this coating is improved by preheating the metal surface to about 100°C before the slurry is applied.

5.6.2 Materials

The materials used to construct lens elements must be such that the equipotential surfaces near the electrodes faithfully follow the contours of the electrode surfaces. These surfaces must be clean, and must withstand periodic cleansing by baking, bead blasting, electropolishing, or harsh chemical action.

The *refractory metals* such as tungsten, tantalum, and particularly molybdenum are probably the best electrode materials. These metals have a low, uniform surface potential, they do not oxidize at ordinary temperatures, and they are bakeable. Refractory metals are hard and rather brittle. Only the wrought material can be machined or formed easily. Stock produced by sintering is very difficult to machine. Electrodes are available in cylindrical and spherical shapes, spun from sheet molybdenum (Bomco, Inc.).

Oxygen-free high-conductivity (OFHC) *copper* (see Section 1.2.5) is often used for electrode fabrication, although Kuyatt used commercial, half-hard copper in some applications. Exposed to air, copper forms a surface oxide, but this oxide is conducting. Copper is bakeable and reasonably machinable.

Stainless steels (see Section 1.2.3) contain domains of different composition, some of which may be ferromagnetic. The magnetic properties of stainless steels vary considerably even between samples of the same net composition, and as a result these materials are unsuitable for use with low-energy charged particles unless they are carefully annealed to remove residual magnetism (see

Section 1.2.3). On the other hand, stainless steels have the advantage of being bakeable, strong, and easy to machine.

Aluminum (see Section 1.2.6) is unsuitable because its surface is rapidly attacked by oxygen in the air to form a hard, insulting layer of oxide. On the other hand, aluminum has a low density, and most aluminum alloys are easy to machine or form by bending, spinning, or rolling. When these properties are important, the surface can be plated with copper or gold to overcome the problem of oxidation.

Electrodes and lenses must be mounted on nonconducting materials. Glass, ceramic, (see Section 1.2.9) or, in some cases, plastic (see Section 1.2.8) can be used for this purpose. Of the ceramic materials, *alumina* is the best insulator and the strongest. Precision-ground rods and balls are commercially available. Alumina circuit board substrate in the form of thin plates laser cut in complex shapes can be obtained quite inexpensively. There are also machinable ceramics from which complicated shapes can be formed (such as MACOR from Corning). As noted in Chapter 3, it is important to be aware that the resistivity of a ceramic is strongly dependent upon temperature. The resistivity of a ceramic material falls by roughly a factor of ten for every 100° (C) increase in temperature.[31] For example, the resistivity of alumina is greater than 10^{17} ohm cm at 100°C and falls below 10^7 ohm cm at 1000°C.

Plastics are machinable and inexpensive. Unfortunately, they are not bakeable. An additional disadvantage is that they contain hydrocarbons that can contaminate the vacuum environment. *Polyimide plastic sheet* (Kapton) is a useful insulator. *Mica* in sheets a few thousandths of an inch thick (*stove mica*) is an excellent insulator.

5.6.3 Lens and Lens-Mount Design

Most electron-optical lenses are created by the fields in the gap between coaxial, cylindrically symmetric electrodes. These electrodes must be designed so that electric fields associated with the lens mounts and the vacuum-container walls do not penetrate the gap between electrodes. The involuted design illustrated in Figure 5.39 ensures that the lens gap is shielded. The step on the shoulder of the lens elements is designed so that the inner gap is of the correct size when the outer gap is adjusted to some standard

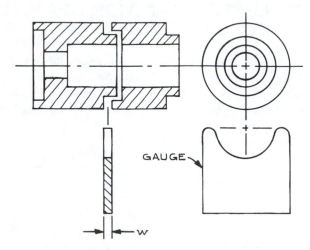

Figure 5.39 Tube lens electrodes are constructed so that external fields cannot penetrate the lens gap. When mounting the lens elements, a gauge is inserted to insure correct spacing at the invisible lens gap.

width. In this way, all of the lenses in a system can be correctly positioned with the aid of a single gauge.

There are two widely used, semikinematic schemes for mounting lens elements. These are the *rod mount* and the *ball mount,* illustrated in Figures 5.40 and 5.41, respectively. In the rod mount, cylindrical lens elements rest on ceramic rods that insulate the elements from one another and from a grounded mounting plate. The critical dimensions are the diameters of the lens elements and the width of the channel in which the rods rest. In order for the lens elements to be coaxial, their outer diameters must be identical. This requirement is met by making all of the elements from a single piece of rod stock that has been carefully turned to a uniform diameter. To ensure that the elements are mounted coaxially, it is then only necessary that the sides of the channel in which the rods rest be parallel. This is easily accomplished in a milling operation. If the vertical position of the lens axis is important, then the width of the channel becomes a critical dimension. Of course, the diameter of the rods is important, but as noted elsewhere (Section 1.2.8), alumina rod, centerless ground to high precision, is commercially available. For maximum strength, the rods should be about 90° apart around the circumference of the lens element.

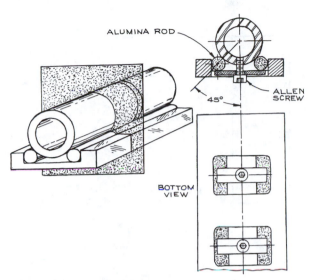

Figure 5.40   Cylindrical lens elements mounted on ceramic rods.

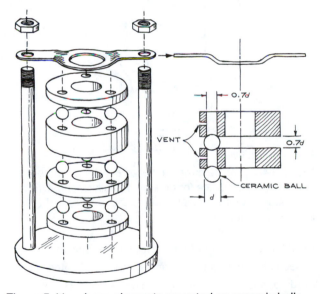

Figure 5.41   Lens elements mounted on ceramic balls.

In the ball-mounting scheme, lens elements are insulated from one another and positioned by ceramic balls that rest in holes near the edges of the lenses. This system is preferred when mounting very thin elements, as

in an aperture lens system. Ideally, only three balls should be used. The critical dimensions are the locations of the holes and their diameter. The holes should be bored, or drilled and reamed, in a milling machine or jig borer using a dividing head or a precision rotary table. For balls of diameter d, the hole diameter should be $d \sin 45°$, in which case the spacing between lens elements will be $d \cos 45°$. A lens system is assembled by stacking lens elements alternating with balls. The stack must be clamped. The clamp should have some spring and should bear on the topmost element only at one or two points near the center of the circle around which the balls are located. A rigid clamp will tend to drive the balls into their holes and may crush the edges of the holes. If the clamp bears too near the edge, there is a danger that the stack may become cocked off axis.

5.6.4 Charged-Particle Detection

For particle energies up to a few tens of keV, there are two widely used detection schemes. The charged particles can be collected on a metal surface and the resultant electrical current measured directly, or they can be detected by collecting the slower secondary electrons that are ejected from a metal surface by impact of the primary particle. Charged particles can also be detected with photographic emulsions, scintillators, and various solid-state devices. Emulsions and scintillators have largely been supplanted by the methods mentioned above, and solid-state devices are only suitable for the detection of high-energy particles (above 30 keV). Table 6.23 lists the properties of common particle detectors.

The collector for the direct detection of a current of charged particles is called a *Faraday cup*. Typical designs are illustrated in Figure 5.42. These simple collectors are connected directly to a current-measuring device and are useful at currents down to the detection limits of modern electrometers—about 10^{-14} A. A properly designed Faraday cup does not permit secondary electrons to escape. For a positive-ion collector, the loss of a secondary electron appears to the current-measuring instrument as an additional ion, while for an electron collector the loss of each secondary electron cancels the effect of an incident primary electron. To prevent the escape of secondary

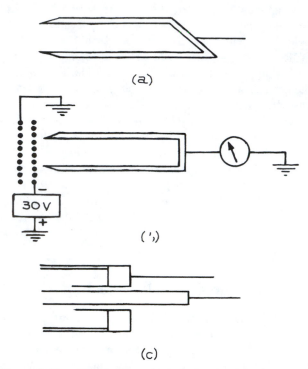

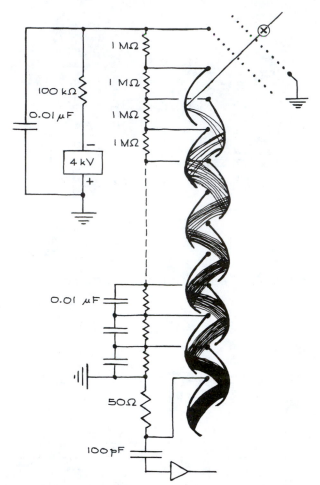

Figure 5.42 Faraday-cup designs. Design (b) has a grid biased to suppress secondary electron emission. Design (c) is a double cup for aligning and focusing a beam.

electrons, the depth of a Faraday cup should be at least five times its diameter. A suppressor aperture or grid biased to about −30 V in front of a Faraday cup effectively prevents the escape of most secondary electrons. A grounded grid in front of the suppressor, as illustrated in Figure 5.42b, prevents field penetration from the suppressor in the direction of the incident current source. When a particle beam must be aligned and sharply focused, a concentric pair of collectors (Figure 5.42c) is useful. With this arrangement, the beam-focusing elements are adjusted to maximize the ratio of current to the inner cup in relation to the current to the outer cup.

Charged-particle currents of less than 10^{-14} A can be detected with an *electron multiplier*. As illustrated in Figure 5.43, these devices consist of a series of electrodes known as *dynodes*. Secondary electrons produced by impact of a particle on the first dynode are accelerated towards the second dynode through a potential drop of 100

Figure 5.43 An electron multiplier. Electrical connections shown are appropriate for detecting positive ions. Grids at the entrance of the multiplier prevent the escape of secondary electrons.

to 300 V. The impact of these electrons results in a number of secondary electrons that are in turn accelerated into the third dynode, and so on. The secondary emission coefficient of the dynodes is typically about 3 electrons per incident particle. The impact of a single particle on a multiplier with n electrodes results in a pulse of about 3^n electrons. Electron multipliers are available with 10 to 20 dynodes to provide gains of 10^6 to 10^9. The dispersion in time of the shower of electrons from the last dynode is of the order of 10 ns. The corresponding current through a 50 Ω measuring resistor produces a pulse of 500 μV to 500 mV that is easily detected with modern pulse-counting electronics. If time resolution is not required, the output current from an electron multiplier can also be measured directly with an electrometer.

The electrodes of an electron multiplier are fabricated of a material that has a high work function—typically a Be-Cu alloy that has been *activated* by some proprietary process. The high work function is desirable because it inhibits thermionic emission that would result in noise pulses at the multiplier output. On the other hand, incident particles must have energies in excess of a few hundred eV to assure efficient secondary emission. When working with lower-energy particles it is necessary to provide an acceleration stage at the multiplier input. For incident particles with energies greater than 300 eV, the detection efficiency of an electron multiplier can exceed 90 percent.

An important variant of the electron multiplier uses a continuous dynode as illustrated in Figure 5.44. These devices are known as *channeltrons*. They are glass tubes with semiconducting inner surfaces. The end-to-end resistance is about 10^9 ohms. In operation, a potential of about 3 kV across the channeltron yields a gain of about 10^8. Channeltrons have the advantage of small size (about 5 cm long), low cost, and ruggedness. Monolithic ceramic channeltrons have recently become available from OptoTechnik. These consist of a ceramic block with an interior serpentine channel. Many configurations are available with entrance aperture as small as a few mm^2 to more than one cm^2. The overall volume ranges from 1 to 5 cm^3. The monolithic ceramic channeltrons are very robust. They have found wide use in rocket-borne instruments where forces at launch can exceed 30 g.

For the detection of positive ions, an electron multiplier is usually operated with the first dynode at a high negative potential and the last dynode near ground potential, as in Figure 5.43. This arrangement provides acceleration of incident ions. For electrons or negative ions, the cathode is usually biased positive by a few hundred volts and the last dynode is at a high positive potential relative to ground, as in Figure 5.44. In this latter configuration, it is only convenient to operate in a pulse-counting mode, since the detection electronics must be electrically insulated from the measuring resistor. Ordinarily, the multiplier output is coupled to the

pulse-counting electronics *via* a high-voltage disc ceramic capacitor of about 100 pF. A problem sometimes encountered with this arrangement is that sparking between dynodes or across dynode resistors, or switching noise, gives rise to large high-frequency transients that are transmitted through the coupling capacitor and damage the electronics. This problem can be cured by coupling the output of the multiplier to the electronics through a transformer that will attenuate large pulses because of saturation. A suitable transformer can be produced by making a ten-turn bifilar winding of well-insulated wire on a ferrite core (e.g. Ferroxcube), as illustrated in Figure 5.45.[32] The number of turns and their positions must be carefully adjusted to assure a proper impedance match between the multiplier and the electronics.

The output current from an electron multiplier is limited by the current available from the resistive voltage divider that establishes the potential of the dynodes. To ensure that the gain of a multiplier is not reduced because of depletion of the dynode charge, the maximum output current should be at least an order of magnitude less than the current drawn by the dynode resistor chain. For channeltrons, the output current should be an order of magnitude less than the current from the high-voltage supply through the (typically 10^9 Ω) resistance of the channel.

The gain of a channeltron is determined not by the absolute size but rather by the ratio of the length of the channel to the diameter. This ratio is about 50 for a conventional channeltron. Very small channeltrons are available in close-packed arrays. Each channel is about 50 μ in diameter and 1 mm long. The arrays, known as *microchannel plates* (MCPs), are available in sizes up to 10 cm on a side. Because the length-to-diameter ratio is only about 20, microchannel plates are usually stacked in pairs to provide easily detected charge pulses. Owing to the small size of the channel, the charge cloud through a channel is temporally much more compact than in a full-size channeltron. The charge cloud is delivered in about 1 ns. With leading-edge discrimination it has been possible to time the arrival of a charged particle at the detector with a resolution of a few hundred picoseconds.

Microchannel plates are used as position-sensitive detectors of electrons and ions at energies up to at least 10 keV. A particle striking the front of the plate triggers an electron avalanche in only one channel. The position of the charge cloud emitted at the output of a channel can be

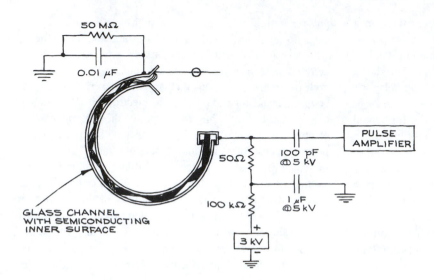

Figure 5.44 A channeltron electron multiplier. Electrical connections shown are for detection of electrons.

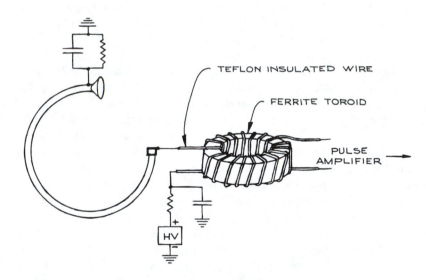

Figure 5.45 A coupling transformer between an electron multiplier and a sensitive pulse amplifier. Properly constructed, a transformer will faithfully transmit signal pulses while attenuating large noise pulses.

located with an accuracy approaching the diameter of the channel. Both one-dimensional and two-dimensional position information can be extracted. A number of methods have been developed to obtain position information.[33] The most straightforward scheme is to provide an array of collectors, each with its own pulse amplifier and counter. This is the most robust method, but expensive and impractical if the full resolution of the microchannel plate is to be realized—a standard 2.5 cm round MCP can potentially provide more than 10^4 pixels. Many position-sensitive detectors employ a resistive strip anode across the back of a stacked pair of MCPs. The arrival of a charge pulse at a point on the anode is detected simultaneously as a pulse at each end or each corner of the anode, and the position is determined from the ratio of the pulse heights.

5.6.5 Magnetic Field Control

A magnetic field may cause aberrations in a charged-particle optical system. The earth's magnetic field is the usual source of problems, but large concentrations of magnetic materials nearby or large currents (such as those associated with a cyclotron) may also produce magnetic fields of sufficient intensity to deflect the trajectories of electrons or ions in an optical system. The earth's magnetic field is about 0.6 gauss directed at an angle of elevation in the north-south plane roughly equal to the local latitude. Whether or not this is a problem depends on the energy and mass of the transmitted charged particles and the dimensions of lens elements and apertures through which the particles must pass. For example, the radius of curvature of the path of a 10 eV electron moving perpendicular to the earth's field is 15 cm. This amounts to a defection of 1.5 mm along a 1 cm path. For a low-energy (<100 eV) electron spectrometer, such considerations may lead to a residual-field tolerance of less than 1 milligauss.

There are two methods used to eliminate magnetic fields. An opposing field can be applied to cancel the offending field throughout the volume occupied by the apparatus or the apparatus can be enclosed in a shield of high permeability ferromagnetic metal that shunts the local field around the enclosure. An electromagnet constructed to make an opposing field is the least expensive—but also the most ungainly—approach. The dimensions of the coils may exceed the dimensions of the apparatus to be shielded by a factor of ten. In addition, without an elaborate feedback control, an electromagnet can only cancel a constant (dipole) field. A magnetic shield need only be large enough to contain the apparatus, but the construction requires expensive fabrication and heat treatment of the shielding alloy that is itself relatively expensive.

Electromagnets for canceling the ambient field are assembled of two or more pairs of current-carrying loops. A continuous solenoid would be impractical since access to the interior volume is limited. Caprari has surveyed optimal geometries for a range of possibilities.[35] The most common arrangement consists of a pair of identical circular magnet coils spaced apart by half the coil diameter. These are referred to as *Helmholtz pairs* or *Helmholtz coils*. The axis of the pair of coils must be oriented parallel to the field that is to be canceled. Three pairs arranged orthogonally can also be used with the current to each adjusted to give a net field that opposes the offending field. The Helmholtz coil geometry is illustrated in Figure 5.46. The field along the axis of a round Helmholtz pair, a distance z above the bottom coil, is given by

$$B = 0.32 \frac{NI}{R} \left\{ \left[1 + \left(\frac{z}{R} \right)^2 \right]^{-3/2} + \left[1 + \left(1 - \frac{z}{R} \right)^2 \right]^{-3/2} \right\} \text{ gauss}$$

where N is the total number of turns, I is in amperes, and R and Z are in cm. For square coils, a similar expression with R replaced by $L/2$ applies. The orientation of the coils and the exact value of the current must be determined with the aid of a magnetometer. The field produced by a Helmholtz pair is of high uniformity only in a small volume midway between the coils and near the axis. For example, a uniformity of 0.1 percent is achieved in a central volume of 2×10^{-2} cubic radii, a uniformity of 1 percent in a volume of 1.5×10^{-1} cubic radii, and a uniformity of 5 percent in a volume of 0.6 cubic radii.[34] The coil dimensions therefore must exceed those of the apparatus to be shielded by at least a factor of ten in order to reduce the magnetic field to a few milligauss. It is also important to recognize that a Helmholtz coil can only cancel the dipole component of the ambient field.

Shielding electron-optical devices from ambient magnetic fields can be accomplished with an enclosure fabricated of special high-permeability nickel-iron alloys.[36] These materials are available as sheet or tubing and thus, as a practical matter, are best formed into simple cylindrical shapes, either with or without end caps depending on requirements of access to the interior. Construction details are important in determining the performance of a magnetic shield. Removable partitions and end caps should be designed to overlap the fixed pieces to which they are attached and they should fit tightly. Permanent joints should be spot-welded lap joints or arc-welded butt joints. Magnetic-shielding alloys only attain their high permeability after annealing in a hydrogen atmosphere following a precisely defined temperature schedule. Mechanical stress degrades the permeability of the material so the annealing is carried out after fabrication. It usually makes sense to have a shield fabricated by a shop that is prepared to do the annealing. The completed shield must be handled carefully. Bending

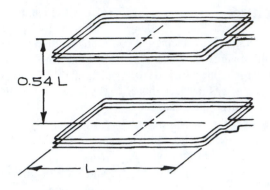

Figure 5.46 The geometry of round and square Helmholtz coils.

or sharp blows degrade the shielding properties as will contact with magnetic materials (such as magnetized tools). The annealed material should not be heated above 400°C. It must be emphasized that magnetic shielding cannot be accomplished casually—a few sheets of μ-metal wrapped around a vacuum system will often do more harm than good.

The performance of a magnetic shield is specified by an attenuation factor A, given by the ratio of the magnetic field H_0 at a point in the absence of the shield to the field strength H_s at that point when it is surrounded by the shield:

$$A = \frac{H_0}{H_s}$$

The calculation of the attenuation factor is generally difficult. Since a shield is usually constructed of sheet and tubing, however, a reasonable estimation can be made from approximations to the attenuation factor for simple geometric shapes. The attenuation of a long cylindrical shield can be approximated as that for an infinitely long cylinder and the attenuation of a closed box can be approximated as that for a sphere. The attenuation factor depends upon the thickness of the shielding material, the overall dimensions of the shield, and the permeability of the shielding material μ. Low-saturation alloys used to shield fields of the order of one gauss have permeabilities of about 40,000. From the formulae below it will be obvious that two or more nested shields of thin material are much preferable to a single thick shield.[37]

For a long cylinder, the attenuation factor is given approximately by

$$A \cong \frac{\mu d}{2R}$$

and for a sphere by

$$A \cong \frac{2\mu d}{3R}$$

where R is the radius of the cylinder or sphere and d is the thickness of the material. For a nested pair of cylinders or spheres of radius R_1 and R_2 spaced apart by a distance $\Delta = |R_1 - R_2|$, the attenuation factor is

$$A \cong A_1 A_2 \frac{2\Delta}{R}$$

where A_1 and A_2 are the attenuation factors for each of the cylinders or each of the spheres alone and R is the mean radius of the two. Note that the attenuation increases with the spacing between the two shields.

A sample calculation illustrates the advantage of multiple shields. A single cylindrical shield, 20 cm in diameter of 1 mm thick material with a permeability of 40,000 gives an attenuation factor of 200. Nested shields 20 cm and 22 cm in diameter of the same material give an attenuation factor of 7000. For shielding from the earth's magnetic field, the double shield reduces the field well into the sub-milligauss range.

An opening in a magnetic shield is often unavoidable. A shield surrounding a vacuum system, for example, must be left open at one end for attachment of the pump and for electrical and mechanical feedthroughs. The external field penetrates into the opening. The attenuation near an opening reaches a value of about two-thirds of the maximum attenuation of the shield at a distance into the shield volume about equal to the mean radius of the opening.[38] A skirt of shielding alloy extending outward from an opening, effectively displacing the opening outward from the shielded volume, helps to alleviate the problem.

CITED REFERENCES

1. K. R. Spangenberg, *Vacuum Tubes,* McGraw-Hill, New York, 1948.

2. J. R. Pierce, *Theory and Design of Electron Beams,* 2nd ed., Van Nostrand, New York, 1954.

3. O. Klemperer and M. E. Barnett, *Electron Optics,* 3rd ed., Cambridge University Press, Cambridge, 1971.

4. V. E. Cosslett, *Introduction to Electron Optics,* Oxford University Press, Oxford, 1946.

5. D. Roy and J. D. Carette, "Design of Electron Spectrometers for Surface Analysis," in *Electron Spectroscopy,* H. Ibach, ed., Topics in Current Physics, chap. 2, Springer, Berlin, 1977.

6. *Methods of Experimental Physics,* Vol. 4A, *Atomic Sources and Detectors,* V. W. Hughes and H. L. Schultz, eds., Academic Press, New York, 1967.

7. E. Harting and F. H. Read, *Electrostatic Lenses,* Elsevier, New York, 1976.

8. (a) F. H. Read, *J. Phys.,* **E2**, 165 (1969); (b) F. H. Read, *J. Phys.,* **E2**, 679 (1969); (c) F. H. Read, A. Adams, and J. R. Soto-Montiel, *J. Phys.,* **E4**, 625 (1971); (d) A. Adams and F. H. Read, *J. Phys.,* **E5**, 150 (1972); (e) A. Adams and F. H. Read, *J. Phys.,* **E5**, 156 (1972).

9. S. Natali, D. DiChio, E. Uva, and C. E. Kuyatt, *Rev. Sci. Instr.,* **43**, 80 (1972); D. DiChio, S. V. Natali, and C. E. Kuyatt, *Rev. Sci. Instr.,* **45**, 559 (1974).

10. D. W. O. Heddle, *Tables of Focal Properties of Three-Element Electrostatic Cylinder Lenses,* J.I.L.A. Report No. 104, University of Colorado, Boulder.

11. R. E. Collins, B. B. Aubrey, P. N. Eisner, and R. J. Celotta, *Rev. Sci. Instr.,* **41**, 1403 (1970).

12. D. DiChio, S. V. Natali, C. E. Kuyatt, and A. Galejs, *Rev. Sci. Instr.,* **45**, 566 (1974).

13. J. A. Simpson, "Electron Guns," in *Methods of Experimental Physics,* Vol. 4A, section 1.5, V. W. Hughes and H. L. Schultz, eds., Academic Press, New York, 1967.

14. An estimation of image expansion at the space charge limit can be obtained from the data of W. Glaser, *Grundlagen der Elektronenoptik,* Springer, Vienna, 1952, p. 75.

15. R. G. Wilson and G. R. Brewer, *Ion Beams,* Wiley, New York, 1973.

16. C. D. Moak, H. E. Banta, J. N. Thurston, J.W. Johnson, and R. F. King, *Rev. Sci. Instr.,* **30**, 694 (1959); M. von Ardenne, *Tabellen der Electronenghysit IonengRysik und Ubermilroskopie,* Deutscher Verlag der Wissenschaften, Berlin, 1956.

17. W. Aberth and J. R. Peterson, *Rev. Sci. Instr.,* **38**, 745 (1967).

18. J. A. Simpson, *Rev. Sci. Instr.,* **32**, 1283 (1961).

19. M. E. Rudd, "Electrostatic Analyzers," in *Low Energy Electron Spectrometry,* by K. D. Sevier, chap. 2, Section 3, Wiley-Interscience, New York, 1972, A. Poulin and D. Roy, *J. Phys.,* **E11**, 35 (1978).

20. T. S. Green and G. A. Proca, *Rev. Sci. Instr.,* **41**, 1409 (1970); G.A. Proca and T.S. Green, *Rev. Sci. Instr.,* **41**, 1778 (1970).

21. V. V. Zashkvara, M. I. Korsunskii, and O. S. Kosmachev, *Soviet Phys. Tech. Phys.,* **11**, 96 (1966); H. Sar-el, *Rev. Sci. Instr.,* **38**, 1210 (1967), and **39**, 533 (1968); J. S. Risley, *Rev. Sci. Instr.,* **43**, 95 (1972).

22. C. E. Kuyatt and J. A. Simpson, *Rev. Sci. Instr.,* **38**, 103 (1967).

23. C. E. Kuyatt, unpublished lecture notes; see also A. J. Williams, III, and J. P. Doering, *J. Chem. Phys.,* **51**, 2859 (1969); J. H. Moore, *J. Chem. Phys.,* **55**, 2760 (1971).

24. R. F. Herzog, *Z. Phys.,* **89**, 447 (1934); **97**, 596 (1935); *Phys. Z.,* **41**, 18 (1940).

25. H. Wollnik and H. Ewald, *Nucl. Instr. Meth.,* **36**, 93 (1965).

26. A. Stamatovic and G. J. Schulz, *Rev. Sci. Instrum.,* **39**, 1752 (1968); A. Stamatovic and G. J. Schulz, *ibid,* **41**, 423 (1970): L. Sanche and G. J. Schulz, *Phys. Rev. A,*

5, 1672 (1972); D. Roy, *Rev. Sci. Instrum.*, **43**, 535 (1972).

27. R. L. Seliger, *J. Appl. Phys.*, **43**, 2352 (1972).

28. E. Segre, *Experimental Nuclear Physics,* Vol. 1, part V, Wiley, New York, 1953.

29. H. Boersch, J. Geiger, and W. Stickel, *Z. Phys.*, **180**, 415 (1964).

30. E. W. Blauth, *Dynamic Mass Spectrometers,* Elsevier, Amsterdam, 1966; P. H. Dawson, *Quadrupole Mass Spectrometer,* Elsevier, Amsterdam, 1976.

31. W. D. Kingery, H. K. Bowen, and D. R. Uhlmann, *Introduction to Ceramics*, 2nd ed., chap. 17, John Wiley & Sons, New York, 1976.

32. J. Millman and H. Taub, *Pulse, Digital, and Switching Waveforms,* Chap. 3, McGraw-Hill, New York, 1965; C. N. Winningstad, *IRE Trans. Nucl. Sci.*, **NS43**, 26 (1959); C. L. Ruthroff, *Proc. IRE*, **47**, 1337 (1959).

33. L. J. Richter and W. Ho, *Rev. Sci. Instrum.*, **57**, 1469 (1986); R. M. Tromp, M. Copel, M. C. Reuter, M. Horn v. Hoegen, J. Speidell, and R. Koudijs, *ibid*, **62**, 2679 (1991).

34. R. K. Cacak and J. R. Craig, *Rev. Sci. Instr.*, **40**, 1468 (1969).

35. R. Caprari, Measurement Science & Technology, **6**, 593 (1995).

36. "The Definitive Guide to Magnetic Shielding", Amuneal Manufacturing Corp., 4737 Darrah Street, Philadelphia, Pa. 19124.

37. V. Schmidt, *Electron Spectrometry of Atoms using Synchrotron Radiation*, Cambridge University Press, Cambridge, (1997), pp. 403–407.

38. W. G. Wadey, *Rev. Sci. Instr.*, **27**, 910 (1956).

COMPUTER SIMULATION PROGRAMS

Simion 3D available from:

Princeton Electronics Systems, Inc.
P.O. Box 8627
Princeton, NJ 08543

Scientific Instrument Services
1027 Old York Rd.
Ringoes, NJ 08551

SciTech International, Inc.
2525 North Elston
Chicago, IL 60647–9939

CPO-2D and CPO-3D available from:

Frank Read and Nick Bowring
Department of Physics and Astronomy
University of Manchester, UK
Email: cpo3d@fs3.ph.man.ac.uk

MANUFACTURERS AND SUPPLIERS

Acheson Colloids Co.
Port Huron, MI 48060
(colloidal graphite)

Amuneal Corp.
4737 Darrah St.
Philadelphia, PA 19124
(215) 535-3000
(magnetic shielding and heat treating)

Bomco, Inc.
Rt. 128
Blackburn Circle
Gloucester, MA 01930
(molybdenum fabrication)

Burle Electro-Optics, Inc. (formerly Galileo)
Galileo Park
Sturbridge, MA 01518
(electron multipliers)

Cliftronics, Inc.
515 Broad St.
Clifton, NJ 07013
(electron guns)

Corning Glass Works
Corning, NY 14830
(machinable ceramic)

H. Cross Co.
363 Park Ave.
Weehawken, NJ 07087
(tungsten wire and ribbon)

Electron Technology
626 Schuyler Ave.
Kearny, NJ 07032
(electron-gun filaments)

Ferroxcube
5083 Kings Highway
Saugerties, NY 12477
(magnet cores)

Industrial Tectonics
P.O. Box 1128
Ann Arbor, MI 48106
(ceramic balls)

K and M Electronics
11 Interstate Drive
West Springfield, MA 01089
(ceramic channel electron multipliers)

Kimble Physics, Inc.
Kimball Hill Rd.
Wilton, NH 03086
(electron and ion guns and electron optics)

McDanel Refractory Porcelain Co.
510 Ninth Ave.
Beaver Falls, PA 15010
(ceramic rod)

Magnetic Metals Co.
Camden, NJ 08101
(magnetic shielding materials and fabrication)

Magnetic Shield Div.
Perfection Mica Co.
740 North Thomas Dr.
Bensenville, IL 60106
(magnetic shielding materials and fabrication)

OptoTechnik
Max-Planck-Str 1
3411 Katlenburg-Lindau
GERMANY
(ceramic channel electron multipliers)

Spectra-Mat, Inc.
Watsonville, CA 95076
(alkali-metal ion sources)

Spruce Pine Mica Co.
100 Mountain Laurel Dr.
Spruce Pine, NC 28777
(mica)

CHAPTER 6

ELECTRONICS

This chapter discusses electronics at a level somewhere between that of a handbook, which consists essentially of charts, tables, and graphs, and a textbook, where the interesting, important, and useful conclusions come only after well-developed discussions with examples. The aim here is a presentation that has sufficient continuity and readability that individual sections can be profitably read without having to refer to preceding sections or other texts. On the other hand, it is important to have useful and frequently referenced material in the form of readily accessible tables, graphs, and diagrams that are sufficiently self-explanatory that very little reference to the text material is necessary. Another important goal is an emphasis on vocabulary. A large amount of jargon in electronics is meaningless to the uninitiated, but when it is necessary to understand the properties of an electronic device from a written technical description, when writing the specifications for electronic equipment, or when talking to an electronics engineer, salesman, or technician, this vocabulary is essential. With this in mind, terms that are not current outside of electronics are italicized.

To be used to best advantage, this chapter should be supplemented with manufacturers' catalogs, data books, applications texts, handbooks, and more specialized texts that treat the topic of interest in depth. Manufacturers of laboratory electronic equipment, discrete devices, and integrated circuits have publications that describe, in clear practical terms, the properties of their products and their applications to a wide variety of tasks. Much of this material is also available on the Internet, and for this reason Internet addresses are given when available.

The material has been organized and written as one explains it to a student or technician coming to work in a laboratory for the first time. The complexity of modern electronics is such that the cut-and-try approach is too inefficient and costly in terms of material and time. There are too many possibilities when connecting devices and multiple-component circuits, it is important to establish a systematic approach based on a limited number of simple, well-understood principles. It is probably not reasonable in the laboratory to expect quick solutions to problems that are entirely outside one's previous experience. The number of really new situations that can arise is limited, however, most problems being variations on a few basic situations. The ability to recognize this and to isolate the source of difficulty comes with experience and mastery of basic principles. When confronted with a new situation involving rack upon rack of equipment, the tendency is to believe that an understanding of how everything works is beyond the capabilities of all but experienced electronics engineers. This is far from the truth. At the operational level, present-day electronics is the most reliable, easy-to-use, and easy-to-understand element of most experiments.

6.1 PRELIMINARIES

6.1.1 Circuit Theory

An understanding of elementary circuit theory and the accompanying vocabulary permits one to reduce complex circuits consisting of many elements to a few essentials, predict the behavior of complex circuits, specify the operation of components, and understand and use data sheets and operation manuals. In routine laboratory work, it is not necessary to be skillful with circuit theory. It is necessary to be able to isolate the basic elements of a circuit and understand their behavior. With that ability, when circuits fail to operate correctly, the causes of the malfunction can be localized and repaired.

Linear circuit theory applies to devices whose output is directly proportional to the applied input. If one increases the current through a resistor by a factor of two, for example, the voltage across it will double. An example of a nonlinear device is a switch that is either open or closed and whose state changes abruptly at a threshold. A nonlinear device can often be treated with linear theory by dividing the response of the device into separate regions over which it behaves in a quasi-linear manner. This is called *linearizing the response curve,* An example is the

piecewise linearization of a diode's current-voltage response as shown in Figure 6.1. The exponentially rising forward current and constant reverse current are represented by straight lines of slope $1/R_f$ and $1/R_r$, which are joined at voltage V_g.

We begin by considering only *passive* linear devices; that is, devices that either dissipate energy (resistors) or store energy in electric (capacitors) or magnetic (inductors, transformers) fields. *Active* devices such as transistors can supply energy to a circuit when appropriately powered by external sources. The analysis of circuits with active devices is based on representations using equivalent circuits consisting of passive devices.

Conventional circuit analysis uses three *lumped* circuit elements—resistors (R), capacitors (C), and inductors (L). This way of analyzing circuits is valid at signal frequencies f for which the wavelength λ is much larger than the physical dimensions of the circuit. Since $\lambda = c/f$, where c is the speed of light, this means that analysis in terms of lumped elements is valid up to frequencies of a few hundred megahertz.

This frequency limitation also excludes waveforms with significant frequency components above a few hundred megahertz, even though the repetition rate of the waveform may be much less. A convenient way to

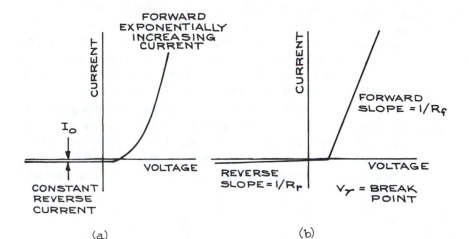

Figure 6.1   (a) Real and (b) piecewise linear representation of the current-voltage characteristics of a diode.

estimate the frequency of the highest-frequency component of a nonsinusoidal waveform is to divide 0.3 by the *rise time* of the waveform, t_r, defined as the time between the 10 percent and 90 percent amplitude points on the leading edge of the waveform. A pulse with a 10 ns rise time, for example, has significant frequency components up to 30 MHz. The *fall time, t_f,* of a waveform is the time between the 10 percent and 90 percent amplitude points on the trailing edge. Figure 6.2 illustrates *rise time* and *fall time*.

Even at low frequencies there are no ideal resistors, capacitors, or inductors. Actual resistors have some capacitance and inductance, while capacitors have resistance and inductance, and inductors have resistance and capacitance. These departures from ideality are largely a matter of construction.

At high frequencies, stray capacitances and inductances become significant, and one commonly speaks of *distributed* parameters in contrast to the *lumped* parameters at low frequencies. Coaxial cable is an example of a type of distributed parameter circuit. The electrical properties of coaxial cable are normally given in terms of resistance and attenuation per unit length as a function of frequency. At high frequencies, the resistance of conductors (even connecting wires) increases due to what is termed *skin effect*. The magnitude of this effect for round cross-section wires is given in Table 6.1 as a function of frequency. High-frequency connections are best made with leads having a large surface area-to-volume ratio, with flat-ribbon geometry being the best.

Conventional circuit theory is based on a few laws, principles, and theorems. In the equations that follow, lowercase letters represent instantaneous values of voltage and current, whereas uppercase letters indicate effective or dc values. It is also convenient to distinguish between root-mean-square (rms), peak-to-peak, and average values of voltage and current for sinusoidally varying voltages. If $v = V\cos \omega t$, where v is the instantaneous value of voltage and V is the peak value, the rms value is $V/\sqrt{2}$, the peak-to-peak value is $2V$, and the average value is clearly zero. This is illustrated in Figure 6.3. Common U.S. line voltage is specified as 110 volts ac, which is the rms value. The peak voltage is 156 volts, so the peak-to-peak voltage is 312 volts.

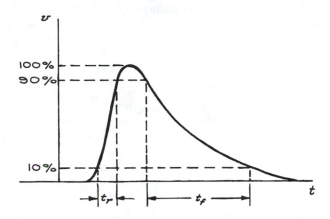

Figure 6.2 Rise time (t_r) and fall time (t_f) of a pulse.

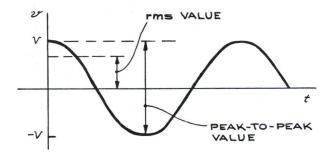

Figure 6.3 Relation of rms to peak-to-peak voltages for a sinusoidal waveform.

Under certain conditions, the rms value is not sufficient for specifying the output of an ac source. A source producing voltage spikes of large amplitude but short duration superimposed on a small sinusoidally varying voltage will have an rms value very close to that without the spikes, but the spikes can have a large effect on circuits connected to the source. When specifying the output of a dc power supply, the magnitude, frequency, and duration of nonsinusoidal waveforms that appear at the output need to be specified as well as the rms value of any ac component of the output.

Laws

(i) Current-voltage relations. For resistors, capacitors, and inductors we have

TABLE 6.1 RATIO OF AC TO DC WIRE RESISTANCE

Wire Gauge	R_{ac}/R_{dc}			
	10^6 Hz	10^7 Hz	10^8 Hz	10^9 Hz
#22	6.9	21.7	69	217
#18	10.9	34.5	109	345
#14	17.6	55.7	176	557
#10	27.6	87.3	276	873

$$v_R = iR, \qquad v_C = \frac{1}{C}\int i\,dt, \qquad v_L = L\frac{di}{dt}$$

respectively, where the resistance R is in ohms; the capacitace C in farads; and the inductance L in henrys.

(ii) Loops and nodes (Kirchhoff's laws).

In these, the sums are algebraic (signs taken into account):

1. Σ (voltage drops around a closed loop) = Σ (voltage sources).

2. Σ (current into a node) = Σ (current out of a node), where a node is a point where two or more elements have a common connection.

Principles

(i) Voltage.

The voltage between any two points in a circuit is equal to the algebraic sum of the voltages produced by each one of the voltage sources separately, replacing all the other sources by their internal resistance.

(ii) Current.

The current past any point in a circuit is the algebraic sum of the currents due to each of the current sources in the circuit, replacing each of the other sources by their internal resistance.

Theorems

(i) Thevenin's theorem.

A real voltage source in a circuit can always be replaced by an ideal voltage source in series with a generalized resistance. An ideal voltage source is one that can maintain a constant voltage across its terminals regardless of the load across it. In other words, an ideal voltage source has zero internal resistance. An automobile battery, with an internal resistance of a few hundredths of an ohm, is a good approximation to an ideal voltage source at currents of a few amperes. Electronically regulated power supplies often have very low effective internal resistances when operated within their voltage and current ratings.

(ii) Norton's theorem.

A real current source in a circuit can always be replaced by an ideal current source shunted by a generalized resistance. An ideal current source is one that supplies a constant current regardless of load—such a source has an infinite internal resistance. Photo-multiplier and electron-multiplier devices provide currents, albeit very low, that are independent of load and approximate ideal current sources.

When connecting a source of current or voltage to a circuit, it is often important to know the internal resistance of the source. This can be determined by first measuring the open-circuit voltage of the source with a high internal-resistance voltmeter and then connecting a variable resistance across the source and adjusting it until the voltmeter reading is one half the open-circuit value. The source resistance is then equal to the value of the variable-resistance setting. If the source has a very high internal resistance, a current measurement can be substituted for the voltage measurement. In this case, the output is shunted with an ammeter and the so-called *short-circuit current* is measured. A variable resistance is then placed in series with the ammeter and adjusted until the current through the ammeter is one half the short-circuit current. The value of the variable resistor at this point is equal to the internal resistance of the source. Analagous measurements can be made for ac sources by using either ac voltmeters and ammeters or an oscilloscope.

6.1.2 Circuit Analysis

For any given source, the choice of representation (Thevenin or Norton) is arbitrary and, in fact, the series resistance in the Thevenin representation is exactly equal to the parallel resistance in the Norton representation. Thevenin's and Norton's theorems simplify the application of the laws and principles discussed above.

The most general method for solving circuit problems is to apply Kirchhoff's laws using the appropriate current-voltage relations for each element in the circuit. This gives rise to one or more linear differential equations, which when solved with the proper boundary conditions give the general solution. This is illustrated for RC circuits in Section 6.1.3.

When dealing with sinusoidal sources of angular frequency ω, circuit analysis can be greatly simplified when only the steady-state solution is required. In this case, circuit capacitances and inductances are replaced by *reactances:*

$$capacitive\ reactance = jX_C, \quad where\ X_C = \frac{-1}{\omega C}$$

$$inductive\ reactance = jX_L, \quad where\ X_L = \omega L$$

$$j = \sqrt{-1}$$

The *impedance* Z of a circuit is obtained by combining reactances and resistances according to the formula $Z = R + j(X_L + X_C)$. These quantities can be represented in the complex plane by vectors (see Figure 6.4). The angle between Z and the real axis is ϕ. By analogy with the *I-V* relations for a pure resistance, X_C is the ratio of the ac voltage across a capacitor to the current through it, X_L is the ratio of the ac voltage across an inductor to the current through it, and Z is the net ratio of ac voltage to current in a circuit composed of resistors, capacitors, and inductors.

The fact that jX_C and jX_L are imaginary means that the voltage and current are 90° out of phase with each other. For a capacitor, the voltage lags the current by 90° while for an inductor the voltage leads the current by 90°.

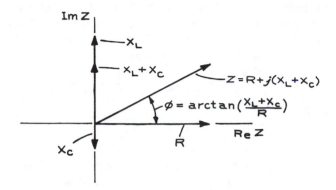

Figure 6.4   Relations between reactance, resistance, impedance, and phase angle.

Another quantity that is occasionally useful in circuit analysis is the *complex admittance Y*, which is the reciprocal of the impedance. The usefulness of the admittance arises in circuits with several parallel branches where the net admittance is the sum of the admittances of the branches.

In carrying out circuit analysis, the following results of the above laws are useful.

Series Circuits

At any instant, the current is the same everywhere in a series circuit, and the algebraic sum of the voltage drops around a circuit equals the algebraic sum of the sources. For circuit elements of impedance Z_1, Z_2,..., Z_N in an N-element series circuit, the total impedance is $Z = Z_1 + Z_2 + ... + Z_N$. If all the elements are resistors, or inductors, or capacitors, the general expression reduces, respectively, to

$$R = R_1 + R_2 + \cdots + R_N$$

$$\frac{1}{C} = \frac{1}{C_1} + \frac{1}{C_2} + \cdots + \frac{1}{C_N}$$

$$L = L_1 + L_2 + \cdots + L_N$$

Parallel Circuits

For circuit elements in a parallel circuit, the voltage drop across each branch is the same while the current through each branch is inversely proportional to the impedance of the branch. The total current through all of the branches is the voltage across the network divided by the equivalent impedance for the network. The equivalent impedance Z and admittance Y for an N-branch parallel circuit are

$$\frac{1}{Z} = \frac{1}{Z_1} + \frac{1}{Z_2} + \cdots + \frac{1}{Z_N}$$

and

$$Y = Y_1 + Y_2 + \cdots + Y_N$$

where $Z_1, Z_2,...,Z_N$ are the impedances of the branches and $Y_1, Y_2,...,Y_N$ are the admittances. In the special cases where all of the circuit elements in the branches are of the same type,

$$\frac{1}{R} = \frac{1}{R_1} + \frac{1}{R_2} + \cdots + \frac{1}{R_N}$$

$$C = C_1 + C_2 + \cdots + C_N$$

$$\frac{1}{L} = \frac{1}{L_1} + \frac{1}{L_2} + \cdots + \frac{1}{L_N}$$

where R, C, and L are the net resistance, capacitance, and inductance of the circuits.

Voltage Dividers

The *voltage divider,* illustrated in Figure 6.5a, is a very common circuit element. The instantaneous voltage across Z_N is $v_{in}[Z_N/(Z_1 + Z_2 + Z_3 + ... + Z_N)]$; that is, the fraction of v_{in} that appears across any circuit element is the impedance of that element divided by the total impedance of the series circuit. Voltage dividers provide an easy way to obtain a variable-voltage output from a fixed-voltage input, but there are limitations. To avoid drawing too much current from the voltage source, the impedance of the voltage divider string should not be too small. If, in the interest of conserving power, however, the impedance is made large, the output impedance of the circuit will be large and v_{out} will depend critically on the load. This can be seen from the Thevenin equivalent of the circuit given in Figure 6.5b, where v_s is the instantaneous voltage of the ideal voltage source. When Z_s is large, the voltage across a load will depend critically on the value of the load. Such *loading* of a divider is to be avoided. For noncritical applications, Z_s should be at most 1/10th of any load.

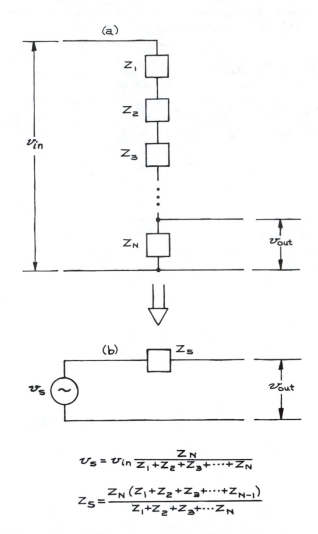

$$v_s = v_{in}\frac{Z_N}{Z_1 + Z_2 + Z_3 + \cdots + Z_N}$$

$$Z_s = \frac{Z_N(Z_1 + Z_2 + Z_3 + \cdots + Z_{N-1})}{Z_1 + Z_2 + Z_3 + \cdots Z_N}$$

Figure 6.5   (a) The voltage divider; (b) the Thevenin equivalent.

Precision, highly linear, multiturn potentiometers are commonly used for position sensing. In this application, the shaft of the potentiometer is coupled mechanically to the moveable element whose position is to be determined, and a stable voltage source is connected across the fixed ends of the potentiometer. The ratio of the voltage from the variable contact of the potentiometer to its end gives the number of turns though which the shaft has rotated.

Equivalent Circuits

Two circuits are *equivalent* if the relationships between the measurable currents and voltages are identical. As has been seen, a circuit with two external terminals can be replaced by its Thevenin or Norton equivalent. Common equivalence transformations for circuits with three terminals *(Miller* and *Y-Δ transformations)* are shown in Figures 6.6 and 6.7. In the first Miller transformation of Figure 6.6, it is necessary to know the ratio of the voltages at nodes 1 and 2; in the second it is necessary to know the ratio of the currents through 1 and 2.

Discussions of active circuits often refer to the *Miller effect.* This is nothing more than the multiplication of Z by $(1 - A_i)$ in the upper left-hand branch, where A_i is the current gain of the active element.

The term *Miller capacitance* refers to the first transformation in the case where Z is purely capacitive.

The transformation results in Z being multiplied by $1/(1 - K)$, where K is the voltage gain with the transformed impedance inserted in the circuit between node 1 and the common or ground node.

The *Y-Δ* transformation allows one to transform a circuit of three elements from a node to a loop configuration and from a loop to a node.

6.1.3 High-Pass and Low-Pass Circuits

Analysis of the *high-pass* and *low-pass* circuits shown in Figure 6.8 illustrates some of the above circuit-analysis principles. The combination of v_s and R_s represents a real voltage source with instantaneous open-circuit voltage v_s and internal resistance R_s. For the high-pass or *differentiating circuit,* the output voltage v_o is across the resistor R; for the low-pass or *integrating circuit,* it is

Figure 6.6   Miller transformations for a circuit with three terminals.

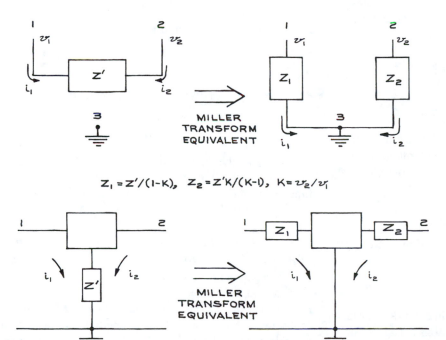

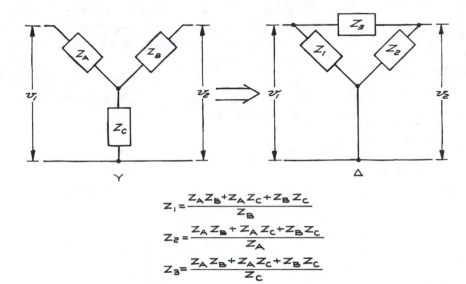

$$Z_1 = \frac{Z_A Z_B + Z_A Z_C + Z_B Z_C}{Z_B}$$

$$Z_2 = \frac{Z_A Z_B + Z_A Z_C + Z_B Z_C}{Z_A}$$

$$Z_3 = \frac{Z_A Z_B + Z_A Z_C + Z_B Z_C}{Z_C}$$

across the capacitor C. Very often the principal features of extremely complex circuits reduce to one of these two circuits, so it is useful to be acquainted with their characteristics.

These circuits can be analyzed by the *differential-equation method*. For either circuit,

$$v_s(t) = iR_s + \frac{1}{C}\int i\,dt + iR$$

This equation uses the fact that the sum of the voltage drops in the circuit equals the sum of the voltage sources and the current is everywhere the same in a series circuit at any instant. Differentiating with respect to time,

$$\frac{dv_s}{dt} = \frac{di}{dt}(R_s + R) + \frac{i}{C}$$

The solution to the homogeneous equation ($dv_s/dt = 0$) is:

$$i = Ae^{-t/R'C}$$

where $R' = R_s + R$ and A is the integration constant determined from the initial conditions. The general solution requires that the functional form of v_s be known. Consider the following three cases:

1.   An ac voltage of amplitude V,

$$v_s = V\cos(\omega t + \phi)$$

2.   A step voltage of amplitude V,

$$v_s = \begin{cases} 0 & for \ \ t < 0 \\ V & for \ \ t > 0 \end{cases}$$

3.   A rectangular pulse of amplitude V and duration T,

$$v_s = \begin{cases} V & for \ \ 0 \le t \le T \\ V & for \ \ t < 0, \ t > T \end{cases}$$

For case 1, the output voltage is sinusoidal at the same frequency as the input voltage. The ratio of v_o to v_s as a function of normalized frequency is shown in Figure 6.9a. At the frequencies $\omega = \omega_H$ and $\omega = \omega_L$ for the two circuits, v_o is $1/\sqrt{2}$ of the maximum value. These frequencies are called the *upper* and *lower corner frequencies*, respectively.

The maximum power that can be delivered to a load is proportional to the square of the output voltage, so that at $\omega = \omega_H$ and $\omega = \omega_L$ the maximum power that the circuits can deliver to a constant load is one-half the maximum possible value. The usual way of expressing this is in terms of decibels (dB), where

$$ratio \ in \ dB = 10\log_{10}\left(\frac{power \ out}{power \ in}\right) = 10\log\left(\frac{v_{out}^2/R_{out}}{v_{in}^2/R_{in}}\right)$$

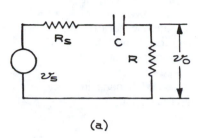 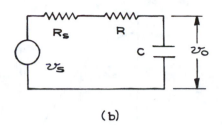

Figure 6.8   (a) High-pass or differentiating circuit; (b) low-pass or integrating circuit.

(a)   (b)

If $R_{out} = R_{in}$, which is often assumed, then (ratio in dB) $= 20\log_{10}[v_{out}/v_{in}]$. When $v_{out}/v_{in} = 1/2$, this is approximately -3, so that -3 dB represents a power reduction of a factor of about two. Since the frequency response of amplifiers, filters, and transducers is routinely given in dB, it is important to keep in mind that the dB scale is logarithmic. Human sensory perception is approximately logarithmic, and a 3 dB change in sound level or light level is barely perceptible.

Because of the reactive element in the circuits (the capacitor), the voltage is not in phase with the current, as illustrated in Figure 6.9b. These plots of phase and log(output voltage) as a function of log(frequency) are called *Bode plots,* after H. W. Bode.[1]

It is often convenient to approximate frequency-response curves by a piecewise linear function. Such idealized Bode plots are shown in Figure 6.10a and b. The *corner frequencies* are where ω/ω_L and $\omega/\omega_H = 1.0$ correspond to the -3 dB points on the unapproximated Bode plots. For most purposes, the simplified curves are an entirely satisfactory representation. From these curves, every tenfold reduction in frequency below ω_L for the high-pass circuit decreases the output voltage by 20 dB, and every twofold reduction decreases it by 6 dB. The low-pass circuit has just the opposite properties: a tenfold increase in frequency above ω_H results in a 20 dB decrease in output voltage, and a twofold increase results in a 6 dB decrease. One often states these facts as *20 dB per decade* and *6 dB per octave.* The linearized phase-response curves are shown in Figure 6.10c. The -3 dB frequencies occur at a phase shift of $-\pi/4$ ($-45°$) for the two circuits.

For the nonrepetitive input voltage waveforms of cases 2 and 3, the output waveforms are given in Figure 6.11.

The output waveforms for the rectangular-wave input function can be used to determine the RC time constants for differentiating and integrating circuits. This is called *square-wave testing.* The RC time constant for the differentiating circuit is obtained by using a square-wave input with a rise time much less than RC and a period much greater than RC. For times small compared with RC, the tilt of the top edge of the output as viewed with a fast rise time oscilloscope (see Figure 6.12) is directly related to RC. The fractional decrease in v_o, in time t_1 is t_1/RC, which can be set equal to $(V - V')/V$ and solved for RC. For the integrating circuit, RC is obtained by measuring the rise time of the output waveform on a fast rise time oscilloscope. Using the definition of the rise time t_r as the time between the 10 percent and 90 percent points on the leading edge of the output waveform, one has the relation

$$t_r = 2.2RC$$

6.1.4 Resonant Circuits

The voltages and currents in circuits with capacitors, inductors, and resistors show oscillatory properties much like those of mechanical oscillators. Electronic circuits have natural frequencies of oscillation and can be critically damped, underdamped, or overdamped, depending on the relations between the values of the circuit parameters. Resonant circuits with ideal capacitors and inductors are of the series or parallel type shown in Figure 6.13. When driven by a sinusoidal input source, the capacitive reactance in the series circuit will cancel the inductive reactance at the precise resonant

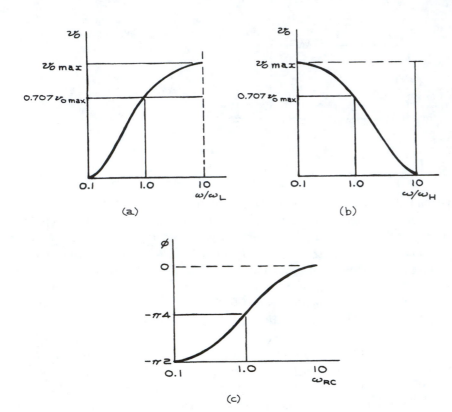

Figure 6.9 Output voltage as a function of frequency for (a) high-pass and (b) low-pass circuits; (c) voltage-current phase relationship in an RC circuit.

frequency ω_0, where $1/C\omega_0 = \omega_0 L$ and $\omega_0 = 1/LC$. At ω_0, the impedance of the series circuit is a minimum and the current through it is a maximum.

For the parallel resonant circuit at low frequencies, the L-branch will have a very low reactance and the current drawn from the source will flow almost entirely through that branch. At high frequencies, the current through the RC branch is limited by the value of R. The total impedance of the parallel circuit is therefore small at low and high frequencies, passing through a maximum at the frequency $\omega_0 = 1/LC$ provided that $R \ll \omega_0 L$. Graphs of the currents in the two circuits as a function of driving frequency are given in Figure 6.14. Real inductors have an associated resistance, which generally must also be taken into account when analyzing circuits.

One measure of the resonance sharpness in the series and parallel circuits is the Q or *quality* of the circuit. For practical purposes, $Q = \omega_0/\Delta\omega$, where $\Delta\omega$ is the full width

at half maximum of the peak or valley. In terms of the circuit parameters, $1/Q = \omega_0 L/R$. This is the ratio of the energy stored (in the capacitor or inductor) to the energy dissipated (in the resistor, per cycle, at resonance. Values of Q as large as 100 can be attained in electrical circuits—mechanical oscillators can attain values as high as 10^6. The phase relationships between voltage and current in series and parallel resonant circuits are shown in Figure 6.15.

The behavior of an LRC circuit upon the application of a step or rectangular input is much like the response of a mechanical system to a sudden impulse. Critically damped, underdamped, and overdamped current flows result.

6.1.5 The Laplace-Transform Method

A very general technique for analyzing circuits for arbitrary input voltage waveforms is the method of

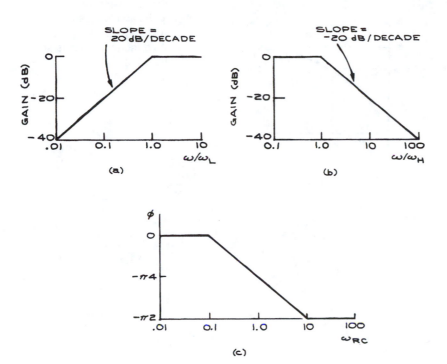

Figure 6.10   Idealized gain response for (a) high-pass and (b) low-pass circuits; (c) idealized phase response.

Laplace transforms. With this method it is possible to use only algebra and lists of transforms—such as those given in Table 6.2—for the solution of differential equations and the evaluation of boundary conditions. The results of the method will be presented without any proofs. The vocabulary of Laplace transforms is used very often in the discussion of circuits and is included in this chapter for that reason.

The method itself is based on an integral transform of the type

$$\bar{f}(s) = \int_0^\infty f(t)e^{-st}dt$$

where $\bar{f}(s)$ is the Laplace transform of $f(t)$, written as $\mathcal{L}[f(t)]$. The function $f(t)$ can involve integrals and differentials. When it is applied to the second-order differential equations that arise in circuit analysis, rather important simplifications occur and results can often be written down by inspection. The Laplace transform of the output voltage $\bar{v}_0(s)$ of a circuit is the Laplace transform

of the input voltage $\bar{v}_i(s)$ times the Laplace transform of the transfer function $\bar{T}(s)$—the *transfer function* being the function relating the output to input. To obtain $T(s)$, the values of all elements in the circuit are replaced by their transform equivalents according to the recipe $R \rightarrow R$, $C \rightarrow 1/sC$, and $L \rightarrow sL$. In the simple case of the voltage divider, the transfer function is the ratio of the impedance of the output-circuit element to the total impedance of the circuit chain. $\bar{T}(s)$ is obtained in exactly the same way using the equivalences for R, C, and L. In general, $\bar{T}(s)$ is in the form of a ratio of two functions of s, $G(s)$ and $H(s)$, which are polynomials in s:

$$\bar{T}(s) = \frac{G(s)}{H(s)}$$

The values of s for which $G(s)$ is 0 are called the *zeros* of $\bar{T}(s)$; the values of s for which $H(s)$ is 0 are called the *poles* of $\bar{T}(s)$. In the most general case, the zeros and poles are complex. The positions of the zeros and poles of $\bar{T}(s)$ in the complex plane give important information on the

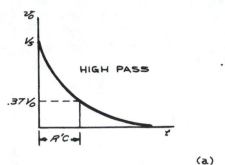

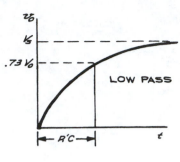

(a)

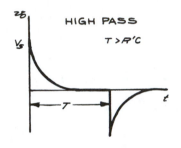

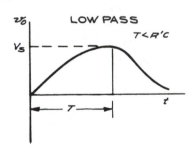

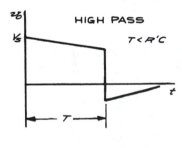

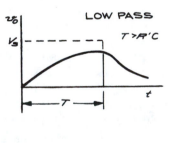

(b)

Figure 6.11 Response of high-pass and low-pass circuits to a step voltage (a) and a rectangular pulse of duration T (b).

properties of the circuit under analysis. When the poles are complex, they occur in pairs, while real-valued poles can occur singly or in pairs. The values of the real and imaginary components of the pole coordinates, usually labeled σ and ω, have important physical meaning. The real component σ is a measure of the damping in the circuit while the imaginary part ω is the natural frequency of oscillation. Negative values of σ give stable circuits in which transient signals all decay to zero with time.

Circuits employing only passive elements behave in this way and are stable. Circuits with active elements can behave in such a way that the output increases with time in response to a transient input signal. Such circuits are unstable and have values of σ greater than zero. They are to be avoided except in the case of oscillators, which must be unstable in order to function.

The Laplace-transform equivalents of the high-pass and low-pass circuits are shown in Figure 6.16. For the high-

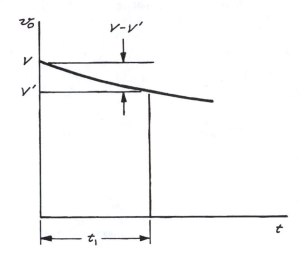

Figure 6.12   Square-wave testing of a high-pass circuit.

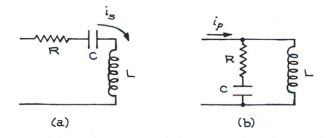

(a)   (b)

Figure 6.13   (a) Series resonant circuit; (b) parallel resonant circuit.

pass circuit, the output voltage across the resistor R for the transformed circuit is

$$\bar{v}_0(s) = \bar{v}_s(s)\frac{sRC}{1+sR'C}$$

where $R' = (R_s + R)$ and $\bar{T}(s)$, which is the transfer function, has a zero at $s = 0$ and a pole at $s = -1/R'C = -\omega_L$. For the low-pass circuit, the output voltage across the capacitor for the transformed circuit is

$$\bar{v}_0(s) = \bar{v}_s(s)\frac{1}{1+sR'C}$$

$T(s)$ has a pole at $s = -1/R'C = -\omega_H$. The steady state frequency and phase response of the circuits are obtained from $T(s)$ by replacing s with $j\omega$. The transfer functions are now, for the low-pass circuit,

$$T(\omega) = \frac{j\omega}{1+j\omega/\omega_L}$$

and

$$T(\omega) = \frac{1}{1+j\omega/\omega_H}$$

for the high-pass circuit.

As seen previously, ω_L and ω_H are the corner frequencies of the circuits. By rationalizing the denominators of the transfer functions, one obtains the phase response. Since there are no inductive elements in the circuits, there is no natural frequency of oscillation. The poles lie on the negative real axis because there are no active elements in the circuit.

6.1.6 RLC circuits

Consider the equivalent circuit and the Laplace transform given in Figure 6.17. To analyze the circuit, consider the parallel combination of R and $1/sC$, which is in series with sL in a voltage-divider configuration:

$$\frac{R/sC}{R+1/sC} = \frac{R}{1+sRC}$$

Thus

$$\bar{v}_0(s) = \bar{v}_i(s)\frac{1}{1+sL/R+s^2LC}$$

The poles of $\bar{T}(s)$ occur at

$$s = \frac{-L/R \pm \sqrt{(L/R)^2 - 4LC}}{2LC}$$

Letting the natural frequency of oscillation of the circuit be $\omega_0 = 1/LC$, we have $Q = R/\omega_0 L$, and the poles can be rewritten as

$$s = \frac{-\omega_0}{2Q} \pm \frac{\omega_0}{2}\sqrt{1/Q^2 - 4}$$

There are three different possibilities for the roots of s:

- $1/Q^2 = 4$: a single real root at $s = -\omega_0/2Q$
- $1/Q^2 - 4 = m^2$ (m real): two real roots at

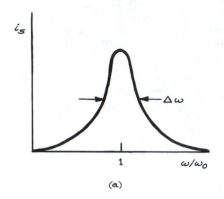 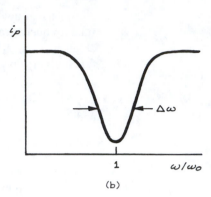

Figure 6.14   Current as a function of frequency for (a) the series resonant circuit and (b) the parallel resonant circuit.

(a)   (b)

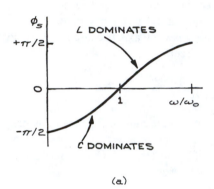 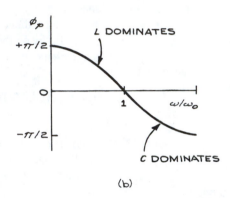

Figure 6.15   Phase relations between voltage and current in (a) the series resonant circuit and (b) the parallel resonant circuit.

(a)   (b)

TABLE 6.2 ELEMENTARY LAPLACE TRANSFORMS

$f(t)$ ($t > 0$)	$f(s)$
$\delta(t)$	1
1	$1/s$
$t^{n-1}/(n-1)!$	$1/s^n$ (n a positive integer)
e^{at}	$\dfrac{1}{s-a}$
$\sin at$	$\dfrac{a}{s^2+a^2}$
$\cos at$	$\dfrac{s}{s^2+a^2}$
$\sinh at$	$\dfrac{a}{s^2-a^2}$

TABLE 6.2 ELEMENTARY LAPLACE TRANSFORMS (CONTINUED)

$f(t)$ $(t > 0)$	$\bar{f}(s)$
$\cosh at$	$\dfrac{s}{s^2 - a^2}$
$\dfrac{1}{2a}\sin at$	$\dfrac{s}{(s^2 + a^2)^2}$
$\dfrac{1}{2a^3}(\sin at - at\cos at)$	$\dfrac{1}{(s^2 + a^2)^2}$
$\dfrac{df(t)}{dt}$	$s\bar{f}(s) - f_0 \left[f_0 = \lim\limits_{t \to 0} f(t) \right]$
$\dfrac{d^2 f(t)}{dt}$	$s\bar{f}(s) - sf_0 - f_1 \left[f_1 = \lim\limits_{t \to 0} \dfrac{df(t)}{dt} \right]$
$\displaystyle\int_0^t f(t')dt'$	$\dfrac{1}{s}\bar{f}(s)$
$\dfrac{1}{a-b}(e^{at} - e^{bt})$	$\dfrac{1}{(s-a)(s-b)}$
$\dfrac{1}{a-b}(ae^{at} - be^{bt})$	$\dfrac{s}{(s-a)(s-b)}$
$\dfrac{1}{a^2}(1 - \cos at)$	$\dfrac{1}{s(s^2 + s^2)}$
$\dfrac{1}{a^3}(at - \sin at)$	$\dfrac{1}{s^2(s^2 + s^2)}$
$\dfrac{1}{ab(b^2 - a^2)}(b\sin at - a\sin bt)$	$\dfrac{1}{(s^2 + a^2)(s^2 + b^2)}$
$\dfrac{1}{b^2 - a^2}(\cos at - \cos bt)$	$\dfrac{s}{(s^2 + a^2)(s^2 + b^2)}$
$\dfrac{1}{\alpha^2 + \beta^2} - \dfrac{e^{-\alpha t}}{\beta\sqrt{\alpha^2 + \beta^2}}\sin[\beta t + \arg(\alpha + i\beta)]$	$\dfrac{1}{s[(s + \alpha)^2 + \beta^2]}$

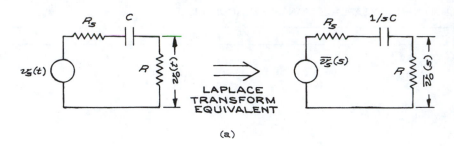

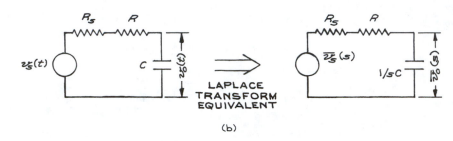

Figure 6.16   Laplace-transform equivalents of (a) the high-pass and (b) the low-pass circuit.

(a)

(b)

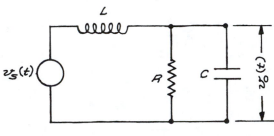

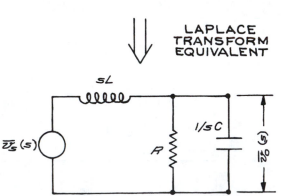

Figure 6.17   An RLC circuit and the Laplace-transform equivalent.

$$s = -\omega_0/2Q \pm \omega_0 m/2$$

- $1/Q^2 - 4 = -m^2$ (m real): two conjugate complex roots at $s = -\omega_0/2Q \pm j\omega_0 m/2$

The magnitude of s in the $\sigma - j\omega$ plane is ω_0; in geometric terms this means that the roots of s are confined to a semicircle of radius ω_0 in the left half of the complex plane (see Figure 6.18).

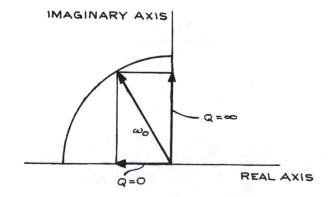

Figure 6.18   Pole trajectory for an RLC circuit.

6.1.7 Transient Response of Resonant Circuits

The stability of a linear system subjected to a driving function depends on the system itself rather than the driving function. For ease of analysis, consider the *RLC* circuit to be driven by a unit step input voltage of the form $v_i(t) = 1$ for $t > 0$. The Laplace transform of the output voltage is then the transform of the input voltage $\bar{v}_i(s)$ multiplied by the transform of the transfer function, $\bar{T}(s)$:

$$\bar{v}_0(s) = \bar{v}_i(s)\bar{T}(s)$$

For the unit step, input $\bar{v}_i(s) = 1/s$ from Table 6.2; therefore

$$\bar{v}_0(s) = \frac{1}{s(1 + sL/R + s^2LC)}$$

To find $v_0(t)$ it is necessary to look up the inverse transform in Table 6.2. The result is

$$v_0(t) = 1 - \frac{e^{-k\omega_0 t}}{\sqrt{1-k^2}}\sin\left[\sqrt{1-k^2}\,\omega_0 t + \arctan\left(\frac{\sqrt{1-k^2}}{k}\right)\right]$$

where $k = 1/2Q$. Critical damping, overdamping, and underdamping correspond to $k = 1$, $k > 1$, and $k < 1$, respectively. The normalized response of the circuit for various values of k is shown in Figure 6.19.

As can be seen, the underdamped waveform overshoots the steady-state value of 1 and oscillates about it with decreasing amplitude. This is sometimes called *ringing*. As damping is increased, the output waveform oscillations decrease in amplitude until $k = 1$ (critical damping) the oscillations are completely gone. Increasing the damping beyond $k = 1$ only increases the time for the waveform to arrive at the steady-state value.

Overshoot, rise time, settling time, and *error band* are terms used to characterize output waveforms. They are illustrated in Figure 6.20. For critical damping, $t_r = 3.3/\omega_0$. By decreasing the damping so that $k < 1$, t_r can be further decreased but at the expense of ringing. A suitable compromise that is often used is $k = 0.707$, under which circumstance the overshoot is 4.3 percent and t_r is reduced to $2.16/\omega_0$. Figure 6.21 is a graph of the percentage overshoot as a function of k.

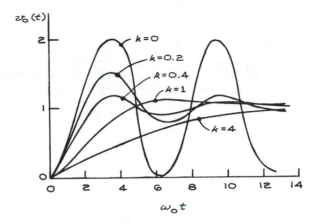

Figure 6.19   Response of a resonant circuit with no damping ($k = 0$), underdamping ($k = 0.2$, 0.4), critical damping ($k = 1$), and overdamping ($k = 4$).

The above analysis is not restricted to passive circuits. Amplifiers with negative feedback can have response functions identical in form to those for *RLC* circuits. The response of such amplifiers to nonrepetitive waveforms is specified using the same terms as for the passive circuit above. In pulse amplifiers, the transient response is the fundamental limitation on the pulse repetition rate. For critically damped amplifiers, the width T of the impulse response and the rise time t_r are approximately related by $T = 1.5t_r$, so that pulses cannot be sent at a rate greater than $1/T$ per second. In this case, T, is generally taken to be the time between the 50 percent points on the output waveform.

For the precision control of temperature in a thermostat, three-term control is often used. With this method, the power applied to the heater element is set equal to a constant background level plus a term proportional to the difference between the set-point temperature T_s and current temperature T, a term proportional to the time integral of $T_s - T$, and a term proportional to the time derivative of $T_s - T$. The resulting differential equation is, under certain circumstances, analogous to that for the *RLC* circuit. Experimental parameters are chosen to obtain the fastest time response consistent with overall system stability. This is discussed in further detail in Section 6.7.9 and in Chapter 7.

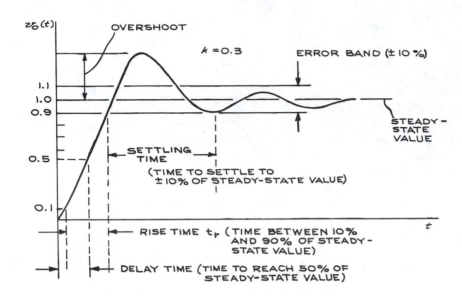

Figure 6.20   Output waveform in an underdamped resonant circuit.

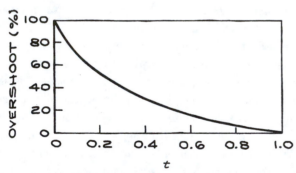

Figure 6.21   Percentage overshoot as a function of damping in a resonant circuit.

6.1.8 Transformers and Mutual Inductance

Transformers consist of primary and secondary windings coupled by a core, which can be composed of magnetic material or air. A voltage is induced in the secondary windings of the transformer when there is a change of current in the primary. The reverse also occurs—this induced voltage in the primary is due to a current change in the secondary. A transformer circuit and the Laplace-transform equivalent are illustrated in Figure 6.22. A dot is associated with each winding of the transformer. The convention is that M, the mutual inductance, is positive if the currents i_1 and i_2 both flow into or out of the dotted ends of the coils, and negative otherwise. The transfer function $\overline{T}(s)$ is obtained by applying Kirchhoff's laws to the two loops:

$$\bar{v}_i(s) = R\bar{i}_1(s) + \frac{\bar{i}_1(s)}{sC} + \bar{i}_1(s)sL_1 - sM\bar{i}_2(s),$$

$$0 = \bar{i}_2(s)sL_2 - \bar{i}_1(s)sM + \bar{i}_2(s)sL_3 + \bar{i}_2(s)R_2,$$

$$\bar{v}_0(s) = \bar{i}_2(s)R_2$$

Eliminating $\bar{i}_1(s)$ and substituting for $\bar{i}_2(s)$ in the third equation gives the result

$$\bar{v}_0(s) = \bar{v}_i(s) \frac{R_2 M C s^2}{\left[cL_1(L_2 + L_3) - M^2 \right] s^3 + \left[R_1(L_1 + L_2) + R_2 L_1 \right] C s^2 + \left[L_2 + L_3 + R_1 R_2 C \right] R_2 s}$$

6.1.9 Compensation

It is often desired to modify the frequency response of a given circuit. Two common methods are the addition of a pole that is much smaller than the other poles of the transfer function and the simultaneous addition of a pole and zero to the transfer function. The addition of a pole is accomplished with a resistor and capacitor in a low-pass configuration at the output of the circuit to be modified, as illustrated in Figure 6.23a. If the original circuit had a pole ω_1 and the introduction of the RC circuit produced a pole at ω_2 where $\omega_2 \ll \omega_1$, the overall response of the circuit would be a 20 dB/decade decrease in gain from ω_2 to ω_1 followed by a 40 dB/decade decrease for frequencies greater than ω_1.

The addition of a pole and zero is accomplished with the circuit in Figure 6.23b. The transfer function for the *pole-zero compensation* network is

$$\frac{R_2 + 1/sC}{R_1 + R_2 + 1/sC} = \frac{1 + sR_2 C}{1 + s(R_1 + R_2)C}$$

with the zero at $s = -1/R_2 C$ and the pole at $s = -1/(R_1 + R_2)C$. If the zero in the transfer function is chosen to cancel the smallest pole of the original circuit, the overall frequency response will be flat up to the frequency corresponding to the new pole. The gain will then decrease by 20 dB/decade with increasing frequency up to the second pole of the original circuit. The gain will then decrease by 40 dB/decade up to the next high-frequency pole, after which the decrease will be 60 dB/decade. The response of the pole-zero-compensated circuit is sharper than that of one with single-pole compensation.

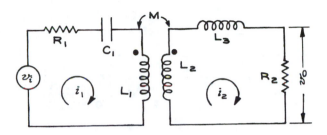

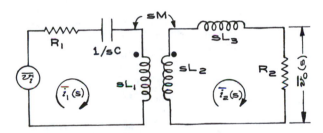

Figure 6.22   A transformer circuit and the Laplace-transform equivalent. M is positive if i_1 and i_2 both flow into or out of the dotted ends of the coil.

6.1.10 Filters

Filters are generally classed as *low-pass, high-pass, band-pass,* and *band-reject.* RC circuits are examples of the first two kinds, while the series and parallel RLC circuits illustrate the last two. A series of band-pass or band-reject filters with pass bands at different but closely spaced frequencies is called a *comb filter.*

The ideal filter would pass unattenuated all frequencies in its pass band while completely rejecting frequencies outside the pass band. The simple circuits we have examined thus far are only approximations to the ideal. There are excellent approximations to the ideal filter that rely on judicious choices of the transfer function poles of

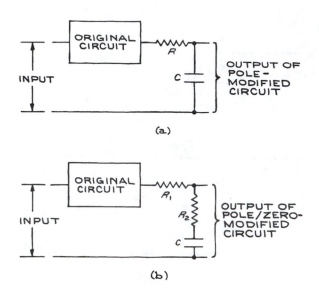

Figure 6.23   Compensation by (a) the addition of a pole and (b) the addition of a pole and a zero to cancel the lowest-frequency pole of the original circuit.

RC and *RLC* circuits. There are four classes of filter design, each of which has its advantages:

1. *Maximally flat* has the flattest amplitude response within the pass band. The *Butterworth* filter is an example.

2. *Equal ripple* has fluctuations in the pass band but greater attenuation in the stop band than the maximally flat filter.

3. *Elliptic* has the maximum rate of attenuation between the pass band and stop band. The *Chebyshev* filter is an example of this type.

4. *Linear phase* has a much less sharp cutoff than the others but maintains an almost linear phase response in the pass band. The *Bessel* filter is an example of this type.

Methods of filter design are well established, and there are many texts and handbooks that provide tables for filter synthesis. Before designing or specifying a filter, it is important to know not only the cutoff and phase properties, but also the transient response and the input and output impedances.

Amplifiers with negative feedback provided by resistors and capacitors can have responses identical to *RLC* circuits. By appropriate choices of the elements in a feedback network, these amplifiers can be made into filters. Especially at low frequencies, where large inductors with low resistance are heavy and expensive, these so-called *active filters* are very useful. There are many texts and handbooks on active filter design. *Lancaster's Active Filter Cookbook* (D. Lancaster, 2nd ed., Butterworth-Heinemann, 1996) is written from a practical point-of-view with a minimum of abstract mathematics. Most active filter designs are based on high-gain integrated-circuit operational amplifiers. With the frequency response of these amplifiers now extending into the megahertz region, active filters are finding increasing application to high-frequency circuits. Stability is an important factor in filter design, especially for large values of *Q*. If drift is to be minimized, highly stable passive components are required.

6.1.11 Computer Aided Circuit Analysis

The *Simulation Program with Integrated-Circuit Emphasis* (SPICE) was originally intended for integrated circuit design where the very large number of circuit elements made traditional circuit analysis slow, cumbersome, and subject to errors. SPICE was first limited to use on mainframe computers with network access, but various versions are now available for personal computers (PCs). One popular PC version of SPICE is PSPICE, now available from Cadence (see Web listing). Because SPICE and PSPICE figure prominently in electrical engineering circuit-analysis courses, they are usually available over university networks. Various versions, including student/faculty and evaluation versions, can be downloaded from the Web. Analysis of digital circuits with SPICE requires detailed information about waveforms and delays within individual units. Digital circuits are increasingly assembled from programmable arrays and the manufacturers of the arrays generally supply the required information with their programming software.

The following procedures should be followed when using circuit-analysis programs:

1. Create a schematic diagram of the circuit
   - Assemble components from the program library
   - Import or create components unavailable in library
   - Assign values to the components
   - Make interconnections

2. Create SPICE/PSPICE file
   - Specify temperature
   - Specify sources
   - Independent
   - Controlled
   - Verify interconnections (nodes)

3. Analysis
   - dc quiescent
   - Transient
   - ac small signal

4. Output
   - Graphs
   - Bode plots
   - Waveforms
   - Numerical tables

5. Adjust circuit to meet design criteria and reanalyze

Analysis programs are extremely useful but cannot replace an understanding of circuit theory. It is unrealistic to believe that one can begin with an approximate circuit and use a program to arrange the values of the components and the configuration until the desired result is attained. Even for a simple circuit of 10 components, the possibilities are so large that an uninformed attempt to arrive at a desired final result cannot be successful. On the other hand, an arbitrary circuit can be analyzed to determine whether it will perform according to the requirements. Most analysis programs come with libraries of common active devices (transistors, diodes, operational amplifiers); however, new devices appear on the market daily and older devices become obsolete so that it is impractical for any single library to contain all possible devices. Adding device specifications to analysis program libraries can be time consuming and it is often more efficient to use available devices with operating parameters close to those of the device in question. This requires additional research, but device specifications are now routinely available on the Web sites of the manufacturers. In some cases, manufacturers provide SPICE files for their devices for loading directly into libraries. An additional benefit of analysis programs is compatibility with schematic capture and printed circuit board layout software. This is discussed in more detail in Section 6.10.3 on printed circuit boards.

6.2 PASSIVE COMPONENTS

In the discussion of passive components, emphasis is placed on choosing the correct type for the function to be performed. It is rarely sufficient to specify the nominal value of a component. This can be seen by looking through the catalog of any electronics components supplier under *resistors* and finding perhaps 50 different resistor types, each with a range of resistance values. The following are some general considerations in the choice of passive components:

1. *Nominal value and tolerance.* An aspect of this is the coding applied to the component. Both color and number codes are used.

2. *Stability.* The nominal value as a function of temperature (the temperature coefficient [*tempco*] is often expressed in percent or in parts per million [ppm]). Age and environmental conditions such as humidity, vibration, or shock also effect component values. On the other hand, components can affect their environment—for example, the outgassing of a capacitor in vacuum or the heating of one component by another one.

3. *Size and shape.* Along with the reduction in size that solid-state electronics has brought, there has been a similar reduction in the size of passive components. Generally, one uses the smallest practical size, but power-density considerations can limit the packing density of power-dissipating components. Also, components come in different shapes with different

lead geometries depending on the intended mounting method. These are discussed in the section on hardware.

4. *Power dissipation and voltage rating.* A resistor has both a power rating and a voltage rating. For very large values of resistance, it is quite possible to operate well within the resistor's power rating yet exceed the voltage rating. The result is electrical breakdown and destruction of the resistor. For moderate values of capacitance, capacitors dissipate very little power and power rating is of no importance. Very high-capacitance capacitors with large effective surface areas, however, have non-negligible shunt conductances. The large currents that flow in such capacitors can cause considerable heat to be generated with potentially fatal results (to the capacitor) .

5. *Noise.* Passive components can introduce noise into an electronic circuit, and one must be particularly aware of this problem in low-level circuits and choose appropriate low-noise components.

6. *Frequency characteristics.* Pure resistors, capacitors, and inductors are practically unrealizable, and the electrical properties of real passive components depend on frequency in a non-ideal way. This is often expressed by an equivalent circuit. Inductors and transformers with magnetic cores, however, are inherently nonlinear and can be treated only approximately in this way.

7. *Derating.* This is an engineering term related to the safety factors to be used when operating components under a variety of conditions. Good design requires that components be operated at no more than 50 percent of the manufacturer's recommended maximum ratings, particularly with respect to power and voltage. At high ambient temperatures components should be derated by even larger amounts. The resulting decrease in stress greatly increases component lifetime. In laboratory equipment, large safety factors should always be used for enhanced reliability.

8. *Cost.* For laboratory applications it is poor practice to economize on passive components by using the least expensive ones that will do the job. Laboratory work is labor intensive. The extra cost of top-grade components is greatly outweighed by the time saved in troubleshooting and repair.

6.2.1 Fixed Resistors and Capacitors

The more common types of fixed and variable resistors and fixed capacitors are listed in Tables 6.3, 6.4, and 6.5. In Table 6.5, the *C* range and *V* range refer to the minimum and maximum values available for all types of a given dielectric. It is not possible to have the maximum capacitance at the maximum voltage. The larger capacitances have lower *working voltages* (WV), while the smaller ones have the higher WV. The temperature coefficients generally apply only to a limited temperature range within the larger operating temperature range.

High component density circuits use *surface mount technology* (SMT) chip resistors and capacitors. These are different from the more traditional wire lead types, both in their configuration and the way they are electrically and mechanically connected to the circuit. Chip resistors and capacitors are common SMT components, but semiconductors and integrated circuits also are manufactured for surface mounting. While high packing density is not often a consideration in laboratory electronics, small component size can be an important advantage for some circuits where stray capacitance and inductance are factors. Surface mount resistors and capacitors have metalized surfaces and are held in place over circuit board pads with solder that connects the SMT metallization and circuit board pads. SMT resistors and capacitors can be little larger than the end of a pencil lead and use special codes to identify their values. The resistor code is a three-digit code (for 2% and greater tolerances) and a tolerance character. The first two digits are the first two significant figures of the resistance and the third digit is the power of ten multiplier. The code 563, for example, is 56 kΩ. For resistances less than 100 Ω the letter R is used to represent a decimal point. For example 33R is a 33 Ω resistor. The tolerance is specified by a single letter at the end of the numeric code.

Tantalum is the most stable of all film-forming materials. Tantalum-foil electrolytics have similar

TABLE 6.3 FIXED RESISTORS

Carbon Composition		Precision Wire-Wound (bobbin)	
R range:	2.7 Ω–22 M Ω	R range:	0.1 Ω–18 M Ω
Power range:	0.1–2 W	Power range:	0.1–2 W
Tolerances.	±5, ±10, ±20%	Tolerances:	±0.05 to ±5%
Temperature range:	−55 to +150°C	Temperature range:	−55 to +145°C
Temperature coefficient:	±0.1%/°C	Temperature coefficient:	±0.0001 to ±0.002%/°C
Voltage coefficient:	±0.03%/V dc	Noise:	Low
Noise:	Highest	Frequency response:	To 25 kHz for high-resistance values unless noninductively wound
Typical capacitance:	0.25 pF, no inductance		
Working voltage:	150 V ($\frac{1}{8}$ W) to 750 V (2 W)		

Precision Carbon Film		Power Wire-Wound	
R range:	1 Ω–100 M Ω	R range:	0.1 Ω–1.3 M Ω
Power range:	0.1–2 W	Power range:	2–225 W
Tolerances.	±0.5, 1, 2%	Tolerances.	±5%, ±10%

Precision Carbon Film		Power Wire-Wound	
Temperature range:	−55 to +165°C	Operating temperature:	−75 to +350°C
Temperature coefficient:	0.02–0.05%/°C.	Maximum rated-power	
Noise:	Low	temperature:	25°C
Frequency response:	To 1 MHz for > 1 kΩ (spinal),	Temperature coefficient:	±0.026%/°C
	to 100 MHz for R < 1 kΩ	Frequency response:	To 25 kHz for high-resistance values unless noninductively wound

Precision Metal Film	
R range:	10 Ω–200 M Ω
Power range:	$\frac{1}{8} - \frac{1}{2}$ W
Tolerances:	±0.1 to ±1%
Temperature range:	−55 to +165°C
Temperature coefficient:	0.0025–0.01%/°C
Noise:	Lowest

characteristics to aluminum-foil electrolytics, but offer superior stability and freedom from leaking. The wet-slug type has the highest volumetric efficiency of any capacitor. Tantalum solid-electrolyte capacitors have much better stability, frequency, and temperature characteristics than liquid-electrolyte types.

The equivalent circuits for fixed resistors and capacitors are given in Figure 6.24. Figure 6.25 illustrates common capacitor types. Coding is given in Tables 6.6, 6.7, and 6.8.

TABLE 6.4 VARIABLE RESISTORS

Resistive Element Material	R Range	Temperature Coefficient (ppm/°C)	Linearity (%)
Carbon composition	50 Ω–0 MΩ	±1000	5.0
Resistance wire	1Ω–00 KΩ	±10	0.25
Conductive plastic	100 Ω–5 MΩ	±100	0.25
Cermet[a]	100 Ω–5 MΩ	±100	0.25

[a] Ceramic-metal hybrid.

6.2.2 Variable Resistors

In electronic equipment, variable resistors are generally called *potentiometers*. More specifically, they can be classed as potentiometers, trimmers, and rheostats. All are three-terminal devices with a fixed terminal at each end of the resistive element and the third terminal attached to the sliding contact or *tap*. Potentiometers are designed for regular movement, trimmers for occasional changes, and rheostats for current limiting in high-power circuits. Some common types of variable resistors are shown in Figure 6.26. There are a number of different resistive-element materials for potentiometers. They are listed in Table 6.4.

General-purpose, single-turn potentiometers come in power ratings from 1/2 to 5 watts and a variety of sizes and tapers (*taper* is the change in resistance as a function of shaft angle). Nonlinear tapers are often used in volume controls, where an approximately logarithmic response is desired. Some designs can be stacked on a common shaft, and it is often also possible to incorporate an on-off switch. Shafts with locking nuts are very useful when it is necessary to fix a given setting. Precision potentiometers can be single-turn or 3-, 5-, or 10-turn types. The resistive elements are wire, conductive plastic, or a ceramic-metal substance called *cernet*. The rotating shaft is usually supported in a bushing, which is mounted with a threaded collar. Power ratings are from 1/4 to 2 watts with a maximum operating temperature of 125°C. Resistance values are from 50 Ω to 200 kΩ, depending on the model, and tolerances are ±1 percent to ±5 percent. Linearity is ±0.25 percent to ±1.0 percent, and temperature coefficients are ±20 ppm/°C. For the 10-turn models, turns-counting dials are available.

Trimmers are used to compensate for the tolerance and variation of fixed components. They are not designed for continuous adjustment, 200 cycles being their design life. The resistance elements are carbon composition, cermet, and wire. They are usually single-turn or have lead screws or worm gears.

6.2.3 Transmission Lines

The most common transmission lines are those with two conductors. A number of common configurations are shown in Figure 6.27. Common VHF flat television antenna lead-in cable is an example of a two-wire transmission line. RG/U cables found in almost all laboratories are examples of coaxial transmission lines—such coaxial cable is also used for UHF and satellite television leads. The wire-and-plane configuration is not common. The ribbon-and-plane configuration occurs very often in printed circuit boards (PCBs), where the ribbon is the metal trace on one side of the board and the plane is a continuous conducting sheet on the opposite side, often called the *ground plane*. A transmission line of this type is called a *microstrip line*. The formulae for the characteristic impedance Z_0 of the geometries in Figure 6.27 are given in Table 6.9.

The properties of a transmission line are given in terms of the series impedance Z and shunt admittance Y per unit length:

$$Z = R + j\omega L$$
$$Y = G + j\omega C$$

where $j = \sqrt{-1}$ and G is the shunt conductance, which is the reciprocal of the shunt resistance per unit length. The characteristic impedance Z_0 is $\sqrt{Z/Y}$. Voltage and current propagate along a transmission line as waves. The

TABLE 6.5 FIXED CAPACITORS

Air	
C range (pF):	<100
Tolerances:	To 0.01%
Stability:	High
Temperature coefficient:	20 ppm/°C
Dissipation factor:	10^{-5}

Mica and Glass	
C range:	1 pF–1 μF
Tolerance:	±1 to ±20%
Operating Temperature:-	–55 to +70, +85, +125, or +150°C depending on construction
Temperature Coefficient:	–20 to +100 ppm/°C
V range:	100 to 8000 WV dc
Dissipation factor:	$(1–7)\times10^{-4}$

Ceramic Disc	
C range:	1 pF–1 μF
Tolerance:	±1 to ±20%
V range:	100–7500 WV dc
Operating temperature:	–55 to + 150°C
Temperature coefficient:	0 ppm/°C for NPO to ±750 ppm/°C for N750
Dissipation factor:	5×10^{-4}

Film			
Dielectric:	Polyester	Polycarbonate	Polystyrene
C range (μF):	0.0001–12 (metalized)	0.0001–50 (metalized)	0.0001–1 (metalized)
V range (WV dc):	50–1600	50–600	50–600
Operating temp. (°C):	–55 to +125	–55 to +125	–55 to +85
Temperature coefficient.:	+250 ppm/°C	Nonmonotonic, ±1% total	–50 to–100 ppm/°C
Tolerances:	±1 to ±20%	±1 to ±10%	±1 to ±10%
Dissipation factor:	10^{-3}	3×10^{-3}	10^{-4}

Electrolytic (polarized or unpolarized)		
Aluminum (general purpose)		
C range (μF):	1–50,000	
Tolerance:	–10 to +75%	
V range:	3–450 WV dc	
Maximum operating temperature:	85°C	
Direct current leakage:	High	
Low temperature stability:	Low	
Internal inductance limits high-frequency performance		
Prone to leaking		
Tantalum (foil and wet slug)		
Electrolyte:	Wet	Dry
C range (μF):	15–2200	0.0047–1000
V range (WV dc):	3–300	2–150
Tolerance:	±10% to –15, +75%	±10 to ±40%

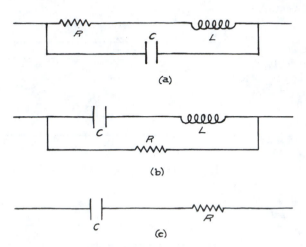

Figure 6.24   Equivalent circuits: (a) resistor; (b) capacitor (the power factor, D, is X_C/R of $1/\omega RC$ and is usually specified at 10^4 to 10^5 Hz, where $\omega L \ll 1/\omega C$; (c) electrolytic capacitor.

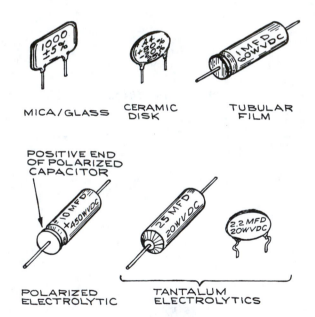

MICA/GLASS   CERAMIC DISK   TUBULAR FILM

POSITIVE END OF POLARIZED CAPACITOR

POLARIZED ELECTROLYTIC   TANTALUM ELECTROLYTICS

Figure 6.25   Some common capacitor types.

properties of the waves are determined by the propagation constant γ and phase velocity v_p. They are given by $\gamma = \sqrt{ZY} = \alpha + j\beta$ and $v_p = \omega/\beta$, where ω is the frequency of the wave. The wavelength in the transmission line is $\lambda = 2\pi v_p/\omega = 2\pi/\beta$. The attenuation of the wave is given by the factor $e^{-\alpha}$ per unit length. These formulae are summarized in Table 6.10.

For most practical transmission lines, the series resistance per unit length R is small compared to the inductive reactance ωL, and the shunt conductance G is small compared to the reciprocal of the capacitance reactance $1/\omega C$. Under these conditions, Z_0 is essentially $\sqrt{L/C}$, α becomes $1/2[R/Z_0 + GZ_0]$, and β becomes $\omega\sqrt{LC}$. The phase velocity is independent of frequency and is the ratio of the speed of light in vacuum to the square root of the dielectric constant of the material separating the conductors. For reasonable values of conductor geometry, Z_0 lies approximately between 50 and 500 Ω. The attenuation a is obtained from the shunt conductance G and series resistance R once Z_0 is known. For most dielectrics, G is very small (for air, G can be taken equal to zero) and R becomes the dominant factor. Because of the *skin effect*, R increases with frequency (see Table 6.1).

The properties of common coaxial cables are given in Table 6.11. As is to be expected, the smaller cables show greater attenuation due to the necessity of using a small-diameter inner conductor to maintain a reasonable ratio of outer conductor diameter to inner conductor diameter. The most common cable has a characteristic impedance of 50 Ω. Twisted pairs of single-conductor wire are used where immunity from noise is important, since unwanted signals induced in both conductors can be canceled electronically by differential amplification.

As a general rule, at least one end of a transmission line should be terminated with a resistance equal to Z_0 to prevent ringing and undesirable reflections if the line is used for sending signals over distances greater than $\lambda/8$. The transmission of pulses with nanosecond rise times over distances of a few centimeters on a printed circuit board, for example, requires termination. In theory, termination of the source or load end of the line is sufficient. In practice, however, it is usually the load end that is terminated. Because 50 Ω cable is so common, electronic instruments designed to operate at high frequencies often have 50 Ω input and output impedances to facilitate connections. Such low-impedance instruments are not suitable for use in combination with lower-

TABLE 6.6 RESISTOR CODING

Color Codes

For general-purpose industrial resistors

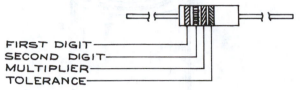

FIRST DIGIT
SECOND DIGIT
MULTIPLIER
TOLERANCE

For general-purpose wire-wound resistors

WIDE FIRST BAND
DENOTES WIREWOUND
RESISTOR

WIDE FIRST BAND
AND BLUE FIFTH
BAND DENOTE FLAME-
PROOF WIREWOUND
RESISTOR

Color	Significant Digit	Multiplier	Tolerance
Black	0	1	—
Brown	1	10	—
Red	2	100	—
Orange	3	1000	—
Yellow	4	10,000	—
Green	5	100,000	—
Blue	6	1,000,000	—
Violet	7	10,000,000	—
Gray	8	—	
White	9	—	
Gold	—	—	±5%
Silver	—	—	±10%
No color	—	—	±20%
	—	—	

Example

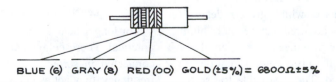

BLUE (6) GRAY (8) RED (00) GOLD (±5%) = 6800Ω±5%

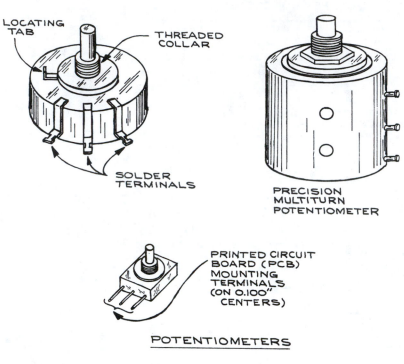

Figure 6.26   Some common types of variable resistors.

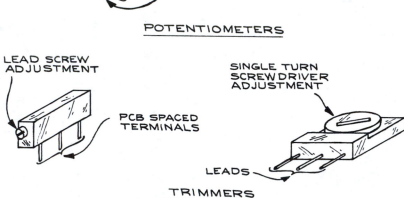

frequency devices, which may be seriously overloaded by 50 Ω

The velocity of propagation v_p of an electrical signal through a transmission line is $1/\sqrt{LC} = Z_0/L = 1/Z_0C$. For 50 Ω cable with a polyethylene or Teflon insulation, v_p is on the order of 2×10^{10} m/sec, which gives a delay of 5 ns/m. For delays up to a few hundred nanoseconds, high-quality, low-loss coaxial cable provides an excellent delay line because no distortion of the propagated signal occurs when it is properly terminated. The best cable to use for critical delay-line applications is rigid or semirigid air-core coaxial cable. With air as the dielectric, the shunt conductance is virtually zero. Semirigid construction assures that the coaxial geometry is maintained within tight tolerances, reducing distortion. Such cable is practical for use at frequencies in the microwave region.

For longer delays, lumped-element delays are the most practical. These delays consist of low-pass filter sections connected in series. For frequencies below their corner frequency, they act as delay lines. The number of sections

TABLE 6.7 CAPACITOR CODING

Molded Mica

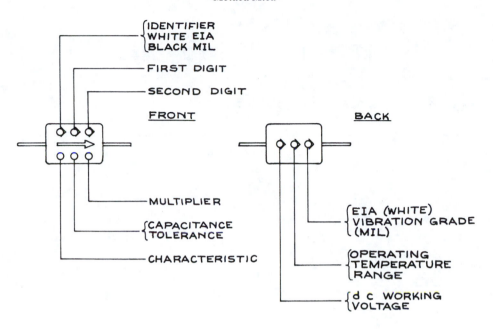

Color	Significant Digits	Multiplier	Capacitance Tolerance (%)	Characteristic	Working Voltage (dc)	Operating Temperature (°C)	EIA/ Vibration
Black	0	1	±20	—	—	−55 to +70	10–55 Hz
Brown	1	10	±1	B	100	—	—
Red	2	100	±2	C	—	−55 to +85	—
Orange	3	1000	—	D	300	—	—
Yellow	4	10,000	—	E	—	−55 to +125	10–2000 Hz
Green	5	—	±5	F	500	—	—
Blue	6	—	—	—	—	−55 to +150	—
Violet	7	—	—	—	—	—	—
Gray	8	—	—	—	—	—	—
White	9	—	—	—	—	—	EIA
Gold	—	—	±0.5	—	1000	—	—
Silver	—	—	±10	—	—	—	—

Equivalences:

10^{-6} F = 1 μF, 1 MFD

10^{-9} F = 0.001 μF, 1000 pF, 1 mF

10^{-12} F = 1 pF, 1 MMFD, 1 μF

TABLE 6.7 CAPACITOR CODING (CONTINUED)

Ceramic Capacitors

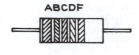

	Six-Dot or Six-Band Code			*Five-Dot or Five-Band Code*
A	Temperature coefficient		A	Temperature coefficient
C	Capacitance		C	Capacitance
D	Capacitance		D	Capacitance
E	Capacitance		E	Capacitance
F	Capacitance tolerance		F	Capacitance tolerance
G	Military code number			

Temperature Characteristics

A	*B*	*Temp. Coeff.*[b]	*A*	*B*	*Temp. Coeff.*[b]	*Ppm/°C*
Gray	Black	Gen. Purpose	Black	—	NP0	0
Orange	Orange	N1500	Brown	—	N030	−30
Yellow	Orange	N2200	Red	—	N080	−80
Green	Orange	N3300	Orange	—	N150	−150
Blue	Orange	N4700	Yellow	—	N220	−220
Red	Violet	P100	Green	—	N330	−330
Green	Blue	P030	Blue	—	N470	−470
Gold	Orange	X5F	Violet	—	N750	−750
Brown	Orange	Z5F	Gold	—	P100	+100
Gold	Yellow	X5P	White	—		
Brown	Yellow	Z5P	Gray	—		
Gold	Blue	X5S				
Brown	Blue	Z5S				
Gold	Gray	X5U				
Brown	Gray	Z5U				

[a]Nominal capacitance code is EIA-RS 198. MIL-SPEC code is not the same. Five- and six-digit codes are both used for radial-lead and axial lead capacitors. Disc capacitors normally have typographical marking but may be color coded.
[b]N designates a negative temperature coefficient; P designates a positive temperature coefficient.

TABLE 6.7 CAPACITOR CODING (CONTINUED)

		Capacitance		
			Nominal Capacitance	
Color	Digit (C&D)	Multiplier (E)	< 10 pF	> 10 pF
Black	0	1	±2.0 pF	±20%
Brown	1	10	±0.1 pF	±1%
Red	2	100	—	±2%
Orange	3	1,000	—	±3%
Yellow	4	10,000	—	+100%, –0%
Green	5	—	±0.5 pF	±5%
Blue	6	—	—	—
Violet	7	—	—	—
Gray	8	0.01	±0.25 pF	+ 80%, –20%
White	9	0.1	±1.0 pF	±10%

TABLE 6.8 SURFACE MOUNT (SMT) CODES

SMT Resistor Tolerance Code	
Letter	Tolerance
D	±0.5%
F	±1.0%
G	±2.0%
J	±5.0%

SMT Capacitor Significant Figures Code[a]

Character	Figures	Character	Figures	Character	Figures
A	1.0	N	3.3	a	2.5
B	1.1	P	3.6	b	3.5
C	1.2	Q	3.9	d	4.0
D	1.3	R	4.3	e	4.5
E	1.5	S	4.7	f	5.0
F	1.6	T	5.1	m	6.0
G	1.8	U	5.6	n	7.0
H	2.0	V	6.2	t	8.0
J	2.2	W	6.8	y	9.0
K	2.4	X	7.5		
L	2.7	Y	8.2		
M	3.0	Z	9.1		

TABLE 6.8 SURFACE MOUNT (SMT) CODES (CONTINUED)

SMT Capacitor Code	
Character	*Multiplier*
0	1
1	10
2	100
3	1,000
4	10,000
5	100,000
6	1,000,000
7	10,000,000
8	100,000,000
9	0.1

[a] SMT capacitors are marked with a two-character code consisting of a letter and number. The letter gives the significant figures in picofarads: the number is the decimal multiplier.
A .01 microfarad capacitor has the code A4 corresponding to $1.0 \times 10^{-6} \times 10^{4}$.

TABLE 6.9 CHARACTERISTIC IMPEDENCE OF TRANSMISSION-LINE GEOMETRIES

Geometry	$Z_0\,(\Omega)$[a]
Two-wire	$\dfrac{138}{\sqrt{\varepsilon}}\ln\dfrac{b}{a}$
Coaxial	$\dfrac{276}{\sqrt{\varepsilon}}\ln\dfrac{b}{a}$
Wire and plane	$\dfrac{60}{\sqrt{\varepsilon}}\ln\dfrac{4h}{d}$
Ribbon and plane or Microstrip line[b]	$\dfrac{87}{\sqrt{\varepsilon+1.41}}\ln\dfrac{5.98h}{0.8w+t}$
Twisted pair (AWG 24–28, 30 turns/foot)	110
Strip line[b]	$\dfrac{60}{\sqrt{\varepsilon}}\ln\dfrac{4b}{0.67\pi w(0.8+t/w)}$

[a] $\sqrt{\varepsilon} = 1.0$ for air and 5.0 for G-10 fiberglass epoxy circuit board.
[b] The thickness of the copper foil is 0.001 in. for 1 oz cladding and 0.002 in. for 2 oz cladding.

n required is related to the ratio of the delay time t_d to the rise time t_r, and is given by $n = 1.5\,t_d/t_r$. The incorporation of multiple taps between the sections allows a choice of delays with a single unit.

For a line length less than a quarter wavelength, a transmission line behaves like a capacitor or an inductor, depending on whether the end of the line is an open circuit or short circuit. A line that is an odd number of quarter-wavelengths long behaves like a parallel resonant *RLC* circuit when short-circuited at one end, and like a series resonant circuit when open-circuited. The Qs of such resonant circuits are much higher than attainable with lumped elements. There are, however, an infinite number of resonances corresponding to frequencies where $v_p/f = (2n + 1)\lambda/4$–where n is an integer.

6.2.4 Coaxial Connectors

Probably more time is spent with connectors than with any other element of electronic hardware. When deciding on the method of connecting various pieces of electronic equipment, it is wise to be aware the kinds of connectors available and choose a family of connectors best suited to the task. In many cases, the choice of connector is dictated by the connectors at the outputs and inputs of the existing equipment.

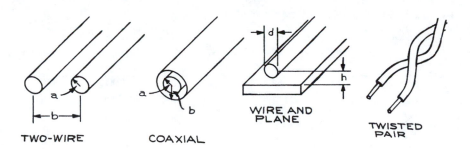

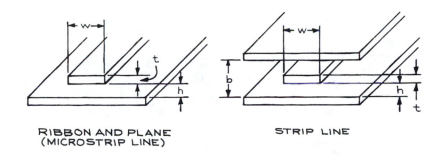

Figure 6.27   Fixed-geometry transmission-line configurations.

TABLE 6.10 GENERAL TRANSMISSION-LINE PROPERTIES

Quantity	Symbol	Relation to Fundamental Parameters
Shunt admittance	Y	$G + j\omega C$
Series impedance	Z	$R + j\omega L$
Characteristic Impedance	Z_0	$\sqrt{Z/Y}$
Wavelength	λ	$2\pi/\beta$
Propagation velocity	v_p	$c/\sqrt{\varepsilon}$
Attenuation constant	α	
Propagation constant	γ	$\sqrt{ZY} = \alpha + j\beta$

The most common connector is currently the BNC. It is a miniature bayonet type, but circuit density has increased enormously with the extensive use of integrated circuits and the BNC connector is now being replaced by a variety of subminiature connectors—among which the LEMO™ has taken the lead for nuclear instrumentation. A list of common coaxial connectors is given in Table 6.12.

In addition to coaxial connector plugs, each series of connectors has a wide variety of jacks and adaptors (generally *plug* refers to the connector attached to the cable or wire, while *jack* refers to the connector to which the plug mates). Jacks for chassis mounting, bulkhead mounting, and printed circuit board mounting are available, as well as hermetically sealed jacks for making coaxial connections through a vacuum wall. *Tees* and

TABLE 6.11 PROPERTIES OF COAXIAL CABLE

Military RG Number	Armor O.D. (in.)	Jacket O.D. (in.)	Jacket Type[a]	Dielectric O.D. (in.)	Dielectric Type[b]	Center Conductor[c]	V.P.[d] (%)	Capacitance (pF/ft.)	Max. RMS Operating Voltage(V)	Nominal Impedance (Ω)
5B	—	0.328	IIa	.181	P	16 S	65.9	28.5	3,000	50
6A/U	—	0.332	IIa	.185	P	21 CW	65.9	20.0	2,700	75
7A/U	—	0.405	I	.285	P	7/21 C	65.9	29.5	5,000	52
8A/U	—	0.405	IIa	.285	P	7/21 C	65.9	29.5	5,000	52
9	—	0.420	II Grey	.280	P	7/21 S	65.9	30.0	5,000	51
9A	—	0.420	II Grey	.280	P	7/21 S	65.9	30.0	5,000	51
9B/U	—	0.420	IIa	.280	P	7/21 S	65.9	30.0	5,000	50
10A/U	0.475	0.405	IIa	.285	P	7/21 C	65.9	29.5	5,000	52
11A	—	0.405	I	.285	P	7/26 TC	65.9	20.5	5,000	75
11A/U	—	0.405	IIa	.285	P	7/26 TC	65.9	20.5	5,000	75
12A/U	0.475	0.405	IIa	.285	P	7/26 TC	65.9	20.5	5,000	75
13	—	0.420	I	.280	P	7/26 TC	65.9	20.5	5,000	74
13A/U	—	0.420	IIa	.280	P	7/26 T	65.9	20.5	5,000	74
14A/U	—	0.545	IIa	.370	P	10 C	65.9	29.5	7,000	52
17A	—	0.870	IIa	.680	P	.188 C	65.9	29.5	11,000	52
17A/U	—	0.870	IIa	.680	P	.188 C	65.9	29.5	11,000	52
18A/U	0.945	0.870	IIa	.680	P	.188 C	65.9	29.5	11,000	52
19A/U	—	1.120	IIa	.910	P	.250 C	65.9	29.5	14,000	52
20A/U	1.195	1.120	IIa	.910	P	.250 C	65.9	29.5	14,000	52
21A	—	0.332	IIa	.185	P	16 N	65.9	29.0	2,700	53
22	—	0.405	I	.285	P	Two 7/.0152 C	65.9	16.0	1,000	95
22B/U	—	0.420	IIa	.285	P	Two 7/.0152 C	65.9	16.0	1,000	95
34B/U	—	0.630	IIa	.460	P	7/.0249 C	65.9	21.5	6,500	75
35B/U	0.945	0.870	IIa	.680	P	.1045 C	65.9	21.0	10,000	75
55B/U	—	0.206	IIIa	.116	P	20 S	65.9	28.5	1,900	53.5
57A/U	—	0.625	IIa	.472	P	Two 7/21 C	65.9	16.0	3,900	95
58/U	—	0.195	I	.116	P	20 C	65.9	28.5	1,900	53.5
58A/U	—	0.195	I	.116	P	19/.0071 TC	65.9	30.0	1,900	50
58C/U	—	0.195	IIa	.116	P	19/.0071 TC	65.9	30.0	1,900	50
59/U	—	0.242	I	.146	P	22 CW	65.9	21.5	2,300	73
59B/U	—	0.242	IIa	.146	P	.023 CW	65.9	21.0	2,300	75
62/U	—	0.242	I	.146	SSP	22 CW	84	13.5	750	93
62A/U	—	0.242	IIa	.146	SSP	22 CW	84	13.5	750	93
62B/U	—	0.242	IIa	.146	SSP	7/32 CW	84	13.5	750	93
63/U	—	0.405	I	.285	SSP	22 CW	84	10.0	1,000	125
63B/U	—	0.405	IIa	.285	SSP	22 CW	84	10.0	1,000	125
71A	—	< 0.250	I	.146	SSP	22 CW	84	13.5	750	93

TABLE 6.11 PROPERTIES OF COAXIAL CABLE (CONTINUED)

Military RG Number	Armor O.D. (in.)	Jacket O.D. (in.)	Jacket Type[a]	Dielectric O.D. (in.)	Dielectric Type[b]	Center Conductor[c]	V.P.[d] (%)	Capacitance (pF/ft.)	Max. RMS Operating Voltage(V)	Nominal Impedance (Ω)
71B/U	—	0.250	IIIa	.146	SSP	22 CW	84	13.5	750	93
74A/U	0.615	0.545	IIa	.370	P	10 C	65.9	29.5	7,000	52
79B/U	0.475	0.405	IIa	.285	SSP	22 CW	84	10.0	1,000	125
87A/U	—	0.425	V	.280	TF	7/20 S	69.5	29.5	5,000	50
108A/U	—	0.235	IIa	.079	P	Two 7/28 TC	68	23.5	1,000	78
111A/U	0.490	0.420	IIa	.285	P	Two 7/.0152 C	65.9	16.0	1,000	95
114/U	—	0.405	I	.285	SSP	33 CW	88	6.5	1,000	185
114A/U	—	0.405	IIa	.285	SSP	33 CW	88	6.5	1,000	185
115	—	0.375	V	.250	'IT	7/21 S	70	29.5	5,000	50
115A/U	—	0.415	V	.250	'IT	7/21 S	70	29.5	4,000	50
116/U	0.475	0.425	V	.280	TF	7/20 S	69.5	29.5	5,000	50
122/U	—	0.160	IIa	.096	P	27/37 TC	65.9	29.5	1,900	50
140/U	—	0.233	V	.146	TF	.025 SCW	69.5	21	2,300	75
141/U	—	0.190	V	.116	TF	.0359 SCW	69.5	28.5	1,900	50
141A/U	—	0.190	V	.116	TF'	.039 SCW	69.5	28.5	1,900	50
142/U	—	0.206	V	.116	TF'	.0359 SCW	69.5	28.5	1,900	50
142/B	—	0.195	IX	.116	TF	.039 SCW	69.5	28.5	1,900	50
143	—	0.325	V	.185	TF	.057 SCW	69.5	28.5	3,000	50
142A/U	—	0.206	V	.116	TF	.039 SCW	69.5	28.5	1,900	50
143A/U	—	0.325	V	.185	TF	.059 SCW	69.5	28.5	3,000	50
149/U	—	0.405	I	.285	P	7/26 TC	65.9	20.5	5,000	75
164/U	—	0.870	IIa	.680	P	.1045 C	65.9	21	10,000	75
174/U	—	0.100	I	.060	P	7/.0063 CW	65.9	30	1,500	50
178B/U	—	0.075	VII	.034	TF	7/38 SCW	69.5	29	1,000	50
179B/U	—	0.105	VII	.063	TF'	7/38 SCW	69.5	19.5	1,200	75
180B/U	—	0.145	VII	.102	TF	7/38 SCW	69.5	15	1,500	95
187/U	—	0.110	VII	.063	TF	7/38 SCW	69.5	19.5	1,200	75
188/U	—	0.110	VII	.060	TF	7/.0067 SCW	69.5	29	1,200	50
195/U	—	0.155	VII	.102	TF	7/38 SCW	69.5	15	1,500	95
196/U	—	0.080	VII	.034	TF	7/38 SCW	69.5	29	1,000	50
209/U	—	0.745	VI	.500	SST	19/.0378 S	84	26.5	3,200	50
210/U	—	0.242	V	.146	SST	22 SCW	84	13.5	750	93
211A/U	—	0.730	V	.620	TF	.190 C	69.5	29.0	7,000	50
212/U	—	0.332	IIa	.185	P	.0556 S	65.9	28.5	3,000	50
213/U	—	0.405	IIa	.285	P	7/.0296 C	65.9	29.5	5,000	50
214/U	—	0.425	IIa	.285	P	7/.0296 S	65.9	30	5,000	50
215/U	0.475	0.405	IIa	.285	P	7/.0296 C	65.9	29.5	5,000	50
216/U	—	0.425	IIa	.285	P	7/26 TC	65.9	20.5	5,000	75

TABLE 6.11 PROPERTIES OF COAXIAL CABLE (CONTINUED)

Military RG Number	Armor O.D. (in.)	Jacket O.D. (in.)	Jacket Type[a]	Dielectric O.D. (in.)	Dielectric Type[b]	Center Conductor[c]	V.P.[d] (%)	Capacitance (pF/ft.)	Max. RMS Operating Voltage(V)	Nominal Impedance (Ω)
217/U	—	'0.545	IIa	.370	P	.106 C	65.9	29.5	7,000	50
218/U	—	0.870	IIa	.680	P	.195 C	65.9	29.5	11,000	50
219/U	0.945	0.870	IIa	.680	P	.195 C	65.9	29.5	11,000	50
220/U	—	1.120	IIa	.910	P	.260 C	65.9	29.5	14,000	50
221/U	1.195	1.120	IIa	.910	P	.260 C	65.9	29.5	14,000	50
222/U	—	0.322	IIa	.185	P	.0556 N	65.9	29	2,700	50
223/U	—	0.216	IIa	.116	P	.035 S	65.9	29.5	1,900	50
225/U	—	0.430	V	.285	TF	7/.0312 S	69.5	29.5	5,000	50
227/U	0.490	0.430	V	.285	TF	7/.0312 S	69.5	29.5	5,000	50
228A/U	0.795	0.730	V	.620	TF	.190 C	69.5	29.0	7,000	50
264/U	—	0.750	PU	.176	P	Four 19/27 C	69.5	40.0	N.A.	40
280/U	—	0.480	IX	.327	TT	9C	80	27	3,000	50
281/U	—	0.750	VI	.500	TT	19/.0378 S	80	27	3,200	50
301/U	—	0.245	IX	.185	TF	7/.0203 K	69.5	28.5	3,000	50
302/U	—	0.206	IX	.146	TF	22 SCW	69.5	21	2,300	75
303/U	—	0.170	IX	.116	TF	.038 SCW	69.5	28.5	1,900	50
304/U	—	0.280	IX	.185	TF	.059 SCW	69.5	28.5	3,000	50
307A/U	—	0.270	IIIa	.146	SSP	19/.0058 S	80.0	17	N.A	75
316/U	—	0.102	IX	.060	TF	7/.0067 SCW	69.5	29	1,200	50

[a] Designation listed at end of this table.
[b] P: polyethylene; SSP: semisolid polyethylene; TF: Teflon; TT: Teflon tape; SST: semisolid Teflon.
[c] Number of strands, gauge (B&S) or O.D. and material are specified. A single solid wire results in the lowest cable attenuation. Stranding increases flexibility. S: silvered copper; C: copper; TC: tinned copper; N: Nichrome; CW: Copperweld; SCW: silvered Copperweld; K: Karma.
[d] 100% × (velocity of propagation)/(velocity in vacuum).

TABLE 6.11 PROPERTIES OF COAXIAL CABLE (CONTINUED)

	Attenuation Ratings for RG/U Cable									
	Nominal Attenuation [dB/(100 ft.)]									
Frequency (MHz):	1.0	10	50	100	200	400	1000	3000	5000	10,000
5, 5A, 5B, 6, 6A, 212	0.26	0.83	1.9	2.7	4.1	5.9	9.8	23.0	32.0	56.0
7	0.18	0.64	1.6	2.4	3.5	5.2	9.0	18.0	25.0	43.0
8, 8A, 10, 10A, 213,215	0.15	0.55	1.3	1.9	2.7	4.1	8.0	16.0	27.0	> 100.0
9, 9A, 9B, 214	0.21	0.66	1.5	2.3	3.3	5.0	8.8	18.0	27.0	45.0
11, 11A, 12, 12A, 13, 13A, 216	0.19	0.66	1.6	2.3	3.3	4.8	7.8	16.5	26.5	> 100.0
14, 14A, 74, 74A, 217, 224	0.12	0.41	1.0	1.4	2.0	3.1	5.5	12.4	19.0	50.0
17, 17A, 18, 18A, 177, 218, 219	0.06	0.24	0.62	0.95	1.5	2.4	4.4	9.5	15.3	> 100.0
19, 19A, 20, 20A, 220, 221	0.04	0.17	0.45	0.69	1.12	1.85	3.6	7.7	11.5	> 100.0

TABLE 6.11 PROPERTIES OF COAXIAL CABLE (CONTINUED)

Attenuation Ratings for RG/U Cable

Frequency (MHz):	Nominal Attenuation [dB/(100 ft.)]									
	1.0	*10*	*50*	*100*	*200*	*400*	*1000*	*3000*	*5000*	*10,000*
21, 21A, 222	1.5	4.4	9.3	13.0	18.0	26.0	43.0	85.0	> 100.0	> 100.0
22, 22B, 111, 111A	0.24	0.80	2.0	3.0	4.5	6.8	12.0	25.0	> 100.0	> 100.0
29	0.32	1.20	2.95	4.4	6.5	9.6	16.2	30.0	44.0	> 100.0
34, 34A, 34B	0.08	0.32	0.85	1.4	2.1	3.3	5.8	16.0	28.0	> 100.0
35, 35A, 35B, 164	0.08	0.58	0.85	1.27	1.95	3.5	8.6	15.5	> 100.0	> 100.0
54, 54A	0.33	0.92	2.15	3.2	4.7	6.8	13.0	25.0	37.0	> 100.0
55, 55A, 55B, 223	0.30	1.2	3.2	4.8	7.0	10.0	16.5	30.5	46.0	> 100.0
57, 57A, 130, 131	0.18	0.65	1.6	2.4	3.5	5.4	9.8	21.0	> 100.0	> 100.0
58, 58B	0.33	1.25	3.15	4.6	6.9	10.5	17.5	37.5	60.0	> 100.0
58A, 58C	0.44	1.4	3.3	4.9	7.4	12.0	24.0	54.0	83.0	> 100.0
59, 59A, 59B	0.33	1.1	2.4	3.4	4.9	7.0	12.0	26.5	42.0	> 100.0
62, 62A, 71, 71A, 71B	0.25	0.85	1.9	2.7	3.8	5.3	8.7	18.5	30.0	83.0
62B	0.31	0.90	2.0	2.9	4.2	6.2	11.0	24.0	38.0	92.0
63, 63B, 79, 79B	0.19	0.52	1.1	1.5	2.3	3.4	5.8	12.0	20.5	44.0
87A, 116, 165, 166, 225, 227	0.18	0.60	1.4	2.1	3.0	4.5	7.6	15.0	21.5	36.5
94	0.15	0.60	1.6	2.2	3.3	5.0	7.0	16.0	25.0	60.0
94A, 226	0.15	0.55	1.2	1.7	2.5	3.5	6.6	15.0	23.0	50.0
108, 108A	0.70	2.3	5.2	7.5	11.0	16.0	26.0	54.0	86.0	> 100.0
114, 114A	0.95	1.3	2.1	2.9	4.4	6.7	11.6	26.0	40.0	65.0
115, IISA, 235	0.17	0.60	1.4	2.0	2.9	4.2	7.0	13.0	20.0	33.0
117, 118, 211, 228	0.09	0.24	0.60	0.90	1.35	2.0	3.5	7.5	12.0	37.0
119, 120	0.12	0.43	1.0	1.5	2.2	3.3	5.5	12.0	17.5	54.0
122	0.40	1.7	4.5	7.0	11.0	16.5	29.0	57.0	87.0	> 100.0
125	0.17	0.50	1.1	1.6	2.3	3.5	6.0	13.5	23.0	> 100.0
140, 141, 141A	0.30	0.90	2.1	3.3	4.7	6.9	13.0	26.0	40.0	90.0
142, 142A, 142B	0.34	1.1	2.7	3.9	5.6	8.0	13.5	27.0	39.0	70.0
143, 143A	0.25	0.85	1.9	2.8	4.0	5.8	9.5	18.0	25.5	52.0
144	0.19	0.60	1.3	1.8	2.6	3.9	7.0	14.0	22.0	50.0
149, 150	0.24	0.88	2.3	3.5	5.4	8.5	16.0	38.0	65.0	> 100.0
161,174	2.3	3.9	6.6	8.9	12.0	17.5	30.0	64.0	99.0	> 100.0
178, 178A, 196	2.6	5.6	10.5	14.0	19.0	28.0	46.0	85.0	> 100.0	> 100.0
179, 179A, 187	3.0	5.3	8.5	10.0	12.5	16.0	24.0	44.0	64.0	> 100.0
180, 180A, 195	2.4	3.3	4.6	5.7	7.6	10.8	17.0	35.0	50.0	88.0
188, 188A	3.1	6.0	9.6	11.4	14.2	16.7	31.0	60.0	82.0	> 100.0
209	0.08	0.27	0.68	1.0	1.6	2.5	4.4	9.5	15.0	48.0
281	0.09	0.32	0.78	1.1	1.7	2.5	4.5	9.0	13.0	24.0

TABLE 6.11 PROPERTIES OF COAXIAL CABLE (CONTINUED)

Jacket Type

Designation	Material	Temperature Limits (°C)
Type I	Black polyvinyl chloride	−40 to +80
Type IIa	Black polyvinyl chloride (noncontaminating):	
	< 1/4 in.	−55 to +80
	> 1/4 in.	−40 to +80
Type IIIa	Black polyethylene	−55 to +85
Type IV	Black synthetic rubber:	
	< 1/4 in.	−55 to +80
	> 1/2 in.	−40 to +80
Type V	Fiberglass	−55 to +250
Type VI	Silicone rubber	−55 to +175
Type VII	Polytetrafluoroethylene (Teflon)	−55 to +200
Type IX	Fluorinated Ethylene Propylene (FEP)	−55 to +200
PU	Polyurethane (Estane)	−60 to +180

TABLE 6.12 COMMON COAXIAL CONNECTOR TYPES

Size Classification	Connector Type	Coupling Method	Cable Size Range (in.)	Maximum Frequency (GHz)	RMS Working Voltage (V)
Medium	UHF	Screw	$\frac{3}{16} - \frac{7}{8}$	0.5	500
	N	Screw			1,000
	C	Bayonet	$\frac{3}{8} - \frac{7}{8}$	11	1,500
	SC	Screw			1,500
	7-mm precision	Sexless Screw	.141, .250, .325, semirigid; RG 214/U, RG 142B/U	18	1,500
Miniature	BNC	Bayonet		4	500
	MHV[a]	Bayonet	$\frac{1}{8} - \frac{3}{8}$	4	5,000
	TNC	Bayonet		11	500
	SHV	Bayonet	0.080-0.420	—	10 000 dc[b]
Subminiature	SMB	Snap on		3	500
	SMC	Screw	$\frac{1}{16} - \frac{1}{8}$	10	500
	SMA	Screw		18	350
	LEMO[c]	Push On	0.100-0.195	4	1 500

Source: Reference 2.

[a] Higher-voltage version of the BNC connector, not mateable with BNC connectors.
[b] NIM high-voltage connector.
[c] 00 shell size. Manufactured in the United States as K-Loc by Kings Electronics Co.

barrels allow one to connect cables together and there are *terminators* that can be attached directly to input or output jacks. In addition, each series has a certain number of interseries adaptors for connection of plugs or jacks to those of another. Some of the different connectors are shown in Figure 6.28. It is advisable to reduce the number of adaptors to a minimum to maintain the transmission-line properties of the connections with as few reflection-producing discontinuities as possible.

A good connector should be easy to fit to the proper cable and should not introduce any discontinuities in the transmission-line properties of the cable. Needless to say, the characteristic impedance of the connector should be identical to that of the cable. The connector should be a strong mechanical fit to the cable, and it should be easy to connect and disconnect from mating connectors.

Any given connector type comes in a variety of styles to accommodate the different cable sizes. Different configurations—such as straight and right angle—are available, and often there are different choices for the mechanical and electrical connections of the connector to the cable, the three most common being a ferrule clamp and screw, crimp, and solder. The procedure for stripping coaxial cable by hand is shown in Figure 6.29a. Shielded lead wire extractors are available for removing insulated lead wire from braided shield. Similar to a syringe, the extractor removes the lead wire from the shield leaving the braid intact. Figure 6.29b shows the assembly of a ferrule-clamp BNC/MHV connector. Crimp-style connectors require the use of a special crimping tool and die. They are stronger, less bulky, more uniform, and electrically more robust than clamp and solder connectors. Crimping avoids the swelling or melting of the cable dielectric that can occur when solder-type connectors are assembled. When attaching connectors to cable it is highly recommended that the manufacturer's cable-cutting procedures be followed so that the inner conductor, insulator, outer conductor (braid or foil), and outer insulator are of the proper length. In this way no discontinuities in the electrical properties of the cable are introduced at the cable-connector interface. Special cutting jigs are available to assure the proper geometry when preparing the cable for insertion intothe connector.

Assembly instructions for several types of coaxial connectors are given in the Chapter 6 appendix.

6.2.5 Relays

A relay is an electrically actuated switch. Relay action can be electromechanical or can be based on solid-state devices. A large number of contact configurations exist.

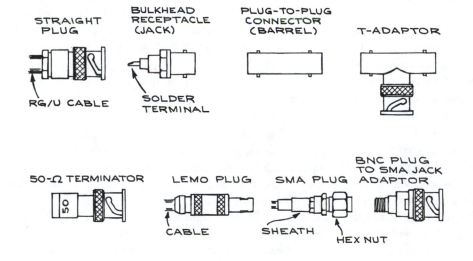

Figure 6.28   Some coaxial connectors.

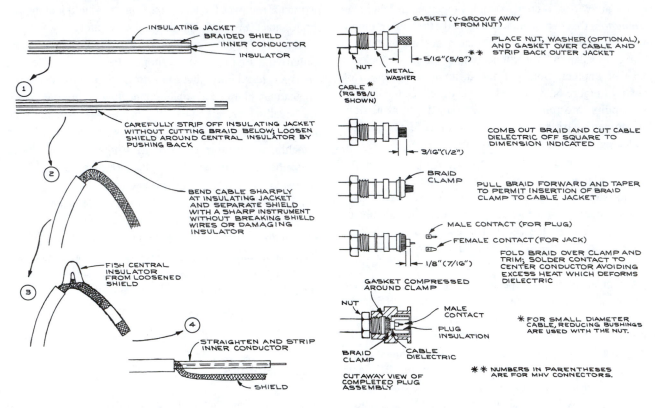

Figure 6.29 (a) Procedure of stripping coaxial cables. (b) Assembly procedure for BNC/MHV ferrule-clamp connector.

Important parameters are the operating voltage and power rating of the relay as well as the contact rating. The speed at which mechanical relays operate is generally in the 10 to 100 ms range. General-purpose electromagnetic relays operate at 6, 12, 24, 48, and 110 V dc and 6, 12, 24, 48, 110, and 220 V ac with coil power ratings of around 1 W (see Figure 6.30a). Such relays have contacts capable of handling from 2 to 10 A. Sensitive mechanical relays use a large number of turns of fine wire on the electromagnet coil. They operate with 28 V dc and require only 1 to 40 mW of coil power. Contacts are rated from 0.5 to 2 A.

A *reed relay* is a glass-encapsulated switch having two flexible thin metal strips, or reeds, as contactors. The switch is placed inside an electromagnetic coil (Figure 6.30b). The relay contacts can be dry or mercury wetted. Standard coil voltages are 6, 12, and 24 V dc, and the coil power is from 50 to 500 mW per reed-switch capsule.

Contacts are rated at 10 mA to 3 A, depending on size. There is also a maximum open-circuit voltage that can be sustained by the opened contacts. The low mass of the reeds makes operating times of 0.2 to 2 ms possible. Small reed relays are made for PCB mounting.

With inductive loads, the fast opening of relay contacts can cause a large voltage to appear across them. Some protective circuits for reducing this problem are illustrated in Figure 6.31.

Solid-state relays consist of a solid-state switching element driven by an appropriate amplifier. Because the input circuit is often optically isolated from the load, these relays introduce little or no noise into the circuit that drives them. The load current determines the switching element. Field-effect transistors (FETs) are used for low-level dc, bipolar junction transistors (BJTs) for intermediate dc currents, and triacs and SCRs for large ac

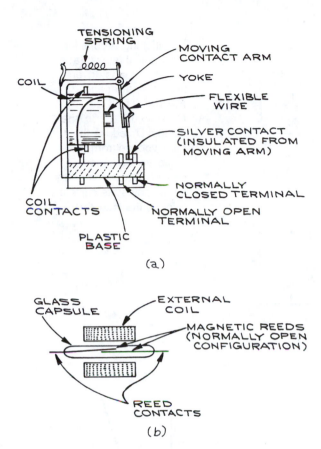

Figure 6.30   Mechanical relays: (a) medium-power relay; (b) reed relay.

and dc currents. Section 6.3 includes a discussion of transistors and solid-state switches. Though there is no contact wear with solid-state relays, they may have high *on* resistance compared with electromechanical devices and require carefully conditioned activating signals. They are much less resistant to overload than are electromechanical devices.

6.3 ACTIVE COMPONENTS

The overall properties of semiconductor diodes and transistors can be understood in terms of the properties of the *p-n* junctions that form them. Without going into solid-

state physics, it is sufficient to know that *p*-type semiconductor material conducts electricity principally through the motion of positive charges (holes), while *n*-type material owes its conductivity mainly to electrons. Both *p*- and *n*-material can be produced from germanium or silicon by the addition of impurity atoms—a procedure called *doping*.

6.3.1 Diodes

Most semiconductor diodes consist of a *p-n* junction as illustrated in Figure 6.32. The schematic representation is shown next to the simplified drawing of the diode. The diode is a unidirectional device: its dc current-voltage (*I-V*) characteristics depend on the polarity of the *anode* with respect to the *cathode*. The *I-V* characteristics of a typical small-signal silicon diode are drawn in Figure 6.33. When the anode is positive with respect to the cathode (*forward-bias*), the diode conducts a large current. When the polarity is reversed (*reverse-bias*), the current is very small. If the reverse voltage is increased beyond the *peak reverse voltage* (PRV), breakdown occurs and the diode conducts strongly again. Except in the case of specially designed Zener diodes, this usually destroys the diode. The forward-bias voltage at which the diode begins to conduct strongly is V_{cutin}. For germanium diodes, V_{cutin} is 0.2 volts, while for silicon diodes it is 0.6 volts. The *I-V* characteristics shown in the graph are accurately represented by the formula

$$I = I_0[e^{V/\eta V_\gamma} - 1]$$

where $I_0 = 10^{-9}$ A for silicon and 10^{-6} A for germanium. I_0 is temperature dependent, increasing at a rate of 11%/°C for Ge and 8%/°C for Si. V_T has the value 0.026 volts at 300 K and is directly proportional to the absolute temperature, while η is a unitless constant equal to 1 for germanium and 2 for silicon.

Diodes can be separated into two types according to the power they can dissipate. *Power diodes* are used in high-current applications, for example, as rectifiers in power supplies. Important specifications for power diodes are the *maximum forward current,* maximum PRV (peak reverse voltage), and *effective forward-bias resistance.* Low-power diodes or *signal diodes* are used to rectify small

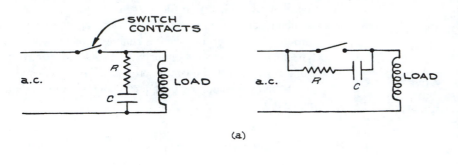

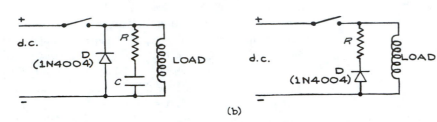

Figure 6.31   Contact-conditioning circuits for mechanical relays: (a) ac (for small loads driven from the power line, $R = 100\ \Omega$ and $C = 0.05\ \mu F$; (b) dc (the diodes, D, reduce the peak voltage from the inductive load; they should be able to handle the steady-state current through the inductor).

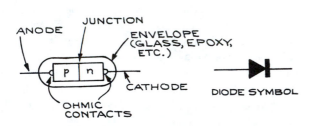

Figure 6.32   A diode and the corresponding symbol.

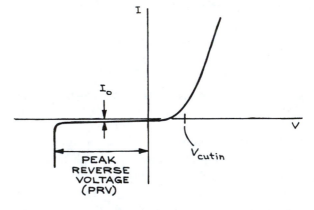

Figure 6.33   Typical current-voltage characteristic of a small-signal silicon diode.

signals, mix two frequencies to produce sum and difference frequencies, and switch low voltages and currents. Besides the current and voltage ratings, the speed with which diodes can be changed from the forward-bias to the reverse-bias condition and back is a factor in high speed switching circuits. Switching speed is related to the effective capacitance of the *p-n* junction. Typical specifications for a power and signal diode are given in Tables 6.13 and 6.14.

Most diodes have a 1N-prefix designation. The two ends of a diode are usually distinguished from each other by a mark—the cathode end having a black band for glass-encapsulated diodes and a white band for black plastic-

encapsulated diodes. Power diodes can dissipate large amounts of heat and are often constructed for ease of mounting to a heat sink for efficient heat dissipation. Some diode configurations are shown in Figure 6.34. In the case of the high-power diode, the cathode is a stud that can be attached directly to a metal heat sink. Generally, fiber or mica washers are used to isolate the cathode

TABLE 6.13 PROPERTIES OF THE 1N914 FAST-SWITCHING DIODE

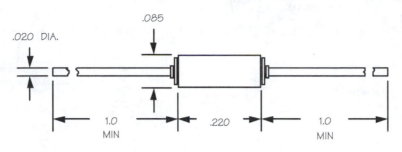

Maximum Ratings	
Peak reverse voltage	75 V
Reverse current	25 nA
Average forward current	75 mA
Peak surge current (1 sec)	500 mA

Electrical Characteristics	
Junction capacitance	4 pF
Reverse recovery time	4–8 ns

TABLE 6.14 PROPERTIES OF 1N4007—1N4007 1-AMPERE SILICON RECTIFIERS

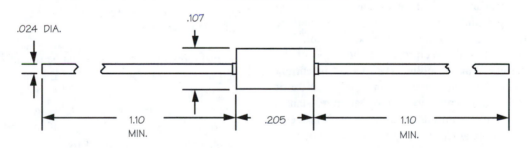

Maximum Ratings						
	4001	4002	4003	4004	4005	4006	4007
Peak reverse voltage (V)	50	100	200	400	600	800	1,000
Continuous reverse voltage (V)	50	100	200	400	600	800	1,000
Average forward current (A)				1			
Peak surge current (1 cycle) (A)				30			
Operating temperature range (°C)				−65 to +175			

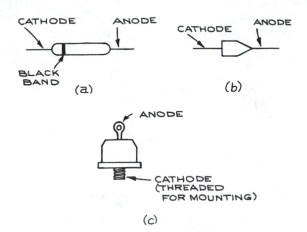

Figure 6.34   Diode case configurations: (a) glass encapsulated signal diode; (b) plastic encapsulated medium-power diode; (c) high-power diode.

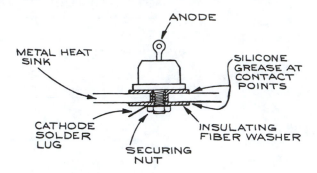

Figure 6.35   Proper power-diode mounting to provide electrical insulation with good heat dissipation.

electrically from the metal heat sink, and thermal conductivity is enhanced with special silicone greases between the washers and heat sink. The mounting technique is shown in Figure 6.35.

Because the current through a diode is a nonlinear function of the voltage across it, the effective resistance changes with diode voltage. This is shown in Figure 6.33, where the reciprocal of the slope of the curve passing through the origin and a point on the *I-V* characteristic curve defines the *static resistance* at that point. The *dynamic resistance* is the reciprocal of the slope of the tangent to the curve at the point. The change in current through a diode for small changes in voltage across it requires knowledge of both static and dynamic resistance. For the purposes of simplified circuit analysis, it is useful to represent the *I-V* characteristic curve of a diode by a so-called *piecewise linear model*. With this model, the forward- and reverse-bias regions are represented by straight lines with the transition occurring at the *cutin* voltage. The piecewise linear curve is shown in Figure 6.36. The slope of the curve at voltages above V_{cutin} is the reciprocal of the average forward resistance. With this model, it is easy to calculate the effect of the diode in a circuit, since it behaves like a resistor in the forward-bias condition with resistance R_f, and in the reverse-bias condition it acts essentially like an open circuit. A

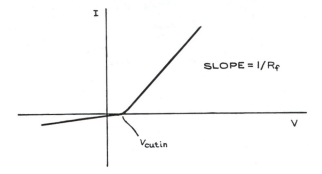

Figure 6.36   Linearized current voltage characteristic of a small signal silicon diode.

graphical representation of this can be obtained by what is called *load-line analysis*. Consider the circuit with an ac voltage source shown in Figure 6.37. The current through R_L can be determined in the following manner: given a value of v_S, the voltage across the diode and the current through it will be determined by the intersection of the straight line with coordinates $(v_S, 0)$ and $(0, v_S/R_L)$, as in Figure 6.38. Different values of v_S will give a family of parallel straight lines. From the intersection points, one constructs a graph of v_L as a function of v_S. This *dynamic curve,* which is shown in Figure 6.39 has much the same appearance as the original *static curve* from which it is derived. If the input waveform is drawn below the horizontal axis, the reflection of it on the dynamic curve will give the output waveform as shown.,

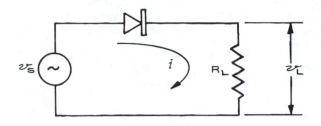

Figure 6.37   Circuit for load-line analysis of a diode.

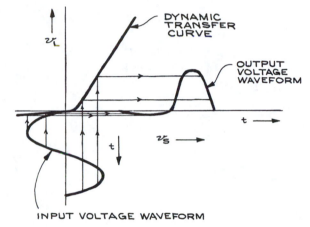

Figure 6.38   Graphical construction of output waveform from input waveform and the dynamic transfer curve.

6.3.2 Transistors

Transistors are three-terminal devices. Any one of the three terminals can be used as an input with a second terminal as output and the third providing the common connection between the input and output circuits.

Bipolar Junction Transistors (BJTs)

These transistors consist of an *emitter,* a *base,* and a *collector.* They are designated *npn* or *pnp,* depending on the materials used for the elements. As can be seen from Figure 6.40, a transistor consists of two back-to-back diodes with the base element common to the two junctions. The major current flow in a transistor is from the emitter to the collector across the base. Relatively little

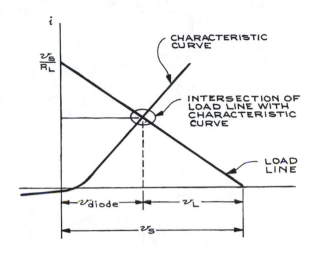

Figure 6.39   Graphical construction for determining the dynamic transfer curve from the load line and the static characteristics curve.

current flows through the base lead. The arrow on the emitter shows the direction of the current flow.

Some common BJT transistor case styles are shown in Figure 6.41. To identify the leads of high frequency transistors, manufacturers' base diagrams should be consulted. With plastic cases, there is a flat on one side of the case to help identify the leads. Leads are sometimes identified by letters, or are in the following order—emitter, base, collector—starting at the flat or tab of sealed metal cans and proceeding clockwise as viewed from the lead side of the transistors. In power transistors, most of the power is dissipated in the collector and the collector terminal is the one with provision for attachment to a heat sink. Standard transistor types have code designations beginning with 2N. Several manufacturers often make the same transistors. Proprietary code designations are also used by manufacturers to identify their products.

Since BJT transistors consist of two *p-n* junctions, a simple ohmmeter test is a good way to check them once the transistor is removed from the circuit. The ohmmeter must provide at least 0.6 V, enough to forward bias a silicon *p-n* junction. The resistance between the emitter and base should be low, on the order of a few hundred ohms or less, when the ohmmeter leads are attached so as

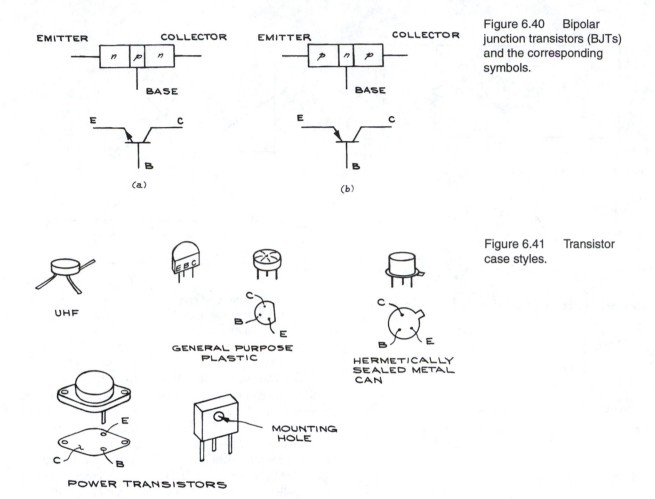

Figure 6.40   Bipolar junction transistors (BJTs) and the corresponding symbols.

Figure 6.41   Transistor case styles.

to forward bias the junction. When the leads are reversed, the resistance should be on the order of megohms. The same should be true of the collector-base junction. The polarity of the ohmmeter leads can be checked with a separate voltmeter.

In normal operation, the emitter-base junction of a transistor is forward biased, while the base-collector junction is reverse biased. The three usual configurations, *common emitter* (CE), *common base* (CB), and *common collector* (CC), are shown schematically in Figure 6.42. The *I-V* characteristics of each configuration consist of two sets of curves, one for the input and the other for the output. The *I-V* curves for a typical general-purpose transistor in the CE configuration, for example, are shown

in Figure 6.43. There is a set of curves rather than a single curve for the input and output circuits because the two circuits interact with each other. When measuring the *I-V* characteristics of either the input or output circuit, the condition of the other one must also therefore be specified. Figure 6.44 shows a circuit for measuring the output characteristics of a CE configuration.

For a fixed series of I_B values, set by adjusting V_{BB}, the current into the collector I_C is measured as a function of V_{CE}. The variation in V_{CE} is obtained by changing V_{CC}. Transistor operation can be understood in terms of the characteristic curves. For a fixed value of I_B, I_C increases rapidly with V_{CE} until a plateau is reached. I_C increases thereafter much more slowly with V_{CE}. In the plateau

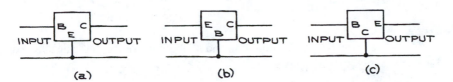

Figure 6.42 (a) Common-emitter (CE), (b) common-base (CB), and (c) common-collector (CC) configurations.

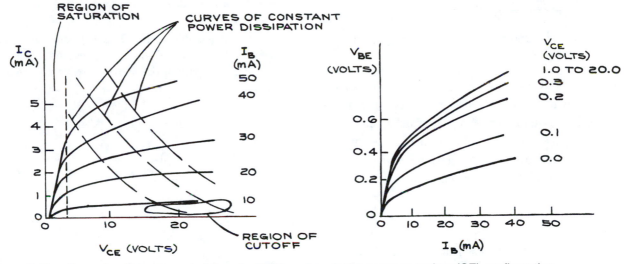

Figure 6.43 Current-voltage characteristics of a BJT transistor in the common emitter (CE) configuration.

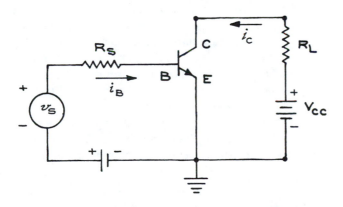

Figure 6.44 Biasing circuit for the BJT common-emitter (CE) configuration.

region, the value of I_C depends on I_B and is almost independent of V_{CE}. This can be made more quantitative by using load-line analysis. Two points are sufficient to establish the load line on the output curves—they are points corresponding to $I_C = 0$ where $V_{CE} = V_{CC}$ and to $V_{CE} = 0$ where $I_C = V_{CC}/R_L$. The output curves with a superimposed load line are shown in Figure 6.45. The intersections of the load line with the characteristic curves establish the operating voltages and currents. It is possible to predict quantitatively how the current through R_L varies with the input base current from these curves. For linear operation, conditions are adjusted so that the transistor operates somewhere in the middle of the characteristic output curves, where they are evenly spaced and parallel for equal base-current increments.

Another way of analyzing transistor operation is by constructing an equivalent circuit with passive elements and sources. Such a circuit can then be analyzed by

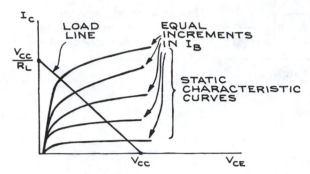

Figure 6.45 Relation between the static characteristic curves and the load line.

standard methods. Because the equivalent circuit contains only linear elements, it can represent the properties of the transistor only within a small region of the characteristic curves about given nominal values of the dc, steady state, or quiescent voltages and currents. Four variables are sufficient to characterize a given configuration—input voltage, input current, output voltage, and output current. The normal convention is to take input voltage and output current as the dependent variables, and input current and output voltage as the independent variables. Lowercase letters apply to instantaneous values of current and voltage and uppercase subscripts indicate that it is the total value of the current or voltage that is under consideration. Mathematically the dependent variables are expressed as

$$v_{Input} = v_{Input}(i_{Input}, v_{Output})$$
$$i_{Output} = i_{Output}(i_{Input}, v_{Output})$$

where the subscripts with the uppercase letters indicate total instantaneous values. For the CE configuration, these equations take the form

$$v_{BE} = v_{BE}(i_B, v_{CE}), \; i_C = i_C(i_B, v_{CE})$$

Expanding V_{BE} and i_C about the quiescent values I_B and V_{CE} gives

$$\Delta v_{BE} = \left.\frac{\partial v_{BE}}{\partial i_B}\right|_{v_{CE}} \Delta i_B + \left.\frac{\partial v_{BE}}{\partial v_{CE}}\right|_{I_B} \Delta v_{CE}$$

$$\Delta i_C = \left.\frac{\partial i_C}{\partial i_B}\right|_{v_{CE}} \Delta i_B + \left.\frac{\partial i_C}{\partial v_{CE}}\right|_{I_B} \Delta v_{CE}$$

The Δ's are now the small differences between the total instantaneous values and the dc quiescent values. By convention, the partial derivatives are represented by lowercase h's:

$$h_{ie} = \left.\frac{\partial v_{BE}}{\partial i_B}\right|_{v_{CE}}, \; h_{re} = \left.\frac{\partial v_{BE}}{\partial v_{CE}}\right|_{I_B},$$

$$h_{fe} = \left.\frac{\partial i_C}{\partial i_B}\right|_{v_{CE}}, \; h_{oe} \; \frac{\partial i_C}{\partial v_{CE}}\right|_{I_B}$$

so that

$$v_{be} = h_{ie}i_b + h_{re}v_{ce}, \; i_c = h_{fe}i_b + h_{oe}v_{ce}$$

The h's are called small-signal *hybrid parameters*. The equations for v_{be} and i_c can be translated into the low-frequency equivalent circuit in Figure 6.46. In physical terms, h_{ie} is the effective emitter-base resistance, (often labeled r_p); h_{fe} is the small signal current gain (ratio of collector current to base current), commonly labeled β; $1/h_{oe}$ is the output resistance, r_o; and h_{re} expresses influence of the output circuit on the input circuit (a small effect that is often neglected). It is necessary to keep in mind that the linear circuit is a good representation of the transistor only for small excursions of the input and output voltages and currents about specified quiescent dc values. The equivalent circuit does not show the biasing network necessary to

Figure 6.46 Hybrid-parameter equivalent circuit.

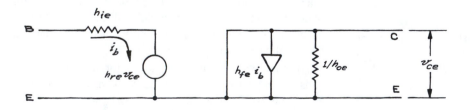

establish the quiescent operating point. Practical circuits using separate power supplies for biasing voltages are shown in Figure 6.47. The output signal is developed across the load resistor R_L, while capacitors C_i and C_o isolate the ac input and output signals from the dc biasing levels. Transistor data sheets give the values of the hybrid parameters for different operating points. The hybrid parameters for the CB and CC configurations can be calculated from those of the CE with standard formulae.[3] The second letter in the subscript of the h's indicates whether the configuration is common emitter, common base, or common collector. Lowercase subscripts indicate small-signal values while uppercase subscripts are for dc, large-signal conditions. By applying the standard methods of circuit analysis to the equivalent circuit, the current, voltage, and power gains can be calculated, as well as the input and output impedances under small-signal conditions. The general properties of the three configurations are given in the graphs of Figure 6.48. Because the charge carriers in the p- and n-materials that make up the transistor cannot respond instantaneously to changes in voltage across the junctions, the hybrid parameter model is strictly valid only at low frequencies. The response of the transistor at high frequencies can be very different, and equivalent-circuit models exist for the high-frequency response. In these models, the effective capacitances between the base-emitter, base-collector, and collector-emitter are represented by capacitors C_{be}, C_{bc}, and C_{ce}.

If the emitter-base junction of a transistor is strongly forward biased (0.8 V for silicon), the current through the transistor will be determined by the external resistance in the circuit. In other words, the resistance between emitter and collector will be low; the voltage between emitter and collector will be on the order of 0.2 V. In this condition, the transistor is said to be *saturated* or *on* and the base current must be sufficient to sustain the current flow from emitter to collector. If the emitter-base junction is reverse biased, no base current will flow and no collector current will flow. Under this condition, the transistor acts as an open circuit and is said to be *cut off*. In the saturated or the cut-off condition, a transistor is similar to a closed or an open switch. Transistor switches can operate at high frequencies, unlike mechanical switches, but the saturated resistance is

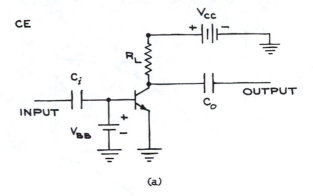

(a)

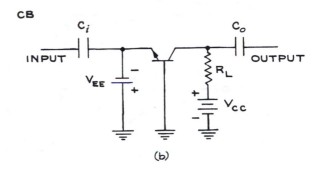

(b)

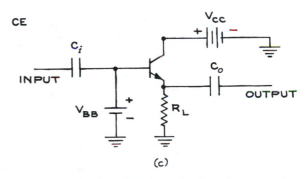

(c)

Figure 6.47   Practical transistor circuits using separate power supplies for biasing voltages.

never zero, and there is always at least a 0.2 V drop from the emitter to the collector.

The operating conditions for bipolar junction transistors are summarized in Table 6.15. They determine whether a

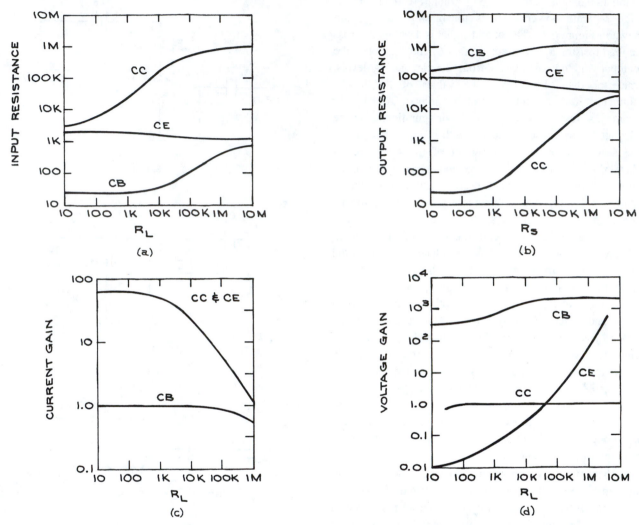

Figure 6.48 Variation of the properties of the CE, CB, and CC transistor configurations with various input parameters: (a) input resistance as a function of load resistance; (b) output resistance as a function of source resistance; (c) current gain as a function of load resistance; (d) voltage gain as a function of load resistance.

transistor is operating in the linear region (*active*), is *saturated*, or is *cut off*. When reading the specifications for a 2N3904 transistor in Tables 6.15 and 6.16, the following conventions are to be noted:

1. Voltages between two terminals are specified by either V or v with two or three subscripts. The first two subscripts indicate the terminals between which the voltage occurs, and the third subscript, when present, indicates the condition of the third terminal. Thus V_{CBO} indicates a dc voltage between collector and base with the emitter open. Subscripts with repeated letters indicate power supply voltages. For example, V_{CC} is the power supply voltage to the collector, and V_{BB} is the power supply voltage to the base.

TABLE 6.15 2N3904 TRANSISTOR PARAMETERS

Parameter	Condition of Transistor		
	Active	Saturated	Cut Off
V_{EB}	0.6 V	0.8 V	< 0.0 V
V_{BC}	5–10 V (reverse bias)	0.6 V (forward bias)	$\cong V_{CC}$ (reverse bias)
V_{EC}	5–10 V	0.2 V	V_{CC}
I_B	$\cong \dfrac{I_C}{h_{FE}}$	$\dfrac{I_C}{h_{FE}}$	$\cong 0$
I_C	$\cong \dfrac{V_{CC} - V_{CE}}{R_L}$	$\dfrac{V_{CC}}{R_L}$	$\cong 0$

2. By convention, all currents are taken to be positive when they flow into the leads of the transistor and negative when they flow out of the leads.

3. Currents are specified by I or i with a subscript indicating the terminal involved. I_B is therefore a dc base current.

Transistors are destroyed if their maximum power capabilities are exceeded, or if, when reverse biasing the junctions, the maximum reverse voltages are exceeded, resulting in electrical breakdown. The maximum power dissipation depends on temperature, and a transistor fixed to a heat sink can dissipate considerably more power than one standing alone in free air. In pulse operation, the maximum steady-state power levels can be momentarily exceeded without damage.

Field-Effect Transistors

The *junction FET* or *JFET* is the simplest *FET*. It has three terminals—a *source*, a *gate*, and a *drain*. An *n*-channel device is shown in Figure 6.49. Current flow is from source to drain with the gate reverse biased with respect to the channel. Because of this reverse-bias, negligible current flows into the gate. The *I-V* characteristics of a small-signal JFET are shown in Figure 6.50. From the curves it can be seen that there is a region in which the drain current decreases linearly with increasing reverse-bias from the gate to source,

independent of the drain-source voltage. This is the region in which the transistor has a linear response. The region $V_{DS} = 0$, I_D increases linearly with V_{DS} for fixed V_{GS}, with the slope equal to the reciprocal of the channel resistance $R_{channel}$. As V_{GS} becomes more negative, $R_{channel}$ increases. Transistor operation in this region is that of a *voltage-controlled resistor.*

The FET can also be used as a switch, with the advantage of no offset voltage from source to drain and almost perfect electrical isolation of the gate signal from the output circuit. Disadvantages are rather high *on* resistances (on the order of hundreds of ohms) and low *off* resistances. Also, switching times for FETs are greater than for BJTs.

The simple structure of FETs results in low noise. As a consequence, they are often used as the active elements in low-level signal amplifiers. FETs with gates insulated by an oxide layer from the conducting channel are called *MOSFETs* (metal oxide-semiconductor FETs). The oxide gate insulation effectively prevents any gate current flow. These FETs can operate in a *depletion mode,* where an effective reverse-bias appears at the surface of the channel, or in an *enhancement mode,* where there is a forward-bias. The channel can also be of *p* or *n* material. The low power consumption, ease of fabrication, and high packing densities have made these devices extremely common in *large-scale integrated circuits* (LSI). Because of the gate insulation, static charge can accumulate on the gate and voltage levels sufficient to punch through the insulation

TABLE 6.16 PROPERTIES OF THE 2N3904 N-P-N SILICON TRANSISTOR

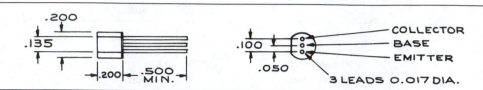

Parameter	Value	Conditions
V_{CBO}	60 V	$I_C = 10\ \mu A$, $I_E = 0$
V_{CEO}	40 V	$I_C = 1\ mA$, $I_B = 0$
V_{EBO}	6 V	$I_E = 10\ \mu A$, $I_C = 0$
h_{FE}	40	$V_{CE} = 1V$, $I_C = 100\ \mu A$
	70	$V_{CE} = 1V$, $I_C = 1\ mA$
V_{BE}	0.65–0.85 V	$I_B = 1\ mA$, $I_C = 10\ mA$
	0.95 V	$I_B = 5\ mA$, $I_C = 50\ mA$
$V_{CE}{}^a$	0.2	$I_B = 1\ mA$, $I_C = 10\ mA$
	0.3	$I_B = 5\ mA$, $I_C = 50\ mA$
h_{ie}	$1\text{--}10\times10^4\ \Omega$	
h_{fe}	100–400	$V_{CE} = 10\ V$, $I_C = 1\ mA$,
h_{re}	$(0.5\text{--}8\times10^{-4})$	$f = 1\ Hz$
h_{oe}	1–40 μmho	
$f_T{}^b$	300 MHz	$V_{CE} = 20\ V$, $I_C = 10\ mA$
$C_{obo}{}^c$	4 pF	$V_{CB} = 5\ V$, $I_E = 0$
$C_{ibo}{}^d$	8 pF	$V_{EB} = 0.5\ V$, $I_C = 0$

Maximum Ratings at 25°C in Free Air:
$V_{CB} = 60\ V$, $V_{CE} = 40\ V$, $V_{EB} = 6\ V$,
I_C (continuous) = 200 mA

Power Dissipation in Free Air:
310 mW at 25°

[a] Saturation.
[b] Transition frequency where $|h_{fe}| = 1$.
[c] Common-base open-circuit output capacitance.
[d] Common-base open-circuit itput capacitance.

are attained. To avoid this, some MOSFET gates are protected with reverse-biased Zener diodes. MOSFET devices are normally packaged in conducting foam or with metal protecting rings that are removed just prior to insertion into the circuit.

Active Loads and Current Sources

Both BJT and JFET transistors can be used as resistors with static resistances of kilohms and dynamic resistances of tens and even hundreds of kilohms. Active loads are commonly used in integrated circuits where resistors are difficult to fabricate. The high dynamic resistance (*dv/di*)

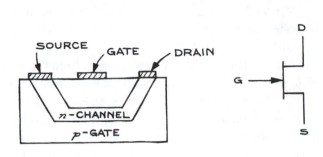

Figure 6.49   Junction field-effect transistor and the corresponding symbol.

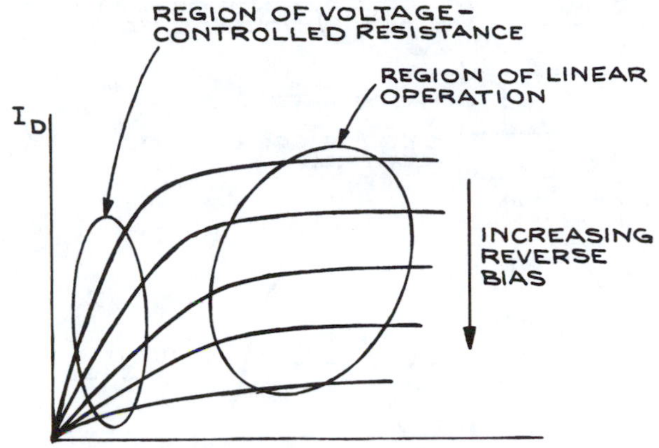

Figure 6.50   Current-voltage characteristics of a small-signal, junction-field-effect transistor (JFET).

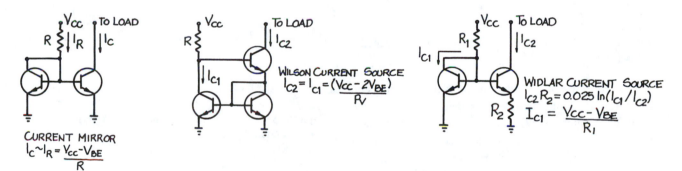

Figure 6.51   Current sources.

of BJTs and FETs in their active region is also used for high output impedance current sources circuits. Examples of current source circuits are given in Figure 6.51

6.3.3 Silicon Controlled Rectifiers

The *silicon-controlled rectifier* (SCR) is a switching device with only two states—*on* and *off*. In the *on* state, the SCR has a low resistance, and some models are capable of passing over 100 A. In the *off* state, the resistance is in the megohm range. The SCR has three terminals: an *anode*, a *cathode*, and a *gate*. The transition

from *off* to *on* occurs when the voltage on the anode is positive with respect to the cathode and a pulse of the correct polarity and magnitude is applied to the gate. Once the SCR is *on*, the gate loses all control over the functioning of the device. The only way to turn it *off* is to reduce the anode-cathode voltage to zero. If a sinusoidal voltage is applied to the SCR anode, the device will be turned *off* once each half cycle. By controlling the point in each positive half cycle when the trigger pulse is applied, the average current through the SCR can be varied over wide limits (see Figure 6.52). This way of regulating the power to a load is very efficient because very little power

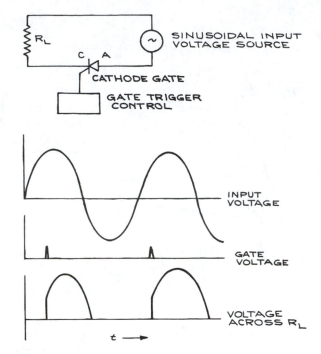

Figure 6.52   Current regulation with a silicon controlled rectifier (SCR).

is dissipated in the SCR. When the SCR is *off*, there is no current through it and the power dissipation is zero. When the current through the SCR is at a maximum, the voltage across it is low, resulting in low power dissipation. Bilateral triggering can be accomplished with a bilateral switch or *triac*. A common application of this kind of power control is the incandescent light dimmer. High-power switching devices like the SCR and triac produce *radio-frequency interference* (rfi) when switched with a voltage across them. *Zero crossing* switching avoids rfi by timing the switching to those instants when there is no voltage across the device. Silicon controlled switches (SCSs) are similar to SCRs, but have the advantage of being able to be turned off with a pulse.

6.3.4 Unijunction Transistors

The *unijunction transistor* (UJT) is a single bar of *n* material to the middle of which *p* material is attached,

forming a *p-n* junction. This is shown schematically in Figure 6.53. The two ends of the bar are bases 1 and 2 (*B*1 and *B*2), while the *p* material is the emitter (*E*). When *B*2 is made positive with respect to *B*1, there will be a potential gradient along the bar and the potential at the point of attachment of the emitter will be a fixed fraction of the potential across the bar. As long as the emitter is less positive than the bar at its point of attachment, the emitter junction will be reversed biased and no current will flow through it. When the emitter potential is increased sufficiently to forward bias the junction, current will flow in the *E-B*1 circuit and V_{EB1} will fall. This is shown in the characteristic curve (see Figure 6.54). A common

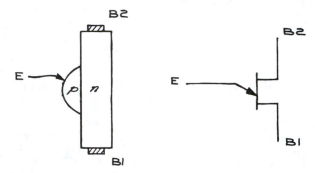

Figure 6.53   Unijunction transistor (UJT) and the corresponding symbol.

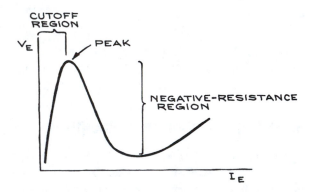

Figure 6.54   Current-voltage characteristic curve for a unijunction transistor showing the region of negative resistance.

application of the UJT is in relaxation oscillators where a capacitor is charged through a resistor to the peak potential of the UJT and then discharged. The UJT is also used in trigger circuits, pulse generators, and frequency dividers. The stable peak voltage, which is a fixed fraction of the interbase voltage, and high pulse-current capability make the UJT very useful for SCR control circuits.

6.3.5 Thyratrons

For fast switching of high currents at high voltages, gas-filled tubes called *thyratrons* are used. They are analogs of SCRs and consist of a heated cathode, an anode, and a grid sealed into a glass, gas-filled envelope. An arc can be struck between the cathode and anode, with the initiation of the arc controlled by the potential on the grid. The grid is usually a cylindrical structure surrounding the anode and cathode from which a baffle or series of baffles with small holes extends between anode and cathode. Since the anode and cathode are almost completely shielded from each other, only a small grid potential is needed to overcome the field at the cathode resulting from the large anode potential. Once the arc has been initiated, the grid loses control over the arc. Grid control is reestablished only when the anode potential falls below the level necessary to sustain the arc. Deuterium-filled thyratrons can switch hundreds of amperes at tens of kilovolts in microseconds or less and are commonly used in the high-voltage discharge sources of lasers. SCRs are often used to supply the trigger pulse to the grid of the thyratron.

6.4 AMPLIFIERS AND PULSE ELECTRONICS

6.4.1 Definition of Terms

In laboratory applications, amplifiers are used to transform low-level signals to levels sufficient for observation and recording or for operation of electronic or electromechanical devices. When choosing an amplifier for a given task, there are a number of considerations that require the definition and explanation of several special terms. The goal of this section is to provide information for the experimentalist who must make decisions regarding the use and specifications of amplifiers.

Amplifiers can be classified in a number of ways, among them the following:

- By input and output variables
- By frequency domain
- By power levels

Each of these classifications will be treated in turn, and any given amplifier will have its place somewhere in each of the classifications. The two most common electrical input and output variables are current and voltage. There are therefore four different possible types of amplifier, corresponding to the two input-variable possibilities and the two output-variable possibilities. These are listed in Table 6.17. The *gain* of an amplifier is the ratio of the output variable to the input variable. This is the equivalent of the transfer function for the passive circuits already considered. For voltage and current amplifiers, the gain is unitless. For the transconductance and transresistance amplifiers, however, it has units of siemens (ohm^{-1} or mho) and ohm, respectively.

TABLE 6.17 TYPES OF AMPLIFIER

Input Variable	Output Variable	Amplifier Type
Voltage	Voltage	Voltage
Voltage	Current	Transconductance
Current	Voltage	Transresistance
Current	Current	Current

There are also amplifiers for which the input variable is the time integral of the current or the charge. Such *charge-sensitive* amplifiers have a voltage as an output, and the gain is expressed in volts per coulomb or inverse farads. Amplifiers with the time derivative of current as an input variable are also possible.

The amplifiers in Table 6.17 have very different input and output properties. If the input is a current, the amplifier should have a low input impedance so as to affect the source as little as possible. On the other hand, if the input is a voltage, the input impedance should be as large as possible to avoid drawing current from the source and changing the source output voltage. The opposite considerations apply to the outputs of the amplifiers. If the output is a current, it is desirable that the output be close to that of an ideal current source—that is, a very high output impedance. If the output is a voltage, the output impedance should be low to approximate an ideal voltage source as closely as possible. These considerations of input and output impedance (see Table 6.18) are of fundamental importance when matching an amplifier to a detector or transducer at the input and a load at the output. If the input device is a current source, one should generally choose a current, transresistance, or

TABLE 6.18 AMPLIFIER IMPEDANCES

Amplifier Type	Input Impedance	Output Impendance
Voltage	High	Low
Transconductance	High	High
Transresistance	Low	Low
Current	Low	High
Charge sensitive	Low	Low

charge-sensitive amplifier; if the input device is a voltage source, a voltage or transconductance amplifier is required. This assumes that amplification with minimum loading of the input circuit is desired. When efficient power transfer is desired, the impedance of the source and load should be identical to the input and output impedances of the amplifier.

Amplifiers can be divided into groups according to the frequency domain in which they are designed to operate (see Table 6.19). Consider the Bode plots of gain and phase shift shown in Figure 6.55 for a capacitance-coupled inverting (output 180° out of phase with input) amplifier. The curves can be understood qualitatively in the following ways—at low frequencies the coupling capacitors (capacitors in series with input and output to block dc levels) reduce the low-frequency gain just as a high-pass filter and at high frequencies the reduced gain of the active devices (transistors) has the effect of shunt capacitances across both the input and output circuits of the amplifier, resulting in behavior similar to that of a low-pass filter. The frequency difference between the corner frequencies (3 dB points of the gain curve) is the *bandwidth* of the amplifier. Because the low corner

TABLE 6.19 AMPLIFIER FREQUENCY RANGES

Amplifier Type	Amplifier Type
dc	0–10 Hz
Audio	10 Hz–10 kHz
rf	100 kHz–1 MHz
Video	30–1000 MHz
VHF	30–300 MHz
UHF	300–1000 MHz
Microwave	1000 MHz–50 GHz

frequency is usually orders of magnitude smaller than the high corner frequency (except in the case of tuned amplifiers that operate over a very small frequency range), the bandwidth can be considered equal to the upper corner frequency. An often used figure of merit for amplifiers is the *gain-bandwidth product* (GBWP), which is the midband gain times the bandwidth. By this criterion, a low-gain, wide-bandwidth amplifier is equivalent to a high-gain, low-bandwidth amplifier.

If it is possible for the output of an amplifier to be coupled back to the input (capacitance coupling is always present to some degree because of stray capacitances) in such a way as to be in phase with the signal at the input, a *positive feedback* or *regenerative* situation will occur, giving rise to oscillations and unstable behavior. This is the origin of the familiar squawking and whistling that occurs in public address systems when the output from the loudspeakers finds its way back to the microphone. The stability of an amplifier against such oscillations is expressed in terms of *gain* and *phase margins*. If the value of the gain of an amplifier (in dB) is negative at frequencies for which the phase of the output is equal to that of the input, oscillations cannot occur. Correspondingly, if the output is out of phase with the input at all frequencies for which the gain is positive, oscillations are prevented. The amount of negative gain at zero phase is the *gain margin;* the phase difference at 0 dB gain is the *phase margin* (see Figure 6.56). The larger these margins, the more stable the amplifier. Wide-band video amplifiers are particularly susceptible to unstable behavior because even small stray capacitances are sufficient to provide a low-impedance path for high frequencies from the output to the input. Because of the

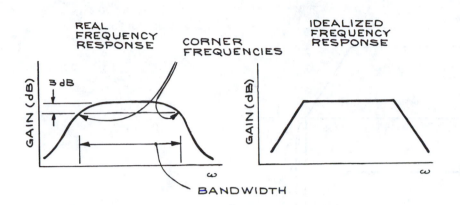

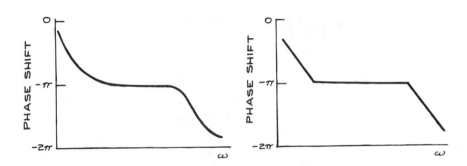

Figure 6.55   Bode plots for a capacitance-coupled inverting amplifier.

large power consumption in the positive feedback (regenerative) mode, prolonged unstable operation of an amplifier can destroy it. This is especially true in the case of fast pulse amplifiers that must have a very large bandwidth in order to accurately amplify pulses with short rise times. Compensation by the addition of poles and zeros to the transfer function can remedy this at the expense of reduced bandwidth.

In classifying amplifiers by power rating, the distinctions are rather arbitrary. Generally, *preamplifiers* are designed to operate with input voltage levels of millivolts or less, input currents of microamperes or less, and input charge of picocoulombs or less. Gain is not as important in preamplifiers as is noise, which limits ultimate sensitivity. In some applications, a unit-gain preamplifier is used as an impedance-matching device. *Power amplifiers* are designed to furnish anywhere from a few to several hundred watts to a matched load. Audio

amplifiers are examples of these. Intermediate between preamplifiers and power amplifiers are amplifiers that operate with input signal levels from preamplifier outputs and produce outputs from a few to several hundred milliwatts. Gains of such amplifiers are high, and noise considerations are much less important than overload recovery. Frequency- and gain-compensating circuits are often incorporated in such amplifiers.

It has been assumed thus far that the output of an amplifier is a linear function of the input at any frequency. Because the active elements (transistors) in solid-state amplifiers are nonlinear, it follows that amplifiers can only operate in a linear way within a limited range of input conditions. For all amplifiers, there is an input signal level beyond which the output signal is severely distorted. Clipping of the output waveform is an obvious form of distortion. Other distortions can be quantified by a Fourier analysis of the output waveform for a sinusoidal input. The

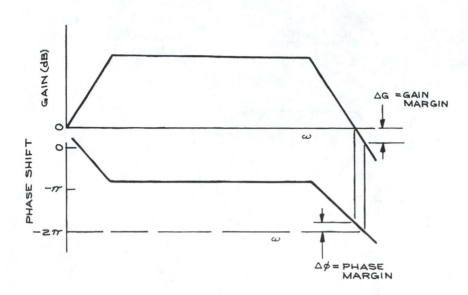

Figure 6.56   Gain and phase margins.

distortion in the output can be described in terms of the coefficients of the harmonics. This *harmonic distortion* is important in critical audio-amplifier applications, but not generally in the laboratory.

The properties of solid-state amplifiers may also be temperature-dependent, and such factors as gain and distortion, though acceptable at room temperature, may become unacceptable at high or low temperatures.

6.4.2 General Transistor-Amplifier Operating Principles

The active elements in solid-state amplifiers can be *n-p-n* or *p-n-p* bipolar junction transistors or field-effect transistors, of which there is a very large variety. FETs can be *n* channel or *p* channel and can be operated in the depletion or the enhancement mode. They can, therefore, have gate structures that are insulated from the conducting channel by an oxide layer (as in MOSFETs), or a direct contact or junction between the gate and conducting channel (as in JFETs).

To use an inherently nonlinear transistor to make a linear amplifier requires great ingenuity. Standard methods involve operating the transistors over a limited range of voltage and current where their nonlinear characteristics can be approximated by linear functions, using compensating circuits to cancel the nonlinearities with devices that have opposite characteristics, or applying negative feedback.

Transistors must be properly biased if they are to work at all. This means supplying, from external sources such as batteries or dc power supply circuits, the correct potential differences across the terminals and the correct currents into them. Even in the absence of an external signal, the transistors are dissipating power. The values of the dc bias (or quiescent) voltages and currents define the *operating point* of the transistor. Signals from an outside source or previous stage are then superimposed on the quiescent values. As long as the signal does not represent too large a deviation from these values, the transistor will operate linearly. If the external signal is sufficiently large however, it can *cut off* the transistor (that is, reduce all the currents through it to zero) or *saturate* it (that is, cause the transistor to act as a short circuit).

Since the bias voltages and currents are dc and the signal voltages and currents are ac, it is possible to separate them. Capacitors or transformers can be used to couple ac signals without disturbing the quiescent dc values. When capacitance coupling is used, the capacitors are called *blocking* capacitors because they block dc voltages and currents. Reasonably sized blocking capacitors usually limit the low-frequency response of amplifiers to greater than 1 Hz (capacitor lead inductance and stray shunt capacitances to ground can also limit the high-frequency response). In many applications the poor transient response of the high-pass circuits formed by the blocking capacitors cannot be tolerated. Direct-coupled amplifiers—where the bias levels of each transistor stage are designed to be compatible with the preceding and succeeding stages—provide high gain to 0 Hz. Because of the need to match active components and resistor ratios very precisely, such amplifiers are often *integrated-circuit* (*IC*) amplifiers, where all of the elements of the circuit are manufactured simultaneously on a single chip. Direct-coupled designs are, in fact, particularly suited to the IC manufacturing process, which favors the production of transistors and diodes but requires quite special techniques for the production of capacitors. *Chopping* is another method of achieving good low-frequency response in amplifiers. With this method, the input signal is chopped at a frequency much higher than the highest characteristic frequency of the input signal. The resulting ac signal can be amplified by conventional ac coupled amplifier stages. Rectification of the output of the final stage restores the input waveform. The advantages of chopper over direct-coupled amplifiers are lower uncompensated input currents and voltages, and comparative freedom from drifts with temperature. These advantages have been largely offset by IC direct-coupled amplifier designs with FET input stages.

Amplifier properties and specifications are best understood by considering as an example the general-purpose 741 IC operational amplifier. The name *operational amplifier* (*op amp*) was originally applied to amplifiers used in analog-computer circuits, which performed a wide variety of mathematical operations. The term is now more generally applied to any high-gain, differential-input amplifier.

Differential-input amplifiers have two input terminals that are electrically isolated from ground and each other. They are called the *noninverting* (+) and *inverting* (–) terminals. The output signal is proportional to the difference between the signals at the two input terminals. *Single-ended* inputs are those with one input terminal and a ground terminal. Most operational amplifier outputs are single-ended—that is, the output is developed between a single output terminal and ground. The ideal operational amplifier has the following characteristics:

- Infinite input impedance
- Zero output impedance
- Infinite voltage gain
- Infinite bandwidth
- Zero output voltage for identical input voltages at the two differential input terminals
- Properties independent of temperature, voltage levels, and frequency

Real operational amplifiers depart from the ideal. Table 6.20 compares the properties of the common 741 IC operational amplifier and the premium OP07. The symbol is shown in Figure 6.57. The supply voltages $V_{CC}+$ and $V_{CC}-$ are applied to the indicated terminals, not to be confused with the signal input terminals. The output is at the apex of the triangle and the terminals marked *null* are for an external potentiometer to cancel the offset voltage in critical applications.

Common operational-amplifier terms include the following:

Common-mode rejection ratio (CMRR). This is the ratio of the difference gain A_D to the common-mode gain A_{CM} (often expressed in dB). The difference gain is the output voltage divided by the algebraic difference of the input voltages. The common mode gain is the output voltage divided by the algebraic sum of the input voltages. For the 741, the 90 dB value of the CMRR means that a one-volt signal at each input terminal will result in a 31 μV output signal. This is shown in the following calculation:

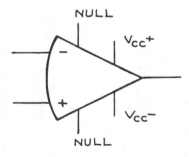

Figure 6.57   Schematic symbol for the operational amplifier, showing input, output, power supply, and nulling terminals.

TABLE 6.20 COMPARISON OF OPERATIONAL-AMPLIFIER PARAMETERS[a]

	741	OP07
CMRR (dB)	90	120
Open-loop dc gain	200,000	500,000
GBWP (MHz)	1.5	0.6
Slew rate (V/μs)	0.5	0.3
Input resistance (MΩ)	2	80
Output resistance (Ω)	75	60
Input offset voltage (mV)	5	0.010
Input offset current (nA)	20	0.3
Input bias current (nA)	80	0.7
Input voltage range (V)	±13	±13
Maximum peak-to-peak output voltage swing (V)	28	25
Power supply rejection ratio (μV/V)	150	5

[a] Typical characteristics at 25°C with power-supply voltages of ±15 V.

$$90dB = 20\log\frac{A_D}{A_{CM}}, \frac{A_D}{A_{CM}} = 10^{4.5} = 3.2 \times 10^4$$

Since the difference voltage is 0.0 V, the output voltage will be the common-mode voltage multiplied by the common-mode gain A_{CM}. For a common-mode voltage of 1.0 V, the output voltage is then $1.0/3.2\times10^4 = 31\times10^{-6}$ V.

Open-loop dc gain. This is the gain in the absence of a connection from output to input—that is, in the absence of any *feedback*. The gain of the amplifier in the presence of a feedback network is called the *closed-loop gain*. With open-loop gains on the order of 200,000, it can be seen that the region of linear operation applies to only very small input signals. At a gain of 200,000, an input signal of +75 μV is sufficient to drive the output to the maximum possible level, which in this case is +15 V from the power supply.

Frequency response. This is given in terms of either the GBWP or the frequency at which the gain equals 0 dB (unit gain). The two methods give nearly the same numbers for a single-pole transfer function. For the 741, the gain has fallen by over five orders of magnitude at 1 MHz from the dc value. One should keep in mind that these frequency-response data are based on so-called *small-signal* conditions, where the input is small enough so that the output stages are operating well within their linear regions.

Slew rate. This gives the time response of the amplifier to large input signals. For this specification the amplifier is connected in a unit-gain (closed-loop) configuration with negative feedback, and an input voltage step is applied. The slew rate is then taken as the ratio of 10 V to time for the output voltage to go from 0 to 10 V. For the 741, this time is 20 μs so the slew rate is 0.5 V/μs.

Input resistance. This is the resistance R_i between the input terminals with one terminal grounded.

Output resistance. This is the resistance R_o (seen by the load) between the output terminal and ground.

Input offset voltage. This is the dc voltage V_{IO} that must be applied across the input terminals to bring the dc output to zero. For the 741, it is 0.8 mV and can be compensated; one would not choose the 741 for the amplification of millivolt dc levels.

Input offset current (I_{IO}). This is the difference between the currents into the input terminals necessary to bring the output voltage to zero.

Input bias current (I_{IB}). This is the average of the currents into the input terminals with the output voltage at zero. It should be noted that these bias currents will flow through circuit elements (generally resistors) attached to the input terminals. If the resistances are not identical, voltage differences will be developed across the input terminals and amplified along with the input signal.

Input voltage range (V_I). This is the maximum allowed input voltage for which proper operation can be maintained.

Maximum peak-to-peak output voltage swing. This is the maximum peak-to-peak output voltage that can be obtained without clipping—about a quiescent output voltage of zero.

Power supply rejection ratio. This is the change in input offset voltage for a given change in power supply voltage. Operational amplifiers are particularly good in this regard and do not require highly regulated power supplies.

Temperature effects. The various offset and bias currents and voltages are temperature dependent. The variations of these parameters with temperature are specified by temperature coefficients.

The 741 is internally compensated with a capacitor at the high-gain second stage to produce a pole in the transfer function at 6 Hz. This ensures stable operation under even 100 percent feedback conditions (of course, this gain in stability reduces the high-frequency response of the amplifier). Two extra terminals are available for connection to an external 10 kΩ potentiometer, the wiper arm of which is connected to the negative power supply. By adjusting the potentiometer, the input offset voltage can be eliminated. Additional features include short-circuit output protection and input overload protection. Single amplifiers come in at least four different packages: an 8-lead metal can, a 14-lead dual in-line package (DIP), an 8-lead mini-DIP, and a 10-lead *Flat Pak.* There is a commercial model—the 741C—that operates over a restricted temperature range, and a military model—the 741A—that operates over an extended temperature range. The 741 sells for about the price of a few 1/4 W composition resistors.

The OP07 has a very low input offset voltage, obtained by trimming thin film resistors in the input circuit. The low offset voltages usually eliminate the need for external null circuits. The OP07 also has low input bias current and high open-loop gain. It is a direct pin-for-pin replacement for the 741. Some examples of more specialized operational amplifiers are given in Table 6.21.

Because of their high gain, wide bandwidth, and small offsets, operational amplifiers with suitable feedback networks can perform a wide range of electronic functions. There are many excellent texts devoted especially to operational-amplifier applications.[4] Here we give only a list of the basic configurations. In all of these it is important to keep in mind the following design rules:

TABLE 6.21 SOME SPECIAL OPERATIONAL AMPLIFIERS

Low offset voltage	OP07
Low bias current	(Burr-Brown) OPA129
High voltage	(Intersil) CA3140 (MOSFET input/BJT output)
Low power consumption	(National) LF441
Single-power-supply operation	(National) LM10
Wide bandwidth	(National) LM318 (Burr-Brown) OPA606
Video preamplifier	733

1. The offset-voltage temperature coefficient must produce voltages much less than the input signal throughout the anticipated temperature range. The offset voltage can be eliminated with an external nulling circuit at any given temperature, but variations with temperature cannot be easily accommodated.

2. The voltages created by the bias currents flowing through resistances attached to the input terminals must be less than the signal voltage. For high output-impedance sources this may create a problem, but it can be solved by introducing an intermediate buffer stage with a high input impedance (so as not to load the source) and a low output impedance (to minimize the effects of input bias current). The temperature coefficient of the offset current must produce offset voltages much less than the signal voltage over the anticipated temperature range. To minimize bias-current effects, the resistances to ground from both input terminals should be the same.

3. The frequency response and compensation must be such that the amplifier will not break into oscillation. The 741 is internally compensated. High-frequency amplifiers lack internal compensation, but have extra terminals for external compensation.

4. The maximum rate of change of a sinusoidal output voltage $V_0 \sin \omega t$ is $V_0 \omega$ If this exceeds the slew rate of the amplifier, distortion will result. Slew rate is directly related to frequency compensation, and high slew rates are obtained with externally compensated amplifiers using the minimum compensation

capacitance consistent with the particular configuration. Data sheets should be consulted.

6.4.3 Operational-Amplifier Circuit Analysis

If ideal behavior can be assumed (as is reasonable for circuits in which the open-loop gain is much larger than the closed-loop gain), only the following two rules need be used to obtain the transfer function of an operational-amplifier circuit as shown schematically in Figure 6.58:

1. The voltage difference across the input terminals is zero.

2. The currents into the input terminals are zero (the inverting terminal is sometimes called the *summing point*).

Applying these to the generalized circuit of Figure 6.58, we see that the voltage at the + input is $v_{i2}Z_3/(Z_2 + Z_3)$, since there is no current flowing into the + input and Z_2, Z_3 form a voltage divider. From Rule 1, the voltage at the −input must also equal $v_{i2}Z_3/(Z_2 + Z_3)$. The current through Z_1 is then the potential difference across it divided by Z_1, or

$$\frac{1}{Z_1}\left[v_{i1} - v_{i2}\frac{Z_3}{(Z_2 + Z_3)}\right]$$

This current must be equal in magnitude and opposite in sign to the current from the output through the feedback element Z_f. This current i_f is given by the potential difference across Z_f divided by Z_f:

$$i_f = \frac{1}{Z_f}\left[v_{out} - v_{i2}\frac{Z_3}{(Z_2 + Z_3)}\right]$$

$$= -\frac{1}{Z_1}\left[v_{i1} - v_{i2}\frac{Z_3}{(Z_2 + Z_3)}\right]$$

Solving for v_{out}, we obtain

$$v_{out} = -v_{i1}\frac{Z_f}{Z_1} + v_{i2}\left(\frac{Z_3}{Z_2 + Z_3}\right)\left(1 + \frac{Z_f}{Z_1}\right)$$

Some useful configurations are illustrated in Figure 6.59. The *summer* circuit (Figure 6.59e) is an elaboration of the inverting amplifier, with the current into the summing point coming from two external sources (v_A and v_B) canceled by the current from the feedback loop, v_{out}/R_f. If $R_A = R_B = R$,

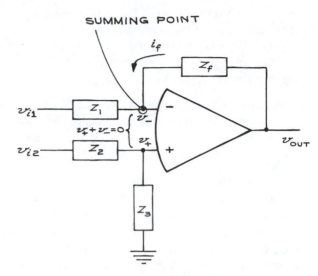

Figure 6.58   The general operational-amplifier circuit. Power supply and nulling terminals are not shown.

then $v_{out} = -(R_f/R)(v_A + v_B)$. If $R_A = R_B$, then v_{out} is equal to the negative of the weighted sum of voltages v_A and v_B—that is,

$$v_{out} = -\left[\left(\frac{R_f}{R_A}\right)v_A + \left(\frac{R_f}{R_B}\right)v_B\right]$$

Any number of external signal sources can be summed in this way.

A sinusoidal input to the *low-pass filter* (Figure 6.59f) gives $v_{out} = (j/\omega RC)v_{i1}$. For a nonsinusoidal input, the current into the summing point from v_{i1} is v_{i1}/R, because the inverting input is at ground (Rule 1). Often one speaks of the inverting input being at *virtual ground*. Although there is no direct connection from the inverting input to ground, the condition in which both inputs are at the same potential results in the inverting input assuming a potential of zero when the noninverting input is at ground potential. The current through the feedback capacitor of this circuit is i_f, and the voltage across the capacitor, v_{out} is equal to $(1/C)\int i_f dt$. Since i_f and v_{i1}/R are equal in magnitude and opposite in sign,

$$v_{out} = (1/C)\int i_f dt = -(1/RC)\int v_{i1} dt$$

and the circuit acts as an *integrator* of the input voltage. Integration times of several minutes or even hours are possible with high quality, low-leakage capacitors and operational amplifiers with small offset voltages and bias currents. For the *high-pass filter* (Figure 6.59g),

$$v_{out} = -RC dv_{i1}/dt$$

and the circuit acts as a differentiator.

The integrator and differentiator circuits are simple forms of active filters. More complex networks involving only capacitors and resistors can be used to obtain poles in the left half of the complex s plane, which in the past could only be obtained with inductors in passive filter networks. The operational amplifier allows the use of reasonable resistor and capacitor values even at frequencies as low as a few hertz. An additional advantage is the high isolation of input from output due to the low output impedance of most operational-amplifier circuits. The limitations of active filters are directly related to the properties of the operational amplifier. Inputs and outputs are usually single-ended and cannot be floated as passive filters can. The input and output voltage ranges are limited, as is the output current. Offset currents and voltages, bias currents, and temperature drifts can all affect active filter performance. A good introduction to active filters is *Active Filter Cookbook* by Don Lancaster.

The circuits in Figure 6.59b through g are examples of the use of operational amplifiers in analog computation. With these circuits, the mathematical operations of addition, subtraction, multiplication by a constant, integration, and differentiation can be performed. Multiplication or division of two voltages is accomplished by a logarithmic amplifier—an adder (for multiplication) or subtractor (for division)—and an exponential amplifier.

Both the *logarithmic* and *exponential* amplifiers rely on the exponential *I-V* characteristics of a *p-n* junction. When this junction is the emitter-base junction of a transistor, the collector current I_C with zero collector-base voltage is

$$I_C = I_0[\exp(qV_{BE}/kT) - 1]$$

where I_0 is a constant for all transistors of a given type and V_{EB} is the emitter-base voltage. Typically, I_0 is 10 to 15 nA for silicon planar transistors.

For $I_C \geq 10^{-8}$ A, the equation reduces to

$$I_C = I_0\exp(qV_{EB}/kT).$$

For the configuration shown in Figure 6.59h,

$$i_1 = v_{i1}/R = I_0\exp(qv_{out}/kT)$$

and $v_{out} = (kT/q)(\ln v_{i1} + \ln RI_0)$. Since k, T, q, R, and I_0 are constants, v_{out} will be proportional to v_{i1}. Practical logarithmic amplifiers can operate over three decades of input voltages. They do, however, need temperature-compensating circuits that generally require the use of matched pairs of transistors. In addition to their arithmetic use, logarithmic amplifiers are frequently used for transforming signal amplitudes that cover several orders of magnitude to a linear scale. Exponential (antilogarithmic) amplifier circuits are logarithmic circuits with the input and feedback circuit elements interchanged.

Logarithmic and *exponential* amplifiers are examples of nonlinear circuits—the output is not linearly related to the input. Operational amplifiers are used extensively in nonlinear circuits. Because the cutin voltage of simple diodes is a few tenths of a volt (0.2 V for Ge and 0.6 V for Si), they cannot be used to rectify millivolt-level ac voltages. A diode in the feedback loop of an operational amplifier (see Figure 6.59i) gives a rectifier with a cutin voltage equal to the diode cutin voltage divided by the open-loop gain of the amplifier. With a slight modification, the rectifier circuit can be converted to a *clamp* (see Figure 6.59j) in which the output follows the input for voltages greater than a reference voltage V_R, but equals V_R for input voltages less than V_R.

Comparator circuits are used to compare an input signal with a reference voltage level. The output is driven to V_{CC^+} or V_{CC^-} depending on whether the input is less than or greater than the reference level (see Figure 6.60a). Using an amplifier in such a saturated mode is very poor practice and practical comparators with internally clamped outputs less than $|V_{CC^+}|$ and $|V_{CC^-}|$ are available. The transition between the two output states can be accelerated by the use of positive feedback (see Figure 6.60b). Such a circuit is called a *Schmitt trigger* and finds wide application in signal conditioning when it is necessary to convert a slowly changing input voltage into an output waveform with a very steep edge. The price paid for the fast transition is *hysteresis*—where the threshold voltage for a positive transition of the output is different from the threshold voltage for a negative transition.

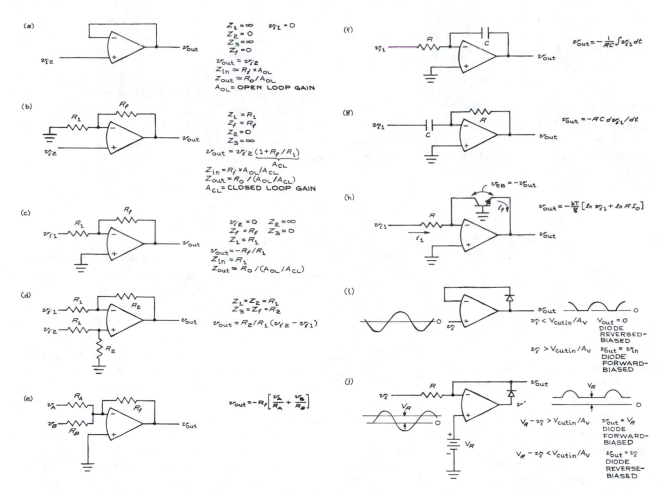

Figure 6.59 Operational-amplifier configurations: (a) follower; (b) follower with gain; (c) inverting amplifier; (d) subtractor; (e) summer; (f) low-pass filter (integrator); (g) high-pass filter (differentiator); (h) logarithmic amplifier; (i) precision rectifier; (j) clamp. The parameters are defined in Figure 6.59.

6.4.4 Instrumentation and Isolation Amplifiers

Instrumentation amplifiers are used for amplifying low-level signals in the presence of large common-mode voltages as, for example, in thermocouple and bridge circuits. The instrumentation amplifier is different from the operational amplifier because the feedback network of the instrumentation amplifier is integral to it and a single external resistor sets the gain. The output of the instrumentation amplifier is usually single-ended with respect to ground and is equal to the gain multiplied by the differential input voltage. Compared to an operational amplifier in the differential configuration with equivalent gain, the instrumentation amplifier has superior common-mode rejection ratio and input impedance.

Instrumentation amplifiers are often made up of three operational amplifiers as shown in Figure 6.61. Amplifiers A1 and A2 together provide a differential gain of $1 + 2R_1/R_G$ and a common-mode gain of unity. Amplifier A3 is a unit-gain differential amplifier. Amplifiers A1 and A2 operate as followers with gain, and resistors R_1 do not affect the input

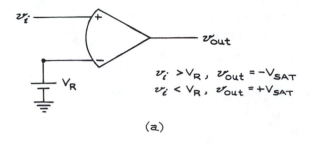

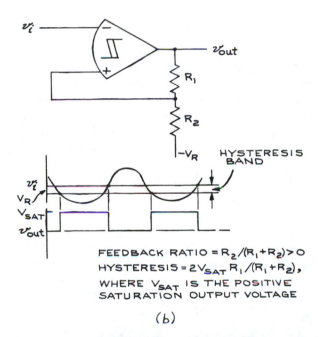

FEEDBACK RATIO $= R_2/(R_1+R_2) > 0$

HYSTERESIS $= 2V_{SAT} R_1/(R_1+R_2)$,

WHERE V_{SAT} IS THE POSITIVE
SATURATION OUTPUT VOLTAGE

(b)

Figure 6.60 (a) Simple comparator ($-V_{sat}$ and $+V_{sat}$ are the minimum and maximum voltages the amplifier can deliver; V_R is a constant reference voltage); (b) Schmitt trigger (comparator with positive feedback).

impedances. Because amplifier A3 is driven by the low output impedances of A1 and A2, the resistors R_0 can have relatively small values in order to optimize CMRR and frequency response while minimizing the effects of input offset currents.

The three-amplifier configuration is limited to common-mode voltages of ± 8 V in most applications. Typical high-performance instrumentation amplifiers are the INA110 (Burr-Brown), LH0038 (National Semiconductor), and

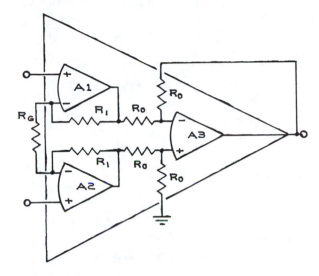

Figure 6.61 Three-amplifier instrumentation amplifier.

AD624 (Analog Devices). Typical specifications for high-performance amplifiers are given in Table 6.22.

Instrumentation amplifiers cannot be used when the common-mode voltage exceeds the power supply voltage. In these cases, an *isolation amplifier* is required. Isolation amplifiers are also useful when the input and output signals of an amplifier are referenced to different common voltage levels. It is often necessary, for example, to impose a low-voltage waveform on a dc high voltage. With an isolation amplifier, the modulation voltage can be derived from a source referenced to ground and the amplifier output can be referenced to the dc high voltage. The internal construction of the amplifier isolates the high voltage from the low-voltage input.

The basic isolation-amplifier configuration is shown in Figure 6.62. For the above example, V_{IN} is the modulation voltage referred to ground ($V_{CM}= 0$), and V_{ISO} is the high voltage modulated by V_{IN}. In cases where V_{CM} exceeds ± 10 volts, the input common terminal is not grounded and the common-mode voltage is referenced to the output common terminal. This effectively transfers V_{CM} across the isolation barrier and allows for V_{CM} voltages up to the maximum isolation rating of the amplifier.

The *isolation-mode rejection ratio* (*IMRR*) and *CMRR* are critical isolation amplifier parameters because of the

TABLE 6.22 TYPICAL HIGH-PERFORMANCE INSTRUMENTATION AMPLIFIERS: SPECIFICATIONS

	INA110	LH0038	AD624
Gain range	1–500	1–2000	1–1000
Gain nonlinearity (%)	0.001–0.01	0.001	0.001
Input impedance Ω	5×10^{12}	5×10^{6}	10^{9}
CMRR (dB)	90–110	120	130
Input offset-voltage drift (μV/°C)	2	0.25	0.25
Input bias current (nA)	0.050	50	50
Bandwidth (MHz)	2.5	2	25
Noise, referred to input (nV/$\sqrt{Hz}$)	10	6	4

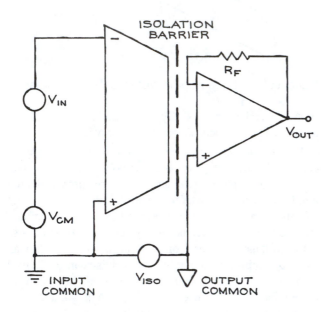

Figure 6.62   Isolation amplifier.

Isolation between input and output is achieved with optical couplers or transformers. The latter are more expensive, but have higher isolation voltage ratings and IMRR. An additional advantage of the transformer-coupled amplifiers is their built-in isolated power supply. The input and output stages of the optically coupled amplifiers must have separate power supplies. This means that the output-stage power supplies must be floated at the isolation voltage. Burr-Brown manufactures a full line of isolation amplifiers for medical and industrial applications. The 3450 and 3656 series are transformer-coupled, while the 3650 and 150100 series are optically coupled. Prices range from tens to hundreds of dollars, depending on specifications. Analog Devices produces a premium transformer-coupled isolation amplifier—the AD295—as well as the less expensive AD202.

6.4.5 Stability and Oscillators

For circuits composed only of passive elements, the poles of the transfer function will always lie on the left-hand side of the complex $s = \sigma + j\omega$ plane—that is, σ will always be less than zero. For circuits with active elements, this is not necessarily the case. If, for example, a fraction of the output signal of an amplifier is returned to the input in phase with the original input signal, an unstable situation will result and the output of the amplifier will increase with time. Analysis of such a circuit will show at least one pole to lie in the positive half of the complex s plane. Unwanted oscillations in amplifiers are due to the regenerative coupling of a fraction of the output signal

large common-mode and isolation voltages present. To calculate the effect of common-mode and isolation voltages on amplifier performance, it is necessary to add the product of IMRR and V_{ISO}, and that of CMRR and V_{CM} to V_{IN}. Typical values of IMRR and CMRR are 140 and 90 dB, respectively. For an IMRR of 140 dB, a 1 kV isolation voltage produces 0.1 mV at the input terminals of the amplifier. Other important parameters specific to isolation amplifiers are the maximum isolation voltage, isolation resistance, and isolation capacitance.

back to the input (see Figure 6.63). Stray capacitances between input and output are sufficient at high frequencies to couple the output signal back to the input. This can be minimized by physically having the output and input as far apart as possible. A copper shield separating the output and input is often also effective, and high-frequency circuits built on printed circuit boards should have generous ground-plane areas.

While the lack of stability is undesirable in amplifiers, sinusoidal *oscillators* depend on such instabilities for their operation. Consider an amplifier with gain A_o (it can be any one of the four already considered). If X_o represents the output signal (a current or voltage) and X_i the input signal (also a current or voltage), $X_o = A_o X_i$. For a simple amplifier $X_i = X_s$ (the source signal).

If a fraction, β, of the output is sent back to the input in such a way as to cancel part of the source signal, the new signal X_i is $X_s - \beta X_o$, and the gain of the amplifier with feedback is the output signal divided by the source signal:

$$A_f = \frac{X_o}{X_s} = \frac{X_o}{X_i + \beta X_o}$$

$$A_f = \frac{A_0}{1 + \beta A_o}$$

For $\beta > 0$, the feedback signal acts to cancel the signal from the source and $A_f < A_o$. This is called *negative* or *degenerative feedback* and is frequently used to stabilize amplifier gain and improve linearity and noise. For $\beta < 0$, the feedback signal adds to the source signal at the input and $A_f > A_o$. This is positive or regenerative feedback and is sometimes used to increase the gain of amplifiers at the expense of stability. If $\beta A_o = -1$, $A_f = \infty$ and the amplifier is unstable, with an output that breaks into oscillation or saturates.

In practice, both A_o and β are frequency-dependent and the response of such feedback amplifiers can be calculated. If only stability criteria are desired, however, it is only necessary to determine whether any poles lie in the right half of the complex plane.

If βA_o can be made equal to -1 for only a single frequency by means of a frequency-selective feedback network, one has an oscillator. There are a large number of frequency-selective circuits based on LC and RC networks. Quartz crystal oscillators are particularly stable

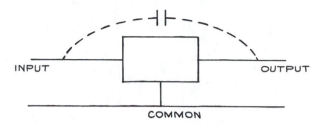

Figure 6.63   Unwanted feedback from stray capacitance.

and free of temperature effects. These devices work by the application of a potential difference across the faces of a quartz crystal, that then becomes deformed and oscillates at a resonant frequency determined by the size and orientation of the crystal planes with respect to the electrodes. Crystal oscillators with Qs of 10^4 to 10^6 at frequencies from tens of kHz to over 100 MHz are commercially available in DIP and SMT configurations.

6.4.6 Detecting and Processing Pulses

There is a large class of experiments, especially in nuclear and particle physics, that deals with the detection and analysis of single events. With the advent of detectors for low-energy photons, electrons, and ions, these techniques have spread to the fields of atomic and molecular physics and physical chemistry. Commercial instrumentation is abundant, but to specify and use it to best advantage requires knowledge of the basic properties of the detectors, amplifiers, and associated processing circuits.

Table 6.23 summarizes the properties of the more common types of detectors. The first part of the table allows one to estimate the output of a given detector for an arbitrary input. A 5 MeV alpha particle depositing all its energy in a silicon detector, for example, will create 1.4×10^6 electron-hole pairs. With a suitable bias across the detector, the charge can be collected in 0.1 to 10 ns. Similar considerations allow one to calculate the outputs of other detectors. The detectors can be represented electrically by the equivalent circuit of Figure 6.64. The choice of preamplifier, amplifier, and shaping and analysis circuits now depends on the information to be obtained from the

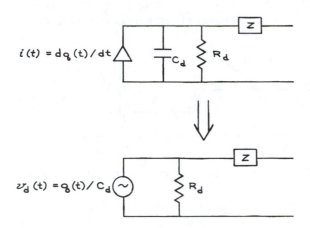

Figure 6.64   Equivalent circuit of a detector.

pulse. Figure 6.65 shows the arrangement of three general categories of experiments—counting, timing, and pulse-height analysis. Each one has different electronics requirements, but before considering each of these systems in detail it is worthwhile to examine the circuits commonly available for amplification and shaping.

Preamplifiers can be voltage-, current-, or charge-sensitive. Voltage preamplifiers are not generally used, because the voltage generated by most detectors depends on the capacitance of the detector (see Figure 6.66), that can change with bias voltage and other circuit parameters. Current preamplifiers have low input impedances and are designed to convert the fast current pulse from a photomultiplier or electron multiplier to a voltage pulse. Their sensitivity is specified as output voltage per input current. Charge-sensitive preamplifiers are the most commonly used preamplifiers because their gain is independent of the capacitance of the detector. Their sensitivity is specified in output voltage per unit input charge or in equivalent output millivolts per megaelectronvolt (MeV) deposited in a given solid-state detector.

A schematic diagram of a combined detector and charge-sensitive preamplifier is shown in Figure 6.66. Neglecting the effect of the coupling capacitor, the Laplace transform of the output voltage $\bar{v}_0(s)$ is

$$\bar{v}_0(s) = \left(\frac{\bar{Q}(s)}{C_f}\right)\frac{1}{1 + 1/R_f C_f s}$$

The function has a pole at $s = -1/R_f C_f$; the Bode plot is given in Figure 6.67. The high-frequency pole at ω_2 is a result of the preamplifier transfer function. The output waveform for a rectangular input pulse of duration τ will be a *tail pulse* of rise time τ, amplitude Q/C_f, and decay time $R_f C_f$.

For very small values of τ, the rise time of the output pulse will be determined by the amplifier itself. Practical amplifiers use high-stability NPO capacitors for C_f with values from 0.1 to 5 pF; the feedback resistor R_f is made as large as is consistent with the signal rate and leakage current. Noise in charge-sensitive preamplifiers is specified in terms of the energy resolution FWHM in keV. This can be converted to an equivalent charge noise with the factors in Table 6.23. To test charge-sensitive preamplifiers, the circuit of Figure 6.68 can be used. An input step voltage applied to the capacitor C_T results in an amount of charge equal to $C_T V_T$ deposited on the input of the preamplifier. For 10^{-15} coulombs with $C_T = 10$ pF, the amplitude of the input voltage pulse must be 10 mV. The 50 Ω resistor in the circuit is for termination purposes. The duration of the test input current will be equal to the rise time of the voltage input.

The output pulses from the preamplifier are generally very different in shape from the input signals. This is the case even with fast current preamplifiers. An important function of the amplifier and shaping circuits is to increase the amplitude of the preamplifier output signal and change its shape to minimize pulse overlap and increase the signal-to-noise ratio. To minimize pulse overlap, the duration of the pulses should be short compared with the average time between them. In more quantitative terms, the rms voltage level of the pulse train to be processed should be much less than the required voltage resolution of the system. If n is the pulse rate and $v(t)$ is the functional form of the pulse, then

$$v_{rms} = \left[n\int_0^\infty v(t)^2 dt\right]^{1/2}$$

For rectangular pulses of unit height and width τ,

$$v_{rms} = n^{1/2}\tau^{1/2}$$

The long-tail pulse from a charge-sensitive preamplifier can be shortened with a differentiating circuit (Figure

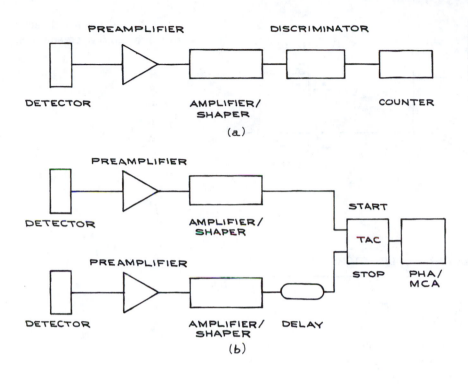

Figure 6.65   Three categories of pulse experiments: (a) counting, (b) coincidence, and (c) pulse-height analysis.

6.69a). The circuit reduces the fall time from $R_f C_f$ to RC. A disadvantage of simple differentiation is undershoot, which appears on the trailing edge of the output pulse. This can be substantially eliminated by modifying the transfer function of the differentiating circuit to include a zero that exactly cancels the pole in the preamplifier transfer function (Figure 6.69b). The Laplace transform of the transfer function of Figure 6.69b is

$$\frac{s + 1/R_2 C_1}{1 + (R_1 + R_2)/R_1 R_2 C_1}$$

If $R_2 C_1$ is set equal to $R_f C_f$ from the preamplifier, undershoot can be eliminated. This technique of *pole-zero compensation* is commonly used in pulse shaping.

For pulse-height analysis, a Gaussian waveform has the best noise characteristics.[5] Such a shape can be obtained from a tail pulse by a single differentiation and an infinite number of cascaded integrations with integration time constants all equal to the differentiation constant. In practice, four integrations give a shape very close to Gaussian. Bipolar pulses can be obtained by double differentiation of a tail pulse with an intermediate integration (Figure 6.70). An important property of the bipolar pulse is that the zero crossing is independent of pulse height and always occurs at a fixed time after its start, determined by the time constant. When used with zero-crossing discriminators, such pulses permit reliable

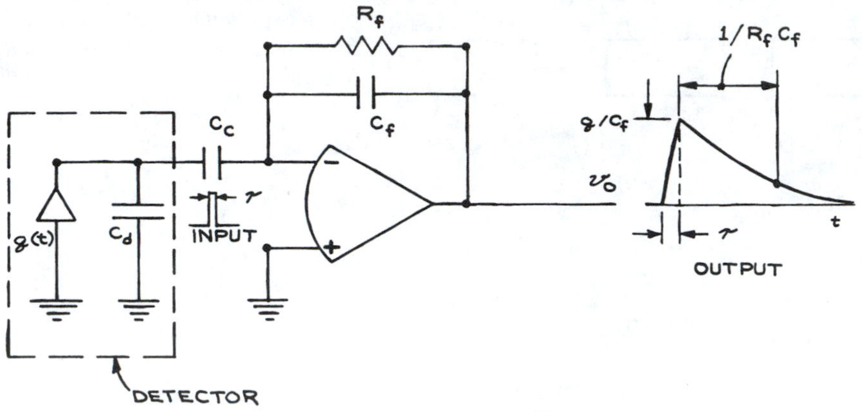

Figure 6.66 Combined detector and charge-sensitive preamplifier.

TABLE 6.23 PARTICLE DETECTORS

Particle Detector	Output Signal Level	Charge-Collection Time

Semiconductor:
  Ge
  Si
	2.8 eV / electron hole pair
	3.5 eV / electron hole pair
		0.1–10 ns[a]

Photomultiplier	10^6–10^7 electrons/photoelectron	1 ns
Electron multiplier	10^6–10^8 electrons/incident electron	0.5–1 ns
Microchannel plate	10^3–10^4 electrons/incident electron	0.1 ns

Scintillator:
  Plastic
  Alkali halide
	3 keV/photon	0.1 – 10 ns
		1 μs

Gas-filled tube	25–35 eV/ion pair	0.01–5 μs

[a]For narrow depletion depth.

timing to be accomplished with long-decay-time pulses. Bipolar pulses, however, have poorer noise characteristics than unipolar Gaussian pulses.

An alternative to *RC* shaping is delay-line shaping. The equivalent of unipolar and bipolar pulses can be obtained with single- and double-delay-line circuits (Figure 6.71). Advantages are the preservation of fast rise times, sharp trailing-edge clipping, and greatly reduced base-line shifts. These advantages are offset by the poor noise properties of the pulses compared with *RC* shaping. Delay-line shaping is mostly confined to timing applications.

The shaping circuits discussed in the previous paragraph are generally incorporated in the variable-gain main amplifier that follows the preamplifier. While noise is a consideration with such amplifiers, it is not as critical as with preamplifiers because of the larger signal levels. Important considerations are linearity and fast overload recovery. To reduce base-line shifts from pulse pileup, main amplifiers often incorporate a *base-line restorer circuit*. The principle of operation is illustrated in Figure 6.72. A noninverting amplifier with gain $\overline{H}(s)$ is placed in the negative-feedback loop of an amplifier with gain $\overline{G}(s)$.

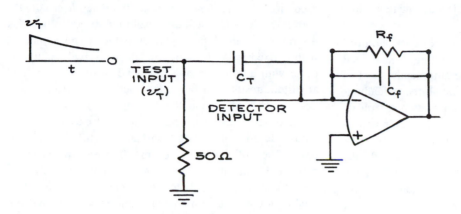

Figure 6.67 Test circuit for charge-sensitive preamplifiers.

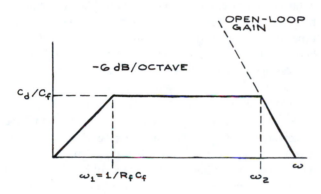

Figure 6.68 Bode plot for a charge-sensitive preamplifier.

The Laplace transform of the transfer function for the system is

$$\frac{\overline{G}(s)}{1 + \overline{G}(s)\overline{H}(s)}$$

If both $\overline{G}(s)$ and $\overline{H}(s)$ are themselves transforms of single-pole transfer functions, given by

$$\overline{G}(s) = \frac{A}{1 + T_1 s}$$

and

$$\overline{H}(s) = \frac{K}{1 + T_2 s}$$

the overall transfer function is

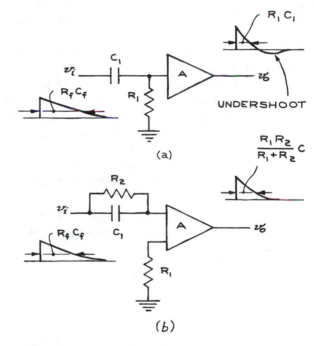

Figure 6.69 Differentiating circuits: (a) without pole-zero compensation; (b) with pole-zero compensation.

$$\frac{A(1 + T_2 s)}{AK + (1 + T_1 s)(1 + T_2 s)}$$

At high frequencies ($s \gg 1$) the transfer function reduces to $\overline{G}(s)$, while for dc ($s = 0$) the function is $1/K$. The overall effect of the circuit is to amplify high frequency-component pulses while attenuating low frequency base-line shifts.

For counting applications, the principal considerations are the pulse rate and discriminator setting. The shaped pulses from the amplifier should be sufficiently short in duration to avoid pileup and base-line shifts. A leading-edge discriminator can be used with a unipolar pulse. Since such units are based on regenerative (positive-feedback) comparator circuits that have hysteresis, the decay of the input pulse must not pass through the recovery voltage level until the triggered state is firmly established. When bipolar pulses are to be counted, zero-crossing discriminators are appropriate.

Timing systems require that the leading-edge information from the detector pulse be preserved as faithfully as possible. Timing errors due to time variations in the processed pulses relative to the input pulse, long-term drift from component aging, and noise must be considered. Wide-band current preamplifiers and fast timing amplifiers with short integrating and differentiating shaping circuits or delay-line shaping should be used. When using wide-band amplifiers, precautions should be taken to electrically terminate all connections properly and avoid stray capacitances that can form positive-feedback paths and produce instability. When the detector signal itself is of sufficient amplitude—as it is for the output of a photomultiplier tube used with a scintillation detector—a timing signal can be taken directly from the detector without further amplification or shaping. Leading-edge and constant-fraction discriminators are most commonly

used for timing. The *constant-fraction discriminator* currently provides the best resolution times. With this discriminator, the input signal is delayed and combined with a fixed fraction of the undelayed signal. A bipolar pulse with a zero crossing independent of rise time and amplitude results can be used to activate a zero-crossing detector. In Figure 6.65b, the delay inserted in the *stop* line of the system is to ensure that the *stop* pulse to the time-to-amplitude converter (TAC) will always follow the *start* pulse. The TAC converts the time difference between the *start* and *stop* input pulses to a single pulse of an amplitude proportional to the time difference. High quality, low-attenuation coaxial cable makes an excellent delay line, and lumped-parameter delay lines are also commercially available. Timing information is usually obtained by processing the output of the TAC with a pulse-height analyzer (PHA). A more direct method is with a time-to-digital converter that converts the time difference between the *start* and *stop* pulses to a digital address for processing. Depending on detector characteristics, time resolution of a few hundred pico-seconds can now be obtained with standard commercial units.

Particle spectroscopy experiments rely on the preservation of the pulse-amplitude information from the detector because the amplitude is a direct measure of the particle energy. Optimum amplitude resolution is obtained by operating the preamplifier and amplifier in their linear ranges, reducing sources of noise and base-line shifts from pileup, and using semi-Gaussian pulse shapes as inputs to the PHA. The function of the PHA is to convert the pulse

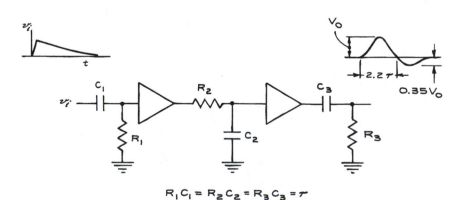

Figure 6.70 Bipolar pulse shaping with CR-RC-CR circuits.

$$R_1 C_1 = R_2 C_2 = R_3 C_3 = \tau$$

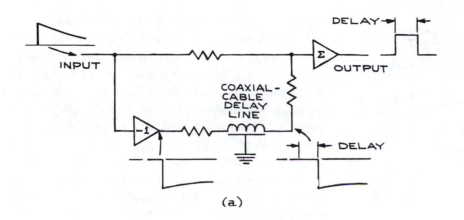

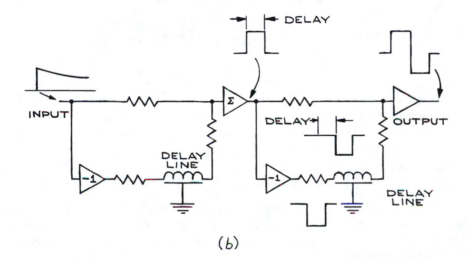

Figure 6.71　(a) Unipolar and (b) bipolar pulse shaping with delay lines.

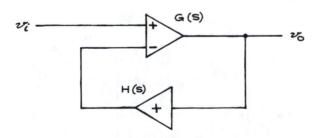

Figure 6.72　A base-line restorer circuit

amplitude to a binary number. This number can then be used to address a memory location in a multi-channel analyzer (MCA) or PC. In operation, each pulse encoded by the PHA increases the contents of the memory location specified by the height (amplitude) of the pulse by one. Nonlinearities in the preamplifier and amplifier can be both differential and integral, illustrated in Figure 6.73. The *differential nonlinearity* is the ratio of amplifier gain at a specified input level to the gain at a reference input level. The *integral nonlinearity* is the maximum vertical deviation of the real gain curve from the ideal straight-line gain curve, expressed as a percentage of maximum output. Such nonlinearities also occur in the circuits of the PHA and must be taken into consideration for high-resolution work.

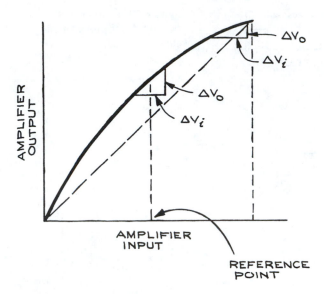

Figure 6.73 Amplifier linearity.

The manufacturers of pulse-processing equipment include application notes and product guides with the catalog descriptions of their instruments. These are generally very helpful and should be read thoroughly before deciding on a system and selecting components.

6.5 POWER SUPPLIES

The design of regulated power supplies follows a standard configuration (see Figure 6.74). Raw ac from the outlet is converted, usually with a transformer, to the desired level. This converted ac is rectified to produce unregulated dc. The unregulated dc then becomes the input to a regulator circuit, whose output is the desired regulated dc voltage.

6.5.1 Power-Supply Specifications

Power supplies are specified by the maximum current, maximum voltage, and maximum power they can deliver. It is not always the case that maximum current can be delivered over the complete voltage range of the supply, nor can that maximum voltage be obtained at all currents. This situation is illustrated in Figure 6.75. Maximum

current can only be obtained over a limited range of voltage without exceeding the power rating of the supply.

The change in output voltage per unit change in input ac voltage is called the *line regulation,* while the change in output voltage per unit change in load or output current is called the *load regulation.* Both are generally specified as a percentage. Many power supplies can be operated in a *constant-current mode* where the output is a current essentially independent of input and output voltage. Regulation specifications then apply to the output current. The load regulation can be translated into an equivalent dynamic-output resistance, since the relative change in load is the negative of the relative change in current ($\Delta R/R = -\Delta I/I$):

$$\frac{\text{load}}{\text{regulation}} = \frac{\Delta V/V}{\Delta R/R} = \frac{\Delta V/V}{\Delta I/I} =$$

$$\frac{|\Delta V/\Delta I|}{V/I} = \frac{R_{\text{dynamic output}}}{R_{\text{load}}}$$

As an example, if a filament power supply produces 3 A at 5 V, and the load regulation is 0.1 percent, then the dynamic output resistance is

$$R_{\text{dynamic output}} = \frac{1 \times 10^{-3} \times 5 \text{volts}}{3 \text{amps}} = 1.67 \times 10^{-3} \Omega$$

Transient response and *recovery time* relate to the ability of the power supply to recover from sudden changes in load or line voltage at the stated operating point. If the line and load fluctuations are rapid, the power

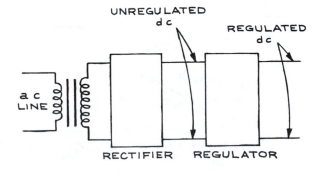

Figure 6.74 Block diagram of the regulator circuit in a power supply.

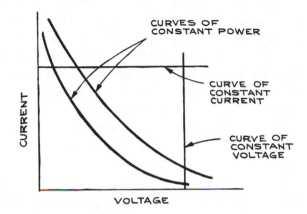

Figure 6.75   Current, voltage, and power relations for a power supply.

supply regulating circuits may not be able to keep up with them and the resulting regulation will be considerably worse than specified. Generally, the better the regulation, the slower the response to changes of line and load conditions. The regulation may also depend on the output power level, a point worth checking in the specifications.

As well as maintaining a constant average dc level, instantaneous values of the output voltage should not deviate appreciably from the average. Deviations come from ripple at twice the line frequency for dissipative regulators, at the switching frequency for switching regulators, and from regulating circuit transients. Specification of the rms ripple gives an indication of the effectiveness of the filtering circuits. It gives no indication, however, of the presence of short duration, large-amplitude voltage spikes on the output. A maximum peak-to-peak noise specification is used to describe this kind of instantaneous voltage deviation. As with regulation, ripple and noise may depend upon the power level.

Even with a constant line voltage and constant load, the output voltage can change if the temperature changes. This is generally indicated by a *temperature-coefficient* specification, which gives the percentage change in output voltage per degree change in temperature about a specified temperature. For laboratory applications, this is usually not a critical specification. When working in extreme environments, however, it may be significant (for highly compact units that depend on forced-air cooling or free-

convection cooling, adequate space must be provided in the mounting to prevent overheating).

Though often not included in the specification list, radio frequency (rf) noise can be an important consideration for supplies used in the vicinity of sensitive wide-band amplifiers and when trying to extract a weak signal from noise. Some types of power supplies produce more rf noise than others because of the design of the regulating circuit. Switching regulators with SCRs are particularly poor in this regard.

Other factors to consider in a power supply are *bipolar operation* (can both the negative and positive terminals be grounded to give positive and negative output voltages, respectively?) and insulation from ground if floating the supply is anticipated. Operation is simplified by front-panel meters indicating both output voltage and current, overload reset switches, and a convenient output terminal configuration and location.

A single output power supply can never be used to supply two voltages, one negative and one positive with respect to ground. When it is necessary to have such voltages as, for example, with operational-amplifiers, two separate supplies must be used.

6.5.2 Regulator Circuits and Programmable Power Supplies

A block diagram of a dissipative regulator circuit in a power supply is illustrated in Figure 6.76. The sensing resistors R_1 and R_2 are across the output of the power supply. A fraction $R_2/(R_1 + R_2)$ of the output voltage is sent to a difference amplifier, where it is compared with a reference voltage, and the amplified difference used to control the *pass element*—which in this case is a transistor in the common-collector configuration. Depending on the output of the difference amplifier, the pass-element output voltage will adjust itself to a value just sufficient to sustain the required input from the difference amplifier. If for some reason the output voltage rises, the output of the difference amplifier will decrease and the pass-element output will decrease. The opposite occurs for a falling output. The series dissipative regulator is inefficient because the difference in voltage between the raw dc input and regulated dc output falls across the pass element that dissipates power equal to this voltage drop times the output current.

More efficient power supplies use circuits to control the *duty cycle* of the pass element. In these circuits, the pass element (a transistor, SCR, or other solid state device) acts as a switch and is either *on* (no voltage drop across it) or *off* (no current through it). The output voltage from such a regulator is

$$V_{out} = V_{in} \frac{t_{on}}{t_{on} + t_{off}}$$

where t_{on} is the time for which the pass element is on and t_{off} is the time for which it is off in each cycle. The optimum switching frequency for switching regulators is between 20 and 100 kHz. At these high frequencies, rfi can cause problems.

In practice, regulator circuits can be very complex, with high-gain difference amplifiers or comparator preregulators, multiple pass elements, overload and overvoltage protection, and temperature compensation. The above description, however, is sufficient for an understanding of remote programmable power supplies. If R_1 and R_2 in Figure 6.76 are replaced by a potentiometer, a change in its setting will result in a change in the regulated output voltage. This is what occurs when one changes the dial settings on the front panel of a variable power supply. The internal potentiometer can clearly be replaced by an external potentiometer (*resistance programming*)—or even more directly—by an external voltage source (*voltage programming*). The addition of such features requires very little modification to the basic power supply regulator circuit, and they are often available at no cost or as low-cost options. Remote programming can be particularly useful in control operations, where a digital code from a computer is changed to a voltage by a digital-to-analog-converter (DAC), and this voltage is used to set the output of a power supply.

Overvoltage protection is an important feature in a power supply. With such protection, one is assured that no failure will result in a high voltage appearing across the output terminals of the supply. *Crowbar* circuits using SCR switches are often used for overvoltage protection. Such circuits short the output terminals of the supply when the output voltage exceeds a preset value. The resulting short circuit current causes a fuse to blow or a circuit breaker to open, shutting down the power supply until the fault that caused the overvoltage is corrected. The speed of response should be included in the specifications because it is essential that the protection circuit respond quickly.

Another built-in safety feature found on power supplies is *foldback current limiting*. When the output current exceeds a specified limit, it is automatically reduced by

Figure 6.76 Block diagram of a regulated power supply.

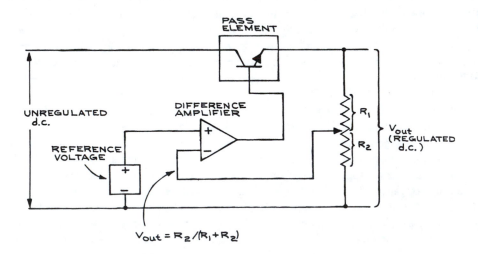

$$V_{out} = R_2/(R_1+R_2)$$

such circuits to a low level and maintained there for a specified time before being reset. Current-limiting circuits are triggered by temperature-sensitive or power-dissipating elements.

In addition to variable power supplies there are a vast number of fixed-voltage power supplies on the market. When working with logic circuits and operational amplifier circuits, only fixed power supply voltages are needed.

Most fixed-voltage supplies incorporated in other units are designed to operate from the ac line, although there are a number of dc-to-dc converters. The least expensive configuration is the card power supply (Figure 6.77a). All components are on a single printed circuit board with edge contacts. The ac input must be brought to one set of contacts and the dc output taken from another. For routine use, the card supply should be mounted in a case or on a rack panel with an on-off switch, a pilot light, a fuse, and output terminals. This requires additional time and expense, which may cancel the initial cost savings from purchasing such a supply.

Frame-mounted supplies (see Figure 6.77b) are similar to the card supplies in that all external connections, switches, and output terminals are missing. The metal frame facilitates mounting, however, and affords some protection to the components. Rather than PCB contacts, frame supplies usually have screw-type barrier connectors.

Fixed-voltage supplies also come in enclosed cases and in encapsulated units suitable for direct connection to a printed circuit board (see Figure 6.77c). Many of the so-called fixed-voltage supplies can be adjusted over a 5 to 10 percent range about the nominal output voltage value. Multiple fixed-voltage units with two and even three voltages are also common. It is often useful to have ±15 V and +5 V available from a single power supply when both linear and digital circuits are used together.

Manufacturers produce a wide range of supplies with different voltages, power ratings, and mechanical configurations. It is therefore rarely worthwhile to construct a power supply. A possible exception lies in the use of integrated-circuit modules. There are two general types: fixed and variable output. The fixed-output units are three-terminal devices—unregulated dc is applied across the input and common terminals, and regulated dc at the appropriate voltage appears between the output and common terminals. A representative unit of this type is the LM109 5 V, 1.5 A

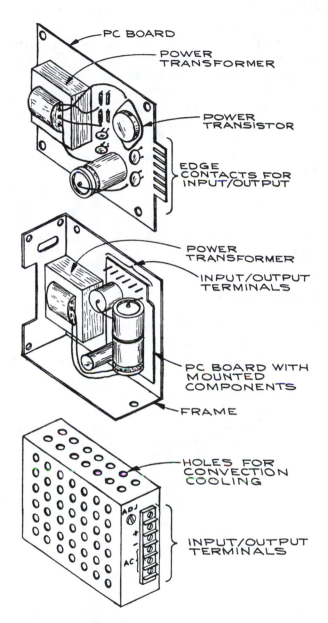

Figure 6.77 Fixed-voltage power supplies; (a) card supply; (b) frame supply; (c) enclosed case.

monolithic regulator. The regulator combines on-chip thermal shutdown with current limiting, an on-chip series pass transistor, and a stable internal voltage reference. Further details about the LM109 and its use in other applications are available from the manufacturer.[6]

The LM105 is an example of a variable voltage integrated-circuit regulator. For this unit, the output is set by the ratio of two external resistances. The maximum output voltage is 40 V, and the maximum output current is 20 mA. When an external pass transistor is added to the basic regulator circuit, the load current can be increased to 500 mA. The LM105 can also be used as a switching regulator. Data and application sheets should be consulted when using such regulators.

Useful power supply circuits with integrated regulators are illustrated in Figure 6.78. Fixed-voltage units are made by several manufacturers at several fixed voltages from 4 to 20 V. Variable units are available for both positive and negative voltages.

6.5.3 Bridges

Bridges are used to measure the electrical properties of a circuit element by comparison with a similar element. The Wheatstone bridge, developed in the early 19th century, compares the potentials from two voltage dividers connected in parallel across a voltage source as shown in Figure 6.79a.

Taking the difference between the potentials at the junction of R_1 and R_2 and R_3 and R_4,

$$V_{\text{out}} = V_{\text{in}} \left[\frac{R_3}{R_1 + R_2} - \frac{R_4}{R_2 + R_4} \right].$$

When

$$V_{\text{out}} = 0, \qquad \frac{R_1}{R_2} = \frac{R_3}{R_4},$$

the null condition. In the standard configuration, R_1 and R_2 are fixed, R_4 is the resistance to be measured, and R_3 is varied until a null or balance condition is achieved. It is important to note that when at null the relation between the resistances is independent of V_{in}. Null-type measurements require a means for varying one element of the bridge, and this is usually accomplished with a system that incorporates feedback.

Bridges in most transducer applications are operated *off-null*, or out of balance. For the simple bridge in Figure 6.79b, three of the resistors are fixed at value R while the

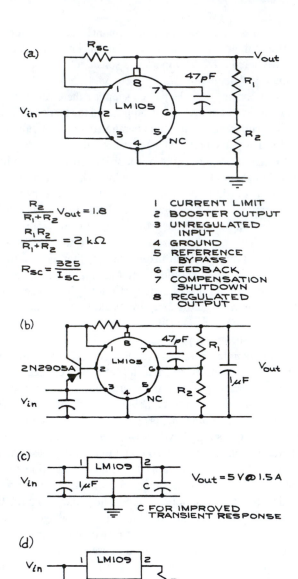

$$\frac{R_2}{R_1 + R_2} V_{\text{out}} = 1.8$$

$$\frac{R_1 R_2}{R_1 + R_2} = 2 \text{ k}\Omega$$

$$R_{sc} = \frac{325}{I_{sc}}$$

1 CURRENT LIMIT
2 BOOSTER OUTPUT
3 UNREGULATED INPUT
4 GROUND
5 REFERENCE BYPASS
6 FEEDBACK
7 COMPENSATION SHUTDOWN
8 REGULATED OUTPUT

Figure 6.78 Examples of circuits with the LM109 fixed-voltage regulator and the LM105 variable-voltage regulator.

one to be measured, R_x, deviates from R by xR—where x is a unitless variable. The output voltage V_{out} is given by

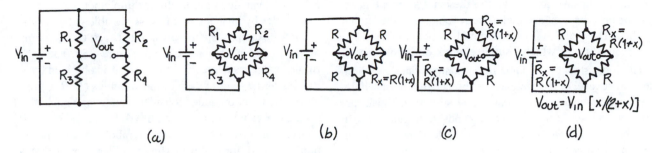

Figure 6.79 (a) Wheatstone bridge; (b) off-null bridge with one variable element; (c) off-null bridge with two variable elements; (d) off-null bridge with a compensating variable element.

$$V_{out} = V_{in}\left[\frac{R}{2R} - \frac{R_x}{R + R_x}\right] = -\frac{V_{in}}{4}\left[\frac{x}{1 + x/2}\right]$$

For $x/2 \ll 1$, $V_{out} \approx -V_{in}(x/4)$.

As an example, let $x = 0.01$, then $V_{out} = -V_{in}(0.01/4)$, so that a 1 percent change in R_x gives a 0.25 percent change in V_{out}. The sensitivity of the bridge, defined as the change in output voltage for a given change in input voltage, is 2.5 mV/V. Bridge sensitivity can be doubled by adding a second identical variable element opposite the first as shown in Figure 6.79c.

While the output of such a configuration is doubled, the nonlinearity remains. The addition of two more variable elements with resistances that change in opposite ways to the original elements results in a bridge with output equal to the fractional change in resistance times the reference voltage. This is shown in Figure 6.79d. Bridges of this type are used with two-element strain gauges and piezo-resistive transducers. The transducers are arranged so that there are complementary changes in resistance for sensors in adjacent bridge arms.

Another way to linearize the output of a bridge circuit is by inserting an operational amplifier to null the bridge by adding a voltage in series with the variable element R_x that is equal in magnitude and opposite in sign to the incremental voltage across R_x. This gives an output that is linear in x for large values of x.

In bridge circuits that are operated out of balance, the output is proportional to the reference voltage V_{in}. A stable reference voltage can be derived from an IC reference source followed by an operational amplifier stage. Analog Devices and Maxim Integrated Products manufacture signal-conditioning power supplies that have programmable outputs and include amplification and filtering.

Because neither side of the bridge from which the voltage is measured is at ground, instrumentation amplifiers are the preferred devices for reading bridge outputs. These amplifiers have the necessary differential inputs, high common-mode rejection ratios, high input impedances, and resistance programmable gain.

6.6 DIGITAL ELECTRONICS

Digital systems are based on circuit elements (usually transistors) operated in a way such that they exist in only one of two states. This is in contrast to analog systems, where the outputs are continuous functions of the input variables. Combinations of two-state, or *binary,* devices can perform arithmetic and logic operations of any degree of complexity.

6.6.1 Binary Counting

With a binary system, it is usual to call the states 0 and 1. Combinations of 0s and 1s can then be used for counting. Some schemes are given in Table 6.24. Note that the *least significant bit* (LSB) or column is 2^0, the next 2^1, and the last or *most significant bit* (MSB) is 2^4. In other words, the binary number system is merely a base-2 system, while the decimal system is base 10. Other systems derived from the binary system are the octal (base 8) and the hexadecimal (base 16) as detailed in Table 6.24. In the binary system, the digit 2 does not appear, and in the octal system 8 does not appear. The hexadecimal system substitutes the letters A, B, C, D, E, and F for the decimal numbers 10, 11, 12, 13, 14, and 15, for which a single symbol does not exist.

The logic state of the logic device output can depend on the voltage level or the presence or absence of a pulse within a given time window. A voltage-level system in which the most positive voltage level corresponds to logical 1 is a *positive-logic* system. One in which the least positive level corresponds to logical 1 is a *negative-logic* system.

TABLE 6.24 SYSTEMS OF NUMERATION

Decimal	Binary	Octal	Hexadecimal
0	0000	00	0
1	0001	01	1
2	0010	02	2
3	0011	03	3
4	0100	04	4
5	0101	05	5
6	0110	06	6
7	0111	07	7
8	1000	10	8
9	1001	11	9
10	1010	12	A
11	1011	13	B
12	1100	14	C
13	1101	15	D
14	1110	16	E
15	1111	17	F

MSB   LSB

6.6.2 Elementary Functions

The basic logic gates, along with the corresponding relationships between input and output variables, are given in Table 6.25 in *truth-table* form.

6.6.3 Boolean Algebra

The rules for manipulating logic expressions based on binary logic were formulated in the nineteenth century by George Boole. The laws and identities are given in Table

6.26. Boolean algebra is useful for implementing complex truth tables by means of the fewest possible gates. Consider, for example, the three-variable (A, B, C), two-function (X, Y) truth table at the top of Figure 6.82. The Boolean algebra expression for X and Y is obtained by looking at the state of each of the variables (A, B, C) for which the function under consideration (X or Y) is in the 1 state. For example, $X = 1$ when $A = 0$, $B = 0$, and $C = 1$. By convention, a bar over a variable symbol indicates negation, so that if $A = 0$, $\overline{A} = 1$. The coincidence of 1s at $\overline{A}$, $\overline{B}$, and C to give $X = 1$ requires an AND operation, which is the first term in the full expression for X in Figure 6.80. The other terms are derived from the truth table in the same way, and they are all connected together by OR operations. After simplification by application of the laws in Table 6.26, the expressions can be implemented with elementary gates. The gates for the function X are shown at the bottom of Figure 6.80.

Another way of simplifying truth tables is by making a logic map called a Karnaugh map. The map for the truth table of Figure 6.80 is shown in Figure 6.81. The squares corresponding to the variables for which X and Y are 1 are shaded. Contiguous shaded squares corresponding to 0 and 1 for a single variable, say C, with the other variables constant, show that the value of the function is independent of the state of C. In the map for Y, therefore, it can be seen that the squares for $\overline{A} \cdot B$ are shaded for C and $\overline{C}$. It therefore follows that the state of Y is not dependent on the variable C for these cases.

6.6.4 Arithmetic Units

After elementary gates, the next simplest logic units are those that perform elementary binary arithmetic operations. Some common arithmetic units are given in Table 6.27. The truth table for the *half adder* follows the rules for binary addition with C the low-order bit of the sum and D the high-order bit. The *full adder* is an elaboration of the half adder with an extra input. When adding two binary numbers of more than one-bit each, provision must be made for the carry operation. This is the function of the C_{n-1} terminal, which accepts the carry bit from the result of adding the preceding lower-order bits in the string (thus the designation C_{n-1}). The results from the addition of the n-

TABLE 6.25 LOGIC GATES

Name	Symbol	Truth Table	Boolean Expression
OR		A B Y 0 0 0 0 1 1 1 0 1 1 1 1	$Y = A + B$
AND (coincidence)		A B Y 0 0 0 0 1 0 1 0 0 1 1 1	$Y = A \bullet B$
NOT (inverter)	Negation Symbol	A Y 0 1 1 0	$Y = \bar{A}$
NOR		A B Y 0 0 1 0 1 0 1 0 0 1 1 0	$Y = \overline{A + B}$
NAND		A B Y 0 0 1 0 1 1 1 0 1 1 1 0	$Y = \overline{A \bullet B}$
XOR (exclusive OR, anti-coincidence)		A B Y 0 0 0 0 1 1 1 0 1 1 1 0	$Y = (B \bullet \bar{A}) + (\bar{B} \bullet A)$
XNOR (exclusive, NOR, equivalence)		A B Y 0 0 1 0 1 0 1 0 0 1 1 1	$Y = \bar{A} \bullet \bar{B} + A \bullet B$

order bits appear at terminals C_n and S_n. The *digital comparator* compares the magnitude of the digital signals at input A_n and B_n and activates output C_n, D_n, or E_n depending on whether $A_n > B_n$, $A_n < B_n$, or $A_n = B_n$. Such elements can be connected together to compare two *n*-bit binary numbers. The parity of a binary number is defined as even (0) if there is an even number of 1s and odd (1) if there is an odd number of 1s. The parity code is often used to check for transmission errors in bit strings. By sending a parity bit with an *n*-bit word and testing the received word against the parity bit, single bit errors can be detected. This

A	B	C	X	Y
0	0	0	0	1
0	0	1	1	0
0	1	0	1	1
0	1	1	0	1
1	0	0	1	0
1	0	1	0	0
1	1	0	0	1
1	1	1	1	0

$$X = (\overline{A} \cdot \overline{B} \cdot C) + (\overline{A} \cdot B \cdot \overline{C}) + (A \cdot \overline{B} \cdot \overline{C}) + (A \cdot B \cdot C)$$
$$Y = (\overline{A} \cdot \overline{B} \cdot \overline{C}) + (\overline{A} \cdot B \cdot \overline{C}) + (\overline{A} \cdot B \cdot C) + (A \cdot B \cdot \overline{C})$$

WHICH SIMPLIFY TO:

EXCLUSIVE OR

$$X = (\overline{A} \cdot \overline{B} + A \cdot B) \cdot C + (\overline{A} \cdot B + A \cdot \overline{B}) \cdot \overline{C}$$
$$Y = \overline{A} \cdot B + (\overline{A} \cdot \overline{B} + A \cdot B) \cdot \overline{C}$$

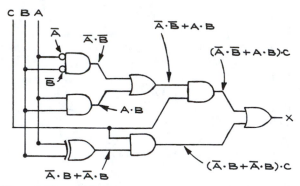

Figure 6.80   Example of a truth table, the equivalent Boolean algebraic expression, and implementation with elementary gates.

is illustrated for a four-bit word with the *parity generator/checker* in Table 6.27.

6.6.5 Data Units

For data acquisition and transmission, *decoders*, *encoders*, *multiplexers*, and *demultiplexers* are commonly used. The *read-only memory* (ROM) is a combination of decoder and encoder and is frequently used to implement complex truth

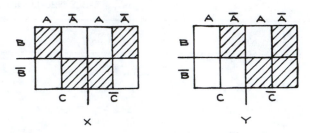

Figure 6.81   Karnaugh map of the truth table of Figure 6.80.

tables. *Programmable* and *erasable* ROMs, called PROMs and EPROMs, can have their internal logic modified for the particular application of the user. Table 6.28 gives the truth tables and input-output terminal arrangements for some simple data units.

6.6.6 Dynamic Systems

The logic units described so far are considered to be *static* because the output levels depend only on the input levels at the time of observation. The previous history of the signals at the different inputs is irrelevant. With *dynamic systems*, output signal levels depend on the history of the signal levels at the input terminals. The truth tables for dynamic systems must specify the previous states of the inputs and outputs in order for the state of a particular output to be fully determined. Subscripts are used to differentiate the time sequence of the output states. Table 6.29 lists a number of *flip-flops* that are the basic building blocks in dynamic systems. Flip-flops can be strung

TABLE 6.26 BOOLEAN ALGEBRA

Laws	Identities
Commutative	
$A + B = B + A$	$A + A = A$
$A \bullet B = B \bullet A$	$A \bullet A = A$
Distributive	$A + 1 = 1$
$A \bullet (B + C) = A \bullet B + B \bullet A$	$A \bullet 1 = A$
DeMorgan	$A + 0 = A$
$\overline{A \bullet B \bullet C} = \overline{A} + \overline{B} + \overline{C}$	$A \bullet 0 = 0$
$\overline{A + B + C} = \overline{A} \bullet \overline{B} \bullet \overline{C}$	

together to make *shift registers* and *counters*. A shift register is a series of *J/K flip-flops* with the Q and $\overline{Q}$ outputs of each stage connected to the J and K inputs of the next stage (see Figure 6.82). Data are entered serially in the first unit, connected as a *D flip-flop,* and move from unit to unit at each clock transition. Shift registers can be used as counters, serial-to-parallel data converters using the parallel outputs, and parallel-to-serial data converters using the parallel inputs.

The basic serial or *ripple* binary counter is shown in Figure 6.83. Three *J/K* flip-flops can count to 2^3 in this arrangement. In the shift-register configuration, eight flip-flops are needed to count to 2^3.

Modulo-*N* counters

($N \neq 2$) can be made by decoding the counter outputs with gates and resetting the flip-flops after N counts. Shift registers and counters are manufactured in a variety of configurations in single 14- and 16-pin DIP and SMT packages.

6.6.7 Digital-to-Analog Conversion

Whenever it is necessary to connect a digital system to an external device that operates in an analog mode, such as a thermocouple (temperature proportional to output voltage), pressure gauge (pressure proportional to output current), or chart recorder (pen deflection proportional to input voltage), converters are needed—either *digital-to-analog* (DAC) or *analog-to-digital* (ADC). Of the two, the DAC is the more fundamental, since many ADCs use DACs in their construction.

A DAC produces an output current or voltage proportional to the magnitude of the digital number at its input terminals (see Figure 6.84). In one mode of operation, the digital inputs to a DAC activate FET analog switches that connect the properly weighted current or voltage from a reference source to the output. In the case of voltage DACs, an operational amplifier in the summing mode adds the weighted input voltages. Electrical DAC specifications to consider are *resolution,* the number of input bits or steps (an eight-bit DAC has 256 steps); *linearity,* the maximum deviation from the best straight line drawn through the graph of output versus digital input; and *accuracy,* the

deviation of the output from that computed on the basis of the digital input. Linearity primarily depends on the resistors in the DAC circuit and the voltage drops across the internal transistor switches. For these reasons, linearity is temperature dependent. Accuracy depends on the factors affecting linearity and also on the reference voltage, either internal or external. In applications where speed of conversion is important, *settling time* must be considered. This is the time necessary for the output to stabilize within ±1/2 of the least significant bit of the final steady-state value. Other considerations are power supply sensitivity, compatibility with input logic levels and output logic levels. If a current DAC is used in an application where the output must be a voltage, a current-to-voltage converter is needed. For low-voltage applications, this can be a resistor. When higher voltages are required, the operational-amplifier circuits shown in Figure 6.85 can be used. The characteristics of the amplifier should not degrade the expected performance of the DAC with respect to linearity, accuracy, or settling time.

A variation of the DAC is the *multiplying DAC.* Since the output of a DAC is directly proportional to the digital input and reference voltage, varying the reference voltage has the effect of multiplying the output by a constant. Digitally modulated waveforms can be easily produced with such devices.

There is a larger variety of ADCs than DACs because the complexity of analog-to-digital conversion lends itself to a variety of techniques. Three ADC schemes are shown in Figure 6.86. With the *counter ADC* (see Figure 6.86a), pulses from an oscillator or *clock* are routed to a counter through control logic, which is activated by the output of a comparator. The counter output goes to a DAC, the output of which is compared with the input signal. When the DAC output equals the input voltage, the comparator changes state and disconnects the clock from the counter. The digital number at the counter outputs is then proportional to the input voltage.

The *dual-slope ADC* (see Figure 6.86b) converts voltage to time by using the input voltage to charge a capacitor to a predetermined voltage and then discharging the capacitor by connecting it to a reference voltage of the opposite polarity. The input voltage is proportional to the ratio of the charge to the discharge times, which is recorded by a counter driven by a clock through control

TABLE 6.27 ARITHMETIC UNITS

Name	Symbol	Truth Table				
		A	B	Sum	C	D
Half adder	A B \ HA	0	0	00	0	0
		0	1	01	0	1
		1	0	01	0	1
		1	1	10	1	0

		A_n	B_n	C_{n-1}	S_n (sum)	C_n (carry)
Full adder	A_n B_n C_{n-1} \ FA \ C_n S_n	0	0	0	0	0
		0	1	0	1	0
		1	0	0	1	0
		1	1	0	0	1
		0	0	1	1	0
		0	1	1	0	1
		1	0	1	0	1
		1	1	1	1	1

		A_n	B_n	C_n ($A_n > B_n$)	D_n ($B_n > A_n$)	E_n ($A_n = B_n$)
Digital comparator	C D \ C_n D_n E_n	0	0	0	0	1
		0	1	0	1	0
		1	0	1	0	0
		1	1	0	0	1

Parity generator/checker

A B C D P \ P'

4-bit word				Parity bits	
A	B	C	D	P	P'
0	0	0	0	0	0
0	0	0	1	1	0
0	0	1	0	1	0
0	0	1	1	0	0
0	1	0	0	1	0
0	1	0	1	0	0
0	1	1	0	0	0
0	1	1	1	1	0
1	0	0	0	1	0
1	0	0	1	0	0
1	0	1	0	0	0
1	0	1	1	1	0
1	1	0	0	0	0
1	1	0	1	1	0
1	1	1	0	1	0
1	1	1	1	0	0

TABLE 6.28 DATA UNITS

Unit	Symbol	Truth Table

Decoder: Activates one of 2^n outputs according to an n-bit code.

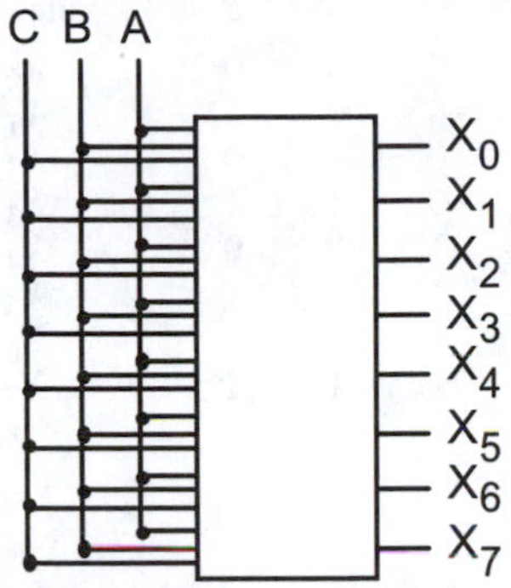

C	B	A	X_0	X_1	X_2	X_3	X_4	X_5	X_6	X_7
0	0	0	1	0	0	0	0	0	0	0
0	0	1	0	1	0	0	0	0	0	0
0	1	0	0	0	1	0	0	0	0	0
0	1	1	0	0	0	1	0	0	0	0
1	0	0	0	0	0	0	1	0	0	0
1	0	1	0	0	0	0	0	1	0	0
1	1	0	0	0	0	0	0	0	1	0
1	1	1	0	0	0	0	0	0	0	1

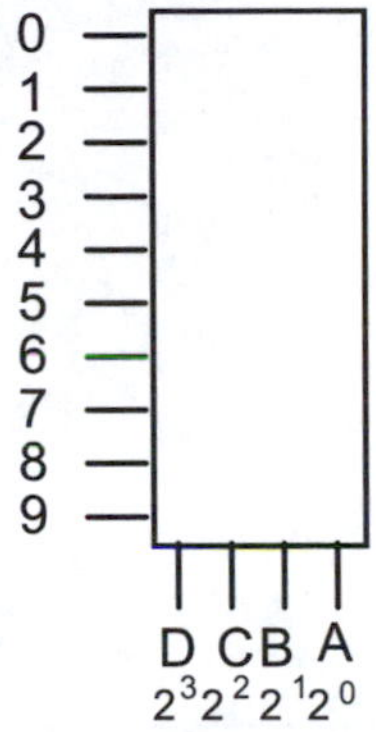

Encoder:

Input	D	C	B	A
0	0	0	0	0
1	0	0	0	1
2	0	0	1	0
3	0	0	1	1
4	0	1	0	0
5	0	1	0	1
6	0	1	1	0
7	0	1	1	1
8	1	0	0	0
9	1	0	0	1

Multiplexer: According to an n-bit code, one of 2^n signal input lines is connected to a single output line.

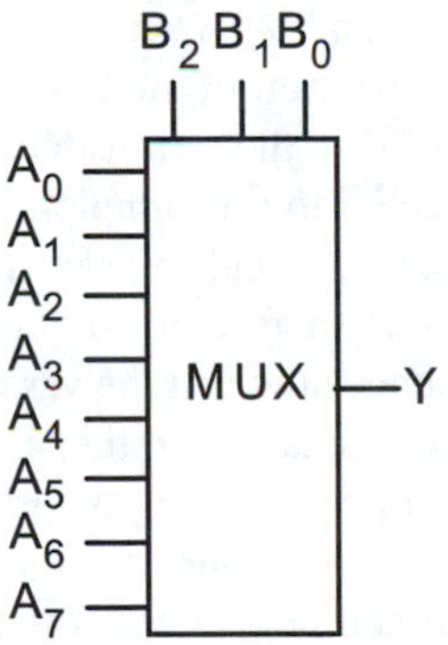

8-to-1 line multiplexer

B_2	B_1	B_0	Y
0	0	0	A_0
0	0	1	A_1
0	1	0	A_2
0	1	1	A_3
1	0	0	A_4
1	0	1	A_5
1	1	0	A_6
1	1	1	A_7

TABLE 6.28 DATA UNITS (CONTINUED)

Unit	Symbol	Truth Table

Unit	Symbol					
Demultiplexer: According to an n-bit code, a single input line is routed to one of 2^n output lines.		C	A	B	Y connected to	
		0	0	0	X_0	
		0	0	1	X_1	
		0	1	0	X_2	
		0	1	1	X_3	
		1	0	0	X_4	
		1	0	1	X_5	
		1	1	0	X_6	
		1	1	1	X_7	

Read-only memory (ROM): Combination of a decoder and encoder to convert an n-bit code to an m-bit code, where n and m are not necessarily equal.

logic. As long as the capacitor, clock, and reference voltage are stable during the conversion interval, the resolution will depend only on the resolution of the capacitor circuit. This method is the one most commonly used in digital multimeters.

The *successive-approximation ADC* (see Figure 6.86c) works like the counter type except that a programmer, rather than a clock and counter, drives the DAC. The programmer starts by activating the most significant bit of the DAC. If the DAC output is greater than the input, the programmer turns the MSB off and the next MSB on, and the comparison is done again. The process of comparison continues to the LSB, after which the conversion is complete. This method is fast and has high resolution. As with DACs, resolution, linearity, accuracy, and settling time are important parameters used for characterizing the operation of a unit.

Modular data-acquisition systems or *data loggers* with an analog multiplexer (MUX), *sample-and-hold* (S/H)

circuit, and ADC are available from many sources. A block diagram of such a system is shown in Figure 6.87. Each of the transducer-amplifier-filter circuits is connected to the sample-and-hold circuit for a fixed time through the multiplexer. The output of the sample-and-hold circuit is then digitally encoded by the ADC. The frequency at which the various inputs are sampled depends on the rate at which the signals are varying. An important theorem of sampling theory states that if a sampled signal contains no Fourier components higher than f_{max}, the signal can be recovered with no distortion if sampled at a rate of $2f_{max}$.

6.6.8 Memories

Large-scale integration (LSI) has made possible the production of high-density, solid-state memories. The typical one-bit memory cell is an R/S flip-flop with supplementary gates shown in Figure 6.88. When the

TABLE 6.29 FLIP-FLOPS

Units	Symbol	Truth Tables			

R/S (reset-set) flip-flop: The basic digital memory unit.

S	R	Q	$\bar{Q}$
0	1	0	1
1	0	1	0
0	0	Not Allowed	
1	1	Hold	

J/K flip-flop: An elaboration of the R/S avoiding the indeterminate 1, 1 and 0, 0 states and with separate preset (Pr) and clear (Cr) inputs, and a clock (Ck) input that must be activated for the outputs to follow the input conditions.

Upon the application of a Ck pulse[a]

J	K	Q_{n+1}	$\bar{Q}_{n+1}$
0	0	Q_n	$\bar{Q}$
0	1	0	1
1	0	1	0
1	1	$\bar{Q}_n$	Q_n

Direct Inputs:

Pr	Cr	Q	$\bar{Q}$
0	0	Disallowed	
0	1	1	0
1	0	0	1
1	1	Normal clocked operation	

D (delay) flip-flop: Whatever is on the input appears at the output after a clock pulse. Can be constructed from a J/K flip-flop by putting an inverter from J to K, with the signal applied to J.

T (toggle) flip-flop: Output changes state at each clock transition. Used as a binary divider. Constructed from a J/K flip-flop by fixing J and K at 1.

[a] The subscript $n + 1$ designates the $(n + 1)$th state as distinct from the previous nth state. When R and S are 0, the output of Q remains constant, equal to its previous value.

address and *write enable* lines are activated, the signal on the *write* line is registered on the Q line. It will remain there as long as the *write enable* line is not subsequently activated. To read the contents of the memory cell, the *address* line has only to be activated. Large-scale memories are composed of millions of such elementary cells. Because the contents of any single cell can be read by activating the appropriate address line, such memories are called random-access memories (RAMs). There are several different ways of arranging the basic memory cells for addressing purposes. For example, a 1024-bit RAM can be arranged as 1024 one-bit words or 256 four-bit words. The addressing scheme in the first case requires 10 address lines, one *data out* line, and one *data in* line. For the 256 × 4 memory, eight address lines are needed as well as four *data in* and four *data out* lines. In addition to the

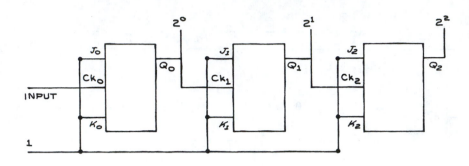

Figure 6.82   J/K flip/flops arranged as a 3-bit register.

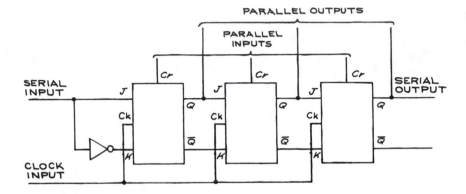

Figure 6.83   J/K flip/flops arranged as an 8-bit ripple binary counter.

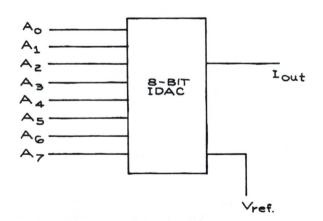

Figure 6.84   An 8-bit current digital-to-analog converter (IDAC).

memory—the CE line is activated when reading or writing data. *Dynamic* RAMs require an additional input, called the *refresh,* to hold the contents of the memory. Refreshing is required continuously—the signal for it is usually supplied from a clock. *Static* RAMs require no such refreshing. Solid-state RAMs are volatile, meaning that their contents are destroyed when the power is interrupted. Battery backup can be provided in applications in which the memory contents must be retained in the event of a power failure.

Important parameters to consider for RAMs are power dissipation, access time, and cycle time. The *access time* is the time required after the activation of the address lines to obtain valid data at the output. The *cycle times* are the times required to complete a read- or write-data procedure. The access and cycle times are related to the power dissipation—generally, the lower the power dissipation, the slower the memory.

Other considerations are power supply voltages and input and output signal levels. The RAM must be able to

address and data lines, a *read/write* ($R/\overline{W}$) line and a *chip enable* (CE) line are needed. The state of the $R/\overline{W}$ line determines whether data are to be read into or out of the

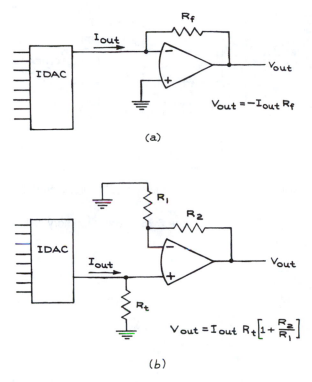

Figure 6.85   Conversion of the current output of an IDAC to a voltage output: (a) inverting; (b) noninverting.

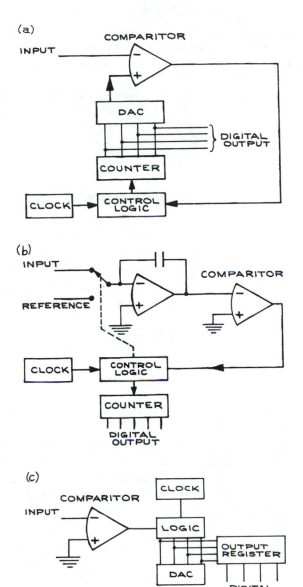

Figure 6.86   Three analog-to-digital converters: (a) counter; (b) dual slope; (c) successive approximation.

accept data from external devices and, in turn, produce output levels capable of driving the circuits connected to its outputs.

ROMs are storage devices with permanent, unalterable patterns of 0s and 1s in the individual cells. ROMs are used extensively in computers for storage of the character generation bit patterns, lookup tables for routine code conversions, and control programs such as those used in calculators.

PROMs are ROMs whose bit pattern can be altered. PROMs are programmed with a PROM *programmer* that permanently *burns* the desired bit pattern into the PROM. The bit pattern can be entered manually (memory location by memory location), serially, or in blocks under the control of a computer connected to the programmer.

Even more useful devices are erasable PROMS, or EPROMS. These are programmed in the usual way, but can also be erased and reprogrammed for another purpose

or in case of errors. Erasing is usually accomplished with ultraviolet light in an EPROM eraser. This can take several minutes. The latest development is EEPROMs, or

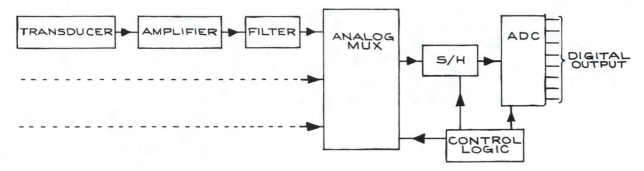

Figure 6.87   Multiple-input data-acquisition system with an analog multiplexer (MUX), sample-and-hold (S/H) circuit, and ADC.

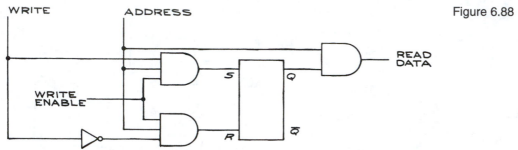

Figure 6.88   Basic memory cell.

electrically erasable PROMs, that can be very quickly erased with electrical signals.

An example of the more common series of EPROMs is the 2700 series, manufactured by INTEL, Advanced Micro Devices, and others. Members of the series are the 2708, 2716, 2732, 2764, 27128, 27256, and 27512. The digits following the 27 give the number of kilobits in the EPROM. For example, the 2732 has 32 kilobits of memory arranged in a 4096×8 pattern—that is, 4 kilobytes. A twelve-bit address ($212 = 4096$) is necessary to address 4 kilobytes. The chip also has 8 output pins, one pin each for the +5 V power, the ground, and the 25 V programming voltage, and a chip-enable pin. The 2732 chip has 24 pins and is programmed by entering 0s into the desired memory locations since all of the bits are 1s upon delivery or after erasure. Programming is accomplished by applying +25 V to the OE/VPP programming-voltage pin, while the address of the byte to be programmed is applied to the appropriate address pins at the same time that the desired eight-bit pattern is placed

on the output pins. When all voltage levels are stable, a 50 ms TTL low-level pulse is applied to the CE/PGM chip enable pin.

PROM programmers are designed to apply the proper voltages in the required sequence to the PROMs. Programmers in kit form cost less than $100. Fully assembled programmers cost from $200 to $500—the more costly ones have the capability to program a wide variety of EPROMs as well as copy and verify EPROM codes. The programmers generally are controlled through the serial port of a personal computer, and software is supplied with the programmer and loaded into the computer. The PROMs are useful in experimental work because they can be programmed to reproduce fairly complex truth tables, eliminating the need for several different logic chips. The truth table of Table 6.32, for example, can be fully implemented with a PROM having as little as 32×4 bits.

The implementation of truth tables with PROMs is inefficient because, in practice, only a limited number of variable combinations are actually used. With a PROM, all possible combinations of all the input variables are represented, resulting in a large bit count most of which is never used. Programmable array logic (PAL), programmable logic arrays (PLAs), and field programmable gate arrays (FPGAs) permit the economical realization of logic expressions with a limited number of inputs and terms. The units perform the functions of a custom IC. Of the three devices, the PLAs and FPGAs are the most flexible. Programmers for these devices are more complicated than PROM programmers and cost, with software, is several hundred dollars for professional models and a few hundred dollars for student models. XILINX, Altera, and Actel are manufacturers of programmable logic devices.

6.6.9 Logic and Function

To illustrate the ideas in the preceding sections on digital logic, we now discuss a method for designing an interlock circuit for a vacuum system. High-vacuum systems often have a vacuum chamber connected to an oil diffusion pump or turbomolecular pump through a gate valve (see Section 3.6.1). The diffusion or turbo-pump has a high-pressure outlet connected to a mechanical rotary pump or diaphragm pump via a foreline and foreline valve. A bypass roughing line with a roughing valve is used for preliminary evacuation of the chamber. Before this is done, however, the foreline valve is closed. When the pressure in the chamber is sufficiently low, the roughing line is closed, the foreline valve is opened, and the chamber is pumped with the diffusion or turbo pump through the gate valve.

To guard against accidents, it is worthwhile to have a system of interlocks that activates valves in the proper sequence and turns the diffusion pump or turbo-pump off in case of an accident. The interlock system consists, first, of *sensors* that continuously sample the variables in the vacuum system, such as cooling-water temperature, cooling-water pressure, foreline pressure, vacuum-chamber pressure, and roughing-line pressure. The security of the system depends on whether the physical property sampled is above or below a predetermined threshold. The second section of the interlock system is the *logic,* which, with combinations of elementary gates, establishes the relationships between the input variables from the sensors and the output functions that control the various vacuum-system operations—such as the opening and closing of the various valves and the application of voltage to the diffusion pump or turbo-pump. Because the output functions only assume one of two values, the interlock system is amenable to binary-logic analysis using either Boolean algebra or Karnaugh maps. Examples of increasing complexity are given below. A way of implementing the system with a *data selector* is also discussed.

The simplest practical interlock system monitors the cooling-water temperature and pressure and turns off the voltage to the diffusion-pump or turbo-pump if the temperature is too high, the pressure is too low, or both. A truth table relating the variables to the function is shown in Table 6.30. The letters and numbers in parentheses are logic symbols assigned to the variables (A, B), function (X), and states of the variables (0,1). The Boolean logic expression for the truth table is

$$X = \bar{A} \bullet \bar{B} = \overline{A + B}$$

This is obtained by noting the condition under which X is in the logical 1 state. Had there been more than one occurrence of 1 for X, the expression would have additional OR terms. The Boolean expression can be implemented with the inverter-and-gate arrangement of Figure 6.89. A bimetallic-switch temperature sensor can provide the temperature variable, and a diaphragm-type pressure sensor can provide the pressure variable. Power to the diffusion pump is controlled by a heavy-duty relay actuated by a solid-state, optically isolated relay driven by the output of the logic circuit. A practical circuit is shown

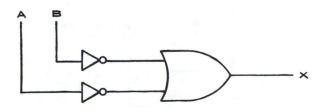

Figure 6.89   Gate arrangement corresponding to Table 6.30.

in Figure 6.90. It is assumed that logical 0 is less than 0.8 V and logical 1 is greater than 2.4 V, consistent with TTL logic. Optical isolators provide an effective interface when controlling power circuits from logic circuits.

An optical isolator contains both an infrared-emitting diode and a photodetector arranged in a single package, so that the light from the diode is efficiently transmitted to the detector through a transparent dielectric medium that provides electrical isolation. The diode-detector pair is shielded from ambient light by opaque encapsulating material. Since only infrared radiation is used to communicate between diode and detector, there is no electrical connection between the two. The communication is also unidirectional because the detector cannot provide a signal that can couple back to the diode. Typical applications for isolators are elimination of common mode input signals and ground loops, level shifting between two circuits, logic control of power circuits, and isolation of logic circuits from noise. The last application commonly

TABLE 6.30 TRUTH TABLE

Variables		Function
Water Pressure (A)	Water Temperature (B)	Diffusion Pump Heater (X)
Low (1	Low (0)	Off (0)
Low (1)	High (1)	Off (0)
High (0)	Low (0)	On (1)
High (0)	High (1)	Off (0)

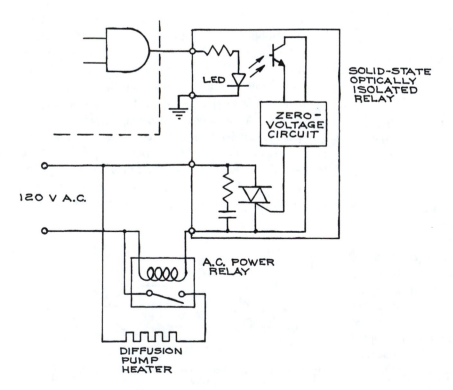

Figure 6.90   Practical circuit for a vacuum-system interlock according to the truth table of Table 6.30.

occurs when using the output logic levels on a computer to control external devices. Infrared emitting diodes and photodetectors are inherently nonlinear devices, but some isolators are specially made for linear analog circuits. Photodiodes, phototransistors, photo-Darlington transistor pairs, and phototriacs are the detector units in isolators. Important parameters to consider when choosing an isolator are the isolation voltage, ratio of output current to light emitting-diode input current, and output power-handling capabilities. Transistor isolators have transfer ratios from 2 to over 50 percent, while the Darlington transistor pairs have ratios from 50 to 500 percent. Isolators with triac outputs require input trigger currents in the 10 to 30 mA range in order to switch 120 V ac at currents up to a few hundred milliamperes. This is sufficient to drive high-power triacs, solenoids, small motors, and mechanical relays. For proper isolator operation, the characteristics of the output stage must be matched to the load. Data sheets should be consulted.

The most common package for low-power isolators is a six-pin DIP. A list of suppliers is given at the end of this chapter.

When switching transformer loads, solenoid valves and contactors, lamp loads, and heaters, output current rating should exceed the load rating by at least a factor of two. Solid-state relays provide isolation, and an important application of solid-state relays is the interface between computers and power devices. Opto 22 makes a line of solid-state relays that can be turned on and off with standard TTL signal levels. The 120D10 can be used to switch up to 10 A of ac current at 120 V, with a maximum voltage drop of 1.6 V. The control circuit and switching circuits are electrically isolated from each other up to 4 kV, and switching of the load occurs at zero voltage in the ac cycle, thereby eliminating switching transients. Because of the isolation between control signal and load, the circuit supplying the control signal is protected from voltage spikes. The relays are 1.2 in. high, 2.4 in. long, and 1.8 in. wide.

A more useful interlock system using three variables and two functions has the truth table of Table 6.31. The variables monitored are water pressure and temperature and diffusion-pump foreline pressure. The functions are the diffusion-pump heater (on or off) and the gate valve (open or closed). The Boolean expression and logic circuit are also given in the table, while the Karnaugh map is

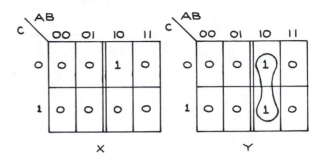

Figure 6.91   Karnaugh map of the truth table of Table 6.31.

shown in Figure 6.91. In the map for Y, the 1s are adjacent, resulting in the simplification that Y is independent of the state of C.

As a final example of logic design, consider increasing the number of variables to five by including the roughing-line pressure and chamber pressure and adding the roughing valve and foreline valve to the functions. With five variables there will be 2^5 (or 32) combinations. Systematic design of the interlock circuit will ensure that none of them is left out. The truth table is given in Table 6.32. Rather than obtain the Boolean expressions for the truth table or make a map, we can use a *data selector* to establish the relationships among the variables and functions. A five-bit data selector has five input address lines, one output line, and 32 data lines. Depending on the five-bit input address, the data on the appropriate input line are transferred to the output line. Implementation of the expression for Y is shown in Figure 6.92a. A four-bit data selector (see Figure 6.92b) can also be used by employing a foldback arrangement. Looking at the A, B, C, and D bits, one sees that there are 16 identical pairs corresponding to E being 1 or 0. For each of those pairs, the function is independent of the state of E, equals E, or equals $\overline{E}$. The 16 input lines of the four-bit data selector will each have 1, E, or $\overline{E}$ attached to them.

Even more complexity can be built into the circuit by having time delays on the functions so that one valve is not opening while the other is closing. Extra functions such as turning on and off and venting the mechanical pump can also be added. This systematic method of analysis ensures that nothing is overlooked. By noting when two functions

TABLE 6.31 THREE-VARIABLE, TWO-FUNCTION INTERLOCK TRUTH TABLE

Truth Table

Variables		*Function*	
Water Temperature (B)	*Foreline Pressure (C)*	*Diffusion-Pump Heater (X)*	*Gate Valve (Y)*
low (0)	low (0)	off (0)	closed (0)
low (0)	high (1)	off (0)	closed (0)
high (1)	low (0)	off (0)	closed (0)
high (1)	high (1)	off (0)	closed (0)
low (0)	low (0)	on (1)	open (1)
low (0)	high (1)	off (0)	open (1)
high (1)	low (0)	off (0)	closed (0)
high (1)	high (1)	off (0)	closed (0)

Boolean Expressions

$$X = A \cdot \bar{B} \cdot \bar{C}$$
$$Y = A \cdot \bar{B} \cdot C + A \cdot \bar{B} \cdot C$$
$$Y = A \cdot \bar{B}(C + \bar{C})$$
$$Y = A \cdot \bar{B}$$

Implementation

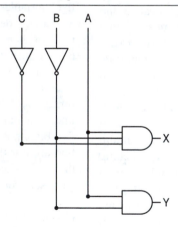

change state upon the change of state of a variable, decisions about sequencing can be made.

6.6.10 Implementing Logic Functions

Implementing functions with transistor gates requires that the 0 and 1 states be associated with voltage or current levels. There are a large number of logic families, all of which have internally consistent 0 and 1 levels, but which generally are not mutually compatible. The choice of a logic family for the implementation of a given function depends on a number of factors—among them are speed, power dissipation, immunity to noise, number of available functions, cost, and compatibility. A summary of the properties of the most common logic families is given in Table 6.33. As can be seen from the table, there is a

TABLE 6.32 FIVE-VARIABLE, FOUR-FUNCTION INTERLOCK TRUTH TABLE

T_{H_2O} (A)	P_{H_2O} (B)	$P_{foreline}$ (C)	$P_{roughing}$ (D)	$P_{chamber}$ (E)	Diffusion Pump (X)	Gate Valve (Y)	Roughing Valve (R)	Foreline Valve (W)
0 (low)	0 (low)	0 (low)	0 (low)	0 (low)	0 (off)	0 (closed)	0 (closed)	0 (closed)
0	0	0	0	1 (high)	0	0	0	0
0	0	0	1 (high)	0	0	0	0	0
0	0	0	1	1	0	0	0	0
0	0	1 (high)	0	0	0	0	0	0
0	0	1	0	1	0	0	0	0
0	0	1	1	0	0	0	0	0
0	0	1	1	1	0	0	0	0
0	1 (high)	0	0	0	1 (on)	1 (open)	0	1 (open)
0	1	0	0	1	0	0	1 (open)	0
0	1	0	1	0	1	1	0	1
0	1	0	1	1	0	0	1	0
0	1	1	0	0	0	0	0	0
0	1	1	0	1	0	0	1	0
0	1	1	1	0	0	0	0	1
0	1	1	1	1	0	0	1	0
1 (high)	0	0	0	0	0	0	0	0
1	0	0	0	1	0	0	0	0
1	0	0	1	0	0	0	0	0
1	0	0	1	1	0	0	0	0
1	0	0	0	0	0	0	0	0
1	0	1	0	1	0	0	0	0
1	0	1	1	0	0	0	0	0
1	0	1	1	1	0	0	0	0
1	1	0	0	0	0	0	0	0
1	1	0	0	1	0	0	0	0
1	1	0	1	0	0	0	0	0
1	1	0	1	1	0	0	0	0
1	1	1	0	0	0	0	0	0
1	1	1	0	1	0	0	0	0
1	1	1	1	0	0	0	0	0
1	1	1	1	1	0	0	0	0

tradeoff between speed and power dissipation—the fastest logic dissipates the most power per gate and the slowest dissipates the least. In an effort to reduce power dissipation while maintaining speed, nominal voltage levels have been reduced to 3.5 and as low as 2.5 V for some logic families from the nominal TTL 5 V. *Noise immunity* is measured by *noise margin,* the meaning of which is illustrated in Figure 6.93. In general, logic families are not directly compatible—translator chips exist for interfamily conversion, however, and some simple conversion circuits are given in Figure 6.94. The abbreviations SSI, MSI, LSI, and VLSI stand for small-scale integration, medium-scale integration, large-scale integration, and very large-scale integration, and are used to specify the number of gates on a chip. SSI is less than 12 gates per chip, MSI is between 12 and 100, LSI is more than 100, and VLSI is more than 1000. Simple logic functions employ SSI and MSI.

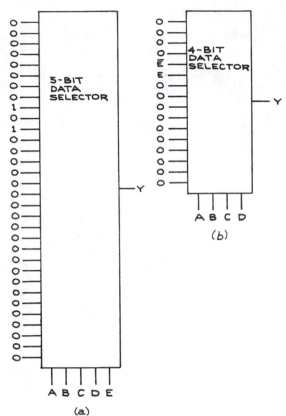

Figure 6.92 Implementation of the Y-function of the truth table of Table 6.31 with (a) a 5-bit data selector and (b) a 4-bit data selector using a fold-back arrangement.

The most common logic family is TTL (transistor-transistor logic) and its variations, but CMOS (complementary symmetry, metal-oxide semiconductor) finds wide application because of its low power consumption and wide range of usable power supply voltages. The TTL line has expanded from the basic 74/54 line to high-performance, high-speed, and low-power consumption series. Emitter-coupled logic (ECL) is only used when the highest speeds are required. The ECL 10,000 series is a slower version of the ECL II series and is not as sensitive to the quality of the interconnections. With the II series, a circuit board with a ground plane must be used and the interconnectors must be carefully laid out to minimize cross-coupling.

There have been important changes in the semiconductor electronics industry in the last few years. These changes reflect new technologies as well as new philosophies in electronics. Whole new families of logic have been born—based either on CMOS or Schottky technologies. The goal is to maximize speed while minimizing power requirements. CMOS is inherently the lowest-power technology, but has suffered in the past from low speed. This has largely been overcome with the HC and HCT logic lines. Both are directly compatible with TTL LS families, but offer greatly improved speed-to-power ratios. Because TTL high levels can be as low as 2.2 V, there can be difficulties driving HC devices from TTL outputs. This is overcome with the HCT family, which is completely compatible with TTL over the full range of TTL operating parameters. It is expected that HC logic will replace the TTL LS family over the next few years. The 4000 D series of CMOS can operate with supply voltages from +5 to +15 V and is used in conjunction with higher-voltage analog circuits. FAST[TM] logic is based on Schottky technology and approaches ECL 10,000 logic in speed, but with much less power consumption. FAST devices bear the designation 54/74F and have the advantage of TTL logic levels combined with the smaller voltage swings and lower noise margins of ECL.[7] FAST is a trademark of Fairchild Camera and Instruments Corporation and is an acronym for Fairchild Advanced Schottky TTL.

Because TTL remains common, some simple TTL circuits are given in Figure 6.95. The mechanical contact conditioner ensures that the output from a mechanical switch will be a single transition rather than a series of voltage spikes normally resulting from the mechanical bouncing of the contacts. TTL gates are either *edge-* or *level-triggered*. Edge triggering refers to the transition between the 0 and 1 levels (positive edge) or 1 and 0 levels (negative edge). Level triggering occurs when the 0 or 1 is attained. Data sheets specify the type of triggering each logic unit requires.

With TTL, as with most logic families, there is the tendency to generate current spikes during switching; this creates power supply noise. To reduce the noise, the power supply must be capacitatively decoupled from the logic circuits. A 0.01 μF ceramic capacitor for each gate and an additional 0.1 μF capacitor for each 20 gates are generally

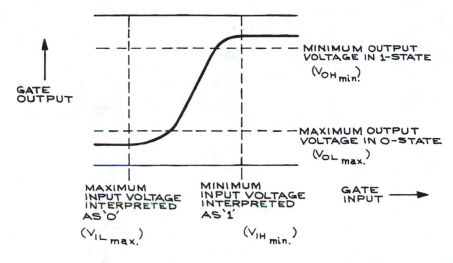

GATE OUTPUT

MINIMUM OUTPUT VOLTAGE IN 1-STATE $(V_{OH_{min}})$

MAXIMUM OUTPUT VOLTAGE IN 0-STATE $(V_{OL_{max}})$

MAXIMUM INPUT VOLTAGE INTERPRETED AS '0' $(V_{IL_{max}})$

MINIMUM INPUT VOLTAGE INTERPRETED AS '1' $(V_{IH_{min}})$

GATE INPUT →

Figure 6.93   Noise margin for 0-state = $(V_{OH_{min}} - V_{IH_{min}})$: noise margin for 1-state $= V_{IL_{max}} - V_{OL_{max}}$.

TABLE 6.33 LOGIC FAMILIES

Logic	Propagation Delay (ns)	Power Dissipation (mW/gate)	0-Level (V)	1-Level (V)	Noise Margin, Low / High (V)	Fanout	Power Supply (V)
TTL							
54/74	10	10	0.2	3.9	0.3/0.7	10	+5
54/74LS	10	2	0.2	3.9	0.3/0.7	20	+5
54/74F	4	4	0.2	3.9	0.3/0.7	33 (50LS)	+5
CMOS							
54/74C	30	0	0.2	4.8	0/7/0.6	> 50 (10LS)	+5
54/74HC	10	0	0.2	4.8	0.7/0.6	> 50 (10LS)	+5
54/74HCT	8	0.010	0.2	4.8	0.7/0.6	> 50 (10LS)	+5
4000 CD						50	
$V_{DD} = 5$ V	115	0.005	0.05	4.95	1.0	50	+5
$V_{DD} = 15$ V	40	0.060	0.05	14.95	2.5		+15
ECL							
10,000	3	24	1.75	0.90	0.3/0.3	10	−5.2
100,000	0.8	40	1.75	0.90	0.3/0.3	10	−4.5

sufficient. Counters and shift registers are especially sensitive to power supply noise and should be decoupled with a 0.1 µF capacitor for every two devices. For optimum performance, unused inputs should be set to 0 by direct connection to ground or 1 by connection to the power supply voltage through a 1 to 10 kΩ resistor.

Some additional features of TTL are three-state outputs and open-collector outputs. A *three-state output* has the standard 0 and 1 levels and also a high-impedance, nonactive state. In this state, the output will assume the level of any active output connected to it. The three-state outputs are designed to be connected together in a common line or bus structure so that, when any single

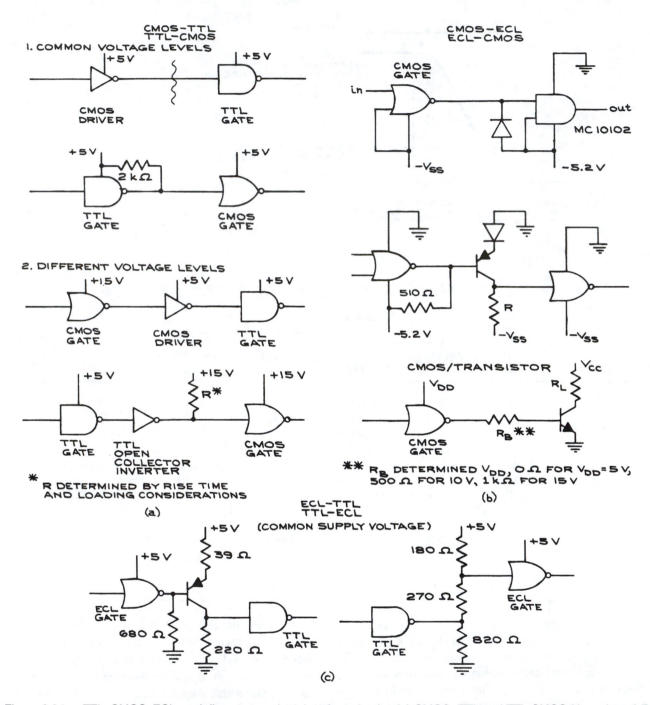

Figure 6.94   TTL, CMOS, ECL, and discrete transistor interface circuits. (a) CMOS -TTL and TTL-CMOS (the value of *R* is determined by rise-time and loading considerations). (From *McMOS Integrated Circuits Data Book*, Motorola, Inc., 1973. Copyright of Motorola, Inc. Used by permission.) (b) CMOS-ECL and ECL-CMOS (*R*$_B$ is determined by *V*$_{DD}$: 0 for *V*$_{DD}$ = 5 V, 500 Ω for 10 V, 1 kΩ for 15 V); (c) ECL-TTL: and TTL-ECL (from *MECL System Design Handbook*, 2nd ed., W.R. Blood Jr., and E. C. Tynan, eds., Motorola, Inc. Used by permission.)

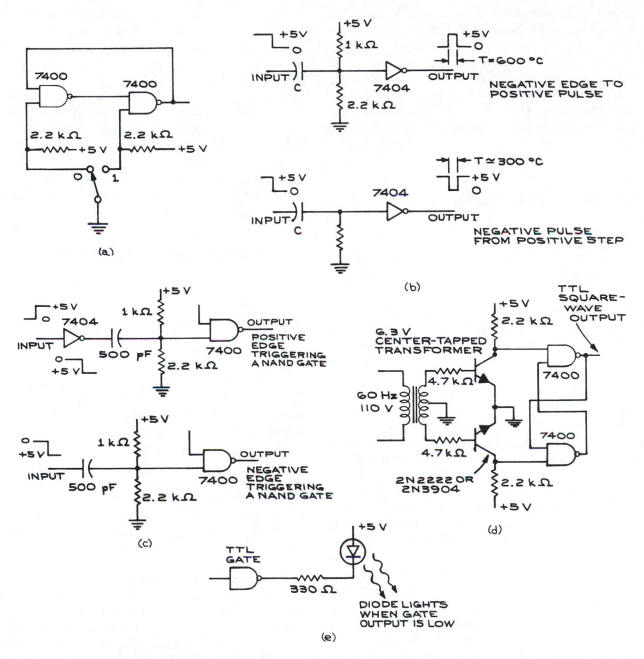

Figure 6.95   TTL signal-conditioning circuits: (a) mechanical-contact conditioner; (b) pulses from step inputs (step duration must exceed output-pulse width); (c) edge triggering; (d) 60 Hz TTL square wave from the power line; (e) TTL to LED.

output is active, all of the others assume its level. The *open-collector output* requires a collector resistor to the power supply for operation. When connected together, open-collector outputs allow one to implement an output AND function (often incorrectly termed wired-OR)—that is, when any output is 0, all become 0. Open-collector outputs have the disadvantage of reduced transition speed and high output impedance in the 1 state. They are usually used as line drivers.

6.7 DATA ACQUISITION

6.7.1 Data Rates

When designing an experiment, an important consideration is the rate at which data are collected. This is necessary because of the need to store and display the data in a form that can be used for subsequent reduction and analysis. It is usually useful to think of an experiment in terms of sampling rate, sampling time, and measurement precision. Consider a spectroscopy experiment where the output of a monochromator is incident on a photomultiplier—the output of which is bursts of electrons that are counted with a counter. The monochromator is stepped at a regular rate and predetermined step size over the wavelength interval of interest. The time spent at each step combined with the number of steps determines the total number of counts recorded. If the total number of signal counts in any interval is C_S and the number of noise counts is C_N, the total number of counts C_T is $C_S + C_N$. The statistical uncertainty in C_T is $\sqrt{C_T}$, and in C_N is $\sqrt{C_N}$. The uncertainty in determining C_S is therefore $\sqrt{C_T + C_N}$, so the relative error is $\sqrt{C_T + C_N}/(C_T - C_N)$. The counts increase linearly with the time of measurement, but the relative uncertainty only decreases as the inverse square root of the time of the measurement. The *data rate* is the number of counts in a measurement interval divided by the interval time.

Bits, bytes, words, and *characters* are all used to specify an amount of information. The number of *bits* necessary to specify a given integer decimal D is given by $N = 3.32\log_{10} D$. If N is a noninteger, it is to be rounded to the next higher integer. If the decimal number is to be signed (that is, + or –),

another bit is added. Eight bits make a *byte,* and two bytes (16 bits) commonly, but not always, make a *word.*

A *character* is the equivalent of one word. If data are to be transferred from the counter to a printer or other display or storage device, the data acceptance rate of the device must match the data output rate of the counter. Data rates are often specified as a *baud rate,* which is the number of times per second there is a transition on the signal line from a 0 to a 1 or from 1 to 0.

For compatibility between two devices, a number of parameters must match in addition to the rate. These include voltage levels, timing, format, code, hand-shaking protocols, and housekeeping procedures.

6.7.2 Voltage Levels and Timing

The most positive and most negative voltages corresponding to the two logic states must match. For example, the TTL logic levels are +3.2 and +0.2 V, while the ECL levels are –0.75 and –1.55 V. Furthermore, the sign of the logic must be identical—that is, positive or negative. *Positive logic* means that the more positive voltage level is taken as a logical 1, and the more negative as logical 0. The reverse is true for *negative logic.* If the signs do not match and everything else is satisfactory, only inversion of the input signal will be required.

Often, a digital input or output circuit is compatible with a logic family of which it is not a member. The 54/74C, HC, and HCT families are *TTL compatible,* for example, which means that the voltage levels are compatible. The sinking and sourcing capabilities may be very different, however, for so-called compatible logic families. *Sinking* has to do with the number of gates that can be attached to the output of a single gate and still maintain reliable operation. With TTL logic, when an output is at logical 0 it must be able to hold all inputs connected to it at logical 0. Since this condition is a consequence of the input transistors being in saturation, the output must be able to sink all of the saturation currents and still keep the inputs at their required levels. A TTL output at logical 1 attached to a TTL input will turn off the input transistor, and very little current will flow in the input circuit. This is the *sourcing* specification. *Fan-in* and *fan-out* are directly related to the sinking and sourcing capabilities of a gate. Fan-in is the number of logic inputs to a gate, while fan-out is the number of logic outputs it can drive. These specifications should be carefully considered when connecting two different logic families. Timing is also important—the required rise and fall times, as well as pulse widths, must be compatible. Timing diagrams show the relationship between signals, and should be consulted. Converter chips that change the signals from one logic family directly to another are readily available.

6.7.3 Format

A *serial format* has all data on a single line in a sequential pattern. This is economical as far as wiring is concerned, but slower than a *parallel format* where there are as many data lines as bits, and all of the data are transferred simultaneously. This is the fastest way to transmit data, but the least economical.

There are also *parallel-serial* formats in which the data are transmitted in sequential parallel codes. An eight-bit byte, for example, could be transmitted with four lines as two four-bit codes. A synchronization code is necessary with such an arrangement to be able to distinguish between the beginning and end of a single-byte string. It is possible to convert from a serial code to a parallel code and vice versa with converters that use shift registers. For serial-to-parallel conversion, data are sent to the registers from one end and then read from the outputs of each stage. For parallel-to-serial conversion, data are put into each register simultaneously and then extracted from the last register sequentially, a bit at a time. Data transmission in one direction over a single line is called *simplex*, in two directions over a single line is called *half duplex*, and in two directions simultaneously, with a single line and two coding frequencies or two separate lines, is *full duplex*. Complete ICs made for data conversion are called universal asynchronous receiver-transmitters (UARTs). They take a wide variety of data formats and convert them to standard codes, one of which is ASCII (American Standard Computer Information Interchange). The ASCII computer code is given in Table 6.34.

TABLE 6.34 ASCII COMPUTER CODE

b_4	b_3	b_2	b_1	0	0	0	0	1	1	1	$1 \leftarrow b_7$
				0	0	1	1	0	0	1	$1 \leftarrow b_6$
$\downarrow$	$\downarrow$	$\downarrow$	$\downarrow$	0	1	0	1	0	1	0	$1 \leftarrow b_5$
0	0	0	0	NUL	DLE	SP	0	@	P	\	p
0	0	0	1	SOH	DC1	!	1	A	Q	a	q
0	0	1	0	STX	DC2	ð	2	B	R	b	r
0	0	1	1	ETX	DC3	#	3	C	S	c	s
0	1	0	0	EOT	DC4	$	4	D	T	d	t
0	1	0	1	ENQ	NAK	%	5	E	U	e	u
0	1	1	0	ACK	SYN	&	6	F	V	f	v
0	1	1	1	BEL	ETB	¢	7	G	W	g	w
1	0	0	0	BS	CAN	(	8	H	X	h	x
1	0	0	1	HT	EM	)	9	I	Y	i	y
1	0	1	0	LF	SUB	*	:	J	Z	j	z
1	0	1	1	VT	ESC	+	;	K	[	k	{
1	1	0	0	FF	FS	,	<	L	\	l		
1	1	0	1	CR	GS	-	=	M	]	m	}
1	1	1	0	SO	RS	.	>	N	^	n	~
1	1	1	1	SI	US	/	?	O	_	o	DEL

Note: This is a 7-bit code to give $2^7 = 128$ different words. Columns 1 and 2 are machine commands such as carriage return (CR), line feed (LF), escape (ESC), etc.

A major difficulty in setting up data handling equipment is the interface. The RS 232C serial interface can handle data rates to 20 kbps—it has a high input impedance (3 to 5 kΩ), uses high voltage (+25 V), and is unbalanced and therefore noise-susceptible. The pin assignments for the standard 25-pin connector used with this interface are given in Figure 6.96. More advanced interfaces such as the RS 485 use balanced circuitry and can handle data rates well in excess of 10 Mbps (megabits per second). There are computer interfaces designed to handle data rates up to hundreds of Mbps.

Modems (modulator-demodulators) are devices for converting a binary code to audio frequencies (300 to 3300 Hz) for transmission over telephone lines, and reconverting the audio frequencies to binary code. They are separate units to which both a telephone receiver and a data terminal are attached, but some modems are incorporated directly into the telephone base.

6.7.4 System Overhead

In all data-transmission schemes there are additional signals that are necessary to assure that the data transfer occurs correctly. These are sometimes called *housekeeping* signals. In the case where such information is exchanged back and forth between the sending unit and the receiving unit, they are called *hand-shaking*. These additional signals contribute to what is called the *system overhead*—that is, additional data capacity which must exist regardless of the useful information transmitted. When data are transferred at a regular rate between two units, they must be kept in step. A *sync* bit is often used to assure synchronization between the two units. When data are transferred at a variable rate that depends on other independent parameters of the system, they are said to be *asynchronous*. In this case, the sending unit must alert the receiving unit of the coming of data, and the receiving unit then should give a *ready to receive* signal, after which the data are transferred. When data transfer is complete, an *end of data* signal is also usually sent. Without these signals, no reliable data transfer can be effected.

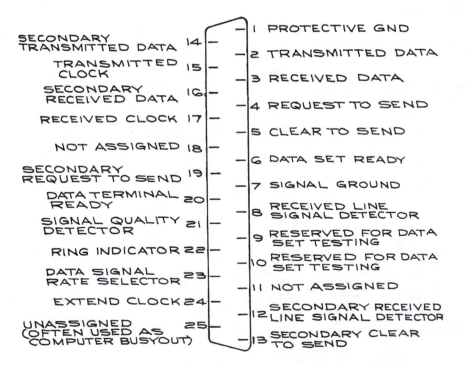

Figure 6.96   RS232C serial interface: 25-contact connector pin assignment.

To ensure the reliability of data transfer and detect errors in the transmission, an extra *parity bit* can be sent. One scheme requires the parity bit to be 0 when the sum of the data bits is even and 1 when the sum is odd. At the receiving unit, data bits are summed and compared with the parity bit—if there is consistency, the data are accepted; if not, they are rejected. A more elaborate way of ensuring integrity is to have the receiving unit echo the received data back to the sender, where they are compared with the original data. Any differences cause the data to be rejected. This system minimizes transmission errors at the cost of additional complexity and lower speed.

Returning to the example of the spectroscopy experiment in Section 6.7.1, if the expected signal produces 10^4 counts per second and the background is 10 counts per second, the statistical error after one second is

$$\frac{\sqrt{10^4 + 10}}{10^4} \cong 1\%$$

If the counting interval is one second, at least 13 bits are needed for the data alone. To be safe, 16 bits should be used, with the extra bits reserved for housekeeping functions. This data transfer rate of two bytes per second is very modest. It is necessary, however, to consider the true rate of data transfer. If the data are held on the output lines of the counter for only 10 ms, the effective rate is 200 bytes per second.

In order to reduce the timing requirements of the counter, a *latch* can be used on the output lines to hold the data for a predetermined amount of time regardless of the subsequent status of the lines. The operation of a four-bit latch is illustrated in Figure 6.97. Data at the input lines to the latch (A_0, A_1, A_2, and A_3) are transferred to the output lines (B_0, B_1, B_2, and B_3) when the control line is activated. In the absence of a start signal on the control line, the output levels cannot change even though the input levels do. In this way, data can be held on the output lines for a time sufficient for them to be processed by the subsequent data-receiving unit. The timing relations among the various signals are important for proper operation of the latch. The control signal must arrive when there are valid data on the input lines.

A data *buffer* serves a purpose similar to a latch. It holds received data for processing by subsequent units or circuits. Printers, which are inherently slow because they

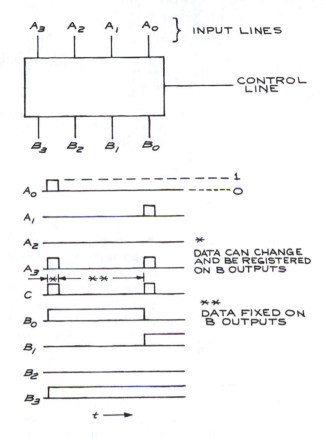

Figure 6.97   A 4-bit latch.

are mechanical devices, often have data buffers that hold the input data until the printing mechanisms can transfer them to paper. The buffer is then ready to accept another batch of data.

6.7.5 Analog Input Signals

The spectroscopy experiment used as an example could be modified by using a detector with an output current proportional to the incident light intensity, as in Figure 6.98. To convert the current to a digital signal requires an ADC. The resolution of the ADC is determined by the number of output bits, and the analog signal must be compatible both in magnitude and in sign with the input requirements of the ADC. If the ADC accepts only a voltage with a full-scale maximum value of 1.0 V and the

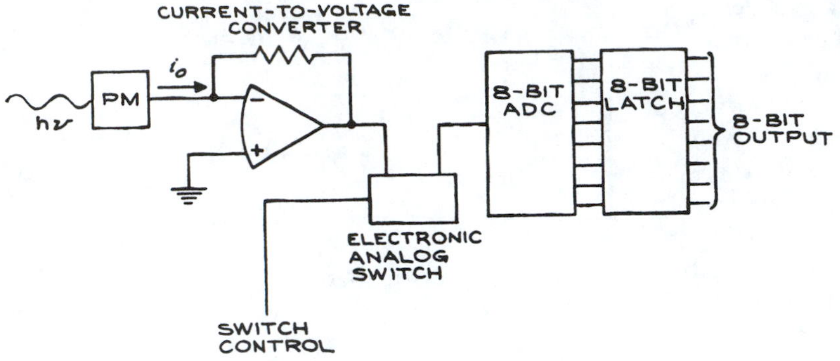

Figure 6.98   Example of a light-measuring system using an ADC with an analog switch at the input to the ADC.

analog signal is a current maximum of 10^{-8} A, conversion from current to voltage can be made using an operational amplifier with a variable-gain element. Response and sampling times are again a consideration. The analog input to the ADC must remain constant long enough for the ADC to respond. This is called the *settling time*. For times less than the settling time, the ADC output will not be a valid reflection of the analog input.

Voltage-to-frequency converters provide an easy way to convert analog signals into digital form. Consider the AD 537 converter manufactured by Analog Devices as an example. Through the selection of two external resistors and a capacitor, frequencies up to 100 kHz can be generated with a maximum input current of 1.00 mA. The circuit for a full-scale input of 1 V and a frequency maximum of 10 KHz is given in Figure 6.99. The combination of the 8.2 kΩ resistor and 5 kΩ potentiometer sets the gain of the circuit. The potentiometer is adjusted so that the output frequency is 10 kHz for a 1 V input. The 20 kΩ potentiometer nulls the offset voltage of the input amplifier. The linearity is 0.05 percent of full scale at 25°C, with a temperature coefficient of +30 ppm of full scale per °C. To minimize noise, a simple low-pass filter is used at the input. The 5 kΩ load resistor R_{out} allows the circuit to drive TTL input stages. The circuit can be used to convert an analog voltage to a digital word by connecting the output to a gated counter. A four-digit decade counter gated for one second will register the input voltage to four significant figures. A similar circuit can be used to convert the output of an analog electrometer to a digital format.

The circuit can also be used to integrate a varying analog signal, as long as the signal variation occurs in a time that is much larger than the reciprocal of the maximum frequency. The area under the peaks in a gas chromatogram, for example, can be recorded by connecting the output from the chromatograph to a voltage-to-frequency converter and using a counter to total the number of output pulses. The scale factor depends on the maximum frequency and full-scale input voltage.

Voltage-to-frequency converters can also be used for frequency-to-voltage conversion. In this configuration, the input can be a TTL pulse train and the output is a voltage. The circuit can be used as an FM demodulator, frequency monitor, or motor speed controller. The AD537 voltage-to-frequency converter, made by Analog Devices, accommodates positive or negative input voltages and negative currents. It operates from a single 5 to 36 V supply and has a maximum full-scale frequency output of 100 kHz.

For the systems under consideration, the output current must not be changing faster than the ADC can respond. When the analog signal is changing too rapidly, a S/H circuit can be used ahead of the ADC. Important parameters of a S/H circuit are the *sampling time* and the *output decay time*. S/H circuits use a capacitor as a memory element, with the sampling time proportional to the capacitance. Because of leakage, capacitors cannot permanently maintain their voltage—fast sampling circuits with small capacitors maintain their output voltages for much less time than slower circuits. The *decay time* is the time for the output to fall to a fixed

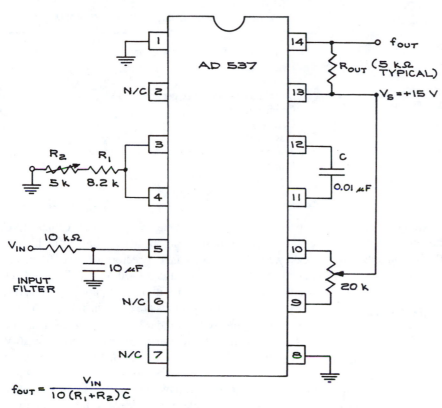

Figure 6.99   Voltage-to-frequency conversion circuit. A 1.00 V input gives a 10 kHz output frequency.

$$f_{OUT} = \frac{V_{IN}}{10\,(R_1 + R_2)\,C}$$

FOR TEMPERATURE STABILITY C TO BE NPO CERAMIC OR POLYSTYRENE

fraction of the original value. A figure of merit for S/H circuits is the ratio of sampling time to decay time. For the simple S/H circuit shown in Figure 6.100, a control signal closes the MOSFET switch, and the 1 μF capacitor is charged with a time constant $R_{on}C$, where R_{on} is the *on* resistance of the switch. Typical values of R_{on} are 10^2 to 10^4 Ω, giving a time constant of 10^{-4} to 10^{-2} s for the sampling time. The decay time depends on capacitor leakage, the *off* resistance of the MOSFET, and the bias currents of the operational amplifier. When high-quality capacitors with polyethylene, polystyrene, polycarbonate, or Teflon dielectrics are used, most of the decay will be due to current through the switch ($\approx 10^{-9}$A) and bias currents (10^{-9} to 10^{-8} A for an average, low bias-current operational amplifier). In this example, the decay will be on the order of 2 to 20 mV/s. Special MOSFET switches

and operational amplifiers with picoampere bias currents can reduce these rates even further.

6.7.6 Multiple Signal Sources: Data Loggers

A multiple-transducer, data-acquisition system is a straightforward extension of the single-transducer system already described. Figure 6.101 shows one possible arrangement. Such systems are useful when it is necessary to monitor and record several different signals at virtually the same time. If the transducers all have analog outputs, the most convenient arrangement is an *analog multiplexer*. The function of the multiplexer is to alternately connect each of the analog signal lines to an ADC circuit and the recording instrument. The switching between the various signal lines is accomplished by separate control lines.

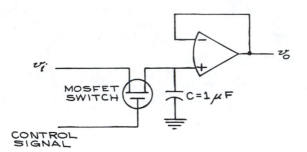

Figure 6.100   A sample-and-hold circuit using a capacitor as memory element.

TABLE 6.35 MODULES AVAILABLE IN NIM

Power supplies
Amplifiers
Discriminators
Delays
Signal generators, pulse generators
Counters/timers
Gates
DACs and ADCs
Analog adders, subtractors, dividers, multipliers
Printers
Signal averagers
Multichannel analyzers

Multiplexing reduces the number of circuit components at the expense of more complex control and synchronization circuits. With the system in Figure 6.101, the data rate can be as much as four times that of a simple system.

6.7.7 Standardized Data-Acquisition Systems

A number of companies manufacture complete data-acquisition units specifically designed for the laboratory. As long as the input signals are compatible and the data rates are within the limits of the unit, they provide a fast and efficient means of constructing a data-acquisition system. Data acquisition from multiple sensors is a common industrial situation and there are several kinds of industrial data loggers. These, however, are designed for reliability rather than flexibility and are generally not suited to laboratory data-acquisition where many different kinds of transducers are used. There are two standardized laboratory data acquisition systems, Nuclear Instrumentation Module (NIM) and CAMAC.

The NIM system consists of individual electronic units that fit into a mounting frame or *bin,* which has a central power supply. The NIM bin is capable of powering several individual units, depending on their size. The bin power supply provides ±24 V, ±12 V, and usually ±6 V. Each full-size bin has 12 slots in which instrument module units can be placed. Units can occupy anywhere from one to six slots. The bin power supplies and individual units have been designed so that a bin can accept the full complement of modules without overloading. A short list of standard NIM modules is given in Table 6.35. Interconnections

between modules are made with coaxial cable. BNC connections are most common, but in some units higher-density, LEMO-type coaxial connectors are used. Input and output specifications are standardized so that NIM modules can be connected together without compatibility concerns. NIM logic level specifications are given in Table 6.36.

Originally the NIM standard voltages were ±12 and ±24 V. As the use of integrated circuits increased, a ±6 V voltage was included in the standard. Modules operating from ±12 and ±24 V supplies, however, are still the most flexible because all bins have these voltages. Where ±6 V is required for a module in a ±12/±24 V bin, it is best to derive it from within the module using the available ±12 V. Use of high-current, 6 V modules in a ±12/±24 V bin should be avoided because the regulation of the ±12 and ±24 V supplies will be degraded by the high current in the common return line. Precise reference voltages should be produced in the modules themselves, since the tolerances of the bin voltages are ±24.00 (±0.7%), ±12.0 (±1.0%), and ±6.00 (±3.0%). The NIM power-connector pin assignments and other details are given in Figure 6.102. The NIM standard logic levels are not compatible directly with common logic families such as TTL or ECL. The power supply voltages are also different from common logic supplies (+5 V) and operational-amplifier supplies (+15 V). Some conversion circuits between NIM, ECL, and TTL levels are given in Figure 6.103.

NIM systems are limited in complexity to hard-wired interconnections between modules. This becomes cumbersome when there are several modules that must be linked together. To circumvent the limitations of NIM systems, the CAMAC system was developed. This system

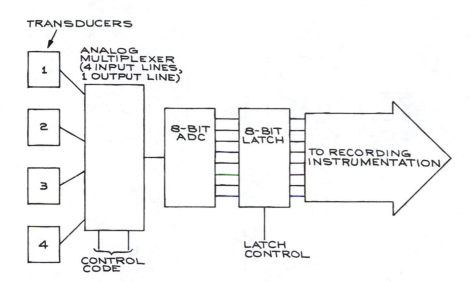

TRANSDUCERS

1

ANALOG
MULTIPLEXER
(4 INPUT LINES,
1 OUTPUT LINE)

2

3

4

8-BIT
ADC

8-BIT
LATCH

TO RECORDING
INSTRUMENTATION

CONTROL
CODE

LATCH
CONTROL

Figure 6.101   A multiple-transducer data-acquisition system using a multiplexer.

retains the use of individual modules that can be plugged into a central power supply unit—in this case, a *crate*. The CAMAC crate, however, has a large number of additional lines besides those for the power.

The additional lines form the *data bus*. The bus has 24 *read* and 24 *write* lines, 4 station *subaddress* lines, and 5-station *function* lines—each station has its own identification line and *look-at-me* (LAM) line. Common to all stations are 2 *timing* lines, 3 *status* lines, and 3 *control* lines. These lines are connected to the control unit (crate controller). Also connected to the controller are the lines from the computer, generally divided into a set of data bits and address bits. The controller interprets the signals from the computer and sends appropriate signals to the individual CAMAC units or *stations*. This is shown schematically in Figure 6.104, with the line designations given in Table 6.37. With each of the units under digital control, it is possible, through the use of a computer connected to the crate controller, to have systems capable of extremely complex logic and control functions. Data can be collected by the computer and stored, displayed, and manipulated to a degree of complexity limited only by the programming language used by the computer.

The CAMAC system offers hardware and software compatibility among individual units. Programming can

be a formidable task because the information transferred between the crate controller and the computer is a binary machine-language code that has no relation to the high-level languages most scientists use for calculations. An alternative is to write the codes in a high-level language, such as FORTRAN, and at appropriate points in the program call a machine-language subroutine that converts the FORTRAN variables to the bit code necessary for control of the crate modules. When done in this way, machine-language programming can be relatively simple.

A CAMAC system can represent a considerable investment in money and time necessary to learn the required programming techniques. The need for a computer to run the system properly—and the computer peripherals such as terminals, printers, and mass storage units—are additional expenses. The advantage of the CAMAC system is the complete flexibility that it provides. As in NIM systems, the individual modules are not costly because there is no need for internal power supplies or complex logic circuits. There are now several manufacturers offering lines of CAMAC crates, controllers, and individual function modules. There are also NIM-CAMAC adapters that permit one to run certain NIM modules under CAMAC control.

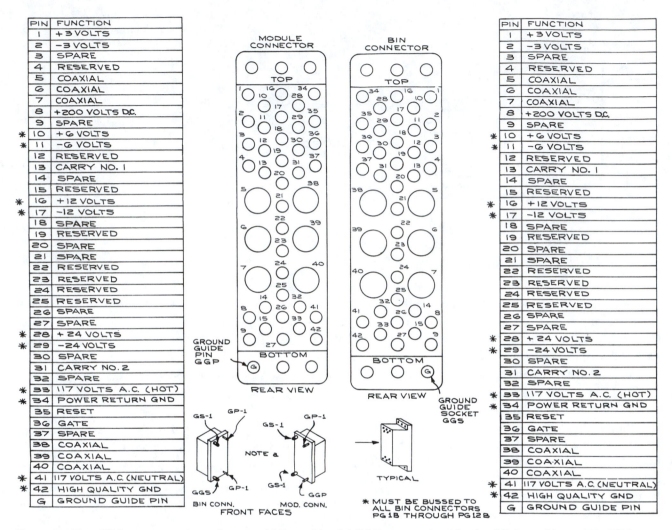

PIN	FUNCTION
1	+3 VOLTS
2	−3 VOLTS
3	SPARE
4	RESERVED
5	COAXIAL
6	COAXIAL
7	COAXIAL
8	+200 VOLTS D.C.
9	SPARE
*10	+6 VOLTS
*11	−6 VOLTS
12	RESERVED
13	CARRY NO. 1
14	SPARE
15	RESERVED
*16	+12 VOLTS
*17	−12 VOLTS
18	SPARE
19	RESERVED
20	SPARE
21	SPARE
22	RESERVED
23	RESERVED
24	RESERVED
25	RESERVED
26	SPARE
27	SPARE
*28	+24 VOLTS
*29	−24 VOLTS
30	SPARE
31	CARRY NO. 2
32	SPARE
*33	117 VOLTS A.C. (HOT)
*34	POWER RETURN GND
35	RESET
36	GATE
37	SPARE
38	COAXIAL
39	COAXIAL
40	COAXIAL
*41	117 VOLTS A.C. (NEUTRAL)
*42	HIGH QUALITY GND
G	GROUND GUIDE PIN

PIN	FUNCTION
1	+3 VOLTS
2	−3 VOLTS
3	SPARE
4	RESERVED
5	COAXIAL
6	COAXIAL
7	COAXIAL
8	+200 VOLTS D.C.
9	SPARE
*10	+6 VOLTS
*11	−6 VOLTS
12	RESERVED
13	CARRY NO. 1
14	SPARE
15	RESERVED
*16	+12 VOLTS
*17	−12 VOLTS
18	SPARE
19	RESERVED
20	SPARE
21	SPARE
22	RESERVED
23	RESERVED
24	RESERVED
25	RESERVED
26	SPARE
27	SPARE
*28	+24 VOLTS
*29	−24 VOLTS
30	SPARE
31	CARRY NO. 2
32	SPARE
*33	117 VOLTS A.C. (HOT)
*34	POWER RETURN GND
35	RESET
36	GATE
37	SPARE
38	COAXIAL
39	COAXIAL
40	COAXIAL
*41	117 VOLTS A.C. (NEUTRAL)
*42	HIGH QUALITY GND
G	GROUND GUIDE PIN

Figure 6.102   NIM connector pin assignment (GP = guide pin, GGP = ground guide pin, GS-1 = guide socket, GGS = ground guide socket).

TABLE 6.36 NIM LOGIC LEVELS

	Output (must deliver)	Input (must deliver)
	Digital Data	
Logic 1	+4 to +12 V	+3 to +12 V
Logic 0	+1 to −2 V	+1.5 to −2 V
	Fast Logic (50 Ω impedance)	
Logic 1	+4 to +12 V	+3 to +12 V
Logic 0	+1 to −2 V	+1.5 to −2 V

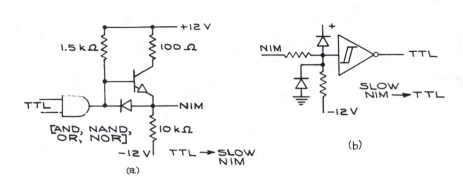

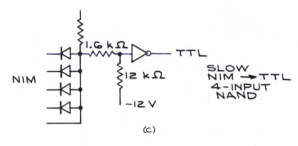

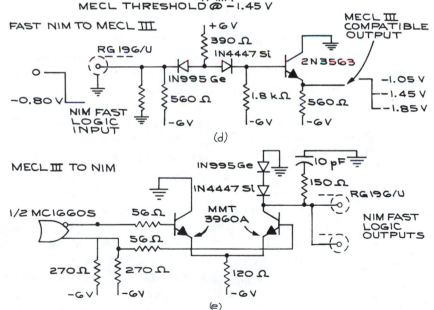

Figure 6.103 (a) NIM-TTL and (b) NIM-ECL conversion circuits (fast NIM levels @ 17 mA: –0.200 V maximum high, –0.600 V minimum low; MECL threshold at –1.45 V). From R. F. Nagel, NIM "Fast Logic Modules Utilizing MECL III Integrated Circuits," *IEEE Trans. Nucl. Sci.*, **NS-19**, 520 (1972). Copyright 1972 IEEE.

TABLE 6.37 CAMAC SIGNAL LINE DESIGNATIONS

Line Code (no. of lines used)	Command	Function
N(1)	Station number	Selects station
A(4)	Sub-address	Selects section of station
F(5)	Function	Defines function
S(2)	Strobe	Controls phase of operation
R(24)	Read	Parallel data to module
W(24)	Write	Parallel data from module
L(1)	Look-at-me	Request for service
B(1)	Busy	Operation in progress
X(1)	Command accepted	Module ready
Q(1)	Response	Status
Z(1)	Initialize	Sets module
I(1)	Inhibit	Disables module
C(1)	Clear	Clears registers
P(5)	Extra lines	
+12V +6V		Power

6.7.8 Control Systems

Control theory establishes a framework that enables one to optimize an active system for accuracy, stability, transient response, resistance to external perturbations, and internal component changes. The method of analysis uses Laplace transforms in a way that is similar to that used in circuit theory (see Section 6.1.5). Control theory is an established engineering subject and there are many excellent texts available. The purpose of this section is to give a brief introduction to the subject and provide general information on how to analyze a system and decide on the proper configuration for accomplishing a given task.

All control systems, whether electrical, mechanical, hydraulic, pneumatic, optical, thermal, acoustical, or electromechanical can be represented in terms of one of the two block diagrams shown in Figure 6.105.

The first system is *open loop*—there is no signal path from the output back to the input. An example is the burner on the top of a stove. Once the position of the burner switch or valve is set, there is no further control over the burner output. The second system is *closed loop*—a portion of the sampled output is sent back to the input and compared with the input signal. An example of a closed loop system is the oven of a stove. The temperature is set by an external input and this input is compared to a signal proportional to the temperature of the oven. If the oven temperature exceeds the set temperature, the oven heat source is turned off, and if the oven temperature is less than the set temperature, the source is turned on.

The standard method for testing an open-loop system is to apply signals at the inputs of each stage, starting with the last and proceeding to the first—verifying the input and output signal at each point. This sequential testing method cannot be used for a closed-loop system because the input to each stage is related to the output of the overall system.

All closed-loop systems have an input, an output, a summing or comparison unit, a controller, the working unit that is controlled, a measuring or sampling unit, and a feedback path. Once mathematical models for each of the units in the system are developed and combined

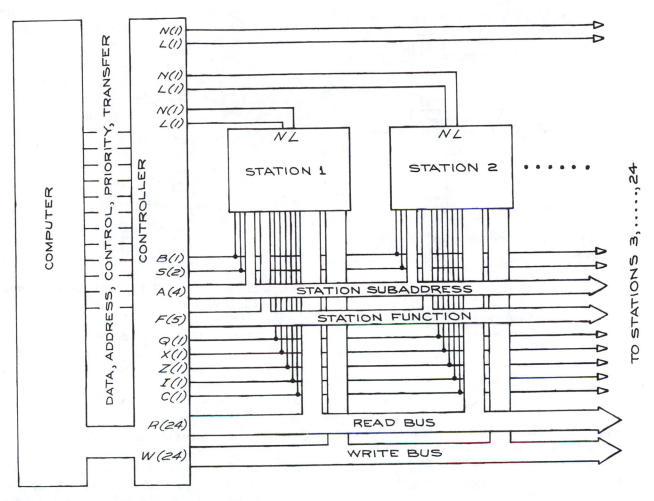

Figure 6.104   The CAMAC bus.

appropriately, the overall performance of the system can be evaluated and optimized by varying the parameters of the model.

Each unit of a control system can be characterized by a transfer or gain function. These functions describe how the input to the unit is transformed by it and converted to the output. The individual transfer functions themselves are, in general, the solutions of ordinary, linear differential equations that are most easily expressed in terms of Laplace transforms (see Section 6.1.5). The full representation of the system in terms of Laplace transforms is particularly easy to assess with respect to stability, accuracy, and transient response.

Consider the general system shown in Figure 6.106 consisting of an input variable $r(t)$, output variable $c(t)$, and transfer functions $g_1(t)$, $g_2(t)$ and $\beta(t)$. The Laplace transforms of these functions are $\overline{R}(s)$, $\overline{G}_1(s)$, $\overline{G}_2(s)$, $\overline{B}(s)$, and $\overline{C}(s)$ respectively. The transfer function for the overall system is

$$g(t) = \frac{g_1(t)g_2(t)}{1 + \beta g_1(t)g_2(t)}$$

with the equivalent Laplace transform

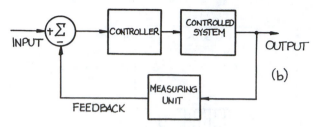

Figure 6.105   Control system block diagram: (a) open loop: (b) closed loop.

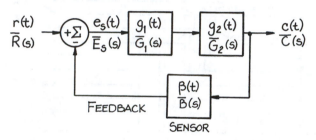

Figure 6.106   Block diagram of a generalized closed-loop control system with schematic representation of the transfer functions and their Laplace transforms.

$$\overline{G}(s) = \frac{\overline{G}_1(s)\overline{G}_2(s)}{1 + \overline{B}(s)\overline{G}_1(s)\overline{G}_2(s)}.$$

$\overline{G}(s)$ can be written as the ratio of two polynomials in s,

$$\overline{G}(s) = \frac{P(s)}{Q(s)}$$

The roots of $P(s)$ give the zeros of the transfer function, and the roots of $Q(s)$ give the poles of the transfer function. The system is stable if the poles of the transfer function lie in the left half of the complex s plane.

System Response

Sensitivity. Sensitivity (S) is a measure of the fractional change in $\overline{G}(s)$ with a fractional change in any

part of the system represented by the variable b. The smaller the S, the less sensitive the system is to changes in any element of the system.

$$S = \frac{\Delta \overline{G}(s)/\overline{G}(s)}{\Delta b/b}$$

As examples, S will be calculated for variations in the controller transfer function $\overline{G}_1(s)$ and for the feedback $\overline{B}(s)$.

Case 1: Relative change in $\overline{G}_1(s)$.

$$S = \frac{1}{1 + \overline{B}(s)\overline{G}_1(s)\overline{G}_2(s)}$$

Case 2: Relative change in $\overline{B}(s)$.

$$S = \frac{-\overline{B}(s)\overline{G}_1(s)\overline{G}_2(s)}{1 + \overline{B}(s)\overline{G}_1(s)\overline{G}_2(s)}$$

For Case 1, S is small when $\overline{B}(s)$, $\overline{G}_1(s)$, $\overline{G}_2(s)$ are large. For Case 2, S is small only if $\overline{B}(s)$, $\overline{G}_1(s)$, $\overline{G}_2(s)$ are small, so a system cannot simultaneously be made insensitive to changes in the transfer function of the controlled unit and transfer function of the feedback element. The solution is to use a high-quality feedback element that remains stable over a wide variety of conditions and increase the transfer functions of the other system elements in order to stabilize the overall system with regard to their changes.

Steady state accuracy. The steady state accuracy of a system $e_{ss}(t)$ is the difference between the input signal $r(t)$ and the closed-loop signal from the output $\beta c(t)$ taken in the limit as $t \rightarrow \infty$. In terms of Laplace transforms, the difference between the input signal and the closed-loop signal from the output is

$$\frac{\overline{R}(s)}{1 + \overline{G}_1(s)\overline{G}_2(s)}$$

In s space, the limit as $t \rightarrow \infty$ is obtained by multiplying

$$\frac{\overline{R}(s)}{1 + \overline{G}_1(s)\overline{G}_2(s)}$$

by s and taking the limit as $s \rightarrow 0$,

$$\overline{E}_{ss} = \lim_{s \rightarrow 0} \frac{s\overline{R}(s)}{1 + \overline{G}_1(s)\overline{G}_2(s)}$$

To evaluate $\overline{E}_{ss}$, it is necessary to specify the signal $\overline{R}(s)$. For a unit step input $\overline{R}(s) = \frac{1}{s}$ (see Table 6.2) so that

$$\bar{E}_{ss} = \lim_{s \to 0} \frac{1}{1 + \bar{G}_1(s)\bar{G}_2(s)}$$

If $\bar{G}_1(s)\bar{G}_2(s)$ is of the form

$$\frac{F(s)}{s^N Q(s)}$$

where $F(s)$ and $Q(s)$ are polynomials in s and $N \geq 1$, E_{ss} will approach zero and the output will approach the input.

Proportional, integral, and differential (PID) control.

A general type of controller is one that provides an output that is the sum of three terms—one proportional to the difference between the reference signal and controlled signal, one proportional to the time integral of the difference, and one proportional to the time differential of the difference. These *PID* controllers use operational amplifiers configured as direct amplifiers, integrators, and differentiators to attain these functions. The Laplace transforms of the three functions give a transfer function that has a form that depends on the details of the operational amplifier circuits. As an example consider the circuit in Figure 6.107. It has the transfer function

$$\bar{G}(s) = \frac{K(1 + sR_1C_1)^2}{sR_1C_1(1 + sR_2C_2)}$$

for

$$\frac{R_1 + R_2}{R_1} = \frac{C_1}{C_2}$$

$\bar{G}(s)$ can be rewritten as the sum of three terms

$$\bar{G}(s) = K\left[1 + \frac{1}{sR_1C_1} + \left(\frac{R_1}{R_1 + R_2}\right)\frac{1 + sR_1C_1}{1 + sR_2C_2}\right]$$

where the first term is the P, the second the I, and the last the D. The unit represented by the transfer function $\bar{H}(s)$ has been added to limit the frequency response and increase the overall stability of the system. Commercial PID controllers allow one to adjust the absolute as well as the relative weights of each of the terms to attain the optimum overall system response. This generally means a response that has the maximum bandwidth consistent with a phase margin of approximately 60°.

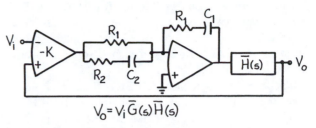

$$V_o = V_i \bar{G}(s)\bar{H}(s)$$

Figure 6.107 PID controller circuit.

6.7.9 Personal Computer (PC) Control of Experiments

Computer operation of experiments is best suited to routine measurements on systems with a high degree of reliability—the latter condition not always being met for ordinary laboratory experiments. It is not worthwhile to add a computer to an experiment that is intended to make a limited number of specialized measurements. Though it may be tedious, it is often much more efficient to take measurements by hand rather than purchase specialized measuring instruments, purchase a computer or a computer interface and software, connect them together, write the necessary program, and put the system into operation. On the other hand, routine measurements in an experiment that continually produces large amounts of data is well suited to computer operation. In any case, before considering computer control it is important to completely understand the experiment. Special attention should be given to the range of variables to be measured and controlled, the response time of the system, and the accuracy and precision of the quantities that are measured and controlled.

Two classes of computer-based measurement and control systems will be discussed. The first is based on the IEEE 488 or GPIB bus and requires measuring and control instruments compatible with the bus as well as the computer, interface circuit, and software. The second type of system is based on an interface circuit that either fits into one of the computer expansion slots or resides in a separate chassis and is connected to the computer through a standard data transmission link like the RS 232 serial interface. With the IEEE 488 system, sensors and transducers are connected to individual instruments, which, in turn, are connected to the IEEE 488 bus. In the second type of system, by contrast, sensors and transducers are directly connected to the interface board via screw terminals. In both systems, some software is needed for the host computer.

It is possible to purchase an IEEE 488 interface card for a PC that will effectively control up to 12 IEEE 488-compatible instruments, provided that there are no severe timing or synchronization requirements. An advantage of the IEEE 488 system is that all signal conditioning and digital-to-analog conversion is done within the individual instruments attached to the bus. The IEEE 488 bus is separate from the bus structure of the control computer, and communication is entirely through the interface. Adapters for converting from the IEEE 488 bus to the RS 232C interface are readily available. IEEE 488 data paths are eight bits (1 byte) wide, and a complete data transmission consists of two serial eight-bit bytes. Each unit connected to the bus is autonomous with its own power supply. Connections to the bus are through standard 24-pin connectors (see Figure 6.108) arranged so that they can be stacked one on top of the other in a daisy chain configuration. There can be up to 15 interconnected devices on a single bus, and the total transmission path length cannot exceed 20 meters. Instruments connected to the bus are designated as *talkers, listeners,* and *controllers.* A tape reader is a talker, for example, while a signal generator is a listener. Some instruments may be able to talk *and* listen, or even (like a variable-range frequency counter) talk, listen, *and* control. The bus itself is passive, and all the circuitry enabling an instrument to talk, listen, or control is contained within the instrument itself. Simple systems built around the bus need not even use a controller but can have just one talker and one listener, as long as the instruments have the built-in interface circuitry that allows their functions to be assigned under some form of local control. The more usual configuration with a personal computer uses interface messages to assign the function of talker or listener to instruments on the bus.

For PC-based IEEE 488 systems, there is much more variation among the software offered by the different manufacturers than among the hardware interface cards. There are three criteria in the choice of software—ease of use, flexibility, and speed. Software that is easy to use is generally slow and cannot be readily adapted to special tasks. For special applications, the software must have many options and a variety of functions. This requires that the experimenter spend additional time organizing and assembling the final program. If programs are to be fast they must be written in assembly language—a time-consuming task.

A simple PC-controlled experiment for measuring the dielectric constant of a liquid as a function of temperature is shown in Figure 6.109. In this experiment, a digital impedance bridge, lock-in amplifier, and digital voltmeter are connected to the computer via an IEEE 488 bus. The dielectric constant and conductivity of the liquid are calculated from the measurement of the capacitance and resistance of a liquid-filled cell. Measurements are made over a range of temperatures, and a digital platinum resistance thermometer measures the thermostat temperature. Programming the IEEE 488 bus usually requires proficiency in a high-level language such as BASIC and complete knowledge of the operation of each instrument on the bus and its device-dependent commands. For the system in Figure 6.109, timing is not a consideration and the measurement sequence can follow the simple flow chart in Figure 6.110.

Implementing the flow chart with the IEEE 488 bus requires knowledge of the characteristics of the instruments on the bus. The thermometer is a *talker*, and the digital voltmeter is a *listener* when its ranges are being set and a *talker* when sending the voltage readings to the computer. To begin, the address of each instrument is set with switches on their back panels. The instruments are wired in daisy-chain fashion to the computer with ribbon cable and the IEEE 488 piggyback plugs. Enough time must be provided in the program for instrument readings to stabilize. This is done with *wait* loops inserted in the program (see Figure 6.111). The input and output data formats for each instrument must also be accommodated in the program so that the commands and data from the computer are correctly read by the instruments and the data from the instruments correctly read and stored by the computer. Even for this relatively simple system, the BASIC program requires several lines of code. For more complex systems with special timing requirements, simple BASIC programs may be too slow or too inflexible. When this is the case, systems with separate software packages are required. Such systems are offered by Keithley Instruments and National Instruments.

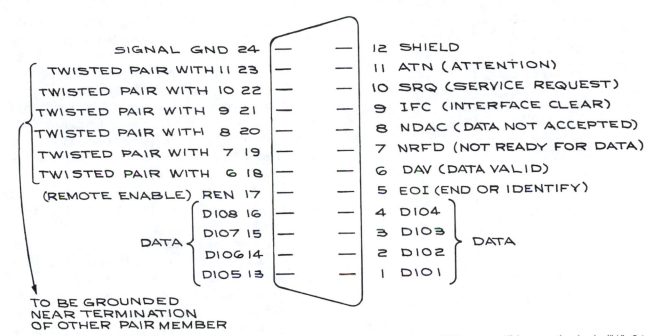

Figure 6.108   IEEE-488 interface-bus connector pin assignment. Logic levels are TTL-compatible negative logic ("1", 0 to 0.4 V; "0", 2.5 to 5.0 V).

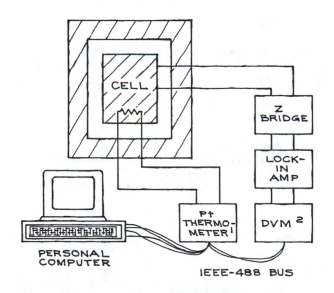

Figure 6.109   A personal-computer-controlled IEEE-488 system: 1, Guildline model 9540 digital thermometer; 2, Hewlett-Packard model 3455A digital voltmeter.

Data acquisition and control systems based on I/O modules and boards that mate directly to a personal computer are justified when there are a large number of standard sensors and the cost of separate IEEE 488 instruments is too great. For an I/O board system, it is necessary to consider

- The number of input channels
- The number of output channels
- Input signals and common-mode levels
- Sampling rates
- Measurement precision
- Computer and software requirements[9]

Interface boards can be programmed directly in BASIC once the addresses of all the I/O ports are known. As with all dynamic systems, timing is crucial and routines must be written to allow time for ADC conversions and data storage. Interpreted BASIC programs are slow—compiled BASIC has a considerable speed advantage. Optimum speed and ease of programming are usually accomplished with software supplied by the manufacturers specifically for their interface boards. Such software has machine

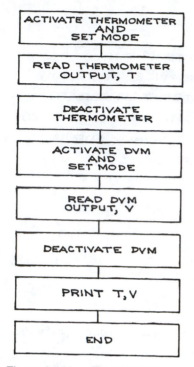

Figure 6.110   Flow chart for measurement sequence in system of Figure 6.109.

language subroutines for initializing the ports and reading and writing data. These routines are designed to be easy to use and can be called from within a BASIC program. The more sophisticated manufacturer-supplied software is menu driven with true foreground-background computer operation and graphic data display with full interfacing to standard spreadsheet packages. Since much more time is spent with software than hardware, it is important that the programming materials supplied by the manufacturer be clear, easy to use, and well documented.

A typical small, commercial, data-acquisition interface circuit is shown in Figure 6.112a. The details of analog input circuits are shown in Figure 6.112b. In this example, there are four single-ended or two differential analog inputs, each connected to a S/H circuit and analog multiplexer. The output of the multiplexer goes to a unit-gain instrumentation amplifier, a programmable-gain amplifier signal-conditioning network, and a second S/H circuit. The output of the second S/H is connected to a twelve-bit ADC, whose output goes to the internal interface bus. A board of this type plugs directly into the expansion slot of a PC. Connections to the input circuits are made through the screw terminals on the board.

```
10  OPEN 7,10            Thermometer assigned device number 10
20  PRINT #7, "CR"       Set thermometer to Celsius, continuous reading
30  FOR I=1 TO 1000: NEXT I   Wait loop
40  INPUT #7, TC$        Read thermometer output
50  FOR I=1 TO 2000: NEXT I   Wait loop
60  TC=VAL(MID$(TC$))*10E-4   Scale the output data
70  CLOSE 7              Deactivate thermometer
80  OPEN 8,22            DVM assigned device number 22
90  A$="F1R7T1M3A1H1D0"  Set DVM to DC volts, auto range, internal
100 PRINT #8,A$          trigger, math "off", autocal "off", high
                         resolution "on", data ready request "off"
110 FOR I=1 TO 2000: NEXT I   Wait loop
120 INPUT #8,B$          Read DVM
130 FOR I=1 TO 5000: NEXT I   Wait loop
140 CLOSE 8              Deactivate DVM
150 VC=VAL(B$)           Convert B$ to a number
160 PRINT "TC=";TC;" VC=";VC   Output data
170 END
```

Figure 6.111   IEEE-488 BASIC program example.

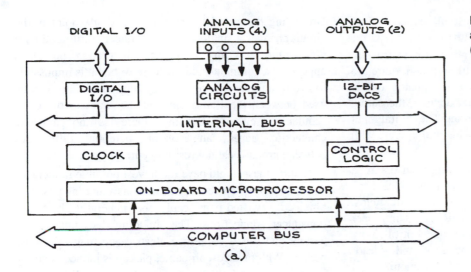

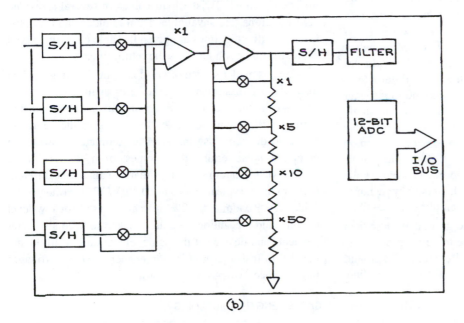

Figure 6.112   (a) Data-acquisition interface; (b) analog input circuits.

S/H circuits capture the analog input signals and hold them long enough for the ADC to convert. The operation of the S/H has already been discussed. Of particular importance are the *acquisition time,* the *aperture time,* the *decay time,* and the *slew rate* (the maximum large-signal change that the output of the S/H circuit can produce).

Input circuits can typically accommodate full-scale input voltages of 20 mV to 10 V. The number of samples per second is from 5000 to 10,000. With the twelve-bit ADC, the system has a maximum resolution of one part in 4000, or 0.025 percent. In practice, this resolution cannot be realized for low-level input signals because of amplifier zero and gain drifts, noise, common-mode error signals, and ADC nonlinearities. These must all be taken into account when assessing the final accuracy of the system.

Important analog output specifications for instrument control are the resolution and the settling time required to achieve the quoted resolution. Since the analog output voltages are the input signals to other devices, the output voltage and current ranges are important. The digital input and output ports are usually 8 bits wide and operate with TTL levels.

The primary concern when connecting sensors to a data-acquisition system is elimination of noise. Noise can be generated on the signal lines from magnetic fields, electric fields, thermal gradients, friction, and improper grounding. These noise sources can be controlled by proper choice of input and output impedances and of lead-wire composition, length, shielding, and placement.

High output-impedance, current-source transducers are less sensitive to magnetically induced noise than low output-impedance, voltage-source transducers. It is important to distinguish between ground and common (or return) lines in transducer circuits. Ground lines do not normally carry current, but return lines do. Return lines carry the signal currents that complete the transducer circuit. Operational amplifiers normally have a single output terminal and terminals for positive and negative power supply voltages. There is no separate common terminal, and the return current must flow in the power-supply lines. For this reason, it is important to bypass the power-supply leads with large, high-frequency ceramic disk capacitors located as close to the amplifier as possible.

When a return line is connected to ground at more than one point in the circuit, it is possible to develop a voltage across the line due to small differences in ground potentials and the finite resistance of the return line. Grounding the signal return line at a single point in the circuit changes a single-ended transducer configuration to *differential*, and any difference in ground potentials appears as a common-mode voltage. When it is impossible to avoid multiple grounds in a system, an isolator can be used between the transducer and the input to the data-acquisition circuit. Because there is no galvanic connection across an isolator, potential differences between grounds cannot result in current flow.

Return lines should have the lowest possible inductance and should overlap the signal lines as much as possible. Shielded, twisted-pair cable is generally best for transducer connections. The shielding does not carry a current and serves to reduce electric and magnetic noise pickup. With twisted-pair cable, pickup is induced equally in both wires and can be canceled by an amplifier with a differential input. Twisted-pair cables in several sizes and configurations are available from Belden and Alpha. Cable-length guidelines are given in Table 6.38. In coaxial cable, the outer conductor is the return line.

A much smaller source of noise is static discharges when insulators of dissimilar materials rub against each other. This is called *triboelectric induction* and can be eliminated with special low-noise cable that has graphite lubricant between dielectric surfaces. The ultimate obstacle to complete noise elimination comes from thermoelectric effects when junctions of dissimilar metals are at different temperatures. Typical values are 0.1 μV/°C. Thermoelectric solder (70% cadmium, 30% tin) can be used for low-level circuits, and precautions should be taken to keep all circuit elements and devices at the same temperature. Carbon and metal-film resistors produce larger thermoelectric voltages than precision wire-wound resistors.

TABLE 6.38 INSTRUMENTATION CABLE LENGTH GUIDELINES

Signal Source	Bandwidth (Hz)	Accuracy (%)	Maximum Cable Length[a]	
			(ft)	(m)
4–20 mA	<10	0.5	1000–5000	300–500
±1– ±10 V	<10	0.5	50–300	15–90
±10 mV– ±1 V	<10	0.5	5-100	1.5–30

[a]Shielded twisted pairs.

Once the sources of noise in the system have been reduced to acceptable levels, input filtering can be used to improve the signal-to-noise ratio. Filtering cannot, however, eliminate the effects of improper wiring.

Example of sensor and control electronics.

In this example, a silicon substrate inside an ultra-high vacuum chamber is to be heated to a predetermined temperature and then maintained at that temperature while an electron gun directs electrons at the surface and a mass spectrometer monitors the masses of the atoms and molecules emitted from the surface. An ion gauge monitors the pressure in the experimental chamber. The data from the experiment are count rates as a function of mass. The rates are recorded and stored for subsequent analysis. A schematic representation of the system is given in Figure 6.113. The system is a mixed digital and analog system with a computer supplying control and data acquisition functions and the heater and thermocouple circuits operating with analog signals. D/A and A/D circuits provide the translation between the digital and analog domains.

It is most convenient to divide the overall system into subsystems and determine the operating conditions for each of them. In the example, the subsystems are

- Substrate/heater
- Thermocouple temperature monitor
- Digital multimeter (DMM)
- Picoammeter current monitor
- Mass spectrometer
- Electron gun
- Ion gauge

The substrate/heater and thermocouple temperature monitor form the temperature-control system that both sends and receives signals to and from the computer. The electron gun receives signals from the computer while the DMM and ion gauge send signals. The mass spectrometer both receives and sends signals. The signals it receives set the mass range and scanning rate. The signals it sends are pulses from the detector.

There are a number of manufacturers of computer interface boards that are specifically designed for the control and monitoring of experiments—among them are National Instruments (manufacturer of LabVIEW), Hewlett-Packard

(Agilent), Keithley Instruments, and Datel. All have catalogs that describe their products in detail and provide very useful information on the general configuration of data-acquisition and control systems. The purpose here is to provide an organizational plan and guidelines.

The first task is to define the operating parameters for each of the subsystems. For the substrate/heater, this means defining the input power necessary to obtain the required temperature of the silicon substrate. Once the power is known, it is possible to define the power supply requirements and its programming input requirements. For this example, the heater is a coil of tungsten wire that has a resistance of 3 Ω at room temperature. The required maximum power was determined to be 81 W (27 V at 3 A). This is supplied by a 300 W power supply with a range of 0 to 30 V at 0 to 10 A. The power supply is voltage programmable with variable gain, so that an input voltage from 0 to 10 V is sufficient to change the output from 0 to 30 V when the internal gain of the power supply is set to 3. The outputs of D/A cards used for power supply control are generally 0 to 1 and 0 to 10 V. When this is not sufficient to drive the output of the power supply to the required maximum level, an intermediate gain stage will be required. The supply that is used is current limiting. For this application, the current-limiting control is set to 3 A so that 3 A are available at all voltage settings. This prevents excessive heating when the filament is cold—a necessary precaution for tungsten that has a temperature coefficient between 4.5 and 8.9×10^{-3} 1/K.

Because the programming voltage to the heater power supply is in the range of 0 to 10 V, a computer interface card with a digital to analog converter is used that has a maximum analog output voltage of 10 V. The input to the card is a twelve-bit code giving a maximum possible resolution of one part in 4096 or 0.025 percent at 30 V output. At 27 V, the resolution is 0.0225 percent, slightly exceeding the maximum available resolution. The thermocouple is a type K NiCr/NiAl, with a working temperature range of −200 to 1300°C. It is non-magnetic and compatible with the ultrahigh vacuum environment where it is located. The thermocouple wires are crimped to a copper lug that is fixed to a stud on the stage holding the silicon substrate. The wires are connected to the leads of a UHV thermocouple electrical feedthrough. The leads on the atmospheric pressure side of the feedthrough are

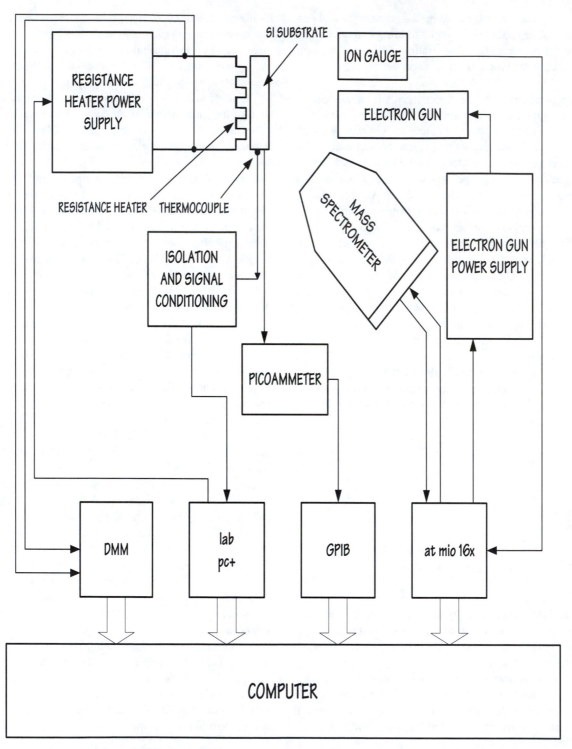

Figure 6.113   System for controlling the temperature of a silicon substrate while monitoring the atoms and molecules emitted from its surface under electron bombardment.

connected to the signal conditioner/isolator unit that converts the output of the thermocouple to an analog voltage that is changed to a temperature in the software that interprets the output from the interface card. A block diagram of the substrate temperature-control system is shown in Figure 6.114.

In this figure, the individual functions are represented by blocks with the Laplace transform of the transfer function for each block written inside the block (see discussion of Laplace transforms in Section 6.1.5). The transforms are obtained from consideration of the differential equations that relate the output of each function to the input. The *compensator* is a proportional/integral type with constants K_1 and K_2 adjusted to attain the optimum response and stability of the circuit (see Section 6.7.8). One detail to be noted is that the output power of the filament amplifier must be linearly related to the difference between the set temperature T_i and the output temperature T_o for a Laplace transform analysis to be used.

Determination of block parameters.

The time response of the D/A, amplifier, filament, thermocouple, amplifier/filter, and A/D are obtained from the specifications of the individual elements. The constants K_9 and K_{10} were chosen so that the input to the A/D in the thermocouple circuit is 0 V at approximately −100°C and 4 V at 1000°C. For the system under consideration,

$$K_3 = 1.2 \times 10^{-5}\,s$$

$$K_4 = 5 \times 10^{-5}\,s$$

$$K_5 = 1 \times 10^{-3}\,s$$

$$K_6 = 1 \times 10^{-3}\,s$$

$$K_7 = 0.3535$$

$$K_8 = 86.67\,s$$

$$K_9 = 0.25$$

$$K_{10} = 1.2 \times 10^{-5}\,s$$

The remaining parameters—thermal resistance R_T, and heat capacity C_T—are characteristic of the experimental arrangement and are determined experimentally from heating and cooling curves and a measurement of the maximum temperature at maximum power input. The assumption is made that the heat loss from the substrate to the surroundings is proportional to the temperature difference between the substrate and surroundings ($T_o - T_s$). The initial slope of the heating curve dT_o/dt (T_o as a function of time for a constant heat input) when the substrate is at the temperature of the surroundings and the heat loss is zero is

$$\frac{dT_o}{dt} = C_T \frac{dQ}{dt}$$

where C_T is the heat capacity of the substrate and dQ/dt is the rate at which heat is added to the substrate. Knowing dQ/dt the input power, and dT_o/dt C_T can be calculated. R_T is correspondingly determined from the asymptotic temperature of the substrate for a constant power input. Assuming heat loss proportional to ($T_o - T_s$), the thermal resistance of the system R is given by $(T_o/T_s)/$ input power. The values of C and R can be checked by a cooling curve measurement. With the substrate initially at $T_o > T_s$ and no heat input, the time for the temperature to fall to $1/e$ of ($T_o - T_s$) is $RC = 300$ s. This is consistent with the values for R_T and C_T of 5.6 K/W and 54 J/K.

Because the time constant for the substrate and load is at least two orders of magnitude larger that any other time constant in the system, it is possible to simplify the circuit to the one shown in Figure 6.115a, in which the overall transfer function of the D/A, amplifier, filament, and feedback transfer function are all set equal to 1. In Figure 6.115b, the substrate, load, and summing junction are replaced by a single transfer function

$$\frac{R_i}{1 + \tau s}$$

where R_i is the effective thermal load equal to $R + T_s/Q_i$. In this case, $\tau = RC = 300$ s, T_s is the temperature of the surroundings, and Q_i is the power input to the substrate at the set temperature T_o. It should be noted that the value of R_i will change with the temperature of the surroundings and the set temperature of the substrate. To maintain the substrate at 500 K, the system requires a heat input of 36 W, giving a value of 13.9 K/W. The overall transfer function for Figure 6.115b, $\overline{T}_o(s)/\overline{T}_i(s)$, is

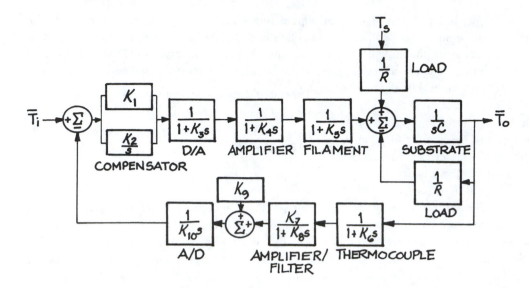

Figure 6.114 Block diagram of the silicon substrate-temperature control system.

$$\bar{T}_o(s)/\bar{T}_i(s) = \frac{13.9(K_1 s + K_2)}{300 s^2 + (13.9 K_1 + 1)s + 13.9 K_2}$$

$$= \frac{\alpha s + 1}{21.6 \beta s^2 + (0.072\beta + \alpha)s + 1}$$

where $\alpha = K_1/K_2$ and $\beta = 1/K_2$. The response and stability of the system are optimized by the appropriate choice of K_1 and K_2. Figure 6.116 shows the frequency dependence of the system transfer function for different values of α and β. This can be done empirically or by using standard control-analysis methods—some of which are discussed in Section 6.7.8 and in more detail in the references at the end of the chapter. Programs such as Mathematica™ and Matlab™ have special packages for control analysis and design.

A DMM card is used to monitor the voltage across the leads of the heater and compare the results with the programmed voltage. The DMM is specified to have a resolution of 5-1/2 digits ±1 in the next to least significant digit. When set on the 10 V range, the card can resolve 1 mV (0.01%)—more than adequate for verifying the output voltage of the power supply considering that the maximum resolution of the DAC is only 0.025 percent.

The at-mio-16x interface board from National Instruments is supplied with software and connectors. It can be installed in any available sixteen-bit expansion slot on a PC. It has 16 single-ended and eight differential analog input channels of 16 bits each, and a maximum sampling rate of 100 ksamples/s. Gains of 1, 2, 5, 10, 20, 50, and 100 are software selectable. The output of the ion gauge is connected to one of the analog input channels. There are two analog output channels driven by sixteen-bit DACs. The two channels are used to control the electron gun and mass spectrometer power supplies. The maximum update rate is 100 ksamples/s and the output software-selectable voltage ranges are 0 to 10 V unipolar, and ±10 V bipolar. The board also has eight digital TTL-compatible I/O lines and three independent sixteen-bit counter/timers. One of the counters is used to register the output counts from the mass spectrometer.

A standard GPIB interface board is used to monitor the output of the picoammeter that registers the current from the electron gun that is collected by the silicon substrate.

6.8 EXTRACTION OF SIGNAL FROM NOISE

6.8.1 Signal-to-Noise Ratio

With many experiments, an important consideration is the minimum detectable signal that can be recovered. This depends on the *signal-to-noise ratio*[10] (SNR or S/N) rather

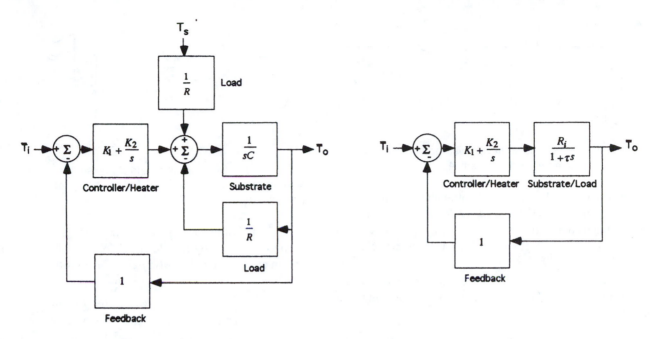

Figure 6.115 (a) Simplified block diagram of the silicon substrate-temperature control system. (b) Block diagram with load and substrate transfer functions combined.

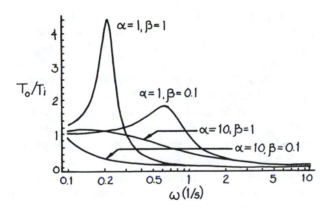

Figure 6.116 Frequency dependence of steady state transfer functions of a range of system parameter values.

than the absolute value of the signal. In counting experiments, where the minimum signal is one event, the uncertainty associated with C_S signal counts is $\sqrt{C_S}$ so that the relative uncertainty is $\sqrt{C_S}/C_S$ or $1/(\sqrt{C_S})$.

In principle, one could attain arbitrarily high precision by counting for long times, but this is often not practical because of changes in the experimental system.

With analog systems, where the signal is a voltage, current, or charge, the SNR depends on the source resistance and capacitance, shot noise, $1/f$ or flicker noise, and amplifier noise. In this discussion, we assume that sources of noise such as rfi do not exist. The noise will be assumed to be confined to the source and preamplifier and to represent the theoretical limit for such a combination.

Consider a detector that is a source of voltage with output impedance R_S, connected to an amplifier with a gain A_0 as shown in Figure 6.117. *Johnson noise* is associated with the random motion of electrons in R_S. Its rms value over a frequency bandwidth Δf in Hz is

$$v_{\text{Johnson}} = \sqrt{4kTR_s\Delta f}$$

where k is Boltzman's constant (1.38×10^{-23} J/K), and T is the absolute temperature of the resistor. Johnson noise is frequency independent—that is, the rms noise per unit bandwidth (noise density) is the same at all frequencies.

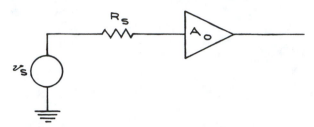

Figure 6.117 Voltage source v_S with output impendance R_s connected to an amplifier of gain A_o.

For this reason, Johnson noise is often called *white noise*. For the purposes of circuit analysis, R_S can be represented by a voltage source giving a voltage $v_{Johnson}$ in series with a noiseless resistor R. The noise current from this source is $i_{Johnson} = v_{Johnson}/R$.

Shot noise has its origin in the discrete nature of the electronic charge. The shot current noise accompanying a dc current I is

$$i_{Shot} = \sqrt{2qi\Delta f}$$

where q is the electronic charge in coulombs. Like Johnson noise, shot noise is frequency independent.

Flicker or *1/f noise* tends to dominate Johnson and shot noise below 100 Hz. The frequency dependence of this noise is of the form $1/f^n$, where n is usually between 0.9 and 1.35. Flicker noise imposes an important limitation on the SNR at low frequencies. For this reason, measurements should, if at all possible, be made at frequencies where white noise dominates.

6.8.2 Optimizing the Signal-to-Noise Ratio

An ideal amplifier of gain A_o can represent a real amplifier with current and voltage noise generators at the input terminal. It is usually assumed that the noise sources are frequency independent. Consider the system in Figure 6.118, where v_n and i_n represent the rms voltage and current noise per $Hz^{1/2}$. The SNR is then the output voltage due to the signal alone divided by the total output voltage, including noise:

$$SNR = \frac{v_s}{\sqrt{4kTR_s + v_n^2 + (i_n R_s)^2}\sqrt{\Delta f}}$$

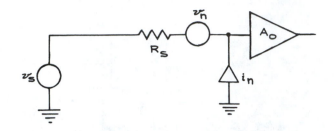

Figure 6.118 Real amplifier and signal source v_s with noise sources explicitly represented by voltage and current generators v_n and i_n.

From this formula it is clear that the SNR can be improved by reducing R_S, Δf, and T. When characterizing noise in an amplifier, the *noise figure* (NF) is often used:

$$NF(dB) = 20\log_{10} \frac{\text{input voltage SNR without amplifier}}{\text{output SNR from amplifier}}$$

An ideal amplifier will have the same SNR at both the input and output terminals. The smallest possible NF is 0, and an NF of 3 dB means that the amplifier has reduced the SNR by two at its output terminals from what it was at the input terminals. Exceedingly good commercial amplifiers can have NFs as low as 0.05 dB over a limited range of frequency and source resistance. To evaluate the importance of the NF, it is necessary to know the input signal level, source resistance, and frequency bandwidth. If the SNR is 10^3 at the input terminals of an amplifier, for example, an amplifier with a NF of 20 will reduce it to 10^2 at the output terminals. This may still be entirely satisfactory for the intended application. Manufacturers of preamplifiers and amplifiers often supply contour plots of constant NF in frequency source-resistance space.

For optimum SNR, it is important to match the source resistance to the amplifier. The optimum value of the source resistor R_{SO} is v_n/i_n. This results in a minimum NF given by

$$NF_{min} = 10\log\left[1 + \frac{2v_n^2}{4kTR_{SO}}\right]$$

where R_{SO} is the optimum value of the source resistance. The source resistance is generally inherent to the source. To transform the source's intrinsic resistance R_S to the optimum resistance R_{SO}, a transformer is used (never an additional series or shunt resistor). If the ratio of secondary to primary turns is α, then $\alpha^2 R_S = R_{SO}$. Although it was stated that the maximum SNR occurs when $R_S = 0$, in practice, all sources have finite resistance and the SNR can be optimized by transformer matching. Photomultipliers, photodiodes, and electron multipliers are best represented (see Figure 6.119) as noiseless constant current sources in parallel with shot-noise and Johnson-noise sources resulting from the addition of a load resistor R_L to the circuit. In this case, the SNR is optimized by using as large a value of R_L as possible and an amplifier with high input impedance and low v_n and i_n. Amplifiers with FET input stages are the best choice. Maximum practical values of R_L are limited by the capacitance C of the input circuit (including cables) and the input impedance of the amplifier. There is no benefit in having R_L greater than the input impedance of the amplifier or so large that the time constant $R_L C$ requires one to work at low frequencies where $1/f$ noise dominates. The justification for using large values of R_L comes from the fact that the voltage signal from a current source is proportional to R_L, while the Johnson noise is proportional to $\sqrt{R_L} R_L$.

An alternative way of specifying amplifier performance is with the *equivalent noise temperature* T_e defined as the necessary increase in temperature of the source resistor to produce the observed noise at the amplifier output—the amplifier being considered noiseless for this purpose—

$$T_e = T(10^{NF/10} - 1)$$

where T is the absolute temperature of the source resistor. Using the example of the amplifier with an NF of 3 dB and a source resistor at 300 K,

$$T_e = 300\text{K}(10^{0.3} - 1) = 300\text{K}$$

In this case, the amplifier introduces an amount of noise equal to the noise from the source resistor.

6.8.3 The Lock-In Amplifier and Gated Integrator or Boxcar

The proper matching of a signal source to an amplifier is the first step to be taken in the recovery of a signal accompanied by noise. Once this has been done correctly,

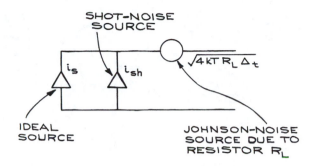

Figure 6.119   Explicit representation of noise sources in a detector.

a number of signal-enhancing techniques can be used to extract the signal.

Because of the difficulty in making dc measurements due to zero drifts, amplifier instabilities, and flicker noise, the signal of interest should, if at all possible, be at a frequency sufficiently high that the dominant noise is white noise. A dc or low-frequency signal can be converted to a higher-frequency by chopping the signal at the higher frequency. It is wise to choose a chopper frequency far from the power-line frequency and its harmonics. Frequencies near known noise frequencies should be avoided.

Signal-enhancing techniques are mainly based on bandwidth reduction. Because the white-noise power per unit bandwidth is constant, reducing the bandwidth will proportionally reduce the noise power. Of course, as the bandwidth goes to zero, the noise and signal power both go to zero as well. Common bandwidth-narrowing circuits are the resonant filter and low-pass filter. For resonant filters, the sharpness of the resonance is given in terms of Q, which is approximately equal to $f_0/\Delta f$, where f_0 is the frequency to which the filter is tuned and Δf is the bandwidth. Practical values of Q for electronic resonant filters vary from 10 to 100. A simple detection system might take the form shown in Figure 6.120. The rectifier converts the ac signal back to dc, so it can be recorded on a chart recorder or read from a meter. With this arrangement, the noise passed by the filter is rectified along with the signal and adds to the recorded dc level.

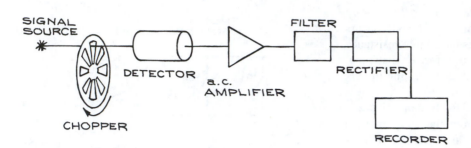

SIGNAL SOURCE

CHOPPER

DETECTOR

a.c. AMPLIFIER

FILTER

RECTIFIER

RECORDER

Figure 6.120   Simple detection and signal-enhancement system.

The effective bandwidth can be substantially narrowed if, in the above arrangement, the ordinary rectifier is replaced by a synchronous rectifier. This kind of rectifier acts as a switch that is opened and closed in synchronization with the chopper. Because the phase relations for each of the noise components are random (a requirement of white noise), they will tend to cancel each other out when averaged—a low-pass filter at the output of the rectifier can be used to do the averaging. The RC time constant of the filter is related to the effective pass band of the system, Δf, by

$$\Delta f = \frac{1}{4RC}$$

for a single-section, low-pass filter, and

$$\Delta f = \frac{1}{4nRC}$$

for n concatenated low-pass filters—each with time constant RC. Quite small values of Δf are possible with such a system. The only limitation is the length of time necessary for the measurement. If Δf is the pass band, the measurement time is approximately $1/\Delta f$ so that it is necessary for the source signal to remain stable over times comparable to $1/\Delta f$. The chopping frequency is normally chosen to be ten times the highest component frequency to be recovered in the source signal.

The system described above is often called a *lock-in amplifier* or *phase-sensitive detector*. The details of the operation of such devices are well documented.[11] As far as signal enhancement is concerned, however, the formulas given above are sufficient for estimating the benefits of such devices. It is good to remember that the SNR is proportional to $1/\sqrt{\Delta f}$, so that narrowing the pass band by

a factor of four increases SNR by a factor of two but increases the time of measurement by a factor of four.

When the signal to be detected has the form of a repetitive low-duty-cycle train of pulses, the lock-in amplifier may not be the best method for signal enhancement. The *duty cycle* is the fraction of time during which the signal of interest is present. With low duty cycles, signal information is available for only a fraction of the total time, while noise is always present. With timing and gating circuits, it is possible to connect the signal line to an RC integrating circuit only during those times when the signal is present. The time constant of the integrator is then chosen to be much larger than the period of the pulse train. The time required for the capacitor to charge to 99 percent of the final voltage level is *4.6RC,* so that five time constants after the first gate opening the capacitor should be charged to within 1 percent of the capacitor's final steady-state value—provided the signal is continuously present. If the signal is present for a fraction γ of the time, the time constant of the integrator must be increased to $5RC/\gamma$. With this system, known as a *gated* or *boxcar integrator,* the SNR is increased only by increasing the time of the measurement. The effective bandwidth of the instrument is $\gamma/4RC$. Table 6.39 compares the important parameters associated with lock-in amplifiers and gated integrators.

6.8.4 Signal Averaging

With a repetitive signal, merely averaging the signal over many cycles can effect improvement in the SNR. Let SNR_0 represent the SNR in one cycle, and N the number

TABLE 6.39 COMPARISON OF LOCK-IN AMPLIFIER AND GATED INTEGRATOR

	Lock-in Amplifier	*Gated Integrator*
Duty cycle, γ	> 0.05	< 0.50
Bandwidth	$1/8\,RC$	$\gamma/4\,RC$
Minimum measurement time	$5\,RC$	$5\,RC/\gamma$
Highest recoverable signal frequency	$1/20\pi\,RC$ for chopping frequency $\geq 1/2\pi\,RC$	$\gamma/20\pi\,RC$ for repetition frequency $\geq 1/2\pi\,RC$
Design notes	1. Determine f_{max}, the highest frequency component of the signal to be recovered. 2. Choose f_s, the chopping frequency, where $f_s = 10 f_{max}$. 3. Choose low-pass filter constants $1/RC = 2\pi f_s$, so as to pass frequency f_{max}. 4. The bandwidth Δf is $1/8\,RC$, and a tuned amplifier with a Q of 10 is sufficient to pass all frequency components of the signal to f_{max}.	1. Required measurement time is $5\,RC/\gamma$. 2. Bandwidth is $\gamma/4\,RC$. 3. Preferred to a lock-in amplifier for signal repletion rate ≤ 10 Hz and low duty cycle.

of cycles over which one averages. The SNR improvement is proportional to $\sqrt{N}$. If each cycle lasts for a time τ, the time T necessary to arrive at a specified SNR is given by

$$T = \left[\frac{SNR_0}{SNR}\right]^2 \tau$$

Lock-in amplifiers and gated integrators are inherently superior to simple signal averaging because of the band-narrowing functions they perform. The bandwidth improvement is only an advantage, however, when the highest frequency component in the signal to be recovered permits a long integrating time.

6.8.5 Waveform Recovery

When it is necessary to know the variation of a signal with time, waveform recovery techniques are needed. The gated integrator can be converted to an instrument for waveform recovery by the inclusion of variable delay and variable gate-width functions. A separate RC network is required for each delay time in such a scheme. A schematic representation of such a system is shown in Figure 6.121. Each separate delay corresponds to a different part of the waveform. The duty cycle for each gate opening is $\tau/T = \gamma$, so that the effective bandwidth is $\gamma/4RC$. An integration time of $5RC/\gamma$ is needed for the charge on each capacitor

to reach a steady-state value. If the waveform has been divided into n parts, the total time of measurement is $5nRC/\gamma$.

Digital schemes for recovering waveforms use ADCs to convert the analog signal level to a digital number, which is then recorded in the appropriate time channel with the aid of delay and gating signals. With such instruments, called *digital signal analyzers,* SNRs are only limited by the length of time of the measurement. The long-term stability of the various components, however, limits the use of very long measuring times. If the absolute value of the waveform is needed, one must record the number of cycles treated by the instrument, and data rates are limited by the conversion speed of the ADC that encodes the amplitude information. An advantage of digital signal analysis is the increased versatility in data manipulation and the production of permanent records on paper or magnetic media. Such data are then available for numerical analysis.

Instruments related to the digital waveform recorders are *transient recorders* and *memory oscilloscopes.* While these offer no gain in SNR over that in the original signal, they do provide outputs that are a record of a single waveform. An inefficient, but commonly used, method of

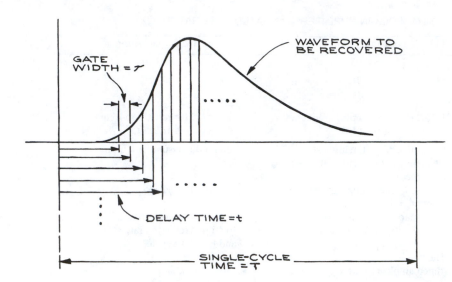

Figure 6.121   Relation between gate width τ, delay time t, and single-cycle time T in waveform recovery. The duty cycle γ is given by τ/T.

SNR enhancement with these instruments is the separate recording of many waveforms and subsequent digital averaging by a computer.

6.8.6 Coincidence and Time Correlation Techniques

There are a few general principles that govern all time correlation measurements. These will be discussed in sufficient detail to be useful in not only constructing experiments where time is a parameter, but also in evaluating the results of time measurements and optimizing the operating parameters.

Correlated events are related in time, and this time relation can be established either with respect to an external clock or to the events themselves. Random or uncorrelated events bear no fixed time relation to each other, on the other hand, their very randomness allows them to be quantified. Consider the passing of cars on a busy street. It is possible to calculate the probability $P_n(n)$ that n cars pass within a given time interval in terms of the average number of cars $\bar{n}$ in the interval where $P_{\bar{n}}(n)$, the Poisson distribution is given by

$$P_{\bar{n}}(n) = \frac{\bar{n}^n}{n!}e^{-\bar{n}}$$

If one finds that 100 cars pass a fixed position on a highway in an hour, for example, then the average number of cars per minute is 100/60. The probability that two cars pass in a minute is given by

$$\frac{0.6^2}{2}e^{-0.6} = 0.10$$

The probability that three times the average number of cars pass per unit time is 0.02. $P_{\bar{n}}(n)$ and the integral of $P_{\bar{n}}(n)$ for $\bar{n} = 1$ are shown in Figure 6.122. It is

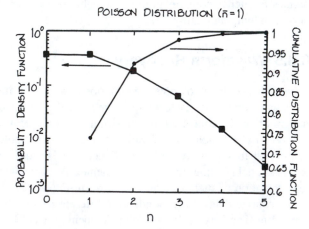

Figure 6.122   Poisson distribution and cumulative Poisson distribution for $\bar{n} = 1$.

worthwhile to note that only a single parameter, the average value $\bar{n}$, is sufficient to define the function. The function is also not symmetric about $\bar{n}$.

This example of passing cars has implications for counting experiments. Consider again the particle counting experiment shown in Figure 6.64a. It consists of a detector, preamplifier, amplifier/shaper, discriminator, and counter. The detector converts the energy of an incoming particle to an electrical signal that is amplified by the preamplifier to a level sufficient to be amplified and shaped by the amplifier. The discriminator converts the signal from the amplifier to a standard electronic pulse of fixed amplitude and time duration, provided that the amplitude of the signal from the amplifier exceeds a set threshold. The counter records the number of pulses from the discriminator for a predetermined period of time. The factors that affect the measurement are counting rate, signal duration, and processing times.

In counting experiments, the instantaneous rate at which particles arrive at the detector can be significantly different from the average rate. In order to assess the rate at which the system can accept data, it is necessary to know how the signals from the detector, preamplifier, amplifier, and discriminator vary with time. If signals arrive at an average rate of 1 kHz at the detector, for example, the average time between the start of each signal is 10^{-3} seconds. If we consider a time interval of 10^{-3} seconds, the average number of signals arriving in that interval is 1—if as many as three pulses can be registered in the 10^{-3} seconds, then 99 percent of all pulses will be counted. To accommodate the case of three pulses in 10^{-3} seconds, it is necessary to recognize that the pulses will be randomly distributed in the 10^{-3} second interval. To register the three randomly distributed pulses in the 10^{-3} second interval requires approximately another factor of three in time resolution. The result is that the system must have the capability of registering events at a 10 kHz uniform rate in order to be able to register randomly arriving events at a 1 kHz average rate with 99 percent efficiency. For 99.9 percent efficiency, the required system bandwidth increases to 100 kHz. Provided that the discriminator and counter are capable of handling the rates, it is then necessary to be sure that the duration of all of the electronic signals are consistent with the required time resolution. For 1 kHz, 10 kHz, and 100 kHz

bandwidths this means signal duration of 0.1, 0.01, and 0.001 ms, respectively. An excellent discussion of the application of statistics to physics experiments is given by Melissinos.[17]

The processing times of the electronic units must also be taken into consideration. There are propagation delays associated with the active devices in the electronics circuits and delays associated with the actual registering of events by the counter. The manufacturers of counters specify processing delays in terms of maximum count rate. A 10 MHz counter may only count at a 10 MHz rate if the input signals arrive at a uniform rate. For randomly arriving signals with a 10 MHz average rate, a system with a 100 MHz bandwidth is required to record 99 percent of the incoming events.

The result that is sought in counting experiments is a *rate* (number of events per unit time). For convenience, we divide the total time of the measurement T into n equal time intervals, each of duration T/n. If there are N_i counts registered in interval i, then the mean or average rate is given by

$$\frac{1}{n} \sum_{i=1}^{n} \frac{N_i}{T/n} = \frac{N_N}{T}$$

where N_N is the total number of counts registered during the course of the experiment. To assess the uncertainty in the overall measured rate, we assume that the individual rate measurements are statistically distributed about the average value. In other words, they arise from statistical fluctuations in the arrival times of the events and not from uncertainties introduced by the measuring instruments.

When the rate measurement is statistically distributed about the mean, the distribution of events can be described by the Poisson distribution, $P_{\bar{n}}(n)$. The uncertainty in the rate is given by the standard deviation and is equal to the square root of the average rate $\sqrt{\bar{n}}$. Most significant is the relative uncertainty

$$\frac{\sqrt{N/t_T}}{N/t_T}.$$

The relative uncertainty decreases as the number of counts per counting interval increases. For a fixed experimental arrangement, increasing the time of the measurement can increase the number of counts. As can be seen from the formula, the relative uncertainty can be made arbitrarily small by lengthening the measurement time. Improvement in relative uncertainty, however, is proportional to the reciprocal of the square root of the measurement time. To reduce the relative uncertainty by a factor of two, it is necessary to increase the measurement time by a factor of four. One soon reaches a point where the fractional improvement in relative uncertainty requires a prohibitively long measurement time.

Coincidence experiments require explicit knowledge of the time correlation between two events. Consider the example of electron impact ionization of an atom in which a single incident electron strikes a target atom or molecule and ejects an electron from it. Because the scattered and ejected electrons arise from the same event, there is a time correlation between their arrival times at the detectors.

In a two-detector coincidence experiment, of which Figure 6.65b is an example, pulses from the two detectors are amplified and then sent to discriminators—the outputs of which are standard rectangular pulses of constant amplitude and duration. The outputs from the two discriminators are then sent to the *start* and *stop* inputs of a TAC or TDC. Even though a single event is responsible for ejected and scattered electrons, the two electrons will not arrive at the detectors simultaneously because of differences in path lengths and electron velocities. Electronic propagation delays and cable delays also contribute to the *start* and *stop* signals not arriving at the inputs of the TAC or TDC at identical times—sometimes the *start* signal arrives first, sometimes the *stop* signal is first. If the signal to the *start* input arrives after the signal to the *stop*, that pair of events will not result in a TAC/TDC output. The result can be a 50 percent reduction in the number of recorded coincidences. To overcome this limitation, a time delay is inserted in the *stop* line between the discriminator and the TAC/TDC. This delay can be a length of coaxial cable (typical delays are on the order of 1 nsec/ft.—see Table 6.11), a lumped delay line, or an electronic delay circuit. The purpose of the delay is to ensure that the stop signal always arrives at the *stop* input of the TAC/TDC after the *start* signal. A perfect coincidence will be recorded at a time difference approximately equal to the delay time. When a time window twice the length of the delay time is used, perfect coincidence will lie at the center of the time window, and it is possible to make an accurate assessment

of the background by considering the regions to either side of the perfect coincidence region. An example of a time spectrum is shown in Figure 6.123.

An alternative to the TAC and PHA/ADC (see Section 6.4.5) is the time-to-digital converter (TDC)—a unit that combines the functions of the TAC and PHA/ADC. There are *start* and *stop* inputs and an output that provides a binary number directly proportional to the time difference between the *stop* and *start* signals. The TDC can be directly connected to a MCA or PC with the appropriate digital interface.

By storing the binary outputs of a PHA/ADC or a TDC in the form of a histogram in the memory of a MCA or PC, analyses can be performed on the data that take full account of the limitations of the individual components of the measuring system. For the preamplifiers and discriminators, time resolution is the principal consideration. For the TAC, resolution and processing time can be critical. The PHA/ADC or TDC provide the link between the analog circuits of the preamplifiers and discriminators, and the digital input of the MCA or PC. The conversion from analog to digital by a PHA/ADC or TDC is not perfectly linear and the deviations from linearity need to be taken into consideration in precise work.

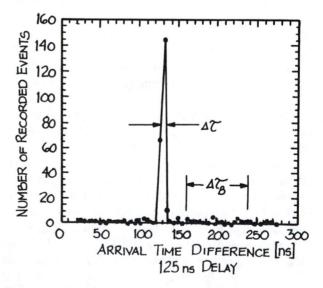

Figure 6.123   Coincidence time spectrum.

To assess the precision of a coincidence measurement, it is necessary to consider not just the event under consideration, but also all other events arriving at the two detectors. The events under study are correlated in time and result in a peak in the time spectrum centered approximately at the delay time. There is also a background level due to events that bear no fixed time relation to each other. If the average rate of the background events in each detector is R_1 and R_2, then the rate that two such events will be recorded within time $\Delta\tau$ is given by R_B, where

$$R_B = R_1 R_2 \Delta\tau$$

Let the rate of the event under study be R_A. It will be proportional to the probability of occurrence of the event p_A, the source function S, and the acceptances and efficiencies of the detectors d_1 and d_2.

$$R_A = p_A S d_1 d_2$$

For the background, each of the rates R_1 and R_2 will be proportional to the source function, the probability for single events p_1 and p_2, and the properties of the individual detectors,

$$R_1 = S p_1 d_1$$
$$R_2 = S p_2 d_2$$

Combining the two expressions for R_1 and R_2

$$R_B = S^2 p_1 p_2 d_1 d_2 \Delta\tau$$

Comparing the expressions for the background rate and the signal rate, one sees that the background increases as the square of the source function while the signal rate is proportional to the source function. The signal-to-background rate R_{AB} is then

$$R_{AB} = \frac{p_A}{p_1 p_2} \frac{1}{S \Delta\tau}$$

It is important to note that the signal is always accompanied by background. We now consider the signal and background after accumulating counts over a time T. For this, it is informative to refer to the time spectrum in Figure 6.123. The total number of counts within an arrival time difference $\Delta\tau$ is N_T and this number is the sum of the signal counts $N_A = R_A T$ and the background counts $N_B = R_B T$:

$$N_T = N_A + N_B.$$

The determination of the background counts must come from an independent measurement, typically made in a region of the time spectrum outside of the signal region, yet representative of the background within the signal region. The uncertainty in the determination of the signal counts δN_A is given by the square root of the uncertainties in the total counts and the background counts:

$$\delta N_A = \sqrt{N_T + N_B}$$

The essential quantity is the relative uncertainty in the signal counts $\delta N_A / N_A$. This is given by

$$\delta N_A / N_A = \sqrt{N_T + N_B} / N_A = \sqrt{N_A + 2N_B} / N_A$$

$$= \sqrt{R_A T + 2 R_B T} / R_A T = \sqrt{\frac{1 + 2/R_{AB}}{R_A T}}$$

Expressing R_A in terms of R_{AB} results in the following formula:

$$\delta N_A / N_A = \sqrt{\frac{(R_{AB} + 2)\Delta\tau}{2}}{\frac{p_A^2}{p_1 p_2} d_1 d_2 T}$$

There are a number of conclusions to be drawn from the above formula. The relative uncertainty can be reduced to an arbitrarily small value by increasing T, but because the relative uncertainty is proportional to $1/\sqrt{T}$, a reduction in relative uncertainty by a factor of two requires a factor of four increase in collection time. Reducing $\Delta\tau$ can also reduce the relative uncertainty. It is understood that $\Delta\tau$ is the smallest time window that includes the entire signal. Using the fastest possible detectors, preamplifiers, and discriminators and minimizing time dispersion in the section of the experiment ahead of the detectors may decrease $\Delta\tau$.

The signal and background rates are not independent, but are coupled through the source function S. As a consequence, the relative uncertainty in the signal decreases with the signal-to-background rate R_{AB}—a somewhat unanticipated result. Dividing $\delta N_A / N_A$ by its value at $R_{AB} = 1$ gives a reduced relative uncertainty $[\delta N_A / N_{AB}]_R$ equal to

$$\sqrt{\frac{(R_{AB} + 2)}{3}}$$

To review this result, consider a case in which there is one signal count and one background count in the time window $\Delta\tau$. The signal-to background ratio is 1, and the relative uncertainty in the signal is

$$\frac{\sqrt{2+1}}{1} = 1.7$$

By increasing the source strength by a factor of 10, the signal will be increased by a factor of 10 and the background by a factor of 100. The signal-to background ratio is now 0.1, but the relative uncertainty in the signal is

$$\frac{\sqrt{110+100}}{10} = 1.45$$

a clear improvement over the larger signal-to-background case.

Another method for reducing the relative uncertainty is to increase the precision of the measurement of background. The above formulae are based on an independent measurement of background over a time window $\Delta\tau$ that is equivalent to the time window within which the signal appears. If a larger time window for the background is used, the uncertainty in the background determination can be correspondingly reduced. Let a time window of width $\Delta\tau_B$ be used for the determination of background, where $\Delta\tau/\Delta\tau_B = \rho < 1$. If the rate at which counts are accumulated in time window $\Delta\tau_B$ is R_{BB}, the background counts to be subtracted from the total counts in time window $\Delta\tau$ become $\rho R_{BB}T = \rho N_{BB}$. The uncertainty in the number of background counts to be subtracted from the total of signal plus background is $\sqrt{\rho^2 N_{BB}}$. The relative uncertainty of the signal is then

$$\sqrt{\frac{1+2\rho^2 N_{BB}/N_A}{N_A}} = \sqrt{\frac{1+2\rho N_B/N_A}{N_A}}$$

The expression for the reduced relative uncertainty is then

$$[\delta N_A/N_{AB}]_R = \sqrt{\frac{(R_{AB}+2\rho)}{3}}$$

The single coincidence measurement can be expanded to a multiple detector arrangement if it is necessary to measure coincidence rates as a function of more than one parameter. This is a much more efficient way of collecting data than having two detectors that are scanned over the range of the parameters. Depending on the signal and background rates, detector outputs can be multiplexed and a detector encoding circuit can be used to identify those detectors responsible for each coincidence signal. If detectors are multiplexed, it is good to remember that the overall count rate is the sum of the rates for all of the detectors. This can be an important consideration when rates become comparable to the reciprocal of system dead time.

It is often the case that signal rates are limited by the processing electronics. Consider a coincidence time spectrum of 100 channels covering 100 ns with a coincidence time window of 10 ns. Assume a signal rate of 1 Hz and a background rate of 1 Hz within the 10 ns window. This background rate implies an uncorrelated event rate of 10 kHz in each of the detectors. To register 99 percent of the incoming events, the dead time of the system can be no larger than 10 ms. The dead time limitation can be substantially reduced with a preprocessing circuit that only accepts events falling within 100 ns of each other (the width of the time spectrum). One way to accomplish this is with a circuit that incorporates delays and gates to only pass signals that fall within a 100 ns time widow. With this circuit, the number of background events that need to be processed each second is reduced from 10,000 to 10, and dead times as long as 10 ms can be accommodated while maintaining a collection efficiency of 99 percent. In a way, this is an extension of the multiparameter processing in which the parameter is the time difference between processed events. Goruganthu et al. discuss a preprocessing circuit.[18]

6.9 GROUNDS AND GROUNDING

6.9.1 Electrical Grounds and Safety

A battery or power supply produces a potential difference between its two output terminals. Only by defining the potential of one of the terminals in some absolute way can the potential of the other terminal be given an absolute value. By convention, the electrical potential of the earth is defined to be zero. When one output terminal of a battery or power supply is connected to the earth, the absolute potential of the other terminal is then equal to the potential difference. Because the surface of the earth is a fairly good conductor of electricity, all points will be at nearly the same potential. A good connection to the earth via a copper rod buried in the ground is the most convenient way of establishing zero potential. Since such connections are not always readily available in the laboratory; one often uses connections to metal water pipes that go underground to define zero potential.

Most connections to the ac line are now made with three-prong plugs and three-conductor wire. The wire is usually color coded in North America—black corresponding to the high-voltage or *hot* side of the line, white to the low-voltage or *neutral* and green to the earth or *ground*. The terminals on ac plugs and receptacles are also coded—the hot terminal is brass, the neutral terminal is chrome plated, and the ground terminal is dyed green. Switches and fuses for use with ac lines should always be placed in the hot line, otherwise the full line voltage will be present at the input even with an open switch or a blown fuse.

The three-wire system is designed for safety. Consider a piece of electrical equipment with a metal case operating from the ac line. Normal practice requires that the case be connected to the green ground wire of the line cord. Should the hot wire contact the case for some reason, a very large current would flow in the hot line and blow a fuse or open a circuit breaker. If the case is not connected to ground, it will assume the potential of the hot line and could give an electric shock to anyone touching it. Of course, if the case were not a conductor, there would be no danger even if the hot line made a connection to it. There are a large number of appliances and hand power tools with so-called *double-insulated* cases that do not require a separate ground connection for safety purposes. Ground-fault (circuit) interrupters (GFI or GFCI) are safety devices that are incorporated in ac power outlets where there is a danger of operators of electrical equipment attached to the outlet touching electrical ground while using the equipment. The GFI senses any current flowing directly to ground and switches off all power to the equipment to minimize electrical shock. GFI electrical outlets are standard in humid environments outdoors and in bathrooms and underground installations.

From this discussion it should be clear that fuses and circuit breakers are important safety elements in electrical circuits. Fuses are generally of either the standard quick-response type or the *slow-blow* type. The former will open whenever its maximum-rated current is exceeded, even momentarily, while the *slow-blow* fuse is designed to sustain momentary current surges in excess of the maximum-rated continuous value. The two types are not interchangeable even though they may have identical maximum continuous-current ratings. Circuit breakers are generally of the electromagnetic-relay or the bimetallic-strip type. They can be reset after opening from an overload.

It is sometimes necessary to *float* or remove all ground connections from a piece of electrical equipment. This is necessary with measuring instruments when a signal between two points in a circuit, neither of which is at ground, is to be measured, amplified, or processed in some way. Power supplies are floated when it is necessary to produce a well-defined potential difference with respect to a nonzero potential. An example is the filament heater supply of a high voltage electron gun. The heater supply is connected across the filament and has one lead at the high voltage. In this configuration, neither heater supply lead is at ground potential. Floating introduces the possibility of additional noise, electrical shock, and electrical breakdown within the circuit with a resulting destruction of components.

At this point it is worthwhile to make the distinction between dc and ac grounds. A *dc ground* connection is a direct low-resistance connection to ground while an *ac ground* connection is one that does not permit dc currents to flow while offering low impedance to ac currents. In Figure 6.124, power supply 1 has the negative terminal and case (G

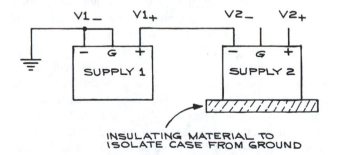

INSULATING MATERIAL TO ISOLATE CASE FROM GROUND

Figure 6.124   Power supply 2 floating on power supply 1. The +, −, and G terminals of supply 2 are assumed to have no direct dc connection to the ground.

terminal) at dc ground. Power supply 2 is floating on power supply 1 and in the absence of any external connections, no current will flow through the terminals of the supplies. Since high-quality power supplies generally have very low output impedances, the ac impedances of the terminals marked $V1_+$, $V2_-$, and $V2_+$ with respect to ground are low, although the potentials at the terminals are not zero. It is good to keep in mind that the impedances to ground are frequency dependent because of the characteristics of the output circuits. At high frequencies, the impedances can become large through the use of components with poor high-frequency characteristics (such as electrolytic capacitors). As a result, transient voltage spikes may not be effectively shunted to ground and may adversely affect the circuits to which the power supplies are connected. A remedy is the use of good high-frequency capacitors of an adequate voltage rating between the terminals $V2_+$, $V1_+$, and ground.

Figure 6.125 illustrates a typical line-operated power supply connected to an ac outlet via a three-wire line cord. Under normal operation, either the negative or the positive terminal is connected to the ground terminal depending on whether a positive or a negative voltage with respect to ground is desired. When floating the power supply, the potential at which the supply is to be floated is applied to either the positive or negative output terminal. This may be satisfactory for levels of as much as a few hundred volts but has its dangers because the internal circuitry is floating while the case is at ground through the green wire in the three-wire line plug. Potentials at least equal to the floating potential plus the potential difference at the output terminals now exist at the terminals of several components in the internal circuitry. Components are often mechanically fixed to the grounded chassis. The electrical insulation between the components and the chassis, as well as that between the primary and secondary windings of the transformer, must be sufficient to sustain the dc voltage difference. In the event of electrical breakdown, components will be destroyed and the shafts of the controls may attain a potential that is equal to the floating potential and thus present a shock hazard. The maximum potential for floating line-powered equipment in this way is generally 600 V.

When floating line-operated equipment at voltages above the rated maximum voltages, it is necessary to completely remove the ground connection to the case. The easiest method is to use a two-terminal to three-terminal ac line-plug converter and not connect the third terminal to ground at the ac outlet. If the case of the instrument is then electrically isolated from all conducting surfaces by placing it in a Plexiglass enclosure, it can be electrically floated at the desired potential without risk of electrical breakdown between the components of the circuit and the case. With the case at a nonzero potential, there is a shock hazard associated with touching it or the controls so that *access must be provided via auxiliary insulated knobs and shafts*. A further difficulty is the possibility of electrical breakdown between the primary and secondary windings of the input transformer. At the primary, the peak voltage values are +155 V, corresponding to the normal peak values of the ac line voltage. The secondary voltage, however, will have a dc component equal to the floating voltage superimposed on the normal ac value. If the resulting dc plus peak-ac secondary voltage exceeds the voltage rating of the transformer insulation, electrical breakdown can occur, destroying the transformer and possibly the rest of the circuit. *Isolation transformers* are commonly used to solve this problem. These transformers have a 1:1 turns ratio and high-voltage insulation between the primary and secondary windings. The primary is connected to the ac line, while the secondary is connected to the primary of the input transformer of the instrument to be floated. In this way, the floating voltage appears across the isolation-transformer windings. The case and controls are still at the floating potential, and precautions similar to those in the former situation must be observed. When specifying an isolation transformer, the power-handling capacity in watts or kilovolt-amperes (KVA) as well as the maximum isolation voltage must be considered.

A substitute for an isolation transformer is two high-voltage filament transformers of the same turns ratio, connected back to back. Such transformers are often used in oscilloscope power supplies. This will produce the necessary 1:1 overall voltage ratio with isolation. Often such transformers are more readily available and less expensive than high-voltage, high-power single-isolation transformers.

When floating instruments that are not directly powered from the ac line, the external power supply (whether it is a battery or line operated) must be floated, too. The

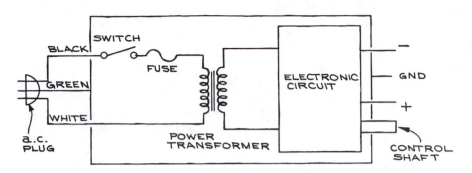

Figure 6.125 Mounting a line-operated power supply so that both output terminals are independent of the ac line ground.

considerations that apply to the floating of line-operated instruments apply here also. Table 6.40 summarizes the important considerations involved when floating power supplies.

6.9.2 Electrical Pickup: Capacitive Effects

It is often necessary to detect and measure electrical signals in the millivolt and even microvolt range. If the signals are accompanied by noise that is generated by external sources, they can be so degraded as to render the measurement useless. The elimination of these external sources—or, if this is not possible, the elimination of their effects—is an important element in the design and construction of electronic measuring, control, and detection systems.

Consider a wire (wire 1) at a potential V_1 in the vicinity of a second wire (wire 2) that is insulated from all conducting surfaces (see Figure 6.126a). Coulomb forces between the charges on the surfaces of the wires will cause polarization of the free charges in wire 2. Because wire 2 is electrically isolated, it will remain electrically neutral. If wire 2 is now connected to ground through a resistance R, a current will momentarily flow due to the presence of the positive charges on wire attracting negative charge to wire 2 from the ground. The equivalent circuit is shown in Figure 6.126b. Typical values of the coupling capacitance are 1 to 100 picofarad (pF). Two twisted #22 cloth-insulated wires, for example, will have a capacitance of 20 pF/ft while the center-wire-to-shield capacitance of RG/58 coaxial cable is 33 pF/ft. If we assume that V_1 is

TABLE 6.40 FLOATING POWER SUPPLIES

A. Normal Operation
  Input-line cord connected to three-terminal ac outlet
  Chassis at ground potential
  + or – output terminal to grounded center terminal for a – or + output voltage
B. Low-voltage floating operation
  Input-line cord to three-terminal ac outlet
  Chassis at ground potential
  + or – terminal to floating potential
  Grounded output terminal unconnected
C. Medium-voltage floating operation
  Input-line cord attached to 3-2 plug adaptor with ground terminal unconnected
  Chassis at floating potential
  + or – output terminal connected to chassis output terminal (at floating potential)
  Case and chassis isolated from all conductors
  Insulating shafts and knobs on all switches and dials
D. High-voltage floating operation
  ac input from isolation transformer
  Instrument case at floating potential
  Case and chassis electrically isolated from all conducting surfaces
  Insulated shafts and knobs on all switches and controls

varying sinusoidally at frequency ω, the rms potential across R is given by

$$v_{1\text{rms}} \frac{R}{R + 1/\omega C} = v_{1\text{rms}} \frac{\omega RC}{1 + \omega RC}$$

(the voltage-divider formula). Thus the rms potential across R increases with R, ω, and C.

In the average laboratory, there are many sources of ac voltage that can couple capacitively to signal lines—open-line sockets, lighting fixtures, and line cords are a few examples. An estimate of the capacitance C_{min}

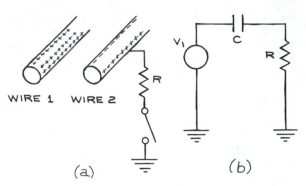

Figure 6.126 (a) Capacitive coupling between two wires; (b) the equivalent electrical circuit with the effect of wire 1 on wire 2 represented by C.

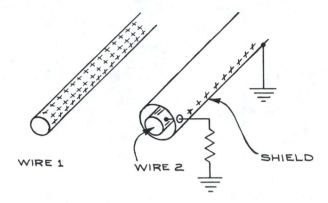

Figure 6.127 Shielding to prevent capacitive coupling.

necessary to induce a 1 mV rms voltage across 1 MΩ from a power-line source can easily be made by setting $V_{1\ rms}$ equal to 120 V, $\omega = 2\pi f = 188$ rad/sec, and neglecting ωRC with respect to 1.

$$C_{min} = \frac{1 \times 10^{-3}\text{volts}}{1.2 \times 10^{2}\text{volts} \times 1.88 \times 10^{2}\text{sec}^{-1} \times 10^{6}\Omega}$$

$$= 0.044 pF$$

This value is easily attained. Such induced voltages can be substantially reduced by the use of shielding. If a grounded shield is placed around wire 2, the situation illustrated in Figure 6.127 will occur. The conducting shield around wire 2 will have a current induced in it when connected to ground, due to the potential on wire 1. This current, however, flows through a low-resistance connection. In addition, and most importantly, any charge on the shield will exist on its surface, and the interior will be field free. In practice, common coaxial braided shield is about 90 percent effective against capacitatively coupled voltages from external sources. Foil shields are even more effective.

6.9.3 Electrical Pickup: Inductive Effects

A changing magnetic field can induce a current in any loop it cuts. The magnitude of the induced current depends on the area of the loop and the time rate of change of the magnetic field. High-resistance circuits are not greatly affected by such inductive effects, but they are important in low-resistance circuits such as the input circuits of

current and pulse amplifiers. Common sources of magnetic-field pickup are transformers, inductors, and wires carrying large ac currents. Effective shielding against low-frequency magnetic fields is accomplished with ferromagnetic enclosures, which act to concentrate the stray magnetic fields within the shield material. The usual practice is to enclose the source of the magnetic fields within such shields. Higher-frequency magnetic fields are best shielded with a copper enclosure. Eddy currents induced in the copper produce counter magnetic fields. A rapidly changing current is a source of time-varying magnetic fields that can themselves induce currents in other circuits. Fast rise time pulses in the low-impedance output circuits of pulse amplifiers can generate rapidly changing magnetic fields, which can induce unwanted currents in any nearby low-impedance input circuit. Once again, the most effective shielding is high conductivity copper sheet or foil.

6.9.4 Electromagnetic Interference and RFI

The high-frequency fluctuation of current or charge in a conductor results in the radiation of part of the energy in the conductor in the form of an electromagnetic wave. Such a wave propagates through space at the speed of light and, when not carrying useful information, is called *radio-frequency interference* (rfi). Sources of such radiation are automobile ignition systems, microwave ovens, electrical discharges, electric motors, electromechanical switches and relays, and electronic switches such as thyratrons, rectifiers, SCRs, and triacs. Power supplies with switching regulators are often sources of rfi.

This interference can be reduced by the use of zero-crossing switching regulators in which the switching only occurs when the voltage across the switch is zero. The most effective shield against rfi is a grounded enclosure of a conducting material, such as copper or aluminum. On the surfaces of such a shield, the electric component of any incident electromagnetic wave is zero and further propagation is not possible. Conducting screens, rather than solid sheet, are often used for rfi shielding because of the savings in weight. When using such screen, it is important that the mesh size be small compared to the wavelength of the highest-frequency component of the rfi. Standards have now been established for rfi emission from certain classes of electrical and electronic equipment. Sensitive, high-frequency-measuring equipment must have good immunity to rfi, and additional rfi shielding often can be purchased as an option for equipment that must operate in an especially noisy environment.

6.9.5 Power-Line-Coupled Noise

Consider the case illustrated in Figure 6.128, where a single supply line powers three circuits. If circuit 1 suddenly requires a large amount of current, circuits 2 and 3 may be affected because of the resistance and inductance of the power-supply line. A large current drawn by circuit 1 can cause a momentary voltage drop at circuits 2 and 3 that can be propagated as noise throughout the circuit. The greater the time rate of change of the current pulse through circuit 1, the larger the effect on the other circuits. This problem is very common with logic circuits, particularly TTL circuits, which have saturating output stages. If a number of gates switch simultaneously, very large transient current spikes can appear on the power line. These transients are then interpreted as a change in logic level by other gates, and transitions occur that are entirely spurious. Standard practice is to use low-resistance, low-inductance power-supply lines to the logic gates and decouple the gates from the power supply with high-

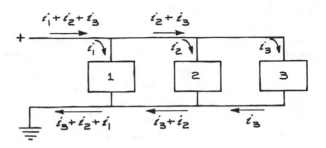

Figure 6.128 Three circuits on a common power line.

frequency capacitors from the power supply terminal of the gate to ground. For TTL, every group of 20 gates must be decoupled from the power supply with a 0.1 µF capacitor, and every group of 100 gates must be decoupled with an additional 0.1 µF tantalum capacitor with good high-frequency characteristics.

Similar situations arise with switched, high-current devices such as temperature-controlled furnaces operated directly from the ac line. Large current transients may affect all equipment plugged into the same line. The only remedy here is to operate the high-current device from a completely separate line circuit.

As an example, consider a series of logic gates to be connected to a power supply by one foot of #22 wire, which has a resistance of 16 Ω/1000 ft and an inductance of 640 µH/1000 ft. If a few gates switch simultaneously, resulting in a current spike of 10 mA with a rise time t_r of 10 ns, the momentary voltage drop along the line will be approximately iZ, where Z is $\sqrt{R^2 + \omega^2 L^2}$. In this case,

$$Z = \frac{}{\sqrt{(16 \times 10^{-3}\,\Omega)^2 + (2\pi \times 3.3 \times 10^7\,\text{sec}^{-1} \times 6.4 \times 10^{-7}\,\text{H})^2}}$$

where $\omega = (2\pi/3)t_r$ has been used in the estimate of the inductive reactance of the wire. The voltage drop is then

$$10^{-2}\,\text{A} \times 133\,\Omega = 1.3\,\text{V}$$

For the TTL logic, this drop could be enough to cause some marginally substandard gates to change state momentarily.

6.9.6 Ground Loops

Instruments are connected to ground in order to have the potentials of those terminals be at 0 V for reference purposes. This, of course, assumes that every ground connection is made directly to the zero potential of the earth. This is often far from the case, since most ground connections are through the third wire of the ac line. Since this line has finite resistance, currents flowing in it will cause potential drops and, depending on the point on the line where a ground connection is made, the potential can be very different from 0 V. When the 0 V references of two instruments that are connected together differ, there is danger of introducing noise into the system. The problem is made all the worse by the fact that the reference points can change their potentials with respect to each other in a way independent of other system parameters. The solution to this problem is to have only a single ground point in the system and connect all of the zero reference points and shields to it. This means removing the ground connections in the power-line cords of all line-operated instruments, and isolating all shields and cases from conducting surfaces that may be connected to ground at points other than the single system ground. All connections to this single system ground should be made as short as possible, with the lowest-resistance conductors available. By having the 0 V reference potential connection and shield connection at the same point in the system, no currents can be induced in the reference line.

To illustrate these ideas, two different ways of connecting a system composed of a signal source, amplifier, and recorder are shown in Figure 6.129. In arrangement (a), each case is separately grounded and additionally connected to the other cases through the outer conductor of the coaxial cable between them. If the three grounds are at different potentials, currents will flow through the coaxial outer shields connecting the devices together. Since the reference-potential line is the shield, it will take on different potentials at different points in the circuit. As these potentials change, the signal levels will change—the result being seen as noise on the signal line.

In arrangement (b), the common reference line and the instrument shields are all attached to ground at a single point—the reference ground. No ground loops can exist, and the reference line remains at the same potential everywhere in the system. Clearly this is a superior arrangement. The implementation may be difficult, however, since it is necessary to connect all of the cases and shields separately to a single ground point. If all of the instruments are mounted in a single rack, each must be electrically isolated from the conducting rack structure. The coaxial-shield connections must also be isolated from the cases, and the ground connection in the ac line cord must be removed from each of the instruments. It is not always necessary to take such precautions with all stages of a system—it may be sufficient to eliminate ground loops from the most sensitive element to which the signal source is connected.

It is worthwhile to understand the origins of ground loop and shielding problems in order to be able to solve them when they arise. A much more extensive treatment is given by Morrison.[12]

The problem of ground loops also occurs when measurements must be made across two electrical terminals, neither of which is at ground potential.[13] Bridge measurements are a common example of this. For floating measurements, instruments that have the reference input terminal electrically separate from the ground or common terminal must be used, otherwise the floating potential (common-mode potential) will be short circuited.

In Figure 6.130, V_S is the source voltage to be measured, v_C is the common-mode voltage, Z_1 is the input impedance of the measuring system, and Z_2 is the impedance from the common of the measuring circuit to the low terminal—normally 10^8 to 10^{10} Ω in parallel with 10^3 to 10^5 pF. The presence of v_C results in currents i_{C1} and i_{C2} flowing through Z_1 and Z_2. This presents no difficulties as long as the resistance of the lines is zero, but if they are not zero, the common-mode currents will result in potential differences at the low input terminal and add to the error of the measurement. Generally, $Z_1 \ll Z_2$—the critical parameter, therefore, is the ratio of the resistance of the low signal line to Z_2. This ratio is frequency dependent because Z_2 is complex—if the source and instrument grounds are not at the same potential, the difference will add to V_C.

The CMRR is a quantity for specifying how well an instrument rejects a common-mode voltage superimposed on the voltage to be measured (the normal mode voltage),

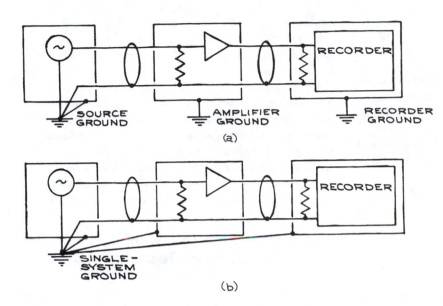

Figure 6.129 System grounding: (a) multiple grounds; (b) single ground to eliminate ground loops.

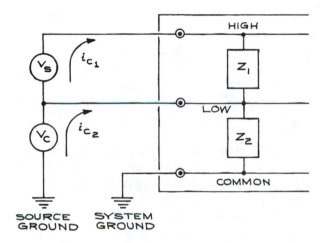

Figure 6.130 Measurement of v_s across two terminals, one of which is not at ground. Current flow through Z_2, the impedance from the low terminal of the measuring instrument to ground, will result in an error voltage v_c.

$$CMRR(dB) = -20\log\frac{v_{NM}}{v_{CM}}$$

where V_{NM} is the common-mode voltage, appearing as a normal-mode voltage, and therefore a source of error. Floatable meters have CMRRs from 80 to 120 dB at dc and 60 to 100 dB at line frequency. When higher values are required, special guarded voltmeters having an additional guard terminal must be used. This terminal, when connected to the low terminal of the source by a low-resistance connection, shunts the common-mode current from the measuring terminals. Such meters offer CMRRs of 160 dB at dc and 140 dB at line frequency.

6.10 HARDWARE AND CONSTRUCTION

It is almost always better to purchase a piece of electronic equipment than to design and construct it. This is certainly the case for power supplies, amplifiers, signal generators, and similar equipment that is mass produced by a large number of companies in a wide variety of models.

When one does, however, decide to construct a piece of electronic equipment because of cost or the unavailability of commercial units, there are a number of well-defined steps to follow:

• Design of the circuit
• Selection of components
• Construction and testing of a breadboard model

- Construction of the final circuit
- Mounting
- Final testing

Except for the simplest circuits, one is well advised to use proven designs that can be found in manufacturers' application books and in books and articles on circuit design. A few such publications are given in the list of references.[14] There are, however, some simple procedures that are well known to electronics engineers (but not often included in textbook examples of circuit design) that can make the difference between a circuit that works and one that does not. These include bandwidth limiting, power supply decoupling, and signal conditioning.

6.10.1 Circuit Diagrams

Two kinds of circuit diagrams are the block diagram and the schematic wiring diagram. The *block diagram* shows the logical layout of the circuit by grouping all elements necessary for single function into a single rectangle. Although simple, such diagrams (when they are well done) are very useful in understanding the general operation of a circuit. The block diagram is much like the flowchart of a computer program.

The *schematic wiring diagram* shows all of the components of the circuit and the connections between them. Figure 6.131 lists symbols used in schematic diagrams. The more complete the schematic, the more useful it is for troubleshooting. A schematic will have the values and ratings of all the components, as well as numbers and color codes when necessary. Good schematics include the values of voltages at critical points in the circuit and drawings of waveforms where appropriate. One generally reads schematic diagrams starting at the upper left with the input and proceeding to the lower right to the output.

A useful addition to the schematic diagram is the *chassis layout diagram,* which shows the physical location of all parts. For ease of drawing, components that appear next to each other on the schematic may be far away from each other on the actual chassis or circuit board.

With increasing use of integrated circuits, it is more and more difficult to identify the functions of different parts of a schematic. This is because the integrated circuits are often only represented by rectangles with pin numbers. To identify their function, it is necessary to consult a manufacturer's catalog. Certain abbreviations common with ICs are given in Tables 6.27 to 6.29. Schematic diagrams also help identify specific components. If there is an operational amplifier on a circuit board with a balance control near it for nulling purposes, it is useful to know where the amplifier is in the circuit and what comes before and after it, before proceeding with adjustments. The location of damaged components can be useful information when deciding on the probable cause of the damage and which other components may also have been affected.

6.10.2 Component Selection and Construction Techniques

When selecting components, the considerations that an electronic engineer finds important may be different from those of a laboratory scientist building only a single example of a circuit. It is always wise to overspecify components—that is, to use components of higher ratings and better quality than a critical cost-effective analysis of the circuit would show to be necessary. Usually the cost of components is a small fraction of total project cost when time is considered.

Once the circuit design has been decided upon and the components assembled, a *breadboard* model can be constructed. This may seem an unnecessary and time-consuming step, but care at this stage will save a great deal of time later. The breadboard circuit should be constructed from a complete, detailed schematic circuit diagram.

There are a number of prototype circuit boards and aids for making breadboard models. Prototype boards use sockets to which components and wires can be connected in a temporary way. The spacing of the sockets is usually on 0.1 inch centers to accommodate the usual 14- and 16-pin dual in-line integrated circuits. The sockets accommodate only a limited range of solid-wire sizes. Outside of this range, the wire is either not held securely or the socket is bent out of shape. Boards of this type are satisfactory for low-frequency circuits and logic circuits using TTL and CMOS. For sensitive high-frequency circuits and ECL logic, the stray capacitance between sockets is a problem, and such circuits must be built on a

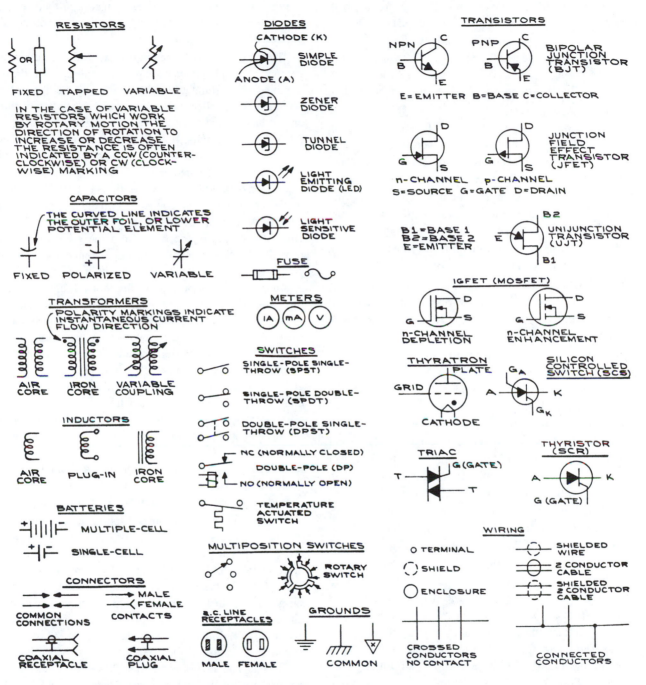

Figure 6.131   Symbols for schematic circuit diagrams.

circuit board with a ground plane if they are to work properly.[15] At this stage, it is necessary that the breadboard circuit follow the schematic in every detail. The mechanical layout should be neat, with interconnections between components as short as possible. This simplifies troubleshooting and avoids pickup problems.

A common difficulty with breadboards is getting signals in and out and applying dc power. Dangling wires and alligator-clip connections invite short circuits and damaged components. Whenever possible, the breadboard should be mounted on a larger structure with terminals and connectors to which semipermanent signal and power supply connections can be made (see Figure 6.132). Connections from these terminals to the breadboard can then be made with short pieces of hookup wire.

The breadboard circuit is fully tested before proceeding to the final construction stage. During testing, the circuit should be operated with the same input and output connections the final circuit will have.

Once the breadboard circuit is functioning properly, the final circuit can be constructed. Depending on the complexity of the circuit, there are a number of construction techniques available. In general, all components are mounted on circuit boards and the boards are fitted into a case or attached to a panel. The power supply can be incorporated into the circuit, or power supply voltages can be brought in from the outside through connectors. For simple circuits involving only a few components and a moderate number of connections, perforated circuit board and push-in terminals can be used (see Figure 6.133a). All connections are individually made and soldered in such circuits. Whenever possible, active components such as ICs and transistors should be used with sockets, and all soldering done with the components removed from the sockets. Perforated circuit board with printed circuit wiring is also convenient for making final circuits (see Figure 6.133b). With this type, all soldering is done on the board and there is no need for separate terminals. Power supply and ground-bus structures are often printed on such boards to facilitate bringing power to the components.

For more complex circuits, custom-printed circuit boards (see Section 6.10.3) or Wire Wrap[TM] boards (see Section 6.10.4) can be used.

The circuit board containing all of the electrical components and a power supply, if required, is typically mounted on a standard 19 in. aluminum rack panel on standoffs. When there is a need for more than two or three circuit boards, the means of mounting and of making the interconnections must be considered more carefully. For multiple circuit boards, *card-cage* mounting—using guides and edge connectors as shown in Figure 6.134—is a common technique. Interconnections between boards are made between the edge connectors on the rear panel of the chassis. A variation of this method uses a *motherboard,* to which all of the connections from the other boards are brought. In this method, the edge connectors provide power to the separate boards and help to mechanically retain the boards in the chassis. Signal leads are brought to the motherboard via multiple pin connectors and flat multiple-conductor cable, using mass termination hardware for increased reliability. A variation of the motherboard configuration includes the edge connectors on the motherboard. An on-off switch, pilot light, and fuse are put on the front panel, as well as input and output connectors and controls. When labeling the panel, it is wise to schematically indicate the electronic functions performed by each section of the circuit with standard symbols. Labels can be applied with the silk-screen technique or with dry transfer letters, India ink, or embossed pressure-sensitive tape. A protective cover should be placed over the circuit (see Figure 6.135).

Construction can be simple. A portable electric drill or drill press is necessary for drilling proper-size holes in panels and covers. A list of electronic-circuit construction tools is given in Table 6.41, and a list of useful hardware is given in Table 6.42. Flathead screws should be used on exterior surfaces, and lock washers should be used under nuts. Standard 1/16 in. circuit board can be cut easily with a hacksaw and drilled with standard high-speed drills. A nibbling tool is useful for making irregular-shaped holes in sheet metal, and a set of chassis punches from 1/2 to 1 1/2 in. speeds up work and makes accurate holes in sheet metal. A tapered hand reamer is very useful for enlarging holes in sheet metal and circuit boards.

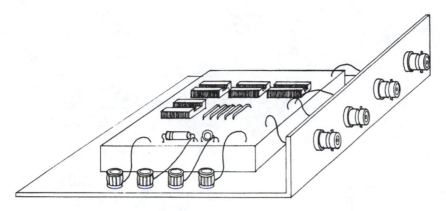

Figure 6.132 A circuit breadboard mounted to allow proper connection of power supply and input and output leads.

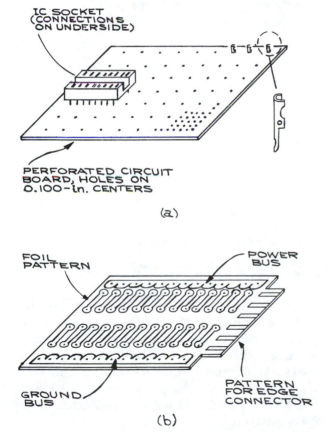

IC SOCKET
(CONNECTIONS
ON UNDERSIDE)

PERFORATED CIRCUIT
BOARD, HOLES ON
0.100-in. CENTERS

(a)

FOIL
PATTERN

POWER
BUS

GROUND
BUS

PATTERN
FOR EDGE
CONNECTOR

(b)

Figure 6.133 Perforated circuit boards for use with (a) pin connections and (b) direct solder connections.

6.10.3 Printed Circuit Boards

The use of PCBs in laboratory electronics is justified when many identical boards are needed and when the electrical properties of the circuit require the kind of controlled geometry that printed circuit boards offer. ECL circuits require microstrip line geometries for interconnections and precise placement of components to achieve fast rise times and at the same time avoid crosstalk among circuit elements. Sensitive low-level amplifier circuits require precise lead placement to eliminate noise pickup. The boards consist of an insulated substrate covered with a metal (usually copper) foil. Some of the common rigid substrates used for copper-clad boards are listed in Table 6.43 along with their codes. Nonrigid boards or *flexboards* are used to fit into irregular confined spaces.

Commercial production of PCBs begins with copper-clad laminated boards, usually epoxy-impregnated fiberglass 0.062 in. thick, from which copper is removed by chemical etching to form the pattern of traces, pads, and busses that are the electrical connections and mounting areas for the individual electronic components and connectors. To increase circuit density, patterns can be formed on both sides of the board with electrical connections between the layers made by plated holes or *vias*. Sandwiches of two and three boards (4 and 6 layers) are common, and current manufacturing capabilities permit as many as forty layers. After fabrication of the board pattern, holes are drilled for component leads, *vias*, and fasteners—a thin layer of solder (solder plate) is flowed

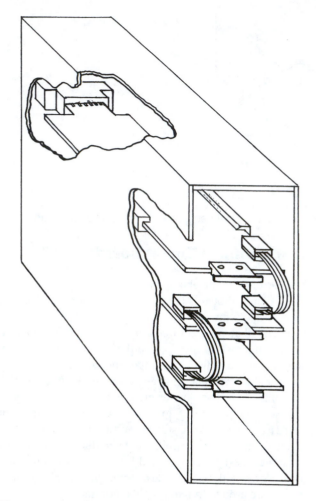

Figure 6.134   Circuit boards mounted in a card cage.

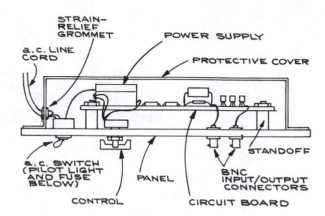

Figure 6.135   Circuit board mounted on a rack panel.

TABLE 6.41 ELECTRONIC-CIRCUIT CONSTRUCTION TOOLS

Nut-drivers
Sheet-metal nibblers
Universal electrician's tool
Needle-nose pliers
Slip-joint pliers
1/2 inch tapered reamer
Wire cutters
Wire strippers
Soldering gun, pencil
Knife
Single-edge razor blades
Solder sucker
Solder wick
Heat gun
Clip-on heat sinks
Chassis punches (5/8, 3/4, 1-1/8, 1-1/4 inch)
Circuit board holder

Note: These tools supplement those listed in Table 1.1.

over the copper and through the *vias* to protect from oxidation and increase the ease of component soldering. Solder mask—a thin, tough, insulating film—is placed over the board, leaving only the solder terminals and pads uncovered. The mask protects the traces, acts as an insulator to prevent short circuits, and prevents solder bridges from occurring when components are soldered to the board. Component outlines and descriptive information are printed on the board using a silkscreen technique.

The circuit patterns are photographically transferred to the boards using photoplots and photo resists. The photoplots are the equivalent of a photographic negative and must contain all of the detail at the required precision of the final circuit board. In the past, photoplots were made by hand with opaque tape and precut pads and pad arrays, but this has been commercially superseded by CAD/CAM (Computer Aided Design/Computer Aided Manufacture) programs that are now widely available for PCs. These programs have two parts—schematic capture and board layout.

Printed circuit boards produced with CAD/CAM software have a number of advantages, even at the single board prototype level:

TABLE 6.42 HARDWARE AND ELECTRONIC EQUIPMENT

Hardware

Screws (flathead and roundhead), nuts, flat washers, lock washers (4-40, 6-32, 10-24)
Solder lugs and terminals
Binding posts
Grommets
Standoffs
Cable clamps
Line cord
Hookup wire
Spaghetti (flexible thin insulating tubing) in assorted sizes
Shrink tubing in assorted sizes
Eyelets
Fuse holders
Indicator lamps
Switches

Electronic Equipment

Signal sources:
   Signal generator
   Pulse generator
   Logic pulse source
VOM/DVM
Oscilloscope
Logic probe
Logic clips
Test leads for VOM and DVM
Oscilloscope probes:
   ×1 probe for low frequencies
   ×10 compensated probe for frequencies above 10 MHz
   Radio-frequency probe for demodulating rf signals

- Uniform electrical and mechanical properties from using standard materials and automated fabrication methods along with electronic and mechanical engineering conventions that are built into the CAD/CAM programs
- Error checking in the schematic capture routines
- Linking the output of the schematic capture program to circuit analysis programs for evaluation of the circuit
- Electronic transmission of the photoplot (Gerber) and drill files
- Competitive pricing and fast turnaround

These advantages have to be weighed against the inconvenience of modifying printed circuit boards that are only meant to be prototypes, and having to acquire and learn the CAD/CAM software. Manufacturers of schematic capture and PCB layout programs are listed at the end of this chapter. Trial versions of their software products are generally available and many offer student and educational institution discounts.

Procedures for producing commercial PCBs using schematic capture and board layout programs are detailed in the following two sections.

Schematic capture. A schematic diagram of the circuit is laid out on the computer using components from the program library. Components not in the library can sometimes be obtained from the manufacturer of the component. The library entry contains mechanical and electrical information about the component, as well as the numbering and designation of the pins and/or terminals. Layout is accomplished by arranging the components on the computer screen with the mouse or by specifying coordinates and then interconnecting the pins of the components. Each connection is a node. From the schematic arrangement, a *netlist* is made. This is a list of all of the components along with their electrical parameters and the nodes to which they are connected. The netlist is the basis of various error-checking routines to verify that inputs and outputs are not grounded and that no component is shorted.

Board layout. The netlist from the schematic capture program is the input to the board-layout program. Layout starts by defining the size and shape of the board and the number and designation of layers. For a two-sided board, one side (the top) is the component side and the other (the bottom) is the wiring side—even though generally there will be wiring on both sides. Components are placed on the board from a stack of outline drawings (*rat nest*) with connections specified by the netlist. The components with their accompanying wires are initially arranged within the board area according to a few simple rules: input and outputs separated by as large a distance as possible, power supply and ground connections as short as possible, a minimum number of crossed wires, and horizontal and vertical routing of wires. Final routing of the connections is done manually for small, uncomplicated boards with only a few components. Autorouting routines are useful for complex boards, but hand routing must still be done in

TABLE 6.43 CIRCUIT BOARD SUBSTRATES

Type of board material	Designation	Application
Paper-base phenolic	XXXP	Hot punching
	XXXPC	Room-temperature punching
Paper-base phenolic	FR-2	Flame-resistant
Paper-base epoxy resin	FR-3	Flame-resistant
Glass-fabric-base epoxy resin	FR-4	General-purpose flame-resistant
	FR-5	Temperature- and flame-resistant
	G-10	General-purpose
	G-11	Temperature-resistant
Glass-fabric-base poly-tetrafluoroethylene resin	GT and GX	Controlled dielectric constant
Polyimide		High dielectric strength

Note: Standard thicknesses, including the copper cladding, range from 1/32 (0.031) to 1/4 (0.250) in., with 1/16 (0.062) in. the most common.

certain circumstances. For small boards, it is possible to skip the schematic capture part of the procedure and go directly to the board-layout program. In this case, parts are placed manually and interconnections made individually. When parts are available in the program library, this can be an efficient method for board-layout, but all error checking is lost with the manual method.

The PCB manufacturer requires computer files to fabricate the boards. These files are routinely transferred electronically over the Internet. The files specify the board layout, board size and shape, hole pattern, solder mask pattern, and silkscreen labels and outlines. It is essential that the files precisely follow the format required by the manufacturer. There is one Gerber file and one aperture file per layer. The Gerber file specifies the interconnections and the aperture file gives the widths of the traces. If solder mask and silkscreen printing are used, each requires a Gerber file and aperture file per layer. Holes are specified by a drill file and a drill list file. For laboratory purposes, the FR-4 grade board is the current standard. The copper-foil cladding is usually cathode-quality, electrolytic copper. The most common thickness is 1.0 oz/ft^2 (35 μm). Turnaround time for double-sided boards is generally three to five days. A rectangular, double-sided board, 3 in. by 5 in., costs less than $100. Per board prices fall substantially for two or more boards.

Components are fixed to PCBs by soldering. Wire lead components are held in place on the board for soldering by inserting the leads through the appropriate holes in the board and bending them outward at 45° so that the

component will not fall out of the board when it is turned over for soldering. After soldering, the leads are clipped short. Surface-mount components are aligned and held in place over the pad pattern while solder is flowed over the terminals and pads.

Although commercial production of PCBs from master artwork has been replaced by fabrication directly from computer files, it is still possible to make a PCB from hand-drawn artwork with kits that are available from electronics suppliers. The kits consist of circuit boards and sensitizers or presensitized boards, developer, ferric chloride etchant, and trays and brushes. The first step is production of the artwork master, which is a precise representation of the final traces that will appear on the board. The master is produced using drafting techniques. To increase speed and accuracy of manual drafting, PCB dry-transfer drafting tapes, socket-hole patterns, and terminal pads are available. The tools needed are a T-square, triangles, an X-Acto™ knife to cut the tape to length, and tweezers for removing the dry transfers from the backing sheet. To avoid dimensional changes with temperature and humidity, the artwork can be prepared on Mylar drafting film rather than ordinary paper. Conventional drafting programs are easier than hand drafting and produce a computer file that can be modified and saved for subsequent use.

The pattern on the master photographic negative is transferred to the circuit board by the use of a *photoresist* sensitizer. Boards can be purchased presensitized, or unsensitized blanks can be coated with a photoresist that is in aerosol or liquid form. A *negative* resist, after exposure to UV light, is unaffected by a subsequent developer solution, whereas UV-exposed, *positive* resist is dissolved by developer. If the master artwork is a positive (that is, the desired foil pattern is represented by opaque areas on a transparent background), a positive resist is required. The sensitized board is exposed to strong light through the full-size master. The board is then placed in a glass tray containing developer. Areas on the board exposed to the light will be dissolved, while those areas protected by the artwork will remain. After full development, the board is rinsed, dried, and placed in an etching solution. It is also possible to laser print or photocopy circuit designs directly onto specially coated transfer paper—available from JDR Microdevices—that is ironed on to a blank PC board. The board is then ready for etching. Several electronics suppliers sell resist pens that are used to draw PCB artwork directly on bare circuit board. Ferric chloride is the most common etchant. It works best when gently heated and agitated. Full etching can take up to one hour per board and should always be done in a glass tray. When etching is complete, the board is rinsed and all etchant residues removed with steel wool or a solvent. A more complete description of the procedure is given in the 1999 edition of the ARRL Handbook for Radio Amateurs. The photoresists have limited shelf life, and the etchant is corrosive.

Transistor and integrated circuit packages for PCB mounting come in a wide variety of forms. Table 6.44 gives the designation of some common packages with representative diagrams.

A technique related to PCBs and based on thick-film technology is used to produce high-density circuit patterns on an alumina (Al_2O_3) substrate. A screen with 200 to 300 lines/in. transfers the desired circuit pattern with a special ink containing suspended metal particles and a binder. When the inked substrate is fired, the metal particles fuse to each other and the substrate, producing a tightly bonded conducting pattern. Components are then soldered to the appropriate pads on the substrate with microsoldering irons or solder paste and carefully controlled hot air. Thick-film, rather than discrete, resistors are often used with this technique because they can be produced by the same printing technique by using a different ink. Small, high-capacitance chip capacitors replace the disc and tubular capacitors used with PCBs. With the thick-film technique, it is often quite easy to incorporate ICs and discrete

TABLE 6.44 TRANSISTOR AND INTEGRATED CIRCUIT PACKAGE ABBREVIATIONS AND OUTLINES

Transistor	
TO-3	Round Can on Diamond Base
TO-5	Round Can
TO-220	Rectangle with Heat Sink
Through Hole IC	
DIP	Dual In-Line Package
PDIP	Plastic In-Line Package
SIL	Single In-Line
SIM	Single In-Line Module
SIP	Single In-Line Package
VIL	Vertical In-Line
ZIP	Zig-Zag In-Line Package
Surface Mount IC	
PSMC	Plastic Surface Mount Component
QSOP	Quarter Size Small Outline Package
SOIC	Small Outline IC
SOL	Small Outline Large
SOP	Small Outline Package
SOT	Small Outline Transistor
SSOP	Small Shrink Outline Package
TQFP	Thin Quad Flat-Pack
TSOP	Thin Small Outline Package
TSSOP	Thin Shrink Small Outline Package
IC Modules	
BGA	Ball Grid Array
DIMM	Dual In-Line Memory Module
JLCC	J-Lead Chip Carrier
LCCC	Leadless Ceramic Chip Carrier
LDCC	Leaded Ceramic Chip Carrier
PGA	Pin Grid Array
PLCC	Plastic J-Lead Chip Carrier
SIMM	Single In-Line Memory Module

components on the same substrate to produce a self-contained hybrid circuit. Such a circuit can be hermetically sealed and has the appearance of a large IC. The advantages of such circuits are small size, excellent electrical

TABLE 6.44 CONTINUED

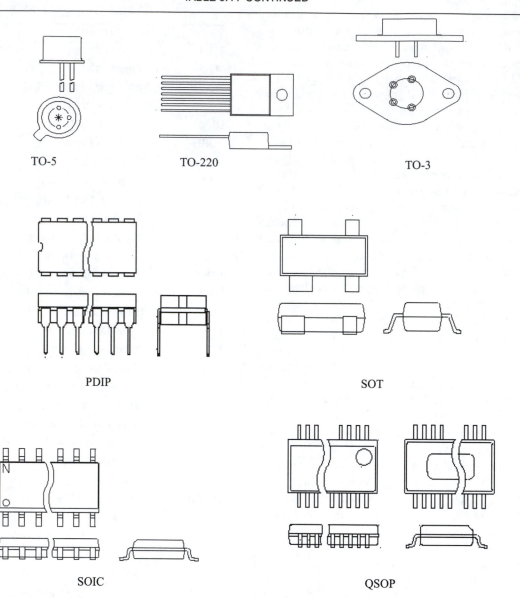

TO-5 TO-220 TO-3

PDIP SOT

SOIC QSOP

TABLE 6.44 CONTINUED

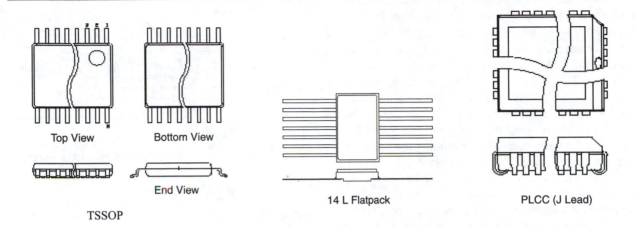

Top View	Bottom View	
End View		
TSSOP	14 L Flatpack	PLCC (J Lead)

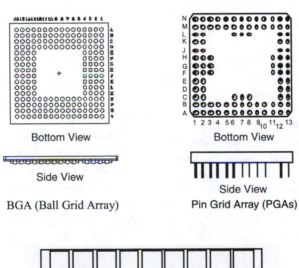

Bottom View

Side View

BGA (Ball Grid Array)

Bottom View

Side View

Pin Grid Array (PGAs)

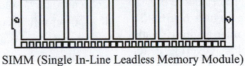

SIMM (Single In-Line Leadless Memory Module)

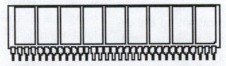

SIP (Single In-Line Leaded Memory Module)

properties, and good mechanical strength. There are a number of companies that produce hybrid circuits starting from schematic diagrams. Because of its specialized nature, thick-film work is best left to such firms.

Solder for electrical and electronic work is an alloy of tin and lead and is specified by the ratio in weight percent of tin to lead. Thus 40/60 solder is 40 percent tin and 60 percent lead by weight. An eutectic alloy with the sharp melting point of 183°C (361°F) is formed when the mixture is 63/37. As one departs from this ratio, the melting point becomes less sharp, extending over a larger temperature range. Fluxes, whose purpose it is to dissolve the oxides on the surfaces of the metals to be joined, are almost always used in soldering. Rosin fluxes are commonly used for electrical and electronic work. Under no circumstances should acid fluxes be used. These are extremely corrosive and can severely damage components, insulation, and the circuit board itself. There are three types of rosin fluxes: R (low activity), RMA (mild activity), and RA (active). The R type is normally used. Solder wire for electronic use generally has flux incorporated in it, though pure flux is also available. The solder should be maintained at 35 to 65°C (60 to 120°F) above its melting point for a time sufficient to completely wet the surface to be joined. Too little heat will result in insufficient wetting of the surfaces and a so-called *cold-solder joint*, which is mechanically and electrically unsatisfactory. Too much heat can destroy components, melt plastic insulation, and burn the circuit board. When soldering heat-sensitive components, a heat sink can be used near the component to protect it. This can be in the form of a pair of long-nose pliers or special tweezers, which are clamped on the lead between the source of heat and component to be protected (see Figure 6.136). To ensure a good joint, the tip of the soldering iron should have a thin bright coat of molten solder covering it, for thermal contact. With time, this coating becomes contaminated with oxides and should be renewed by wiping the surface and reapplying fresh solder. This is called *tinning* the iron. The properly tinned hot tip is then brought in contact with the surfaces to be soldered, and only when they have reached the temperature necessary to melt the solder should solder be applied to them. A common mistake is to melt the solder on the surface of the iron and hope it will flow onto the surfaces to be soldered.

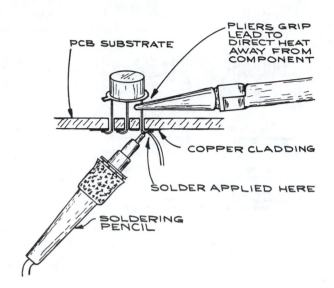

Figure 6.136 Pliers used as a heat sink. A heat sink is required when soldering to the leads of sensitive components.

Since the molten solder flows toward the hottest point, this method leads to wasted solder and cold-solder joints. Poor solder joints also result from insufficient heat and from contaminated surfaces, which resist being wetted by the solder even in the presence of flux. When soldering is complete, the flux should be removed with a commercial flux solvent or isopropyl alcohol.

Soldering surface mount components require that solder flow over the circuit board pad and component contact. This is illustrated in Figure 6.137. For small chip resistors and capacitors, soldering is best done with the aid of a low-power microscope.

Soldering irons are generally of two types, the soldering gun (Figure 6.138a) and the single-element soldering pencil (Figure 6.138b). Guns come in single- and dual-watt models from 100 to 325 W. They have the advantage of being able to be turned on and off quickly, but are generally too powerful and bulky for PCB work. Soldering pencils come in wattages from 12 to 75 W with a large variety of tip shapes. Thermostatic control and stands with tip-cleaning sponges are often used with them. For best results, soldering-iron tips should be cleaned and tinned regularly,

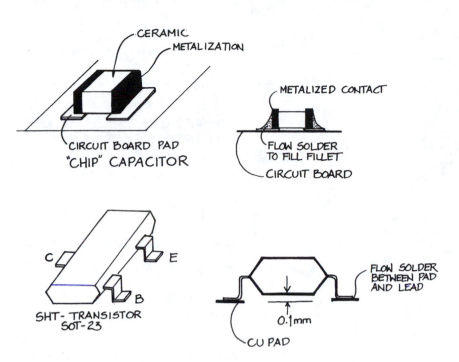

Figure 6.137   Soldering surface mount (SMT) components.

and the heating elements should be checked for good electrical, mechanical, and thermal contact to the tips.

It is often necessary to remove a component from a PCB and replace it. One method that does not require desoldering the old component is pictured in Figure 6.139. This technique does not work with ICs and multilead components, and desoldering from the board is necessary. Three common desoldering methods are solder wick, solder aspiration, and heat and pull. *Solder wick* is copper braid that withdraws solder from a joint by capillary action when heated and placed on top of the solder to be removed. Solder can also be removed by using a sucking tool to aspirate the solder after melting it with a soldering iron. The tool can be a rubber bulb with a heat-resistant nonmetallic tip, or a triggered spring-loaded syringe. Special heat-and-pull soldering irons are needed to remove multilead components such as ICs. These irons have tips shaped to heat all leads simultaneously. After removal of the component, the holes must be cleaned of residual solder before proceeding with the insertion of another component. Solder removal by heating followed by shaking or blowing should be avoided. It results in solder splashes that can cause short circuits.

PCB tracks that have been damaged by overheating or mechanical stresses can be removed with a knife and replaced with new sections soldered directly to undamaged foil. When pads become detached from the substrate, they can be replaced with new ones anchored to the board with swaged eyelets.

In the preceding discussion of PCBs, low-frequency applications have been assumed. If the board is made for high-frequency applications, a great deal more care is required in component placement and lead geometry. For fast logic circuits using ECL ICs, microstrip line geometries are recommended.[15]

6.10.4 Wire Wrap™ Boards

Boards made by the Gardner-Denver Company under the trade name Wire Wrap™ have sockets spaced at intervals corresponding to the spacing of the leads on integrated circuits. The sockets have long, square cross-section posts, so that when the socket is swayged into a substrate material, the post extends on the back side. The length of the post determines how many separate connections can be made to it. Three- and four-layer length posts are the most

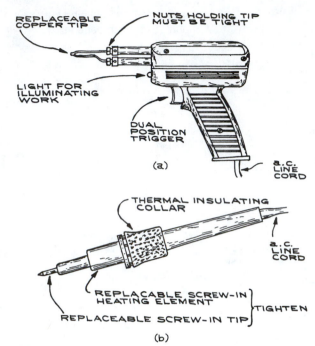

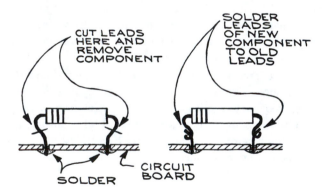

Figure 6.138   (a) Soldering gun; (b) single-element soldering pencil.

Figure 6.139   Replacing a component on a printed circuit board.

common (see Figure 6.140). A special wrapping tool is used to make connections to the pins. The tool can be manual, line operated, battery operated, or air operated. The hand tool is entirely adequate for circuits with up to 10 ICs. As the wire is wrapped around the post, the high pressures generated between the wire and the sharp

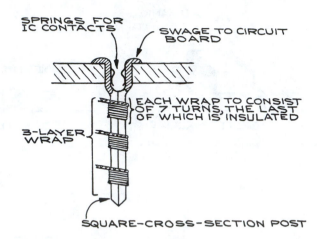

Figure 6.140   Wire Wrap™ post.

corners of the post form a kind of cold weld, which has good electrical and mechanical properties. The advantages of such a system are as follows:

- No solder connections to cause heat damage to components and circuit board
- Electrical components that plug into sockets and are therefore easily removed for replacement or testing
- Higher component density than obtainable with double-sided circuit boards because wire crossings are possible without short circuits
- Ease of removing and remaking connections (an unwrapping tool that is a companion to the wrapping tool allows one to remove connections—however, if the connection that one wants to remove is not the highest of a stack, one must remove the upper ones to get to it)
- No artwork, photographic reproduction, masking, or etching required
- Reliability
- Speed of fabrication

In practice, one uses single-conductor wire specially made for wrapping. The three most common gauges are AWG 30 (0.25 mm), 28 (0.32 mm), and 26 (0.40 mm). Different wrap and unwrap tools are used for each gauge. One can obtain the wire in rolls or precut and prestripped lengths. If one uses rolls, it is necessary to have a wire stripper to remove 1 in. of insulation at either end of the length of wire to be wrapped. Thermal strippers ensure that the conductor will not be nicked when stripped, but special mechanical strippers that do an adequate job can also be obtained inexpensively.

The usual Wire Wrap™ board consists of a matrix of sockets with spacing corresponding to the spacing of the terminals on the normal 14- or 16-pin IC. After the placement of ICs on the board has been decided upon, a Wire Wrap™ list is made, which indicates the pins between which connections are to be made. Each pin is identified by a letter and number code identifying the column and row of the pin (see Figure 6.141b). Once the list is complete, all of the wrapping can be done at one time. It is common for Wire Wrap™ boards to have a ground plane and a power supply voltage plane or bus. The Wire Wrap™ posts can be connected to these planes at intervals of 14 to 16 pins. A solder bridge can sometimes be made between the appropriate pin and the exposed ground plane. Connections to the plane from outside the board can be made with a screw-and-lug connection to tie points on the board. Signal connections are usually via ribbon connectors to DIP sockets.

Discrete components are used with mounting platforms or, if sufficient space is available, they can be wrapped directly to the posts. The platforms have pins that fit the socket holes in the board, and the discrete components are soldered to the posts or forked terminals on the top of the platform (see Figure 6.142). When a circuit with only a few ICs must be constructed, an inexpensive method is to use individual Wire Wrap™ sockets and glue them to perforated circuit board with 0.100 in. hole-spacing. This is much less expensive than commercial boards and entirely adequate for low-frequency applications. A disadvantage of Wire Wrap™ boards is their high initial cost, but they can be reused.

Initial verification of the connections on a board is done by removing all components, attaching the ground and power-supply leads, and measuring the voltage at each pin. They should all be consistent with the Wire Wrap™ list.

6.10.5 Wires and Cables

There are many types of electrical wires used in the laboratory. Selection of the correct type is important for correctly functioning equipment. Wire should be selected according to its voltage and current rating for routine low-frequency operation. For applications involving high frequencies and low signal levels, more care must be taken. The use of multiple-conductor cables can simplify wiring, and some knowledge of the kinds of multiple conductor configurations is useful. Table 6.45 lists the diameter, allowable current, and resistance per 1000 feet of B&S-gauge insulated copper wire. Table 6.46 lists the electrical properties of common thermoplastics used for insulation.

TABLE 6.45 CURRENT-CARRYING CAPACITIES OF COPPER-INSULATED WIRE

B&S Gauge	Diameter (in.)	Allowable Current[a] (A)	Resistance per 1000 ft[b] (Ω)
8	0.128	50	0.628
10	0.102	30	0.999
12	0.081	25	1.588
14	0.064	20	2.525
16	0.051	10	4.016
18	0.040	5	6.385
20	0.032	3.2	10.15
22	0.025	2.0	16.14
24	0.020	1.25	25.67
26	0.016	0.80	40.81
28	0.013	0.53	64.90
30	0.010	0.31	103.2

[a] For rubber-insulated wires the allowable current should be reduced by 30%.
[b] At 20°C (68°F).

The most common wire is ac-line cord—the type used for connection to the ac outlet. Twin-conductor *zip cord* should be avoided in laboratory applications because of its limited resistance to mechanical stresses. Three-conductor color-coded line cord is best suited for the laboratory. The double insulation of the wires provides extra mechanical and electrical protection. The standard code is black for hot, white for neutral, and green for ground, and should be observed at both the plug and the chassis end of the cord. To minimize damage to the line cord at the chassis end, strain relief and mechanical protection of the insulation should be provided. This is often accomplished with a single strain-relief grommet or combination plastic

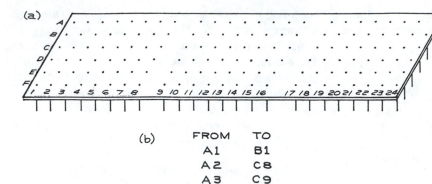

(a)

(b)

FROM	TO
A 1	B 1
A 2	C 8
A 3	C 9
A 4	F 1
A 5	F 3
A 6	F 2
⋮	⋮

Figure 6.141 (a) Wire Wrap™ board; (b) an example of a wrap list.

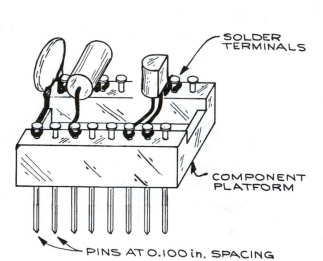

Figure 6.142 Platform for mounting discrete components. The platform then plugs into the Wire Wrap™ board.

grommet and clamp, as in Figure 6.143. Line cord is almost always stranded to enhance flexibility. If the wire is to be used in high-current-carrying applications, care should be taken not to cut the strands when stripping, since that will limit the capacity of the wire and cause heating at the stripped end. Sometimes the individual strands are insulated with varnish. When making electrical connections with wire of this kind, the insulation must be removed with fine emery paper or steel wool.

Another type of wire used for ac lines is the solid-conductor #12 or #14 gauge wire for power distribution. The most common are two- and three-conductor Romex, which is PVC-insulated, and two-conductor BX, which has a metal-armored outer covering. Both types can be routed within walls or outside, in either metal or PVC conduit. Because of the heavy gauge of the conductors, connections are made to screw terminals or with *wire nuts* when splices are involved. The installation of ac lines generally requires a professional electrician who knows the conventions, regulations (codes), and procedures for fusing and connecting to existing power lines.

For electronic circuits in which the voltages are less than a few hundred volts and the currents are a few amperes, one uses hookup wire. This can be solid conductor or multistranded. The solid-conductor wire is easier to use when making connections because it does not fray, but it lacks the flexibility of multistranded wire. Hookup wire is generally *tinned*—that is, coated with a tin-lead alloy to enhance solderability. The insulation is usually PVC, polyethylene, or Teflon.

When stripping the wire, care should be taken to avoid nicking the conductor of the solid wire and cutting strands of the multistranded wire. There is a large variety of wire strippers on the market (see Figure 6.144), but none is foolproof and all require a certain amount of skill if the

TABLE 6.46 ELECTRICAL PROPERTIES OF THERMOPLASTICS

Material	Trade Name	Volume Resistivity (Ω cm)	Dielectric Strength (V/mil)	Power Factor at 60 Hz	Characteristics
ABS	Lustran	10^{15}–10^{17}	300–450	.003–.007	Tough, with average overall electrical properties
Acetals	Delrin	10^{14}	500	.004–.005	Strong with good electrical properties to 125°C
Acrylics	Lucite, Plexiglas, Perspex	$> 10^{14}$	450–480	.04–.05	Resistant to arcing
Fluorocarbons:					
CTFE	Kel-F	10^{18}	450	.015	Excellent electrical properties, some cold flow
FEP	Teflon FEP	$> 10^{18}$	500	.0002	Properties similar to TFE, good to 400°F
TFE	Teflon TFE	$> 10^{18}$	400	$< .0001$	One of the best electrical materials to 300 to 500°F; cold flow
Polyamides	Nylon	10^{14}–10^{15}	300–400	.04–.6	Good general electrical properties; absorbs water
Polyamide-imides and polyimides	Vespel, Kapton	10^{16}–10^{17}	400	.002–.003	Useful operating temperatures from 400 to 700°F; excellent electrical properties
Polycarbonates	Lexan	10^{16}	410	.0001–.0005	Good electrical, excellent mechanical properties; low water absorption
Polyethylene and polypropylenes	–	10^{15}–10^{18}	450–1000	.0001–.006	Good electrical, weak mechanical and thermal properties
Polyethylene terephthalates	Mylar	$> 10^{16}$	500–710	.0003	Tough, excellent dielectric properties
PVC	Saran	10^{11}–10^{16}	300–1100	.01–.15	Low cost; general purpose; average electrical properties

(a)
PLACE BUSHING ON CORD

(b)
PRESS TOGETHER WITH PLIERS

(c)
SNAP INTO HOLE

Figure 6.143   Mounting live cord at the chassis to relieve strain on the electrical connections. Mounting live cord at the chassis to relieve strain on the electrical connections.

conductor is to remain undamaged. Thermal strippers that melt the insulation locally before it is withdrawn from the wire work very well but are not portable. For Teflon-insulated wire, thermal strippers are very useful because the slippery material is difficult to grip.

When working with stranded wire, it is recommended that the ends be tinned. It should be remembered, however, that solder will destroy the flexibility of multistrand wire so the tinning should be confined to a short length at the end. For coding purposes, the insulation of hookup wire comes in many colors and combinations of colors. Color coding is a very useful way to avoid wiring errors and makes subsequent troubleshooting easier.

Test-prod wire is highly flexible multistrand wire with rubber insulation rated at a few thousand volts. Such wire comes with red or black insulation and is used with multimeters and in high-voltage circuitry.

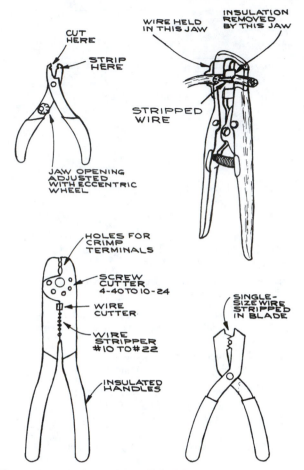

Figure 6.144   Wire strippers.

TABLE 6.47 TYPICAL HIGH-VOLTAGE TEFLON-INSULATED CABLE

dc Voltage Rating[a] (kV)	Conductor	Cable Diameter (in.)
10	19/36[b]	0.077
15		0.097
20		0.117
25		0.137
40		0.197
50		0.237

[a] ac voltage rating is typically 20–25% of dc
[b] 19 strands of #36 wire.

For voltages from a few kilovolts to several tens of kilovolts, special high-voltage cable must be used. The most common jacket materials are silicone rubber, Teflon, and Kapton. Silicone-rubber insulation is very flexible but has only modest dielectric breakdown strength and volume resistivity, so that high-voltage cables made with it have large diameters. Kapton, a polyimide, has high breakdown strength and volume resistivity, and cables made with it are relatively small in diameter. The principal disadvantage of Kapton is its stiffness. The properties of Teflon are intermediate between those of silicone rubber and Kapton, and it is usually the best compromise. Typical properties of high-voltage Teflon cable are given in Table 6.47. One source of high-voltage cable is the ignition cables used in automobiles. The silicone-rubber-insulated stainless-steel or copper stranded-conductor type found in high-performance cars can be used to 50 kV. The more common carbon-filament conductor cable used to minimize rfi is not suitable.

Magnet wire is varnish-insulated, solid-conductor wire used for winding electromagnets, transformers, and inductors. The thin insulation means that one can pack a large number of turns in a given volume. To make electrical connections with such wire, the varnish must be removed with emery paper or steel wool. This magnet wire is not a suitable substitute for hookup wire. When using it for electromagnets, one should be aware of the possible buildup of heat around the inner windings, which can destroy the insulation and create short circuits. The varnish insulation is rated for voltage break down and heat resistance.

For high-current electromagnets, rectangular cross-section wire is used because the high surface-area-to-volume ratio facilitates convection cooling. In applications that require conduction cooling, hollow-core wire is used, through which cooling water can be circulated.

In addition to two- and three-conductor line cord, there are many other multiconductor, round-cable configurations. The conductors can be solid or stranded. The wires are color coded for identification. For low-level, low-frequency signals, the wires can be enclosed in a single shield, as is done for microphone cable. Configurations where there is individual shielding of single wires also are available. Shielding can be in the form of a braid or a foil—if it is foil, a *drain wire* is also included, which is electrically connected to the shield and to which connections can be made. For high-frequency applications there is coaxial cable or 300 Ω, parallel-conductor cable with transmission-line properties. These are discussed in the sections on coaxial cable and connectors (Sections 6.2.3 and 6.2.4, respectively).

Flat multiconductor cable is very useful when a large number of wires are needed as with computer and data interfaces. There are many different arrangements of the wires within the cable. The simplest uses round parallel conductors. There are also flat conductors and alternating round and flat conductors. Flat conductors minimize interconductor capacitance and interference among signals. Twisted pairs of wires are used for transmission lines because constant impedance can be maintained and the wires are flexible and easy to route. An alternating twisted pair-straight geometry is often used for data transmission. The great advantage of flat cable is that mass termination insulation displacement connectors can be used. These connectors are designed so that the cable is clamped and electrical contact is made to each of the wires in a single operation with a special tool. No stripping, soldering, or crimping is required. The reliability of such connections is excellent.

6.10.6 Connectors

Probably the best-known connector combination is the *binding post* and *banana plug* or *tip plug* (see Figure 6.145). Binding posts have the advantage of permitting a wide variety of conductors and conductor terminations to be made to them in a rapid, reliable semipermanent way. They have the disadvantage of being bulky and highly susceptible to the pickup and radiation of electromagnetic energy. When mounting binding posts on a chassis, the mounting hole should be made large enough to accommodate the shoulders on the insulating sleeves. In this way, the central conductor is kept well away from the chassis. The standard spacing between posts is 0.75 inches. This corresponds to the spacing of twin-conductor banana plugs (see Figure 6.146), which are quite common and useful, especially when it is necessary to connect coaxial cable to binding posts. The twin plug usually has one terminal marked *ground*, to which the shield of coaxial cable is attached.

Alligator clips (see Figure 6.147), although useful with test leads, should not be used in any permanent or even semipermanent installation. The type of clip into which one inserts a banana plug is more useful than a clip requiring solder or screw connections. Flexible insulating sleeves that fit over the clips prevent short circuits and are always used when the clip is at the end of a test lead.

Phone plugs (see Figure 6.148) come in several different sizes and provide one means for the termination and connection of coaxial cable carrying low-frequency signals. Because this type of plug is polarized (that is, the connections can only be made in one way), it is sometimes used for low-voltage, low-current dc power supply connections. There are many variations on the basic plug jack design. There are plugs that can accommodate three or more separate connections, and there are jacks that remain shorted until the insertion of the plug.

Insulated *barrier strips* (see Figure 6.149) with screw terminals are a good way of making semipermanent connections for dc applications and are often found on the rear panels of power supplies.

For terminating the ends of wires, there are a variety of terminals that are attached to the wire by soldering or crimping (see Figure 6.150). The terminal used should be of the correct size with respect to both the wire used and the opening at the end. Quick-connect, friction-type, push-on terminals are very useful in dc applications when the connection must be repeatedly made and broken. Such terminals are often also found on mechanical relays, circuit breakers, and mechanical switches. Crimp connectors are color coded according to the wire-size range they can accommodate. It is important to use the proper crimp-tool opening—color coded on many tools—to avoid too loose a crimp (too large a hole) or a severed wire (too small a hole). Three different types of crimping tool are shown in Figure 6.151.

When soldering connectors, only enough heat to make a good joint should be used, since too much heat will melt the insulation. Some type of *third hand* is helpful in this regard—a small vise or an alligator clip at the end of a heavy piece of solid copper wire anchored to a metal base.

A very large part of the electronics-hardware industry is devoted to the manufacture of connectors. For laboratory applications, only a few of the most common ones will be described, and some guidelines for connector selection

(a)

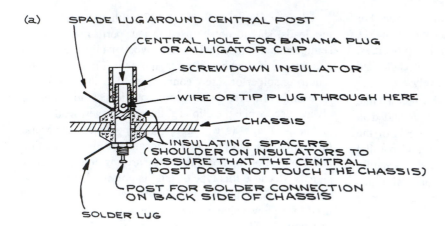

SPADE LUG AROUND CENTRAL POST

CENTRAL HOLE FOR BANANA PLUG OR ALLIGATOR CLIP

SCREW DOWN INSULATOR

WIRE OR TIP PLUG THROUGH HERE

CHASSIS

INSULATING SPACERS (SHOULDER ON INSULATORS TO ASSURE THAT THE CENTRAL POST DOES NOT TOUCH THE CHASSIS)

POST FOR SOLDER CONNECTION ON BACK SIDE OF CHASSIS

SOLDER LUG

Figure 6.145
(a) Binding post;
(b) banana plug.

(b)

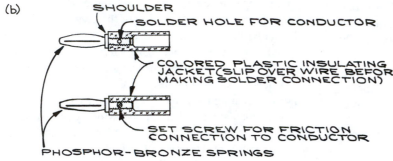

SHOULDER

SOLDER HOLE FOR CONDUCTOR

COLORED PLASTIC INSULATING JACKET (SLIP OVER WIRE BEFORE MAKING SOLDER CONNECTION)

SET SCREW FOR FRICTION CONNECTION TO CONDUCTOR

PHOSPHOR-BRONZE SPRINGS

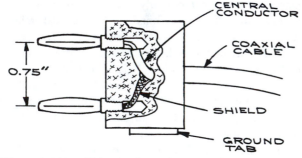

CENTRAL CONDUCTOR

COAXIAL CABLE

SHIELD

GROUND TAB

0.75"

Figure 6.146   Twin-conductor banana plug.

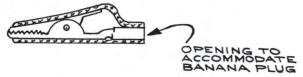

OPENING TO ACCOMMODATE BANANA PLUG

Figure 6.147   Alligator clip (insulated).

will be given. Electrical contacts between the wires and connector pins can be made by soldering or crimping. For some connectors, the pins must be removed to make the electrical connection and then inserted into the connector block. Removal of the pin then requires a special tool. A common power connector using crimped sockets and pins is shown in Figure 6.152. High-density connectors using crimped terminals and integral holding latches are shown in Figure 6.153.

D-connectors derive their name from the cross-sectional shape of the connector body (see Figure 6.154). The connector can have up to 50 contacts and connection is made in a variety of ways; by soldering the conductors into a hollow recess at the back of the contact (solder pot) with crimp pins and sockets, and with ribbon cable mass termination contacts. Such connectors can be attached directly to a panel with the correct cutout. When fitted with a shell (either plastic or metal) incorporating a cable

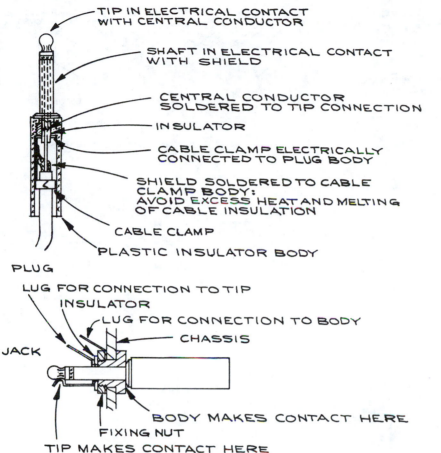

Figure 6.148   Phone plug.

TIP IN ELECTRICAL CONTACT WITH CENTRAL CONDUCTOR

SHAFT IN ELECTRICAL CONTACT WITH SHIELD

CENTRAL CONDUCTOR SOLDERED TO TIP CONNECTION

INSULATOR

CABLE CLAMP ELECTRICALLY CONNECTED TO PLUG BODY

SHIELD SOLDERED TO CABLE CLAMP BODY: AVOID EXCESS HEAT AND MELTING OF CABLE INSULATION

CABLE CLAMP

PLASTIC INSULATOR BODY

PLUG

LUG FOR CONNECTION TO TIP

INSULATOR

LUG FOR CONNECTION TO BODY

CHASSIS

JACK

BODY MAKES CONTACT HERE

FIXING NUT

TIP MAKES CONTACT HERE

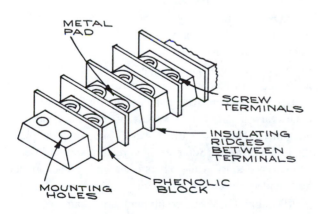

METAL PAD

SCREW TERMINALS

INSULATING RIDGES BETWEEN TERMINALS

MOUNTING HOLES

PHENOLIC BLOCK

Figure 6.149   Barrier strip.

clamp, they become plugs. Locking accessories are also available for securing the plug to the mating jack, and the D-shape assures correct mating. Double-density connectors of this type are also made, and there exists a wide variety of configurations with high-voltage, high-current, and coaxial contacts in addition to the standard single pins. Because pins are very closely spaced, *shrink tubing* is routinely used over the solder connections. In this case, the tubing serves as a strain relief as well as an insulating function. Packages of shrink tubing with an assortment of sizes are convenient. If the tubing is too small in diameter, it will split upon being shrunk—if too large, it will not fit the enclosed wire tightly. Shrinking should be done with a heat gun rather than a soldering

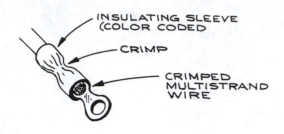

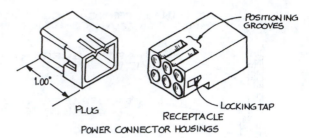

Pin and socket connector housings for 0.093" diameter crimp terminals

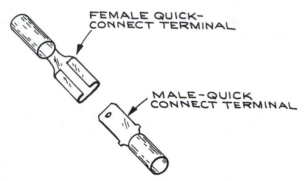

Figure 6.150   Wire terminals.

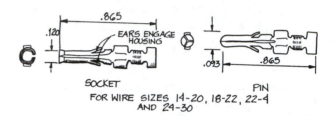

Figure 6.152   Multiple contact connector, crimp terminals, insertion/extraction tool.

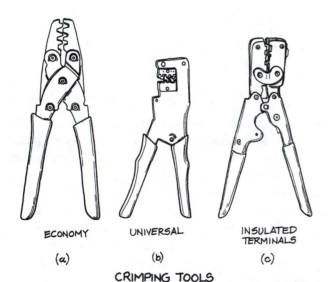

CRIMPING TOOLS

Figure 6.151   Crimping tools.

iron. The usual RS232C interface uses a 25-pin, subminiature D-connector.

Threaded circular connectors (see Figure 6.155) are based on designs originally used by the military for aircraft. They employ separate removable pins and a threaded mating collar to make a mechanically secure union between male and female connectors. There is a broad range of sizes, styles, and pin arrangements, and for this reason these connectors are used on a variety of laboratory electronic equipment. There are various ways of retaining the pins in the insulator block. Some of them require special insertion and removal tools, while others use a second backup block held in place inside the connector.

NIM and CAMAC equipment are examples of the connection of a modular electronic unit to a main chassis. The basic connector is a rectangular insulator block, which

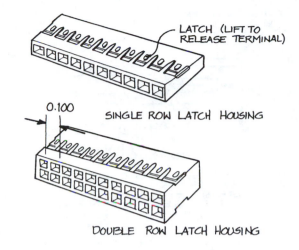

SINGLE ROW LATCH HOUSING

0.100

DOUBLE ROW LATCH HOUSING

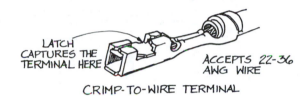

CRIMP-TO-WIRE TERMINAL

MULTIPLE CONDUCTOR INTERCONNECTIONS

Figure 6.153   Multiple terminal connector showing the crimped terminal.

holds contact pins maintained in place by one-way spring action. Connection to the pins is made by crimping. The pins can be readily inserted into the block, but an extraction tool is required to remove them. When the connector has a large number of pins, considerable force is required for mating and threaded screw jacks are often used. Extra pins can be purchased separately and shrouds are available to convert the connector to a plug.

No high-voltage connector is entirely satisfactory, and for this reason they should be avoided and permanent connections with ceramic feedthroughs or standoffs made whenever possible. European-type phenolic sparkplug connectors make acceptable connectors when used with ignition cable—a drilled-out Teflon-insulated rf coaxial connector can also be used. These are illustrated in Figure 6.156.

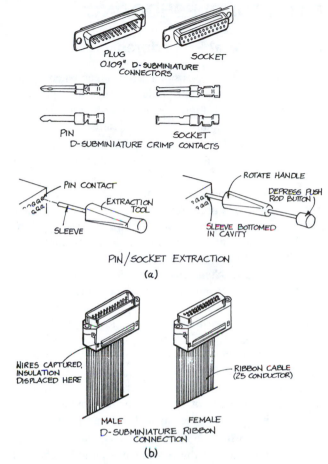

Figure 6.154   D-subminiature connectors (a) crimped contacts (b) ribbon cable termination.

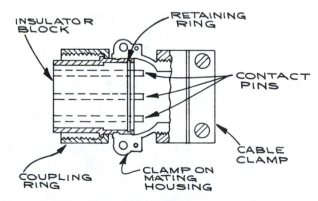

Figure 6.155   Threaded military-type connector.

6.11 TROUBLESHOOTING

6.11.1 General Procedures

There are a number of steps that can be taken when a piece of electronic equipment fails to operate properly, depending on the complexity of the equipment and the availability of test equipment and diagrams.

The worst symptoms are usually the easiest to treat. An experienced TV repairman once noted that 95 percent of all malfunctions in TV sets could be found and corrected in less than one half hour. For the other five percent, the best solution was to sell the customer a new set. This also applies to laboratory equipment. With increasing use of highly reliable integrated circuits and resistance and capacitance modules, the most unreliable elements of a circuit are the mechanical parts (such as switches, dials, and fans) and the connectors—especially the connectors. When an amplifier or power supply no longer works, therefore, one's first impulse should not be to take out a screwdriver and wrench and begin disassembling the chassis. Instead, one needs to be sure that it is plugged in, that the on-off switch is functioning, that all fuses are intact, that the cooling fans are operating, and that the air filters are clean and not blocked. Switch contacts can be cleaned with spray cleaners made for the purpose.

If these measures produce no results, the various dial and switch settings of the instrument should be checked. Often what is interpreted as a malfunction is merely an incorrect setting, causing the instrument to operate in an unexpected mode.

If these simple actions are not effective, the next course of action depends on the complexity of the equipment and the availability of circuit diagrams. To repair a complex circuit without circuit diagrams is almost impossible. If they are not at hand, write to the manufacturer for the necessary diagrams and troubleshooting procedures.

There are, of course, certain visual checks that one can perform without knowledge of the circuit. Charred resistors and an overheated circuit board are readily apparent, as are loose and dangling wires. One difficulty with merely replacing a damaged component, however, is that only the symptom may be treated. The charred resistor, for example, may be due to a failure in another component.

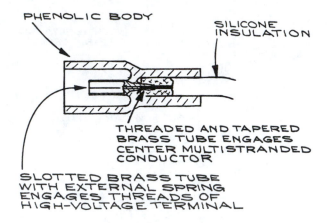

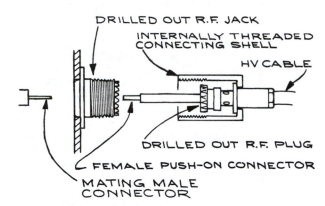

Figure 6.156 High-voltage connectors made from spark-plug and rf connectors.

Troubleshooting is most effective and efficient when a complete set of circuit diagrams, schematic drawings, chassis diagrams, and functional block diagrams are available. These are often accompanied by a troubleshooting guide, which lists the most frequently encountered malfunctions, the symptoms, and the remedies. The area in which the malfunction is occurring can often be localized by studying the functional block diagram of the unit. As an example, consider a counter-timer that responds normally to input signals but gives erratic readings when counting for preset time intervals, and erratic times when in the preset-counts mode. In this case, the gating circuit and the time base are obviously suspect. One should begin by looking at that part of the circuit.

Good schematic diagrams include waveforms and voltage levels at all critical points in the circuit. These should be checked first. Well-designed circuit boards often have the dc voltage levels printed directly at the appropriate test points and components identified by a printed letter-and-number code that corresponds to an identical code on the schematic diagram.

A common difficulty is gaining access to components and connections in the restricted space between boards in a multiple circuit-board assembly. *Extender boards* (see Figure 6.157) are circuit boards arranged in such a way that they mate to the circuit-board connector at one end and the circuit board itself at the other end. The extender board is of sufficient length to place the board in a position where the components and circuit interconnections are easily accessible.

Signal injection is a common method for isolating a circuit fault. With this technique, an appropriate signal is injected at the input of the suspected circuit element, and the output is monitored. This works well for linear circuits such as amplifiers, but becomes quite complex for digital circuits where it may be necessary to stimulate several input terminals and monitor several output terminals simultaneously—and often in synchronization with a clock signal. Logic pursers and probes, as well as clips, are useful for small-scale testing. They are shown in Figure 6.158. For more complex circuits, it is necessary to capture the signals and store them for subsequent analysis. There are logic analyzers made expressly for this purpose. They are expensive, however, and only worthwhile in situations where a large amount of digital circuit troubleshooting is done—this is more the domain of the electronic technician than the experimental scientist. Often it is much more economical to replace all of the suspected ICs in a given section (especially if they are easily removable and in sockets) than to test each one individually.

Some common sources of faults in electronic equipment are the following:

- Electrolytic capacitors in general
- Pass transistors in the output circuit of power supplies
- Input transistors in the input circuits of amplifiers and preamplifiers
- Mechanical switches and potentiometers

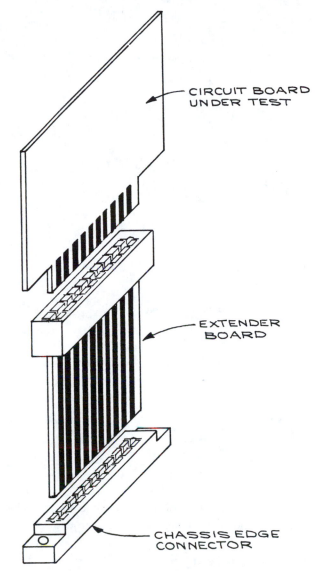

Figure 6.157   Use of the extender board for troubleshooting.

Repeated failure of input transistors in low-level preamplifier circuits can often be cured by placing two diodes such as 1N914s across the input, as shown in Figure 6.159. As long as the input signal does not exceed a few tenths of a volt, both diodes remain nonconducting, allowing the unattenuated signal to pass. Should the signal

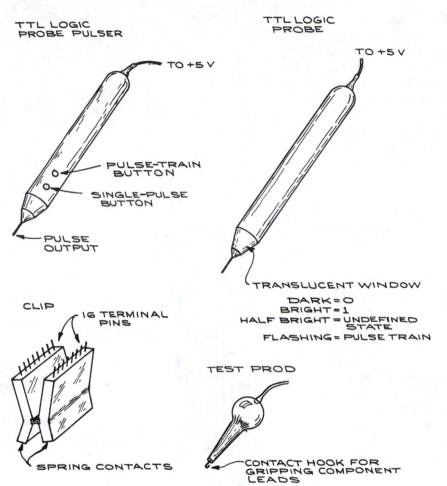

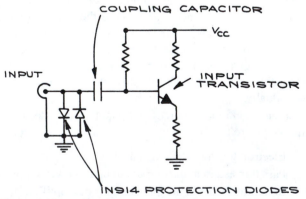

Figure 6.158   Logic pulser, probe, clip, and test prod.

exceed 0.6 V—positive or negative—one or the other diode will conduct, shunting the signal to ground.

6.11.2 Identifying Parts

One can often localize the malfunction to a circuit component but may not know enough about the component to be able to replace it. The parts list may give the equipment manufacturer's component code rather than the standard code of the component manufacturer. When this is the case, it is necessary to identify the component from the code printed on it. Most ICs, transistors, and other components have a four-digit date code giving the date of manufacture. The first two digits are the year and

Figure 6.159   Use of protection diodes at a low level input.

the next two are the week of the year. Thus, 7712 is the 12th week (last week in April) in 1977. The other codes are the manufacturer's component designation. To help identify the manufacturer, a list of logos can be found at the following Web sites:

http//www.icmaster.com/LogosA-M.asp

http//www.icmaster.com/LogosN-Z.asp

The site *http//www.elektronikforum.de/ic-id* has links to *icmaster* and others. Semiconductor manufacturer homepages and data sheets are given at *http//www.bgs.nu/ sdw/p.html*. A list of the prefix codes of semiconductor manufacturers is given in Table 6.48. Once the

TABLE 6.48 SEMICONDUCTOR INTEGRATED-CIRCUIT CODE PREFIXES

Company	Prefix[a]
Analog Devices	AD
Advanced Micro Devices	Am
General Instrument	AY, GIC, GP
Intel	C, I
TRW	CA, TDC, MPY, CMP, DAC, MAT, OP
Precision Monolithics	PM, REF, SSS
National Semiconductor	DM, LF, LFT, LH, LM, NH
Fairchild	F, μ A, μ L, Unx
Ferranti	FSS, ZLD
GE	GEL
Motorola	HEP, MC, MCC, MCM, MFC, MM, MWM
Intersil	ICH, ICL, ICM, IM
ITT	ITT, MIC
Siliconix	L, LD
Fugitsu	MB
Mostek	MK
Plessey	MN, SL, SP
Signetics	N, NE, S, SE, SP
Raytheon	R, RAY, RC, RM
Texas Instruments	SN, TMS
Sprague	ULN, ULS
Westinghouse	WC, WM
Hewlett-Packard	5082-$nnnn$

[a]x = number; n = letter.

manufacturer is known, the appropriate data book or Web site can be consulted. One should replace components with caution, paying attention to the package type and temperature range. The three standard ranges are commercial (0 to 70°C), industrial (–25 to +85°C), and military (–55 to +125°C). Sometimes specially selected or matched components are used. Replacement with off-the-shelf units may not work in this case.

CITED REFERENCES

1. H. W. Bode, *Network Analysis and Feedback Amplifier Design,* Van Nostrand, Princeton, N.J., 1945.

2. Electronic Design, **24**, 63 (1976).

3. J. Millman and C. C. Halkias, *Integrated Electronics: Analog and Digital Circuits and Systems,* McGraw-Hill, New York, 1972, pp. 244–245.

4. J. G. Graeme, *Operational Amplifiers, Design and Application,* G. E. Tobey and L. P. Huelsman, eds., McGraw-Hill, New York, 1971; *Designing with Operational Amplifiers,* McGraw-Hill, New York, 1977.

5. E. Fairstein and J. Hahn, "Nuclear Pulse Amplifiers—Fundamentals and Design Practice," *Nucleonics*, **23**, No. 7, 56 (1965); ibid., No. 9, 81 (1965); ibid., No. 11, 50 (1965); ibid., **24**, No. 1, 54 (1966); ibid., No. 3, 68 (1966).

6. *Voltage Regulator Handbook,* National Semiconductor Corporation, Santa Clara, Calif.

7. *FAST Data Manual,* Signetics Corp., Sunnyvale, Calif., 1984.

8. *Standard Nuclear Instrument Modules,* adopted by AEC Committee on Nuclear Instrument Modules, U.S. Government Publication TID-20893 (Rev. 3).

9. P. H. Garrett, *Analog I/O Design,* Reston Publishing Co., Reston, Va., 1981.

10. S. Letzter and N. Webster, "Noise in Amplifiers," *IEEE Spectrum*, August, 1970, pp. 67–75.

11. John C. Fisher, "Lock in the Devil, Educe Him or Take Him for the Last Ride in a Boxcar?," *Tek Talk, Princeton Applied Research,* **6**, No. 1.

12. R. Morrison, *Grounding and Shielding Techniques in Instrumentation,* Wiley, New York, 1967.

13. *Floating Measurements and Guarding,* Application Note 123, Hewlett-Packard, 1970.

14. *Circuits for Electronics Engineers,* S. Weber, ed., Electronics Book Series, McGraw-Hill, New York, 1977; *Circuit Design Idea Handbook,* W. Furlow, ed., Cahner's Books, Boston, 1974; *Electronics Circuit Designer's Casebook,* Electronics, New York; *Signetics Analog Manual, Applications, Specifications,* Signetics Corporation, Sunnyvale, Calif.; *Linear Applications Handbook, Vols. 1* and *2,* National Semiconductor Corporation, Santa Clara, Calif.

15. W. R. Blood, Jr., *MECL Applications Handbook,* 2nd ed., Motorola Semiconductor Products, 1972.

16. D. Lindsey, *The Design and Drafting of Printed Circuits,* 2nd ed., Bishop Graphics, Westlake Village, Calif., 1984.

17. A. C. Melissinos, *Experiments in Modern Physics*, Academic Press, New York, 1966. chap. 10.

18. R. R. Goruganthu, M. A. Coplan, J. H. Moore, J. A. Tossell, (e,2e) momentum spectroscopic study of the interaction of –CH3 and –CF3 groups with the carbon-carbon triple bond, *J. Chem. Phys.*, **89**, 25 (1988).

19. *Transducer Interfacing Handbook,* D. H. Sheingold, ed., Analog Devices, Norwood, Mass., 1980.

GENERAL REFERENCES

CAMAC and IEEE-488 (GPIB or HP-IB)

CAMAC: A Modular Instrumentation System for Data Handling, chap. 4–6. ESONE Committee, Report EUR 4100, 1972.

CAMAC Tutorial Issue, *IEEE Trans. Nucl. Sci.*, **NS-20**, No. 2 (April 1973).

D. Horelick and R. S. Larsen, CAMAC: A Modular Standard," *IEEE Spectrum*, April 1976, p. 50.

HP-IB General Information, Hewlett-Packard Publication No. 5952–0058.

IEEE Standard 488, available from the IEEE Standards Office, 345 E. 47th St., New York, NY 10017.

Specifications for the CAMAC Serial Highway and Serial Crate Controller Type 62, chap. 14. Report EUR 6100e, Commission of the European Communities, Greel, Belgium, 1976.

Circuit Theory

P. Grivet, *The Physics of Transmission Lines at High and Very High Frequencies,* Academic Press, New York, 1970.

H. V. Malmstadt, C. G. Enke, and S. R. Crouch with G. Horlick, *Electronic Measurements for Scientists,* Benjamin, Menlo Park, Calif., 1974.

J. Millman and A. Grabel, *Microelectronics,* 2nd ed., McGraw-Hill, New York, 1987.

C. J. Savant, Jr., *Fundamentals of the Laplace Transform,* McGraw-Hill, New York, 1962.

A. I. Zverev, *Handbook of Filter Synthesis,* Wiley, New York, 1967.

Printed Circuit Board Software Schematic Capture and Board Layout

Electronics Workbench
908 Niagra Falls Blvd., Ste. 068
North Tonawanda, New York 14120-2060
(800) 263-5552
FAX: (416) 977-1818
http://www.electronicsworkbench.com

PADS Software Inc.
165 Forest St.
Marlborough, MA 01752
(508) 485-4300
FAX: (508) 485-7171
http://www.pads.com

Ivex Design International
15232 NW Greenbrier Pkwy.
Beaverton, OR 97006
(503) 531-3555
FAX: (503) 629-4907
(WinDraft, WinBoard, IvexView, IvexSpice)

Cadence Design Systems
2655 Seely Rd.
San Jose, CA 95134
(800) 746-6223
FAX: (408) 943-0513
http://www.cadence.com

Mentor Graphics Corporation
8005 S.W. Boeckman Rd.
Wilsonville, OR 97070
(800) 547-3000
http://www.mentor.com/pcb/

P-CAD
12348 High Bluff Dr.
San Diego, CA 92130
(858) 350-3000
FAX: (858) 350-3001
http://www.pcad.com

Protel Technologies, Inc.
5252 North Edgewood Dr. Ste. 175
Provo, UT 84604
(800) 544-4186
FAX: (801) 224-0558
http://www.protel.com

Printed Circuit Board Fabricators

Golden Gate Graphics
2408 East Nichols Cr.
Centennial, CO 80122
http://www.goldengategraphics.com

South Bay Circuits
210 Hillsdale Ave.
San Jose, CA 95136-1392
http://sbcinc.com

Advanced Circuits
21100 East 33rd Dr.
Aurora CO 80011
http://www.advancedcircuits.com

Control Analysis and Design

R. C. Dorf, *Modern Control Systems*, 6th ed., Addison-Wesley, New York, 1992.

T. E. Fortmann and K. L. Hitz, *An Introduction to Linear Control Systems, Marcel Dekker*, 1977.

G. F. Franklin, D. J. Powell, and A. Emami-Naeini, *Feedback Control of Dynamic Systems*, 2nd ed., Addison-Wesley, New York, 1991.

B. C. Kuo, *Automatic Control Systems*, 6th ed., Prentice-Hall, N.J., 1990.

K. Ogata, *Modern Control Engineering*, 2nd ed., Prentice-Hall, N.J., 1991.

C. L. Phillips and R. D. Harbor, *Feedback Control Systems*, 2nd ed., Prentice-Hall, N.J., 1991.

Components

M. Grossman, "Focus on rf Connectors," Electronic Design, **24**, No. 11, 60 (1976).

C. A. Harper, "To Compare Electrical Insulators," Electronic Design, **24**, No. 11, 72 (1976).

Handbook of Components for Electronics, C. A. Harper, ed., McGraw-Hill, New York, 1977.

T. H. Jones, *Electronic Components Handbook,* Reston Publishing, Reston, Va., 1978.

"Selecting Capacitors Properly," Electronic Design, **13**, 66 (1977).

Data Books

D.A.T.A. Books, D.A.T.A., San Diego, Calif. Volumes include *Optoelectronics, Digital I.C., Linear I.C., Transistor, Diode, Thyristor, Power Semiconductor, Interface I.C.*

MECL System Design Handbook, 2nd ed., W. R. Book, Jr., and E. C. Tynan, Jr., eds., Motorola Semiconductor Products, 1972.

Motorola Semiconductor Data Library: Vol. 1, *EIA Type Numbers to 1N5000 and 2N5000;* Vol. 2, *Discrete Products, EIA Type Numbers 1N5000 and up, 2N5000 and up, 3N...and 4N...;* Vol. 3, *Discrete Products, Motorola Non-Registered Type Numbers;* Vol. 4, *MECL Integrated Circuits.*

National Semiconductor Data Books: Linear, TTL, CMOS, MOSFET, Memory, Discrete, National Semiconductor Corporation, Santa Clara, Calif.

Optoelectronics Designer's Catalog, Hewlett-Packard, 1984.

Signetics Data Manual; Logic, Memories, Interface, Analog, Microprocessor, Military, Signetics, Sunnyvale, Calif.

The TTL Data Book for Design Engineers, 1st ed., Texas Instruments, 1973.

Voltage Regulator Handbook, National Semiconductor Corporation, Santa Clara, Calif.

Handbooks

The Electronics Handbook (The Electrical Engineering Handbook Series), J. C. Whitaker, ed., CRC Press/IEEE Press, 1996.

ARRL Handbook, 77th ed., American Radio Relay League, Newington, Conn., 2000.

Electronics Designers' Handbook, 2nd ed., L. J. Giacoletto, ed., McGraw-Hill, New York, 1977.

T. D. S. Hamilton, *Handbook of Linear Integrated Electronics for Research,* McGraw-Hill, New York, 1977.

ITT Engineering Staff, *Reference Data for Radio Engineers,* 6th ed., Sams, Indianapolis, 1979.

Radio Engineering Handbook, K. Henney, ed., McGraw-Hill, New York, 1959.

Master Catalogs

Electronic Design's Gold Book: Master Catalog and Directory of Suppliers to Electronics Manufacturers, 7th ed., Hayden, Rochelle Park, N.J., 1980.

Electronic Engineers Master, Vols. 1 and 2, 23rd ed., United Technical Publications, Garden City, N.J., 1980.

Noise

S. Letzter and N. Webster, "Noise in Amplifiers," *IEEE Spectrum*, August 1970, pp. 67–75.

A. V. D. Ziel, "Noise in Solid-State Devices and Lasers", Proc. IEEE, **58**, No. 8, 1178 (1970).

Particle and Radiation Detection

Electronics for Nuclear Particle Analysis, L. J. Herbst, ed., Oxford University Press, London, 1970.

Electro-Optics Handbook: A Compendium of Useful Information and Technical Data, RCA Defense Electronics Products, Aerospace Systems Division, Burlington, Mass., 1968.

Glenn F. Knoll, *Radiation Detection and Measurement,* Wiley, New York, 1979.

Optoelectronics Designer's Catalog, Hewlett-Packard, 1980.

Modular Pulse-Processing Electronics, ORTEC, 801 South Illinois Avenue, Oak Ridge, TN 37831-0895, 2001.

Practical Electronics

Design Techniques for Electronics Engineers, Electronics Book Series, McGraw-Hill, New York, 1977.

P. Horowitz and W. Hill, *The Art of Electronics,* Cambridge University Press, 2nd ed., Cambridge, 1989.

D. Lancaster, *CMOS Cookbook,* Sams, Indianapolis, 1977.

D. Lancaster, *TTL Cookbook,* Sams, Indianapolis, 1974.

Trade Publications

Digital Design, Benwill, Boston: computers, peripherals, systems; monthly.

Electronic Component News, Chilton Col, P.O. Box 2010, Radnor, PA 19089; monthly.

Electronic Design, Hyden Publishing Co., Inc., a subsidiary of VNU, 10 Mulholland Dr., Hasbrouck Heights, NJ 07604; monthly.

Electronic Engineering Times, CMP Publications, Manhasset, N.Y.; biweekly tabloid.

Instrument and Apparatus News (IAN), Chilton, Radnor, Pa.: instruments, industrial controls, digital systems; monthly.

Troubleshooting

R. E. Gasperini, *Digital Trouble Shooting,* Hayden, Rochelle Part, NJ, 1976.

Techniques of Digital Trouble Shooting, Hewlett-Packard Application Note 163-1.

Electronic Measurements

Low Level Measurements: Precision DC Current, Voltage, and Resistance Measurements, 5th ed., Keithly Instruments.

6.12 MANUFACTURERS AND SUPPLIERS

Breadboards and Circuit Boards

Tyco Printed Circuit Group
710 North 600 West
Logan, Utah 84321
(435) 753-4700
FAX: (435) 753-7544
http://www.tpcg-logan.com/

Associated Electronics
9325 Progress Pkwy.
Box 540

Mentor, OH 44060
(216) 354-2101

E&L Instruments, Inc.
70 Fulton Terrace
New Haven, CT 06512
(203) 624-3103

Global Specialties
Mail Order Division
P.O. Box 1405
New Haven, CT 06505
(800) 345-6251

Vector Electronic Co., Inc.
12460 Gladstone Ave.
Dept. EP, Box 4336
Sylmar, CA 91342
(818) 365-9661

Coaxial Connectors

Allied Amphenol Products
4300 Commerce Ct.
Lisle, IL 60532
(312) 983-3500

AMP Inc.
Box 3608
Harrisburg, PA 17105
(717) 564-0100

ITT Cannon
10550 Talbert Ave.
P.O. Box 8040
Fountain Valley, CA 92708
(714) 964-7400

Kings Electronics Co., Inc.
40 Marbledale Rd.
Tuckahoe, NY 10707
(914) 793-5000

Omni Spectra, Inc.
Microwave Connector Division
21 Continental Blvd.

Merrimack, NH 03057
(603) 424-4111

Pasternack Enterprises
Coaxial Products Division
P.O. Box 16759
Irving, CA 92713
(714) 261-1920

TRW Connector Division
1501 Morse Ave.
Elk Grove Village, IL 60007
(312) 981-6000

Lock-In Amplifiers

EG&G Princeton Applied Research
P.O. Box 2565
Princeton, NJ 08543
(609) 452-2111

Evans Electronics
P.O. Box 5055
Berkeley, CA 94705
(415) 653-3083

Ithaco
735 W. Clinton St.
Box 818
Ithaca, NY 14850

Stanford Research Systems, Inc.
460 California Ave.
Palo Alto, CA 94306
(415) 324-3790

Mail-Order Suppliers of Components

Digi-Key Corp.
701 Brooks Avenue South
Thief River Falls, MN 56701
(800) 344-4539 or (218) 681-6674
FAX: (218) 681-3380
Email: webmaster@digikey.com
http://www.digikey.com

Jameco
1355 Shoreway Rd.
Belmont, CA 94002
(800) 831-4242
FAX: (800) 237-6948
http://www.jameco.com

JDR Microdevices
1859 South 10th St.
San Jose, CA 95112-4108
(800) 538-5000
FAX: (408) 494-1420
http://www.jdr.com

Mouser Electronics
(800) 346-6873
http://www.mouser.com

Mouser Electronics West
11433 Woodside Avenue
Santee, CA 92071-4795
FAX: (619) 449-6041

Mouser Electronics Central
958 North Main St.
Mansfield, TX 76063-4827
FAX: (817) 483-0931

Mouser Electronics East
12 Emery Avenue
Randolph, NJ 07869-1362
FAX: (973) 328-7120

Multiple-Pin Connectors/Sockets

Mill-Max Mfg. Corp
190 Pine Hollow Rd., P.O. Box 300
Oyster Bay, NY 11771
(516) 922-6000
FAX: (516) 922-9253
Email: techserv@mill-max.com
http://www.mill-max.com

AMP Special Industries
P.O. Box 1776

Paoli, PA 19301
(215) 647-1000

ITT Cannon
10550 Talbert Ave.
P.O. Box 8040
Fountain Valley, CA 92708
(714) 964-7400

Nuclear Electronics (NIM)

Berkeley Nucleonics Corp.
1198 Tenth St.
Berkeley, CA 94710
(415) 527-1121

Canberra Industries, Inc.
One State St.
Meriden, CT 06450
(203) 238-2351

EG&G Ortec
100 Midland Rd.
Oak Ridge, TN 37831
(615) 482-4411

Harshaw Filtrol Partnership
6801 Cochran Rd.
Solon, OH 44139
(216) 248-7400

Le Croy
700 S. Main St.
Spring Valley, NY 10977
(914) 425-2000

Mech-Tronics Nuclear
1701 N. 25th Ave.
Melrose Park, IL 60160
(312) 344-0202

Philips Scientific
305 Island Rd.
Mahwah, NJ 07430
(201) 934-8015

Tennelec, Inc.
601 Oak Ridge Tpke.
P.O. Box 2560
Oak Ridge, TN 37831
(615) 483-8405

Tracor Northern
2551 W. Belthne
Middleton, WI 53562
(608) 831-6511

Operational Amplifiers and Analog Integrated Circuits

AMPTEK Inc. (Detectors, Preamplifiers)
6 De Angelo Dr.
Bedford, MA 01730-2204
(781) 275-2242
FAX: (781) 275-3470
Email: Sales@amptek.com
http://www.amptek.com

Analog Devices, Inc.
One Technology Way
P.O. Box 9106
Norwood, MA 02062
(617) 329-4700

Burr-Brown Research Corp.
International Airport Industrial Park
P.O. Box 11400
Tucson, AZ 85734
(602) 746-1111

G.E. Datel Systems, Inc.
11 Cabot Blvd.
Mansfield, MA 02048
(617) 339-9341

Harris Semiconductor
Box 883
Melbourn, FL 32901
(305) 724-7400

Hybrid Systems Corp.
87 Second Ave.

Burlington, MA 01803
(617) 272-1522

Power Supplies

Astrodyne (DC/DC and Switching Power Supplies)
300 Myles Standish Blvd.
Taunton, MA 02780
(800) 823-8080
FAX: (508) 823-8181
http://www.astrodyne.com

Avansys (DC to DC Converters)
2248 N. First St.
San Jose, CA 95131
(408) 943-8100
FAX: (408) 943-8106
http://www.Avansys.com/products.htm

Bertan Associates, Inc.
121 New South Rd.
Hicksville, NY 11801
(516) 433-3110

Data Translation (Data Acquisition Boards)
100 Locke Dr.
Marlboro MA 01752-1192
(800) 525-8528
Email: Info@datx.com
http://www.datatranslation.com

Deltron, Inc.
Wissahickon Ave.
North Wales, PA 19454
(215) 699-9261

EMCO High Voltage Corporation
11126 Ridge Rd.
Sutter Creek, CA 95685
(209) 223-3626
FAX: (209) 223-2779
http://www.emcohighvoltage.com

Glassman High Voltage, Inc.
P.O. Box 317

124 West Main St.
High Bridge, NJ 08829-0317
(908) 638-3800
FAX: (908) 638-3700
http://www.glassmanhv.com

John Fluke Mfg. Co.
Box C9090, MS250C
Everett, WA 98206
(206) 356-5400

Kepco
131-38 Sanford Ave.
Flushing, NY 11352
(212) 461-7000

Lambda Electronics
515 Broad Hollow Rd.
Melville, NY 11747
(516) 694-4200

Pico Electronics, Inc.
143 Sparks Avenue
Pelham, NY 10803-1837
(800) 431-1064
FAX: (914) 738-8225
Email: info@picolelectronics.com
http://www.picoelectronics.com

Power/Mate Co.
514 S. River St.
Hackensack, NJ 07601
(201) 440-3100

Sola
1717 Busse Rd.
Elk Grove Village, IL 60007
(312) 439-2800

Sorenson Co.
676 E. Island Pond Rd.
Manchester, NH 03103
(603) 668-4500

Printed Circuit Board Graphics Materials

Bishop Graphics, Inc.
5388 Sterling Center
Westlake Village, CA 91359
(213) 991-2600

Suppliers of Materials for PCB Fabrication

The Datak Corp.
65 71st St.
Guttenberg, NJ 07093
(201) 869-2200

G C Electronics
Household International
1801 Morgan St.
P.O. Box 1209
Rockford, IL 61105
(815) 968-9661

JAMECO Electronics
1355 Shoreway Rd.
Belmont, CA 94002
(415) 592-8097

Kepro Circuit Systems
630 Axminster Dr.
Fenton, MO 63026
(314) 343-1630

Vector Electronic Company
12460 Gladstone Ave.
Dept. EP, Box 4336
Sylmar, CA 91342
(818) 365-9661

Suppliers of FETs, Analog Switches

Vishay Siliconix
One Greenwich Place
Shelton, Connecticut 06484
(203) 452-5664
FAX: (203) 452-5670
http://www.vishay.com/

Intersil
(888) 468-3774
http://www.intersil.com/

Suppliers of Optical Isolators, Solid-State Relays

Clairex Electronics
560 S. 3rd Ave.
Mt. Vernon, NY 10550
(914) 664-6602

General Electric Co., Power Electronics
Semiconductor Department
Electronics Park
Bldg. 7
Syracuse, NY 13221
(315) 456-3606

Hewlett-Packard Co.
1820 Embarcadero Rd.
Palo Alto, CA 94303

Honeywell Optoelectronics
830 E. Arapaho Dr.
Richardson, TX 75081
(214) 234-4271

ITT Electro-Optical Products
ITT Corporation
7635 Plantation Rd.
Roanoke, VA 24019
(703) 563-0371

NEC Electronics, Inc.
401 Ellis St.
Mountain View, CA 94039
(415) 960-6000

Opto 22
43044 Business Park Dr.
Temecula, CA 92590
(909) 695-3017 or (800) 321-OPTO
FAX: (909) 695-3017
http://www.opto22.com

Texas Instruments Optoelectronics
P.O. Box 225012, MS12
Dallas, TX 75265
(214) 995-3821

Suppliers of Personal-Computer-Based Data Acquisition and Control Equipment

American Data Acquisition Corporation (PCI A/D Boards)
70 Tower Office Park, Woburn, MA 01801
(781) 935-3200
FAX: (781) 938-6553
Email: info@adac.com
http://www.adac.com

Analog Devices, Inc.
One Technology Way
P.O. Box 9106
Norwood, MA 02062
(617) 329-4700

Burr-Brown Corporation
P.O. Box 11400
Tucson, AZ 85734
(520) 746-1111
(800) 548-6132
FAX: (520)741-3895
http://www.burr-brown.com/

Capital Equipment Corp.
Burlington, MA 01803

ComputerBoards, Inc. (Data Acquisition, Control, GPIB Interface Boards)
16 Commerce Blvd.
Middleboro, MA 02346
(508) 946-5100
FAX: (508) 946-9500
http://www.computerboards.com

Data Translation, Inc.
100 Locke Dr.
Marlboro, MA 01752-1192
(800) 525-8528
FAX: (508) 481-8620

Email: info@datx.com
http://www.datatranslation.com

G.E./Datel Systems, Inc.
11 Cabot Blvd.
Mansfield, MA 02048
(617) 339-9341
Hewlett-Packard Co.
1820 Embarcadero Rd.
Palo Alto, CA 94303

Interactive Microware, Inc.
P.O. Box 139
State College, PA 16804
(814) 238-8294

John Fluke Mfg. Co., Inc.
P.O. Box C9090, MS250C
Everett, WA 98206
(206) 356-5400

Keithley Instruments, Inc.
Data Acquisition and Control Division
28775 Aurora Rd.
Cleveland, OH 44139
(216) 248-0400

National Instruments
12109 Technology Blvd.
Austin, TX 78727-6204
(512) 250-9119

Wavetek San Diego, Inc.
9405 Balboa Ave.
San Diego, CA 91123
(619) 279-2200

Circuit Analysis Software

Spectrum Software (Microcap software)
1021 S. Wolfe Rd.
Sunnyvale, CA 94086
(408) 738-4387
FAX: (408) 738-4702
www.spectrum-soft.com

Cadence Corporation (PSpice software)
555 River Oaks Pkwy.
San Jose, CA 95134
(408) 943-1234
FAX: (409) 943-0513
www.pspice.com/
(See also Printed Circuit Board Software)

Suppliers of PROM Programmers and Erasers

APROTEK
1071-A Avenida Acaso
Camarillo, CA 93010
(805) 987-2454

B and C Microsystems
355 West Olive Ave.
Sunnyvale, CA 94086
(408) 730-5511

BP Microsystems
10681 Haddington #190
Houston, TX 77043
(713) 461-9430

DATA I/O
10525 Willows Rd., N.E.
P.O. Box 97046
Redmond, WA 98073

DATARASE, available from
JDR Micro Devices
110 Knowles Dr.
Los Gatos, CA 95030
(408) 866-6200

GTEK
P.O. Drawer 1346
399 Hwy. 90
Bay St. Louis, MS 39520
(601) 467-8048

JAMECO Electronics
1355 Shoreway Rd.

Belmont, CA 94002
(415) 592-8097

Wire Wrap Boards

Augat, Inc.
33 Perry Ave.
P.O. Box 779
Attleboro, MA 02703

Robinson-Nugent, Inc.
800 East 8th St.
New Albany, IN 47150
(812) 945-0211

Wire Wrap Tools

Gardner-Denver Co.
Pneutronics Division
P.O. Box 88
Reed City, MI 49677

OK Industries, Inc.
3455 Conner St.
Bronx, NY 10475
(212) 994-6600

Hardware

Conec (Connectors)
(919) 460-8800
FAX: (919) 460-0141
http://www.conec.com

Bud Industries, Inc. (Enclosures)
(440) 946-3200
FAX: (440) 951-4015
http://www.BUDIND.com

Probe Master (Test Probes)
215 Denny Way
El Cajon, CA 92020
(800) 772-1519
Email: sales@probemaster.com

L-Com (Cables and Connectors)
45 Beechwood Dr.
North Andover, MA
(800) 350-5266

FAX: (978) 589-9484
http://www.L-com.com

Techni-Tool (Tools, Supplies)
5 Apollo Rd.
P.O. Box 368
Plymouth Meeting, PA 19462-0368
(800) 832-4866
FAX: (610) 828-5623
Email: sales@techni-tool.com

Contact East, Inc. (Products for Assembling, Testing and Repairing)
335 Willow St.
North Andover, MA 01845-5995
(800) 225-5370
FAX: (800) 743-8141
http://www.contacteast.com

MICRODOT®

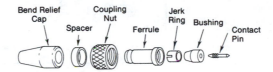

Bend Relief Cap · Spacer · Coupling Nut · Ferrule · Jerk Ring · Bushing · Contact Pin

Coaxial Cable RG-196/U

STEP 1 . . . Slide bend relief cap, spacer, coupling nut and ferrule over cable.

Cable   Braid   1/2"

STEP 2 . . . Strip cable to the dimension shown.

Cable   Braid

STEP 3 . . . Place jerk ring over shield, slotted end of ring should butt against cable jacket. Squeeze the ring snugly around the shield with a slight taper at the slotted end. Be careful not to squeeze the ring out of round.

Cut Here

STEP 4 . . . Comb out braid and cut it close to the jerk ring as shown.

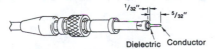

1/32"   5/32"

Dielectric   Conductor

STEP 5 . . . Strip dielectric from end of conductor 5/32" and to 1/32" from the jerk ring (be careful not to nick conductor). If conductor is stranded it must be tinned.

STEP 6 . . . Slip bushing over conductor and dielectric until bushing is against braid.

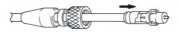

STEP 7 . . . Hold ferrule in one hand, with other hand push on the bushing until jerk ring and part of bushing start inside the ferrule.

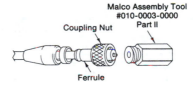

Coupling Nut   Malco Assembly Tool #010-0003-0000 Part II

Ferrule

STEP 8 . . . Slide coupling nut over ferrule. Then screw coupling nut to Malco Tool, Part II.

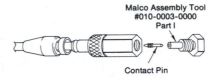

Malco Assembly Tool #010-0003-0000 Part I

Contact Pin

STEP 9 . . . Insert contact pin with barbed end out, in small hole at the end of Malco Tool, Part I, and screw into Part II as far as possible, thus forcing contact pin completely into bushing.

STEP 10 . . . Slide the bend relief cap up and over the end of the ferrule until it snaps into the groove. Remove Malco Tools, Part I and Part II.

To obtain the Malco Assembly Tools contact:
Malco, 306 Pasadena Ave., South Pasadena, CA 91030,
Phone (818) 799-9171

SMB

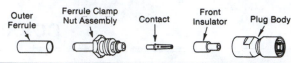

Outer Ferrule | Ferrule Clamp Nut Assembly | Contact | Front Insulator | Plug Body

Coaxial Cable RG-174/U

13/64"   11/64"

31/64"

STEP 1 . . . Slide outer ferrule over cable. Strip cable jacket, braid, and dielectric to the dimensions shown. Make all cuts sharp and square. IMPORTANT: Do not nick braid, dielectric, or center conductor. Tin center conductor, avoid excessive heat.

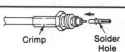

STEP 2 . . . Slightly flare out end of cable braid as shown. DO NOT comb out braid. Install ferrule clamp nut assembly onto cable so that the ferrule portion slides under braid, and insulator inside the assembly butts against cable dielectric.

Crimp   Solder Hole

STEP 3 . . . Slide outer ferrule over and up against nut. Make sure no slack exists in braid. Crimp outer ferrule with Ceramaseal's Crimp Tool #880A2840-04 or Thomas & Betts tool #WT-400, keeping cable dielectric bottomed out inside. Install center contact and solder in place. Use only rosin or non-corrosive flux. Do not get solder on the outside of the contact.

STEP 4 . . . Insert front insulator over contact.

STEP 5 . . . Screw plug body onto prepared cable termination and wrench-tighten by holding the ferrule clamp nut stationary while rotating the plug body.

Finished assembly.

BNC

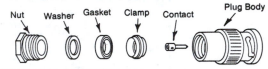

Nut   Washer   Gasket   Clamp   Contact   Plug Body

Coaxial Cables RG-58/U RG-140/U RG-141/U

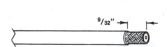

9/32"

STEP 1 . . . Cut jacket and strip to dimension shown.

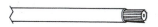

STEP 2 . . . Comb out braid and taper toward the center.

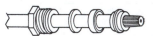

STEP 3 . . . Place the nut, washer, and gasket over the cable, then slide the clamp on over the braid so that the inner shoulder fits against the end of the cable jacket.

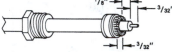

1/8"   3/32"

3/32"

STEP 4 . . . With the clamp in place, fold back braid as shown and trim 3/32" from the end. Trim dielectric to the dimensions shown.

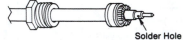

Solder Hole

STEP 5 . . . Slip contact in place so it butts against the dielectric and solder in place. Remove excess solder from outside of contact. Be sure cable dielectric is not heated excessively and swollen so as to prevent dielectric from entering into connector body.

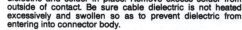

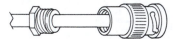

STEP 6 . . . Push assembly into body as far as it will go. Slide nut into body and screw in place with wrench until tight. For this operation, hold cable and shell rigid and rotate nut.

All dimensions are in inches

To obtain the Thomas & Betts Crimp Tool contact: Thomas & Betts Corp., 920-T Route 202, Raritan, NJ 08869, Phone (201) 685-1600, Telex 833190

N

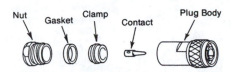

Nut Gasket Clamp Contact Plug Body

Coaxial Cables
RG-8/U
RG-9/U
RG-213/U
RG-214/U
RG-225/U

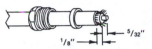

9/32″

STEP 1 . . . Place nut and gasket with "V" groove toward clamp, over cable and cut jacket to the dimension shown.

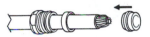

1/8″ 5/32″

STEP 2 . . . Comb out braid and fold out. Cut off cable dielectric to the dimensions shown.

STEP 3 . . . Pull braid wires forward and taper toward conductor. Place clamp over braid so that the inner shoulder fits against the end of the cable jacket.

Solder Hole

STEP 4 . . . Fold back braid as shown, trim to proper length and form over clamp as shown. Solder contact to center conductor.

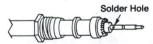

STEP 5 . . . Insert cable and parts into connector body. Make sure sharp edge of clamp seats properly in gasket. Tighten nut.

SMA

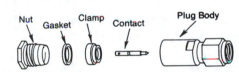

Nut Gasket Clamp Contact Plug Body

Coaxial Cables
RG-55/U
RG-58/U
RG-141/U
RG-142/U
RG-223/U

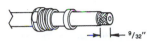

17/64″

STEP 1 . . . Place nut and gasket over cable and cut jacket to dimension shown.

1/8″

STEP 4 . . . Trim dielectric 1/8″ from the end of the cable. Do not nick center conductor.

STEP 2 . . . Comb out braid and taper forward toward the conductor.

Solder Hole

STEP 5 . . . Solder contact in place so as to be seated squarely against dielectric. Clean all surfaces thoroughly.

STEP 3 . . . Place clamp over braid and push back against jacket. Fold braid back against clamp and trim as necessary so that wires do not touch shoulder of clamp.

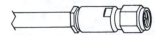

STEP 6 . . . Thread connector assembly onto prepared cable assembly. Tighten to 20-25 in./lbs. torque.

All dimensions are in inches

TRUE TRIAX

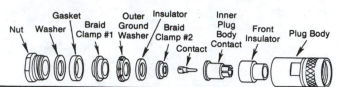

Nut | Washer | Gasket | Braid Clamp #1 | Outer Ground Washer | Insulator | Braid Clamp #2 | Inner Plug Body Contact | Front Insulator | Plug Body

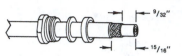

Coaxial Cables
RG-8/U
RG-9/U
RG-10/U

STEP 1 . . . Slide nut, washer and gasket over cable. Cut off outside jacket to 15/16" dimension shown. Make a clean cut, being very careful not to nick braid. Cut outer braid to 19/32" dimension as shown.

STEP 2 . . . Slide first braid clamp over braid up to outer jacket. Fold first braid back over clamp, making sure braid is evenly distributed over the surface of clamp. Trim second jacket to 15/32" dimension as shown.

STEP 3 . . . Trim second braid to dimension 5/16" as shown. Slide on outer ground washer, insulator and second braid clamp over inner braid.

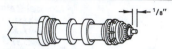

STEP 4 . . . Fold second braid back over braid clamp, again making sure that braid is evenly distributed over the surface of the clamp. Trim inner dielectric to the 1/8" dimension as shown.

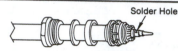

Solder Hole

STEP 5 . . . Tin the inside hole of the contact. Tin wire and insert into contact and solder. Remove any excess solder. Be sure cable dielectric is not heated excessively and swollen so as to prevent dielectric from entering the plug body.

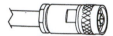

STEP 5 . . . Place front insulator and inner plug body contact into back of plug body and push into proper place. Insert cable-contact assembly into plug body. Screw nut into body with wrench until moderately tight.

MHV

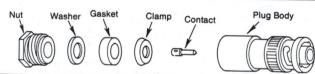

Nut | Washer | Gasket | Clamp | Contact | Plug Body

Coaxial Cable
RG-59/U

STEP 1 . . . Slide nut, washer and gasket (in improved version, with "V" groove toward clamp) over jacket, and cut off jacket 19/32" from end of cable as shown.

STEP 2 . . . Comb out braid and pull braid wires forward and taper toward conductor.

STEP 3 . . . Place clamp over braid and push back against cable jacket. Fold back braid wires as shown, trim to proper length and evenly form over clamp as shown. Cut dielectric to 1/8" dimension shown. Tin exposed conductor using minimum amount of heat.

Solder Hole

STEP 4 . . . Solder contact to conductor. Remove excess flux and solder from outside of contact.

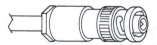

STEP 5 . . . Insert prepared cable termination into plug body. (In improved version make sure sharp edge of clamp seats properly in gasket.) Tighten nut moderately, holding plug body stationary.

All dimensions are in inches

SHV, 5 KV

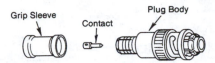

Grip Sleeve   Contact   Plug Body

Coaxial Cable RG-59/U

7/32"   5/32"

15/16"

STEP 1 . . . Slide the grip sleeve over the cable. Trim the end of the cable to the dimensions shown.

Solder Hole

STEP 2 . . . Tin end of the cable and inside of contact. Solder the contact to the center conductor. Flare out the braid without fraying.

STEP 3 . . . Push the assembled contact into the plug body until the dielectric bottoms against the shoulder within the plug body. At this point, the body grip fingers should be under the flared braid as shown.

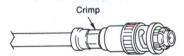

Crimp

STEP 4 . . . Slide the grip sleeve forward against the plug body, over the braid and crimp as shown. Use Kings Hand Crimp Tool #KTH-1000 and Crimp Die #KTH-2062.

BSHV, 7.5 KV

Nut   Clamp   Rear Insulator   Front Insulator with Contact Attached   Plug Body

Coaxial Cable RG-59/U

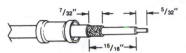

47/64"

STEP 1 . . . Slide nut and clamp over cable. Cut the jacket to the dimension shown. Do not nick braid.

3/16"

STEP 2 . . . Comb out braid and fold back against jacket. Cut dielectric to the dimension shown, do not nick conductor.

Crimp

STEP 3 . . . Carefully slide rear insulator over cable just so it covers the braid. Insert the conductor into the crimp end of the front insulator and crimp as shown.

STEP 4 . . . Carefully slide the rear insulator up against the front insulator as shown. Fold braid out and slide clamp up to meet the braid. Fold the braid back over the clamp and trim so as to fit around the clamp as shown.

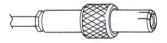

STEP 5 . . . Slide prepared cable end into plug body and tighten nut moderately.

All dimensions are in inches

To obtain the Kings Hand Crimp Tool contact:
Kings Electronics Co. Inc., 40 Marblehead Road, Tuckahoe, N.Y. 10707,
Phone (914) 793-5000, Fax (914) 793-5092

SHV, 15 KV

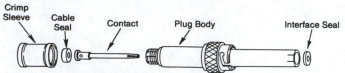

Crimp Sleeve · Cable Seal · Contact · Plug Body · Interface Seal

NOTE: Before assembly be sure you can distinguish between the cable seal and the interface seal. The cable seal has a .09" hole and is .10" thick, however the interface seal has a .13" hole and is .07" thick. Do not interchange.

Coaxial Cable RG-8/U

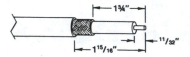

STEP 1 . . . Slide the crimp sleeve over the cable. Trim the end of the cable to the dimensions shown and tin the exposed conductor.

Solder Hole

STEP 2 . . . Slip cable seal over center conductor then slide contact over conductor. Push contact against cable seal and maintain this slight pressure while soldering contact in place through solder hole.

STEP 3 . . . Flare out braid without fraying the ends. Insert prepared cable assembly into the plug body carefully. DO NOT pinch or otherwise damage cable seal. Guide braid smoothly over the splined crimp area of the body until the contact shoulder butts against the inner insulator.

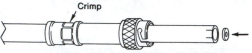

Crimp

STEP 4 . . . Slide the grip sleeve into position and crimp as shown. Use Ceramaseal's Crimping Tool #880A2840-02 or Thomas & Betts Crimping Tool #WT-540 with Crimping Die #5452. Make sure braid does not extend beyond the end of the grip sleeve. Insert interface seal into the open end of the plug. Carefully slide the interface seal over contact until it bottoms evenly around the contact.

HN

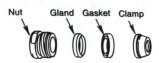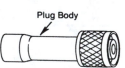

Nut · Gland · Gasket · Clamp · Sleeve · Rear Insulator · Contact · Plug Body

STEP 1 . . . Cut end of cable even. Strip off vinyl jacket to the dimension shown.

STEP 2 . . . Comb out braid and cut dielectric to the dimension shown.

STEP 3 . . . Taper braid wires forward and slide nut and gland onto jacket. Make certain knife edge of gland is toward end of cable. Then slide gasket onto jacket with "V" groove toward gland. Clamp is now pushed over braid so that the internal shoulder butts flush against the cable jacket.

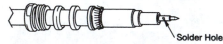

STEP 4 . . . Fold braid back over clamp and trim. Tin exposed center conductor using minimum amount of heat. Slide sleeve and rear insulator over cable dielectric.

Solder Hole

STEP 5 . . . Solder contact to center conductor. Rear insulator must seat against cable dielectric and contact shoulder must be flush with insulator face as shown. Coat cable dielectric and insulator mating surfaces with Amphenol #53-307 Silicone Compound or equal to achieve 5 KV peak rating under operating conditions.

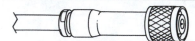

STEP 6 . . . Slide prepared cable termination carefully into body. Be sure knife edge of gland remains in groove of gasket. Tighten nut with wrench, holding body stationary. Gasket should be cut in half during tightening.

All dimensions are in inches

CHAPTER 7

MEASUREMENT AND CONTROL OF TEMPERATURE[*]

There are two levels of concern with temperature in scientific experiments. One is the control of temperature to achieve some secondary (but essential) aim in the apparatus. Examples are the use of a cold trap in a vacuum line and the use of heaters and coolants in a distillation column. For such needs, the temperature needs only to be known and kept constant to within a few kelvins. In the second case, the measurement of the dependence of physical parameters on temperature is a primary aim of the experiment. For example, a physicist can learn a lot about the nature of a material by measuring its heat capacity as a function of temperature. An organic chemist studies the energetics of a chemical reaction by measuring its rate of reaction as a function of temperature. For these experiments, the temperature must be varied over a range and controlled at any point in that range, at resolutions better than 1 K.

Sometimes the temperature must be known *accurately;* that is, the measurement must be closely related to the International Temperature Scale (ITS). For example, accurate measurements are necessary if the new data are to be compared to and used with other measurements on the system being studied. If measurements of the density of water are to be combined with measurements of its kinematic viscosity to calculate its shear viscosity as a function of temperature, then the temperature must be measured with the same accuracy in both the density

measurements and the kinematic viscosity measurements. For some purposes, the *precision* is more important than the accuracy. Such is the case when it is the derivative of some property with respect to temperature that is of interest. The changes in temperature need to be measured with very fine resolution, but the accuracy (the exact temperature on the international scale) is not as important. A physicist wishing to check a theoretical prediction for the thermal expansion of a superconducting solid will be more concerned with precision than with accuracy, even though he or she will have some requirements on the accuracy.

In this chapter we will describe how to measure and control the temperature for simple needs, usually meaning a resolution of 1 K. We will also describe more precise and accurate methods for temperature measurement and control—methods which achieve a resolution of 0.001 K and better. Since we must be able to measure the temperature in order to control it, we will first discuss thermometry. For both measurement and control, we will focus on the temperature range from 10 K to about 1800 K, and on methods that are practical for most laboratories. We will discuss some techniques which are valid at high temperatures, such as thermocouple thermometry, but we will not discuss techniques peculiar to very high temperatures (>1300 K), such as optical pyrometry.[1,2] Nor will we discuss techniques peculiar to very low temperatures (<1 K), such as vapor-pressure thermometry[3] or magnetic thermometry.[4] We will not discuss gas thermometry, which is rarely of value except in standards laboratories.[5,6]

---
[*] By Sandra C. Greer, Department of Chemical Engineering and Department of Chemistry and Biochemistry, University of Maryland at College Park.

7.1 THE MEASUREMENT OF TEMPERATURE

A thermometer can be any system with a measurable property which depends strongly on temperature. We will discuss here the *working* thermometers useful in most laboratories, as opposed to *standard* thermometers, used in calibration laboratories.[7–9]

7.1.1 Expansion Thermometers

Expansion thermometers use the volume of the thermometric system as the thermometric property. For example, the common mercury thermometer uses the change in volume (as reflected in the change in length) of a column of mercury confined to a cylindrical capillary tube. There exist other kinds of expansion thermometers—depending, for example, on the thermal expansion of a metal—that have industrial applications.[8] We will limit ourselves to the *liquid-in-glass thermometers* used in scientific research.[9]

Such thermometers are limited in range by the freezing and boiling points of the liquids used and by the softening point of the glass in which the liquid is contained. Liquids other than mercury can be used—for example, pentane can be used to 143 K. Design considerations also include the choice of glass and the relative dimensions of the parts of the thermometer.[9] Thermometers can be designed for high sensitivity in a particular range by including a section of larger diameter—a contraction chamber—in the liquid column so that the liquid expands in that volume before it reaches the range of interest (see Figure 7.1). The Beckmann differential thermometer is a mercury thermometer with a mercury reservoir that allows the amount of mercury in the stem—and thus the range of the thermometer—to be adjusted.[7]

There are a number of difficulties encountered when trying to obtain high accuracy and precision with liquid-in-glass thermometers.[7–9] Good thermal contact between the thermometer and the system of interest is difficult to achieve for systems other than fluids. The thermometer can be sensitive to pressure changes. If the thermometer is heated to its highest temperatures, it will require days to

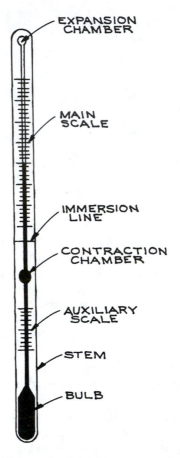

Figure 7.1   Mercury-in-glass thermometer. The auxiliary scale is designed to allow calibration at the ice point. The contraction chamber permits high sensitivity (a small capillary diameter) without having a very long stem. The immersion line indicates the depth to which the thermometer was immersed in the calibration bath. The expansion chamber prevents the buildup of pressure and avoids breakage on overheating.

return to its original calibration because of hysteresis in the glass.

In truth, a liquid-in-glass thermometer is seldom the best choice for thermometry in the modern laboratory. In addition to the problems of obtaining good accuracy and precision, liquid-in-glass thermometers do not lend themselves readily to computer automation of the measurement process. The best uses of liquid-in-glass thermometers are for approximate measurements (±0.1 K) that need not be automated, or as laboratory standards for the calibration to 0.01 K of other thermometers.

Mercury thermometers designed to be accurate and stable to about 0.01 K over temperature ranges of 3 to 5 K are commercially available (see "Manufacturers and Suppliers" at the end of this chapter). Care must be taken in using them as standards. Mount the mercury thermometers in a stirred bath with a controlled temperature (see Section 7.2). Be careful when reading the meniscus position to avoid parallax error—a telescope makes this easier. Use the thermometer at the same immersion depth at which it was originally calibrated—a difference in immersion depth will necessitate stem corrections.[7,9] Tap the thermometer before each measurement to avoid sticking of the mercury column to the stem wall. Monitor the stability of the thermometer by regularly measuring the ice point (see Section 7.1.6). A final hint: If the mercury in a thermometer becomes separated, it can be rejoined by cooling the thermometer bulb with dry ice shavings until all the mercury is back in the bulb.[9]

7.1.2 Thermocouples

Thermocouples use the voltage that develops between two junctions of two different metals when the junctions are at different temperatures as the thermometric property.[9–11] When a difference in electron density exists between two points in a metallic conductor, a voltage develops between those two points. Such a difference in electron density can arise even in a conductor made of a homogeneous metal because the electron density varies when the temperature varies. Thus a temperature gradient across a metallic material causes a voltage to develop across that material. If we try to measure this voltage by attaching leads made of the same metal, then the voltage difference will also develop in the leads and the lead voltages will cancel.[12] If the leads are of a different metal, as shown in Figure 7.2a, then the difference in the thermally induced voltages between the two metals will cause a net voltage between the junctions—the Seebeck effect. This thermoelectric voltage, or *thermal emf*, depends upon the particular choice of metals and upon the difference in temperature

between the junctions, and depends only slightly on the pressure. It is important to realize that the Seebeck effect is a bulk property of metals in a temperature gradient. It is not a property of the junctions themselves. It is not a *contact potential*, and treating it as such can lead to mistakes in the use of thermocouples.[13]

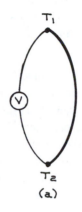

(a)

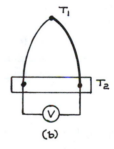

(b)

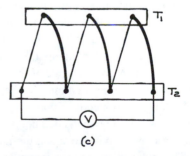

(c)

Figure 7.2   Thermocouple configurations. Lines of different thicknesses represent wires of different metals or alloys. T_1 and T_2 are two different temperatures. V is a voltmeter.

The connection of two wires of two different metals at junctions with two different temperatures, as in Figure 7.2a, is called a *thermocouple*. The voltage developed across the thermocouple is a measure of the difference in temperature $(T_1 - T_2)$. The relationship between the thermoelectric voltage and the temperature difference, while nearly a direct proportionality, cannot be predicted, but must be measured for a given pair of metals. The nature of the Seebeck effect leads to several principles useful in the design of such thermocouple thermometers:[8,14,15]

1. *The law of homogeneous metals:* A thermocouple requires two different metals. If the two wires are of the same material, then no net thermal voltage will develop, even in the presence of a temperature gradient. Voltmeter lead wires, as shown in Figure 7.2b, will therefore not contribute to the measured voltage if they are made of the same metal, and if their ends (at T_2 and at the voltmeter) are at the same temperatures so that they are under the same temperature gradient. On the other hand, inhomogeneities due to strains or impurities in the metal conductor will be, in effect, regions of a different metal. They will act as mini-thermocouples in the presence of a temperature gradient, generating spurious thermal voltages and leading to inaccuracies in the thermometry.

2. *The law of intermediate metals:* A thermocouple requires two different temperatures. If there is no difference in temperature between the ends of a pair of wires, then no net thermal voltage will develop between them, even if the wires are of different metals. In Figure 7.2a, for example, if $T_1 = T_2$, no voltage develops. A consequence of this law is that another wire may be inserted into the thermocouple circuit, as long as both ends of that wire are at a common temperature, which need not be either of the temperatures involved in the actual thermocouple (T_1 or T_2). This means that the thermocouple junctions may be made with any kind of solder or any other method of attachment, since the junction itself is all at the same temperature.

3. *The law of successive metals:* If thermocouples are made of pairs of metal A with metal B, metal B with metal C, and metal A with metal C, then the voltages developed between a given pair of junction temperatures will be related as $V_{AC} = V_{AB} + V_{BC}$. This equation allows the calculation of tables of thermocouple voltages for any pair of metals A and C if the tables are already available for A and C with respect to a reference metal B.

4. *The law of successive temperatures:* If a given thermocouple develops a voltage V_a between T_1 and T_2 and a voltage V_b between T_2 and T_3, then it will develop a voltage $V_c = V_a + V_b$ between T_1 and T_3. It follows that only the junction temperatures matter and that intermediate temperatures along homogeneous wires do not matter. It also follows that if thermocouple tables are available in which voltages are given for each temperature T_1 with respect to a reference temperature T_2, then voltages may be calculated with respect to a different reference temperature T_3 if the voltage V_b is known.

Thermocouple junctions may be placed in series in order to amplify the voltage signal. Such a *thermopile* is shown in Figure 7.2c, in which there are three identical thermocouples in series, producing (since voltages in series add) a voltage three times that of a single thermocouple.

Since any two metals or alloys can form a thermocouple, there are a great many possible thermocouple pairs. The choice of a particular pair will depend upon its sensitivity in the temperature range of interest and upon such considerations as corrosion resistance. Several thermocouple pairs have come into common use—their properties are well known, and good-quality wire is readily available. Table 7.1 gives the most common thermocouple types and their properties. Types C, D, and G are all various combinations of tungsten and rhenium alloys and are valuable at high temperatures because these metals do not readily vaporize. Type E is common in low-temperature applications because the thermal conductivities of the materials are low. Type J has the advantage of high sensitivity and the disadvantage that iron is readily oxidized. Type K has a wide range of usefulness, low thermal conductivity, and good resistance to oxidation, but cannot be used in a reducing atmosphere above 1100 K and shows instability in the calibration due to a phase transition in Chromel at 570 K. Type N is a relatively new thermocouple, intended as a much more stable substitute for Type K. Type S has low sensitivity, but has high stability and thus is useful as a standard. Type T, probably used most often, has the advantage that one material is copper, from which the voltmeter leads are also likely to be made. The chance of extraneous thermal voltages is therefore reduced. On the other hand, copper oxidizes above 620 K and has a very high thermal conductivity (see Table 7.4 in Section 7.2.2 below). The last two Chromel thermocouples in Table 7.1, one with gold-iron and one with copper-iron alloys, are used at cryogenic temperatures, along with types E, K, and T.

The construction of reliable thermocouple thermometers requires some care. Use good-quality, annealed thermocouple wire. Each wire can be individually tested for homogeneity by connecting its ends to a voltmeter and passing the wire slowly through liquid nitrogen. Wires that do not show any measurable thermoelectric voltage are suitable for use.[16] Use the largest-diameter wire that you can tolerate—large wire is more likely to be homogeneous and is less susceptible to strains. Make the junctions mechanically sound and

electrically and thermally continuous. In principle, thermocouple wires can be twisted or clamped to make a junction.[9] In practice, better results are obtained if they are soldered or welded (see Section 1.3). For work below 443 K, Type T can be joined with ordinary tin-lead solder, using a rosin flux. All can be soldered with low melting tin-silver solder. For work at higher temperatures and for better mechanical strength, all can be brazed with a high-melting silver alloy (see Section 1.3.2). All but Type T can be spot welded.[17] All can be welded with a torch. In this welding process, a reducing flame of oxygen with either natural gas or acetylene is used to melt the two wires together into a firm bead. Ready-made thermocouples, thermopiles, and thermopile arrays can be purchased from SensArray, InstruLab, Omega, et al.

Make good thermal contact between the junctions and the points at which the temperatures are to be measured. Junctions may be mechanically attached to solids using a thermally conducting grease, or bonded with a thermally conducting adhesive (Cotronics, Omega, Wakefield). Thermocouples are often placed in protective Pyrex or ceramic tubes (usually in a thermally conducting oil), but such arrangements increase the response time. Protect the wires from stresses. The wires may be electrically

Table 7.1 PROPERTIES OF COMMON THERMOCOUPLES

Type[a]	Materials	Useful Range (K)	Sensitivity ($\mu V/K$) 20 K	300 K
C, D, G	Tungsten/Rhenium	32–2600	—	20
E	Chromel[b]/Constantan[c]	10–700	8.5	61
J	Iron/Constantan[c]	60–1000	—	52
K	Chromel[b]/Alumel[d]	10–1500	4.1	41
N	Nicrosil/Nisil[e]	10–1500	3	26
S	Platinum/PtRh	220–1800	—	6
T	Copper/Constantan[c]	20–700	4.6	41
	Chromel[b]/AuFe	10–1000	15	20
	Chromel[b]/CuFe	10–300	>11	—

[a] The type designation is that of the Instrument Society of America.

[b] Chromel is an alloy of nickel with 10% chromium.

[c] Constantan is an alloy of copper with 43% nickel.

[d] Alumel is an alloy of nickel with 2% aluminum, 2% manganese, and 1% silicon.

[e] Nicrosil is a alloy of nickel, chromium, and silicon; Nisil is an alloy of nickel, magnesium, and silicon.

insulated with Teflon (to 553 K) or Kapton (to 700 K)—fiberglass or ceramic is required for higher temperatures. Thermocouples produce low-level dc voltages, which require particular care in measurement.[18,19] Appropriate procedures should be followed for shielding and grounding in the measurement circuitry (see Section 6.9). Commercial thermocouple IC amplifiers are available (Analog Devices, Linear Technology) for $15 to $50.[11] Sometimes (e. g., for control purposes) it is desirable to linearize the output signal, and such circuits have been designed.[11] A digital voltmeter of appropriate sensitivity suffices for the measurement of the thermocouple voltage.

Thermocouples measure the temperature differences between junctions. This property is advantageous when a difference is just what is required, as in some temperature-control circuits. If, instead, an absolute temperature measurement is required, then one junction (the *reference junction*) must be kept at a constant, known temperature. For example, let T_2 in Figure 7.2 be the reference junction. The absolute temperature at T_1 is then obtained by combining the known T_2 and the difference $(T_1 - T_2)$ as obtained from the thermocouple voltage (see later in this section). One way to do this is to keep the reference junction at 273.15 K by immersing it in a slurry of ice and water (all made from distilled water). The reference temperature can also be established with commercially available (but expensive) fixed point cells (Hart, Omega).[9]

Two less expensive methods are making an isothermal block, or using an electronic compensation circuit. An isothermal block consists of a material of high thermal conductivity (see Table 7.4) onto which the reference junctions and an independent thermometer are mounted with good thermal contact and surrounded by thermal insulation (see Table 7.4). The independent thermometer is then used to establish the reference temperature. This is useful even though it requires an independent thermometer because that thermometer only needs to function near room temperature, but then permits thermocouple measurements far from room temperature.

An electronic compensation circuit or *electronic ice point* is an electronic circuit which measures the temperature at the reference junction and then adds a voltage to the thermocouple circuit that compensates for the fact that the reference junction is not actually at 273.15 K.[11,20] Complete devices are available (Omega)

for about $100. Analog Devices and Linear Technology make integrated circuits (AD594, AD595, LT1025) that include compensator and amplifier for $15 to $40. Circuits can also be made from components.[19] Reference junctions made from ice baths or other fixed point baths allow temperatures to be measured with higher accuracy (± 0.01 K) than do isothermal blocks or electronic ice points (± 0.4 to ± 1 K).

In order to convert the thermocouple voltage to a temperature, it is necessary to use conversion tables or to calibrate directly. The National Institute of Standards and Technology has published extensive tables of thermocouple voltages.[21] A calibration using the tables is likely to be accurate only to a few degrees, due to variations among wires. For more accurate measurements (0.1 K), the thermocouple must be compared with a standard thermometer—usually a platinum resistance thermometer (see Section 7.1.3)—over the temperature range of interest. Voltage-temperature points from either the tables or from a direct calibration can be fitted by a polynomial:

$$T = a_0 + a_1 V + a_2 V^2 + \dots \qquad (7.1)$$

where T is the temperature, V is the voltage, and enough terms are included to describe the data. The calibration determines the accuracy of the temperature measurement on the ITS (see Section 7.1.6). The precision of temperature measurements with thermocouples is considerably better than the accuracy because temperature differences can be resolved to 0.001 K using thermopiles and sensitive amplifiers/voltmeters.

The advantages of thermocouples as thermometers are: they are simple and inexpensive, thermopiles can have a resolution of 0.001 K, they can be used over a very broad range of temperature (see Table 7.1), they are small, and they respond quickly. The disadvantages are: the signals are small and hard to measure, accuracy is hard to attain and maintain, and a reference junction is necessary for absolute temperature measurements.

7.1.3 Resistance Thermometers

Resistance thermometers use the dependence of electrical resistance on temperature as the thermometric property.

There are two main kinds of resistance temperature detectors (*RTDs*)—metal RTDs and thermistors.

Metal resistance thermometers consist of wire coils or thin films of metals such as nickel or platinum, together with the instrumentation needed to measure the electrical resistance. Platinum is common because it is resistant to corrosion and its electrical resistance is nearly linearly dependent on temperature.[9,22] Indeed, the platinum resistance thermometer (PRT) is the standard device for the ITS of 1990 between 13.8033 K and 961.78 K (see Section 7.1.4).[23] As the temperature increases, the increase in kinetic energy of the platinum atoms leads to an increase in the electrical resistivity of about 0.4 percent/K. Some typical platinum resistance thermometers are shown in Figure 7.3. Usually, at room temperature, PRTs have resistances of 25 or 100 Ω, but thin-film PRTs can have resistances of 1000 to 2000 Ω. The resistance as a function of temperature behaves as (see Figure 7.4)

$$R = R_0 + R_1 T + R_2 T^2 + \dots \qquad (7.2)$$

where the T^2 term is important for measurement with a resolution of millikelvins. With care in the resistor construction and in the measurement circuitry, a resolution of 0.1 mK can be achieved with platinum resistance thermometers over a very broad range of temperatures (see Figure 7.4 and Table 7.2).

Thin *surface mount* foil resistance thermometers made from platinum, nickel-iron, and copper are available (Minco). For cryogenic temperatures, copper, constantan, manganin, ruthenium oxide, and rhodium-iron make useful resistance thermometers.[3,24]

Thermistors are small beads of semiconducting material (metal oxides).[25] While electrons flow freely in a metal, in a semiconductor the electrons must first be excited into a conducting state. This makes a thermistor different from a resistance thermometer in that the electrical resistance is much more sensitive to temperature change, the resistance decreases as temperature increases, and the resistance is not linear in its temperature dependence. The resistance of a thermistor decreases at a rate of about 4 percent/K, which is ten times more sensitivity than that of a PRT. The best empirical function for a thermistor is (see Figure 7.4)

$$1/T = A + B(\ln R) + C(\ln R)^3 + \dots \qquad (7.3)$$

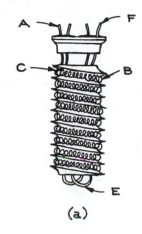

(a)

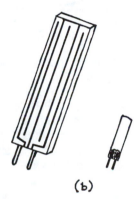

(b)

Figure 7.3   Platinum resistance thermometers. (a) Standard, high-resolution model made by Minco Products, Inc., in which the platinum is wound as a wire coil on a ceramic base. Two leads (A) pass through the lid and attach to one end of the coil (B) at C. At the other end, two leads are again attached (E) and pass through the center of the base to the lid at F. The assembly is then enclosed in an inert atmosphere. This PRT is 3/8 in. long. (b) Film RTDs such as those made by Omega Engineering, Inc. These are called *thick* and *thin* film RTDs, repectively. They can be very small—1/8 in. or less in length.

where *A*, *B*, and *C* are empirical constants and in which higher-order terms of odd power may be required for a broad temperature range or for high resolution.[8,26] If a linearized resistance is required, then linearizing circuits can be made or bought.[19] For example, the YSI model 4800 ($75) linearizes to ±0.1 K from 273 K to 373 K.

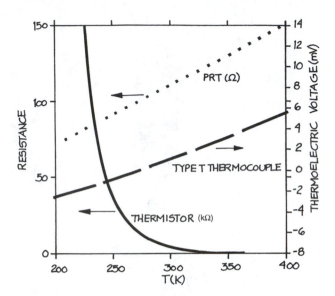

Figure 7.4   The left ordinate gives the resistance as a function of temperature *T* for a thermistor (solid line) that has a room temperature resistance of about 4 kΩ and for a platinum resistance thermometer, PRT, (dotted line) that has a room temperature resistance of about 100 Ω. Note that the scales differ by a factor of 1000. The right ordinate gives the thermoelectric voltage referenced to 273 K of a Type T copper/constantan thermocouple (dashed line) as a function of temperature.

Typical thermistor resistances are 1 to 10 kΩ. Thermistors readily resolve a millikelvin and can be made to resolve a microkelvin.[27] The disadvantage is that their resistance rapidly decreases with increasing temperature and thus their usefulness decreases above about 373 K. Thermistors designed for use up to 523 K, however, are available (YSI). Interchangeable thermistors (matched as well as ±0.05 K over 50 K) and ultra-stable thermistors are also available (YSI, Thermometrics, Betatherm).

Thin surface-mount thermistors are available from YSI and Thermometrics. Special thermistors for radio frequency field environments are available (BSD Medical Corp.) and are known as *Vitek probes*; they require special circuits to measure their high (5 MΩ) resistances,[28] and are fragile and fail if their probes are kinked or bent.

For cryogenic temperatures, there are a number of special semiconductor resistance thermometers available,

including doped germanium, carbon (graphite), and carbon-in-glass.[3,24] The carbon resistors have long been used for cryogenic thermometry, but suffer from considerable instability on temperature cycling; the carbon-in-glass resistors are somewhat more stable. Table 7.2 gives typical ranges and sensitivities for these thermometers.

Complete resistance thermometry systems are commercially available for low-resolution (0.1–1 K) measurements (Barnant, Hart, Omega, Techne, Yokagawa) and for high-resolution measurements to 1 mK (Guildline, Hart, Instrulab, YSI, Lake Shore), 0.1 mK (Julius Peters) or even 0.01 mK (Hart). Nevertheless, it is useful to understand the considerations in constructing the resistance-measuring equipment so that one may take advantage of equipment one already has, or make a thermometer for a special need, such as for very high resolution in a particular temperature range.

The first consideration for a resistance thermometer is its stability. The stability requirements depend on the resolution required and determine the price of the resistor. Thermistors and RTDs have been developed which drift less than 5 mK per year (YSI, Thermometrics). Mechanical stresses caused by mounting (e. g., use of hard adhesives) can cause drifts in resistance thermometers.

The second consideration for a resistance thermometer is its *self-heating*—the current in the resistor must be sufficiently low that the resistor itself is not heated above the temperature of its surroundings. The *dissipation factor* of the resistor specifies the self-heating. For example, a typical dissipation factor for a thermistor is 4 µW/mK. This means that a current resulting in 4 µW of power will heat the thermistor by 1 mK. To obtain a resolution of 1 mK with a 4 kΩ thermistor, the current is thus limited to 3×10^{-5} A. For a 25 Ω PRT, the current limit is about 3 mA for a resolution of 1 mK.

The third consideration is the possibility of thermal emfs in the thermometer circuit. These are the very effects that make thermocouple thermometers possible (see Section 7.1.2), but as extraneous voltages they can interfere with resistance thermometry when the resistance is determined by measuring a voltage. Thermal emfs may be evaluated by reversing the current in the resistance circuit. The effect will add for one current direction and

TABLE 7.2 PROPERTIES OF COMMON RESISTANCE THERMOMETERS

			Sensitivity ($\mu V/K$)	
Type[a]	Materials	Useful Range (K)	20 K	300 K
Conductor	Platinum	20–1,000	10	0.1
	Rhodium-iron[b]	0.5–300	0.05	5
	Ruthenium oxide	0.04–40	5[c]	—
Semiconductor	Thermistor	77–600	—	0.01
	Germanium[b]	1–30	0.5	—
	Carbon	0.04–300	1	10^3
	Carbon-in-glass	1–100	0.75	—
Diode	Silicon[b]	0.01–300	20	15
	Gallium arsenide[b]	0.01–300	15	100

[a] Approximate resolution under typical conditions.

[b] Not recommended for use in magnetic field.[24]

[c] Stability at 4.2 K.

subtract for the other. If such effects are present, they may be corrected by averaging the two readings or by using an ac resistance-measuring method.

The fourth consideration is the effect of the resistor leads. If the resistance of the lead wires enters into the resistance measurement, then that resistance and its temperature dependence can cause an error in the thermometry. Using short wires of large diameter can minimize lead effects. They can be eliminated by using *four-wire measurement* techniques—techniques in which two wires carry the current through the resistor and two other wires are used to measure the voltage across it. If the voltage is measured with a high-impedance device, then no significant current is carried in the voltage leads, no significant voltage drop occurs across those leads, and there is no lead error.

The design of precision resistance-measuring circuits is an art that changes as technology advances.[4,7,9,15,29] We will not discuss the details of such designs here, but refer the reader to the literature[4,30] and to the manufacturers of resistance instruments (Guildline, Hart, Linear Research, and various manufacturers of digital multimeters). In Figure 7.5, we present a simple circuit that requires only a digital voltmeter, a mercury battery, and a standard resistor for the assembly of a resistance thermometer with a resolution of 1 mK. For a 4 kΩ thermistor, a standard resistor of 50 kΩ is required to keep the current below the

level of self-heating. The standard resistor must be stable with time and temperature. Vishay Resistor Products makes such stable resistors. A mercury battery is a cheap and stable voltage source. The resistance R_T will indicate the temperature. R_T equals V_T/I, where V_T is the voltage drop across R_T and I is the current through R_T. The current I is the same in R_S, thus I equals V_S/R_S, where V_S is the voltage drop across the standard resistor and R_S is its resistance. Thus

$$R_T = R_S \, (V_T/V_S) \qquad (7.4)$$

If R_S is known, then only V_T and V_S need to be measured to obtain R_T. R_S need not be known exactly, since it is a constant. The ratio $R = V_T/V_S$ can be used as the thermometric variable when the thermometer is calibrated, and Equation 7.3 can be written with this ratio replacing R. In order to measure temperature to 1 mK, the voltage drops of 0.1 to 1 V need to be measured to six significant figures; some voltmeters will directly measure the ratio R.

7.1.4 Integrated Circuit Thermometers

A number of integrated circuit thermometers have been developed in recent years. They produce voltages or currents that are linear (within about 1 K) in temperature,[19] or even digital outputs.[31] Typical sensitivities are 1 μA/K and 10 mV/K, leading to typical

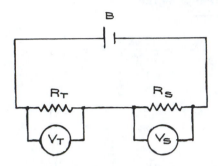

Figure 7.5   A simple resistance thermometry circuit: a series arrangement of a thermistor or a platinum resistance thermometer (R_T), a standard resistor (R_S), and a 1.35 V mercury battery (B).

of 223 to 423 K. IC thermometers therefore may be useful for some purposes. They are used as the independent thermometers in the thermocouple reference junction circuits discussed earlier, and they have many uses in general manufacturing. They are not yet capable of refined temperature measurements.

Most versions have the temperature variation of a silicon junction potential or some other silicon circuit topology as the working principle.[32] Examples are the LM335 (+10 mV/K, $1), the LM35 (+10 mV/K, $1),[33] the AD590 (+1 μA/K, $2), and the AD7416 (10-bit temperature-to-digit converter, $1). Silicon and gallium arsenide diodes use the voltage drop across the diode at constant current as the thermometric property—they are sensitive and stable (see Table 7.2). There are also temperature sensors (e. g., AD22100, +22.5 mV/K, $1) that have the temperature dependence of a thin metal film as the working principle, so that they are *IC resistance thermometers*. Analog Devices and National Semiconductor describe their wide selection of devices on their web pages (see Manufacturers and Suppliers).

7.1.5 Comparison of Thermometers

Mercury thermometers are cheap and easy to use, but are limited in precision (0.01 K) and in means of automation. Thermocouples are cheap, are useful over a wide range, have good precision (mK), and can be automated. They must, however, be designed with attention to the minimization of spurious voltages. Platinum resistance thermometers offer broad range and very good precision (0.1 mK), at higher cost. Thermistors allow extremely high precision (1 μK), but for a limited range in temperature. Resistance thermometers are readily automated. Integrated circuit thermometers offer simple, easily automated temperature measurement when low resolution of 0.1 to 1 K is all that is required.

Thermometry is a very broad subject and there is much that we have not been able to include here. We have not discussed techniques peculiar to very high temperatures, such as optical pyrometry.[1] We have also not discussed techniques peculiar to very low temperatures (<1 K), such as vapor-pressure thermometry or magnetic thermometry.[30,34–36] Several kinds of thermocouples and resistance thermometers, however, are useful at both high and low temperatures. Gas thermometry is rarely of value in other than standards laboratories.[5,6] Fiber optic thermometry,[5] especially fiber optic fluorescence thermometry,[37,38] is emerging as important for the remote sensing of temperature.

7.1.6 Thermometer Calibration

Ideally, the working scientist would like to have temperature measurements which are on the fundamental, thermodynamic temperature scale. Such measurements can be achieved only in standards laboratories, with very expensive and very complex thermometers such as ideal-gas thermometers.[39] The thermodynamic temperature is conveyed to the working laboratory by means of the International Temperature Scale, which was last redefined in 1990 (ITS-90).[23] The ITS approximates the thermodynamic temperature scale as closely as is technically and practically possible by defining the temperatures for a set of fixed points and by specifying the means for interpolation between those fixed points. *Fixed points* are temperatures that can always be reproduced, usually the temperatures at which phase transitions occur. The ITS is maintained in the United States at the National Institute of Standards and Technology (NIST). Differences between the thermodynamic temperature scale and ITS-90 are 0.020 K at 373 K and 0.150 K at 903 K.[23]

Other laboratories may send a thermometer (usually a platinum resistance thermometer) to NIST for calibration on the ITS-90, and then use that *working standard* to calibrate still other thermometers. Such a calibration can cost $1000 or more. Most companies which make and sell thermometers maintain such secondary standards, with which they can perform calibrations traceable to NIST.

For most scientific laboratories, the best procedure is to acquire a thermometer calibrated to the accuracy needed for that laboratory, either by purchasing a calibrated thermometer or by sending a thermometer to NIST for calibration. Once calibrated, the thermometer should be regularly checked for shifts or drifts in its calibration by measuring a fixed point. The most reliable of such fixed points is the triple point of water, which can be reproduced to better than 0.001 K. Triple point cells are commercially available (Hart). However, triple point measurements are not made without difficulty.[9] It is much easier to use the ice point as the fixed point. The temperature at which ice and water coexist at one atmosphere of pressure is 273.15 K and can be reproduced in most laboratories to 0.002 K. Distilled water is first degassed by several cycles of freezing it, evacuating the gas above it, and then remelting it. The ice point can then be achieved by mixing distilled, degassed water with small pieces of ice made from distilled, degassed water, and then continuously stirring the resultant slurry while the temperature measurement is made. NIST[40] sells fourteeen other phase transition fixed point cells ranging from –235 to +1235 K, including the tin freezing point (505.078 K) and the mercury triple point (234.316 K), and also sells six superconductive fixed-point devices for temperatures between 0.5 and 9.2 K. The prices range from $100 to $1000 per cell. In addition, many fixed-point cells are now commercially available (Hart).

Other thermometers can be compared with the working standard by placing both in a temperature-controlled environment (see Section 7.2; also see calibration systems from Cole-Parmer), close together, and under the same conditions at which the working standard was calibrated. The temperature is then varied over the range of interest, and measurements are made on the working standard and on the thermometer under calibration. The temperature is taken from the working standard, and an appropriate equation (see previous sections) is fitted to the thermometric property of the second thermometer as a function of temperature. Keep in mind E. B. Wilson's admonition that "An experimenter of experience would as soon use calibrations carried out by others as he [or she] would use a stranger's toothbrush."[41]

7.2 THE CONTROL OF TEMPERATURE

We will first consider cases for which the temperature is to be controlled at a single fixed value. Usually such cases require only rough control, to a degree or two. For example, we may want a trap that will condense one vapor, but not others, in a vacuum system. Precise control at fixed temperatures would be required for thermometer calibrations with fixed points (see Section 7.1.6). We will also consider the control of temperature over a range, within which any value may be selected. We will discuss how to achieve such control at various levels of precision.

7.2.1 Temperature Control at Fixed Temperatures

Control at fixed temperatures is most easily achieved at points of phase equilibrium.[7,8] For example, the point at which the solid, liquid, and gas phases of a pure substance coexist (the triple point) has a completely determined pressure and a completely determined temperature. Triple points thus make excellent temperature reference points, with accuracies of about 0.001 K. As discussed in Section 7.1.5, several triple points are used to define the ITS. However, since triple points can be difficult to achieve and to maintain, they are not appropriate for temperature control in most laboratories.

The temperatures of vaporization, sublimation, and melting of pure substances are fully determined when the pressure is fixed. These phase transitions are readily achieved and maintained, and are therefore very useful for temperature control. The *eutectic temperature,* the temperature in a two-component mixture at which a liquid solution and both pure solids coexist at a fixed pressure, can also serve as a temperature control point. For all these phase equilibria, the accuracy and stability of the temperature depend on the accuracy and stability of the pressure, as well as the purity of the materials used.

Melting-point baths are easily made and therefore are in common use. The making of an ice-point bath is discussed in Section 7.1.5. Melting-point baths using organic liquids or aqueous solutions, with temperatures ranging from 113 to 286 K, can be prepared.[42-44] These *slush baths* are made by cooling the organic liquid with either dry ice or liquid nitrogen until the bath liquid begins to freeze and a slush of solid and liquid is formed. The slush should be contained in a Dewar flask and frequently stirred. Care must be taken in using liquid nitrogen, which can condense air and possibly create explosive mixtures.

Some useful phase-transition baths are given in Table 7.3; Coyne gives a useful table of melting points for aqueous solutions.[44]

7.2.2 Temperature Control at Variable Temperatures

Control at variable temperatures requires that the temperature be measured by a control thermometer and compared with the desired temperature (the *set point, T_S*). The difference in temperature from the set point indicates whether the system needs to be heated or cooled. If the control thermometer provides an electrical signal, then that signal can be used to drive a heater or a cooler, and thus to control the temperature.

Figure 7.6 shows the main elements of such a *thermostat*. The thermostat will include the control thermometer, from which an electrical signal goes to the signal conditioner, where the signal is compared with the signal expected at the set point, conditioned further in ways to be described below, and then amplified to drive the thermostat heater or cooler. Although it is possible to control a cooling device such as Peltier junction,[45] it is usually easier to control a heating element. The thermostat may therefore include a cooler operating at a constant power to cool the system to a temperature below the set point, from which the set point is reached by means of the control heater. On the other hand, the thermostat may require an auxiliary heater, operating at constant power, to raise the temperature to near the set point. The control heater then achieves the set point from this temperature.

The thermostat will also include a measurement thermometer, independent of the control thermometer.

The *thermal circuit* may be modeled in analogy to an electrical circuit.[46-50] The flow of heat is analogous to a flow of electrical current. There is a *thermal resistance R* associated with each element. For a wall of thickness d, area A, and thermal conductivity k, R is given by

$$R = d/Ak \qquad (7.5)$$

Table 7.4 gives values of the thermal conductivity for various materials used in thermostat construction. In analogy with Ohm's law for current flow, the heat flow J induced by a temperature difference ΔT across the wall is

$$J = \Delta T/R \qquad (7.6)$$

The *thermal capacitance C* of a thermostat element is its specific heat, values of which are also given in Table 7.4. The *thermal time constant* for an element is RC. The analysis of such thermal circuits can, in principle, be made in the same way as that for electrical circuits.[47,48,50]

For temperature control to ± 0.1 K or greater, the controller can simply turn the control on or off, depending on whether the thermostat is too hot or too cold. This *on-off control* will, of course, produce large oscillations around the set point. For better temperature control, the controller must be a feedback controller (see Section 6.7.9).[29,45,47]

First, the thermostat will require a constant *background* heating. The auxiliary heater should do most of the background heating, rather than the control heater, but the control heater could have a level, P_B, of background power.

Second, the controller will provide a power P_P to the heater proportional to the deviation from the desired set point:

$$P_P = G(T_S - T) \qquad (7.7)$$

where G is the *proportional gain* and T is the current temperature. As G is increased, a critical value will be reached at which temperature oscillations will occur; i.e., the control will revert to the on-off mode. The total heater power P is then

$$P = P_B + G(T_S - T) \qquad (7.8)$$

TABLE 7.3 FIXED TEMPERATURES FROM PHASE TRANSITIONS [a]

Temperature (K)	(°C)	System	Phase Transition	Precision (K)
356	183	67% tin, 33% lead	Eutectic	2
273	100.0	Water	Vaporization	0.1
302.92	29.77	Gallium	Melting	0.05
286	13	p-Xylene	Melting	1
273	0.0	Water	Melting	0.002
263	−10	Diethylene glycol	Melting	1
252	−21	23% sodium chloride/77% water	Eutectic	2
243	−30	Bromobenzene	Melting	1
232	−41	Acetonitrile	Melting	1
221	−52	Benzyl acetate	Melting	1
210	−63	Chloroform	Melting	1
200	−73	Trichloroethylene	Melting	1
195	−78	Carbon dioxide	Sublimation	1
189	−84	Ethyl acetate	Melting	1
182	−91	Heptane	Melting	1
175	−98	Methanol	Melting	1
166	−107	Isooctane	Melting	1
157	−116	Ethanol	Melting	1
147	−126	Methylcyclohexane	Melting	1
142	−131	Pentane	Melting	1
132	−141	1,5-Hexadiene	Melting	1
113	−160	Isopentane	Melting	1
77	−196	Nitrogen	Vaporization	1

[a] See references 7, 23, 42, 43.

[b] Percentages are given as weight per cent.

[c] The precision/accuracy is that to be expected without extraordinary attention to purity and pressure control.

Third, the need for background power is likely to change slowly, since the heat losses to the environment will change as the temperature of the thermostat is varied and as the temperature of the environment changes. The controller can adjust for this changing requirement by providing to the heater a power related to the integral over time of the deviation from the set point. The total heater power then becomes

$$P = P_B + G(T_S - T) + R\int(T_S - T)dt \qquad (7.9)$$

This *integral control* is also called *reset control,* and R is the reset constant. The combination of proportional control and integral control is called *two-term control* or *PI control.*

Sometimes another control term, proportional to the derivative of the temperature with time, is included:

$$P = P_B + G(T_S - T) + R\int(T_S - T)dt + D(dT/dt) \qquad (7.10)$$

The derivative term alters the response when there are fast changes in temperature. Control using proportional, integral, and derivative terms is called *three-term control* or *PID control.* Because laboratory thermostats rarely change temperature rapidly, derivative control is not usually needed.

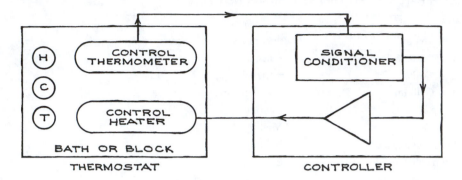

Figure 7.6   The elements of a thermostat. The controller uses the signal from the control thermometer to drive the control heater. The thermostat can also include an auxiliary heater (H), a cooling element (C), and a measurement thermometer (T).

Detailed analysis of a thermostat and its controller is rarely possible, because the parameters R and C are not well known, the thermal resistances at interfaces (e.g., between the heater and the thermostat) are very hard to assess, and effects enter due to radiation and convection. Nonetheless, the study of model circuits has led to some generalizations about thermostat design that are of practical use.[47,48]

Heating is done by using materials of high conductivity and by distributing the heaters. For fluid baths (see next section), the heat is distributed by stirring. The bath or block should have a long time constant with respect to the exchange of heat with the environment. Conductive and convective losses are reduced by using insulating materials or by enclosing the thermostat in a vacuum. Radiation can be reduced by using reflective materials. The heater should have a short time constant for transmitting heat to the bath. It should therefore be of high-conductivity material and should be in good contact with the block or bath fluid. The control thermometer should have a very short time constant for response to the bath or block temperature. It should therefore be small and should be in good contact with the block or fluid. The proportional gain should be set at about half the value at which oscillations begin. The reset constant should equal the thermal resistance between the thermostat bath or block and the environment, multiplied by the thermal capacitance of the bath or block.

The detailed design of analog signal conditioning control circuits is outside the scope of this chapter. Circuits for PID controllers using operational amplifiers

are available,[27,51–53] and Sloman et al. have reviewed the recent literature.[45] The previous discussion on thermometry noted that it can be helpful for control circuits to have a measurement thermometer that produces an output linear in temperature, and included references to such circuits and devices. For a proportional control at the level of ±0.1 to ±1 K, the AD22100 voltage output temperature sensor is particularly attractive, since it operates over 223 to 423 K, has a linearized output, and costs about $1.[54] The AD590 has even been used for control to ±0.1 mK near room temperature.[55] There are also IC controllers now available. The Analog Devices TMP01, for example, is a sensor and comparator with "a control signal from one of two outputs when the device is either above or below a specific temperature range."[56] Wavelength Electronics sells control ICs and complete instruments that control to ±3 mK using the AD590 or LM355 sensors.

A variety of complete temperature control instruments that operate on a signal from an analog thermometer are commercially available. Simple on-off or proportional controllers designed to control to 0.1 to 2.0 K cost $40 to $1000 (Minco, Omega, Cole-Parmer). Instruments are available with PI control (e. g. Tronac, ±0.0003 K) or PID control (e. g. Minco, ±0.1 K, $150; LakeShore, ±0.001 K, $2000). These are analog controllers, some of which can be controlled by computer, and some of which even have LabView software drivers (National Instruments).

It is also possible to convert the analog thermometer signal to a digital signal, operate on that signal by

TABLE 7.4 PROPERTIES OF THERMOSTAT MATERIALS AT 298 K

Conductor	Thermal Conductivity (W/cm K)	Specific Heat (J/g K)	Density (g/cm³)	Emissivity (polished)
Aluminum	2.4	0.90	2.7	0.05
Brass	1.2	0.38	8.5	0.03
Constantan	0.22	0.4	8.6	—
Copper	4.0	0.39	8.9	0.02
Manganin	0.21	0.41	8.5	—
Nickel	0.91	0.44	8.9	0.05
Platinum	0.69	0.13	21.5	0.04
Stainless steel	0.14	0.4	7.8	0.1

Insulator	Thermal Conductivity (W/cm K)	Specific Heat (J/g K)	Density (g/cm³)	Useful Temperature Range (K)
Air (1 atm)	0.00025	1.0	0.0012	—
Alumina	0.06[a]	1.0	4.0	3 to 2200
Bakelite	0.0023	1.6	1.3	3 to 390
Fiberglass	0.00002–0.004	0.8	0.06	3 to 1100
Macor	0.0017	0.75	2.5	3 to 1300
Magnesia	0.07[a]	1.2	3.6	3 to 2700
Nylon	0.0024	1.7	1.1	3 to 420
Pyrex	0.001	0.8	2.2	3 to 1100
Styrofoam	0.00003–0.0003	1.3	0.05	3 to 350
Teflon TFE	0.004	1.0	2.1	3 to 550

[a] At about 1500 K.

computer, then convert back to analog heater power.[45,57,58] A digital control system can linearize thermometer output in the software, by using fits to Equations 7.1 through 7.3, rather than by using hardware linearization. Such a digital controller is especially convenient as a part of a computerized data collection system, to change the temperature at programmed time or temperature intervals. The digital controllers commercially available now are all intended for cryogenic temperatures, but can be used at higher temperatures (about 450 K) with resolutions as good as ±1 mK (see Scientific Instruments, Inc., and Lakeshore).

Stirred liquid baths, in which the temperature of a working liquid is controlled, are the most common type of thermostat because they are simple, versatile, and inexpensive. Commercial controller-stirrer combinations that can control baths to ±0.1 K are available for a few hundred dollars. A number of commercial baths are available that control to ±0.01 to ±0.03 K in ranges between 80 and 100 K. Hart Scientific and Guildline make precision baths stable at levels of ±5 mK to ±0.2 mK, in ranges between 80 and 300 K.

Often the researcher prefers to construct a bath—to save money, to achieve better control, or to provide a particular configuration. In fact, there is no commercial bath with control to ±1 mK or better and with which the experimenter can view the experiment. If you need such a bath (for example, for phase separation studies), you must make it. Figure 7.7 shows a typical fluid bath designed to control to ±1 mK or better. The bath vessel can be any large container—36 L cylindrical Pyrex jars are available that can be used over a large temperature range and are transparent. Radiative heat losses can be reduced with a layer of aluminum foil or reflective plastic such as is used for camping blankets, placed next to the vessel. Fiberglass batting of the kind used to insulate buildings is easily wrapped around the vessel to insulate it. Styrofoam sheets about 3 in. thick can be cut to fit around the vessel. Styrofoam can also be formed *in situ* if a mold is placed around the vessel (Read Plastics). The bath is supported on three sections of glass or plastic tubing that are filled with insulation. Much of the heat loss or gain of the bath is through its top, so it is important that the top be well insulated. An insulated lid, cut to allow the various components to be inserted, is one solution. Figure 7.7 shows an insulating layer of hollow 3/4" O.D. polypropylene balls instead. The balls insulate while allowing easy access to the bath, but can be used only below about 400 K.

The choice of a liquid for the bath depends on temperature range, cost, and convenience. Table 7.5 gives the properties of some common bath liquids. Sometimes an oil will be preferred because the electrical conductivity of aqueous solutions will make electrical connections within the bath awkward and unsafe. It is worth noting that silicone oils can be ordered with a rust inhibitor added, but those inhibitors turn brown and make the bath opaque.

The control heater, as discussed above, should be made from material of high thermal conductivity. A commercial heater with a copper sheath and a power of about 150 to 250 W is satisfactory (Omega Engineering). An ordinary 60 W light bulb is an easy and cheap control heater. A power cord is soldered to the bulb and the connections electrically insulated with a silicone adhesive sealant. The background heater need not respond quickly, and can therefore be a glass or stainless steel immersion heater (Omega Engineering). Its power will depend on the temperature range, the heat capacity of the bath, and the heat losses of the bath. A manually variable transformer can drive the background heater.

There are several ways of providing a cooling element to the bath. Figure 7.7 shows a coil of copper tubing, through which cold liquid is circulated from an external source. That source can be a commercial refrigerated bath with temperature control of a few degrees. Cold liquid can also be obtained by inserting a commercial immersion cooler into a container of fluid (e.g., diethylene glycol) and pumping that liquid through the bath-cooling coil. An immersion cooler can also be inserted directly into the stirred liquid bath, but these coolers are not designed to work at elevated temperatures and such a practice will eventually damage them.

The bath stirrer is an ordinary laboratory stirrer of 1/20 or 1/10 hp. The shaft is of stainless steel to reduce heat transfer from the motor. Stirrers can transmit electrical noise, which can interfere with the controller and thermometers, so it may be necessary to provide the stirrer with an electrical shield.

The configuration of the bath elements is crucial. Figure 7.7 shows the heaters placed near the stirrers so that the heat is readily distributed. The cooling coil is itself somewhat distributed. The point is to arrange the elements so that the stirrer will distribute the hot and cold liquids and prevent temperature gradients.

Some safety precautions are in order. The possibility that the bath will overheat can be precluded by installing a device which will turn off all power if a set temperature is exceeded. Omega, Cole-Parmer, and Instruments for Research and Industry sell such devices for $85 to $300, but much cheaper (<$10) devices (thermal switches) are available from Newark Electronics. A pan made of sheet metal or plywood, sealed at the joints, and placed under the bath, will contain the bath liquid should the vessel break. Above 400 K, the bath liquids may fume and require the installation of a fume hood. Of course, one should never put hands or tools into a fluid bath when any of its components are powered.

Enclosure thermostats control the temperature of an experimental block or stage held in an environment of vacuum or of stagnant air. Such thermostats, rather than stirred liquid thermostats, are used if state-of-the-art temperature control (±0.01 mK) is required,[27,51,59] if the temperature range needed is above or below that of available bath liquids, or if the experimental apparatus cannot be immersed in a stirred liquid. For example, optical experiments are difficult in stirred baths because bubbles and debris get into the optical path. For many experiments, the vibrations that are inevitable in a stirred

TABLE 7.5 PROPERTIES OF BATH LIQUIDS AT 298 K

Liquid	Useful Temperature Range (°C)	(K)	Viscosity (cP)	Thermal Conductivity (W/cm K)	Density (g/cm³)	Specific Heat (J/g K)	Price Range ($/L)
Silicone oils:[a]							
200	−40 to 180	233 to 450	1	0.001	0.96	1.7	20
510	−50 to 180	223 to 450	50	0.001	0.62	1.7	20
550	−40 to 200	233 to 470	100	0.001	1.1	1.7	20
710	−10 to 230	263 to 500	500	0.001	1.1	1.7	50
Hydrocarbon oils:[b]							
74971	25 to 160	298 to 430	~ 10	~ 0.001	0.8	2	20
74972	25 to 150	298 to 420	~ 10	~ 0.001	0.8	2	20
74974	25 to 110	298 to 380	~ 10	~ 0.001	0.8	2	20
Water	5 to 90	278 to 360	1.0	0.0060	1.0	4.2	0
Ethylene glycol:							
100%	−11 to 180	260 to 450	32	0.0026	1.1	2.4	10
50 wt% water	−33 to 105	240 to 380	3.7	—	1.1	3.4	5
Ethanol:							
100%	−117 to 60	160 to 330	1.0	0.0017	0.79	2.5	4
59 wt% water	−30 to 75	240 to 350	2.4	—	0.92	4.0	2
Methanol:							
100%	−90 to 60	180 to 330	0.55	0.0014	0.80	2.6	5
50 wt% water	−50 to 65	220 to 340	1.8	—	0.92	3.6	2

[a] Dow Corning Corp.
[b] Precision Scientific, Inc.

bath cannot be tolerated. Much of the work on enclosure thermostats is in the literature on cryogenics (*cryostats*)[3,4,60] and on calorimetry.[61,62]

In an enclosure thermostat, the temperature is controlled by means of concentric layers that are in sequence thermally insulating, thermally conducting, and thermally controlled.[63] Insulating layers are usually vacuum regions. Conducting layers are made of metal. A controlled layer, that must be conducting, has a heater and a thermometer connected to a controller (see above). The innermost conducting layer is the experimental stage or block. The result is that the alternating layers of conductors and insulators attenuate thermal gradients and fluctuations, making them smaller and smaller as the experimental stage is approached. There are two possible design approaches for such a thermostat. The first is to maximize the heat capacities of the layers in order to reduce the thermal gradients.[59] The disadvantage of this approach is that the time constants are very long, and thus changes in temperature are very slow. The second approach is to maximize the number of stages in order to attenuate

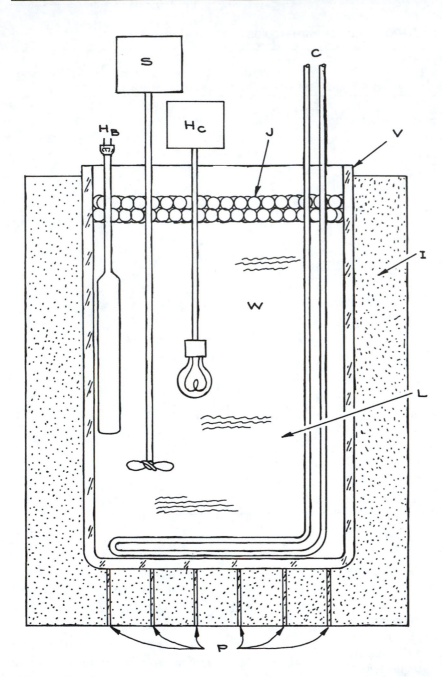

Figure 7.7   A stirred liquid thermostat. The bath elements are: C, a coil through which cold liquid flows; H_B, a commercial blade heater serving as a background heater; H_C, a 60 W light bulb serving as the control heater; I, styrofoam insulation; J, polypropylene balls serving as the top insulation; L, the bath liquid; P, plastic tubes supporting the vessel; S, the stirrer; V, a Pyrex jar serving as the bath vessel; and W, the workspace. The control thermometer and the measurement thermometer are not shown.

gradients.[27] This approach has the advantage of short time constants, but is limited by practical size considerations.

The only commercially available enclosure thermostats are cryostats for low temperature regimes,[4] so it is useful to discuss designs for higher temperatures and for more general purposes. Figure 7.8 shows a configuration for a simple enclosure that will control temperature to 1 mK or better. There are four conducting layers and three insulating layers (all vacuum) in this thermostat. This thermostat can operate only at temperatures above that of its outermost layer. That layer must therefore be provided with a means of achieving the required temperature (usually about 0.5 K below the temperature at the experimental stage). For temperatures up to 20 K above room temperature, the outer layer can be left at room temperature. For temperatures below room temperature, the outer layer can be immersed in a cold bath or can have copper tubing soldered around it, through which cold liquid can be circulated. For example, the vacuum container could be immersed in liquid nitrogen to operate between 77 and 127 K. Likewise, for temperatures more than 20 K above room temperature, a hot bath or hot circulating liquid can be used.

For some designs, it may be possible to use thermoelectric coolers (TEC or Peltier devices)[33,45,64] or miniature refrigerators to provide cooling below ambient. A TEC is the inverse of a thermocouple. For a thermocouple, a temperature difference between junctions induces a voltage, but for a TEC, a current through junctions induces a temperature difference. A TEC device has the advantage that it can be either a heater or a cooler, depending on the direction of current through it; it has the disadvantage that in the cooling mode, the heat removed from the system must be removed by some kind of heat exchanger (flowing air or water). One TEC can pump up to 125 W, and they can be cascaded for increased cooling. Devices cost $10 to $100 and are made by Marlowe, Melcor, and Ferrotec. The Ferrotec web page has a valuable introduction and design guide.[65] Miniature refrigerators (MMR, starting at $1690) operate on the principle of Joule-Thomson cooling. They require flowing gas as the heat exchanger and, depending on the choice of gas, can operate at temperatures between 83 and 273 K (303 K with two-stages) with a cooling capacity of 0.5 to 10 W/cm^2 per stage and control of ± 0.1 K.

Of course, the choice of solders, insulating materials, etc., will depend on the temperature range of operation. Copper is the best choice for conducting layers (see Table 7.4), but aluminum is not bad and is much lighter. Conducting stages of low emissivity will further increase the isolation between stages (Table 7.4).

While Figure 7.8 shows an evacuated thermostat, it is often convenient for the insulating layer to be air. In fact,

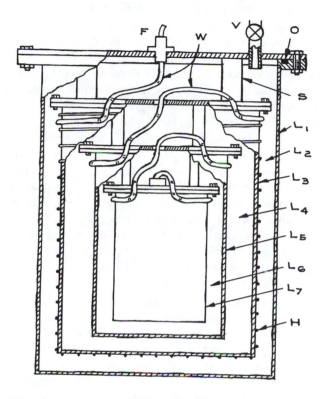

Figure 7.8 A simple enclosure thermostat. The thermostat is made of seven layers. The outer layer (L_1) is a vacuum container made of thermally conducting metal and held at a temperature slightly below the desired set point. The lid is sealed with the O-ring seal at O, and the can is evacuated by means of valve V. Layers L_2, L_4, and L_6 are insulating vacuum layers. Layer L_3 is a thermally conducting controlled layer, wound with heater H and fitted with a control thermometer (not shown). Layer L_5 is a conducting layer. Layer L_7 is the experimental stage of block. Supports (S) are made of an insulator such as nylon. Wires (W) to the thermometers, heaters, and experimental transducers enter through the feedthrough (F) and are thermally anchored on each successive stage.

the thermal conductivity of air is not significantly decreased as the pressure decreases until the pressure reaches about 10^{-4} torr.[3] Thus air at 1 atm is just as good an insulator as an evacuated space, unless low pressures can be achieved. At low temperatures, a disadvantage of air is that water will condense out of it. A disadvantage of a vacuum is that vibrations are transmitted from the pump to the thermostat.

The heater in Figure 7.8 is a wire, wound symmetrically and bifilarly (to reduce inductive effects) on the outside of the controlled layer. The heater is shown as simply glued to the outside of the layer, but better thermal contact is obtained if the wire is glued into grooves machined into the metal. Manganin wire is good for heaters because it has a high electrical resistivity that does not change significantly with temperature (MWS Wire Industries). Foil heaters, which can be glued on where needed, are often useful for the end plates (Omega, Minco). It is important that the heater power be distributed as evenly as possible over the controlled layer. The total heater power must be calculated from the estimated heat losses for the thermostat. A thermostat like that in Figure 7.8, 18 cm high and 18 cm in diameter and operating between 273 and 313 K, required a heater of 315 Ω operated at 0 to 15 V.[66]

Because the heat distribution is entirely passive, attention must be given to all asymmetrical heat losses that could lead to temperature gradients.[3] All support pieces connecting one layer to another must be of an insulating material (such as nylon or glass) and should present a minimum cross section for heat transfer. All wires entering the thermostat should be of the smallest possible diameter. All wires must be *thermally anchored* at each stage. They must be wrapped around the stage in good thermal contact with it, so that they will come to the temperature of that stage. Since windows in a thermostat easily cause gradients, they must be kept small and measures taken to compensate for losses through them.

Cryostats or Dewars for low temperatures are commercially available, and Richardson et al. give a full discussion of criteria and suppliers.[4]

CITED REFERENCES

1. *Precision Measurement and Calibration: Selected NBS Papers on Temperature*, J. F. Swindells, ed., Vol. 2, U. S. Government Printing Office, Washington, D.C., 1968.

2. J. T. Yates, *Experimental Innovations in Surface Science,* 3rd ed., Springer-Verlag, New York, 1998.

3. G. K. White, *Experimental Techniques in Low-Temperature Physics*, Clarendon Press, Oxford, 1979.

4. *Experimental Techniques in Condensed Matter Physics at Low Temperatures*, R. C. Richardson and E. N. Smith, eds, Addison-Wesley, New York, 1988.

5. P. R. N. Childs, J. R. Greenwood, and C. A. Long, *Rev. Sci. Instrum.* **71,** 2959 (2000).

6. D. A. Astrov, L. B. Beliansky, Y. A. Dedikov, S. P. Polunin, and A. A. Zakharov, *Metrologia* **26,** 151 (1989).

7. T. J. Quinn, *Temperature*, 2nd ed., Academic Press, New York, 1990.

8. J. F. Schooley, *Thermometry*, CRC Press, Inc., Boca Raton, Fla., 1986.

9. J. V. Nicholas and D. R. White, *Traceable Temperatures: An Introduction to Temperature Measurement and Calibration,* John Wiley, New York, 1994.

10. R. D. Barnard, *Thermoelectricity in Metals and Alloys*, Halsted Press, John Wiley, New York, 1972.

11. Linear Technology, *Thermocouple Measurement*, *http://www.linear-tech.com/cgi-bin/ parser?action=pdf&var=an28.pdf* (1988).

12. D. D. Pollack, *Thermoelectricity: Theory, Thermometry, Tool; ASTM Special Publication 852*, American Society for Testing and Materials, Philadelphia, 1985.

13. R. P. Reed, *Measurement and Control News* **30,** 137 (1996).

14. W. F. Roeser, *J. Appl. Physics* **11,** 388 (1940).

15. *Temperature: Its Measurement and Control in Science and Industry*, Vol. 5, J. F. Schooley, ed., American Institute of Physics, New York, 1982.

16. C. A. Mossman, J. L. Horton, and R. L. Anderson, in *Temperature: Its Measurement and Control in Science and Industry*, Vol. 5, J. F. Schooley, ed, American Institute of Physics, New York, 1982, p. 923.

17. R. A. Secco and R. F. Tucker, *Rev. Sci. Instrum.* **63,** 5485 (1992).

18. *Low Level Measurements,* 2nd ed., J. Yeager and M. A. Hrusch-Tupta, eds., Keithley Instruments, Cleveland, Ohio, 1998.

19. P. Horowitz and W. Hill, *The Art of Electronics,* Cambridge University Press, Cambridge, 1989,

20. Hewlett-Packard, *Practical Temperature Measurements: Application Note 290,* http://www.tm.agilent.com/data/downloads/eng/tmo/Notes/pdf/Data_acq_AN290_pract_meas.pdf (1997).

21. G. W. Burns, M. G. Scroger, G. F. Strouse, M. C. Croarkin, and W. F. Guthrie, *Temperature-Electromotive Force Reference Functions and Tables for the Letter-Designated Thermocouple Types Based on the ITS-90, Natl. Inst. of Stand. and Technol., Monograph 175,* U. S. Government Printing Office, Washington, D.C., 1993.

22. J. L. Riddle, G. T. Furukawa, and H. H. Plumb, *Platinum Resistance Thermometry, NBS Monograph 126,* U. S. Government Printing Office, Washington, D.C., 1973.

23. H. Preston-Thomas, *Metrologia* **2** and **107** (1990).

24. *Temperature Measurement and Control,* Lake Shore Cryotronics, Westerville, Ohio, 1999.

25. Betatherm, *http://www.betatherm.com/comprona.htm.*

26. Thermometrics, *NTC Thermistors,* http://www.thermometrics.com/assets/images/ntcnotes.pdf.

27. R. B. Kopelman, Ph.D. dissertation, University of Maryland at College Park, 1983.

28. R. R. Bowman, *IEEE Trans. MTT* **24,** 43 (1976).

29. T. D. McGee, *Principles and Methods of Temperature Measurement,* John Wiley & Sons, New York, 1988.

30. X. G. Feng, S. Zelakiewicz, and T. J. Gramila, *Rev. Sci. Instrum.* **70,** 2365 (1999).

31. Analog Devices, *http://www.analog.com/support/standard_linear/selection_guides/temp_sensors.pdf.*

32. A. Devices, *Interfacing the AD22100 Temperature Sensor to a Low Cost Single-chip Microcontroller,* http://www.analog.com/techsupt/application_notes/AN-395.pdf.

33. X. Zhu, E. Krochmann, and J. Chen, *Rev. Sci. Instrum.* **63,** 1999 (1992).

34. R. J. Soulen and W. E. Fogle, *Physics Today,* 36 (1997).

35. G. Schuster, D. Hechtfischer, and B. Fellmuth, *Repts. Prog. Phys.* **57,** 187 (1994).

36. L. G. Rubin, *Cryogenics* **37,** 341 (1997).

37. K. T. V. Grattan and Z. Y. Zhang, *Fiber Optic Fluorescence Thermometry,* Chapman and Hall, London, 1995.

38. F. Izaguirre, G. Csanky, and G. F. Hawkins, *Rev. Sci. Instrum.* **62,** 1916 (1991).

39. L. A. Guildner and W. Thomas, in *Temperature: Its Measurement and Control in Science and Industry,* Vol. 5, J. F. Schooley, ed., American Institute of Physics, New York, 1982, p. 9.

40. *NIST Standard Reference Materials Catalog,* N. M. Trahey, ed., U. S. Government Printing Office, Washington, D.C., 1998–1999.

41. E. B. Wilson, *An Introduction to Scientific Research,* McGraw-Hill, New York, 1952.

42. A. M. Phipps and D. N. Hume, *J. Chem. Ed.* **45,** 664 (1968).

43. R. E. Rondeau, *J. Chem. Eng. Data* **11,** 124 (1966).

44. G. S. Coyne, *The Laboratory Companion: A Practical Guide to Materials, Equipment, and Technique,* John Wiley, New York, 1997.

45. A. W. Sloman, P. Buggs, J. Malloy, and D. Stewart, *Meas. Sci. Tech.* **7,** 1653 (1996).

46. E. D. West, Constant Temperature Baths, in *Treatise in Analytical Chemistry,* I. M. Kolthoff and P. J. Elving, eds., John Wiley, New York, 1967.

47. M. Kutz, *Temperature Control,* John Wiley, New York, 1968.

48. M. V. Swaay, *J. Chem. Ed.* **46,** A515 (1969).

49. W. K. Roots, *Fundamentals of Temperature Control*, Academic Press, New York, 1969.

50. E. M. Forgan, *Cryogenics* **14**, 207 (1974).

51. D. Sarid and D. S. Cannell, *Rev. Sci. Instrum.* **45**, 1082 (1974).

52. L. Bruschi, R. Storti, and G. Torzo, *Rev. Sci. Instrum.* **56**, 427 (1985).

53. M. A. Rubio, L. Conde, and E. Riande, *Rev. Sci. Instrum.* **59**, 2041 (1988).

54. Analog Devices, *Voltage Output Temperature Sensor with Signal Conditioning*, http://products.analog.com/products/info.asp?product=AD22100 (1994).

55. R. D. Esman and D. L. Rode, *Rev. Sci. Instrum.* **54**, 1368 (1983).

56. Analog Devices, *Low Power, Programmable Temperature Controller*, http://products.analog.com/products/info.asp?product=TMP01 (1995).

57. R. B. Strem, B. K. Das, and S. C. Greer, *Rev. Sci. Instrum.* **52**, 1705 (1981).

58. P. Cofrancesco, U. Ruffina, M. Villa, P. Grossi, and R. Scattolini, *Rev. Sci. Instrum.* **62**, 1311 (1991).

59. B. J. Thysse, *J. Chem. Phys.* **74**, 4678 (1981).

60. A. C. Rose-Innes, *Low Temperature Techniques*. The English Universities Press, Inc., London, 1964.

61. J. P. McCullough and D. W. Scott, *Experimental Thermodynamics*, Vol. I, Butterworth's, London, 1969.

62. J. L. Hemmerich, J.-C. Loos, A. Miller, and P. Milverton, *Rev. Sci. Instrum.* **67**, 3877 (1996).

63. T. M. Sporton, *J. Phys. E. Sci. Instrum.* **5**, 317 (1972).

64. A. Kojima, C. Ishii, K. Tozaki, S. Matsuda, T. Nakayama, N. Tsuda, Y. Yoshimura, and H. Iwasaki, *Rev. Sci. Instrum.* **68**, 2301 (1997).

65. Ferrotec-America, *http://www.ferrotec-america.com/*.

66. J. L. Tveekrem, Ph.D. dissertation, University of Maryland at College Park, 1986.

GENERAL REFERENCES

P. R. N. Childs, J. R. Greenwood, and C. A. Long, "Review of Temperature Measurement," *Rev. Sci, Instrum.* **71**, 2959–2978 (2000).

Hewlett-Packard, *Practical Temperature Measurements: Application Note*, http://www.tm.agilent.com/data/downloads/eng/tmo/Notes/pdf/Data_acq_AN290_pract_meas.pdf

P. A. Kinsie, *Thermocouple Temperature Measurement*, Wiley Interscience, New York, 1973.

Thomas D. McGee, *Principles and Methods of Temperature Measurement*, John Wiley and Sons, New York, 1988.

L. Michalski, K. Eckersdorf, and J. McGhee, *Temperature Measurement*, John Wiley & Sons, New York, 1991.

National Institute of Standards and Technology, Gaithersburg, MD, Thermometry Group, *http://www.nist.gov/cstl/div836/836.05/greenbal/Biblio.htm*

J. V. Nicholas and D. R. White, *Traceable Temperatures: An Introduction to Temperature Measurement and Calibration*, John Wiley & Sons, New York, 1994.

G. R. Peacock, *Temperature Sensors*, *http://www.temperatures.com/*.

Frank Pobell, *Matter and Methods at Low Temperatures*, Springer, Berlin, 1996.

Daniel D. Pollack, *Thermocouples: Theory and Properties*, CRC Press, Boca Raton, Fla., 1991.

Robert C. Richardson and Eric N. Smith, eds., *Experimental Techniques in Condensed Matter Physics at Low Temperatures*, Addison-Wesley, New York, 1988.

T. J. Quinn, *Temperature,* 2nd ed., Academic, New York, 1990,

J. L. Riddle, G. T. Furukawa, and H. H. Plumb, *Platinum Resistance Thermometry, NBS Monograph 40, U.S. National Bureau of Standards,* U.S. Government Printing Office, Washington, D.C., 1979.

J. F. Schooley, ed., *Temperature: Its Measurement and Control in Science and Industry,* Vol. 5, American Institute of Physics, New York, 1982. (See also other volumes in this series.)

J. F. Schooley, *Thermometry,* CRC Press, Boca Raton, Fla., 1986.

J. F. Swindells, ed., *Precision Measurement and Calibration*, NBS Special Publication 300, Vol. 2, U. S. Government Printing Office, Washington, D.C., 1968.

J. A. Wise, *Liquid-in-Glass Thermometry, NBS Monograph 150*, U.S. Government Printing Office, Washington, D.C. 1976.

MANUFACTURERS AND SUPPLIERS

Ace Glass, Inc.
P.O. Box 688
1430 North West Blvd.
Vineland, NJ 08360
(800) 223-4524
http://www.aceglass.com/
(stirrers, temperature controllers, heaters)

Analog Devices, Inc.
Building Three
Three Technology Way,
Norwood, MA
(781) 329-4700
http://www.analogdevices.com/
(integrated circuits for thermocouple amplifiers and compensators; good application notes)

Barnant Company
28W092 Commercial Ave.
Barrington, IL 60010
(800) 637-3739
http://www.barnant.com/
(digital thermometers, resistance probes, temperature controllers)

Betatherm Corp.
910 Turnpike Rd.
Shrewsbury, MA 01545
(508) 842-0516
http://www.betatherm.com
(precision thermistors, chip and disc thermistors, custom probes)

Brinkmann Instruments, Inc.
One Cantiague Rd.
P. O. Box 1019
Westbury, NY 11590-0207
(800) 645-3050
http://www.brinkmann.com/
(baths, chillers)

Brooklyn Thermometer Co., Inc.
90 Verdi St.
Farmingdale, NY 11735
(800) 241-6316
http://www.brooklynthermometer.com/
(mercury thermometers)

BSD Medical Corporation
2188 West 2200 South
Salt Lake City, UT 84119-1326
(801) 972-5555
http://www.bsdmc.com/
(vitek probes for thermometry in rf fields)

Burr-Brown Corporation
PO Box 11400
Tucson, AZ 85734
(800) 548-6132
http://burr-brown.com
(diode thermometers)

Cole-Parmer Instrument Co.
625 East Bunker Court
Vernon Hills, IL 60061-1844
(800) 323-4340
http://www.coleparmer.com/
(hollow plastic balls, simple temperature baths and controllers, bath fluids)

Cotronics Corp.
3379 Shore Parkway
Brooklyn, NY 11235
(718) 646-7996
http://www.cotronics.com/
(high-temperature materials and adhesives)

DCC Corporation
7300 North Crescent Boulevard
Pennsauken, NJ 08110
(856) 662-7272
http://www.dcccorporation.com/
(thermocouple welders, data loggers)

Dow Corning Corp.
Midland, MI 48640
(517) 496-4000
http://www.dowcorning.com/
(silicone oils, silicone adhesive sealant)

Ferrotec America Corp.
40 Simon St.
Nashua, NH 03060
(603) 883-9800
http://www.ferrotec-america.com/
(peltier devices)

FTS Systems, Inc.
P.O. Box 158
Route 209
Stone Ridge, NY 12484-0158
http://www.ftssystems.com/
(800) 824-0400
(coolers, circulators, baths, cold traps)

Guildline Instruments, Inc.
103 Commerce St.
Suite 160
Lake Mary, FL 32746
(800) 232-4536
http://www.guildline.com/
(precision digital platinum resistance thermometers, resistance bridges, baths)

Haake, Inc.
53 W. Century Rd.

Paramus, N.J 07652
(800) 631-1369
http://www.haake-usa.com/
(temperature controlled baths and circulators, bath fluids)

Hart Scientific
799 E. Utah Valley Drive
American fork, UT 84003-9775
(800) 438-4278
http://www.hartscientific.com/
(standard and precision platinum resistance thermometers, precision digital thermometers, fixed point cells, precision temperature-controlled baths, precision liquid-in-glass thermometers, bath fluids, controllers, ac resistance bridge)

Instrulab, Inc.
205 Lamar St.
P.O. Box 98
Dayton, OH 45404-0098
(800) 241-2241
http://www.instrulab.com/
(precision digital thermometers, precision resistance thermometer probes, thermocouple assemblies and wire)

Instruments for Research and Industry
P. O. Box 159G
Cheltenham, PA 19012
(215) 379-3333
http://www.i-2-r.com/
(over-temperature protectors, simple controllers)

Julabo USA, Inc.
Allentown, PA 18103
(800)-4-JULABO
http://www.julabo.com
(circulators)

Julius Peters GMBH Berlin
Stromstrassse 30
1000 Berlin 21
Phone 395 12 16
(digital thermometer with 0.1 mK resolution)

LakeShore Cryotronics, Inc.
64 East Walnut St.

Westerville, OH 43081-2399
(614) 891-2243
http://www.lakeshore.com/
(cryogenic thermometers, digital controllers, platinum sensors, thermocouple wire)

Linear Research, Inc.
5231 Cushman Place, Suite 21
San Diego, CA 92110
(619) 299-0719
http://www.linearresearch.com/
(precision ac resistance bridges)

Linear Technology Corp.
1630-T McCarthy Blvd.
Milpitas, CA 95035-7487
(408) 432-1900
http://www.linear-tech.com
(integrated circuits for thermocouple amplifiers and compensators; good application notes)

Marlow Industries, Inc.
10451 Vista Park Rd.
Dallas, TX 75238-1645
(214) 340-4900
http://www.marlow.com
(thermoelectric coolers)

Measurements Group Inc.
P.O. Box 27777
Raleigh, NC 27611
(919) 365-3800
http://www.measurementsgroup.com/
(software for temperature transducers)

Melcor
1040 Spruce St.
Trenton, NJ 08648
(609) 393-4178
http://www.melcor.com/
(thermoelectric coolers, thermal greases and epoxies, computer programs for cooler analysis and control)

Minco Products, Inc.
7300 Commerce Lane
Minneapolis, MN 55432

(612) 571-3121
http://www.minco.com/
(platinum resistance thermometers, foil heaters, thermocouple assemblies, controllers for +/-0.1 to +/-1 K)

MMR Technologies, Inc.
1400 North Shoreline Blvd.
Suite A-5
Mountain View, CA 94043-1346
(650) 962-9620
http://www.mmr.com/
(microminiature refrigerators, controllers, measurement systems)

MWS Wire Industries
31200 Cedar Valley Drive
Westlake Village, CA 91362
(818) 991-8553
http://www.mwswire.com/
(heater and other specialty wires)

National Institute of Standards and Technology
Calibration Program
Gaithersburg, MD 20899
(301) 975-2002
http://www.nist.gov/public_affairs/labs2.htm
(thermometer calibrations)

National Instruments
11500 N. Mopac Expressway
Austin, TX 78759-3504
(512) 794-0100
http://www.natinst.com
(LabView instrument control software)

National Semiconductor
2900 Semiconductor Drive
P.O. Box 58090
Santa Clara, CA 95050
(408) 721-5000
http://www.national.com/
(integrated circuit temperature sensors)

Neslab Instruments, Inc.
P.O. Box 1178
Portsmouth, NH 03802

(800) 4NE-SLAB
http://www.neslabinc.com/
(immersion coolers, temperature-controlled baths and circulators, hollow plastic balls, algicide)

Newark Electronics
4801 N. Ravenswood
Chicago, IL 60640
(773) 784-5100
http://www.newark.com and *http://www.farnell,com*
(thermal switches and thermostats)

Omega Engineering Co., Inc.
One Omega Dr.
P. O. Box 4047
Stamford, CT 06907-0047
(800) 848-4286
http://www.omega.com/
(thermocouple products, resistance thermometers, simple controllers, mercury thermometers, thermally conductive epoxy)

PolyScience Corp.
6600 West Touhy Ave.
Niles, IL 60714
(800) 229-7569
http://www.polyscience.com/
(temperature-controlled baths and circulators)

Read Plastics, Inc.
12331 Wilkins Ave.
Rockville, MD 20852
(301) 881-7900
(polystyrene blocks and foam mix)

Rosemount, Inc.
12001 Technology Drive
Eden Prairie, MN 55344
(800) 949-9307
http://www.rosemount.com/
(platinum resistance thermometers, thermocouple assemblies)

Scientific Instruments, Inc.
4400 West Tiffany Dr.
Mangonia Park
West Palm Beach, FL 3407

(407) 881-8500
http://www.scientificinstruments.com/
(cryogenic thermometers and controllers)

SensArray Corporation
3410 Garrett Drive
Santa Clara, CA 95054
(408) 727-4656
http://www.sensarray.com/
(thermocouple arrays)

Sensor Scientific, Inc.
6 Kings Bridge Rd.
Fairfield, NJ 07004
(973) 227-7790
http://www.sensorsci.com/
(thermistors)

Techne, Inc.
743-T Alexander Rd.
Princeton, NJ 08540-6328
(800) 225-9243
http://www.techneusa.com/
(temperature-controlled baths and circulators, digital thermometers, plastic balls, bath fluids)

Tek-Temp Instruments, Inc.
410 Magnolia Ave.
Croydon, PA 19021
(888) 835-8367
http://www.tek-tempinstruments.com/
(recirculating chillers)

Thermo Electric Co., Inc.
109-T 5th St.
Saddle Brook, NJ 07663 USA
(888) 834-8367
http://www.thermo-electric-online.com/
(thermocouple products)

Thermometrics
808-T U.S. Highway 1
Edison, NJ 08817
(732) 287-2870
http://www.thermometrics.com/
(ultrastable precision thermistors)

Tronac, Inc.
1804 S. Columbus Lane
Orem, UT 84057
(801) 224-1131
(precision temperature controllers)

Vishay Intertechnology
63 Lincoln Highway
Malvern, PA 19355-2120
(610) 644-1300
http://www.vishay.com/
Rep: TMI Delaware Valley
5930 Iron Frame Way
Columbia, MD 21044
(410) 964-5420
(stable resistors)

VWR Scientific
1310 Goshen Parkway
West Chester, PA 19380
(800) 932-5000
http://www.vwrsp.com/
(heaters, stirrers, circulators, large Pyrex jars)

Wakefield Engineering
100 Cummings Center, Suite 157H

Beverly, MA 01915
(781) 406-3000
http://www.wakefield.com/
(thermally conducting adhesives and lubricants)

Wavelength Electronics
P. O. Box 865
Bozeman, MT 59771
(406) 687-4910
http://www.wavelengthelectronics.com/tcc.asp
(temperature control components and instruments)

YSI, Inc.
1725-T Brannum Lane
Yellow Springs, OH 45387
(800) 747-5367
http://www.ysi.com/
(precision thermistors, platinum resistance thermometers, digital thermometers, calibrations)

Yokogawa Corporation of America
2 Dart Rd.
Newan, GA 30265-1040
(770) 253-7000
http://www.yca.com/
(simple digital thermometers)

RECOMMENDED VALUES OF PHYSICAL CONSTANTS AND CONVERSION FACTORS

Quantity	Symbol	Value (with S.D. Uncertainty)
Speed of light in a vacuum	c	2.99792458×10^8 m s^{-1} (exact)
Permeability of a vacuum	μ_0	$4\pi \times 10^{-7}$ H m^{-1}
Permittivity of a vacuum, $1/\mu_0 c^2$	ϵ_0	$8.854187817\ldots \times 10^{-12}$ F m^{-1}
Elementary charge (of proton)	e	$1.60217733(49) \times 10^{-19}$ C
Gravitational constant	G	$6.67259(85) \times 10^{-11}$ N m^2 kg^{-2}
Unified atomic mass constant	m_u	$1.6605402(10) \times 10^{-27}$ kg
Rest mass:		
of electron	m_e	$9.1093897(54) \times 10^{-31}$ kg
	m_e/m_u	$5.48579903(13) \times 10^{-4}$
of proton	m_p	$1.6726231(10) \times 10^{-27}$ kg
	m_p/m_u	$1.007276470(12)$
	m_p/m_e	$1836.152701(37)$
of neutron	m_n	$1.6749286(10) \times 10^{-27}$ kg
	m_n/m_u	$1.008664904(14)$
Energy equivalence of rest mass:		
of electron, $m_e c^2/(e/C)$		$0.51099906(15)$ MeV
of proton, $m_p c^2/(e/C)$		$938.27231(28)$ MeV
of neutron, $m_n c^2/(e/C)$		$939.56563(28)$ MeV
Planck constant	h	$6.6260755(40) \times 10^{-34}$ J s
$h/2\pi$	$\hbar$	$1.05457266(63) \times 10^{-34}$ J s
Rydberg constant, $\mu_0^2 m_e e^4 c^3/8h^3$	R_∞	$1.0973731534(13) \times 10^7$ m^{-1}
Fine-structure constant, $\mu_0 e^2 c/2h$	α	$7.29735308(33) \times 10^{-3}$
$(\mu_0 e^2 c/2h)^{-1}$	α^{-1}	$137.0359895(61)$
Hartree energy, $2R_\infty hc$	E_h	$4.3597482(26) \times 10^{-18}$ J
Bohr radius, $h^2/(\pi \mu_0 c^2 m_e e^2)$	a_0	$5.29177249(24) \times 10^{-11}$ m
Electron radius, $\mu_0 e^2/(4\pi m_e)$	r_e	$2.81794092(38) \times 10^{-15}$ m
Compton wavelength:		
of electron, $h/m_e c$	λ_C	$2.42631058(22) \times 10^{-12}$ m
of proton, $h/m_p c$	$\lambda_{C,p}$	$1.32141002(12) \times 10^{-15}$ m
of neutron, $h/m_n c$	$\lambda_{C,n}$	$1.31959110(12) \times 10^{-15}$ m
Avogadro constant	L, N_A	$6.0221367(36) \times 10^{23}$ mol^{-1}
Faraday constant	F	$9.6485309(29) \times 10^4$ C mol^{-1}
Molar gas constant	R	$8.314510(70)$ J K^{-1} mol^{-1}
Boltzmann constant, R/L	k	$1.380658(12) \times 10^{-23}$ J K^{-1}
Stefan-Boltzmann constant	σ	$5.67051(19) \times 10^{-8}$ W m^{-2} K^{-4}
First radiation constant, $2\pi hc^2$	c_1	$3.7417749(22) \times 10^{-16}$ J m^2 s^{-1}
Second radiation constant, hc/k	c_2	$1.438769(12) \times 10^{-2}$ m K
Bohr magneton, $e\hbar/2m_e$	μ_B	$9.2740154(31) \times 10^{-24}$ J T^{-1}
Nuclear magneton, $e\hbar/2m_p$	μ_N	$5.0507866(17) \times 10^{-27}$ J T^{-1}
Gyromagnetic ratio of proton	γ_p	$2.67522128(81) \times 10^8$ s^{-1} T^{-1}
	$\gamma_p/2\pi$	$4.2577469(13) \times 10^7$ Hz T^{-1}
(proton in pure H$_2$O)[a]	γ_p'	$2.67515255(81) \times 10^8$ s^{-1} T^{-1}
	$\gamma_p'/2\pi$	$4.2576375(13) \times 10^7$ Hz T^{-1}
Magnetic flux quantum, $h/2e$	Φ_0	$2.06783461(61) \times 10^{-15}$ V s
Magnetic moment of electron	μ_e	$9.2847701(31) \times 10^{-24}$ J T^{-1}
in terms of Bohr magneton	μ_e/μ_B	$1.001159652193(10)$
Magnetic moment of proton	μ_p	$1.41060761(47) \times 10^{-26}$ J T^{-1}
in terms of Bohr magneton	μ_p/μ_B	$1.521032202(15) \times 10^{-3}$
(proton in pure H$_2$O)[a]	μ_p'/μ_B	$1.520993129(17) \times 10^{-3}$
in terms of nuclear magneton	μ_p/μ_N	$2.792847386(63)$

[a] As measured in a spherical sample of pure water at 25°C with the diamagnetic shielding effect $1 - \mu_p'/\mu_p = 25.689(15) \times 10^{-6}$.

INDEX

G

inteferometers 216
integral control 593
integral nonlinearity 471
integrated circuit thermometers 589–590
integrated circuits 120
integrated-circuit (IC) amplifiers 457
integrating circuits 405
integrators 460
intensity 141
interface boards 513
interference fits 39
interferometers 192
    design considerations for 297
    double-beam 294–295, 297
    Fabry-Perot 292–295
    Mach-Zehnder 296
    Michelson 294–295
    Twyman-Green 236, 295
interlock system 489
intermediate metals, law of 584
internal resistance 402
internal threads 25
International Temperature Scale (ITS) 581, 590
intrinsic photoconductive detectors 313
Invar alloys 13
inverted images 360
inverted magnetrons 91
inverting terminals 457
ion gauges 517
ion pumps 103
ion sources 374–376
ion-beam devices
    lens-mount design 388–389
    materials for 387–388
    vacuum requirements 387
ionization 89–91
ionizations gauges 89–91
iridium 90

iron 17
iron alloys. See steels
irradiance 235
isolation amplifiers 463
isolation transformers 532
isolation-mode rejection ratio (IMRR) 463–464
isooctane 592
isopentane 592
isopropyl alcohols 219
isotropic radiators 235
ITS (International Temperature Scale) 581, 590
I-V curves 444

J

J/K flip-flops 481, 485
jack 431
Janos Technology 224
jets 117
jeweler's screwdrivers 2
JFETs 456
Johnson noise 521
joining of materials
    by adhesives 31–32
    by demountable fasteners 25–31
    permanent 22–25
joints 66–67
Joule-Thomson cooling 599
journal bearings 50
JP Sercel Associates 261

K

Kapton 20, 554
    as electrical insulator 106
    insulating thermocouples with 586
    as lens material 388

W